Industrial Waste

Also of Interest

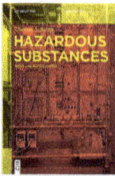

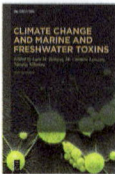

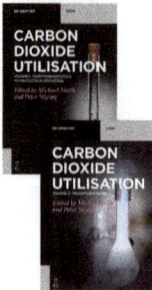

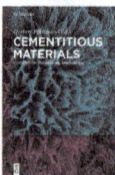

Industrial Waste

Characterization, Modification and Applications
of Residues

Edited by
Herbert Pöllmann

DE GRUYTER

Editor
Prof. Dr. Dr. Herbert Pöllmann
Martin-Luther-University Halle
Nat. Fak.III
Von-Seckendorff-Platz 3
06120 Halle
herbert.poellmann@geo.uni-halle.de

ISBN 978-3-11-067486-6
e-ISBN (PDF) 978-3-11-067494-1
e-ISBN (EPUB) 978-3-11-067502-3

Library of Congress Control Number: 2021933488

Bibliographic information published by the Deutsche Nationalbibliothek
The Deutsche Nationalbibliothek lists this publication in the Deutsche Nationalbibliografie;
detailed bibliographic data are available on the Internet at http://dnb.dnb.de.

© 2021 Walter de Gruyter GmbH, Berlin/Boston
Cover image: ZWL (Zentrum für Werkstoffanalytik)
Typesetting: Integra Software Services Pvt. Ltd.
Printing and binding: CPI books GmbH, Leck

www.degruyter.com

Contents

Part 3: Use and application of industrial residues

Part 4: **Residues from mining**

List of Authors

Jørgen Skibsted
Aarhus University, Department of Chemistry
and Interdisciplinary Nanoscience Center
(iNANO), Langelandsgade 140, DK-8000
Aarhus C

Sabrina Galluccio
Institute of Geological sciences, mineralogy/
geochemistry, Martin-Luther-University Halle,
Halle (Saale), Germany, sabrina.
galluccio@geo.uni-halle.de

Herbert Pöllmann
Institute of Geological sciences, mineralogy/
geochemistry, Martin-Luther-University Halle,
Von Seckendorffplatz 3
06120 Halle,
Halle (Saale), Germany

Katrin Schollbach
Eindhoven University of Technology
PO Box 5135600
MB Eindhoven
Netherlands

Sieger van der Laan
Tata Steel Europe
Postbus 10.0001970 CA IJmuiden
Netherlands

Herbert Pöllmann
Martin Luther University Halle
Institute for Geological Sciences, Mineralogy/
Geochemistry,
Von Seckendorffplatz3
06120 Halle (Saale) Germany

R. Kilian
University of Halle/Saale, Institute
for geological Sciences, Applied
mineralogy

Stefan Stöber
Institute for Geological Sciences, Mineralogy/
Geochemistry, Martin-Luther-University Halle
06120 Halle (Saale) Germany
email.stefan.stoeber@geo.uni-halle.de

Herbert Pöllmann
Martin Luther University Halle
Institute for Geological Sciences,
Mineralogy/Geochemistry,
06120 Halle (Saale), Germany

Andreas Ehrenberg
FEhS – Institut für Baustoff-Forschung e.V.,
Bliersheimer Str. 62,
47229 Duisburg-Rheinhausen,
Germany

Thomas A. Bier
TU Bergakademie Freiberg, IKFVW, Leipziger
Straße 28, 09599 Freiberg
Germany

Eva Kränzlein
Cluster of Excellence Engineering of
Advanced Materials (EAM)
Friedrich-Alexander-Universität
Erlangen-Nürnberg (FAU)
Cauerstr. 3, 91058 Erlangen, Germany

Elsa Qoku
TU Bergakademie Freiberg, IKFVW, Leipziger
Straße 28, 09599 Freiberg

Sandra Waida
TU Bergakademie Freiberg, IKFVW, Leipziger
Straße 28, 09599 Freiberg

Chubaakum Pongener
1Advanced Materials Group, Materials
Science and Technology Division, CSIR-NEIST
Jorhat, Assam. 785001, India.
E-mail: chubaakum@gmail.com

https://doi.org/10.1515/9783110674941-203

Koweteu Sekhamo
2Department of Geology, Patkai Christian
College, Dimapur, Nagaland, India -797115

Rajib Lochan Goswamee
Material Science Division
NEIST (RRL) Jorhat
ASSAM-785 006
INDIA

N.B.Singh
Department of Chemistry and Biochemistry,
SBSR & RTDC
Sharda University, Greater Noida, India
Email: nbsingh43@gmail.com

Mirja Illikainen
University of Oulu
Faculty of Technology
Erkki Koiso-Kanttilan katu, door V2
FI-90014 University of Oulu
Finland

Jenni Kiventerä
University of Oulu, Fibre and Particle
Engineering Research Unit, P.O.Box 4300,
90014 University of Oulu, FINLAND
mirja.illikainen@oulu.fi; +358 40 588 5904

Thomas Neumann
SCHWENK Zement GmbH & Co.
KGLaudenbacher Weg5
97753 Karlstadt
Germany

Kai Tandon
Section Mineralogy, Petrology &
Geochemistry,
Department of Earth and Environmental
Sciences, Ludwig-Maximilians-Universität
München, Theresienstr. 41, 80333 Munich,
Germany

Soraya Heuss-Aßbichler
Section Mineralogy, Petrology &
Geochemistry,
Department of Earth and Environmental
Sciences, Ludwig-Maximilians-Universität
München, Theresienstr. 41, 80333 Munich,
Germany

Iphigenia Anagnostopoulos
Section Mineralogy, Petrology &
Geochemistry,
Department of Earth and Environmental
Sciences, Ludwig-Maximilians-Universität
München, Theresienstr. 41, 80333 Munich,
Germany

Soraya Heuss-Aßbichler
Section Mineralogy, Petrology & Geochemistry,
Department of Earth and Environmental
Sciences, Ludwig-Maximilians-Universität
München, Theresienstr. 41, 80333 Munich,
Germany

Susmita Sarmah (1)
ICAR NBSS-LUP (NER CENTER)
Chemical Laboratory
Jamuguri Road, Rowriah 1
785004 Jorhat/Assam
India

Jitu Saikia (2)
Pandit Deendayal Upadhyaya
Adarsha Mahavidyalaya, Ratowa, Biswanath
Department of Chemistry
Gingia 2
784184 Biswanath
Assam/India

Pinky Saikia (3)
Joya Gogoi College
Department of Chemistry
Khumtai 3
785619 Golaghat
Assam/India

Champa Gogoi (4)
C N B College, Bokakhat
A T Road 4
785612 Golaghat
Assam, India

Rajib Lochan Goswamee
Advanced Materials Group, Materials Science
and Technology Division Council of Scientific
and Industrial Research – North East Institute
of Science and Technology (formerly RRL
Jorhat) Jorhat, Assam India 785006
Email- goswamirl@neist.res.in,
rajibgoswamee@yahoo.com
Phone 0091-9435352686, 0091-376-2370081

PRESENT ADDRESSES -
1ICAR NBSS-LUP (NER CENTER) Jamuguri
Road, Rowriah, Jorhat 785004, Assam, India
2,3Dept of Chem., Jorhat Institute of Science &
Technol., Sotai, Jorhat 785010, Assam India
4Dept of Chem., C N B College – Bokakhat
785612, Golaghat, Assam India

Andreas Kamradt
Economic Geology & Petrology Research Unit
Institute of Geosciences and Geography
Martin Luther University Halle Wittenberg
von-Seckendorff-Platz 3, D-06120 Halle,
Germany

Tim Rödel
Martin Luther University
Institute of Geosciences and Geography
Von Seckendorffplatz 3
06120 Halle
Germany

Stefan Kiefer
Friedrich Schiller University
Institute of Geosciences
Burgweg 11
07749 Jena
Germany

Gregor Borg
Martin Luther University
Institute of Geosciences and Geography
Von Seckendorffplatz 3
06120 Halle
Germany

Bruno A. M. Figueira
Universidade Federal do Oeste do Pará,
Programa de Pós Graduação em Sociedade,
Ambiente e Qualidade de Vida, Santarém-
Pará, Brazil.

Kássia L. L Marinho
Universidade Federal do Oeste do Pará,
Programa de Pós Graduação em Sociedade,
Ambiente e Qualidade de Vida, Santarém-
Pará, Brazil.

Dorsan S. Moraes
Universidade Federal do Pará, Instituto de
Geociências, Belém-Pará, Belém-PA, Brazil.

Oscar J. C. Fernandez
Instituto Federal do Pará, Programa de Pós
Graduação em Engenharia de Materiais,
Belém-PA, Brazil.

Marcondes L. da Costa
Universidade Federal do Pará, Programa de
Pós-Graduação em Geologia e Geoquímica,
Belém-Pará, Brazil
Email: mlc@ufpa.br

Leonardo Boiadeiro Ayres Negrão
Institute of Geosciences and Geography,
Mineralogy/Geochemistry, Martin
Luther University Halle-Wittenberg,
Von-Seckendorff-Platz 3, 06120 Halle,
Germany

Marcondes Lima da Costa
Institute of Geosciences, Universidade
Federal do Pará, Belém-PA, Brazil
Email : mlc@ufpa.br

Herbert Pöllmann
Martin Luther University Halle
Institute for Geological Sciences,
Mineralogy/Geochemistry,
Von Seckendorffplatz3
06120 Halle (Saale) Germany
negrao@gmail.com

Part 1: **Measurement and properties**

Jørgen Skibsted

Chapter 1
Characterization of supplementary cementitious materials and their quantification in cement blends by solid-state NMR

Abstract: Waste materials such as fly ashes, slags, silica fume and recycled glass are often used as supplementary cementitious materials (SCMs) in blended cements, where they can partly replace Portland clinker and contribute to a reduction in the environmental footprint of concrete. The SCMs are generally less crystalline or amorphous compounds rich in silica and/or alumina. Their pozzolanic reactivity is partly ascribed to the amorphous structure, which complicates their analysis and quantification in hydrated cement blends using a range of conventional analytical tools. In this context, solid-state NMR represents an important analytical tool, as it allows quantification of crystalline and amorphous components in an equal manner. This chapter illustrates the potential of solid-state NMR in studies of pozzolanic waste products in blended Portland cements including silica fume, fly ashes, slags and glasses. A general description of the quantitative solid-state NMR approach is given and the advantages and limitations of the method in studies of waste materials and blended cements are discussed. It is shown that solid-state NMR can provide accurate measures for the degree of SCM reaction for the different SCMs along with the reaction degrees of the main phases in Portland cement. Moreover, it is illustrated that valuable information about the principal hydration product, the calcium-alumino-silicate-hydrate (C-(A)-S-H) phase, and the impact of SCMs on its composition and structure can be derived from ^{27}Al and ^{29}Si NMR spectra. Finally, it is shown that similar solid-state NMR approaches can be used to follow the carbonation process in end-of-life concrete exposed to enforced carbonation.

Keywords: NMR spectroscopy, Supplementary Cementitious Materials, Blended cements, Quantification, Reactivity

Acknowledgements: PhD Søren L. Poulsen and PhD Shuai Nie are acknowledged for providing some of the experimental results shown in the figures of this chapter.

Jørgen Skibsted, Department of Chemistry and Interdisciplinary Nanoscience Center (iNANO), Aarhus University, Langelandsgade 140, DK-8000 Aarhus C, Denmark

https://doi.org/10.1515/9783110674941-001

1 Introduction

Supplementary cementitious materials (SCMs) embrace industrial waste products, natural pozzolans and minerals that exhibit hydraulic or pozzolanic properties [1–3]. Under alkaline conditions or in composite Portland cement blends, they will partly react and form hydrated phases of cementitious value that contribute to the strength development. As a result of the high demands for cement-based materials globally, CO_2 emissions associated with Portland cement production represent a major challenge for the cement industry, as it currently contributes with 6–7% of the anthropogenic CO_2 emissions [4]. Although, several approaches have been identified to lower CO_2 emissions from the concrete sector, the most direct and widely used approach is the partly replacement of Portland clinker with SCMs. Thus, a significant amount of research has been performed during the past two decades focusing on the reactivity of SCMs [1–3] and on the optimization of Portland cement – SCM blends in terms of clinker replacement, strength development and durability of the composite cements [5–10]. Currently, the most common SCMs World-wide are limestone ($CaCO_3$), fly ashes, blast furnace slags, silica fume (SiO_2), and calcined clays. However, a range of other waste products, available locally or in smaller amounts, such a rice husk ashes, ground waste glass, paper pulp and red mud, can also be used as SCMs in composite cements. Apart from limestone, most of these SCMs are amorphous materials rich in silica and/or alumina and often with glass-like phases as their most reactive components [1–3].

The amorphous nature of SCMs, which can also include domains with different composition and structure, complicates their structural characterization and quantification in blended systems, since they cannot be directly detected by X-ray diffraction techniques that probe crystallographic long-range order. Powder X-ray diffraction combined with Rietveld analysis represents one of the most prominent tools to analyze Portland cement-based materials. Recent progress has been made by the Rietveld method, using internal standards to determine the total amount of amorphous phases [11], and by the combination of Rietveld and profile summations methods, the so-called PONCKS method, for quantification of fly ashes and slags in cement blends [12, 13]. The composition and amount of SCMs may also be obtained by image analysis of electron microscopy micrographs with element mapping (e.g. SEM-EDS), where improvements recently have been demonstrated in the analysis of fly ashes [14, 15]. Other direct methods for determination of the degree of SCM reaction in blended cements include solid-state NMR spectroscopy (*vide infra*) and selective dissolution techniques [16, 17], where the latter utilizes that hydrate phases and unreacted clinker can be dissolved in specific solutions leaving the unreacted SCM as residue. However, significant amounts of clinkers and hydrate phases may appear in the residue [18, 19], making selective dissolution approaches less reliable. The SCM reaction in blended systems can also be estimated by indirect methods such as calorimetry, chemical shrinkage, and measurements of bound water or calcium hydroxide consumption by thermogravimetric methods. The principal direct and indirect techniques for determination of the amount of SCM in blended cement at a certain hydration time have been compared in an earlier review [20], which concluded that no unique technique for this type of analysis is presently available. Each of the techniques mentioned above are suitable for specific SCMs and cement blends but exhibit limitations from a broader perspective of applications. This suggests that the most reliable description of a given SCM – Portland cement system is achieved by multiple analyses with complementary tools, eventually combined with thermodynamic modelling approaches.

Solid-state NMR spectroscopy is an element specific analytical tool, where the detected resonances depend on local structural order for a given NMR spin nucleus and its coupling with nearby nuclear spins. A principal advantage of the technique is that it probes crystalline and amorphous phases in an equal manner, which has paved the way for its wide range of applications in cement chemistry [21, 22] and materials science in general [23]. The basics of solid-state NMR with magic-angle spinning (MAS) and its applications to cement-based materials have been described in two recent reviews [24, 25], whereas this chapter attempts to cover the potential of solid-state NMR in studies of waste materials with focus on SCMs used in blended cements.

2 ^{27}Al and ^{29}Si NMR studies of anhydrous SCMs

2.1 Supplementary cementitious materials

The applicability of solid-state NMR to characterize the structure of waste materials and their reactivity in cement blends will be illustrated for silica fume, slags, fly ashes, glasses and carbonated recycled concrete. Silica fume is a waste product from the ferrosilicon industry and generally it has a bulk SiO_2 content above 90 wt%. It consists of amorphous silica with a particle size in the nanometer range and is thereby much smaller than cement grains. Slags or ground granulated blast furnace slags (GGBFS) are a by-product from the manufacture of pig iron, where the silicate-rich liquid in the blast furnace is isolated, cooled by granulation, and ground to a particle size in the μm-range similar to the size of cement clinkers. GGBFS is almost completely amorphous, consisting mainly of calcium-magnesium alumino-silicate glass phases with small amounts of Na, K, Ti, Mn, and Fe present. Fly ashes or coal fly ashes are the waste product from coal combustion power plants. They consist primarily of alumino-silicate rich glassy phases with small fractions of crystalline components such as quartz, iron oxides, mullite, lime and periclase. Glasses exhibit also pozzolanic properties and they can be produced from readily available mineral sources such as clay, limestone, quartz, and alkali feldspars, as a potential replacement for slags and fly ashes in future blended cements. Alternatively waste glasses can be used, as they are also available on large scales globally [26]. This includes wastes of bottle and packaging glasses, which are generally composed of silicate-rich soda-lime glasses, although the high alkali content may be a concern from durability perspectives. Finally, carbonated recycled concrete represents a new type of SCM [27, 28], which can significantly contribute to a reduction of the CO_2 footprint associated with cement production, a more circular use of concrete, and raw materials savings as outlined recently [29, 30]. The carbonation of fines of hydrated cement paste can easily be followed by solid-state NMR, which will be illustrated in the final section of this chapter.

2.2 ^{29}Si NMR

The SCMs briefly described in section 2.1 are all dominated by amorphous phases, rich in silica or alumina and silica. Thus, these SCMs can conveniently and with good sensitivity be studied by ^{27}Al and ^{29}Si MAS NMR. ^{29}Si has a nuclear spin of $I = \frac{1}{2}$, a natural abundance of 4.7% and normally small ^{29}Si chemical shift anisotropies in cement related compounds [31]. This implies that single-pulse ^{29}Si NMR spectra can generally be acquired with moderate spinning frequencies ($\nu_R \approx 5\text{–}10$ kHz), where the intensities of the different silicate species will be gathered in centerband peaks with Gaussian-like lineshapes and with negligible intensities in the spinning sidebands. Thus, the intensities can be directly related to the molar fractions in the sample for the individual peak positions that reflect differences in ^{29}Si chemical environments. For silicate minerals and cement phases, silicon is almost exclusively present in tetrahedral coordination (SiO_4 sites), which gives chemical shifts from approx. –60 ppm to –125 ppm, relative to neat tetramethylsilane, $Si(CH_3)_4(l)$. However, a few minerals contain Si in octahedral coordination, which resonate from approx. –180 ppm to –210 ppm [23]. Five-fold coordinated Si sites with chemical shifts around –150 ppm have also been reported for silicate glasses produced under high pressure [32]. The ^{29}Si chemical shifts for silicate sites mainly reflect the degree of condensation of the SiO_4 tetrahedra (Q^n, $n = 0\text{–}4$), where an increase in condensation (n) results in a decrease in chemical shift by 5–10 ppm [33]. Thus, isolated SiO_4 tetrahedra (Q^0) resonate from approx. –60 to –80 ppm, whereas fully condensed SiO_4 sites (Q^4) exhibit chemical shifts from approx. –100 to –125 ppm. Moreover, the replacement of a SiO_4 site by an AlO_4 tetrahedron in the network results in a shift towards higher frequency of about 2–6 ppm [34]. For example, this is observed when the bridging Si sites of the silicate chains in the calcium-silicate-hydrate (C-S-H) structure is substituted by an AlO_4 tetrahedron [35], since this changes the ^{29}Si chemical shift of the neighboring silicon from around –84.5 ppm (pairing Q^2 site) to –81.5 ppm (pairing Q^2(1Al) site). This effect can be more pronounced for glasses, which can include structural sites where up to four Si sites are replaced by Al (*i.e.*, Q^4(mAl), $m = 0\text{–}4$). For calcium aluminosilicate glasses, the different levels of Si replacement (m) results in chemical shifts from approx. –70 ppm to –120 ppm, where the different types of SiO_4 environments (Q^n(mAl), $n = 0\text{–}4$, $m = 0\text{–}n$) cannot be resolved in the ^{29}Si NMR spectra as a result of the amorphous nature of the glasses. However, modelling of the ^{29}Si resonance lineshape and frequency range can in such cases provide information about the distribution of $Q^n(m\text{Al})$ sites in the glasses.

The ^{29}Si NMR spectra in Figure 1.1 of an ordinary Portland cement (oPc), silica fume ($SiO_2 = 98.5$ wt%), a fly ash with low iron content (0.43 wt%) and a slag with a high Al content (20.0 wt%) and a high fraction of amorphous material (94 wt%) show the characteristic frequency ranges for the silicate species in these samples. The spectrum of the oPc contains resonances in the Q^0 range (−68 to −78 ppm) from the alite and belite phases, where a deconvolution of the spectrum, using sub-spectra for alite and belite [36], allows determination of the quantities of these phases (*i.e.*, 62.4 ± 2.3 wt% alite and 16.4 ± 1.4 wt% belite) from their relative intensities and the bulk content of SiO_2 from XRF analysis [37]. Silica fume shows a broad resonance from approx. −100 to −125 ppm, which is ascribed solely to $Q^4(0\text{Al})$ units, where the substantial resonance broadening reflects the amorphous nature and small particle size of SiO_2 in this material. However, the silica fume resonance will not overlap with the resonances from alite, belite and the C-S-H hydration products, which facilitates its quantification in ^{29}Si NMR spectra of Portland cement – silica fume blends [38, 39]. The ^{29}Si NMR spectrum of the fly ash (72.2 wt% SiO_2, 24.2 wt% Al_2O_3, and 0.1 wt% CaO) shows several broad overlapping resonances, covering a spectral range from −90 to −125 ppm, reflecting that a range of different $Q^n(m\text{Al})$ environments are present in the sample. The high-frequency part (roughly −90 to −100 ppm) may reflect the presence of a mullite-like phase ($Al_{4+2x}Si_{2-2x}O_{10-x}\square_x$; $0.2 \leq x \leq 0.6$; $\square$ represents an oxygen vacancy) [40] while the low-frequency part can be assigned to a glassy aluminosilicate phase dominated by $Q^4(0\text{Al})$, $Q^4(1\text{Al})$ and $Q^4(2\text{Al})$ sites. Only the resonances in the high-frequency part of the spectrum will slightly overlap with the ^{29}Si resonances of the C-S-H phase in hydrated Portland cement – fly ash blends. However, spectral overlap may be a severe problem in ^{29}Si NMR studies of Portland cement – slag blends, since the broadened resonance observed for the slag (Figure 1.1) covers a range from −65 to −90 ppm, which is the same spectral region as found for the alite, belite and C-S-H phases of hydrated Portland cement. The spectrum of the slag closely resembles spectra of synthesized calcium-alumino-silicate glasses [41] and magnesium-substituted calcium-alumino-silicate glasses [42], in accordance with its composition (34.6 wt% SiO_2, 20.0 wt% Al_2O_3, 32.5 wt% CaO, and 9.2 wt% MgO) and reflecting that these types of glasses constitute the dominating amorphous component in the slag.

2.3 ^{27}Al NMR

As a half-integer spin nucleus, ^{27}Al ($I = 5/2$) NMR spectra are strongly affected by the quadrupole interaction between the spin nucleus and the surrounding electric field gradients. This interaction cannot be averaged out solely by magic-angle spinning,

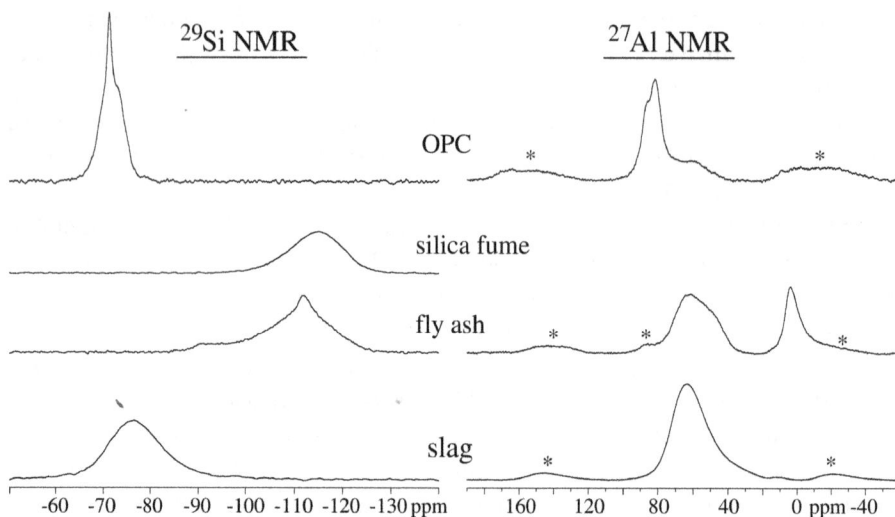

Figure 1.1: ^{29}Si and ^{27}Al MAS NMR spectra of anhydrous samples of an ordinary Portland cement (OPC), silica fume, a fly ash (FA-1, Figure 1.4) with a low iron content, and a slag. The ^{29}Si NMR spectra are acquired at 9.39 T using a spinning frequency of $v_R = 12.0$ kHz whereas the ^{27}Al NMR spectra were obtained at 14.09 T using $v_R = 13.0$ kHz. The asterisks indicate spinning sidebands.

which can result in spectra with rather complicated line shapes that deviate significantly from simple Gaussian or Lorentzian peaks as observed in ^{29}Si NMR spectra. Most ^{27}Al NMR studies of cementitious systems focus on the ^{27}Al central ($m = \frac{1}{2} \leftrightarrow m = -\frac{1}{2}$) transition [24], which is not affected by the first-order quadrupolar interaction but only the second-order term. The second-order quadrupolar interaction is inversely proportional to the magnetic field strength (B_0), implying that the linewidth decrease with increasing B_0 fields. Thus, improvements in spectral resolution of ^{27}Al resonances from different aluminate sites are often achieved by single-pulse ^{27}Al MAS NMR experiments performed at high magnetic fields (~14.1–23.5 T) using high spinning frequencies (~$v_R > 10$ kHz). The ^{27}Al isotope has a natural abundance of 100% and generally very short ^{27}Al spin-lattice relaxation times, which implies that ^{27}Al NMR spectra often can be acquired in rather short time.

The principal source of structural information is derived from the isotropic ^{27}Al chemical shifts, which clearly allow differentiation of coordination states as shown in a pioneering study for aluminum in tetrahedral and octahedral coordination [43]. For aluminate species (AlO$_x$) in minerals, zeolites and cementitious systems, the states of tetrahedral, five-fold and octahedral coordination are distinguished by ^{27}Al chemical shifts of roughly 85–50 ppm, 40–25 ppm and 20 to −10 ppm, respectively. Variations within these regions depend mainly on cations in the second coordination sphere, the bonding network and local geometries. For cementitious systems, the anhydrous phases generally have Al in tetrahedral coordination whereas the hydration products contain octahedrally coordinated Al [24]. This provides a direct mean to follow hydration reactions by ^{27}Al MAS NMR, as used in several studies and demonstrated for both calcium aluminate [44] and Portland cements [45].

The ^{27}Al NMR spectra of the anhydrous ordinary Portland cement, fly ash and slag (Figure 1.1) show clearly that the resonance frequencies reflect the Al coordination state. The spectrum of the ordinary Portland cement contains two rather narrow resonances at 81 and 86 ppm, which originate from tetrahedral Al guest-ions incorporated in the alite ('Ca$_3$SiO$_5$') and belite ('Ca$_2$SiO$_4$') phases, respectively [46]. These resonances are superimposed on a broader peak (80–30 ppm), which corresponds to the tetrahedral Al sites in the tricalcium aluminate (Ca$_3$Al$_2$O$_6$) phase. The fly ash includes most likely two different AlO$_4$ species, as seen by the main tetrahedral peak at 59 ppm with a clear shoulder at 49 ppm, along with a single resonance (2 ppm) corresponding to octahedral aluminum. The fly ash contains 24.3 wt% Al$_2$O$_3$, and analysis of the two distinct centerbands indicates that 68 and 32% of the aluminum are present as AlO$_4$ and AlO$_6$ species, respectively. The ^{27}Al NMR spectrum of the slag (Figure 1.1) is dominated by a resonance at 66 ppm, which is within the typical chemical shift region for AlO$_4$ sites in a network bonded to four SiO$_4$ tetrahedra (i.e. Al(-O-Si)$_4$ sites). In addition, a very small fraction (<1%) of octahedrally coordinated Al is seen (peak at 11 ppm), and it cannot be excluded that the slag also contains a small fraction of five-fold coordinated aluminum, which may account for the shoulder to the AlO$_4$ resonance at roughly 35 ppm. This AlO$_5$ resonance may potentially be resolved in ^{27}Al NMR spectra acquired at higher magnetic fields, as shown recently for synthesized magnesium-substituted calcium aluminosilicate glasses [42].

2.4 Experimental considerations of quantitative NMR

Quantification of structurally different components in SCMs and of SCM reactivity in pozzolanic tests and Portland cement blends from solid-state NMR spectra requires that quantitatively reliable spectra are achieved with good signal-to-noise (S/N) ratios, the latter being crucial for the quality and precision of the data analysis/simulations of the acquired spectra. Such spectra are generally obtained by the single-pulse NMR experiment, where the response from the sample ('free induction decay' – FID) is acquired immediately after the excitation pulse. When the nuclear spins are at thermal equilibrium again, the experiment can be repeated to improve the S/N ratio, and these repetitions are often performed +1000 times to achieve the desired S/N. The repetition rate (d_1) is governed by the spins with the longest spin-lattice relaxation time (T_1), where full relaxation is considered to be achieved for $d_1 > 5T_1$ (i.e. > 99.3% of the magnetization is at equilibrium), when a 90° flip angle is used for the excitation pulse. Lower values for d_1 can also be used when combined with shorter pulse flip angles (α). The optimum flip angle is the so-called Ernst angle (α_E): $\cos\alpha_E = \exp(-(d_1 + a_t)/T_1)$, where a$_t$ is the acquisition time for the FID (the condition $d_1 \gg a_t$ is often fulfilled for solids). For a single-pulse experiment with a given flip angle, the simplest and fastest approach to secure full relaxation between each repetition is to acquire an array of experiments (with relative few scans) with increasing repetition time (d_1). The repetition delay is then chosen as the lowest value of d_1 where steady-state intensities are observed for all resonances. A quantitatively reliable spectrum is then achieved with this d_1 value along with a suitable number of repetitions to give the desired S/N ratio. Alternatively, the T_1 values for the individual sites can be determined from inversion-recovery or saturation recovery experiments (see ref [24]. for further description). Although, these experiments can be rather time-consuming, the T_1 values can in some cases provide additional information that may assist spectral assignments of the resonances or provide indications on the level of paramagnetic impurities in the studied samples. For SCMs and Portland cement systems, the spin-lattice relaxation is often governed by paramagnetic ions (*e.g.*, Fe^{2+}, Fe^{3+}, Cr^{3+}, Ni^{2+}, Cu^{2+}) present as impurity ions in the compounds or as separate phases in the near vicinity. The presence of these ions results in a very efficient relaxation mechanism caused by dipolar interactions between the nuclear spin and the spin of the unpaired electrons, since the gyromagnetic ratio (γ) of the electrons is about three orders of magnitude larger than the γ-values of the nuclear spins.

For several mineral and cementitious samples, the presence of small amounts of paramagnetic ions, incorporated in the crystal lattices as impurities, results in that ^{29}Si NMR spectra can be obtained in a decent time despite its rather low natural abundance. As an example, the T_1 relaxation time for Si in synthesized belite (β-Ca_2SiO_4) may exceed several hundred seconds, whereas is has been reported to be $T_1 = 3.6$ s in a white Portland cement (0.3 wt% Fe_2O_3) and $T_1 = 0.5$ s in an ordinary Portland cement (2.4 wt% Fe_2O_3) [36]. The faster relaxation rate for belite in the ordinary Portland cement reflects the higher content of paramagnetic impurities and the presence of a larger amount of the ferrite phase. The T_1 values are slightly magnetic-field dependent, and the T_1 values for Si in belite are generally longer than the corresponding values for alite in Portland cements [36, 37]. Thus, quantitative ^{29}Si NMR spectra of hydrated Portland cements can often be acquired with relaxation delays of 10–30 s (90° excitation pulse), depending of the cement type, and good S/N ratios are generally achieved for overnight experiments under these conditions. Other components in blended cements may exhibit longer relaxation times, for example silica fume [47] and synthesized glasses [41] with very small amounts of paramagnetic impurities.

Higher concentrations of paramagnetic impurities or iron (roughly > 5 wt% Fe_2O_3) will result in a significant line broadening and signal loss due to the efficient relaxation, which may prevent the use of solid-state NMR for such materials. This may be a problem for oil-well cement, with large amounts of the ferrite phase, fly ashes and certain clay minerals which can include substantial amounts of Fe^{2+}/Fe^{3+} ions in their layer structure. Moreover, care should be exercised when introducing samples with high concentrations of ferro- or paramagnetic ions into the NMR magnet, as the strong interactions between these ions and the magnetic field may harm the NMR probe during sample spinning or in the worst case result in the sample being stocked in the magnet. For pulverized fuel ashes, containing 8.6 wt% Fe_2O_3, a procedure has been established for partly removal of magnetic particles by sieving and repeatedly isolation of magnetic particles by a hand magnet [48]. This procedure had no significant impact on the content of major oxides but it was found to be effective in reducing the paramagnetic line broadening in the ^{29}Si NMR spectra.

As a half-integer spin quadrupole nucleus ($I = 5/2$), the observed intensities in single-pulse ^{27}Al NMR spectra may also depend on the size of the quadrupole coupling interaction, which can be quite different for different aluminate species and structural sites [23, 24]. The quadrupole interaction is described by the quadrupole coupling constant, C_Q, and the associated asymmetry parameter (η_Q), where C_Q is proportional to the principal element of the electric field gradient tensor (*i.e.*, the distortion of the electronic environment). Single-pulse experiments incrementing the length of the excitation pulse (pulse angle) will show an excitation profile for the central transition ($m = 1/2 \leftrightarrow m = -1/2$) that depends on the ^{27}Al quadrupole coupling parameters, the magnetic field strength and the radio-frequency (*rf*) field strength of the excitation pulse ($\nu_{RF} = \gamma B_1/2\pi$, where B_1 is the *rf* field strength). A pure sinusoidal intensity variation is observed for very small or vanishing quadrupole couplings ($\nu_{RF} \gg C_Q$), corresponding to the response from a liquid sample, where maximum intensity is observed for a 90° pulse (τ_{90}^{liq}). For strong quadrupole interactions, corresponding to the condition $\nu_{RF} \ll C_Q$, maximum intensity is observed for $\tau_{max} = \tau_{90}^{liq}/(I + 1/2)$, *i.e.*, a 30° pulse for a $I = 5/2$ nucleus. However, for short excitation pulses there is a regime where the intensity is almost independent of the size of the quadrupole interaction [49], which corresponds to pulse angles fulfilling the condition $\tau \leq \pi/[2(I + 1/2)]$. Thus, the application of pulse angles below this value (*e.g.*, a 30° pulse for $I = 5/2$) will result in spectra with quantitatively reliable central-transition intensities for sites experiencing different quadrupole interactions, when the requirement of full spin-lattice relaxation is also met. ^{27}Al NMR experiments typically employ *rf* field strengths in the range $\gamma B_1/2\pi = 50$–100 kHz and thus quantitatively reliable spectra are obtained with pulse widths below 1.6–0.8 µs. The presence of the quadrupole interaction for ^{27}Al also contributes to the relaxation of magnetization and thus, ^{27}Al NMR spectra can most often be acquired with very short relaxation delays ($d_1 \sim 0.5$–4 s). This fact in combination with the 100% natural abundance of the ^{27}Al isotope implies that ^{27}Al NMR spectra in general can be obtained in short time (0.5–2 hours) even for Portland cements or SCMs with low Al_2O_3 contents (~1–2 wt% Al_2O_3).

3 ^{27}Al and ^{29}Si NMR studies of SCMs in binary Portland cement blends

3.1 ^{29}Si NMR studies of silica fume blends

^{29}Si MAS NMR spectra of four different anhydrous white Portland cement (wPc) – silica fume blends are shown in Figure 1.2 and illustrate that the observed signal intensities for the alite/belite resonances (Q^0 peaks at −65 to −78 ppm) and silica fume (Q^4(0Al) at −100 to −125 ppm) are proportional to the molar fraction of silicon in these phases. Thus, spectral integration or simulation of these spectra can give the relative molar fractions of alite, belite and silica fume, which can be converted into relative mass units (wt%) if the bulk SiO_2 content is available from other types of chemical analysis (*e.g.* XRF analysis). As a result of its high reactivity and high SiO_2 content, resulting in a fast consumption of portlandite ($Ca(OH)_2$) from the hydrating cement, Portland cements with silica fume contain typically rather low quantities (5–10 wt%) of silica fume. ^{29}Si NMR spectra following the hydration from 1 to 90 days for a wPc with 10 wt% silica fume added are also included in Figure 1.2 and show that silica fume and mainly the alite phase of the wPc are consumed during hydration. The consumption of silicate species from all anhydrous phases result in resonances in the range −78 to −90 ppm, corresponding to an aluminum-substituted calcium-silicate-hydrate phase (C-(A)-S-H). For hydrated Portland cement blends, ^{29}Si NMR generally allows distinction of three different silicate species from the C-(A)-S-H phase; the Q^1 resonance from silicate dimers and chain end groups at roughly − 79 ppm, the Q^2 resonance from silicate chain sites at −83 to −86 ppm, and a partly resolved Q^2(1Al) peak at roughly −81 ppm from chain silicate sites connected to an AlO_4 tetrahedron located as a bridging chain site of the C-(A)-S-H structure. Since the ^{29}Si NMR resonances of the C-(A)-S-H phase do not overlap with the broad peak from silica fume, the degree of silica fume reaction (DoR_{SF}) can be determined from spectral integration or simulations of the spectra before and after hydration, using the relationship:

$$DoR(t) = 1 - \frac{I(t)}{I(t=0)} \qquad (1.1)$$

Here, $I(t = 0)$ and $I(t)$ are the relative intensities for the silica fume peak before and after hydration for the time t, respectively, *i.e.*, $I(t) = I_{SF}(t)/I_{tot}(t)$ where I_{SF} and I_{tot} are the intensity for silica fume and the total intensity of the spectrum. An advantage of determination of DoR by this approach is that it relies solely on relative intensities and thereby does not require an internal or external intensity reference or knowledge on actual sample weight or water content, the principal assumption being that no silicate species are removed from the blend during hydration.

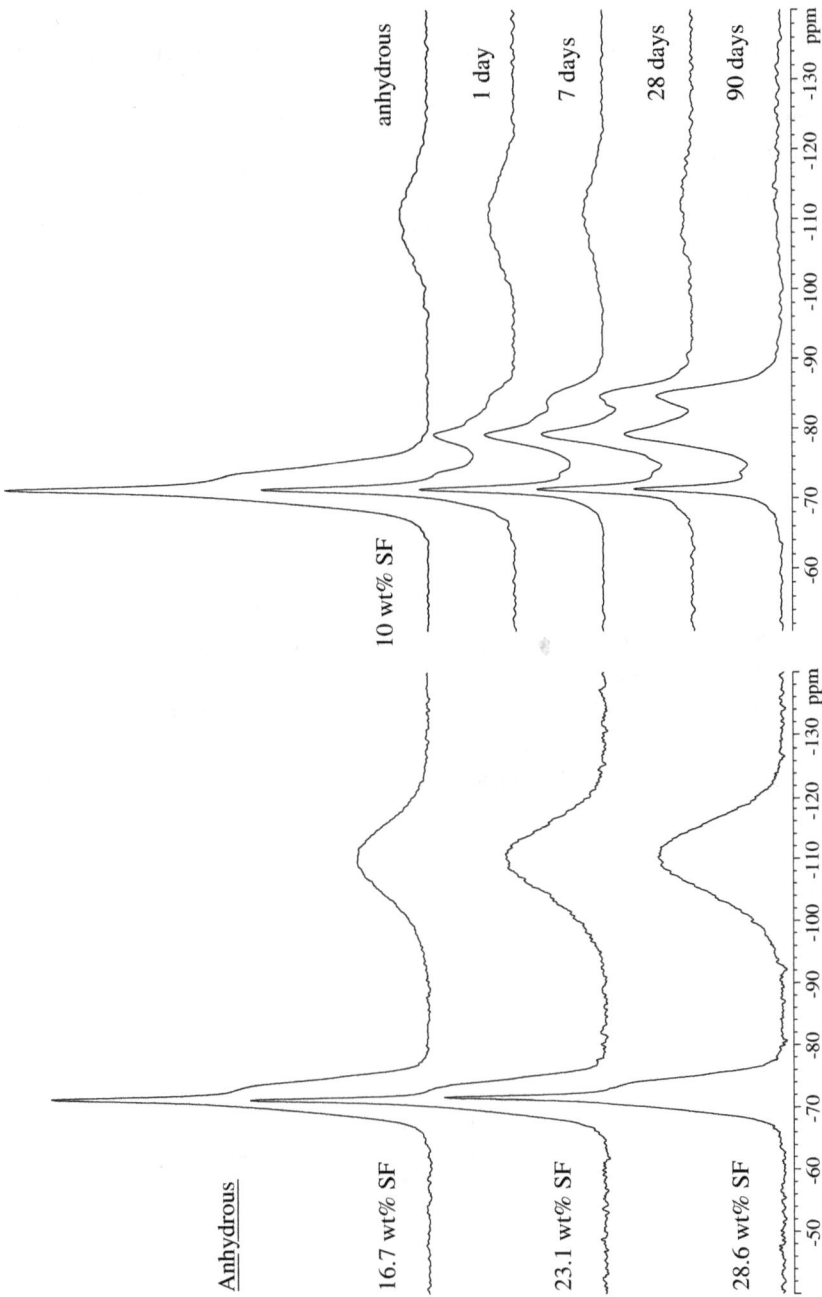

Figure 1.2: ^{29}Si MAS NMR spectra (9.39 T, $v_R = 12.0$ kHz) of anhydrous white Portland cement – silica fume (SF) blends including different quantities of silica fume and of hydrated samples (1–90 days) for the blend including 10 wt% silica fume. Modified after [50].

Further information can be obtained from a full simulation analysis of the spectra for the hydrated wPc – silica fume samples (Figure 1.2), using simulated sub-spectra for alite, belite, silica fume and the Q^1, Q^2(1Al) and Q^2 resonances of the C-(A)-S-H phase. The relative intensities from such simulations provide the *DoR* values for alite, belite and silica fume, using eq. (1.1), and information about the average alumino-silicate chain length (*CL*), the pure-silicate chain length (*CL*$_{Si}$), and the fraction of tetra-hedrally coordinated Al (*Al*$_{IV}$/*Si*) incorporated in the C-(A)-S-H structure [51–53], using the equations:

$$CL = \frac{2\left(Q^1 + Q^2 + \frac{3}{2}Q^2(1Al)\right)}{Q^1} \tag{1.2}$$

$$CL_{Si} = \frac{2(Q^1 + Q^2 + Q^2(1Al))}{Q^1 + Q^2(1Al)} \tag{1.3}$$

$$\frac{Al_{IV}}{Si} = \frac{Q^2(1Al)}{2(Q^1 + Q^2(1Al) + Q^2)} \tag{1.4}$$

where Q^1, Q^2(1Al) and Q^2 denote the relative intensities for these peaks from the C-A-S-H phase. The result of such simulations are shown in Figure 1.3, where the *DoR*$_{SF}$ values reveal almost complete silica fume reaction after 90 days of hydration for the wPc blend incorporating 10 wt% silica fume. Moreover, comparison of the average alumino-silicate chain lengths for the silica fume blend with the corresponding values for the same wPc without silica fume added shows consistently larger values of *CL* for

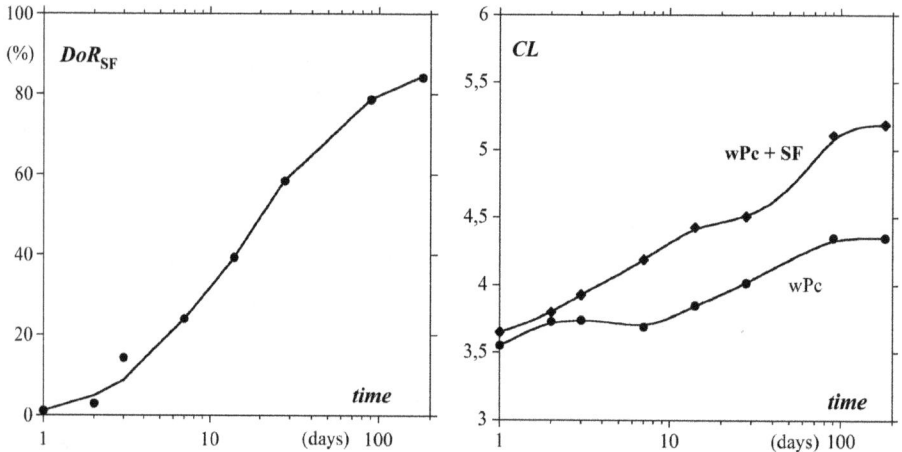

Figure 1.3: Left: Degree of silica fume reaction (*DoR*$_{SF}$) for a white Portland cement (wPc) – silica fume (SF) blend (90:10 w/w) as function of hydration time. Right: Comparison of the average aluminosilicate chain length (*CL*) of the C-(A)-S-H phase for the wPc + SF blend and a pure wPc as function of hydration time. All data are derived from simulations of ^{29}Si NMR spectra.

the silica fume blend. An increase in *CL* indicates a decalcification of the C-(A)-S-H phase, following the relationships between *CL* and the Ca/Si ratio observed by ^{29}Si NMR for synthesized C-S-H samples [54–56]. Thereby the results in Figure 1.3 suggest that silica fume not only consumes the available portlandite from the hydrating cement but its reaction also results in a C-(A)-S-H phase with a lower Ca/Si ratio than Ca/Si = 1.75, typically found for the C-S-H phase in mature Portland cement pastes [57]. This observation agrees well with a recent ^{1}H NMR relaxometry study [58] of a similar wPc – silica fume system, where mass balance calculations indicated the ratio Ca/(Si + Al) = 1.33 for the C-(A)-S-H phase after hydration for 28 days.

3.2 NMR studies of blends with fly ash

Solid-state NMR studies of cement blends with fly ashes are often complicated by the fact that fly ashes contain an appreciable amount of iron (bulk Fe_2O_3 > 5 wt%), which leads to line-broadening of the resonances and intensity loss, as a result of the strong dipolar interaction between the unpaired electron spins from Fe^{2+}/Fe^{3+} and the ^{27}Al and ^{29}Si spins of the principal phases in the fly ash. This is illustrated in Figure 1.4 by ^{29}Si NMR spectra of three different fly ashes with bulk Fe_2O_3 contents of 0.4 wt% (FA-1), 5.7 wt% (FA-2) and 8.3 wt% (FA-3). The low iron content in FA-1 results in a small degree of paramagnetic line-broadening, as seen by the line width of the centerband peaks and the low intensities of the first- and second-order spinning sidebands. The centerband covers a spectral range from −85 ppm to −125 ppm (Figure 1.1), and distinct peak regions for the Si sites in the mullite component and the Q^4(0Al), Q^4(1Al) and Q^4 (2Al) sites of the glassy phase can be identified (*c.f.* section 2.2). However, the centerbands for FA2- and FA-3 are much broader, covering a range from approx. −75 ppm to −130 ppm, and without any distinct features for specific SiO_4 environments. Moreover, a significant part of the intensity is present in the first- to fourth-order spinning sidebands observed on each side of the centerband resonances. A determination of the total intensities for the centerband and spinning sidebands in the ^{29}Si NMR spectra of the three fly ashes gives the values 1.0: 0.80: 0.80 for FA-1: FA-2: FA-3 when the intensities are normalized to the bulk SiO_2 content in the individual samples and the total intensity for FA-1. These data indicates that 20% of the expected ^{29}Si NMR intensity is not observed, as a result of a very efficient relaxation for ^{29}Si spins with un-paired electrons in their nearest vicinity. This intensity loss combined with the broadened centerband lineshapes and the significant fraction of signal intensity in the spinning sidebands complicate the analysis of fly ashes with typical iron contents. Moreover, it questions whether reliable quantitative data on fly ash reaction can be obtained from solid-state NMR experiments of fly ash – cement blends or alkali-activated systems where fly ash is a principal component.

    ^{29}Si NMR spectra similar to those for the anhydrous fly ashes, FA-2 and FA-3 (Figure 1.4), have been reported in the literature [59–64] for coal fly ashes containing

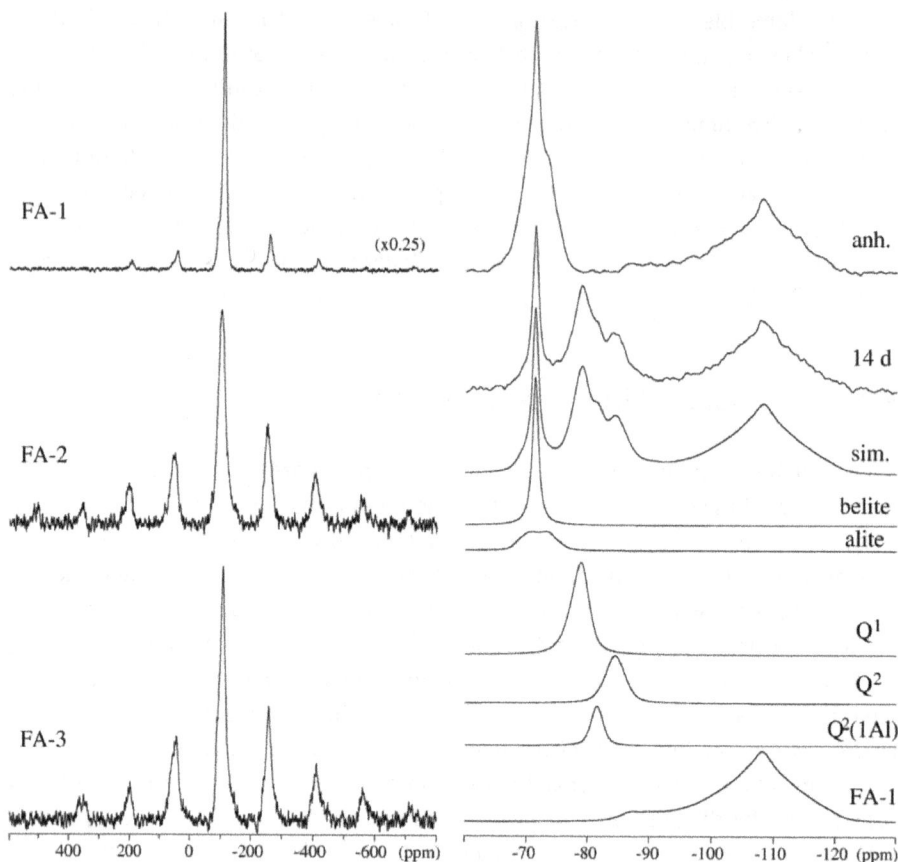

Figure 1.4: Left: ^{29}Si MAS NMR spectra (9.39 T, $v_R = 12.0$ kHz) of three anhydrous fly ash samples containing 0.43, 5.72 and 8.3 wt% Fe_2O_3 for FA-1, FA-2 and FA-2, respectively. Right: ^{29}Si MAS NMR spectra (9.39 T, $v_R = 12.0$ kHz) for a wPc – FA-1 blend (70:30 w/w) before hydration (anh.) and after 14 days of hydration. The spectra below show the optimized simulation of the experimental spectrum after hydration for 14 days along with sub-spectra of the individual components, constituting the overall simulation. Modified after [50].

4.2–8.1 wt% Fe_2O_3. The centerband in these spectra covers typically a range from −80 to −125 ppm with some partly resolved spectral features that reflect the presence of mullite, a glassy aluminosilicate phase and in some cases minor contents of quartz. In several studies these broad centerband peaks are simulated using a number (5–9) Gaussian-shaped peaks to account for the overall line shape. However, care should be exercised in the assignment of these peaks to distinct $Q^n(mAl)$ units, since the contribution to the overall line shape from the paramagnetic broadening cannot be unraveled from the spectra. Most previous studies relate to alkali-activated materials, and for some alkali-activated systems [59–61] a number of distinct peaks emerge in the spectra after alkali-activation, reflecting the formation of a zeolitic sodium-alumino-silicate-hydrate (N-A-S-H) phase. The improved spectral resolution after alkali-activation may reflect a higher degree of ordering of the SiO_4 environments in the N-A-S-H phase or more likely a lower degree of Fe-incorporation in this reaction product.

For the fly ash with low Fe_2O_3 content (FA-1), only the low-intensity part of the centerband line shape overlaps partly with the resonances from the C-(A)-S-H hydration product. Thus, the degree of fly ash reaction can be determined for FA-1 in cement blends by deconvolution of the ^{29}Si NMR spectra before and after hydration, as illustrated in Figure 1.4 for a blend of wPc (70 wt%) and FA-1 (30 wt%) hydrated for 14 days. A series of samples following the hydration of this blend for one to 180 days have been examined by ^{29}Si NMR [50] and a visual inspection of these spectra indicates that the intensity distribution in the resonances from fly ash does not change as the hydration proceeds. Thus, it is assumed that the fly ash dissolves congruently, implying that its intensity contribution to the spectra of the hydrated blends can be simulated with a sub-spectrum that matches the intensity distribution in the spectrum of the anhydrous fly-ash. Such a sub-spectrum is constructed by simulation of the fly-ash spectrum using a number of Gaussian/Lorentzian line shapes with similar linewidths (in the present case 9 peaks with linewidths of *FWHM* = 3.8–7.3 ppm). Least-squares fitting of this sub-spectrum along with the sub-spectrum for alite and Gaussian/Lorentzian peaks for belite and the Q^1, $Q^2(1Al)$ and Q^2 peaks of the C-(A)-S-H phase allows determination of the degree of reaction (*DoR*) for alite, belite and fly ash from the simulated relative intensities and eq. (1.1). For the blend hydrated for 14 days, this gives DoR_{Alite} = 85.4%, DoR_{Belite} = 12.3 wt%, and DoR_{FA} = 0.5%, showing that the fly ash is non-reactive during the early hydration. However, DoR_{FA} values of 18–20% are observed after 90 and 180 days of hydration [50], indicating a low degree of pozzolanic reaction for FA-1 in this blend type. Similar low degrees of fly ash reaction have been reported for binary Portland cement – fly ash blends from selective dissolution experiments and back scattered electron microscopy image analysis investigations [18].

Information about fly ash reaction can principally also be obtained from ^{27}Al NMR. However, fly ashes include aluminum in both tetrahedral and octahedral coordination (Figure 1.1), which for Portland cement blends complicates the analysis since the principal aluminum containing hydration products (the AFt and AFm phases) all contain Al in octahedral coordination. These phases exhibit ^{27}Al chemical shifts in a narrow range (5–14 ppm [24]) which will overlap strongly with the octahedral resonance from anhydrous fly ash. Moreover, the spectral resolution in the ^{27}Al NMR spectra of fly ashes may also suffer from severe line-broadening from couplings with unpaired electrons and strong spinning sideband intensities as seen in previous studies [62, 63, 65, 66]. In some cases, this may even prevent a clear separation of the tetrahedral and octahedral centerband resonances from their first-order spinning sidebands [62, 63, 66], even at high magnetic fields [65]. In alkali-activation of fly ashes with 8 M NaOH solution, the formation of the N-A-S-H phase has been clearly observed [61], as this phase solely contains Al in tetrahedral coordination in a rather well defined environment. This results in a narrow tetrahedral resonance in the ^{27}Al NMR spectra, clearly distinct from the unreacted fly ash resonances.

3.3 ^{27}Al and ^{29}Si NMR studies of slag blends

The principal component in ground granulated blast furnace slags is most often a magnesium-calcium aluminosilicate glass phase, where calcium and magnesium charge-balance the tetrahedral Al sites and contribute to the depolymerization of the aluminosilicate network. This principal phase includes a number of different $Q^n(mAl)$ environments, resulting in broad and rather featureless resonances in ^{29}Si NMR spectra that may cover a range from approx. −70 ppm to −120 ppm, where the width and peak maxima mainly depend on the Al/Si ratio and the degree of depolymerization of the aluminosilicate framework. Thus, the slag peak overlaps strongly with resonances from the anhydrous cement, as indicated in Figure 1.1, and from the curing and hydration products, as seen in several studies of alkali-activated slags systems [67–71] and hydrated Portland cement – slags blends [72–76]. This overlap may be so severe that a reliable estimation of the degree of slag reaction hardly can be obtained from simulations of the ^{29}Si NMR spectra. To address this problem, ^{29}Si NMR has earlier been combined with selective dissolution approaches, where the calcium silicate phases were dissolved by treatment in a solution of ethylene diamine tetra acetic acid ($Na_2H_2[EDTA]$) with triethanolamine ($(CH_3CH_2)_3N$), leaving a solid residue assigned to unreacted slag [72]. The ^{29}Si NMR spectrum of the residue was then used to deconvolute the spectrum of the hydrated cement – slag blends. However, the line shape deviated somewhat from the corresponding spectrum of the anhydrous slag, which was ascribed to incongruent dissolution of the slag. A similar approach, using a sub-spectrum of a dissolution residue, has been employed in a study of alkali-activated slags [69], where it was found that the most convincing result from deconvolution of the ^{29}Si NMR spectra was achieved using the spectrum of the anhydrous slag as the sub-spectrum. This result suggests the absence of preferential dissolution of specific Si sites on the surface of the slag grains during curing in an alkaline environment, as least at the scale detectable by ^{29}Si NMR. This is supported by more recent studies of the dissolution of synthesized glasses in alkaline conditions (pH = 13), which show a congruent dissolution of aluminate and silicate species in blast furnace slag-like glasses [77, 78].

An improved distinction of the slag in cement blends can be achieved by ^{27}Al NMR, since the centerband resonance from the slag is often dominated by an asymmetric but rather well defined resonance from Al in tetrahedral coordination, as illustrated in Figure 1.1. This distinction may be improved at high magnetic fields, utilizing the increase in chemical shift dispersion and reduction in second-order quadrupolar broadening. Resonances in the same region will be present from the tricalcium aluminate phase in Portland cements [45]. However, this phase has

typically fully reacted after a few days of hydration and thus, the centerband resonance from the slag will only partly overlap with resonances from the four- and
five-fold coordinated Al sites present in the C-(A)-S-H phase. A determination of the
slag reaction can be achieved by subtraction of the spectrum of the anhydrous slag
from the spectrum of the hydrated Portland cement – slag blend in such a manner that
the contribution from the anhydrous slag will be removed from the spectrum of the
hydrated cement blend. This is illustrated in Figure 1.5, where the subtraction provides
a measure for the intensity in the slag resonance in the hydrated cement ($I^S_{PCS}(t)$).
Using this intensity along with the corresponding intensity for the spectrum of the anhydrous slag ($I^S(0)$), measured on the same absolute intensity scale, allows calculation
of the degree of slag reaction ($DoR_S(t)$) from the expression:

$$DoR_S(t) = \left(1 - \frac{I^S_{PCS}(t)m_S(1-LOI(0))}{I^S(0)m_{PCS}(1-LOI(t))F_{PCS}}\right) \tag{1.5}$$

Here, m_S and m_{PCS} are the analyzed sample weights of slag and hydrated Portland cement – slag blends in the ^{27}Al NMR experiments, respectively, and $LOI(0)$ and $LOI(t)$
are the loss of ignitions of the cement blend before and after hydration, determined by
thermogravimetric analysis or loss of ignition experiments at 1000 °C. The factor, F_{PCS},
is the fraction of slag in the anhydrous cement blend. Ideally, the mix ratio can be

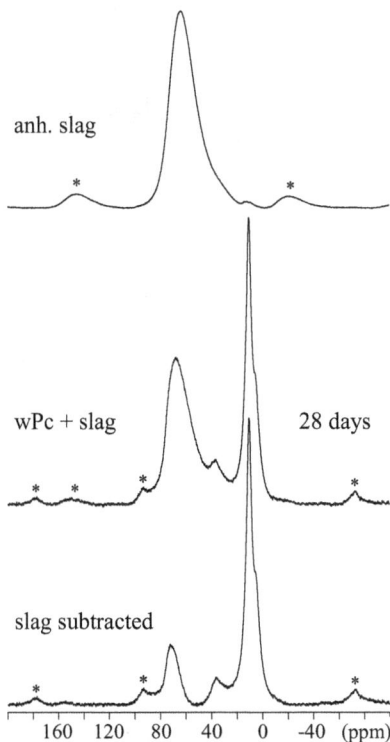

Figure 1.5: ^{27}Al MAS NMR spectra (14.09 T,
$v_R = 13.0$ kHz) of anhydrous slag and a white Portland
cement – slag blend (60:40 w/w) hydrated for
28 days. The lower spectrum shows a scaled
difference plot, where the anhydrous slag component
has been subtracted using the spectrum of the
anhydrous slag. The asterisks indicate spinning
sidebands.

used for F_{PCS}, however, a better value may be achieved by the analysis of an ^{27}Al NMR spectrum of the anhydrous cement using eq. (1.5) by adjustment of F_{PCS} to $DoR_S(0) = 0$. The ^{27}Al NMR spectra of the hydrated cement in Figure 1.5 correspond to a white Portland cement – slag blend (60:40 wt%) hydrated for 28 days. Analysis of the sample, using the procedure described above gives $DoR_S(28\ d) = 56.9\%$, which is in very good agreement the degree of slag reaction determined by backscattered electron image analysis (BSE/IA) combined with SEM – EDS mapping (55.4%) for the same cement blend [19]. Moreover, for other Portland cement – slag systems, this type of ^{27}Al NMR analysis has provided $DoR_S(t)$ values that agree well results from isothermal calorimetry and chemical shrinkage measurements for the same samples [20]. The ^{27}Al NMR spectrum, where the slag contribution has been subtracted, provides details about the fractions of four-, five- and six-fold coordinated Al in the C-(A)-S-H phase by the resonances at 72, 35, and 5 ppm, respectively. The octahedral resonance at 5 ppm overlaps with a strong peak at 9 ppm, which originates from non-resolved resonances from the AFm phases, in particular monosulphate and hydroxy-hydrotalcite ($Mg_4Al_2(OH)_{14} \cdot 3H_2O$), the latter incorporating magnesium released from the slag reaction.

3.4 ^{27}Al and ^{29}Si NMR studies of glasses

Calcium aluminosilicate (CAS) glasses have been studied widely by the cement community, since their compositions and chemical properties resemble those of the principal components in fly ashes and slags [3]. Several of these investigations have focused on synthesized CAS glasses or partially substituted CAS glasses and included solid-state ^{27}Al and ^{29}Si NMR studies, contributing with information about their basic structure [79–87] and reactivity in cement blends [41, 42, 88, 89]. Generally, ^{29}Si NMR spectra of CAS glasses show a broad, nearly Gaussian-like peak, which reflects the disordered nature of the glass structure and the presence of a range of different $Q^n(mAl)$ environments. This is illustrated for three CAS glasses in Figure 1.6, which are synthesized with the same molar Al/Si ratio (Al/Si = 0.36) but different contents of CaO (37.9, 22.5 and 10.5 mol% for CAS1 – CAS3, respectively [42]). This corresponds to typical compositions for the glass phase in slags (CAS1) and fly ashes (CAS2 and CAS3). The shift in the center of gravity (δ^{cg}) for the ^{29}Si resonance towards lower frequency with decreasing CaO content reflects an increasing polymerization of the glass structure (δ^{cg} = −84.4, −93.7, and −100.3 ppm for CAS1 – CAS3, respectively). A partly replacement of Ca by Mg for these glasses does not significantly affect the network structure and SiO_4 polymerization, which has been documented by a nearly linear correlation between δ^{cg} and the MgO + CaO content for magnesium-substituted CAS glasses [42].

The corresponding ^{27}Al NMR spectra (Figure 1.6) are dominated by a resonance with center of gravity at approx. 55 ppm originating from AlO_4 sites of the type Al(−O−Si)$_4$, in agreement with earlier studies [79, 81–85]. However, indications of Al in five- and six

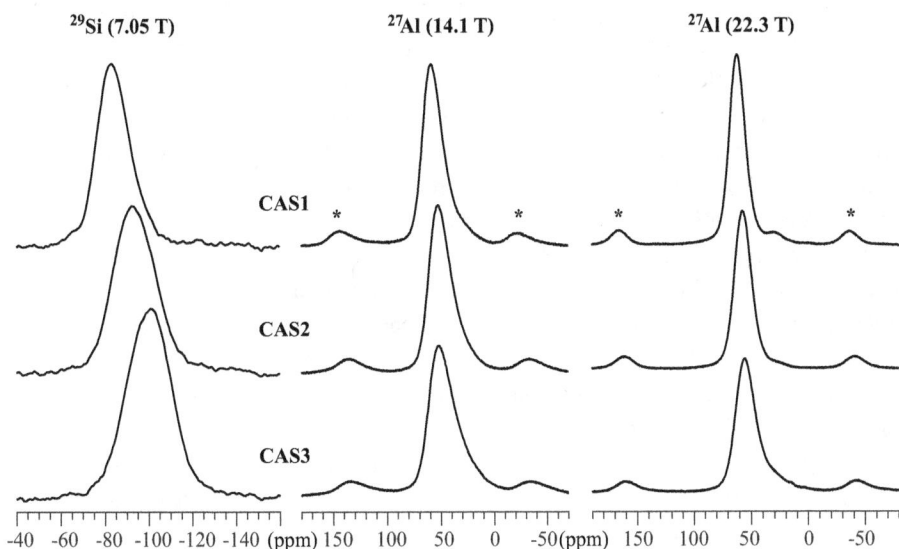

Figure 1.6: ^{29}Si and ^{27}Al MAS NMR spectra of three CAS glasses synthesized with the molar ratio Al/Si = 0.36 and different contents of CaO (37.9, 22.5, and 10.5 mol% CaO for CAS1 – CAS3, respectively). The ^{29}Si NMR spectra (7.05 T) employed a spinning speed of v_R = 7.0 kHz and a 30-s relaxation delay. The ^{27}Al NMR spectra were obtained at 14.1 T and 22.3 T using 2-s relaxation delays and spinning speeds of 13.0 kHz and 25.0 kHz, respectively. Spinning sidebands are indicated in the first row spectra by asterisks. Modified after [42].

fold coordination are also seen by minor peaks/shoulders in the range 40 to 0 ppm. This is most clearly apparent in the ^{27}Al NMR spectra acquired at the very high magnetic field (22.3 T), in particular for the CAS1 glass, where a centerband resonance at approx. 35 ppm is clearly observed for five-fold coordinated Al. The ^{27}Al NMR spectra at 14.1 T and 22.3 T also illustrate the reduction in linewidth (*FWHM*) obtained at high magnetic field, reflecting the inverse proportionality of the second-order quadrupolar interaction with the magnetic field strength. For the CAS1 glass, *FWHM* = 22.4 ppm and *FWHM* = 16.1 ppm is obtained at 14.1 and 22.3 T for the AlO$_4$ centerband, which shows a change in linewidth which is slightly higher than the ideal factor of 0.63 on going from 14.1 to 23.3 T in the case of pure second-order quadrupolar broadening. The higher ratio reflects that the lineshape is also affected by a distribution in ^{27}Al isotropic chemical shifts and quadrupole coupling tensors elements as a result of the amorphous nature of the glass. Lineshapes of this type can be simulated using the Czjzek model [90], which considers a distribution in electric field gradient (EFG) tensors (or quadrupole tensor elements) caused by structural disorder within the nearest coordination spheres of the nuclear site. This approach has been successfully applied in the analysis of ^{27}Al NMR spectra for a number of different materials. In the present case it also gives a reliable distinction between the contributions to the centerband lineshape from Al in four-, five- and six-fold coordination. For example, simulation by this approach of the ^{27}Al NMR spectrum for the CAS1 glass at 22.3 T gives the molar fractions of 91.2% Al_{IV}, 8.3% Al_V, and 0.5% Al_{VI}, whereas the corresponding numbers for CAS3 are 81.0% Al_{IV}, 17.0% Al_V, and 2.0% Al_{VI} [42]. These values indicate that the fraction of five-fold Al increases with increasing network polymerization, in agreement with recent studies of similar CAS glasses [41]. Different structural models for the glass network structure have been examined by mutual simulations of the ^{29}Si resonances for CAS glasses with different compositions [41, 80, 87, 91, 92] using overlapping resonances for the different types of $Q^n(mAl)$ sites, restrictions on linewidths and chemical shifts for the individual components and the distribution of Al_{IV}, Al_V, and Al_{VI} from ^{27}Al NMR spectra. For example, by this approach it has been shown the network AlO_4 units are not randomly distributed $Al(-O-Si)_4$ sites in the glass network but form a quasi-heterogeneous intermediate-range order which is highly polymerized with alternating SiO_4 and $(\frac{1}{2}CaO)AlO_4$ units [85].

The degree of glass reaction in cement blends can be determined using either ^{29}Si or ^{27}Al NMR and the approach described above for the analysis of Portland cement slags blends. For example, this has utilized to investigate sodium-substituted CAS glasses in blends with Portland cement (30:70 wt%), which revealed degrees of glass reaction above 50% after 90 days hydration [82]. The glasses were prepared by melting natural minerals (clays, quartz, and limestone) and thus such glasses could represent a valuable alternative to slags and fly ashes as SCMs in blended cements, when these two common SCMs will be sourced out for environmentally reasons. Glass reaction under alkaline environments has also been examined by ^{27}Al and ^{29}Si NMR for glass samples exposed to Chapelle or Frattini-like tests [41, 42, 93]. This procedure is illustrated in Figure 1.7 for a CAS glass (18.4 mol% CaO, 18.4 mol% Al_2O_3 and 63.2 mol% SiO_2), where 0.40 g of the 20–40 µm fraction of the ground glass sample and 1.20 g $Ca(OH)_2$ were mixed with 50.0 mL of de-ionized water [42]. The lime-saturated mixture was cured in a sealed container for 7 days at 40 °C with continuous stirring, and the suspension was then filtered and dried over silica gel at room temperature. Figure 1.7 shows the ^{27}Al and ^{29}Si NMR spectra of the anhydrous glass as well as of the solid residue from the pozzolanic test, where the latter show resonances from unreacted glass as well as the hydration products formed by reaction of the glass with $Ca(OH)_2$. The Al_{IV} resonance of the anhydrous glass is clearly identified in the ^{27}Al NMR spectrum of the solid residue and thus the subtraction of the anhydrous component can be performed with good precision. This subtraction allows calculation of the degree of glass reaction, when the total and subtracted intensities for weighed samples are evaluated. Moreover, the calculation include corrections for the amounts of bound water, $Ca(OH)_2$ and $CaCO_3$ present in the solid phase, as determined from TGA experiments on the solid residue. From the ^{27}Al NMR spectra in Figure 1.7, a degree of glass reaction of $DoR_{Al} = 64.5\%$ is obtained for the CAS glass. The ^{29}Si NMR spectra can be analyzed in a similar manner, providing an independent measure on the degree of slag reaction. However, a severe overlap of ^{29}Si resonances is observed between the anhydrous glass peak and the resonances from the C-(A)-S-H phase, which for some glass

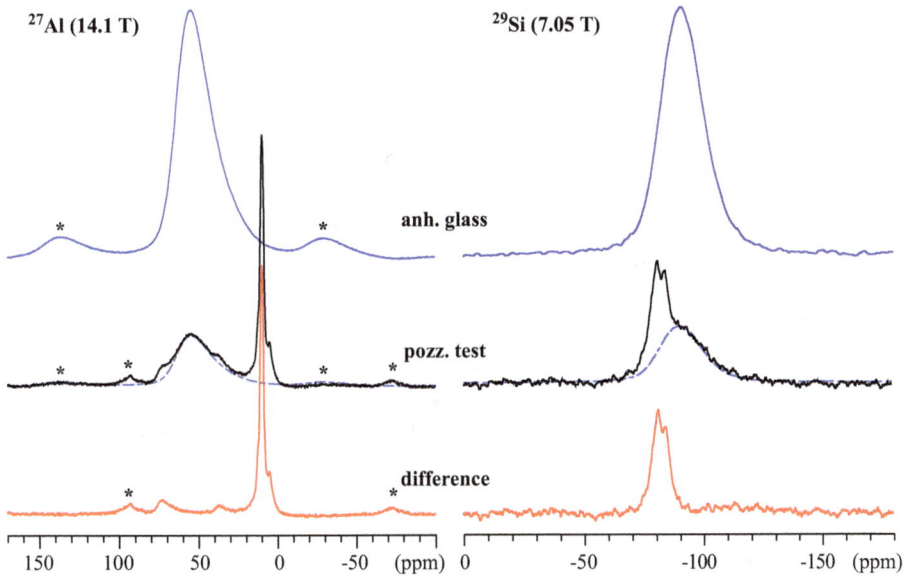

Figure 1.7: ^{27}Al (14.09 T) and ^{29}Si (7.05 T) MAS NMR spectra of a synthesized CAS glass (18.4 mol% CaO, 18.4 mol% Al_2O_3, and 63.2 mol% SiO_2) before and after pozzolanic testing (see text). The degree of glass reaction is obtained by subtraction of scaled versions of the spectra of the anhydrous glass (dashed blue lines in middle part) from the spectra after the pozzolanic test. This gives the difference spectra shown in the lower row. Asterisks indicate spinning sidebands. Modified after [42].

compositions, in particular depolymerized glass structures, may prevent a reliable construction of the difference spectrum. In the present case (Figure 1.7), the intensities for the anhydrous glass and the subtracted spectrum result in a degree of glass reaction of $DoR_{Si} = 67.8\%$, which is similar to the value obtained from ^{27}Al NMR. However, the ^{27}Al NMR spectra show the most clear reflection of the glass peak and thus, the ^{27}Al NMR spectra will provide the most precise value for the glass reaction, as also found in two recent studies of the pozzolanic reactivity of CAS glasses [41, 42]. Alternatively, the ^{29}Si NMR spectra can be analyzed using the DoR value from ^{27}Al NMR to construct the difference spectrum. This spectrum includes only resonances from the C-(A)-S-H phase and can be used to determine the characteristic CL, CL_{Si} and Al_{IV}/Si parameters (eqs. (1.2)–(1.4)) for this phase. For example, these parameters have provided useful information in a study of the pozzolanic reactivity of the clay mineral montmorillonite exposed to heat treatment at different temperatures [94]. Here, a clear increase in the fraction of Al incorporated in the C-(A)-S-H phase was observed for the temperatures with optimum pozzolanic reactivity, suggesting that high reactivity is associated with an intermediate structure of calcined montmorillonite from which Al^{3+} species are easily dissolved.

3.5 Carbonated recycled concrete as an SCM

Demolished end-of-life concrete represents a large source of waste material, which cannot be used one-on-one as replacement of natural aggregate (sand, gravel and stone) in concrete, mainly because of its inferior material properties such as water absorption which affects the fresh concrete workability and leads to shrinkage. Thus, recycled concrete is mainly down-cycled as road base material or only partially used as aggregate in low-grade concrete [95]. Recent attention to the large CO_2 emissions associated with cement production and new carbon capture utilization (CCU) technologies has initiated research in carbonation of fines derived from end-of-life concrete for CO_2 sequestration [29, 30]. A significant uptake of CO_2 can be achieved by full carbonation of the Ca-bearing phases in hydrated cement, producing a material that can be used for stable landfills or as a pozzolan in new cement formulations. Thereby, these technologies can contribute to a circular utilization of concrete and provide approaches for chemical bonding of CO_2 on a geological time scale.

A promising technique is wet carbonation of the concrete fines under alkaline conditions, which has been demonstrated for hydrated ordinary Portland cement (CEM-I) [96] and blended cements incorporating slag (CEM-III) [97]. Under these conditions at ambient temperature and pressure and in a gas stream containing 10% CO_2, portlandite, the C-(A)-S-H phase, ettringite, monosulfate (AFm) as well as remains of the anhydrous alite and belite clinker phases will carbonate and form $CaCO_3$ and an amorphous alumina-silica gel as the principal phases [96, 97]. The wet carbonation processes of the hydrated cement paste can easily be followed by ^{29}Si and ^{27}Al NMR as illustrated in Figure 1.8. The ^{29}Si NMR spectrum before carbonation is dominated by the Q^1 and Q^2 resonances from the C-(A)-S-H phase (−79 ppm to −85 ppm) with minor peaks from reminiscences of alite and belite, as seen by the intensity in the −68 to −76 ppm region. Upon wet carbonation the intensities of these resonances gradually decrease and produces a broad resonance in the range −80 ppm to −120 ppm, representing an amorphous alumina-silica gel, which is composed of a range of different $Q^4(mAl)$ and $Q^3(nAl)$ silicate species. The spectra following the wet carbonation from 2.9 to 43 min. show that the intensity of the Q^1 peak decreases prior to the consumption of the Q^2 silicate species, which reflects a decalcification of the C-(A)-S-H phase. This is in agreement with earlier studies of the carbonation of synthesized C-S-H phases [56], which show that this phase decalcifies until a Ca/Si ratio about 0.67 is reached after which it decomposes into $CaCO_3$ and silica gel. The ^{27}Al NMR spectrum before wet carbonation includes resonances from four- and six-fold coordinated Al from the C-(A)-S-H phase (70 and 5 ppm) as well as from ettringite (13.4 ppm) and the AFm phases (10.6 ppm). During the initial wet carbonation, a re-wetting of the dried sample takes place which can explain the more clear reflection of the ettringite peak after 2.9 min of carbonation. After longer carbonation times, a new resonance at 60.9 ppm is observed which originates from Al-(O-Si)$_4$ sites in the amorphous alumina-silica phase. The resonance is formed at the expense of aluminum in the C-(A)-S-H

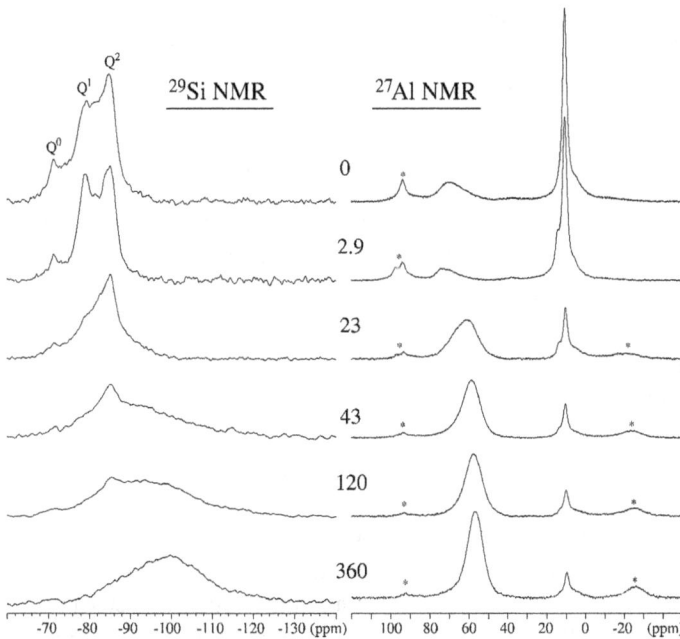

Figure 1.8: ^{29}Si and ^{27}Al MAS NMR spectra following the wet carbonation of a hydrated CEM-I paste with time (the carbonation time given in min.). The ^{29}Si NMR spectra were acquired at 7.05 T using a spinning speed of 7.0 kHz and a 15-s relaxation delay whereas the ^{27}Al NMR spectra were obtained at 14.09 T, using a spinning speed of 13.0 kHz and a 2-s relaxation delay. Asterisks indicate spinning sidebands. The CEM-I was hydrated for 3 months at 40 °C using a water/cement ratio of 0.4, where after it is was dried at 105 °C and ground to a fine powder. Modified after [93].

phase as well as in ettringite and monosulphate (AFm), and after 360 min. of carbonation only a small amount of the AFm phase remains. The linewidth of the Al(4) peak decreases and shifts slightly to lower frequency (56.8 ppm) with carbonation time, which may reflect an ordering of the AlO$_4$ environments in the alumina-silica gel, probably as a consequence of a continued polymerization of the silicate network in this phase. The clear reflection of the amorphous alumina-silica gel in the ^{27}Al NMR spectra is expected to provide the basis for quantification of the pozzolanic reactivity of this phase in new composite cements where carbonated concrete fines are used as SCMs, employing the same approach as describe above for slags and glasses.

4 Conclusions and outlook

A principal advantage of solid-state NMR spectroscopy in studies of waste materials and their use in composite cements is its ability to detect and quantify amorphous and crystalline phases in an equal manner. However, limitations of the technique applies to studies of materials with high contents of paramagnetic elements (*e.g.* Fe, Cr, Co, and Mn species) such as commonly found in fly ashes. The detected NMR parameters (*e.g.* chemical shifts and quadrupole coupling parameters) are most sensitive to local structural features as defined by the first to third coordination spheres of the NMR nucleus. This makes solid-state NMR and important supplement to X-ray diffraction techniques that probes long-range order and crystalline phases.

Research in applications of waste materials as supplementary cementitious materials (SCMs) in composite cements is increasing, reflecting the global challenges that the cement industry faces for development of more sustainable concrete. The SCMs with highest potential are less-crystalline or amorphous phases and NMR approaches will definitely play an important role in the exploration of the structure and pozzolanic reactivity of these new SCMs.

The NMR methods described in this chapter have focused on ^{27}Al and ^{29}Si NMR and the extraction of structural and quantitative information from these spectra for common waste materials and Portland cement – SCM blends. However, the NMR periodic table provides a wealth of other spin nuclei which can be used in the exploration of local structure and pozzolanic reactivity for a range of other industrial waste materials. For example ^{11}B, ^{19}F, ^{23}Na ^{25}Mg and ^{31}P have been utilized in other areas of cement chemistry and all these elements can studied by NMR without isotropic enrichment. Finally, it should be emphasized that the structural and compositional complexity of most waste materials implies that there is no unique analytical tool that can provide all desired details on structure, composition and reactivity that in needed to explore their chemical and physical properties. Thus, the best description will always be obtained by using a multitude of techniques that complement each other by the pieces of information they provide. In this context, solid-state NMR should be considered as one of the important methods in the tool box of microstructural characterization techniques.

References

[1]   Lothenbach B, Scrivener K, Hooton RD. Supplementary cementitious materials. Cem Concr Res. 2011, 41, 1244–1256.
[2]   Snellings R, Mertens G, Elsen J. Supplementary Cementitious Materials Rev Mineral Geochem. 2012, 74, 211–278.
[3]   Skibsted J, Snellings R, Reactivity of supplementary cementitious materials (SCMs) in cement blends. Cem Concr Res. 2019, 124 105799.
[4]   Schneider M. The cement industry on the way to a low-carbon future. Cem Concr Res. 2019, 124, 105792.
[5]   Gartner E. Industrially interesting approaches for "low-CO$_2$" cements. Cem Concr Res. 2004, 34, 1489–1498.

[6]   Damtoft JS, Lukasik J, Herfort D, Sorrentino D, Gartner EM. Sustainable development and climate change initiatives. Cem Concr Res. 2008, 38, 115–127.

[7]   Schneider M, Romer M, Tschudin M, Bolio H. Sustainable cement production – present and future. Cem Concr Res. 2011, 41, 642–650.

[8]   Siddique R, Khan MI. Supplementary cementing materials. Springer Science and Business Media, Berlin, 2011.

[9]   Juenger MCG, Siddique R. Recent advances in understanding the role of supplementary cementitious materials in concrete. Cem Concr Res. 2015, 78, 71–80.

[10] Juenger MCG, Snellings R, Bernal SA, Supplementary cementitious materials: new sources, characterization, and performance insights. Cem Concr Res. 2019, 122, 257–273.

[11] Westphal T, Füllmann T, Pöllmann H. Rietveld quantification of amorphous portions with an internal standard – mathematical consequences of the experimental approach. Powder Diffr. 2009, 24, 239–243.

[12] Snellings R, Salze A, Scrivener KL. Use of X-ray diffraction to quantify amorphous supplementary cementitious materials in anhydrous and hydrated blended cements. Cem Concr Res. 2014, 64, 89–98.

[13] Stetsko YP, Shanahan N, Deford H, Zayed A. Quantification of supplementary cementitious content in blended Portland cement using an iterative Rietveld – PONKCS technique. J Appl Cryst. 2017, 50, 498–507.

[14] Durdzinski PT, Dunant CF, Ben Haha M, Scrivener KL. A new quantification method based on SEM-EDS to assess the fly ash composition and study the reaction of its individual components in hydrating cement paste. Cem Concr Res. 2015, 73, 111–122.

[15] Pfingsten J, Rickert J, Lipus K. Estimation of the content of ground granulated blast furnace slag and different pozzolanas in hardened cement. Constr Build Mater. 2018, 165, 931–938.

[16] Ohsawa S, Asaga K, Goto S, Daimon M. Quantitative determination of fly ash in the hydrated fly ash – $CaSO_4 \cdot 2H_2O$ – $Ca(OH)_2$ system. Cem Concr Res. 1985, 15, 357–366.

[17] Luke K, Glasser FP. Selective dissolution of hydrated blast furnace slag cements. Cem Concr Res. 1987, 17, 273–282.

[18] Ben Haha M, De Weerdt K, Lothenbach B. Quantification of the degree of reaction of fly ash. Cem Concr Res. 2010, 40, 1620–1629.

[19] Kocaba V, Gallucci E, Scrivener KL, Methods for determination of degree of reaction of slag in blended cement pastes. Cem Concr Res. 2012, 42, 511–525.

[20] Scrivener KL, Lothenbach B, De Belie N, Gruyaert E, Skibsted J, Snelllings R, Vollpracht A. TC 238-SCM: hydration and microstructure of concrete with SCMs. State of the art on methods to determine the degree of reaction of SCMs. Mater Struct. 2015, 48, 835–862.

[21] Colombet P, Grimmer A-R, Zanni H, Sozzanni P. Nuclear magnetic resonance spectroscopy of cement-based materials. Berlin, Germany, Springer-Verlag Berlin Heidelberg, 1998.

[22] Skibsted J, Hall C. Characterization of cement minerals, cements and their reaction products at the atomic and nano scale. Cem Concr Res. 2008, 38, 205–225.

[23] Mackenzie MJD, Smith ME. Multinuclear solid-state NMR of inorganic materials. Oxford, UK, Pergamon, 2002.

[24] Skibsted J. High-resolution solid-state nuclear magnetic resonance spectroscopy of Portland cement-based systems. In: Scrivener K, Snellings R, Lothenbach B., eds. A practical guide to microstructural analysis of cementitious materials. Boca Raton, FL, USA, CRC Press, 2016, 213–286.

[25] Walkley B, Provis JL. Solid-state nuclear magnetic resonance spectroscopy of cements. Mater Today Adv. 2019, 1, 100007.

[26] Shi C, Zheng K. A review on the use of waste glasses in the production of cement and concrete. Res Conserv Recy. 2007, 52, 234–247.

[27] Lu B, Shi C, Zhang J, Wang J. Effects of carbonated hardened cement paste powder on hydration and microstructure of Portland cement, Constr Build Mater. 2018, 186, 699–708.

[28] Zajac M, Skocek J, Durdzinski P, Bullerjahn F, Skibsted J, Ben Haha M. Effect of carbonated cement paste on composite cement hydration and performance. Cem Concr Res. 2020, 134, 106090.

[29] Andersson R, Stripple H, Gustafsson T, Ljungkrantz C. Carbonation as a method to improve climate performance for cement based material. Cem Concr Res. 2019, 124, 105819.

[30] Skocek, J, Zajac M, Ben Haha M. Carbon capture and utilization by mineralization of cement pastes derived from recycled concrete. Sci Reports. 2020, 10, 5614.

[31] Hansen MR, Jakobsen HJ, Skibsted J, ^{29}Si chemical shift anisotropies in calcium silicates from high-field ^{29}Si MAS NMR spectroscopy. Inorg Chem. 2003, 42, 2368–2377.

[32] Stebbins JF, Poe BT. Pentacoordinate silicon in high-pressure crystalline and glassy phases of calcium disilicate ($CaSi_2O_5$). Geophys Res. Lett. 1999, 26, 2521–2523.

[33] Lippmaa E, Mägi M, Samoson A, Engelhardt G, Grimmer A-R. Structural studies of silicates by solid-state high-resolution ^{29}Si NMR. J Am Chem Soc. 1980, 102, 4889–4893.

[34] Mägi M, Lippmaa E, Samoson A, Engelhardt G, Grimmer A-R. Solid-state high-resolution silicon-29 chemical shifts in silicates. J Phys Chem. 1984, 88, 1518–1522.

[35] Richardson, IG, Brough AR, Brydson R, Groves GW, Dobson CM. Location of aluminum in substituted calcium silicate hydrate (C-S-H) gels as determined by ^{29}Si and ^{27}Al NMR and EELS. J Am Ceram Soc. 1993, 76, 2285.

[36] Skibsted J, Jakobsen HJ, Hall C. Quantification of Calcium Silicate Phases in Portland Cements by ^{29}Si MAS NMR Spectroscopy. J Chem Soc Faraday Trans. 1995, 91, 4423–4430.

[37] Poulsen SL, Kocaba V, Le Saoût G, Jakobsen HJ, Scrivener KL, Skibsted J. Improved quantification of alite and belite in anhydrous Portland cements by ^{29}Si MAS NMR: effects of paramagnetic ions, Solid State Nucl Magn Reson. 2009, 36, 32–44.

[38] Hjorth J, Skibsted J, Jakobsen HJ. ^{29}Si MAS NMR studies of Portland cement components and effects of microsilica on the hydration reaction. Cem Concr Res. 1988, 18, 789–798.

[39] Justnes H, Meland I, Bjoergum JO, Krane JA. ^{29}Si MAS NMR study of the pozzolanic activity of condensed silica fume and the hydration of di- and tricalcium silicates. Adv Cem Res. 1990, 3, 111–116.

[40] Merwin, LH, Sebald, A, Rager H, Schneider H. ^{29}Si and ^{27}Al MAS NMR spectroscopy of mullite. Phys Chem Miner. 1991, 18, 47–52.

[41] Kucharczyk S, Zajac M, Stabler C, Thomsen RM, Ben Haha M, Skibsted J, Deja J. S. Structure and reactivity of synthetic $CaO-Al_2O_3-SiO_2$ glasses. Cem Concr Res. 2019, 120, 77–91.

[42] Nie S, Thomsen RM, Skibsted J. Impact of Mg substitution on the structure and pozzolanic reactivity of calcium aluminosilicate ($CaO-Al_2O_3-SiO_2$) glasses. Cem Concr Res. 2020 138, 106231.

[43] Müller D, Gessner W, Behrens HJ, Scheler G. Determination of the aluminum coordination in aluminum-oxygen compounds by solid-state high-resolution Al-27 NMR. Chem Phys Lett. 1981, 79, 59–62.

[44] Müller D, Rettel A, Gessner W, Scheler G. An application of solid-state magic-angle spinning 27Al NMR to the study of cement hydration. J Magn Reson. 1984, 57, 152–156.

[45] Skibsted J, Henderson E, Jakobsen HJ. Characterization of calcium aluminate phases in cements by ^{27}Al MAS NMR spectroscopy. Inorg Chem. 1993, 32, 1013–1027.

[46] Skibsted J, Hall C, Jakobsen HJ. Direct observation of aluminum guest ions in the silicate phases of cement minerals by ^{27}Al MAS NMR spectroscopy. J Chem Soc Faraday Trans. 1994, 90, 2095–2098.

[47] Hilbig H, Köhler, FH, Schießl P. Quantitative ^{29}Si MAS NMR spectroscopy of cement and silica fume containing paramagnetic ions. Cem Concr Res. 2006, 36, 326–329.

[48] Richardson IG, Girão AV, Taylor R, Jia, S. Hydration of water- and alkali-activated white Portland cement pastes and blends with low-calcium pulverized fuel ash. Cem Concr Res. 2016, 83, 1–18.

[49] Samoson A, Lippmaa E. Excitation phenomena and line intensities in high-resolution NMR powder spectra of half-integer quadrupolar nuclei. Phys Rev. 1983, B28, 6567–6570.

[50] Poulsen SL. Methodologies for measuring the degree of reaction in Portland cement blends with supplementary cementitious materials by ^{29}Si and ^{27}Al MAS NMR spectroscopy. PhD thesis, Aarhus University, Denmark, 2009.

[51] Richardson IG, Brough AR, Groves GW, Dobson CM. The characterization of hardened alkali-activated blast-blast furnace slag pastes and the nature of the calcium silicate hydrate (C-S-H) phase. Cem Concr Res. 1994, 24, 813–829.

[52] Richardson IG, Groves GW. The structure of the calcium silicate hydrate phases present in hardened pastes of white Portland cement/blast-furnace slag blends. J. Mater. Sci. 1997, 32, 4793–4802.

[53] Andersen MD, Jakobsen HJ, Skibsted J. Characterization of white Portland cement hydration and the C-S-H structure in the presence of sodium aluminate by ^{27}Al and ^{29}Si MAS NMR spectroscopy. Cem Concr Res. 2004, 34, 857–868.

[54] Richardson IG, Model structures for C-(A)-S-H(I). Acta Crystallogr Sect B. 2014, 70, 903–923.

[55] Haas J, Nonat A. From C–S–H to C–A–S–H: Experimental study and thermodynamic modelling. Cem Concr Res. 2015, 68, 124–138.

[56] Sevelsted TF, Skibsted J. Carbonation of C-S-H and C-A-S-H samples studied by ^{13}C, ^{27}Al and ^{29}Si MAS NMR spectroscopy. Cem Concr Res. 2015, 71, 56–65.

[57] Richardson IG. The nature of C-S-H in hardened cements. Cem Concr Res. 1999, 29, 1131–1147.

[58] Muller ACA, Scrivener KL, Skibsted J, Gajewics AM, McDonald PJ. Influence of silica fume on the microstructure of cement pastes: New insights from ^{1}H NMR relaxometry. Cem Concr Res. 2015, 74, 116–125.

[59] Palomo A, Alonso A, Fernandez-Jimenez A, Sobrados I, Sanz J. Alkaline Activation of Fly Ashes: NMR Study of the Reaction Products. J Am Ceram Soc. 2004, 87, 1141–1145.

[60] Criado M, Fernandez-Jimenez A, Palomo A, Sobrados I, Sanz J. Effect of the SiO_2/Na_2O ratio on the alkali activation of fly ash. Part II: ^{29}Si MAS-NMR Survey. Microp Mesop Mater. 2008, 109, 525–534.

[61] Ruiz-Santaquiteria C, Skibsted J, Fernandez-Jimenez A, Palomo A. Alkaline solution/binder ratio as determining factor in the alkaline activation of aluminosilicates. Cem Concr Res. 2012, 42, 1242–1251.

[62] Bernal SA, Provis JL, Walkley B, San Nicolas R, Gehman JD, Brice DG, Kilcullen AR, Duxson P, van Deventer JSJ. Gel nanostructure in alkali-activated binders based on slag and fly ash, and effects of accelerated carbonation. Cem Concr Res. 2013, 53, 127–144.

[63] Peng Z, Vance K, Dakhane A, Marzke R, Neithalath N. Microstructural and ^{29}Si MAS NMR spectroscopic evaluations of alkali cationic effects on fly ash activation. Cem Concr Comp. 2015, 57 34–43.

[64] Gao X, Yu QL, Brouwers HJH. Apply ^{29}Si, ^{27}Al MAS NMR and selective dissolution in identifying the reaction degree of alkali activated slag-fly ash composites. Ceramics Int. 2017, 43, 12408–12419.

[65] Sankar K, Sutrisno A, Kriven WM. Slag-fly ash and slag-metakaolin binders: Part II – Properties of precursors and NMR study of poorly ordered phases. J Am Ceram Soc. 2019, 102, 3204–3227.

[66] Gupta R, Tomar AS, Mishra D, Sanghi SK. Multinuclear MAS NMR characterization of fly-ash-based advanced sodium aluminosilicate geopolymer: exploring solid-state reactions. ChemistrySelect 2020, 5, 4920–4927.

[67] Wang S-D, Scrivener KL. ^{29}Si and ^{27}Al NMR study of alkali-activated slag, Cem Concr Res. 2003, 33, 769–774.

[68] Hilbig H, Buchwald A. The effect of activator concentration on reaction degree and structure formation of alkali-activated ground granulated blast furnace slag. J Mater Sci. 2006, 41, 6488–6491

[69] Le Saout G, Ben Haha M, Winnefeld F, Lothenbach B. Hydration degree of alkali activated slags. J Am Ceram Soc. 2012, 94, 4541–4547.

[70] Park S, Abate SY, Lee HK, Kim H-K. On the quantification of degrees of reaction and hydration of sodium silicate-activated slag cements. Mater Struct. 2020, 53, 65.

[71] Walkley B, San Nicolas R, Sanic M-A, Bernal SA, van Deventer JSJ, Provis JL. Structural evolution of synthetic alkali-activated CaO-MgO-Na$_2$O-Al$_2$O$_3$-SiO$_2$ materials is influenced by Mg content. Cem Concr Res. 2017, 99, 155–171.

[72] Dyson HM, Richardson IG, Brough AR. A combined ^{29}Si MAS NMR and selective dissolution technique for the quantitative evaluation of hydrated blast furnace slag cement blends. J Am Ceram Soc. 2007, 90, 598–602.

[73] Brunet F, Charpentier T, Chao CN, Peycelon H, Nonat A. Characterization by Solid-State NMR and selective dissolution techniques of anhydrous and hydrated CEM V cement pastes. Cem Concr Res. 2010, 40, 208–19.

[74] Taylor R, Richardson IG, Brydson RMD, Composition and microstructure of 20-year-old ordinary Portland cement-ground granulated blast-furnace slag blends containing 0 to 100% slag. Cem Concr Res. 2010, 40, 971–83.

[75] Mendes A, Gates WP, Sanjayan JG, Collins F. NMR, XRD, IR and synchrotron NEXAFS spectroscopic studies of OPC and OPC/slag cement paste hydrates. Mater Struct. 2011, 44, 1773–1791.

[76] Lothenbach B, Le Saout G, Ben Haha M, Figi R, Wieland E. Hydration of a low-alkali CEM III/B-SiO$_2$ cement (LAC). Cem Concr Res. 2012, 42, 410–423.

[77] Snellings R, Solution-controlled dissolution of supplementary cementitious material glasses at pH 13: the effect of solution composition on glass dissolution rates. J Am Ceram Soc. 2013, 96, 2467–2475.

[78] Newlands KC, Foss M, Matchei T, Skibsted J, Macphee DE. Early stage dissolution characteristics of alumino silicate glasses with blast furnace slag- and fly-ash-like compositions. J Am Ceram Soc. 2017, 100, 1941–1955.

[79] Merzbacher CI, Sherriff BL, Hartman JS, White WB. A high-resolution ^{29}Si and ^{27}Al NMR study of alkaline earth aluminosilicate glasses, J Non-Cryst Solids. 1990, 124, 194–206.

[80] Thomsen RM, Skibsted J, Yue Y. The charge-balancing role of calcium and alkali ions in per-alkaline aluminosilicate glasses. J Phys Chem B. 2018, 122, 3184–3195.

[81] Lee SK, Stebbins, JF. The structure of aluminosilicate glasses: high-resolution ^{17}O and ^{27}Al MAS MQMAS NMR study. J Phys Chem B. 2000, 104, 4091–4100.

[82] Stebbins JF, Kroecker S, Lee SK, Kiczenski TJ. Quantification of five- and six-coordinated aluminum ions in aluminosilicate and fluoride-containing glasses by high-field, high-resolution ^{27}Al NMR. J Non-Cryst solids. 2000, 275, 1–6.

[83] Neuville DR, Cormier L, Massiot D. Al coordination and speciation in calcium aluminosilicate glasses: Effects of composition determined by ^{27}Al MQ-MAS NMR and Raman spectroscopy, Chem Geol. 2006, 229, 173–185.

[84] Neuville DR, Cormier L, Montouillout V, Massiot D. Local Al site distribution in aluminosilicate glasses by ^{27}Al MQMAS NMR. J Non-Cryst solids. 2007, 353, 180–184.

[85] Moesgaard M, Keding R, Skibsted J, Yue Y. Evidence of intermediate-range order heterogeneity in calcium aluminosilicate glasses. Chem Mater. 2010, 22, 4471–4483.

[86] Thompson LM, Stebbins JF. Non-stoichiometric non-bridging oxygens and five-coordinated aluminum in alkaline earth aluminosilicate glasses: Effect of modifier cation size. J Non-Cryst solids. 2012, 358, 1783–1789.

[87] Eden M. ^{27}Al NMR Studies of Aluminosilicate Glasses, Ann Rep NMR Spec. 2015, 86, 237–331.

[88] Moesgaard M, Herfort D, Skibsted J, Yue Y. Calcium alumino-silicate glasses as supplementary cementitious materials. Glass Technol: Eur J Glass Sci Techn A. 2010, 51, 183–190.

[89] Moesgaard M, Poulsen SL, Herfort D, Steenberg M, Kirkegaard LF, Skibsted J, Yue Y. Hydration of blended Portland cements containing calcium- aluminosilicate glass powder. J Am Ceram Soc. 2012, 95, 403–409.

[90] d'Espinose de Lacaillerie J-B, Fretigny C, Massiot D. MAS NMR spectra of quadrupolar nuclei in disordered solids: the Czjzek model. J Magn Reson. 2008, 192, 244–251.

[91] Murdoch JB, Stebbins JF, Carmichael ISE. High-resolution ^{29}Si NMR study of silicate and aluminosilicate glasses: the effect of network-modifying cations, Am Mineral. 1985, 70, 332–343.

[92] Lee SK, Stebbins JF. The degree of aluminum avoidance in aluminosilicate glasses. Am Mineral. 1999, 84, 937–945.

[93] B.S. EN, 196-5, Methods of Testing Cement–Part 5: Pozzolanicity Test for Pozzolanic Cements, Eur. Comm. Stand. (CEN), Brussels, 2011.

[94] Garg N, Skibsted J. Thermal activation of a pure montmorillonite clay and its reactivity in cementitious systems, J Phys Chem C. 2014, 118, 11464–11477.

[95] Müller C. Closing the loop: what type of concrete-reuse is the most sustainable option? European Cement Research Academy, ECRA, Düsseldorf, 2015.

[96] Zajac M, Skibsted J, Skocek J, Durdzinski P, Bullerjahn F, Ben Haha M. Phase assemblage and microstructure of cement paste subjected to enforced, wet carbonation, Cem Concr Res. 2020, 130, 105990.

[97] Zajac M, Skibsted J, Durdzinski P, Bullerjahn F, Skocek J, Ben Haha M. Kinetics of enforced carbonation of cement paste. Cem Concr Res. 2020, 131, 106013.

Sabrina Galluccio, Herbert Pöllmann

Chapter 2
Mineralogical quantification of cements, wastes and supplementary cementitious materials

Abstract: The reduction of CO_2 during the production process of cementitious materials plays a very important role, as enormous amounts of carbon dioxide are produced during the decarbonisation process of limestone. For this reason, many carbonate-free raw materials are investigated as replacement and supplementary cementitious materials coming from industry, the processing industry and also from primary industries, but natural materials are also investigated. These inorganic-based materials consist of many different cristalline and amorphous components. In order to understand their reaction behaviour and the overall compositions, different mixtures of supplementary cementitious materials (metakaolin, fly ash, limestone, slags) and cements are under investigation. All these different and very complex mixtures have to be characterized in order to ensure the optimum quality of these cementitious mixtures. For these mineralogical quantifications, different determination methods were applied to quantify the different mineralogical phases. These methods include the determination of amorphous contents, the Rietveld Analysis, the Partial Least Squares Regression (PLSR) method, the Cluster Analysis with Principal Component Analysis (PCA) and the Partial Or Not Known Crystal Structure method (PONKCS).

Keywords: mineral quantification, PLSR, PONKCS, Rietveld Method, Cluster Analysis, XRD

1 Introduction

Supplementary cementitious materials result from different sources in nature and also from industrial production. These materials came back into focus because their use can make a significant contribution to CO_2 reduction in the production of cementitious materials. The aspect of reducing industrial residues through recycling is also essential. Since the addition of supplementary cementitous materials leads

Sabrina Galluccio, Institute of Geological sciences, mineralogy/geochemistry,
Martin-Luther-University Halle, Halle (Saale), Germany, e-mail: sabrina.galluccio@geo.uni-halle.de
Herbert Pöllmann, Institute of Geological sciences, mineralogy/geochemistry,
Martin-Luther-University Halle, Halle (Saale), Germany

https://doi.org/10.1515/9783110674941-002

to different composite cements, it is highly necessary to determine the reaction behaviour and their part during the hydration of the interground or interblended components [1–3]. The various naturally occuring materials from volcanic and sedimentary sources such as clays, pozzolanic materials are available in large high quantities and can be used [4–19]. Due to their formation the natural materials differ in their mineralogical and chemical composition and should be optimized for this application. Therefore, clays are often thermally treated to be activated, while other products are used without treatment with their original composition. Natural and industrial pozzolanes also contain large amounts of amorphous phases, which should be determined as they are partially reactive [20, 21]. Their reaction behaviour is often caused by their amorphous contents, as these materials do not contain typical cement minerals such as calcium silicates and calcium aluminates.

Other basic sources of mineral additions come from artificial sources and are mainly based on industrial by-products from ore processing (slags) or from energy production, like fly ash and bottom ash. As these materials have different compositions and thermal histories, their chemical and mineralogical phase relations must also be determined. Some of these supplementary cement constituents are known for a long time, like latent hydraulic slags, others have to be investigated and characterized newly.

As these components vary in their reaction behaviour, it is very likely that their mineralogical quantification refers to their different properties and can be determined by different methods:

1. Rietveld method of these complex systems, including determination of amorphous contents [22–26]
2. Set up and application of a referenced file using PONKCS (**P**artial **O**r **N**ot **K**nown **C**rystal **S**tructure) method for further quantification [27–32]
3. Cluster Analysis including the various 2–4 component systems and summarizing similar phase contents and properties [33–42]
4. Set up and determination of phases and contents by PLSR method (**P**artial **L**east **S**quares **R**efinement) [43–52]

2 Materials and methods

2.1 Materials

Using an ordinary Portland cement (OPC, CEM I 42.5 R), binary mixtures were prepared with the amorphous materials puzzolan, copper slag, granulated blast furnace slag and metakaolin. The chemical composition of the materials can be seen in Figure 2.1. The mixtures with OPC were made by additions of the different supplementary materials in 5 or 10 weight percent steps (5%, 10%, 20%, 30%, 40%, 50%,

60%, 70%, 80%, 90%, 95%). In a second step binary mixtures were prepared in a similar manner of Calcium Aluminate cement (CAC) with anhydrite, metakaolin and calciumcarbonate. In a third step binary mixtures were prepared of anhydrite with gypsum and calcium carbonate and of gypsum with calcium carbonate. Finally ternary mixtures were prepared of OPC with CAC and metakaolin.

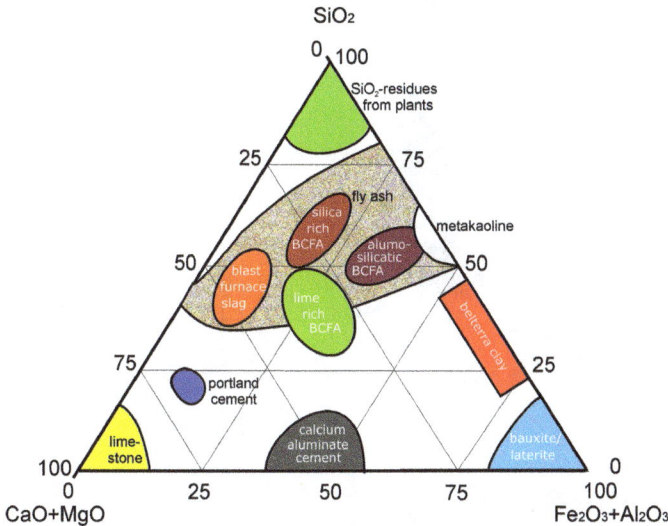

Figure 2.1: Chemical composition of OPC, by-products and major supplementary cementitious materials groups in the ternary diagramm $SiO_2 - CaO + MgO - Fe_2O_3 + Al_2O_3$ (wt.% based).

2.2 Powder X-ray diffraction

Powder X-ray diffraction analyses were performed using a PANalytical X'Pert[3] powder diffractometer equipped with a PIXcel[1D] Detector and a copper anode (CuKα_1 λ = 1,5418 Å, 45 kV, 40 mA), positioned in the θ/θ-Bragg-Brentano-Geometry. The measured 2θ angle range was 5 to 70° with a step width of 0.0131 °2θ and counting time per step of 20.4 s. The samples were prepared by the backloading method in a standard sample carrier (diameter: 27 mm) and measured with 15 mm beam mask, 0.04 rad soller slits and fixed slits (0.25° and 0.5°). The measured data were afterwards treated for interpretation and quantification using the High score suite programs [28]. The X-ray diffractograms of OPC, CAC, metakaolin and granulated blast furnace slag are shown in Figures 2.2–2.5.

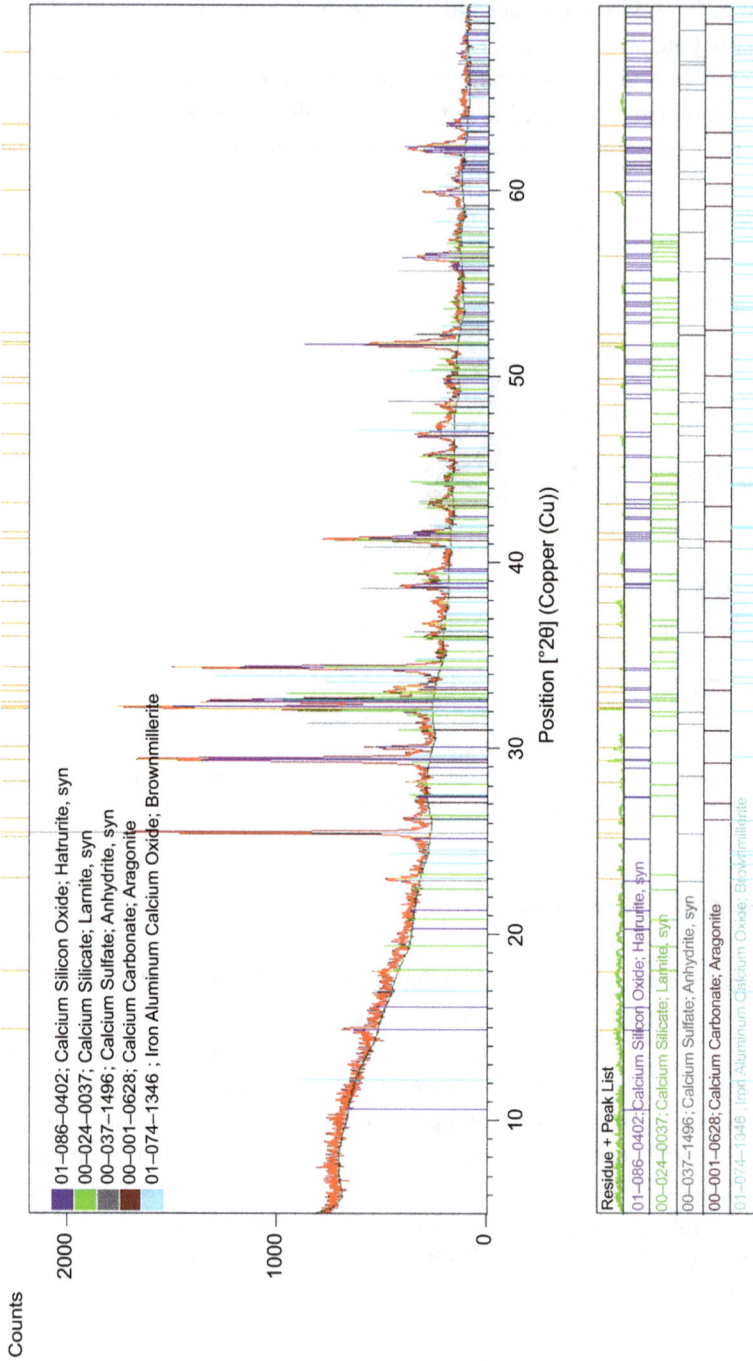

Figure 2.2: Mineralogical composition of OPC (CEM I 42.5 R): alite, belite, anhydrite, aragonite, brownmillerite.

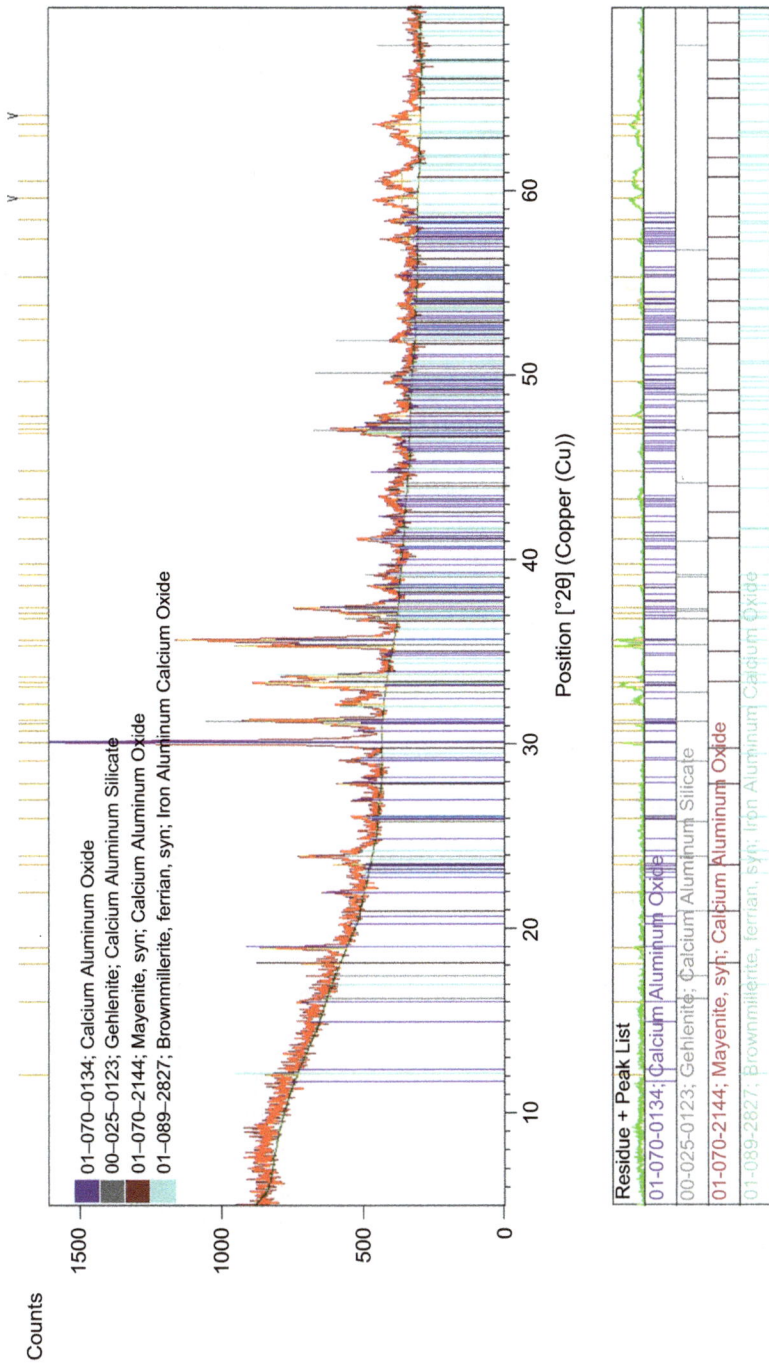

Figure 2.3: Mineralogical composition of CAC: calciumaluminate, gehlenite, mayenite, brownmillerite.

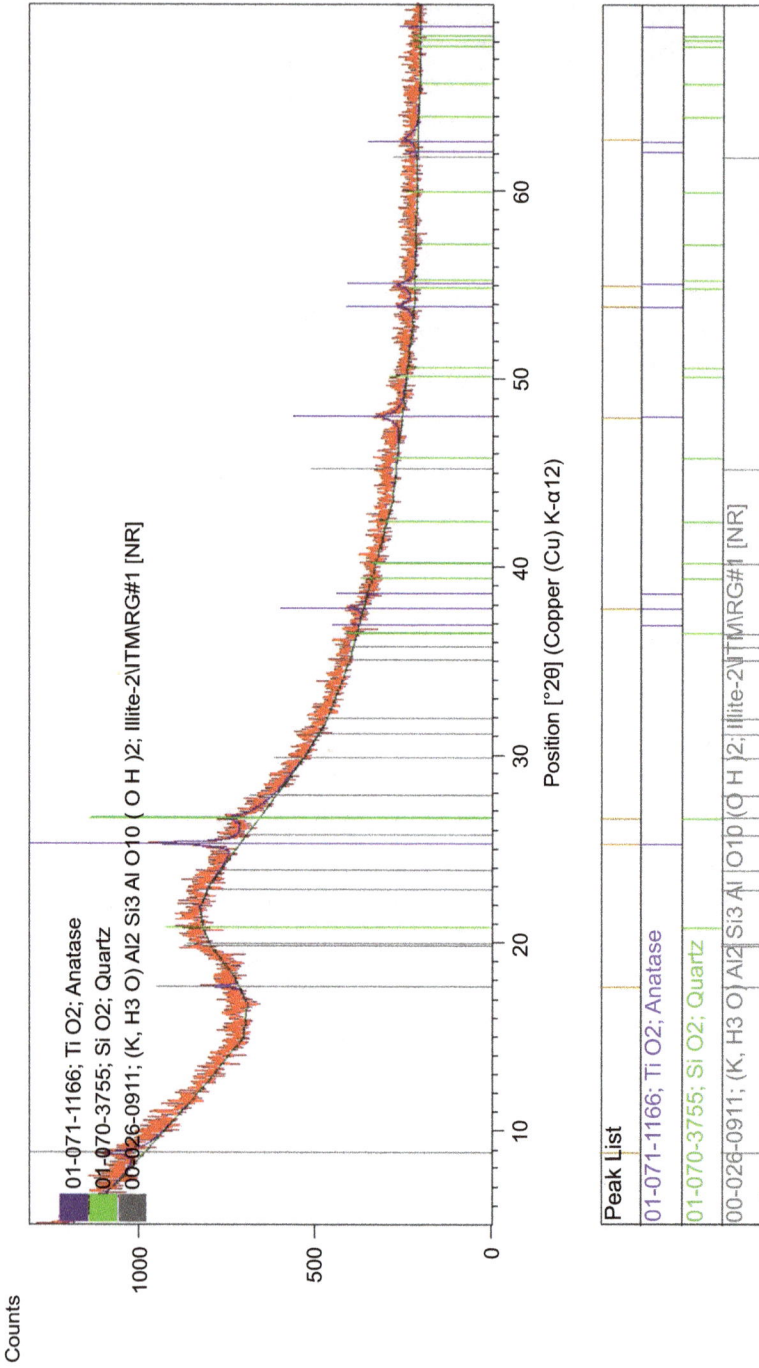

Figure 2.4: Mineralogical composition of metakaolin: anatase, quartz, illite.

Counts

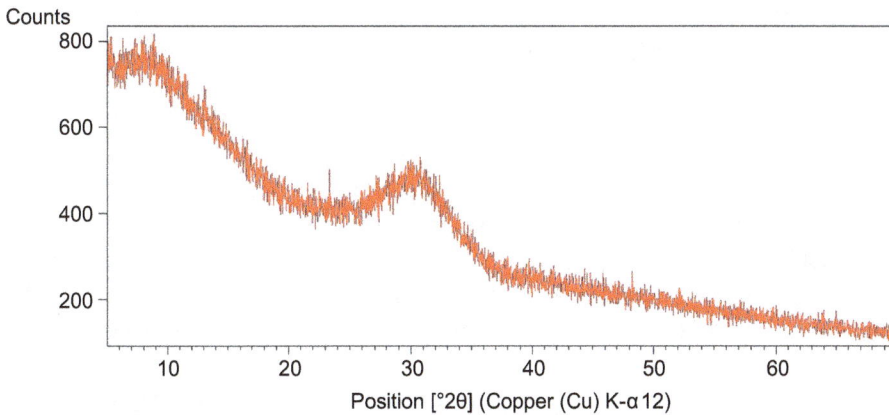

Figure 2.5: X-ray diffractogram of granulated blast furnace slag showing only amorpous components.

2.3 Rietveld method

The Rietveld method [22, 23] was used for the quantification of the total phase composition including the amorphous content [24, 25]. For the analysis the 4.8 version of High-Score Plus from Malvern Panalytical was used [26]. Scale factors, zero-shift and cell and peak shape parameters were refined. The peak shapes were fitted by pseudo-Voigt functions. A rutile standard (10%) (Kronos 2900-TiO$_2$, supplied by KRONOS TITAN GmbH) was added to the samples to determine the amorphous content. The crystallographic information files were taken from the Inorganic Crystal Structure Database.

2.4 Partial Or Not Known Crystal Structure

Scarlett & Madsen [28] developed a direct method to quantify phases with **P**artial **O**r **N**o **K**nown **C**rystal **S**tructure (PONKCS), which can be used to determine the amorphous content of supplementary cementitious materials (SCM). The method can also separate the individual portions of the amorphous portion of cement and SCMs in the total amorphous content of the mixture [29–31]. Snellings identified a detection limit of 5–10% for the amorphous content in this method [32]. With the PONKCS method an hkl file is created by fitting a random cell and symmetry to an unknown crystal structure like for example an amorphous phase. After calibrating with another method like Rietveld by fitting the pseudo formula mass, the hkl file can be used for quantitative determinations of samples with similar phase contents without establishing the use of further admixtures with internal standards. Therefore, the actually observed phase peak intensity is used, instead of the data of a crystal structure, that could be based on slightly different substances, that do not really represent the actual occuring phase.

2.5 Cluster analysis

With the Cluster Analysis closely related scans are sorted automatically into separate clusters. Thus samples with similar phase contents and properties are divided into groups. It thus simplifies the analysis of large amounts of data which can be quantified, even in complex mixtures within seconds after the XRD measurement. These quantifications even can be automated very easily, which can be of high interest for industrial applications. A dendrogram provides a graphical representation of the result of an agglomerating hierarchical Cluster Analysis. Principal Component Analysis (PCA) was also performed using the statistical data. The PCA can be performed as an independent method to visualize the quality of clustering [33–42].

2.6 Partial least square regression

Since Wolds work [46] on the partial least squares, the Partial Least Square Regression (PLSR) method is widely used. This method is well applicable if the data is strongly collinear and the number of variables is larger than observations, which is common for analytical data such as XRD patterns [26, 47–49]. A prediction of hidden information from the raw data is offered, which may be of great interest for industrial process control [26, 50–52]. The PLSR method can be used for a fast and reliable quality control of several sample sets [30, 50]. For this purpose, a calibration curve with several admixtures must first be prepared. With a statistical method a regression is performed with relation to the principal component of the sample. The linear regression model is found by projecting the predicted variables and the observable variables to a new space area. The calculated regression model can then be applied to unknown samples for quantifying the principal component very quickly and easily without further admixtures or calculation fittings.

3 Results and discussion

3.1 Rietveld method

The Rietveld Method was applied on the granulated blast furnace slag already shown in Figure 2.5. The sharp reflections belong to rutile, the used internal standard (Figure 2.6). The blast furnace slag consists only of amorphous components.

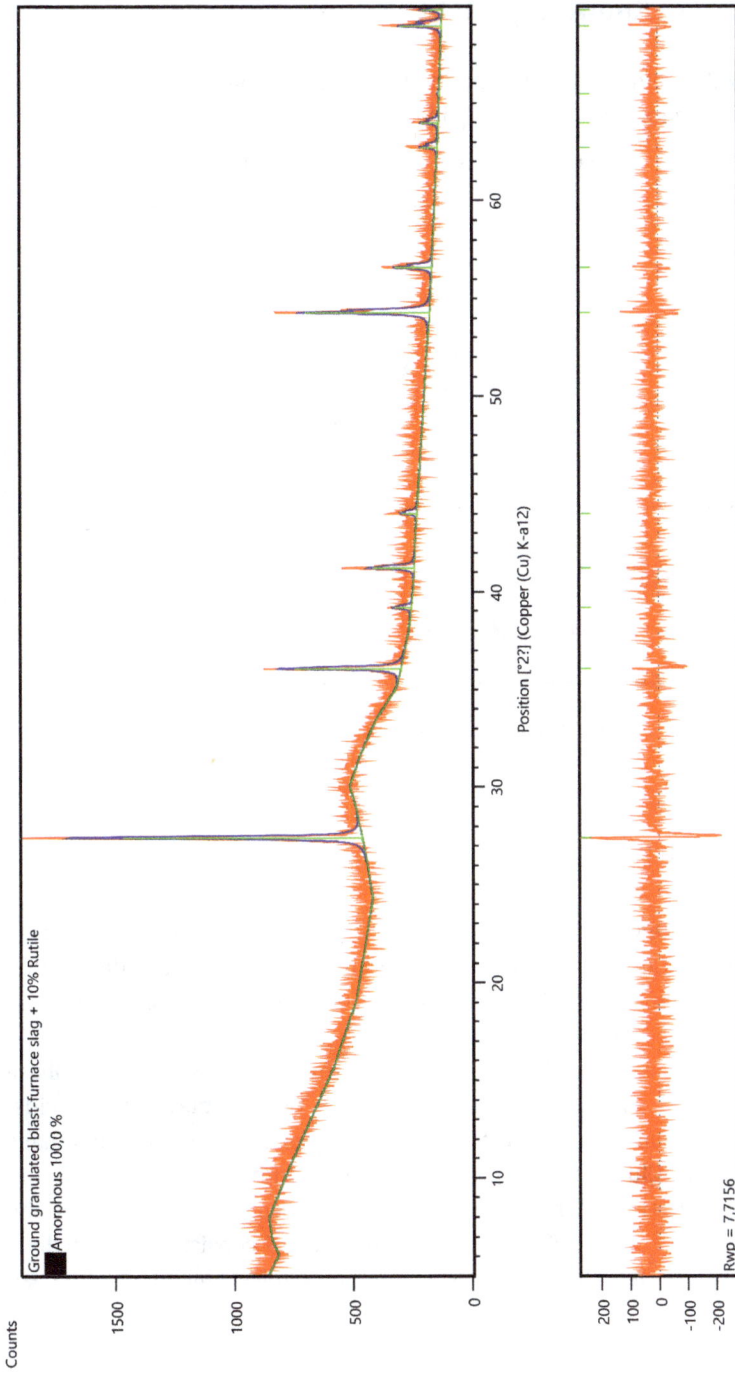

Figure 2.6: Rietveld Analysis of granulated blast furnace slag with 10 wt.-% rutile as internal standard.

3.2 Partial Or Not Known Crystal Structure

Using the PONKCS method, the amorphous humps of the previously investigated granulated blast furnace slag can be described by two created hkl-files (Figure 2.7).

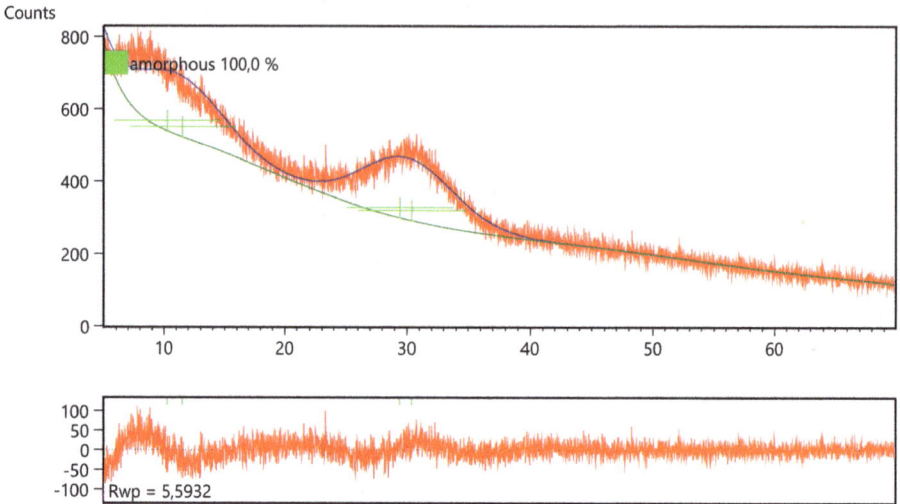

Figure 2.7: Hkl-file of the amorphous humps of granulated blast furnace slag created by using the PONKCS method.

3.3 Cluster analysis

Thirteen mixtures of OPC (shown in Figure 2.2) with granulated blast furnace slag were analysed by the Cluster Analysis. A dendogram can be used to decide whether a more detailed classification with more clusters or a broader classification with fewer clusters should be calculated (Figure 2.8).

These 13 mixtures were divided by the PCA into five groups with similar phase compositions and thus similar properties (Figure 2.9). Two samples were not clustered due to their unique compositon. The view in Figure 2.10 shows the thumbnails of the diffractograms. Cluster no. 1 consists of mixtures with high OPC content and low slag content. Cluster no. 2 shows mixtures with reversed composition: high slag content and low OPC content. The clusters no. 3, 4 and 5 consist of intermediate mixtures and the pure OPC and the pure slag are the samples which have not been clustered.

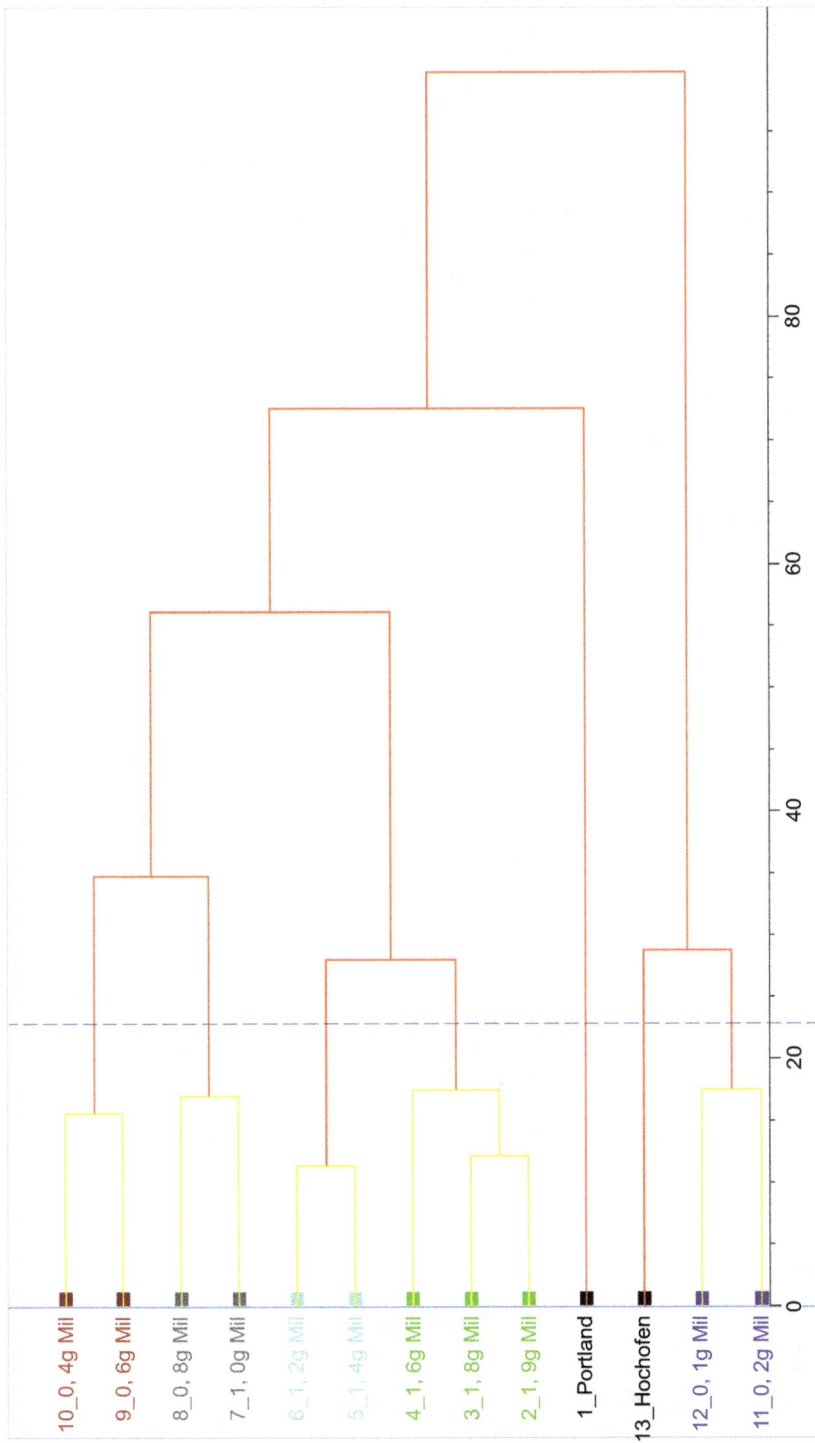

Figure 2.8: Dendrogram of OPC with granulated blast furnace slag.

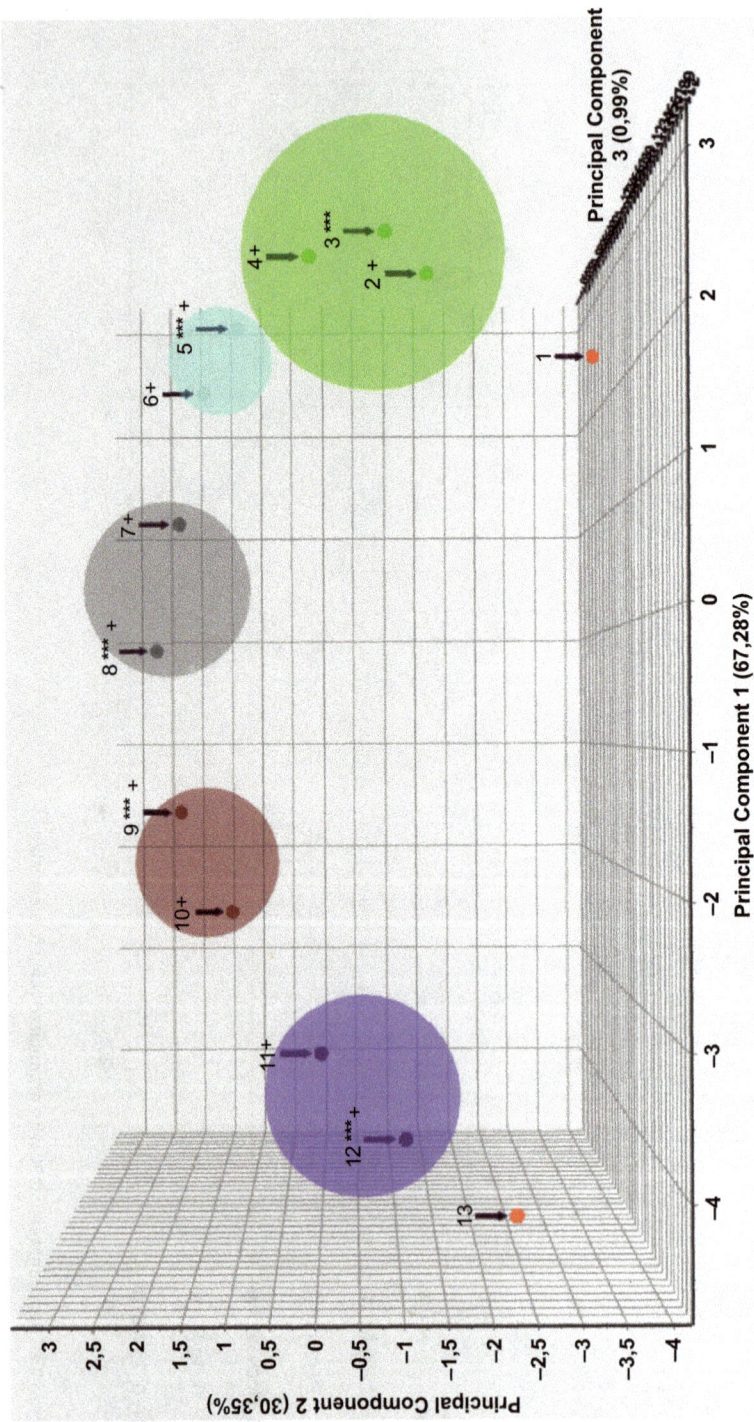

Figure 2.9: Thirteen mixtures of OPC with granulated blast furnace slag were sorted into different clusters by the Principal Component Analysis.

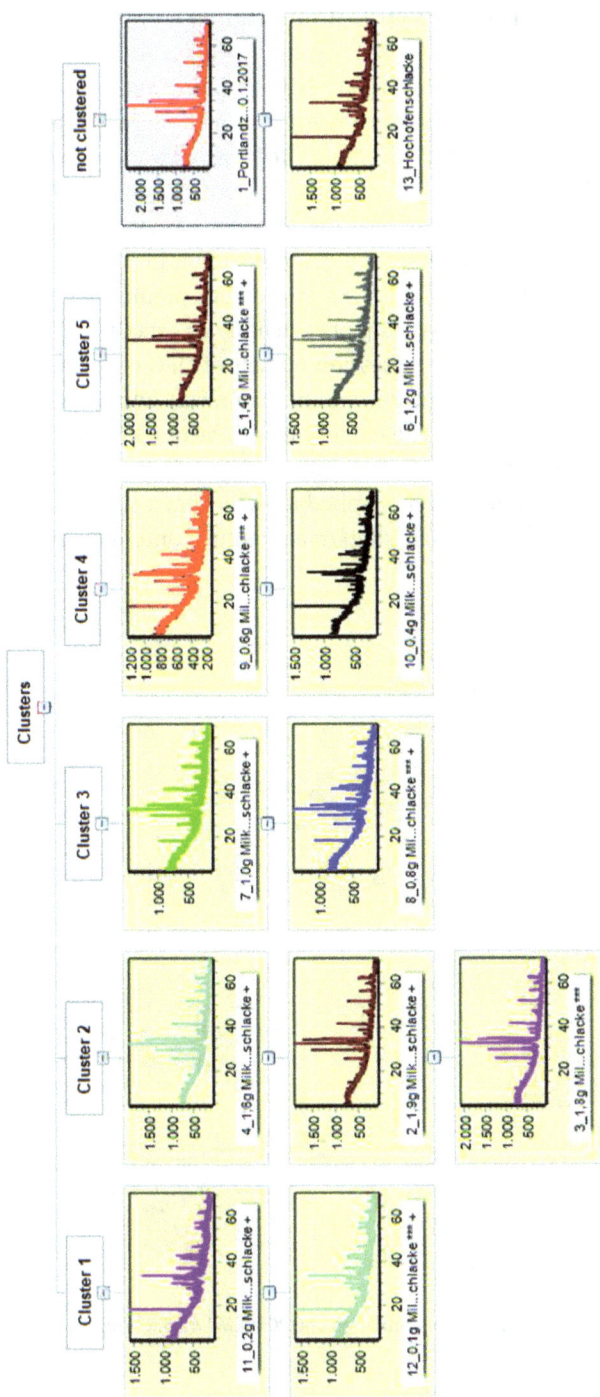

Figure 2.10: Thumbnails of the clusters of OPC with granulated blast furnace slag calculated by the Cluster Analysis.

3.4 PLSR method applied on binary OPC mixtures

Binary mixtures were prepared of OPC with puzzolan, copper slag, granulated blast furnace slag and metakaolin. From these mixtures regression models were created using the PLSR method (Figures 2.11–2.14). The linear regression was performed with relation to OPC as main component. Thus, the OPC content rises in the direction of the right upper side and the content of the other mixing component rises in the opposite direction in the diagram. The blue squares mark the initial weights in weight percentages and the red squares show the OPC content predicted by the model. The two squares should therefore lie on top of each other as exactly as possible in a good regression line, as it is in the regression models of OPC with the two slags. Non-matching predicted and weighed OPC contents can result from weighing or preparation mistakes (Figures 2.11, 2.14). The model can also be quickly rewritten to use the other mixing component as the main component. Depending on the used main component the regression model can be applied to determine the unknown OPC/ puzzolan/ copper slag/ blast furnace slag/ metakaolin content in other samples.

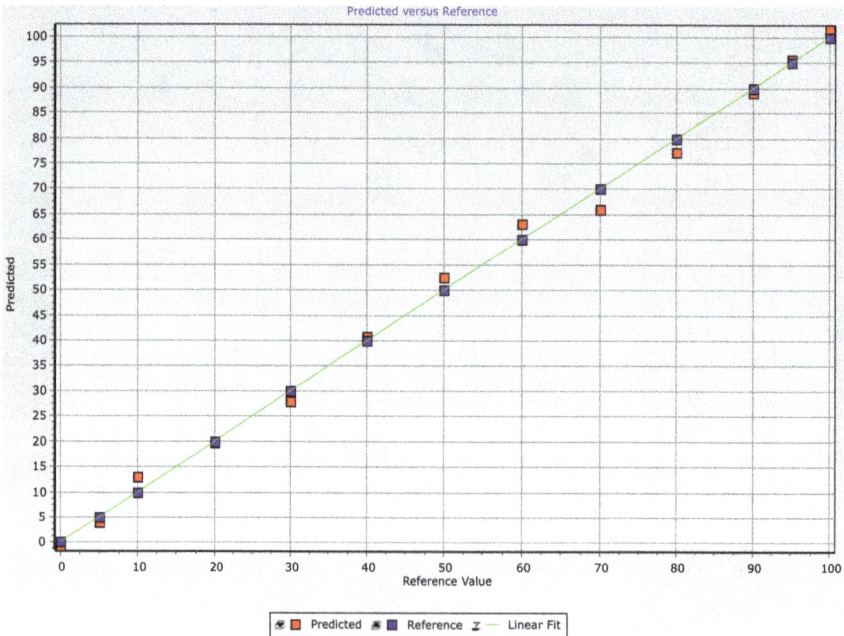

Figure 2.11: Partial Least Squares Regression of binary mixtures of OPC (CEM I 42.5 R) and puzzolan.

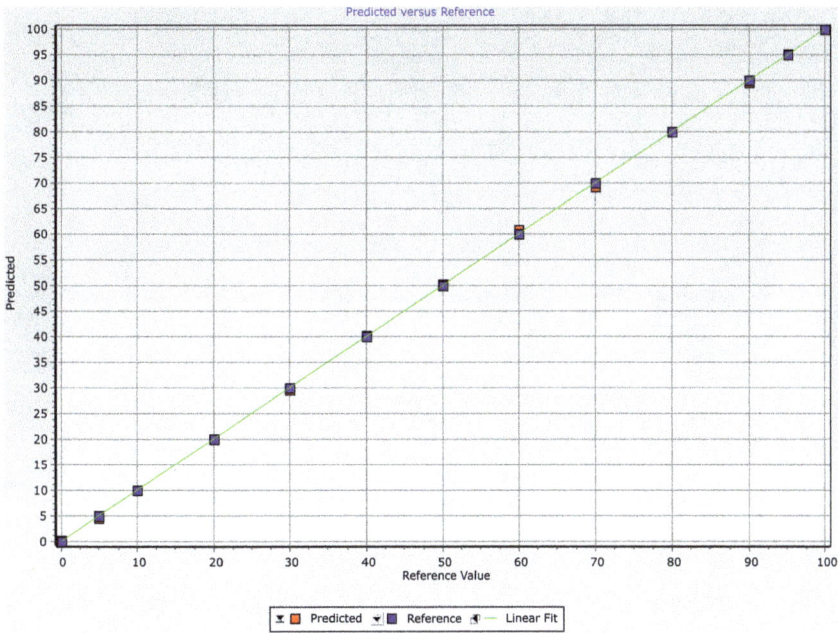

Figure 2.12: Partial Least Squares Regression of binary mixtures of OPC (CEM I 42.5 R) and copper slag.

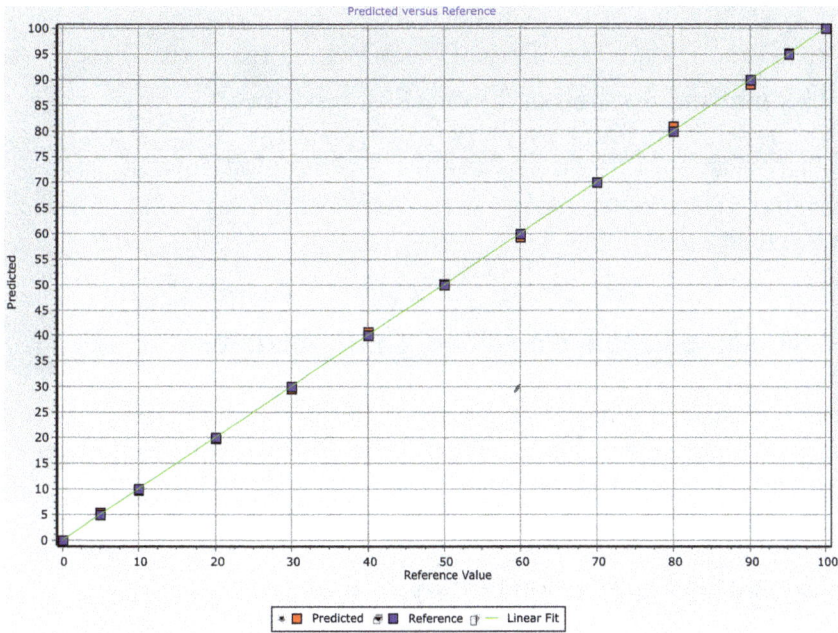

Figure 2.13: Partial Least Squares Regression of binary mixtures of OPC (CEM I 42.5 R) and granulated blast furnace slag.

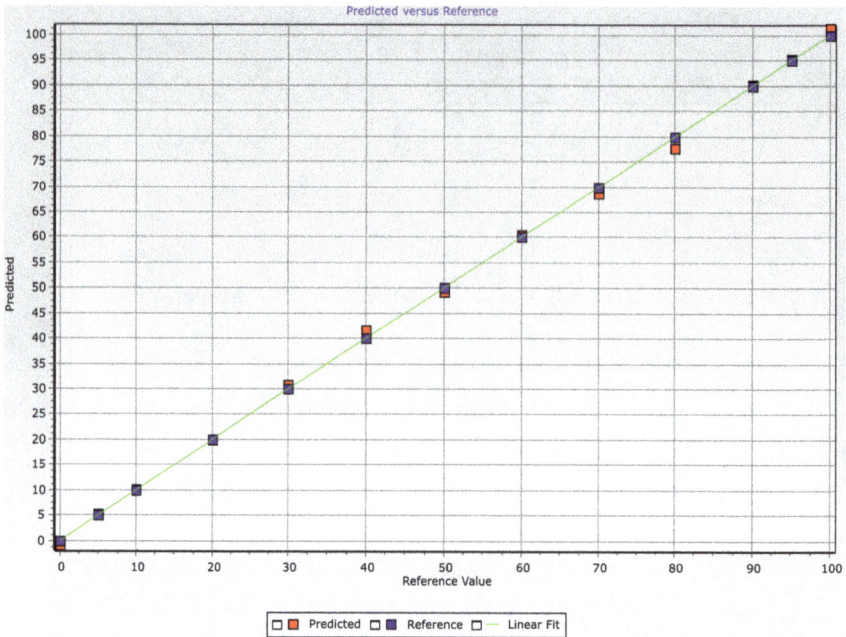

Figure 2.14: Partial Least Squares Regression of binary mixtures of OPC (CEM I 42.5 R) and metakaolin.

3.5 PLSR method applied on binary CAC mixtures

In a second step binary mixtures were prepared of CAC with anhydrite and metakaolin. From these mixtures regression models were created using the PLSR method (Figures 2.15 and 2.16).

3.6 PLSR method applied on binary CaSO$_4$ mixtures

In a third step binary mixtures were prepared of anhydrite with calcium carbonate, anhydrite with gypsum and gypsum with calcium cabonate. From these mixtures regression models were created using the PLSR method (Figures 2.17–2.19).

3.7 PLSR method applied on ternary mixtures

Finally ternary mixtures were prepared of OPC with CAC and metakaolin. From these mixtures regression models were created using the PLSR method (Figure 2.20).

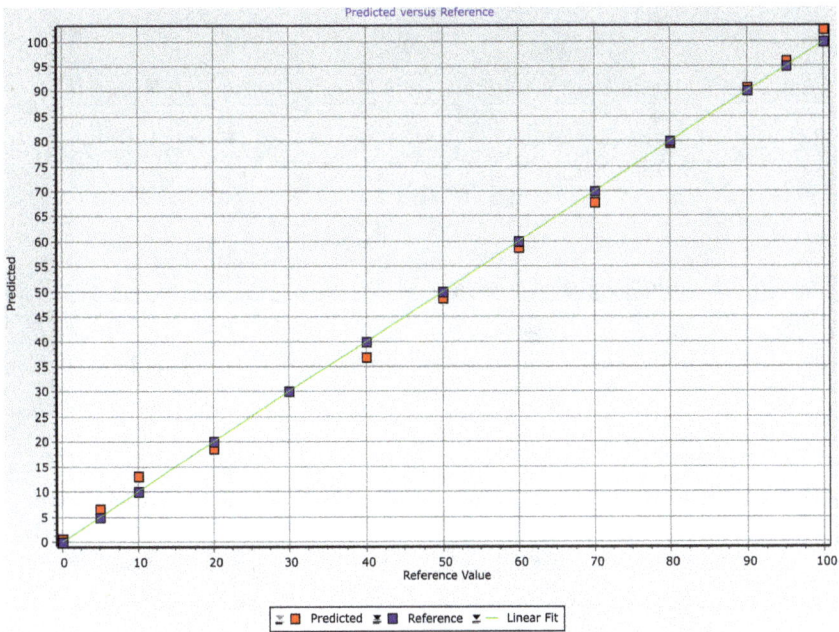

Figure 2.15: Partial Least Squares Regression of binary mixtures of CAC and anhydrite.

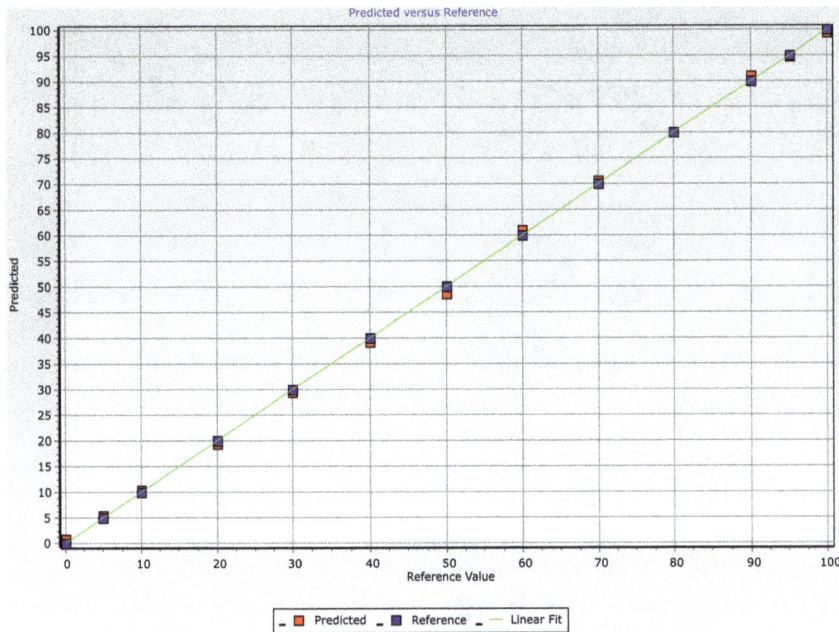

Figure 2.16: Partial Least Squares Regression of binary mixtures of CAC and metakaolin.

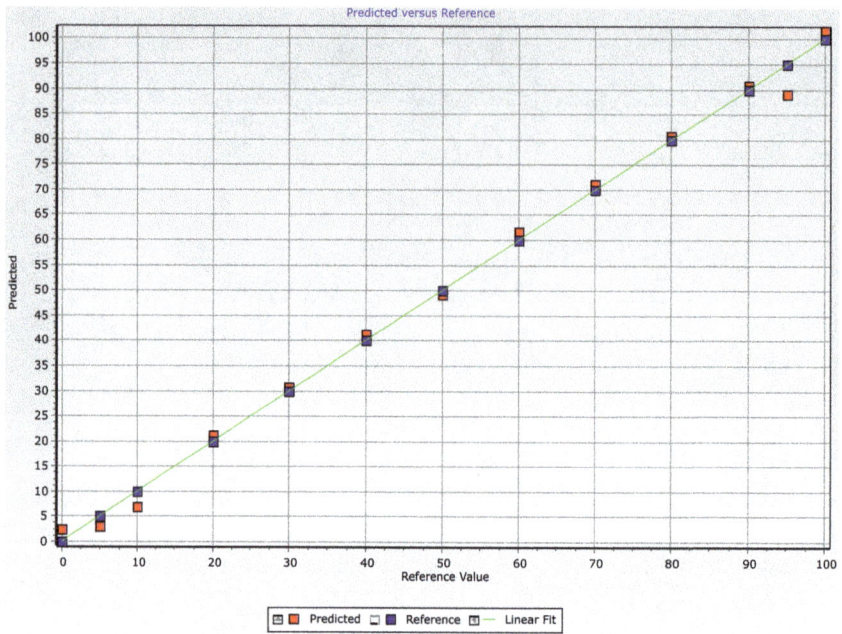

Figure 2.17: Partial Least Squares Regression of binary mixtures of anhydrite and calcium carbonate.

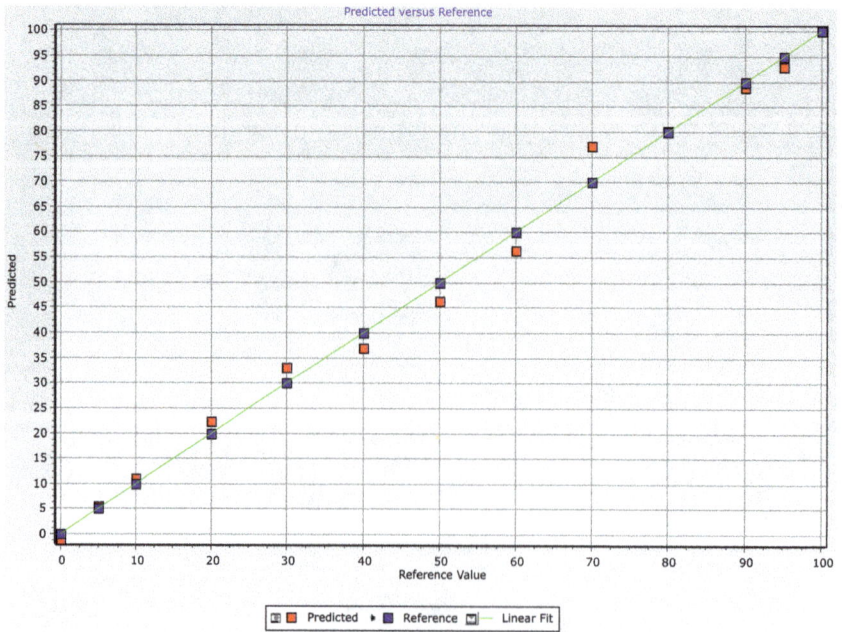

Figure 2.18: Partial Least Squares Regression of binary mixtures of anhydrite and gypsum.

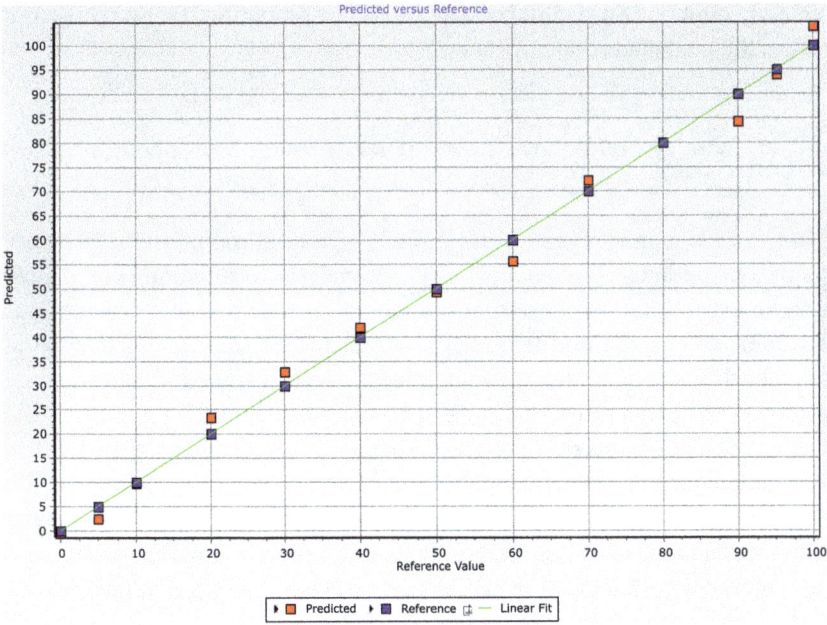

Figure 2.19: Partial Least Squares Regression of binary mixtures of gypsum and calcium carbonate.

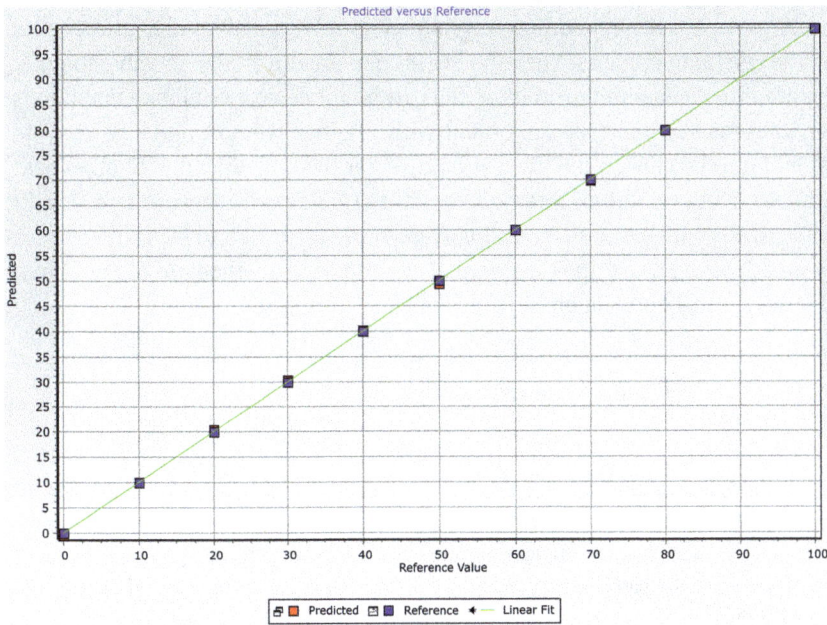

Figure 2.20: Partial Least Squares Regression of ternary mixtures of OPC (CEM I 42.5 R), CAC and metakaolin.

Although more effort is required to homogenise ternary mixtures, the weighed content matches the predicted content very well.

4 Conclusions

The Cluster and PLSR analysis can be used easily to control compositions of binary and ternary cementitious mixtures. For the PLSR analysis, a calibration curve must be created for every different phase composition and mixtures in that cementitious materials field. This is slightly time-consuming at the beginning, but in case of many samples during a quantitative evaluation, the very rapid quantitative method can then easily be automatized. Once the regression model has been created, the quantification of other samples is very fast. Therefore it is mainly used in applications with many samples available as it is the case for industrial operations. The Cluster Analysis simplifies the evaluation and sorting of large data sets, also under the aspect of different properties of the relevant mixtures. Apart from the XRD scans, no further information of admixtures is required. After the set up of a quantitative model this type of quantification can be automized easily.

The quantification by the Rietveld Analysis method is more complex, but gives more detailed information, such as the total phase composition, including the amorphous content. For the determination of the amorphous content, a crystalline internal standard must be added. In addition, crystallographic information files are required to refine the crystalline phases. With the PONKCS method it is also possible to determine the amorphous content or in case, data from unknown cristalline compounds. However, this method must be calibrated with another method like e.g. Rietveld. After calibration, the created hkl file can be used for the quantification of other samples without having to add an internal standard. Therefore it is also interesting for rapid applications with many samples involved. While Rietveld and PONKCS methods are only for advanced users, the Cluster analysis and PLSR method are easier to handle, but provide excellent data on samples very fast.

References

[1]   Galluccio S, Pöllmann H. Usage of secondary raw materials (residual bauxite, laterite and residual kaolin) for the synthesis of CSA-based cements, DMG-conference, Jena, CD, 2014.

[2]   Galluccio S, Beirau T, Pöllmann H. Maximization of the reuse of industrial residues for the production of eco-friendly CSA-belite clinker. Construction and Building Materials 2019, 208, 250–257.

[3]   Galluccio S, Pöllmann H. (2019): Quantifications of cements composed of OPC, Calcined Clays, Pozzolanes and Limestone. 3rd International Conference on Calcined Clays for Sustainable Concrete, Neu-Delhi, 2, 135–153.

[4]   Barata MS, Angélica RS, Pöllmann H, Costa ML. The use of wastes derived from kaolinite processing industries as a pozzolanic material for high-performance mortar and concretes. Eur J Min 17, 2005, 1, 10.

[5]   Barthel M, Rübner K, Kühne HC, Rogge A, Dehn F. From waste materials to products in the cement industry. In: Advances in cement research. ICE publishing, 2016, 1–11.

[6]   Costa ML, Moraes EL. Mineralogy, geochemistry and genesis of kaolins from the Amazon region. Mineralium Deposita 1998, 33, 283–297.

[7]   Costa ML, Souza DJL, Angélica RS. The contribution of lateritization processes to the formation of the kaolin deposits from eastern Amazon. Journal of South American Earth Sciences 2009, 27, 219–234.

[8]   Korndörfer J, Pöllmann H, Costa ML, Angélica RS. Mine tailings from the lateritic gold mine Igarape Bahia (Carajas region) – a basis for active hydraulic binders. Economic Geology ZAG SH1, 2000, 187–192.

[9]   Maia AAB, Saldanha E, Angélica RS, Souza CAG, Neves RF. Utilização de rejeito de caulim da Amazônia na síntese da zeólita A. Cerâmica 2007, 53, 319–324.

[10]   Maia AAB, Angélica RS, Neves RF. Estabilidade Térmica da Zeólita A – Sintetizada a partir de um Rejeito de Caulim da Amazônia. Cerâmica 2008, 54, 345–350.

[11]   Maia AA, Angélica RS, Neves RF, Pöllmann H, Straub C, Saalwächter K. Use of 29Si and 27Al MAS NMR to study thermal activation of kaolinites from Brazilian Amazon kaolin wastes. Appl Clay Science 2013, 189–196.

[12]   Mestdagh MM, Vievoye L, Ilerbillon AJ. Iron in kaolinite: II The relationship between kaolinite crystallinity and iron content. Clay Minerals 1980, 15, 1–13.

[13]   Pöllmann H, Da Costa ML, Angélica RS. Sustainable secondary resources from brazilian kaolin deposits for the production of calcined clays. In: Calcined clays for sustainable concrete. Springer, 2015, 21–26.

[14]   Scrivener K, Favier A. Calcined clays for sustainable concrete. Springer, 2015.

[15]   Setzer F. Investigations on supplementary cementitious materials, Univ. Halle, Mineralogy, (Accessed March 22, 2020, at http://www.geologie.uni-halle.de/igw/mingeo/mingeo.html)

[16]   Siddique R, Cachim P. Waste and supplementary cementitious materials in concrete. 1st ed., Woodhead Publishing, 2018.

[17]   Snellings R, Mertens G, Elsen J. Supplementary Cementitious Materials. In: – Broekmans, MATM, Pöllmann, H. ed. Applied Mineralogy in Cement & Concrete: Chantilly, Mineralogical Society of America, 2012, 211–278.

[18]   Kakali G, Perraki T, Tsivilis S, Badogiannis E. Thermal treatment of kaolin: the effect of mineralogy on the pozzolanic activity. Applied Clay Science 2001, 20, 73–80.

[19]   Scrivener K, Vanderley MJ, Gartner EM. Eco-efficient cements: Potential economically viable solutions for a low-CO_2, cement-based materials industry, UNEP, 2016.

[20]   Dinnebier RE, Kern A. Quantification of amorphous phases – theory, PXRD -13 Workshop, Bad Herrenalb, may 2015.

[21]   Madsen IC, Scarlett NVY, Kern A. Description and survey of methodologies for the determination of amorphous content via X-ray powder diffraction. Z Kristallogr 2011, 226.

[22]   Rietveld HM. Line profiles of neutron powder-diffraction peaks for structure refinement. Acta Crystallogr 1967, 22, 151–152.

[23]   Rietveld HM. A Profile Refinement Method for Nuclear and Magnetic Structures. J Appl Crystallogr 1969, 2, 65–71.

[24]   Whitfield PS, Mitchell LD. Quantitative Rietveld analysis of the amorphous content in cements and clinkers. J Mater Sci, 2003, 38, 4415–442.

[25] Westphal T, Füllmann T, Pöllmann H. Rietveld quantification of amorphous portions with an internal standard – Mathematical consequences of the experimental approach. Powd Diff 2009, 24, 239–243.

[26] Degen T, Sadki M, Bron E, König U, Nénert G. The Highscore suite. Pow Diff 2014, 29, 13–18.

[27] Walenta G, Füllmann T. Advances in quantitative XRD analysis for clinker, cements and cementitious additions. Adv X-ray Anal, 2004, 47, 287–296.

[28] Scarlett NVY, Madsen IC. Quantification of phases with partial or no known crystal structure. Powd Diff 2006, 21, 4, 278–284.

[29] Stetsko YP, Shanahan N, Deford H, Zayed A. Quantification of supplementary cementitious content in blended Portland cement using an iterative Rietveld–PONKCS technique. J Appl Cryst 2017, 50, 2, 498–507.

[30] Snellings R, Salze A, Scrivener KL. Use of X-ray diffraction to quantify amorphous supplementary cementitious materials in anhydrous and hydrated blended cements. Cem Concr Res 2014, 64, 89–98.

[31] Singh GVPB, Subramaniam KVL. Quantitative XRD analysis of binary blends of siliceous fly ash and hydrated cement. J Mater Civ Eng 2016, 28.

[32] Snellings R. Assessing, Understanding and Unlocking Supplementary Cementitious Materials. RILEM Tech Lett 2016, 1, 50–55.

[33] König U, Degen T. Cluster analysis of XRD data for ore evaluation. Accuracy in Powder Diffraction IV NIST, Gaithersburg MD, USA April 22–25, 2013.

[34] Liao B, Chen J. The application of cluster analysis in X-ray diffraction phase analysis. J Appl Cryst 1992, 25, 336–339.

[35] Gilmore CJ, Barr G, Paisley J. High-throughput powder diffraction. I. A new approach to qualitative and quantitative powder diffraction pattern analysis using full pattern profiles. J Appl Crystallogr 2004, 37, 231–242.

[36] Barr G, Dong W, Gilmore CJ. High-throughput powder diffraction. II. Applications of clustering methods and multivariate data analysis. J Appl Crystallogr 2004, 37, 243–252.

[37] Lohninger H. Data Analysis. Berlin, New York, Tokyo, Springer-Verlag, 1999.

[38] Lance GN, Williams WT. A general theory of classification sorting strategies 1., Hierarchical systems. Comp J 1966, 9, 373–380.

[39] Kelley LA, Gardner SP, Sutcliffe MJ. An automated approach for clustering an ensemble of NMR-derived protein structures into conformationally-related subfamilies. Protein Engineering 1996, 9, 1063–1065.

[40] Mardia KV, Kent JT, Bibby JM. Multivariate analysis. London, Academic Press, 1979.

[41] Rousseeuw JP. Silhouettes: a graphical aid to the interpretation and validation of cluster analysis. J Comp Appl Math 1987, 20, 53–65.

[42] Sato M, Sato Y, Jain LC. Fuzzy Clustering Models and Applications, Studies in Fuzziness and Soft Computing, vol. 9, New York, Springer group, 1997.

[43] Beckers D, Degen T, König U. Analysis of hidden information – PLSR on XRD raw data. Acta Cryst. Section A, 2014, 70.

[44] Degen T, König U, Norberg N. PLSR as a new XRD method for downstream processing of ores: – Case study: Fe^{2+} determination in iron ore sinter. Powder Diffraction 2014, 29.

[45] König U, Norberg N. Partial Least Square Regression (PLSR) – A new alternative XRD method for process control in aluminium industries, Conference paper, 2014.

[46] Wold H. Estimation of principal components and related models by iterative least squares. In: Krishnaiah PR, ed. Multivariate Analysis, New York, USA, Academic Press, 1966, 391–420.

[47] Wold S. et al., The collinearity problem in linear regression, the partial least squares approach to generalized inverses, SIAM J Sci Stat Comput 1984, 5, 735–743.

[48] Wold S, Sjöström M, Eriksson L. PLS-regression: a basic tool of chemometrics. Chemometrics and Intelligent Laboratory Systems 2001, 58, 109–130.

[49] Melo CCA, Paz SPA, Estimation of AvAl$_2$O$_3$ and RxSiO$_2$ by Partial Least Square Regression (PLSR) on XRD data: A Case Study Using Low Grade Bauxites. Travaux 47, Proceedings of the 36th International ICSOBA Conference, Belem, Brazil, 29 October – 1 November, 2018.

[50] Wold S et al. Multivariate Data Analysis in Chemistry. In: Kowalski BR, ed. Chemometrics: mathematics and statistics in chemistry. The Netherlands, Reidel Publishing Company, 1984, 17–95.

[51] König U, Norberg N, Process Control in Aluminium Industry – News in the XRD Tool Box. Proceedings of 35th International ICSOBA Conference, Travaux No. 46, Hamburg, Germany, October 2017, 2–5.

[52] Webster NAS et al., Predicting iron ore sinter strength through partial least square regression (PLSR) analysis of X-ray diffraction patterns. Powder Diffraction 2017, 32, 66–69.

Katrin Schollbach, Sieger van der Laan

Chapter 3
Microstructure analysis with quantitative phase mapping using SEM-EDS and Phase Recognition and Characterization (PARC) Software: applied to steelmaking slag

Abstract: The amount, spatial distribution, and composition of phases determine the physical and chemical properties of a material and reveals information about its processing, its reactivity, or its changes due to corrosion and weathering. The characterization challenge, especially with heterogeneous multiphase materials, is to acquire quantitative microstructural data that is relevant for material properties on the macroscale. This often requires microscopy of extensive surface areas and yields very large datasets for X-ray microanalysis (Spectral Images) obtained with Scanning Electron Microscopy (SEM-EDS).

The PARC (PhAse Recognition and Characterization) approach is a solution for dealing with such large Spectral Image datasets and can generate new insights about a material and lay a foundation for understanding it by close visualization of its phase distribution. In this chapter, PARC is applied to converter slag and its partial alteration products from hydration and carbonation. Converter slag, synonymous with basic oxygen furnace (BOF) slag and Linz-Donawitz (LD) slag, is a by-product of steelmaking, globally produced at about 0.2 Billion tons annually, and forms when hot metal from the blast furnace is converted into raw steel. Unlike blast furnace slag itself, converter slag is fully crystalline and not widely used as a building material. It contains C2S (Ca_2SiO_4,) as the most common phase, followed by Mg-wuestite ((Fe,Mg)O) and C2(A, F) ($Ca_2(Fe,Al)_2O_5$) and shows no pozzolanic reactivity but instead is slightly hydraulic and easily carbonated. Analysis of unreacted converter slag as well as hydrated and carbonated systems are discussed to demonstrate how the quantitative evaluation of microstructure with PARC can be comprehensively integrated with bulk analytical observations from quantitative X-Ray diffraction, X-Ray fluorescence, thermogravimetric analysis, and density/porosimetry.

Acknowledgements: Corrie van Hoek for Electron Microscopy, Stefan Melzer for QXRD, Gang Liu, Winnie Franco-Santos, Jonathan Zepper and Anna Kaja for sample material and exemplary data. Frank van der Does for sample preparation. Jonathan Zepper for statistical sampling techniques.

Katrin Schollbach, Sieger van der Laan, Tata Steel, R&D, Microstructure & Surface Characterization (MSC), IJmuiden, The Netherlands; Department of the Built Environment, Eindhoven University of Technology, Eindhoven, The Netherlands

https://doi.org/10.1515/9783110674941-003

Keywords: Microstructure, X-ray microanalysis, Scanning Electron Microscopy, Converter Slag, PARC

1 Introduction

1.1 Microstructure analysis

Microstructure analysis forms the basis for understanding any material. The amount, spatial distribution, and composition of phases determine physical and chemical properties. The microstructure reveals information about the processing of a material, its reactivity, or changes due to corrosion and weathering. The challenge in microstructure analysis is to acquire quantitative information that is relevant for material properties on the macro-scale. This is easier for homogeneous single-phase materials but more difficult with heterogeneous multiphase materials and may require microscopy of extensive surface areas. Microstructure analysis at 1 μm resolution, which is the resolution limit for standard optical microscopy and typically used for X-ray microanalysis on Scanning Electron Microscopy (SEM), yields very large datasets if substantial, representative areas are to be evaluated (10^6 data points for 1 mm^2). Fortunately, the handling and storage of large datasets have become standard fare for most labs, however, for the evaluation few tools are at our disposal, often custom development is required. In this chapter, the PARC (PhAse Recognition and Characterisation) approach will be discussed that was developed to handle large microanalysis datasets. And it is demonstrated how the quantitative evaluation of such datasets allows a comprehensive integration of microstructure with bulk analytical observations, like those obtained from quantitative X-Ray diffraction (QXRD), X-Ray fluorescence (XRF), thermogravimetric analysis (TGA), and He-pycnometry, or Mercury Porosimetry (MIP).

1.2 SEM-EDS microanalysis

One of the most powerful methods to investigate microstructure is Scanning Electron Microscopy (SEM) in combination with Energy-dispersive X-ray spectroscopy (EDS). Typically, a "point and shoot" approach is taken (Figure 3.1), where the composition of single points is measured via EDS. The signal detected with EDS is generated from an area of about 1μm^2, depending on the composition and the accelerating voltage used [1]. In order to determine phase distribution on a larger scale, elemental mapping can be done by measuring each point of a surface with EDS (Figure 3.1) and generating a spectral imaging (SI) dataset (Figure 3.2). Modern detectors allow for the mapping of large areas in a short time however it is still challenging to determine the spatial distribution and quantity of phases. For example a

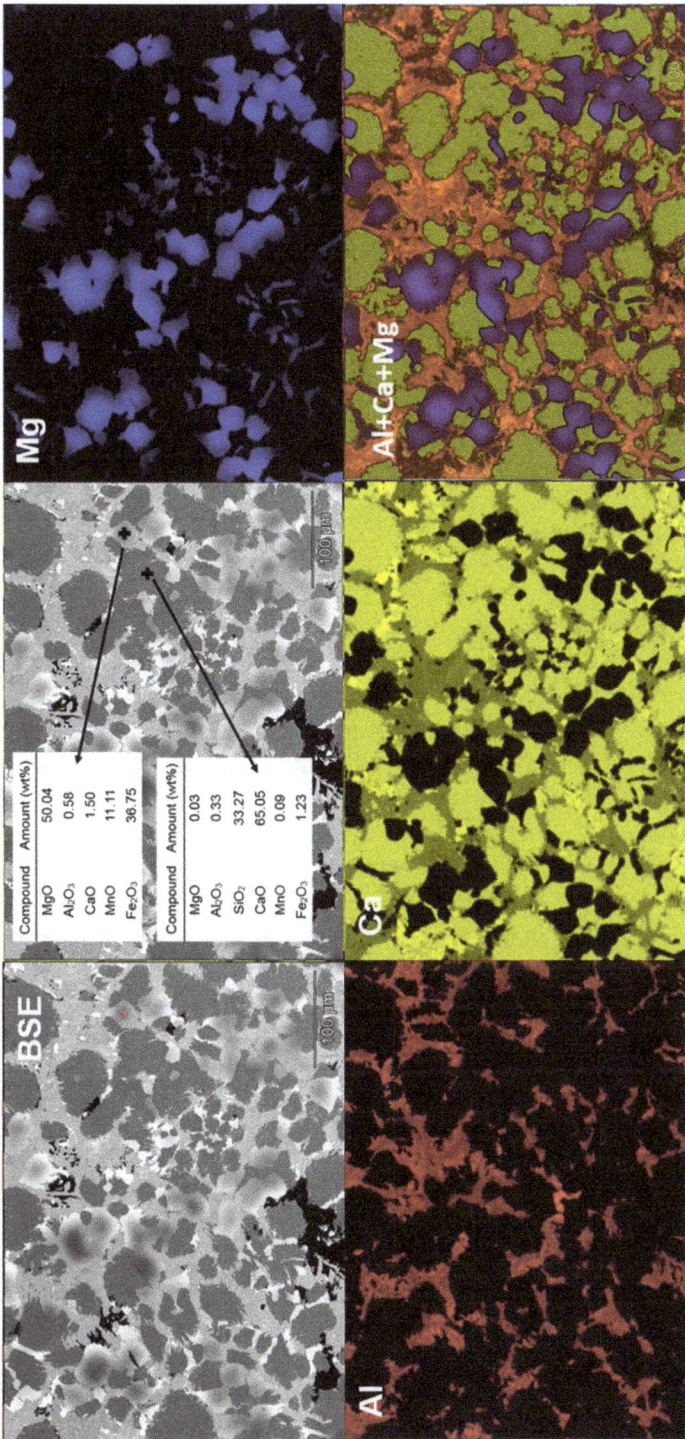

Compound	Amount (wt%)
MgO	50.04
Al₂O₃	0.58
CaO	1.50
MnO	11.11
Fe₂O₃	36.75

Compound	Amount (wt%)
MgO	0.03
Al₂O₃	0.33
SiO₂	33.27
CaO	65.05
MnO	0.09
Fe₂O₃	1.23

Figure 3.1: Point and shoot of converter slag, element maps for Mg, Al and Ca, as well as the complementary element overlay map.

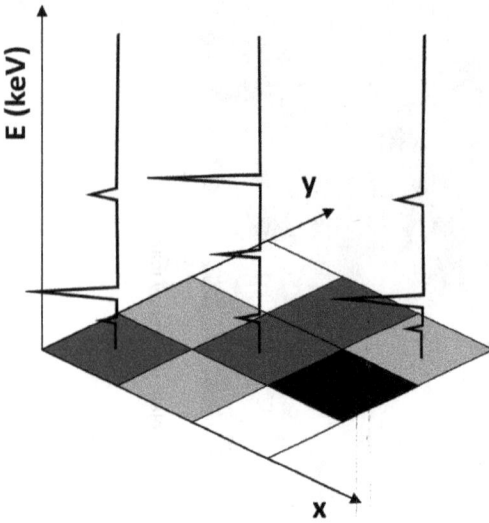

Figure 3.2: Schematic of an SI image set.

simple overlay of different element maps can only yield unique information if no more than three different elements of interest are present as there are three primary colors and all other colors are derived from them. In the example shown in Figure 3.1, the mix of the Ca and Al signal results in an orange phase. An additional element therefore cannot use the color orange, or it will be impossible to attribute the color correctly. Of course it is still possible to use element ratios and other information to generate quantitative phase maps, where each phase is assigned a unique color, but the process has to be repeated for each new SI data set and is usually not automated [2]. Another approach is to compare the EDS spectra of each measured pixel and group them according to composition [3], which is a much more powerful approach. For this purpose, commercial software routines are offered by the leading manufacturers of EDS microanalytical systems. These can be based on Principle Component Analysis, a statistical approach, or comparison to library phase spectra, or intelligently combining overlay maps. All three approaches have drawbacks; principle components are not the same as phases, and in complex multi-component materials that do not represent a phase equilibrium state, dozens of phases can be present, more than can be statistically meaningful extracted as principle components. Likewise, with library spectra, thousands of reference spectra are required to cover any range of phases that may occur when studying a diversity of materials, or contrary, with few references, only a very limited range of materials can be covered. Intelligently combining overlay maps comes closest to what would be expected from a phase mapping routine. Another issue is the difficulty in creating large phase maps by stitching multiple image fields, maintaining the compositional phase definition throughout and applying the same color scheme in the maps, as well as compiling summed information. The automation of SI image interpretation is thereby difficult, and the true potential of SI data is not utilized.

PhAse Recognition and Characterization (PARC) Software was developed at Tata Steel [4] to solve this problem and extract the maximum information from an SI data set with great flexibility including the automatic detection of phases, their quantity, and composition, and many user-determined refinement parameters.

In this chapter, the PARC approach to the analysis of SI data sets is explained and illustrated using converter steel slag as an example. We will show the primary solidification mineralogy of the slag and the subsequent changes during hydration and carbonation and how PARC data can be integrated with observations on bulk chemical composition (XRF), phase amounts (QXRD) volatile content (TGA) and density (He-pycnometry/MIP).

1.3 Some background on converter slag

Converter slag, also called basic oxygen furnace (BOF) slag or Linz-Donawitz (LD) slag, is a byproduct of steelmaking when pig iron is converted into raw steel [5]. Between 80–120 kg are produced per ton of steel. With a global steel production of 1.9 Billion (10^9) tons, this amounts to around 0.2 Billion tons of converter slag annually [6]. The slag is produced batch-wise during the refinement of hot metal from the Blast Furnace, and each batch has its own composition depending on the desired composition of the steel, the blast furnace operation parameters, and the fluxes (lime, dolime) used to create the slag. When the liquid slag is removed from the converter at the end of the refining process and transferred into a slag ladle it has a temperature between 1600 and 1750°C. It immediately forms a solid top layer that insulates the remaining liquid slag. The slag is transported to the slag yard where it is poured into pits and cooled by water spraying. This means the outer layers cool down very quickly, but the bulk can take up to 24 h.

Unlike blast furnace slag (BFS), which is generated during the production of pig iron [7], converter slag is not widely used as a building material. Its bulk composition is very different, and more variable, with a much higher Fe_2O_3 and lower Al_2O_3 and SiO_2 content than BFS (Table 3.1). This results in the mineralogical composition that is given in Table 3.2 [4, 8]. C_2S (Ca_2SiO_4) is the most common phase, followed by Mg-wuestite ((Fe,Mg)O) and $C_2(A,F)$ ($Ca_2(Fe,Al)_2O_5$). Minor amounts of Magnetite (Fe_3O_4), C_3S (Ca_3SiO_5), and CaO can also be present. This is the reason why converter slag shows no pozzolanic reactivity but is instead slightly hydraulic. It has led researchers to investigate the application as a cement replacement material [9–11]. However, the reported reactivity is very low, and the hydration behavior poorly understood, which has prevented widespread use. Due to its high strength and density converter slag can also be used as an aggregate in concrete, but the presence of free lime often causes problems with expansion [12, 13] and so the slag is mostly used for backfilling if it is not directly stockpiled or landfilled [14].

Table 3.1: Typical composition ranges for converter slag and blast furnace slag.

Oxide (wt%)	CaO	Fe_2O_3	SiO_2	MgO	MnO	Al_2O_3	P_2O_5	TiO_2
Converter Slag	35–50	15–25	10–20	5–10	2–7	1–5	0.5–2	0–2
Blast Furnace Slag	30–50	0.5–2.5	25–40	5–10	0–1	5–15	0–1	0–1

Table 3.2: Typical mineralogy of converter slag.

Formula	Name	Amount (wt%)
β-C_2S	Larnite/Alite	}30–50
α'_H-C_2S	Bredigite	
$C_2(A,F)$	Srebrodolskite	15–25
(Fe,Mg)O	Magnesiowuestite	20–30
Fe_3O_4	Magnetite	0–10
CaO	Free lime	0–10
C_3S	Hatrurite/Belite	0–5
MgO	Periclase	0–2

A thorough understanding of the physical and chemical characteristics of a by-product is the first requirement for its qualification as raw material for new applications. And quantitative analysis of microstructure and phase composition is the first step towards developing this understanding. Based on the knowledge of phase composition and microstructure, the material's potential reactivity can be anticipated (e.g. in binder systems or during weathering) and answers given to questions such as: Are there slag phases that will react, and if so under which conditions (pH, temperature, etc.)? And what are potential reaction products that form, given the reactivity of the slag's phase constituents?

Many potential strategies could be used to increase the reactivity of converter slag and turn it into a building material, such as mechanical or alkali activation, granulation, or even chemical modification of the liquid slag. However, in order to understand what is going on, a careful characterization of the starting material is required.

1.4 Representative sampling and analysis of a heterogeneous material like converter slag

The converter slag, produced in batches of 20–35 tons, is accumulated until enough slag is present for efficient metal-removal processing, commonly several hundreds of batches. During the processing of the slag, mixing is achieved to a certain extent, however, to obtain a slag sample representative of all batches, many subsamples need to be taken from the processing stream and combined. There are several protocols describing representative sampling of heterogeneous aggregates e.g. EN 932-1, ASTM D75 / D75M-19 [15, 16]. It is important to carefully assess the variability in mineralogy and microstructure before applying slag as a secondary raw material, especially in case of high-end applications such as cement replacement. In this respect, quality control is just as essential as it is for cement itself. The sequence of steps to follow for characterization is given in Table 3.3. Following the AP04 [16] protocol a sampling survey of the 0–25 mm fraction of the processed slag yields 32 kg of material, that is broken and split to obtain a sample for chemical analysis and QXRD, and a sample of the 1–2 mm fraction for microscopy. The 1–2 mm grain fraction is ideal for this purpose, because 1) with ~800 grain in one 5 cm round polished mount (Figure 3.3), statistically many individual batches should be represented [17], 2) with microstructural features at the scale of 100 μm, individual grains of this size can still give a fair representation of the diversity in microstructures, 3) in smaller size fractions, grains are overrepresented that show effects of water-cooling, carbonation and oxidation as these processes tend to produce smaller grains. For microstructure analysis, the slag grains are mounted in epoxy-resin and polished. Care needs to be taken since the slag is reactive with water and CO_2, and water-free polishing is required using an alcohol-based polishing suspension. Next, an optical microscopy image is acquired of the entire sample mount surface at low magnification (Figure 3.3) (objective Epiplan NEOFLUAR 5.0x/0.13 with ~2 μm resolution and 1 μm pixel size). Such images (~14 GB raw data, 1GB as JPEG) are sufficient for basic Image Analysis to resolve e.g., C_3S phenocrysts and free lime quantitatively in the accumulated slag batch in terms of numbers of grains with these features. On a subset of the grains, SEM-analysis is performed.

Table 3.3: Sequence of steps for characterization of heterogeneous raw materials like converter slag.

Step	Purpose	Sample preparation
1	Representative sampling of slag	AP04 protocol 32 kg in 32 subsamples from heap [17]
2	Representative subsample	Crushing and sample splitting
3	Chemical analysis (XRF)	Grinding analytical fine (125 um)
4	Leaching (ICP-OES) one batch leaching (EN 12457-2) or column test (NEN 7343)	One batch test: Sieving/Crushing below 4 mm Column test: same, 95% below 4 mm
6	Phase analysis (XRD followed by Rietveld quantification)	Grinding to 15 μm with internal standard (Si)
7	Microstructure analyses (Reflected light microscopy and SEM-EDS)	Mounting in epoxy resin, grinding and water-free polishing down to 0.3 μm, C-coating for SEM-EDS

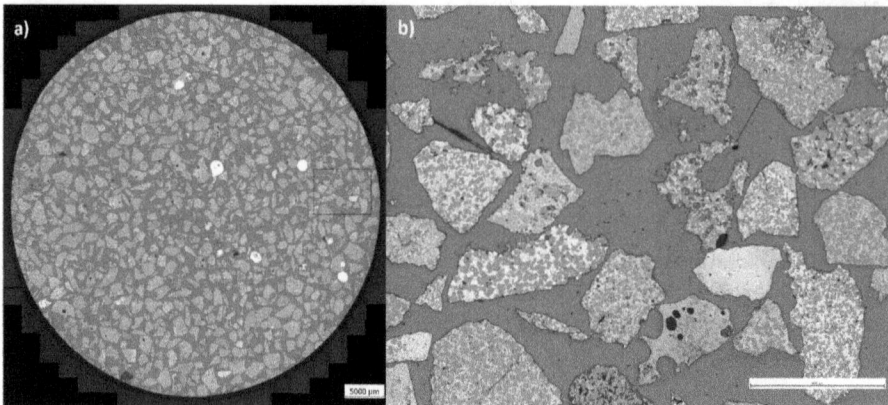

Figure 3.3: a) Polished mount of converter slag grains for optical microscopy and PARC analysis b) Grain mount detail.

The analytical approach with SEM and EDS Spectral Imaging is essentially the same for all ceramic, geologic and slag materials. Being inherently non-conductive, the samples are carbon coated prior to SEM analysis. Back Scatter (BSE) and Spectral Images are typically collected at 15kV for several adjacent fields (as many as 10×10). Each field has a resolution of 1024×768 (BSE) or 512×384 pixels (SI), for a maximal spatial resolution, of 0.5 and 1 µm respectively. A lower resolution can be chosen if the material consists of large homogeneous single-phase domains, or if a large area needs to be imaged. A higher resolution is less useful because the sample volume analysed will start overlapping between adjacent pixels because there is a minimum volume of about 1 µm from which the EDS signal is generated. With lower acceleration voltage the step size can be decreased somewhat (typically 0.7 micron at 10 kV), because the size of the minimum excitation volume decreases.

In our lab data are collected using SEMs with dual EDS detectors roughly at 180° opposite positions on the SEM column. Instruments with a single detector can be used equally well, but acquisition times will be longer. We make use of two FEG-SEM instruments, a JEOL 7001F equipped with a Thermo-Fischer Scientific microanalysis system with two SDD/EDS detectors of 30 mm^2 each, and a ZEISS Gemini450 with an Oxford microanalysis system with two SDD/EDS detectors of 170 mm^2 each. The acquisition time for the Spectral Image is dependent on the count rate, i.e. the size of the EDS detector and beam current, and the desired spectral resolution. Using the Thermo Fischer Scientific system with beam conditions of 15 kV and 10nA beam current a Spectral Image for one field is acquired in 30 min. With the Oxford system the acquisition can be completed in 5 min.

2 Phase identification from Spectral Image data – application of PARC

2.1 General working of PARC (Phase Recognition and Characterization)

PARC converts Spectral Image data into phase distribution maps. The SI data set contains the EDS spectrum for each pixel of an image field, that is 512 × 384 pixel and their corresponding spectra (Figure 3.2). The PARC software groups these pixels based on the occurrence of peaks above a user defined intensity in the EDS spectrum (Figure 3.4). The spectrum below 0.9 keV is excluded. Figure 3.5 shows a list of such groups, the amount of pixel that belong to it and the peaks they are based on.

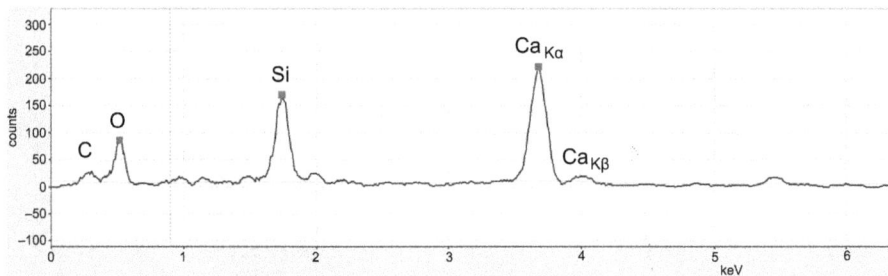

Figure 3.4: Example of a single pixel spectrum corresponding to a PARC group generically labelled {Si, Ca}. The peaks above the threshold are marked red. Energies below 0.9 keV are cut-off (O and C peaks).

The 0.9 keV cut-off means that only elements with a higher atomic number than $Z = 10$ (Ne and higher) are used for the PARC grouping (see Section 2.2). The PARC pixel sorting results in groups of unique peak/element combinations that fall into one of three categories:
1) 'empty pixels' with no peaks above the 0.9 keV cut-off and the minimum peak intensity
2) 'Embedding pixels' selected using user defined energy filtering to separate the mounting medium (usually epoxy) from the sample material
3) PARC groups of unique element peak combinations

Conventionally PARC groups are named with the elements that define the group using the notation {element1, element2, etc}. Each pixel is assigned a color based on the group it belongs to and shown in the phase map. Groups are merged whose only difference is the absence/presence of a Kβ peak above the threshold, as they belong to the same phase. An example are Group 3 and 6 in Figure 3.5. After this a list of groups is

id	color	show	p1	p2	p3	p4	p5	cnt	net cnt	cnt %	net %	label	spect...
0		✓						0		0.00		unclassified	
1		✓						2798		1.42		empty spectra	1.00
2		✓						0		0.00		embedding	1.00
3		✓	Si-KA1	Ca-KA1				88381	88381	44.95	45.60		1.00
4		✓	Mg-KA1	Fe-KA1				37004	37004	18.82	19.09		1.00
5		✓	Al-KA1	Ca-KA1				19002	19002	9.66	9.80		1.00
6		✓	Si-KA1	Ca-KA1	Ca-KB1			15786	15786	8.03	8.15		1.00
7		✓	Mg-KA1	Ca-KA1	Fe-KA1			9541	9541	4.85	4.92		1.00
8		✓	Mg-KA1	Si-KA1	Ca-KA1			4982	4982	2.53	2.57		1.00
9		✓	Al-KA1	Si-KA1	Ca-KA1			3511	3511	1.79	1.81		1.00
10		✓	Mg-KA1	Si-KA1	Ca-KA1	Fe-KA1		3507	3507	1.78	1.81		1.00
11		✓	Al-KA1	Ca-KA1	Fe-KA1			3096	3096	1.57	1.60		1.00
12		✓	Mg-KA1					2870	2870	1.46	1.48		1.00
13		✓	Si-KA1	P-KA1	Ca-KA1			1171	1171	0.60	0.60		1.00
14		✓	Ca-KA1					939	939	0.48	0.48		1.00
15		✓	Ca-KA1	Fe-KA1				805	805	0.41	0.42		1.00
16		✓	Al-KA1	Ca-KA1	Ti-KA1			456	456	0.23	0.24		1.00
17		✓	Mg-KA1	Ca-KA1				356	356	0.18	0.18		1.00
18		✓	Si-KA1	Ca-KA1	Fe-KA1			342	342	0.17	0.18		1.00
19		✓	Al-KA1	Ca-KA1	Ca-KB1			313	313	0.16	0.16		1.00
20		✓	Si-KA1	P-KA1	Ca-KA1	Ca-KB1		292	292	0.15	0.15		1.00
21		✓	Mg-KA1	Al-KA1	Ca-KA1	Fe-KA1		238	238	0.12	0.12		1.00
22		✓	Al-KA1	Si-KA1	Ca-KA1	Ca-KB1		195	195	0.10	0.10		1.00
23		✓	Ca-KA1	Ti-KA1				133	133	0.07	0.07		1.00
24		✓	Mg-KA1	Si-KA1	Ca-KA1	Ca-KB1		90	90	0.05	0.05		1.00
25		✓	Ca-KA1	Ca-KB1				78	78	0.04	0.04		1.00
26		✓	Mg-KA1	Al-KA1	Ca-KA1			72	72	0.04	0.04		1.00
27		✓	Si-KA1	S-KA1	Ca-KA1			72	72	0.04	0.04		1.00
28		✓	Al-KA1	Si-KA1	Ca-KA1	Fe-KA1		70	70	0.04	0.04		1.00
29		✓	Fe-KA1					68	68	0.03	0.04		1.00
30		✓	Mg-KA1	Si-KA1	Fe-KA1			56	56	0.03	0.03		1.00
31		✓	Si-KA1	Ca-KA1	Ti-KA1			48	48	0.02	0.02		1.00
32		✓	Al-KA1	Ca-KA1	Ca-KB1	Fe-KA1		48	48	0.02	0.02		1.00
33		✓	Mg-KA1	Mn-KA1	Fe-KA1			38	38	0.02	0.02		1.00
34		✓	S-KA1	Ca-KA1				31	31	0.02	0.02		1.00
35		✓	S-KA1	Ca-KA1	Fe-KA1			23	23	0.01	0.01		1.00
36		✓	Al-KA1	Si-KA1	Ca-KA1	Ti-KA1		19	19	0.01	0.01		1.00
37		✓	Mg-KA1	Si-KA1				18	18	0.01	0.01		1.00
38		✓	Al-KA1	S-KA1	Ca-KA1			16	16	0.01	0.01		1.00
39		✓	Si-KA1	S-KA1	Ca-KA1	Ca-KB1		15	15	0.01	0.01		1.00
40		✓	Mg-KA1	Ca-KA1	Mn-KA1	Fe-KA1		14	14	0.01	0.01		1.00
41		✓	Al-KA1	S-KA1	Ca-KA1	Fe-KA1		14	14	0.01	0.01		1.00

Figure 3.5: Shortened list of PARC groups after automatic grouping of pixels contained in one SI field of converter slag. The full list includes 74 different groups.

generated and further refinement steps can be taken by merging groups that belong to the same mineral phase.

Often, with compositional variability in phases, they become split up into multiple different PARC groups when different elements reach a peak-height above threshold. In converter slag this is frequently the case for $C_2(A,F)$ and $(Mg,Fe)O$. $(Mg,Fe)O$ for example also contains Mn and can get split up into the groups {Mg}, {Mg,Fe}, {Fe} and {Fe,Mn}. For this reason, PARC groups can be merged if so desired.

Additionally, each group should be checked for the occurrence of different phases within the same group, which can happen if the phases contain the same element peaks, but with different peak intensities. An example in converter slag are C_2S and C_3S that will both get sorted into a {Si,Ca} group. These different phases can be separated using density plots of element vs element intensity (or mathematical expressions of peak intensity combinations). The number of pixels with the same intensities is indicated with a color scale (from violet for low abundance, to red, for high abundance) (Figures 3.6 and 3.7). A single cluster of pixels usually indicates the presence of only one phase (Figure 3.6), while two or more distinct clusters indicate

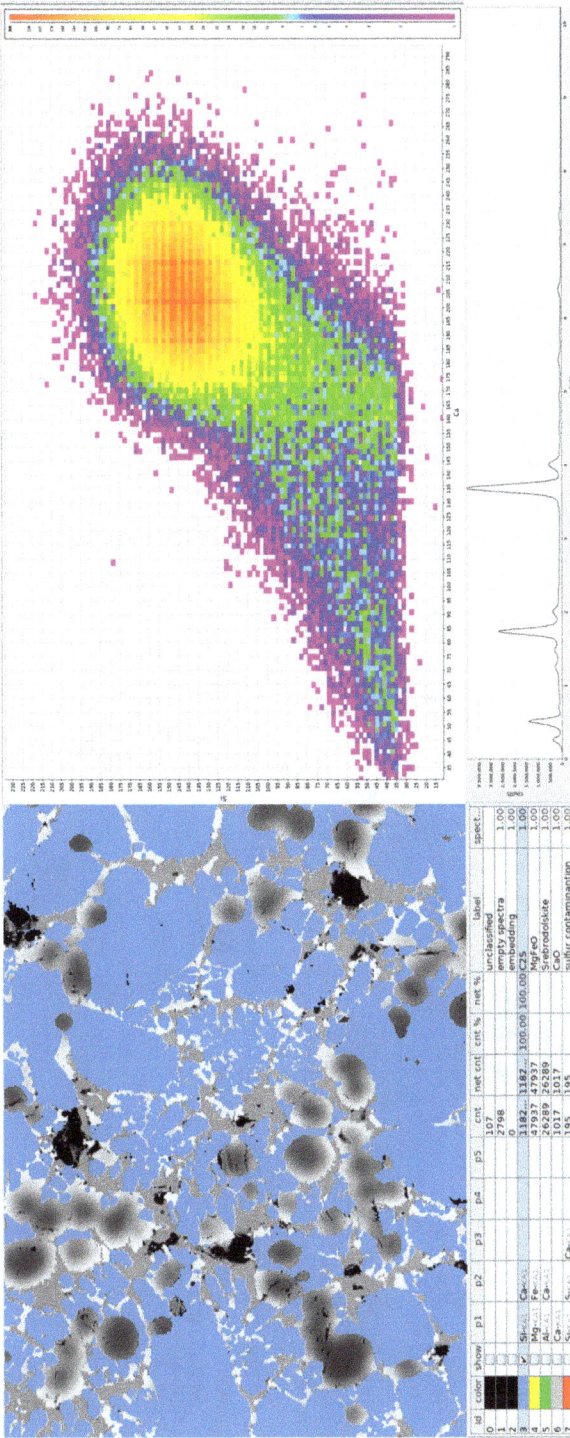

Figure 3.6: From top left: Location of pixels belonging to {Si,Ca} PARC group, corresponding to C₂S, in BSE image. Top right: A Ca-Si density plot of all pixel belonging to the {Si,Ca} group. Lower right: The sum spectrum of all C₂S pixels representing the average phase composition of this image field. Lower left: The PARC groups table with the {Si,Ca} group selected showing that 118.000 pixels are assigned to this group.

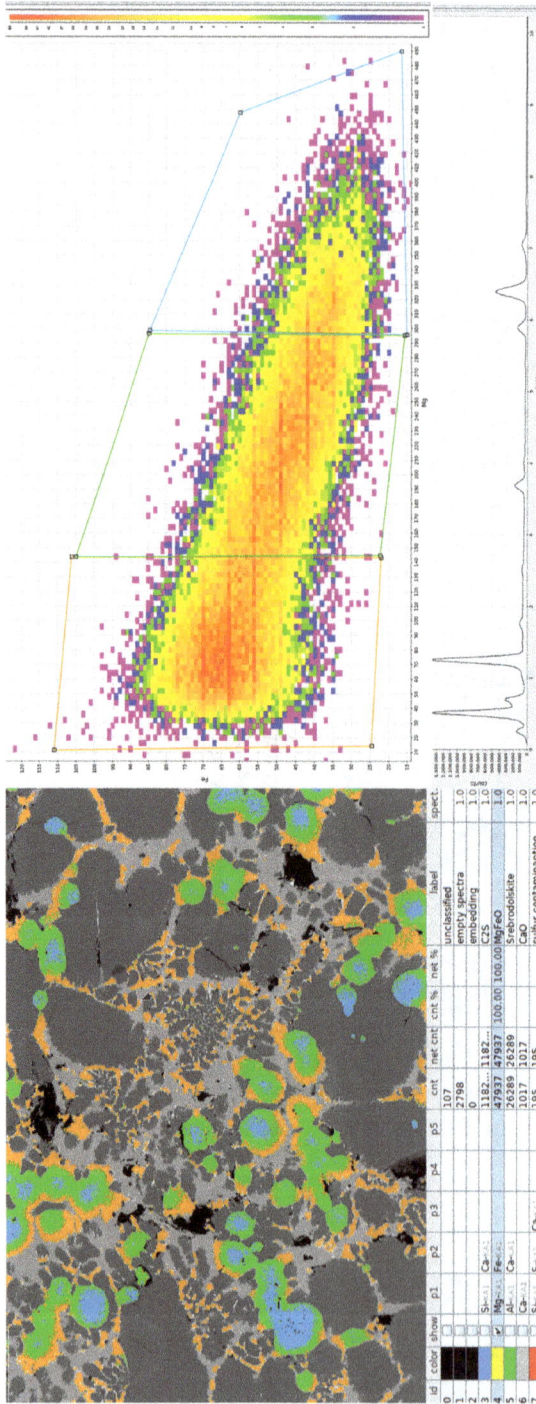

Figure 3.7: From top left: Location of pixels belonging to {Mg,Fe} PARC group, corresponding to (Mg,Fe)O, in BSE image. Top right: A Mg-Fe density plot of all pixel belonging to the {Mg,Fe} group separated into 3 compositional ranges. The polygon colors correspond to the phase-colors in the BSE image. This indicates that (Mg,Fe)O cores are Mg-rich and Fe-rich compositions are found as rims and in the quench-crystallized ground mass. Lower right: The sum spectrum of all (Mg,Fe)O pixels representing the average phase composition of this image field. Lower left: The PARC groups table with the {Mg,Fe} group selected showing that 48.000 pixels are assigned to this group.

the presence of several phases that should be separated. If the pixels of a density plot are evenly spread out over a larger area it is usually evidence for a solid solution (Figure 3.7). The phases defined this way can be stored as a model and applied to different SI fields of the same sample as well as to other samples.

Once this model is established (Figure 3.8) the composition of each PARC group can be determined by exporting the sum spectrum of all pixels belonging to the group and quantifying it. Marginal pixels which are bordering on other PARC phases in an image are excluded in the quantification by applying a so-called erosion, because their spectrum likely is contaminated by signal from the adjacent phase (Figures 3.9 and 3.10). Pure spectra, quantified standard less using Z-A-F correction, yield compositions that can be used to identify a phase and usually corresponds well with the known phase stoichiometry. The PARC groups should therefore be regarded as phases, defined based on composition.

color	show	p1	p2	p3	p4	p5	cnt	net cnt	cnt %	net %	label	spect...
	✔						3088		1.57		unclassified	
	✔						2		0.00		empty spectra	1.00
	✔						0		0.00		embedding	1.00
	✔	Si-KA1	Ca-KA1				1182...	1182...	60.13	61.09	C2S	1.00
	✔	Mg-KA1	Fe-KA1				47937	47937	24.38	24.77	MgFeO	1.00
	✔	Al-KA1	Ca-KA1				26273	26273	13.36	13.58	Srebrodolskite	1.00
	✔	Ca-KA1					960	960	0.49	0.50	CaO	1.00
	✔	Si-KA1	S-KA1	Ca-KA1			126	126	0.06	0.07	sulfur contaminantion	1.00

Figure 3.8: List of PARC phases after the final refinement step.

The PARC model can also be applied to SI-fields that were acquired in a "tiled" mode to cover a large sample area and the results are stitched together.

2.2 Limitations of EDS

2.2.1 Light elements

Light elements (<0.9 keV) analyses are unreliable with EDS since they overlap with the first series transition element's (Ti to Cu) L-lines, that are very commonly present in samples. The light-element signal is also strongly affected by topography and porosity. The signal from light elements can still be used for developing the PARC Phase-model in theory and allows to make distinctions for phases that only differ in light elements. For example, portlandite ($Ca(OH)_2$), calcite ($CaCO_3$) and lime (CaO), all belong to the same PARC group {Ca}, but all differ in their oxygen content. Likewise, Hematite (Fe_2O_3), Magnetite (Fe_3O_4), Wuestite (FeO) and Goethite (FeOOH) are all grouped together as {Fe}. However, if such phase separation is attempted based on oxygen intensity, exceedingly long acquisition times are required. For this reason, we prefer to use complementary techniques, such as Reflected light microscopy

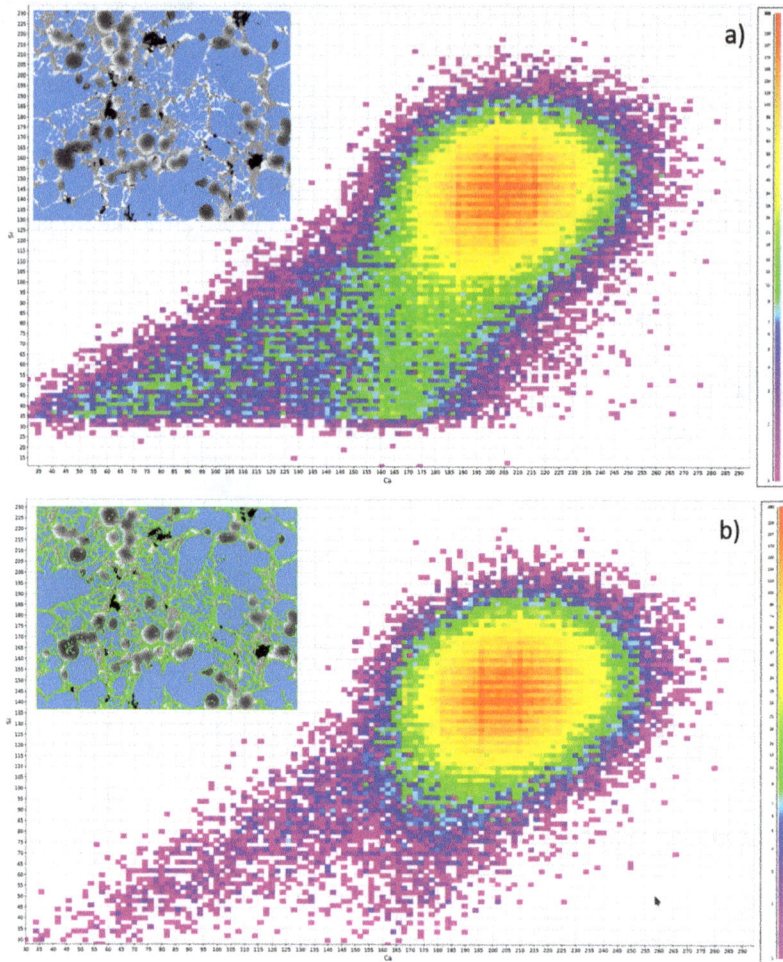

Figure 3.9: a) Density plot of the {Si,Ca} PARC group without erosion. The inset shows the location of the pixels belonging to this group. b) Density plot of the {Si,Ca} PARC group after erosion was applied. The inset shows the location of the pixels belonging to the group as well as the pixels that were removed during erosion.

to distinguish between the various Fe-oxides or lime, portlandite and calcite. Although quantitative analysis of oxygen is not performed, the knowledge that all elements in slag occur in oxidic form is the basis for the common practice of reporting chemical compositions as element oxides, as we do in this chapter.

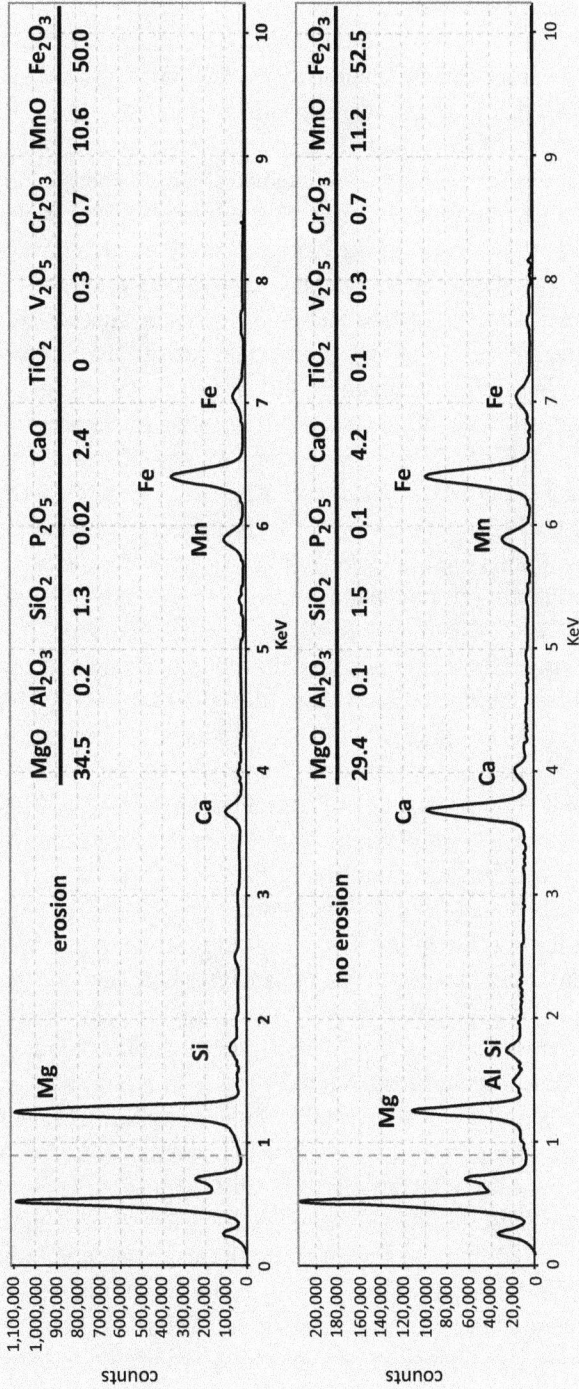

Figure 3.10: Sum spectrum and chemical composition of the {Mg,Fe} group with and without erosion.

2.2.2 Spatial resolution and mixed spectra

The spatial chemical resolution that can be obtained with EDS analysis is determined by the applied acceleration voltage of the electron beam (typically 15 kV are used), the average atomic number of the material composition and its density [1]. The effective spatial resolution obtained for a phase of known chemical composition density and section thickness can be modeled from first principles using Monte Carlo simulation of the electron trajectories in the material [18]. We only consider polished mounts or thin sections (30 μm) for optical microscopy here and not thin films (<0.1 μm) as used for Transmission Electron Microscopy. This is a useful exercise to determine the minimum size of intergrown phases that can still be resolved at the chosen conditions of spectral data acquisition. In Figure 3.11 the size of the excitation volume (=lateral and depth chemical resolution) is shown for tobermorite $(Ca_5Si_6O_{16}(OH)_2 \cdot 4H_2O)$ with 15 keV and 7 keV acceleration voltage (SEM settings) and a density of 2.4 g/cm^3 and a hypothetical density of 1.82 g/cm^3. This density can be thought of as tobermorite with 25% nanoporosity. This is a good approximation for C-S-H gel that commonly contains high porosity and has a composition close to tobermorite. The Monte Carlo simulations model the electron paths and their energy loss with material penetration. As long as electrons still have sufficient energy, they can generate characteristic X-rays. For a tobermorite composition of normal density (2.4 g/cm^3) at 15 keV most emitted characteristic X-rays, (thus detectable with EDS) are derived from penetration depths of 1 μm, at 7 keV from depths of 0.3 μm. For the lower density tobermorite (1.82 g/cm^3) most emitted X-rays will be derived from depths to 1.5–2 μm at 15 keV and at 7 keV from depths of around 0.4 μm (Figure 3.12). Considering the symmetry of the excitation volume from which the X-rays derive, the lateral resolution is similar to these depth values.

   Considering the major improvement in spatial resolution from 1–2 μm at 15 keV to 0.3–0.4 μm at 7 keV it is tempting to acquire SI datasets at low acceleration voltage, since a step-size of 0.4 μm would have no overlap in analyses between adjacent points. However, the low acceleration voltage comes with a penalty in X-ray yield at the EDS detector. As can be seen in Figure 3.12, at 7 keV the emitted intensity for Ca drops by 1/3 to 1/2 and for heavier elements the intensity loss would become even more severe. For this reason, it is not practical to acquire SI datasets at low acceleration voltages, because longer counting times are required, and common elements of the first transition series (e.g. Fe) cannot be detected with their K-lines. In general, lowering the acceleration voltage limits the energy spectral range available for element detection, concerning all K, L, and M lines. In summary, for the acceleration voltage of 15 keV a spatial compositional resolution of ~1 μm is achieved with adequate intensity for the common elements and reasonable measurement times, however for different phases intergrown at a smaller scale only an average composition of the intergrowth will be obtained. To obtain a higher resolution, thin films (~100 nm) can

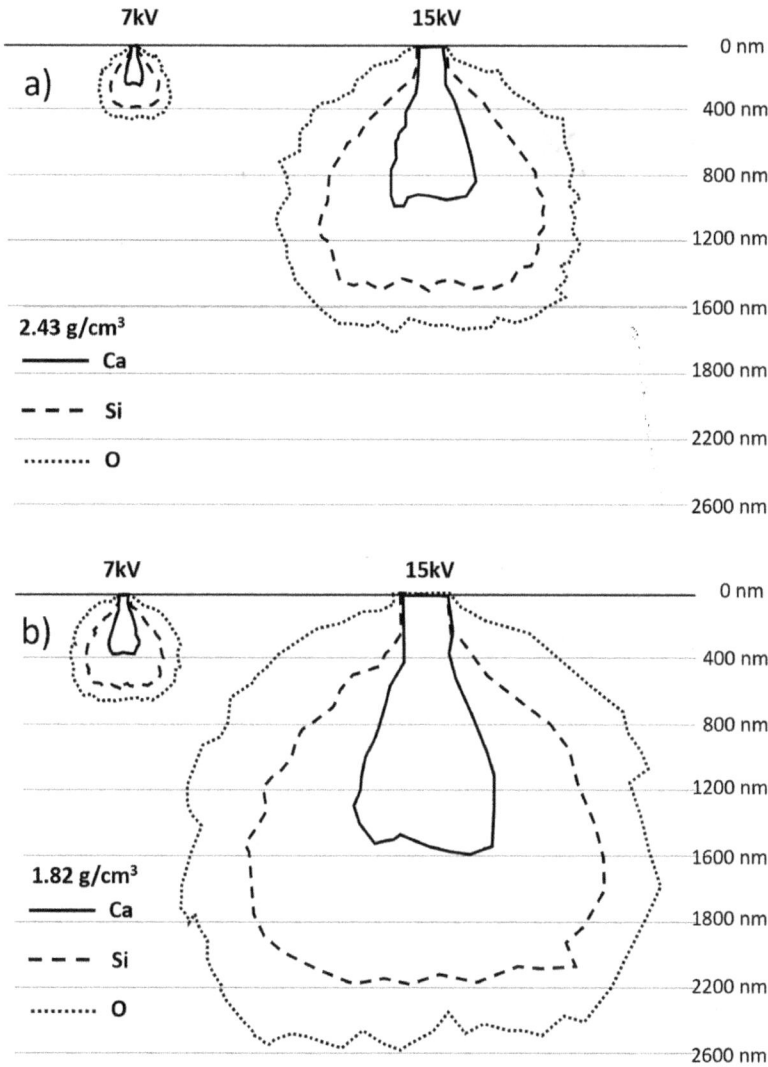

Figure 3.11: Monte Carlo Simulation of the excitation volume in Tobermorite using a 7 and 15 kV acceleration voltage and a density of 2.43 g/cm³ and 1.82 g/cm³.

be used, effectively removing most of the excitation volume as is common practice in Transmission Electron Microscopy (TEM) but will not be discussed here further.

For the PARC approach this has the following consequences. A finely intergrown domain in a sample material can be split up in individual PARC groups based on differences in peak intensities, if the individual phases are still resolved. If the intergrowth is resolved, but very small, a pure phase spectrum cannot be obtained because the pixel-erosion step will leave no spectrum. However, un-eroded spectra can be obtained and quantified. They will not yield a reliable phase composition but can still be useful.

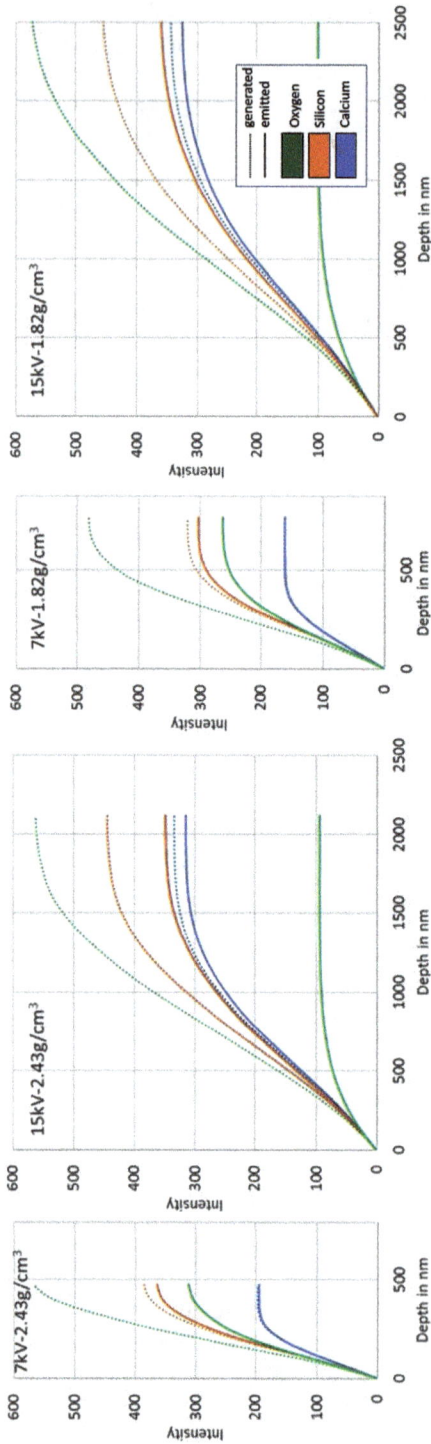

Figure 3.12: X-ray yields of the Monte Carlo simulation in Figure 3.11.

If the intergrowth is beyond resolution, it is indistinguishable whether a PARC group spectrum belongs to an intergrowth or a single phase (Figure 3.13). Compositions of finely intergrown phases that are not spatially resolved, such as cement-hydration products are still useful to obtain. They allow mass balance calculations to be performed on reaction product phases as will be shown in Section 3.3.

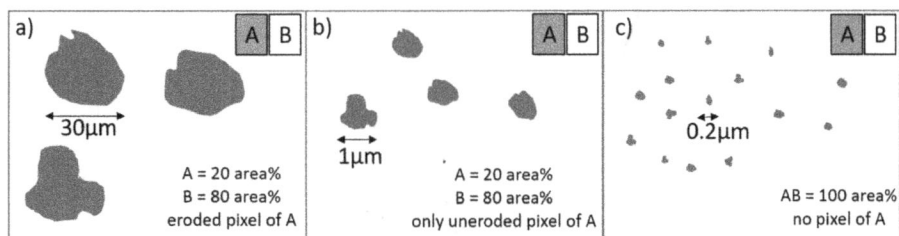

Figure 3.13: Effect of phase intergrowth size on PARC resolution a) Phase A is large enough that erosion can be applied to gain a clean spectrum b) Phase A is large enough to be resolved with EDS but erosion cannot be applied c) Phase is too small to be resolved.

2.3 PARC sensitivity to rare phases

Rare peak combinations occurring only in a few pixels are faithfully picked up in the PARC grouping. These rare phases could stem from sample contamination or be genuinely occurring rare phases associated with the material under investigation. For example, we recently discovered that converter slag can contain rare sulfur rich pixels (PARC group {FeS}), an interesting finding in itself, indicating that a sulfide phase can occur in such an oxidic slag (Figure 3.14). The {FeS} group can be found as last phase to crystallize with the eutectic phase assemblage of $(Mg,Fe)O$, $C_2(A,F)$ and Vanadium-rich C_2S. It is also an example of a phase where obtaining a clean spectrum using erosion is not possible and the resulting composition, expressed as oxide components, (46.4 wt% Fe_2O_3, 23.9 wt% SO_3, 23.5 wt% CaO and 6.2 wt% other oxides) is heavily influenced by the surrounding phases. The presence of large amounts of sulfur is nevertheless clear. Checking with optical microscopy and an EDX spot analysis, the presence of a sulfide phase was confirmed. The resulting composition of 44.1 wt% Fe, 24.1 wt% Mn, 1.6 wt% Ca and 30.0 wt% S is not stoichiometric and not surprisingly, deviates from the composition of the {FeS} group which is based on many more sulfide bearing pixels.

2.4 PARC sensitivity to low-concentration elements

Minor elements that occur at low concentration in a material will not be used to create PARC groups when their concentration always stays below the peak threshold value

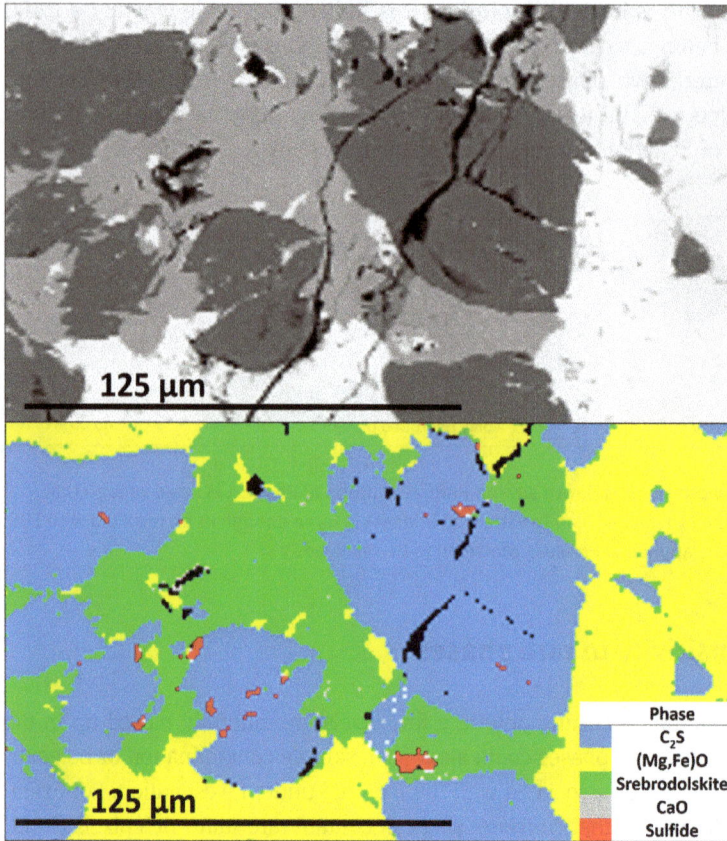

Figure 3.14: BSE image (top) and PARC phase map (bottom) of converter slag. The composition of the sulfide phase is 44.1 wt% Fe, 24.1 wt% Mn, 1.6 wt% Ca and 30.0 wt% S.

used for the grouping. Nonetheless, minor (<1 wt%) components of phases can be found in the PARC phase compositions. Especially the minor elements of abundant phases can be measured very accurately. The sum spectra of such abundant phases will have exceedingly high peak intensities (as high as millions of counts) and low concentration elements are then detectable above background. As an example, a converter slag associated with a steel batch containing 560 ppm Mo was investigated. It contained about 160 ppm Mo itself based on bulk chemical analysis (Table 3.4) and PARC analysis was performed to determine which slag phase hosts the Mo.

From the PARC phase map, the phase spectra were derived for all phases. It was found that only the $C_2(A,F)$ PARC phase contained Mo above the background level (Figure 3.15). Quantifying the sum spectrum for Mo-bearing $C_2(A,F)$ the Mo-content was determined to be 1300 ppm. Note that Nb is also present in $C_2(A,F)$.

Comparing the Mo content of $C_2(A,F)$ to the analysed bulk composition, knowing that $C_2(A,F)$ makes up 15–20 wt% of the slag, gives reasonable agreement. 15 wt% C_2

Table 3.4: Bulk chemical composition of converter slag associated with Mo-bearing steel.

Major Oxides	Al_2O_3	CaO	Fe_2O_3	K_2O	MgO	MnO	P_2O5	SiO_2	TiO_2
(wt%)	0.756	48.2	26.7	0.012	6.75	3.21	1.22	13.2	0.764
Minor Elements	Ba	Cr	Cu	Mo	Ni	Pb	V	Zn	LOI
(wt%)	0.0042	0.162	0.0011	0.0162	0.0059	0.0017	0.179	0.0047	2.3

(A,F) containing 1300 ppm Mo would imply 195 ppm for the bulk composition of the slag (and 260 ppm for 20 wt%), while the measured value is 160 ppm (Table 3.4). Minor element analysis with EDS should be treated with care, because it is not always clear to what extent Z-A-F correction procedures and background fitting are appropriate for quantification.

3 PARC application to hydration and carbonation of converter Slag

3.1 Hydration of converter slag

In order to use converter slag as a cement replacement material it is necessary to understand its reactivity. In a series of experiments, we looked at the hydration reactions of finely ground converter slag (D50 = 19 μm) using just water and slag at a ratio of 0.24 [19]. Only the PARC results are given here as an example (Figure 3.16) (Table 3.5), but the investigation also involved QXRD, XRF, TG and strength testing. The results show that the converter steel slag is indeed reactive under these conditions. The hydration product covers an area of 38.1 % and contains CaO, SiO_2 and Fe_2O_3 as the major oxides. XRD analysis shows that while the hydration product is largely amorphous, crystalline phases such as hydrogarnet, specifically $Ca_3Fe_2(SiO_4)_x$ $(OH)_{4(3-x)}$ are formed too. The converter slag system contains little Al_2O_3, so Fe-hydrogarnet (hydroandradite) is formed instead of the Al-analogue [20]. Due to the sub-micron size the hydrogarnet cannot be distinguished in the amorphous hydration product (Section 2.2.2) and the composition in Table 3.5 reflects the average composition of all hydration products.

A Ca-rich phase $(CaO/Ca(OH)_2/CaCO_3)$ that is likely a mix of $Ca(OH)_2$ and $CaCO_3$ is also present but its exact nature cannot be determined without the help of other methods (Section 2.2.1). The unreacted slag minerals $(C_2S, (Mg,Fe)O$ and $C_2(A,F))$ make up 57.6 area% in total. Due to their identical chemical composition it is not possible to differentiate between α'_H-C_2S and β-C_2S, but the C_2S in converter slag contains a significant amount of P_2O_5 and V_2O_5 that sets it apart from the C_2S in

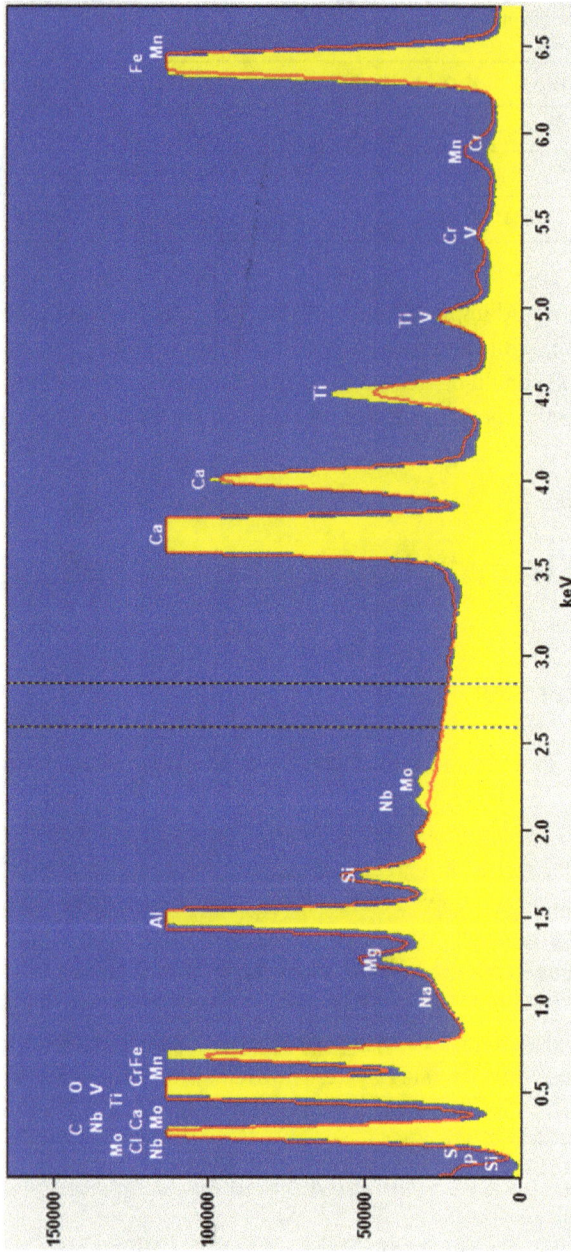

Figure 3.15: Sum Spectrum of $C_2(A,F)$ in standard Mo-free slag (red line) compared to Mo-bearing slag (yellow).

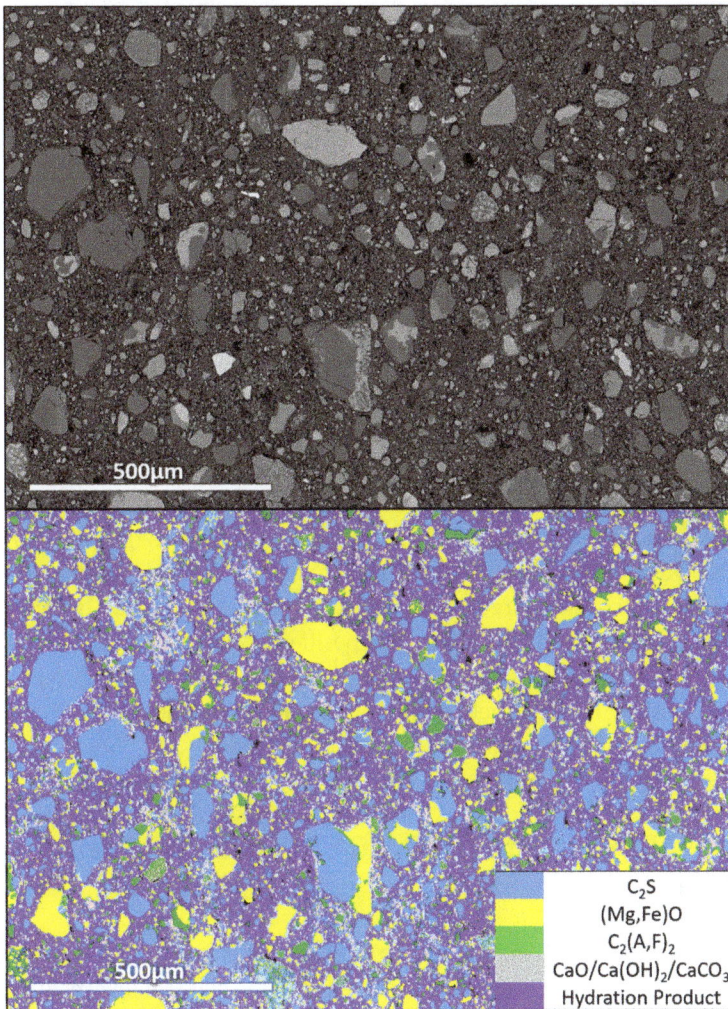

Figure 3.16: BSE image (top) and PARC phase map (bottom) of hydrated converter slag.

Table 3.5: Area% and composition of phases present in hydrated converter slag.

Phases	Area%	MgO	Al$_2$O$_3$	SiO$_2$	P$_2$O$_5$	CaO	TiO$_2$	V$_2$O$_5$	Cr$_2$O$_3$	MnO	Fe$_2$O$_3$
C$_2$S	32.8	0.28	0.53	28.21	3.38	61.39	1.19	1.12	1.22	0.09	2.17
(Mg,Fe)O	14.9	23.76	0.20	0.76	0.02	2.53	0.03	0.12	0.47	11.71	60.33
C$_2$(A,F)	9.9	0.64	11.15	1.83	0.08	42.33	5.12	1.51	0.39	1.24	35.10
CaO/Ca(OH)$_2$/CaCO$_3$	4.3	1.21	0.87	3.93	0.35	83.15	0.62	0.40	0.18	1.53	6.88
Hydration product	38.1	3.51	2.57	19.58	2.00	46.25	1.56	1.34	0.49	2.56	18.06

cement clinker. The Cr_2O_3 content is largely an artefact as a result of the Si + Ca sum peak in the C_2S spectrum overlapping with the Cr Kα emission line [21]. The (Mg,Fe)O also contains a significant amount of MnO (11.7 wt% on average).

3.2 Carbonation of converter slag

A binder can be formed by hydrating a material but also by carbonating it. So we looked at the carbonation of converter slag in mortars that were composed of Portland cement and slag at a ratio of 1 [22]. Typically, a dense microstructure is desired in concrete to achieve high strengths. However, for carbonation CO_2 needs to diffuse into the mortar, so a certain amount of open porosity is necessary, and part of this study was to figure out an optimal particle size distribution for the gas transport. Figure 3.17 and Table 3.6 show PARC results for a mortar that was part of this study. It is clear from the grey scale image alone that carbonation is quite effective in creating a very dense microstructure. This carbonation product is richer in SiO_2 and Al_2O_3 than the hydration product discussed in Section 3.1, due to the contribution of the Portland cement. It consists largely of calcite and amorphous phases, but they are too small to be resolved with EDX and the composition of the carbonation phase represents the average composition of several phases similar to the reaction product in hydrated converter slag. The relatively high V_2O_5 and P_2O_5 content also indicates that the C_2S in converter slag carbonates and releases these oxides.

3.3 Evaluating PARC data with QXRD, TGA and bulk chemical composition

In both the hydration and the carbonation study the challenge is to combine the data from various techniques to arrive at one consistent picture of the material investigated. A complete dataset ideally consists of 1) PARC phase amounts and compositions, 2) QXRD phase amounts and phase densities, 3) TGA volatile content and 4) Sample density (apparent and true density) 5) bulk chemical composition.

The first challenge is to reconcile the phase amounts and compositions from PARC with the bulk chemical composition of the material, e.g. from XRF analysis. A second challenge is to make the comparison between PARC and QXRD. How to perform such a comparison is the subject of this section using hydrated steel slag as an example (Figure 3.16). A word of caution about the data we show here, the values in the datasets are based on a combination of several different samples and have been simplified for the purpose of this example.

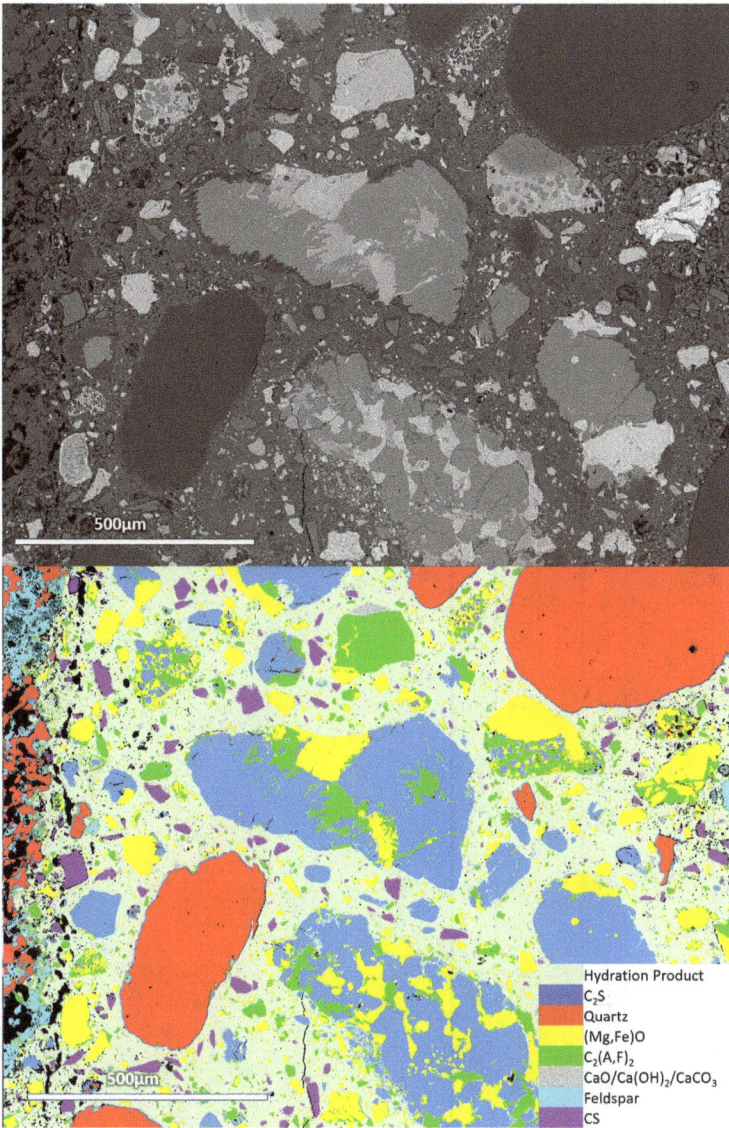

Figure 3.17: BSE image (left) and PARC phase map (right) of carbonated steel slag.

3.3.1 PARC phase identification from stoichiometry

For reconstruction of the bulk chemical composition of the material from PARC data it is critical to know the density of the PARC phases. In order to do this, we need to *identify* the PARC phases first, based on their composition and phase stoichiometry. PARC phase stoichiometry can be evaluated in the same manner as any phase

Table 3.6: Area% and composition of phases present in carbonated converter slag.

Phase	Area%	Na_2O	MgO	Al_2O_3	SiO_2	P_2O_5	SO_3	K_2O	CaO	TiO_2	V_2O_5	Cr_2O_3	MnO	Fe_2O_3
Carbonation product	45.2	2.0	2.2	2.3	25.5	2.5	0.6	0.1	53.1	1.4	1.1	0.5	1.0	7.8
$CaO/Ca(OH)_2/CaCO_3$	1.3	0.27	1.5	0.4	3.4	1.1	0.7	0.0	85.5	0.6	0.3	0.1	2.8	3.3
Feldspar	2.1	1.31	0.0	16.7	64.3	0.4	0.6	11.8	3.0	0.1	0.0	0.2	0.1	1.6
Quartz	15	0.19	0.0	0.0	98.0	0.0	1.5	0.0	0.0	0.0	0.0	0.0	0.0	0.2
CS	2.8	0.51	10.5	12.9	33.2	0.3	2.9	0.4	35.4	1.6	0.1	0.4	0.5	1.3
$C_2(A,F)$	8.3	0	0.7	7.0	3.7	0.2	0.1	0.0	39.4	4.7	1.4	0.5	1.9	40.3
C_2S	13.6	0.15	0.1	0.6	29.3	3.1	0.3	0.0	61.6	1.1	1.1	1.3	0.0	1.5
$(Mg,Fe)O$	11.3	0.13	18.5	0.3	2.9	0.2	0.1	0.0	5.9	0.1	0.2	0.5	13.4	57.6

composition obtained with other techniques. In Table 3.7, an exemplary data set is used to show how phase stoichiometry is determined.

The measured PARC oxide composition is given in weight%. Using the Mol-mass of the oxide components the molar composition for each phase is calculated and normalized to the expected number of cations per formulae unit. A normalization to the amount of oxygen in the formula is also possible [23]. For C_2S this would be 4 oxygen or 3 cations per formulae unit (p.f.u.). If properly done, the occurrence of solid solutions should be taken into account. C_2S (3 cations p.f.u.) forms a solid solution with C_3P (5 cations p.f.u.), while $C_4(A,F)_2$ (8 cations p.f.u.) forms a solid solution with CT (2 cations p.f.u.). Note that in Table 3.7 we have opted to qualitatively treat the phase stoichiometry regarding solid solution effects. So, instead of splitting the analysis into the solid solution end-member compositions, the analyses have been normalized to the dominant end-member stoichiometry, of 3 cations p.f.u. for C_2S and 8 cations p.f.u. for $C_4(A,F)_2$. Next, the cations that are likely to occupy the same structural position in the crystal lattice are grouped together. The oxy-anion forming cations: P, V, Si, Ti are grouped, trivalent cations Al, Fe^{3+}, Cr, divalent small cations Mg, Mn, Fe^{2+}, and the large (earth) alkali cations Ca, Na, K. For multivalent elements, like Fe occurring as Fe^{2+} and Fe^{3+}, the likely valency for the specific host phase is chosen. Minor elements in the PARC analyses, e.g. S and Cl, are omitted from the mineral formulae if they occupy no known site in the crystal structures. From Table 3.7 it is clear that the PARC groups {Mg,Fe}, {Si,Ca} and {Ca,Fe} can be identified respectively as wuestite (Mg,Fe)O, some C_2S polymorph(s) and Srebrodolskite $Ca_2(A,F)$. The {Ca} group as lime, portlandite or calcite, and the {Si,Ca,Fe} group as a reaction product phase.

3.3.2 PARC phase density

PARC phases are measured as area% in the SEM images, which directly corresponds to vol% [24, 25]. If their densities are known, the area% can be recalculated to wt% and compared with QXRD or XRF results. The easiest way to determine the density of a PARC phase after it has been identified is to simply look it up in the literature [26], but many phases, especially those containing solid solutions have a density that can be quite variable. A better alternative is to use the densities that are automatically calculated during QXRD.

The density of the {Ca} phase poses a problem, however, because it could correspond to CaO, $Ca(OH)_2$, or $CaCO_3$ and all three can be present in a cementitious system containing converter slag. XRD data indeed confirms that CaO and $Ca(OH)_2$ are present in our example so the weighted density of both phases can be used (Table 3.8). This was also done for other PARC groups such as {Mg,Fe}. The amount of water bound in {Ca} can also be determined using the TGA mass-loss corresponding to $Ca(OH)_2$. The TGA value for this sample is 0.71 wt% H_2O equivalent to 3 wt% $Ca(OH)_2$ (Table 3.9). Since $CaCO_3$ is present the same consideration is made for CO_2.

Table 3.7: Stoichiometry of PARC phases with allocated volatiles to lime portlandite and reaction product from TGA [12] and reconstructed bulk composition of slag.

Mol mass	PARC group	{Mg,Fe}	{Ca,Fe}	{Ca,Si}	{Ca}	{Ca, Si, Fe}	Bulk Composition	
	Phase	Wuestite +Magnetite	Srebrodolskite	Larnite + alpha'C_2S	f-C/CH/Cc	Reaction Product	calc. PARC	analyzed XRF
	Formula	$(Mg,Mn,Fe)O$	$Ca_2(Fe^{3+}, Al, Ti, V)_2O_5$	$Ca_2(Si, V, Fe)O_4$-$Ca_3P_2O_8$ ss.	$(Ca,Fe)O$			
g/mol	Oxide (wt%)							
62	Na_2O	0.03	0.15	0.14	0.04	0	0.08	
40.3	MgO	25.99	0.75	0.17	0.26	5.35	8.46	8.0
40.3	MgO	25.99	0.75	0.17	0.26	5.35	8.46	8.0
101	Al_2O_3	0.16	10.44	0.61	0.25	2.83	2.34	2.3
60	SiO_3	0.54	2.11	29.04	0.94	16.95	13.41	13.4
142	P_2O_5	0.02	0.16	3.33	0.31	1.93	1.52	1.6
	SO_3	0	0.08	0.17	0.17	0.23	0.12	
	Cl	0	0.02	0.02	0.03	0.07	0.03	
94.2	K_2O	0	0.04	0.03	0.08	0.03	0.03	
56	CaO	1.90	42.38	61.83	94.89	42.37	40.07	38.0

80	TiO$_2$	0.03	5.37	1.00	0.57	1.67	1.47	1.4
181.8	V$_2$O$_5$	0.10	1.48	1.07	0.05	1.24	0.85	1.0
152	Cr$_2$O$_3$	0.55	0.34	0.05†	0.03	0.43	0.31	0.3
71	MnO	11.94	1.34	0.03	0.46	3.46	4.25	4.2
159.8	Fe$_2$O$_3$(total)	58.74	35.33	1.73	1.92	23.44	26.82	28.0
159.8	Fe$_2$O$_3$	0	35.33	1.73	0	23.44	10.62	
71.8	FeO	52.85	0	0	1.73	0	14.58	
18	H$_2$O allocated	–	–	–	14.7	21.3		5.34
44	CO$_2$ allocated	–	–	–	–	4.5		0.97

(continued)

Table 3.7 (continued)

Mol mass / PARC group	{Mg,Fe}	{Ca,Fe}	{Ca,Si}	{Ca}	{Ca, Si, Fe}	Bulk Composition calc. PARC	analyzed XRF
Phase	Wuestite +Magnetite	Srebrodolskite	Larnite + alpha'C$_2$S	f-C/CH/Cc	Reaction Product		
Formula	(Mg,Mn,Fe)O	Ca$_2$(Fe^{3+}, Al, Ti, V)$_2$O$_5$	Ca$_2$(Si, V, Fe)O$_4$–Ca$_3$P$_2$O$_8$ ss.	(Ca,Fe)O			
Fe^{2+}/(Fe^{2+}+Fe^{3+})	1	0	0	1	0	99.8	
Cations p.f.u. Stoichiometry	1	8	3	1	8		
Si^{4+}	0.006	0.179	0.852	0.009	1.380		
P^{5+}	0.000	0.011	0.021	0.002	0.133		
V^{4+}	0.001	0.083	0.083	0.000	0.067		
Ti^{4+}	0.000	0.341	0.022	0.004	0.102		
Al$_{3+}$	0.002	1.051	0.021	0.003	0.274		
Fe^{3+}	0.000	2.247	0.038	0.000	1.433		
Cr^{3+}	0.005	0.023	0.000	0.000	0		
Mg^{2+}	0.402	0.095	0.007	0.004	0.648		
Fe^{2+}	0.459	0.000	0.000	0.014	0.000		
Mn^{2+}	0.105	0.096	0.001	0.004	0.238		
Ca^{2+}	0.021	3.846	1.944	0.959	3.695		
Na$^+$	0.001	0.025	0.008	0.001	0.000		
K$^+$	0.000	0.004	0.001	0.001	0.003		
H$_2$O	—	—	—	0.415		6.876	
CO$_2$			—			0.592	

Site totals (bracketed sums):

Site	{Mg,Fe}	{Ca,Fe}	{Ca,Si}	{Ca}	{Ca, Si, Fe}
Si+P+V+Ti	0.01	0.61	0.98	0.02	1.7
Al+Fe^{3+}+Cr		3.32	0.06		1.7
Mg+Fe^{2+}+Mn	0.99	0.19	0.01	0.98	0.9
Ca+Na+K	0.00	3.87	1.95	0.00	3.7

The density of an amorphous reaction product is usually not something that can be looked up or measured directly. Such phases also may contain sub-micron porosity, which is included in the Area%/Vol% as determined with SEM/PARC. A common example for a phase that is generally both amorphous and highly porous is C-S-H gel, a hydration product of Portland cement. Its nano-porosity is variable, depending on age, water/solid ratio of the cement as well as the composition of the reaction products themselves.

However, if all the densities of PARC phases corresponding to crystalline phases are known, the density of the reaction product including its nano-porosity can be calculated by difference from the bulk density, following:

$$\rho^{bulk} = \sum_{\varphi=1}^{n} \frac{Area\%_{\varphi xtal}}{100\%} \cdot \rho_{\varphi xtal} + \frac{Area\%_{react.\,prod.}}{100\%} \cdot \rho_{react.\,prod.}$$

Where ρ_{bulk} represents the measured value for the bulk material density, n is the number of phases, $Area\%_{\varphi Xtal}$ $\rho_{\varphi Xtal}$ and $Area\%_{react.prod}$, $\rho_{react.prod.}$ represent the $Volume\%$ and $density$ of the crystalline PARC phases (φ_{xtal}) and the reaction product, respectively. The expression states that each phase volume contributes to the total density of a sample and that this sum value should be equal to the bulk density. In Table 3.8 the results are presented for the example PARC data set.

The calculated value, by difference, is 3.28 g/cm^3 and is very sensitive to errors in the underlying data, which all end up in the density value for the reaction product {Ca, Si, Fe}. Alternatively a density for the amorphous component is assumed and a PARC based density for the sample material is independently calculated. We have used 2.6 g/cm^3 resulting in a material density of 3.44 g/cm^3 (Table 3.8) based on PARC, compared to 3.63 g/cm^3 from XRD.

3.3.3 PARC data consistency with bulk composition

Once each PARC phase has been identified and the density is known, their contribution in the bulk chemical composition can be calculated and then summed up. This makes it possible to compare the PARC results to bulk chemical analysis such as XRF. However, it should be realized that the PARC phase compositions do not include volatiles, since EDS measurements cannot easily be used to determine water or carbonate contents. So, the calculated bulk composition is volatile-free and can be compared to an also volatile-free XRF bulk composition. This summation can be expressed as:

$$^{bulk}_{i}m_{Ox} = \sum_{\varphi=1}^{n} \frac{Area\%}{100\%} \cdot \rho_{\varphi} \cdot {}^{\varphi}_{i}m_{Ox}$$

Table 3.8: Calculating Vol% from PARC phases into wt%.

PARC group Phase Formula	{Mg,Fe} Wuestite + Magnetite (Mg,Mn,Fe)O	{Ca,Fe} Srebrodolskite $Ca_2(Fe^{3+}, Al, Ti, V)_2 O_5$	{Ca,Si} Larnite + alpha'C2S $Ca_2(Si, V, Fe)O_4$- $Ca_3P_2O_8$ ss.	{Ca} f-C/CH/Cc CaO	{Ca, Si, Fe} Reaction Product	Total PARC
Density XRD (g/cm³) 3.63 Area%	18.4	12.8	33.4	6.6	28.8	100.0
Group Density †	5.14	3.79	3.27	2.59	2.59 3.28*	3.44
Norm. wt% phase	27.4	14.1	31.8	5.0	21.7	100.0

† densities based on XRD Rietveld

* PARC effective density of reaction product calculated by difference from XRD total density

Where, $^{bulk}_i m_{Ox}$ represent the total content of an oxide (in wt%) in the bulk chemical composition $^{\varphi}_i m_{Ox}$ represents the wt% of an oxide in phase φ, ρ_φ is the density of the phase φ, while n is the number of PARC phases.

The total of the PARC phases in Area% (Vol_{total}) in the example of Table 3.8 amounts to a value lower than 100 %, when disregarding a contamination phase, or some pixels that could not be assigned to a proper phase. This does not affect the validity of the mass-based calculations and is unconnected to the sum total of PARC phase masses. The total mass being 100%, is caused by the chosen normalization to 100% for the phase weight fractions in PARC. Choosing to normalize to 100 wt%, or to 100 wt% minus wt% of volatiles (TGA), are equally valid choices. However, when summing up fractional contributions of each oxide to derive the bulk chemical sample composition (for comparison to an independent bulk chemical analysis e.g., from XRF), the total mass of the bulk chemical composition should arrive at exactly the value to which was chosen to normalize the bulk composition (Table 3.7). Leaving out untrustworthy components (e.g., minor SO_3, Cl and Cr_2O_3 from C_2S), the bulk total will become correspondingly lower.

3.3.4 Combining TGA data with PARC phase compositions

Table 3.9: Hydrated converter slag volatile content in wt% from TGA as well the $CaCO_3$ content based on H_2O loss.

$H_2O < 400\ °C$	H_2O in $Ca(OH)_2$	CO_2 in $CaCO_3$	$Ca(OH)_2$ based on H_2O
4.61 wt%	0.97 wt%	0.73 wt%	3 wt%

Mass loss in TGA measurements can be attributed to the volatile species H_2O and CO_2, based on gas-analysis, and on association with the decomposition temperatures of phases. In Table 3.9 the mass losses for our example material are presented.

The volatiles measured with TGA up to a temperature of 400 °C were allocated to the reaction product, PARC phase {Si,Ca,Fe}. It should be kept in mind that TGA measures the volatiles in the bulk sample, but these volatiles are only present in a small number of phases, because the original slag minerals are known to be anhydrous. That means that the 0.71 wt% portlandite water (TGA) is contained in 5 wt% of the PARC {Ca} phase (Table 3.8), which in turn corresponds to a H_2O content of 14.7 wt% in the {Ca} phase. Similarly, the 4.61 wt% H_2O-content and 0.97 wt% CO_2-content of the bulk sample is contained in ~21.7 wt% reaction product and works out to 21.3 wt% H_2O and 4.5 wt% CO_2 in the {Si,Ca,Fe} phase. It should be clear that changing the assumptions on phase densities affects the mass proportion of the reaction product in the balance calculated using the equation, and therefore also affects the concentrations of the allocated volatiles in the reaction product.

3.3.5 Comparison of PARC phase amounts to QXRD

XRD data is indispensable to identify the phases in unknown materials and for optimal consistency of a Rietveld quantification, the PARC phase compositions (mol fractions) can be used to define the site occupancy of existing crystal structures [27, 28]. As indicated in Table 3.7, the Al/Fe^{3+} ratio of the PARC reaction product phase is used for the hydrotalcite ($Mg_6Al_2[(OH)_{16}|CO_3] \cdot 4H_2O$) and hydrogarnets ($Ca_3Fe_2(SiO_4)_x(OH)_{4(3-x)}$) in Table 3.10.

The results of QXRD can be compared to the PARC phase amounts recalculated to wt%, however, for a good comparison the advantages and disadvantages of each method need to be considered. In general, differences in crystal structures are picked up by XRD, while differences in chemistry are seen in PARC. This means that it is easy to detect different amorphous phases using PARC (as long as their grain size is above the resolution limit) while this is not possible with QXRD unless methods like PONCKS are used [29], that come with their own limitations. In return it is often possible to detect crystalline phases with XRD that are below resolution for PARC. Many phases measured by QXRD directly match a PARC phase and phase amounts can be compared if the results from PARC are converted into wt%. This is the case for $C_4(A,F)_2$ for example, which directly matches the PARC group {Ca,Fe}. The PARC group {Si,Ca} however contains the mineral phases β-C_2S, α'_H-C_2S and C_3S, because for all three Ca and Si are the only two elements with peaks above the threshold in the EDS spectrum. So, they need to be summed up for a comparison (Table 3.10). The same calculation can be performed for the {Ca} group with the corresponding Ca-phases in XRD (CaO, $Ca(OH)_2$, $CaCO_3$).

QXRD also tells us that our reaction product is largely amorphous but that it contains hydrotalcite ($Mg_6Al_2[(OH)_{16}|CO_3] \cdot 4H_2O$) and hydrogarnet ($Ca_3Fe_2(SiO_4)_x$ $(OH)_{4(3-x)}$) (Table 3.10) [20]. These crystalline reaction products are very small and intergrown with the amorphous reaction products so they cannot be seen with PARC and are part of the reaction product phase {Si,Ca,Fe}. For volatile-free converter slag the PARC and QXRD results generally match well [4]. For samples including volatiles, such as hydrated or carbonated samples, the story becomes a bit more complicated, because CO_2 and H_2O are not detected via EDX and these phases tend to contain micro porosity, which makes it more difficult to apply the correct density (Section 3.3.2).

Fortunately, TGA measurements provide information on the total volatiles in a sample, that can be compared to volatiles from the crystalline volatile-bearing phases as quantified with QXRD (Table 3.10). In general, and in the example of Table 3.9, we find more $Ca(OH)_2$ in the TGA measurement than detected with QXRD. Also, the amount of {Ca}-phase in PARC typically exceeds the amount which is found with

Table 3.10: Rietveld phase amounts compared to PARC.

Phases	Formula	XRD wt%	2σ error‡	Phase Density"	H2O contr. Bulk	CO2 contr. bulk	PARC group	XRD wt%
Total		100		3.64*	1.90	0.82		100
Amorphous		23.0	2.8	2.6†				
Fe-Katoite	$Ca_3Fe_2^{3+}(OH)_{12}$	6.0	0.4	2.9	1.52		{Ca,Si,Fe}	29.4
Hydrotalcite	$Mg_6(Fe^{3+},Al)_2(CO_3)(OH)_{16}\cdot4(H_2O)$	0.4	1.0	2.1	0.13	0.03		
Calcite	$CaCO_3$	1.8	0.3	2.7		0.79		
Portlandite	$Ca(OH)_2$	1.0	0.1	2.2	0.24			
Lime	CaO	0.2	0.1	3.3			{Ca}	3.0
Hatrurite C3S	Ca_3SiO_5	1.9	0.5	3.2				
α C2S	Ca_2SiO_4	2.0	0.3	3.3				
α' C2S	Ca_2SiO_4	9.7	0.7	3.3				
Larnite β C2S	Ca_2SiO_4	15.8	1	3.3			{Ca,Si}	29.4
Fe-Perovskite	$Ca(Ti,Fe)O_3$	0.4	0.5	4.0			{Ca,Fe}	14.4
Srebrodolskite	$Ca_2Fe_2O_5$	14.0	0.7	3.8				
Iron metal	Fe-metal	0	0.1	7.9				
Magnetite	Fe_3O_4	10.5	0.4	5.1				
Mg-Wuestite	$Mg(Fe)O$	6.8	1.4	4.6				
Fe-Wuestite	$Fe(Mg)O$	6.6	1.3	5.7			{Mg,Fe}	23.9

(continued)

Table 3.10 (continued)

Phases	Fe-Wuestite	Mg-Wuestite	Magnetite	Iron metal	Srebrodolskite	Fe-Perovskite	Larnite β C₂S	α' C₂S	α C₂S	Hatrurite C₃S	Lime	Portlandite	Calcite	Hydrotalcite	Fe-Katoite	Amorphous	Total
Group Density	5.14				3.79		3.27				2.59			2.6			3.44 ?
PARC wt%	27.4				14.1		31.8				5.0			21.7			100

† estimated density for amorphous content
* calculated material density from the fractional contributions of XRD phases
" phase densities based on XRD-Rietveld unit cell
‡ representative errors for Rietveld fit
? calculated material density from fractional contributions of PARC groups

QXRD. Determining an accurate $Ca(OH)_2$ content with TGA can be difficult because weight loss in the temperature range of $Ca(OH)_2$ decomposition can be influenced by the decomposition of other hydration products such as C-S-H or in our case hydrotalcite $(Mg_6Al_2[(OH)_{16}|CO_3] \cdot 4H_2O)$ and hydrogarnet $(Ca_3(Fe)_2(SiO_4)_x(OH)_{4(3-x)})$. More detailed information about that topic can be found in Scrivener et al [30].

Lastly, the crystalline product phases as identified with XRD should reside in the PARC product phase area. From QXRD it appears that 20 % of the product phase is crystalline, consisting of hydrotalcite $(Mg_6Al_2[(OH)_{16}|CO_3] \cdot 4H_2O)$ and hydrogarnets $(Ca_3Fe_2(SiO_4)_x(OH)_{4(3-x)})$, and approximately 80 % is amorphous. Note that the XRD measurements are not affected by sub-micron porosity the way the area% in PARC are. The PARC reaction product amounts to 21.7 wt% and the XRD to 29.4 wt% (of which 6.4 wt% is crystalline phases). The weakest links in these calculations are: 1) the determination of the amorphous content with XRD 2) the density determination, by difference, of the reaction product based on the bulk density (Section 3.3.2). The amorphous content in our samples using XRD is derived by the internal standard method:

$$wt\%_{am} = (1 - r\%_{St}/wt\%_{St})/(100 - r\%_{St})*10^4,$$

where $r\%_{st}$ is amount of standard material added and $wt\%_{St}$ the proportion of standard material derived by the Rietveld method. The accuracy of the amorphous content is a result of the absolute error made by obtaining the amount of the standard material in the samples using the Rietveld method. It is reasonable to assume that the maximal absolute error is about 1 wt% when adding 10 wt% of standard. Consequently, according to the formula the detection of less than 10 wt% of amorphous material is not reliable. Zhao et al. [31] give a lower limit of 15 wt% amorphous material, above which the accuracy significantly improves.

The advice is to look carefully at the Rietveld error values and to be aware that 0.1 wt% difference in measured internal standard value works out to ~1 wt% difference in amorphous content. Therefore, the XRD-amorphous content, taking the error on the internal standard into account, could equally well be only half its value.

Once the volatiles from TGA are assigned to the known crystalline, volatile bearing phases, the remaining volatiles can be attributed to the amorphous reaction product. The composition of the amorphous reaction product can be analyzed further, when the crystalline hydration products as determined via XRD are subtracted from the {Si,Ca,Fe} PARC phase. The remaining phase is the average oxide composition of the amorphous hydration product. Assuming that the amorphous reaction product phases are equivalent in composition to their crystalline counterparts, Franco-Santos et al. [32] found that a fourth phase, a CSH-phase, was needed to complement the reaction product composition. The amount of this phase and its average composition can be calculated by subtracting the crystalline reaction products, whose amounts are known from XRD.

4 Outlook – Spectral Imaging (PARC) future development and integration with optical microscopy

To combine quantitative bulk material characteristics (chemical composition, phase amounts and density) with spatially resolved data (microscopy, phase compositions) is a challenging undertaking, and requires that detailed attention is payed to checks for internal consistency. A weakness of the current approach is that the density of the phase that is calculated by difference from the bulk density (Section 3.3.2) combines all errors of the other phase densities, much like the amorphous content determined with QXRD contains all errors of the other refined phases. The only independent consistency check is the comparison of the calculated PARC bulk composition to an independent bulk analysis (XRF) (Section 3.3.3). The problem derives from the use of normalized 100% EDS analysis, which excludes H_2O and CO_2 and is caused by the practice of performing standardless analysis. Switching to calibrated analysis, as is common practice in Electron Microprobe analysis, would lower the analytical total of the oxides and allow the quantification of volatiles. However, it is unclear how nanoporosity would affect analytical totals in calibrated analysis.

Another major advancement would be possible by integration of large area SEM/PARC images with images based on light optical microscopy to examine the same area. To correlate information, it is important to get perfect matching within the resolution of the EDX (typically 1 µm) over the entire area. Image distortion differences between the two techniques lie far outside an acceptable 1 µm range. With single image fields (400×500 µm^2) the correlation can be obtained manually by defining corresponding features (pixels) between the images, subsequently calculating and applying the required correction. BigWarp [33], a plugin for FIJI, is such a tool for manual, interactive, landmark-based deformable image alignment [34]. For large stitched images, corresponding to many fields, such corrections can be defined routinely and is subject of ongoing development in our lab with the goal to achieve automated image alignment. Once light optical microscopy images can be aligned with SEM this also opens the possibility to integrate Raman and FTIR data. The combination of Raman-, or FTIR-microscopy with SEM/EDS data is a promising prospect as it could help to gain more information about amorphous hydration products among other things. The mapping could also be combined with TEM to gain more insight into the composition of very small or finely intergrown phases. Of course, for all microscopy methods corrections will be required for image correlation.

There is also potential to integrate MIP data with PARC. MIP gives volume fractions of pore sizes, but the spatial distribution of pore sizes in a sample is not known. However, missing mass in calibrated EDS spectra should be indicative about volatiles as well as pores.

References

[1] J. I. Goldstein, D. E. Newbury, J. R. Michael, N. W. M. Ritchie, J. H. J. Scott and D. C. Joy, Scanning electron microscopy and x-ray microanalysis, 2017.

[2] R. T. Chancey, P. Stutzman, M. C. G. Juenger and D. W. Fowler, "Comprehensive phase characterization of crystalline and amorphous phases of a Class F fly ash," *Cement and Concrete Research*, vol. 40, p. 146–156, 1 2010.

[3] P. T. Durdziński, C. F. Dunant, M. B. Haha and K. L. Scrivener, "A new quantification method based on SEM-EDS to assess fly ash composition and study the reaction of its individual components in hydrating cement paste," *Cement and Concrete Research*, vol. 73, p. 111–122, 7 2015.

[4] C. van Hoek, J. Small and S. van der Laan, " Large-Area Phase Mapping Using P h A se R ecognition and C haracterization (PARC) Software," *Microscopy Today*, vol. 24, p. 12–21, 2016.

[5] H. Jalkanen and L. Holappa, "Converter Steelmaking," in *Treatise on Process Metallurgy*, S. Seetharaman, Ed., Boston, Elsevier, 2014, p. 223–270.

[6] World Steel Association, "Steel industry co-products – worldsteel position paper," 2018.

[7] Y. Yang, K. Raipala and L. Holappa, "Chapter 1.1 – Ironmaking," in *Treatise on Process Metallurgy*, S. Seetharaman, Ed., Boston, Elsevier, 2014, p. 2–88.

[8] J. Waligora, D. Bulteel, P. Degrugilliers, D. Damidot, J. L. L. Potdevin and M. Measson, "Chemical and mineralogical characterizations of LD converter steel slags: A multi-analytical techniques approach," *Materials Characterization*, vol. 61, p. 39–48, 1 2010.

[9] Q. Wang, P. Yan, J. Yang and B. Zhang, "Influence of steel slag on mechanical properties and durability of concrete," *Construction and Building Materials*, vol. 47, p. 1414–1420, 10 2013.

[10] S. Kourounis, S. Tsivilis, P. E. E. Tsakiridis, G. D. D. Papadimitriou and Z. Tsibouki, "Properties and hydration of blended cements with steelmaking slag," *Cement and Concrete Research*, vol. 37, p. 815–822, 6 2007.

[11] J. N. Murphy, T. R. Meadowcroft and P. V. Barr, "Enhancement of the cementitious properties of steelmaking slag," *Canadian Metallurgical Quarterly*, vol. 36, p. 315–331, 1997.

[12] L. M. Juckes, "The volume stability of modern steelmaking slags," *Mineral Processing and Extractive Metallurgy*, vol. 112, p. 177–197, 2003.

[13] G. Wang, "Determination of the expansion force of coarse steel slag aggregate," *Construction and Building Materials*, vol. 24, p. 1961–1966, 2010.

[14] H. Cheng, H. Yi, J. Wang, Y. Wan, G. Xu, H. Chen, H. Cheng, J. Wang, Y. Wan, H. Chen, H. Yi, J. Wang, Y. Wan, G. Xu and H. Chen, "An Overview of Utilization of Steel Slag," *Procedia Environmental Sciences*, vol. 16, p. 791–801, 2012.

[15] CDOT, "Colorado Procedure Standard Practice for Sampling of Aggregates," p. 3–6, 2017.

[16] F. I. f. Q. A. o. S. Management, *ACCREDITATION PROGRAMME BUILDING MATERIALS DECREE SECTION: SAMPLE PRE-TREATMENT AP04 – V Contents of Sample Pre-Treatment (V)*, 2005.

[17] J. C. O. Zepper, F. van Breemen, J. Small, K. Schollbach and S. R. van der Laan, "Quality monitoring and representative sampling of Converter slag from steel production for application in building materials," *in preparation*.

[18] D. Drouin, *Casino – Monte Carlo Simulations of electron trajectories in Solids*, 2016.

[19] A. Kaja, S. Schollbach, S. Melzer, S. R. van der Laan and H. J. H. Brouwers, "Hydration of potassium citrate-activated BOF slag," *Cement and Concrete Research*, vol. 140, p. 106291, 2020.

[20] B. Z. Dilnesa, B. Lothenbach, G. Renaudin, A. Wichser and D. Kulik, "Synthesis and characterization of hydrogarnet $Ca_3(Al_xFe_{1-x})_2(SiO_4)_y(OH)_{4(3-y)}$," *Cement and Concrete Research*, vol. 59, p. 96–111, 5 2014.

[21] K. Schollbach, M. J. Ahmed and S. R. Laan, "The mineralogy of air granulated converter slag," *International Journal of Ceramic Engineering & Science*, vol. 3, p. 21–36, 1 2021.

[22] G. Liu, K. Schollbach, S. van der Laan, P. Tang, M. V. A. Florea and H. J. H. Brouwers, "Recycling and utilization of high volume converter steel slag into CO2 activated mortars – The role of slag particle size," *Resources, Conservation and Recycling*, vol. 160, p. 104883, 2020.

[23] G. T. R. Droop, " A general equation for estimating Fe 3+ concentrations in ferromagnesian silicates and oxides from microprobe analyses, using stoichiometric criteria," *Mineralogical Magazine*, vol. 51, p. 431–435, 1987.

[24] M. D. Higgins, Quantitative Textural Measurements in Igneous and Metamorphic Petrology, Cambridge: Cambridge University Press, 2006.

[25] E. E. Underwood, "Quantitative Stereology for Microstructural Analysis," in *Microstructural Analysis*, 1973.

[26] Mineralogical Society of America, *Minerology database*.

[27] R. T. Downs and M. Hall-Wallace, "The American Mineralogist crystal structure database," *American Mineralogist*, 2003.

[28] M. Hellenbrandt, "The inorganic crystal structure database (ICSD) – Present and future," in *Crystallography Reviews*, 2004.

[29] N. V. Y. Scarlett and I. C. Madsen, "Quantification of phases with partial or no known crystal structures," *Powder Diffraction*, vol. 21, p. 278–284, 2006.

[30] K. Scrivener, R. Snellings and B. Lothenbach, A Practical Guide to Microstructural Analysis of Cementitious Materials, 2018.

[31] P. Zhao, L. Lu, X. Liu, A. G. De La Torre and X. Cheng, "Error analysis and correction for quantitative phase analysis based on rietveld-internal standard method: Whether the minor phases can be ignored?" *Crystals*, vol. 8, p. 1–11, 2018.

[32] W. Franco Santos, K. Schollbach, S. R. van der Laan and H. J. H. Brouwers, "Quantification of the hydration in converter slag," *submitted*, 2020.

[33] J. A. Bogovic, P. Hanslovsky, A. Wong and S. Saalfeld, "Robust registration of calcium images by learned contrast synthesis," in *Proceedings – International Symposium on Biomedical Imaging*, 2016.

[34] C. A. Schneider, W. S. Rasband and K. W. Eliceiri, *NIH Image to ImageJ: 25 years of image analysis*, 2012.

H. Pöllmann, R. Kilian

Chapter 4
The use of µXRF in the characterization of industrial wastes and pozzolanes

Abstract: A wide variety of natural pozzolanes and industrial slags and ashes were analysed using micro X-ray fluorescence spectroscopy (µXRF) and results were compared to chemical analysis by bulk X-ray fluorescence spectroscopy (XRF) and phase determination by powder X-ray diffraction (XRD). Using the integral µXRF spectrum, results are overall comparable to the XRF results. µXRF maps were segmented based on the spectral characteristics using simple clustering algorithms and as far as phase determination was possible from the chemical composition, the derived modal composition can be compared to XRD. An advantage of µXRF is clearly that the analysis incorporates the microstructural aspect of the material, e.g. phase intergrowth, determination of particle sizes or porosity amongst others, which cannot be captured by bulk analytical methods require destruction of the sample material.

Based on mineralogical and chemical analyses the pozzolanity of various natural and industrial materials are characterized and described. Different methods and their results for volcanic ashes, bottom ashes from coal combustion, slags from waste incineration and metal ore processing are given for comparison

Keywords: µXRF, pozzolanity, XRD, XRF, slag, microstructure, volcanic ash

1 Introduction

Large quantities of industrial residues are produced year by year in different industries. As iron and steel industry slags are used for many years in cement production, many of these products from other industry can also be used for partial cement replacement. Due to the political pressure coming for the necessity of CO_2-reduction in the course of cement production these residues can help to get rid of industrial wastes and also in reduction of carbon dioxide during cement production. They can be used as part of the raw material for clinker production, but also as additional materials to produce composite cements. But also replacement of limestone by CO_2-free natural/

Acknowledgements: Thanks are due to ZWL/Lauf, especially Dipl.-Ing. S. Winter and Dr. J. Göske for help and providing images of different pozzolanic materials by SEM investigations.

H. Pöllmann, R. Kilian, University of Halle/Saale, Institute of geological Sciences, Applied mineralogy

https://doi.org/10.1515/9783110674941-004

industrial raw materials or increase of reactivity of clinker minerals leads to a reduced amount of cement materials in cement application. Reductions in process conditions (reduced temperatures, other phases, mineralizers, grinding aids, total efficiency of processes) or clinker minerals with reduced CO_2-output-different compositions. Other possibilities may be activation of slowly reacting cements by introducing highly reactive clinker minerals or the production of clinker minerals with increased specific surface and increased reactivity. Also activation of clinker minerals or cement clinkering using mineralizers. Grinding aids have been proved to reduce energy consumption. Very new is a future development of the replacement of hydration reaction by carbonation process.

Different wastes have already been shown that they are useful in the cement industry, all tons of wastes coming from these industries are not enough to fulfill these necessities. Therefore there must also be some other natural materials which can fulfill these requirements as supplementary cementitious materials. Mainly clays and metaclays, but also zeolites, trass or laterites are coming in the focus nowadays, because they fulfill the necessary requirements, as they do not produce CO_2 and they are available everywhere.

But also very fine grained volcanic ashes can be very useful, as there use has been known since a long time as pozzolanic materials. Famous are the Roman buildings made from natural pozzolanic materials. Even the name pozzolanes can be derived from the Pozzuoli-area in main land Italy close to Naples. In literature [1–75] different artificial and natural materials are described and their potential as supplementary cementitious or cement replacement material are investigated.

In Figure 4.1 typical supplementary cementitious materials from industrial and natural sources with mixtures of OPC are summarized.

Many occurrences worldwide close to ancient or nowadays volcanoes form large deposits of these pozzolanes. As we do know typical parameters of cements and pozzolanes they can be summarized as follows (Figure 4.2).

Nevertheless, as there exist many different supplementary materials, it is essential to provide and use characterization techniques to determine the different properties. Hydration reactions are often due to the amorphous contents, therefore some determination techniques including the amorphous contents will be described.

Typical quantitative results of pozzolanic materials and various mixtures, including the amorphous content and with different cements were described by methods of PLSR and PONCKS and also clustering [chapter 2].

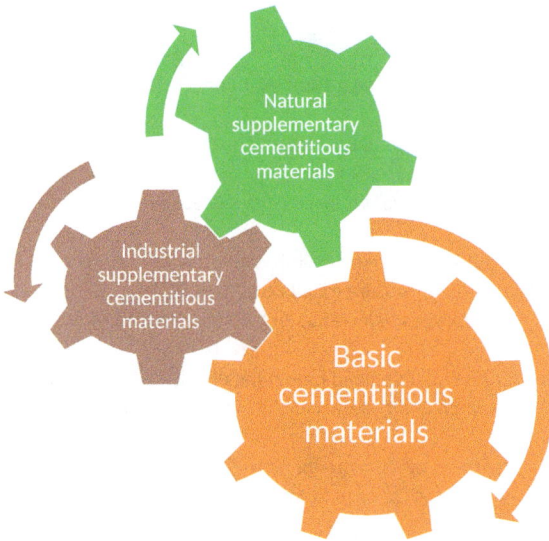

Figure 4.1: Composite cements made of OPC and mixtures with supplementary cementitious materials.

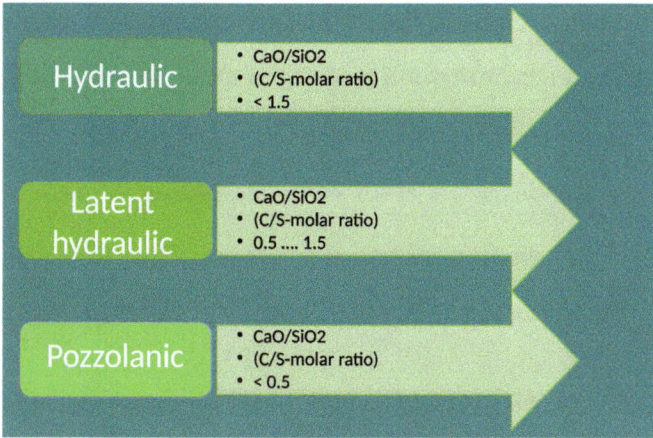

Figure 4.2: Parameters for the definition of pozzolanic material.

Literature

[1]   Ahmadi B, Shekarchi M (2010) Use of natural zeolite as a supplementary cementitious material. Cem Concr Comp 32:134–141

[2]   Akman MS, Mazlum F, Esenli F (1992) Comparative study of natural pozzolans used in blended cement production. ACI Special Publication 132:471–494

[3]   Alexander KM (1960) Reactivity of ultrafine powders produced from siliceous rocks. J Am Concr 157:557–569

[4]   Ambroise J, Gniewek J, Dejean J, Péra J (1987) Hydration of synthetic pozzolanic binders obtained by thermal activation of montmorillonite. Ceram Bull 66:1731–1733

[5]   Ambroise J, Maximilien S, Péra J (1994) Properties of metakaolin blended cements. Adv Cem Based Mater 1:161–168

[6]   Ambroise J, Murat M, Péra J (1985) Hydration reaction and hardening of calcined clays and related minerals. V. Extension of the research and general conclusions. Cem Concr Res 15: 261–268

[7]   Asbridge AH, Page CL, Page MM (2002) Effects of metakaolin, water/binder ratio and interfacial transition zones on the microhardness of cement mortars. Cem Concr Res 32: 1365–1369

[8]   Ayub M, Yusuf M, Beg A, Faruqi FA (1988) Pozzolanic properties of burnt clays. Pakistan J Sci Ind R 31:1–5

[9]   Badogiannis E, Kakali G, Dimopoulou G, Chaniotakis E, Tsivilis S (2005) Metakaolin as a main cement constituent. Exploitation of poor Greek kaolins. Cem Concr Comp 27:197–203

[10] Bakolas A, Aggelakopoulou E, Moropoulou A, Anagnostopoulou S (2006) Evaluation of pozzolanic activity and physico-mechanical characteristics in metakaolin-lime pastes. J Therm Anal Calorim 84:157–163

[11] Baronio G, Binda L (1997) Study of the pozzolanicity of some bricks and clays. Constr Build Mater 11:41–46

[12] Benezet JC, Benhassaine A (1999) Grinding and pozzolanic reactivity of quartz powders. Powder Technol 105:167–171

[13] Biricik H, Aköz F, Berktay I, Tulgar AN (1999) Study of pozzolanic properties of wheat straw ash. Cem Concr Res 29:637–643

[14] Bougara A, Lynsdale C, Milestone NB (2010) Reactivity and performance of blast furnace slags of different origin. Cem Concr Comp 32:319–324

[15] Brough AR, Dobson CM, Richardson IG, Groves GW (1995) A study of the pozzolanic reaction by solid-state 29Si nuclear magnetic resonance using selective isotopie enrichment. J Mater Sci 30:1671–1678

[16] Caputo D, Liguori B, Colella C (2008) Some advances in understanding the pozzolanic activity of zeolites: The effect of zeolite structure. Cem Concr Comp 30:455–462

[17] Javdar A, Yetgin Ş (2007) Availability of tuffs from northeast of Turkey as natural pozzolan on cement, some chemical and mechanical relationships. Constr Build Mater 21:2066–2071

[18] Cheriaf M, Cavalcante Rocha J, Péra J (1999) Pozzolanic properties of pulverized coal combustion bottom ash. Cem Concr Res 29:1387–1391

[19] Chusilp N, Jaturapitakkul C, Kiattikomol K (2009) Utilization of bagasse ash as a pozzolanic material in concrete. Constr Build Mater 23:3352–3358

[20] Colella C, de Gennaro M, Aiello R (2001) Use of zeolitic tuff in the building industry. Rev Mineral Geochem 45:551–587

[21] Coles DG, Ragaini RC, Ondov JM, Fisher GL, Silberman D, Prentice BA (1979) Chemical studies of stack fly ash from a coal-fired power plant. Environ Sci Technol 13: 455–459

[22] Cook DJ (1986a) Natural pozzolanas. In: Cement Replacement Materials. Swamy RN (ed) Surrey University Press, London, p 1–39

[23] Cook DJ (1986b) Calcined clay, shale and other soils. In: Cement Replacement Materials. Swamy RN (ed) Surrey University Press, London, p 40–72

[24] Cook DJ (1986c) Rice husk ash. In: Cement Replacement Materials. Swamy RN (ed) Surrey University Press, London, p 171–196

[25] Cordeiro GC, Filho RDT, Fairbairn EMR (2009b) Use of ultrafine rice husk ash with high-carbon content as pozzolan in high performance concrete. Mater Struct 42:983–992

[26] Cordeiro GC, Filho RDT, Tavares LM, Fairbairn EMR (2008) Pozzolanic activity and filler effect of sugar cane bagasse ash in Portland cement and lime mortars. Cem Concr Res 30:410–418

[27] Cordeiro GC, Filho RDT, Tavares LM, Fairbairn EMR (2009a) Ultrafine grinding of sugar cane bagasse ash for application as pozzolanic admixture in concrete. Cem Concr Res 39:110–115

[28] Davraz M, Giindiiz L (2005) Engineering properties of amorphous silica as a new natural pozzolan for use in concrete. Cem Concr Res 35:1251–1261

[29] Day RL, Shi C (1994) Influence of the fineness of pozzolan on the strength of lime natural-pozzolan cement pastes. Cem Concr Res 24:1485–1491

[30] De Sensale GR (2006) Strength development of concrete with rice-husk ash. Cem Concr Comp 28:158–160

[31] De Silva PS, Glasser FP (1992) Pozzolanic activation of metakaolin. Adv Cement Res 4:167–178

[32] Degirmenci N, Yilmaz A (2009) Use of diatomite as partial replacement for Portland cement in cement mortars.Constr Build Mater 23:284–288

[33] Dihr RK (1986) Pulverised-fuel ash. In: Cement Replacement Materials. Swamy RN (ed) Surrey University Press, London, p 197–255

[34] Douglas E, Elola A, Malhotra VM (1990) Characterisation of ground granulated blast furnace slag and fly ashes and their hydration in Portland cement blends. Cem Concr Aggr 12:38–46

[35] Erdem TK, Meral J, Tokyay M, Erdogan TY (2007) Use of perlite as a pozzolanic addition in producing blended cements. Cem Concr Comp 29:13–21

[36] Erdogdu K, Tiirker P (1998) Effects of fly ash particle size on strength of Portland cement fly ash mortars. Cem Concr Res 28: 1217–1222

[37] Escalante JI, Gomez LY, Johal KK, Mendoza G, Mancha H, Mendez J (2001). Reactivity of blast furnace slag in Portland cement blends hydrated under different conditions. Cem Concr Res 31:1403–1409

[38] Feng NQ, Chan SYN, He ZS, Tsang MKC (1997) Shale ash concrete. Cem Concr Res 27:279–291

[39] Fernandez R, Martirena F, Scrivener KL (2011) The origin of the pozzolanic activity of calcined clay minerals: A comparison between kaolinite, illite and montmorillonite. Cem Concr Res 41: 113–122

[40] Ferreira C, Ribeiro A, Ottosen L (2003) Possible applications for municipal solid waste fly ash. J Hazard Mater B96:201–216

[41] Frias M, Rodriguez O, Vegas I, Vigil R (2008) Properties of calcined clay waste and its influence on blended cement behavior. J Am Ceram Soc 91:1226–1230

[42] Gieré R, Carleton LE, Gregory RL (2003) Micro- and nanochemistry of fly ash from a coal-fired power plant. Am Mineral 88:1853–1865

[43] Goto S, Fujimori H, Tsuda T, Ooshiro K, Yamamoto S, Ioku K (2007) Structure analysis of glass of CaO-Al2O3-SiO2 system by means of NMR. Proc 12th Int Congr Chem Cement Concrete.

[44] Hanna KM, Afify A (1974) Evaluation of the activity of pozzolanic materials. J Appl Chem Biotechn 24:751–757

[45] He C, Mackovicky E, Osbaeck B (1996) Thermal treatment and pozzolanic activity of Na- and Ca-montmorillonite. Appl Clay Sci 10:351–368

[46] He C, Mackovicky E, Osbaeck B (2000) Thermal stability and pozzolanic activity of raw and calcined mixedlayer mica/smectite. Appl Clay Sci 17:141–161

[47] He C, Makovicky E, Osbaeck B (1995) Thermal stability and pozzolanic activity of raw and calcined illite. Appl Clay Sci 9:337–354

[48] He C, Osbaeck B, Makovicky E (1995) Pozzolanic reactions of six principal clay minerals: activation, reactivity assessment and technological effects. Cem Concr Res 25:1691–1702

[49] Hooton RD, Emery JJ (1983) Glass content determination and strength development predictions for vitrified blast furnace slag. ACI Special Publication 79:943–962

[50] Ish-Shalom M, Bentur A, Grinberg T (1980) Cementing properties of oil-shale ash: I. Effect of burning method and temperature. Cem Concr Res 10:799–807

[51] James J, Rao MS (1986) Reactivity of rice husk ash. Cem Concr Res 16:296–302

[52] Johansson S, Andersen PJ (1990) Pozzolanic activity of calcined moler clay. Cem Concr Res 20: 447–452

[53] Kakali G, Perraki T, Tsivilis S, Badogiannis E (2001) Thermal treatment of kaolin: the effect of mineralogy on the pozzolanic activity. Appl Clay Sci 20:73–80

[54] Kirby CS, Rimstidt JD (1993) Mineralogy and surface properties of municipal solid waste ash. Environ Sci Technol 27:652–660

[55] Lavat AE, Trezza MA, Poggi M (2009) Characterization of ceramic roof tile wastes as pozzolanic admixture. Waste Manage 29:1666–1674

[56] Liebig E, Althaus E (1998) Pozzolanic activity of volcanic tuff and suevite: effects of calcination. Cem Concr Res 28:567–575

[57] Ludwig U, Schwiete HE (1963) Lime combination and new formations in the trass-lime reactions. Zem-Kalk-Gips 10:421–431

[58] Massazza F (1974) Chemistry of pozzolanic additions and mixed cements. Proceedings of the 6th International Congress on the Chemistry of Cement, 1–65

[59] Massazza F (2001) Pozzolana and pozzolanic cements. In: Lea's Chemistry of Cement and Concrete. Hewlett PC (ed) Butterworth-Heinemann, Oxford, p 471–636

[60] Massazza F (2002) Properties and applications of natural pozzolanas. In: Structure and Performance of Cements, 2nd edition. Bensted J, Barnes P (eds.) Spon Press, London, p 326–352

[61] Mielenz RC, White LP, Glantz OJ (1950) Effect of calcination on natural pozzolanas. Symp on Use of Pozzolanic Materials in Mortars and Concrete. ASTM Special Technical Publication 99: 43–92

[62] Nair DG, Fraaij A, Klaassen AAK, Kentgens APM (2008) A structural investigation relating pozzolanic activity of rice husk ashes. Cem Concr Res 38:861–869

[63] Papadakis VG (1999) Effect of fly ash on Portland cement systems Part I. Low-calcium fly ash. Cem Concr Res 29:1727–1736

[64] Péra J, Ambroise J, Messi A (1998) Pozzolanic activity of calcined laterite. Silicates Industriés 7–8:87–106

[65] Roy DM, Arjunan P, Silsbee MR (2001) Effect of silica fume, metakaolin, and low-calcium fly ash on chemical resistance of concrete. Cem Concr Res 31:1809–1813

[66] Sabir BB, Wild S, Bai J (2001) Metakaolin and calcined clays as pozzolans for concrete: a review. Cem Concr Comp 23:441–454

[67] Shi C (2001) An overview on the activation of reactivity of natural pozzolans. Can J Civil Eng 28:778–786

[68] Shi C, Day RL (1995) Microstructure and reactivity of natural pozzolans, fly ash and blast furnace slag, Proceedings of the 17th International Conference on Cement Microscopy 150–161

[69] Snellings R, Mertens G, Elsen J (2010b) Calorimetric evolution of the pozzolanic reaction of zeolites. J. Therm. Anal. Calorim. 101:97–105

[70] Snellings, R., Mertens, G. & Elsen, J.: Supplementary cementitious materials, Reviews in mineralogy 74, 211–278, (2012)

[71] Swamy RN (1986) Cement Replacement Materials. Surrey University Press, London.

[72] Vassilev SV, Menendez R, Alvarez D, Diaz-Somoano M, Martinez-Tarazona MR (2003): Phase-mineral and chemical composition of coal fly ashes as a basis for their multicomponent utilization. 1. Characterization of feed coals and fly ashes. Fuel 82:1793–1811

[73] Vassilev SV, Menendez R, Diaz-Somoano M, Martinez-Tarazona MR (2004): Phase-mineral and chemical composition of coal fly ashes as a basis for their multicomponent utilization. 2. Characterization of ceramic cenosphere and salt concentrates. Fuel 83:585–603

[74] Watt JD, Thorne DJ (1965) Composition and pozzolanic properties of pulverised-fuel ashes, Part 1–2. J Appl Chem 15:585–604

[75] Yu LH, Ou H, Lee LL (2003) Investigation on pozzolanic effect of perlite in concrete. Cement Concr Res 33:73–76

2 Introduction of used analytical techniques with special emphasis to µXRF

Different techniques, chemical analysis, thermal analysis, XRD, XRF, µXRF, SEM micro-analysis and imaging, pozzolanity etc. with special emphasis on µXRF for some detailed investigations.

2.1 Micro X-ray fluorescence spetroscopy (µXRF) and data processing

Micro x-ray fluorescence spectroscopy uses an x-ray source exciting a small spot on a sample surface which allows to record the fluorescence spectrum as a function of spatial sample coordinates ([2.1, 2.2] for an overview see [2.3]). Excitation of a small spot can be achieved by focusing the beam through a polycapillary lens ([2.7 2.8] for an overview see [2.6]). We used a µXRF (Bruker Tornado M4 using software version 1.6.0.320), equipped with a combination of a 60 mm 2 silicon drift detector and a scintillator detector, using a Ag X-ray tube with a maximum output of 30 W (at 50 kV, 600 nA). X-rays are focused to a ~ 20 µm spot using a polycapillary lens under which samples are scanned through a movable stage. The spot size of 20 µm needs to be considered as a lower limit and assumes that the sample surface is perfectly in focus. The actual size may increase as a function of sample topography. Usable stage speeds for scanning vary in the range of 4 to 0.5 mm/s, but can be as high as 10 mm/s (Figure 4.3). Full spectra (0–40 keV at a resolution of 10 eV) are read out at constant intervals (interval/scan speed), with a resulting pixel size of typically 20 µm which defines the x,y coordinates of the hyperspectral map. Sample intervals smaller than the spot size (supersampling) are possible and under certain conditions may yield a higher spatial resolution – however at the cost of a much increased mapping time (Figure 4.4). Figure 4.4 displays the count distributions for Fe-Kα and Al-K for different stage speeds (time/pixel) and read-out intervals (pixel size). Discrete peaks in the distribution indicate a clearer distinction between areas of different phases with less noise in the count maps. These properties are an indicator for the ability of perform a successful phase segmentation based on count maps and clustering or thresholding techniques. While supersampling usually does not help in noise suppression, the exact positions of phase boundaries can be approximated more precisely (Figure 4.4 d) at the cost of lower gradients. However, a true higher spatial resolution can only be gained if the point-spread function is available for spatial deconvolution, since the excitation volume may exceed the spot size up to several times. Usually, the spatial resolution depends on the analyzed material, in particular the density contrast of neighboring phases and the weight of the element of interest (Figures 4.5, 4.6). While the size and position of two highly absorbing phases can be determined more accurately, heavy elements in a light matrix show an excitation volume up to > 100 µm width. In addition, the unknown extend of the phase at depth may add to the width of the possible phase boundary region (Figure 4.5, profile 1, Fe-signal). For light elements, especially in phases next to a highly absorbing phase, the boundary between phases can be determined more precisely (Figure 4.5, profile 2/3). Shape, size and composition of particles however can only be determined in an adequate way if particles have diameters larger several 100 µm.

Element maps in units of counts per second (cps) are derived from the hyperspectral maps by subtracting the background and deconvolution of each point with the spectra of all detected elements. Point analysis on materials with a matrix of moderate to high density can be calibrated to yield quantitative results ([2.5]). While composition in weight percent could be calculated for each pixel of the maps (Figure 4.6), given the short measurement time for each point, binning of pixels is usually required to obtain usable spectra for quantification. Computing compositions of larger maps is currently not feasible due to software restrictions. Accordingly, the spectral noise, the differing sizes of excitation volumes, differences in x-ray absorption of neighboring phases and the variable continuation of the phase at depth results in non-stoichiometric or simply non-existent compositions in the sample plane and especially in phase boundary areas (Figure 4.7).

For simple phase assemblages (e.g. 4 major components), multichannel images can be created from individual element maps to visualize individual phases. To derive phase maps, pores, cracks and other surface irregularities were masked using a mask derived from a minimum threshold on the average obtained from all normalized count maps. Subsequently stacks of the masked elements maps were segmented using a k-means clustering algorithm ([2.4]) implemented in imageJ (https://imagej.net) using the ijptoolkit (https://github.com/ij-plugins/ijp-toolkit). Every pixel presents a point a multidimensional space, with the number of dimensions equivalent to the number of element maps used. A phase with a constant composition is ideally presented by a single point in this element space. Since phase compositions may be slightly variable, measurements are noisy and phase boundary areas represent mixed measurements, realistically a single phase represents a point cloud in the element space and the k-means algorithm allows to find an optimal solution for the separation of a predefined number of point clouds. After segmenting, pseudo-phases at particle boundaries are merged into the matrix for high density particles and into the particles in the case of a high density matrix. Phase maps were used to derived modal composition in terms of volume fractions V_a/V (which are equal to area fractions A_a/A) of identified components. The relative error is given by sqrt($[(\sigma_a/\mu_a)^2 + 1]/N_a$) where N_a is the number, μ_a the arithmetic mean and σ_a the standard deviation of phase areas A_a of phase a ([2.9]).

2.2 Sample preparation

For the µXRF analyses, solid material was saw cut and eventually ground to an even surface. In the case of loose material (e.g. unconsolidated ashes), sample material was embedded in epoxy resin, saw cut and ground if necessary. For both preparation methods, particles of fine grained material may show a certain range in composition, depending on the position of the cross sectional area on the sample surface with respect to the particle extend below the surface. This effect is more pronounced in granular material dispersed in epoxy resin due to the low absorption of the resin. In parts, this problem can be overcome by using high quality thin sections of constant thickness, thick enough to absorb the Si and Ca signal emitted from the glass slides. Resin soaked samples cannot be analyzed for chlorine content.

Histograms reveal that the fast maps have higher noise level and a limited dynamic range compared to slower maps. Whether the increased noise level poses a problem for phase segmentation obviously depends on the size of particles and the composition. In general, if maps of heavier elements can be used and phases are several pixels in diameter, fast maps are of sufficient quality for segmentation (Figure 4.3d).

In general, histograms of element counts can be used to approximately indicate how much a phase can be separated from other phases or background, given the volume proportion is sufficiently large. Clearly separable peaks are equivalent to phases which are easier to separate in comparison with broad areas of overlapping peaks, which either related to poorly spatially resolved phases or phases with similar element counts and noisy data. In the case of phases assemblages which are only composed of light elements, in any case longer measurement times need to be accepted.

Supersampling, sampling at spatial intervals smaller than the spot size, provides a more smoothly appearing element map. However, for a true gain in spatial

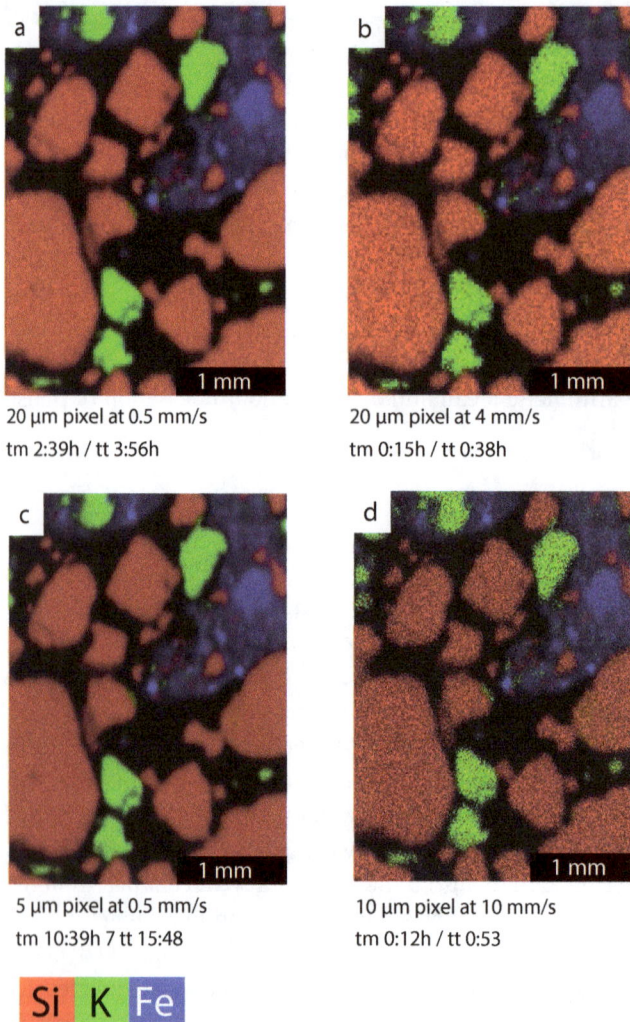

Figure 4.3: Composite RGB images using the counts for Si, K and Fe in the red, green and blue channel, respectively. tm is measurement time, tt is total time needed by the machine to acquire the data. Images are cropped from entire map of 10 by 9 mm. Slag from special waste incineration (4 a2). a) 20 μm pixel size acquired at the slowest possible speed compared to fast acquisition speed (b). Supersampling at slowest speed and 5 μm pixels (c) and at a very high acquisition speed and 10 μm pixels. tm is measurement time (beam on sample) and tt total time required to acquire the map.

resolution, i.e. locating phase boundaries, a spatial deconvolution for each element as a function of matrix absorption would have to be performed. A result at a comparable noise level comes at a large time penalty (Figure 4.4) and on average, no gain in location is achieved (Figure 4.5).

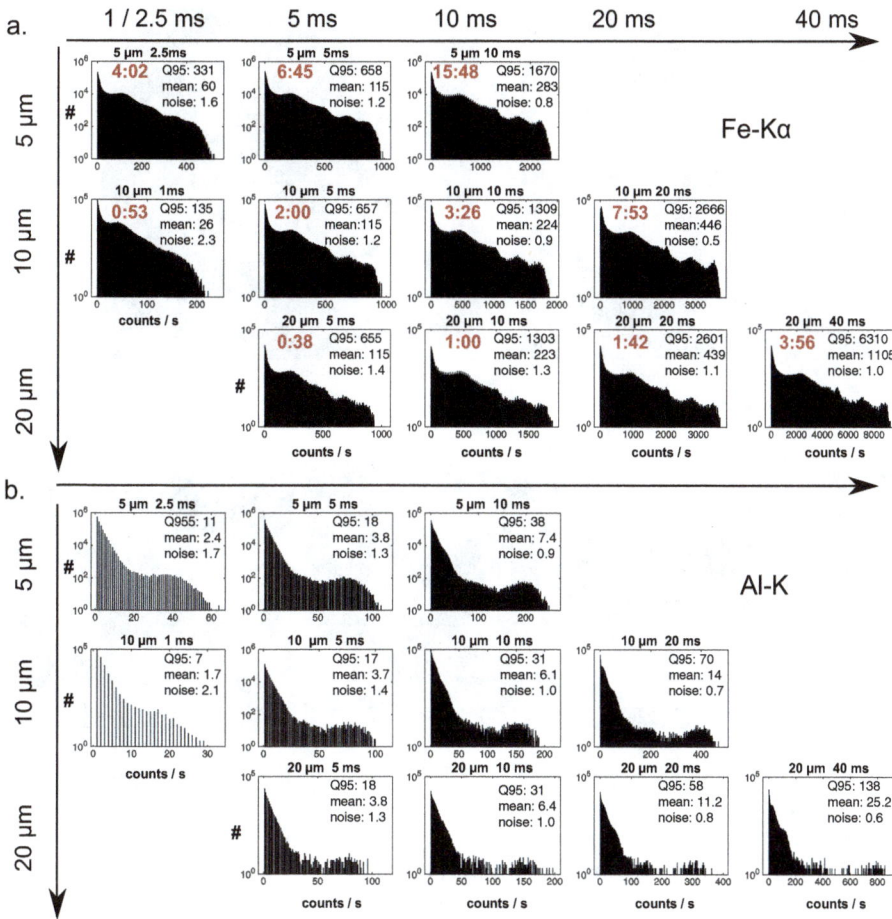

Figure 4.4: Compilation of µXRF mappings for different pixel sizes, given on the y-axis (5,10, 20 µm) and stage velocities in the range of 0.5 mm/s (slowest possible) to 10 mm/s, resulting in the times in ms /pixel given on the x-axis. a) Histograms of counts/s for Fe-Kα and b) Al-K in logarithmic frequencies. The maps has a size of 10 by 9 mm size (slag of special waste incineration) and total mapping times are indicated in (a) in h:mm in red. The amount of counts for each pixel is given as the 0.95 quantile (Q95) as well as the mean counts within each histogram. Noise for each element map is given as the norm of the difference of the measured map with the median filtered (1 pixel neighborhood) map and decreases with decreasing pixel size and increasing times / pixel.

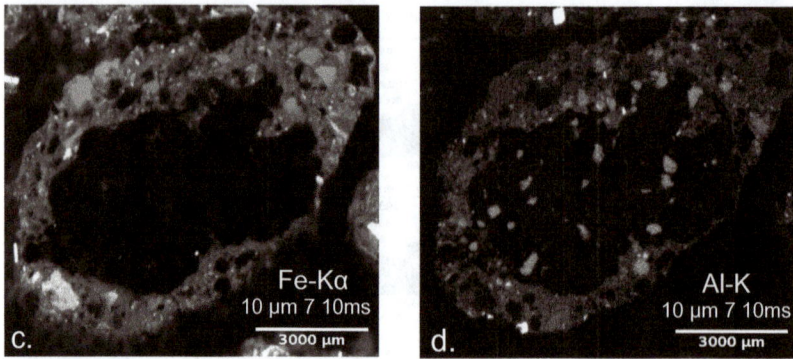

Figure 4.4 (continued)

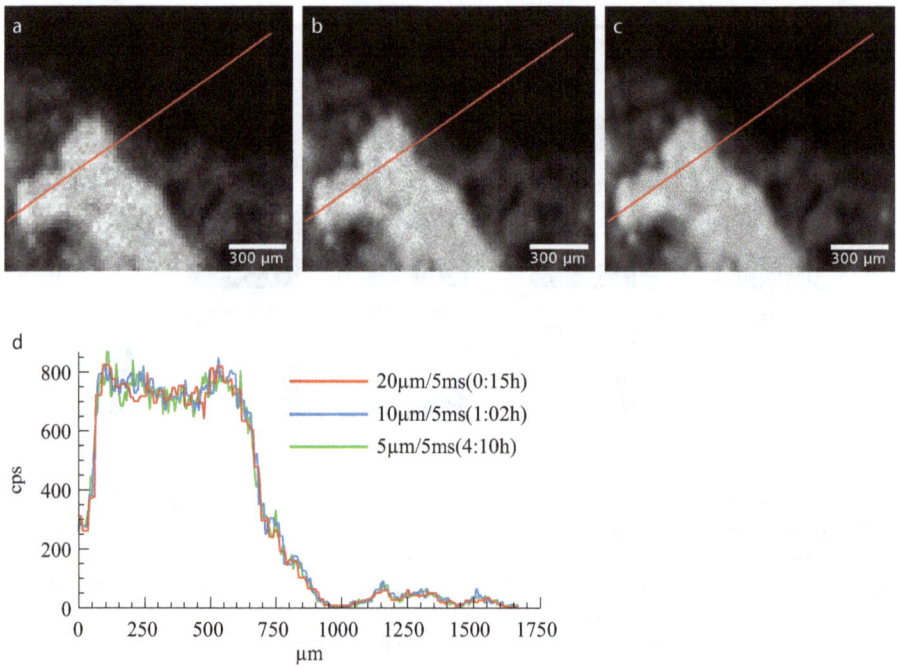

Figure 4.5: Effect of supersampling. Fe-maps with 20 μm (a), 10 μm (b) and 5 μm(c) pixel size using the same integration time over each pixel. d) Spatial profile of Fe-counts along line. Slag from special waste incineration (4 a2).

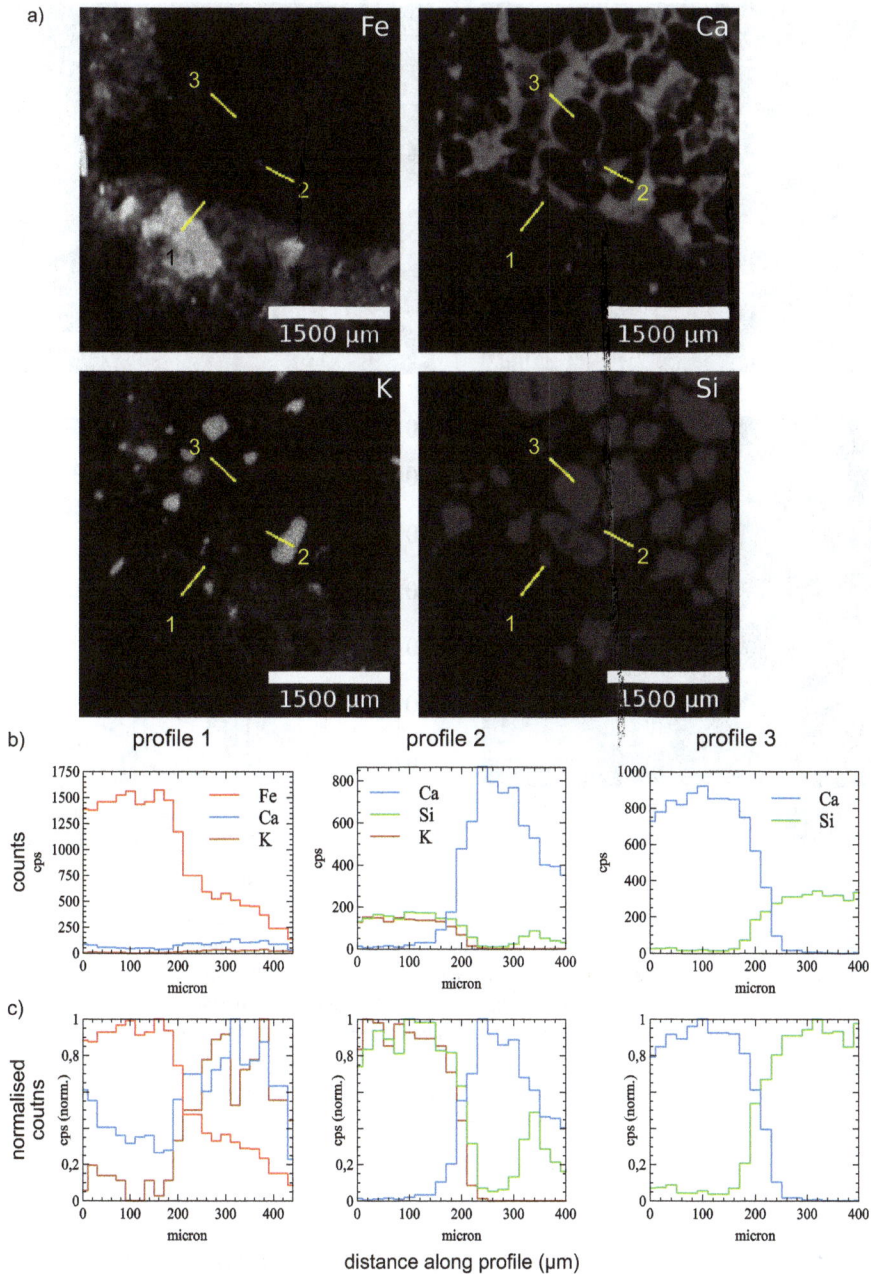

Figure 4.6: Spatial resolution as a function of element and phase. a) Maps of counts (in counts/s) for Fe, Ca, Si and K. Yellow traces indicate profile lines starting at the number. b) Counts plotted along profile line indicated in (a). In profile 1, Fe, Ca and K counts are evaluated, in profile 2 Ca, Si and K and in profile 3 Ca and Si. c) Same profiles as shown in (b) but counts for each element normalized to 1. Slag from special waste incineration (4 a2).

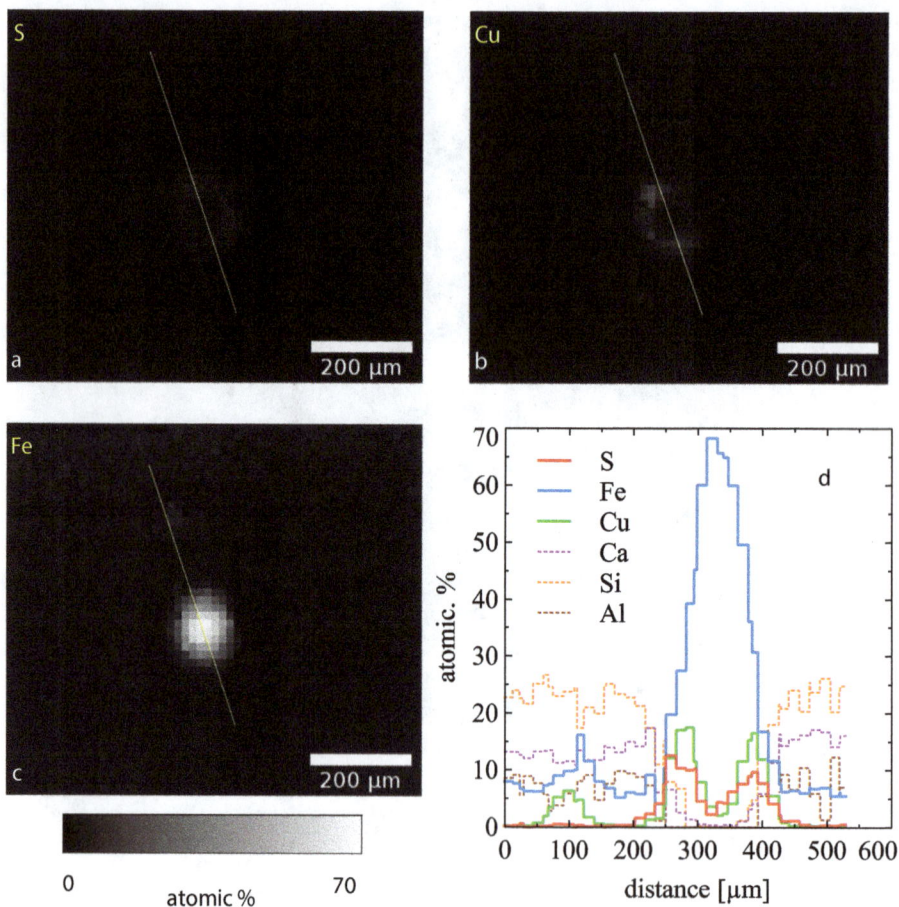

Figure 4.7: Spatial resolution of the μXRF. (a,b,c) maps calculated for atomic percent of a Fe-droplet from the copper slag of Helbra/Mansfeld (see 4a.3) showing a Cu-S rim. The map was acquired at a steps of 15 μm. d) profile plot along the line indicated in (a,b,c) showing the gradual decrease of the apparent Fe content from the center of the droplet. Given that the exact position with respect to the surface can never be known, the size cannot be exactly determined. The lack of a concentration plateau suggests that the activation volume is at least the size of the Fe-droplet.

Literature

[1]   Carpenter, D.A., Taylor, M.A., Holcombe, C.E., 1989, Applications of Laboratory X-ray Microprobe to Materials Analysis, Adv. X-Ray Anal.32, 115.
[2]   Engström, P., Larsson, S., Rindby, A., Stocklassa, B., 1989, A 200 μm X-ray microbeam spectrometer, Nucl. Instrum. Methods B 36, 222–226, doi:10.1016/0168-583X(89)90588-0
[3]   Haschke, M., 2014, Laboratory micro-x-ray fluorescence spectroscopy: Instrumentation and applications, Springer series in surface sciences, volume 55, Springer.

[4]   Jain, A. K., Dubes, R. C., 1988, Algorithms for Clustering Data, Prentice-Hall.
[5]   Nieuwenhuis, M.: Chemische Analytik von Geomaterialien mit der y-RFA: Möglichkeiten und Grenzen, MSc-Arbeit Göttingen, (2017)
[6]   MacDonald, C., 2010, Focusing Polycapillary Optics and Their Applications, X-Ray Optics and Instrumentation, vol 2010, ID 867049,p. 17,doi:10.1155/2010/867049
[7]   Marton, L., 1966, X ray fiber optics, Applied Physics Letters, vol. 9,no. 5, pp. 194 195.
[8]   Mosher, D., Stephanakis, S.J., 1976, X-ray "light pipes", Applied Physics Letters, vol. 29, no. 2, pp. 105–107.
[9]   Underwood, E.E., 1970, Quantitative Stereology. Addison-Wesley.

3 Description of natural supplementary cementitious materials

The following different natural and artificial pozzolanes were under investigation and are characterized using several different techniques and determinations of properties.

3.1 Natural pozzolanes from different sources

3.1.1 Indonesian volcanic ash from Kelimutu mountain

The different natural rock materials coming from various occurrences are described from their origin in [1.3 a1–6.3.a1]. Their mineralogical and chemical parameters are given in Figure 4.8 and Tables 4.1 and 4.2.

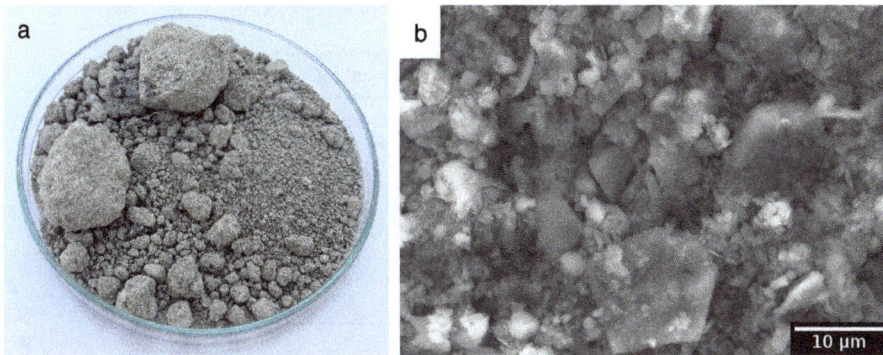

Figure 4.8: a) Sample of the ash from Kelimutu. b) SEM/SE image showing Tridymite crystals among others.

Table 4.1: Chemical composition of "Kelimutu" mountain ash. µXRF data is normalized from a total content of 69%, analyzed area is ~330 mm².

Chemistry	mass%	mass% (av. µXRF)
LOI%	6,6	–
Na_2O (%)	1,0	3.9
MgO (%)	1,3	–
Al_2O_3 (%)	6,7	13.1
SiO_2 (%)	77,1	72.6
P_2O_5 (%)	0,1	–
SO_3 (%)	0,0	1.6
K_2O (%)	1,0	5.5
CaO (%)	1,1	1.0
TiO_2 (%)	1,1	0.4
V_2O_5 (%)	0,0	–
Cr_2O_3 (%)	0,0	–
Mn_2O_3 (%)	0,0	–
MnO (%)	–	0.0
Fe_2O_3 (%)	2,4	–
FeO (%)	–	1.5

Table 4.2: Mineralogical Composition of "Kelimutu" Mountain ash by XRD.

Primary phases	Chemical formula
Quarz	SiO_2
Tridymite	SiO_2
Cristobalite	SiO_2
Na-Alunite	$(Na,Ca)_{1-x}Al_3(SO_4)_2(OH)_6$
Glass	

A few larger and cohesive ash particles were soaked in epoxy resin, mounted, cut and ground evenly. The total analyzed area used for the bulk composition obtained from the sum of the µXRF spectra corresponds to the large particles while out-of-focus background is not incorporated. Clasts themselves are fine grained, angular and show a heterogeneous internal composition (Figure 4.9). Compositionally different areas are themselves angular and embedded in a non-resolvable matrix. Clasts show a very broad range in size and are loosely bound, forming a porous framework. Pore space and cement is compositionally also variable across areas of a few mm, most pores are however filled with epoxy resin, hence are assumed to be empty in the original material.

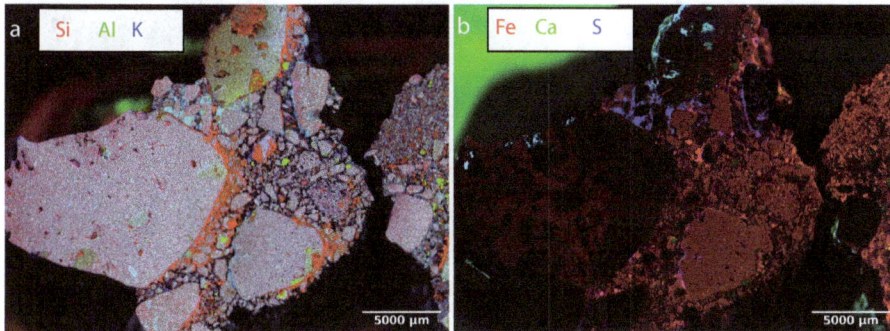

Figure 4.9: µXRF composite maps of "Kelimutu" Mountain ash. a) Si-Al-K and b) Fe-Ca-S composite map obtained at 15 µm pixel size at 0.625 mm/s. Angular clasts of fine grained material with minute compositional differences are bound by a heterogeneous matrix (orange and turquoise (Na-Alunite and an unknown phase) areas between pale particles in (a)). The dark-lilac areas between particles in (a) are mostly porosity. A Ca-S phase is found in certain pores and clast exterior (turquoise color in (b)). The heterogeneous nature of the clasts is easily identified in the Fe/Ca data.

Literature

[1. 3 a1] Global Volcanism Program (2013b) „Kelimutu" (264140) in Volcanoes of the World, v. 4.6.0. Venzke, E (ed.). Smithsonian Institution.

[2. 3 a1] Global Volcanism Program, (2013). „Kelimutu" *in* Volcanoes of the World, v. 4.9.1 (17 Sep 2020). Venzke, E (ed.). Smithsonian Institution. Downloaded 06 Dec 2020 (https://volcano.si.edu/volcano.cfm?vn=264140). https://doi.org/10.5479/si.GVP.VOTW4-2013

[3. 3 a1] Murphy, S., Rouwet, D. and Wright, R.: Color and temperature of the crater lakes at „Kelimutu" volcanoes through time, Bull. Of volcanology 80, 12,017

[4. 3 a1] Pasternack GB, Varekamp JC (1994) The geochemistry of the Keli Mutu crater lakes, Flores, Indonesia. Geochem J 28: 243–262. https://doi.org/10.2343/geochemj.28.243

[5. 3 a1] H. Subagyo, and B.H. Prasetyo, *Indonesian Jour. of Agricultural Science*, 4(1), 2003, 1–11

[6. 3 a1] Van Goreel, Regional geology, Bibliography of the geology of Indonesia and surrounding areas, pdf online, (2013)

3 a2 Indonesian pozzolane from Blue beach/Flores/Indonesia

The pozzolanes material from island of Flores/Indonesia is described by [1–3 a2–5. 3a 2]. The relevant chemical and mineralogical data are summarized in Figure 4.10 and Tables 4.3 and 4.4.

Figure 4.10: a) Volcanic ash from Blue Beach/Flores. b) SEM/SE image of lathy crystals in matrix.

Table 4.3: Chemical composition of pozzolane from Flores/ Indonesia. μXRF data is normalized from a total content of 65%, analyzed area is 700 mm².

Chemistry	mass%	mass% (av. μXRF)
LOI%	9,4	–
Na$_2$O (%)	1,8	1.9
MgO (%)	3,0	2.9
Al$_2$O$_3$ (%)	15,4	17.1
SiO$_2$ (%)	56,7	62.0
P$_2$O$_5$ (%)	0,1	–
SO$_3$ (%)	0,0	0.1
K$_2$O (%)	1,9	2.3
CaO (%)	3,7	4.3
TiO$_2$ (%)	0,6	0.6
V$_2$O$_5$ (%)	0,0	0.0

Table 4.3 (continued)

Chemistry	mass%	mass% (av. µXRF)
Cr_2O_3 (%)	0,0	–
Mn_2O_3 (%)	0,1	–
MnO (%)	–	0.1
Fe_2O_3 (%)	6,7	–
FeO (%)	–	8.3

Table 4.4: Mineralogy of pozzolanic ash from Blue Beach/Flores/Indonesia.

primary phases	chemical formula
Plagioclase	$(Ca,Na)(Al,Si)_2Si_2O_8$
Heulandite	$(Ca,Na,K)_3[Al_3Si_9O_{24}].7.8H_2O$
Quarz	SiO_2
Faujasite	$Ca_{39.36}(Al_{96}Si_{96}O_{384})(H_2O)_{60.64}$

The composite element maps were obtained from a saw-cut slab of the sample. A homogeneous distribution of primary phases (Plagioclase, K-feldspar, Clinopyroxene), a homogeneously distributed Ti-Fe phase (Ilmenite, Ulvöspinel or similar) as well as irregularly occurring veins of an almost pure manganese phase (Manganite, Pyrolusite, Hausmannite or simmilar) can be recognized. The exact nature of some phases cannot be determined unequivocally from the µXRF count maps, such as the turquoise phases in Figure 4.11 b/c, which may amount to a plagioclase or Faujasite. The composition of selected areas within the larger ones of those grains suggest that those are still plagioclase (~An84). Most mottled Ca- and Fe-rich grains may amount to variably altered primary phases. In difference to the phases determined by XRD, µXRF scans find a K-bearing phase, compatible with a potassium feldspar as well as the Fe-Ti phase (magenta-pink in Figure 4.11 c/d) and the heterogeneous distribution of the manganese phase.

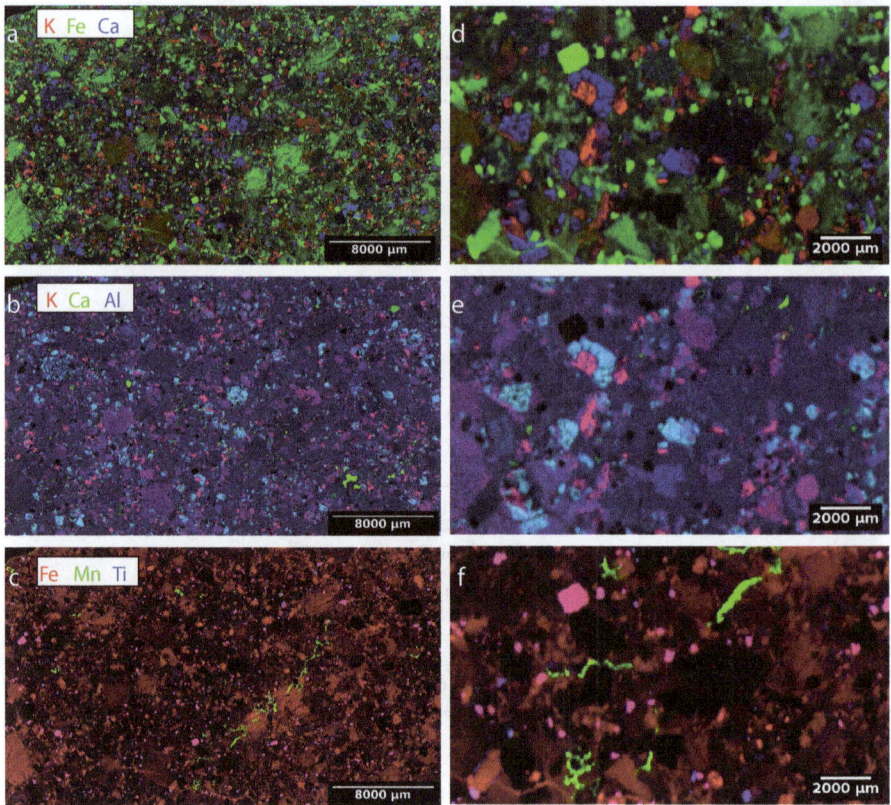

Figure 4.11: µXRF composite maps of pozzolane Blue Beach/Flores/Indonesia. a) K-Fe-Ca, b) K-Ca-Al and c) Fe-Mn-Ti composite maps. d,e,f) show a close-up of the corresponding maps to the left.

Literature

[1. 3 a2] Barber et al. 1981. The geology and tectonics of Eastern Indonesia: Review of the Seatar workshop, 9–14 July 1979, Bandung, Indonesia. In the geology and tectonics of Eastern Indonesia, GRDC, Special Publication, No. 2, 1981, pp. 7–28. Cardwell and Isack, 1981 Special publication GRDC.

[2. 3 a2] Katili, J.A., 1975. Volcanism and plate tectonics in the indonesian island arc. Tectonophysics, 26, 165–188

[3. 3 a2] Matsuda, K., Sriwana, T., Primulyana, S. and Futagoishi, M.: Chemical and isotopic studies of well discharge fluid of the Mataloko geothermal field, Flores, Indonesia, Bull. Geol. Surv. Japan, 53, (2002), 343–353

[4. 3 a2] Muraoka, H., Nasution, A., Simanjuntak, J., Dwipa, S., Takahashi M:; Takahashi H., Matsuda, K., Sueyoshi, Y.: Geology and geothermal systems in the Bajawa Rift Zone, Flores, Eastern Indonesia, Proc.World Geothermal congress, Antalya, 24/25.4. 2005.

[5.3 a2]   Muraoka, H., Nasution, A., Urai, M., Takahashi, M. and Takashima, I.: Geochemistry of volcanic rocks in the Bajawa geothermal field, central Flores, Indonesia, Bull. Geol. Surv. Japan, 53, (2002c), 147–159

3 a3 Krakatau mountain/Indonesia – volcanic slag (2017)

Typical rock samples of the old New Krakatau volcano were included and the mineralogy and chemistry described in Figure 4.12 a,b and Tables 4.5 and 4.6. Relevant literature on the composition and formation of rocks is summarized in [1. 3 a3–13. 3 a3].

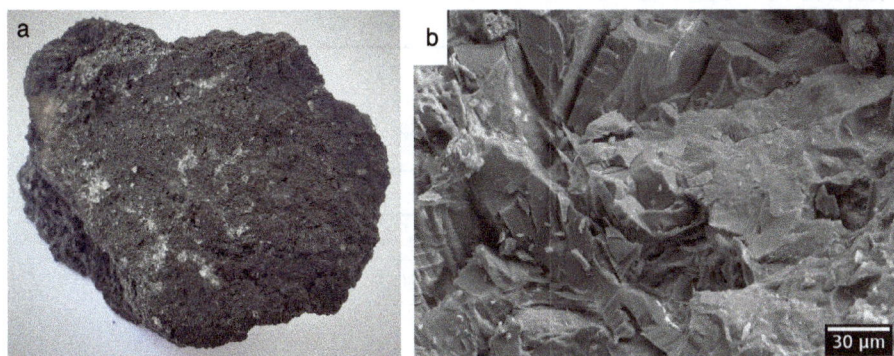

Figure 4.12: a) Rock sample of Krakatau mountain/Indonesia. b) SEM/SE image of plagioclase in rock matrix.

Table 4.5: Chemistry of Krakatau mountain rock. µXRF data normalized from 70%, analyzed area is 1056 mm^2.

Chemistry	mass%	mass% (av. µXRF)
LOI%	0,0	–
Na$_2$O (%)	3,4	5,3
MgO (%)	3,2	2,4
Al$_2$O$_3$ (%)	17,5	19,7
SiO$_2$ (%)	55,1	52,6
P$_2$O$_5$ (%)	0,4	0,1
SO$_3$ (%)	0,0	0,2
K$_2$O (%)	1,8	1,0
CaO (%)	7,5	7,9

Table 4.5 (continued)

Chemistry	mass%	mass% (av. µXRF)
TiO_2 (%)	1,1	1,1
V_2O_5 (%)	0,0	0,0
Cr_2O_3 (%)	0,0	–
Mn_2O_3 (%)	0,2	–
MnO (%)	–	0,2
Fe_2O_3 (%)	8,8	–
FeO (%)	–	8,7

Table 4.6: Phase compositions from XRD measurements.

primary phases	chemical formula
Plagioclase	$Na_{0.499}Ca_{0.491}[Al_{1.488}Si_{2.506}O_8]$
Pyroxene	$Ca_{0.05}Mg_{0.9}Fe_{1.05}[Si_2O_6]$
Cristobalite	SiO_2
Augite	$Ca(Fe,Mg)[Si_2O_6]$
Magnesioferrite	$MgFe_2O_4$

One of the old slags and ashes before 2019 eruptions, is described here as an example for pozzolanic material which were recently replaced by a new outbreak of the volcano bringing new pozzolanic materials:

Table 4.7: Phases and phase proportions identified from µXRF data corresponding to Figure 4.13b. (Ab: Albite, An: Anorthite, Or: Orthoclase, En: Enstatite, Fs: Ferrosilite, Wo: Wollastonite, Usp: Ulvoespinel, Mag: Magnetite).

phase	volume fraction%	abs. error%
Plagioclase ($Ab_{46}An_{53}Or_1$ and $Ab_{75}An_{25}$)	24.6	1.8
Clinopyroxene (Augite)	2.3	0.4
Orthopyroxene ($En_{83}Fs_{14}Wo_3$)	3.4	0.3
Spinel ($Usp_{40}Mag_{60}$)	1.0	0.1
Matrix	57.5	na
Holes	11.2	0.7

For the slag of the Krakatau volcano, µXRF count maps (Figure 4.17a) and (Table 4.7) have been used to derive a phase map (Figure 4.17b), discriminating the major contributing phases. The phase map allows to derive volume fractions and their errors of phases, matrix (here, everything below the spatial resolution of the µXRF) and holes in the sample surface. Under the assumption of a careful sample preparation, holes in the saw-cut and flat-ground specimen surface, would amount to macro porosity. Differences between the µXRF and XRD derived phases are the absence of Cristobalite as well as the presence of a Clinopyroxene (Augite) as well as a Ti-

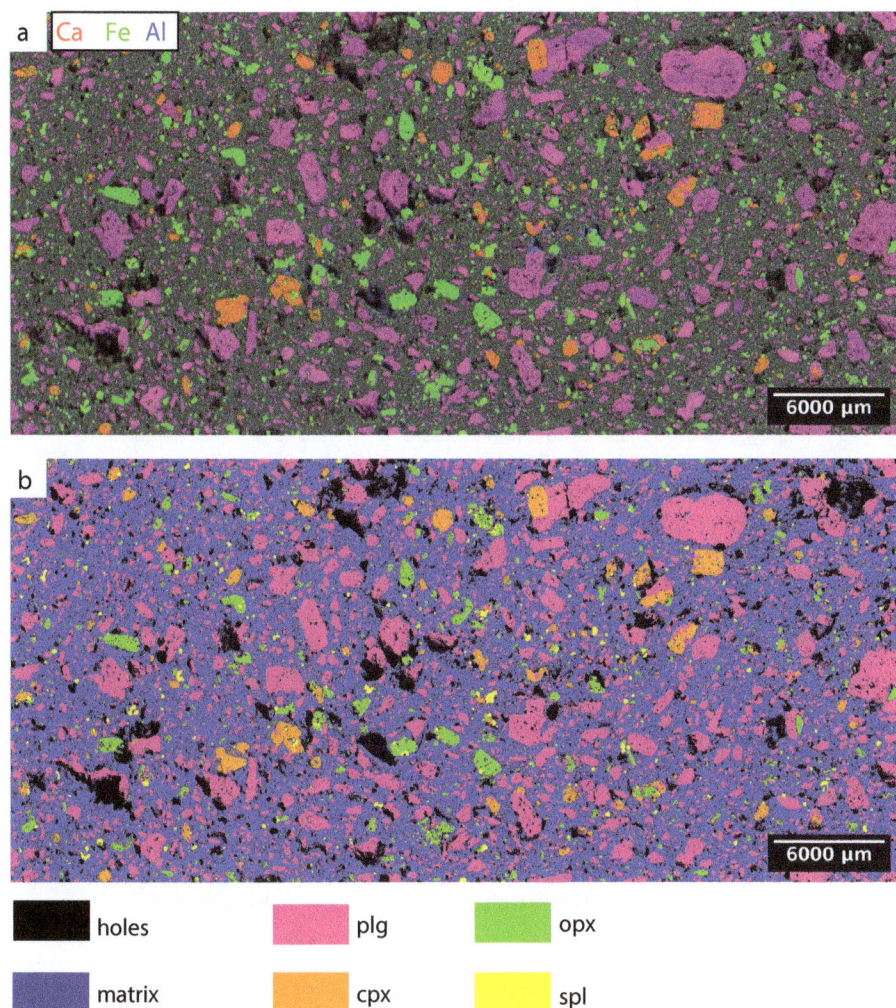

Figure 4.13: µXRF map of Krakatau Mountain slag. a) Ca-Fe-Al composite map obtained from saw-cut slab. b) Segmented phase map using kmeans clustering obtained from Si, Fe, Al, Ca, K, Ti maps.

bearing phase, most likely a spinel, in the µXRF data. The absence of Cristobalite might indicate that the phase occurs at a grain size below the spatial resolution of the µXRF. The difference between $Usp_{40}Mag_{60}$ and Magnesioferrite in the lattice parameter a is less than 1% which is usually insufficient to allow an exact characterization by XRD.

Literature

[1. 3 a3]   Abdurrachman, M., Sri Widiyantoro, S. Priadi, B. and Taufik Ismail: Geochemistry and Structure of Krakatoa Volcano in the Sunda Strait, Indonesia, geosciencesAbdurrachman, M. Insights from Pb and O isotopes into along-arc variations in subduction inputs and crustal assimilation for volcanic rocks in Java, Sunda arc, Indonesia. Geochim. Cosmochim. Acta, 2014,139,205–226.

[2. 3 a3]   Abdurrachman, M. Geology and Petrology of Quaternary Papandayan Volcano and Genetic Relationship of Volcanic Rocks from the Triangular Volcanic Complex around Bandung Basin, West Java, Indonesia, Unpublished Ph.D. Thesis, Akita University, Japan, 2012.

[3. 3 a3]   Gardner, M.F.; Troll, V.R.; Gamble, J.A.; Gertisser, R.; Hart, G.L.; Ellam, R.M.; Harris, C.; Wolff, J.A. Crustal differentiation processes at Krakatau Volcano, Indonesia. J. Petrol. 2013,54, 149–182. Hall, R. Indonesia, Geology. In Encyclopedia of Islands;

[4. 3 a3]   Gillespie, R., Clague, D.A., Eds.;University of California Press: Oakland, CA, USA, 2008.

[5. 3 a3]   Handley, H.K., Blichert-Toft, J., Gertisser, R., Macpherson, C.G., Turner, S., Zaennudin, A., 5. Abdurrachman, M. (2014). Insights from Pb and O isotopes into along-arc variations in subduction inputs and crustal assimilation for volcanic rocks in Java, Sunda arc, Indonesia. Geochimica et Cosmochimica Acta 139, 205–266, (2014)

[6. 3 a3]   Judd JW (1889) The earlier eruptions of Krakatau. Nature 40:365–366

[7. 3 a3]   Mandeville CW, Carey S, Sigurdsson H (1996b) Sedimentology of the Krakatau 1883 submarine pyroclastic deposits. Bull Volcanol96:512–529

[8. 3 a3]   Nishimura S, Harjono H (1992) The Krakatau Islands: the geotectonic setting. Geo Journal 28:87–98

[9. 3 a3]   Stehn, C.E. The geology and volcanism of the Krakatau group. In Guidebook for the 4th Pacific Science Congress; University of Michigan: Ann Arbor, MI, USA, 1929; pp. 1–55.

[10. 3 a3] Turner, S.; Foden, J.U. Th and Ra disequilibria, Sr, Nd, and Pb isotope and trace element variations in Sunda arc lavas: Predominance of a subducted sediment component. Contributions Mineral. Petrol. 2001,142, 43–57.

[11. 3 a3] Thornton, I.: Krakatau, the destruction and reassembly of an island ecosystem, paperback, (1997) Volcano Discovery. Krakatoa Volcano. Available online.

[12. 3 a3] Winchester, S.: The day the world exploded, Krakatoa, (2004)

[13. 3 a3] Zen, M.T., Hadikusumo, D. (1964) Recent changes in the KrakatauVolcano. Bull Volcanol 27:259–268

3 a4 Pozzolanic material from South America-Peru

A typical hardened pozzolanic material was investigated from South America and the relevant mineralogical and chemical data are given in Figure 4.14 and Tables 4.8–4.10.

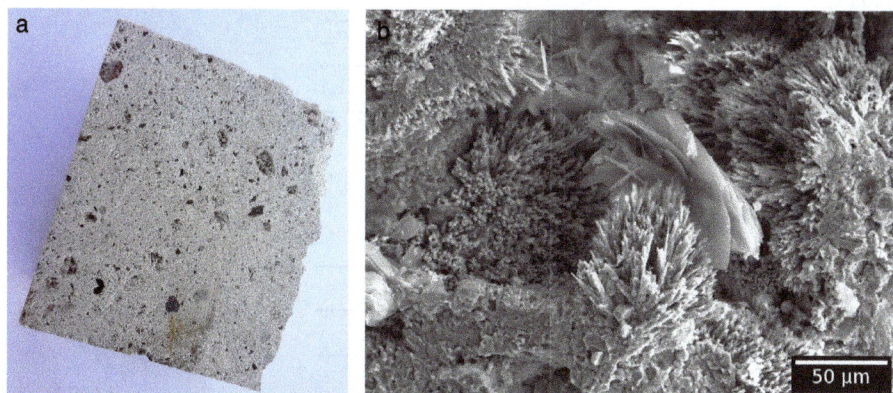

Figure 4.14: a) Macroscopic view of pozzolane sample from Peru. Sample edge roughly 8 cm long. b) SEM/SE image of pozzolane sample from Peru.

Table 4.8: Chemical analysis of Peruvian pozzolane. µXRF data normalized from 66%. The analyzed area is 2100 mm².

Chemistry	mass%	mass% (av. µXRF)
LOI%	0,8	–
Na$_2$O (%)	4,3	5.6
MgO (%)	0,4	0.0
Al$_2$O$_3$ (%)	13,9	14.6
SiO$_2$ (%)	72,2	72.2
P$_2$O$_5$ (%)	0,1	0.0
SO$_3$ (%)	0,0	0.1
K$_2$O (%)	4,0	4.8
CaO (%)	1,1	1.0
TiO$_2$ (%)	0,2	0.3
V$_2$O$_5$ (%)	0,0	–
Cr$_2$O$_3$ (%)	0,0	0.0

Table 4.8 (continued)

Chemistry	mass%	mass% (av. µXRF)
Mn_2O_3 (%)	0,0	–
MnO (%)	–	0.1
Fe_2O_3 (%)	1,5	–
FeO (%)	–	1.1

Table 4.9: Mineralogy of Peruvian pozzolanes from XRD measurements.

primary phases	chemical formula
Sanidine	$K(Al,Fe)[Si_3O_8]$
Plagioclase	$(Ca,Na)[Al_2Si_2O_8]$
Cristobalite	SiO_2
Augite	$Ca(Fe,Mg)[Si_2O_6]$
Magnesioferrite	$MgFe_2O_4$
Muscovite	$(K,Na)_{0.75}(Al,Mg)_2[(Si,Al)_4O_{10}](OH,O)_2$
Glass	

Table 4.10: Phases and phase proportions derived from kmeans-based segmentation of µXRF data. Note that holes correspond to resolvable porosity; pores smaller < 100 µm have a high chance to remain undetected. Especially the volume of matrix 2 will contain significant porosity.

Phase	volume%	error%
Plagioclase	3.3	0.2
Amphibole	2.5	0.1
Orthopyroxene	0.17	0.03
Clinopyroxene	0.06	0.02
Spinel	1.1	0.1
Matrix 1	1.3	0.4
Matrix 2	73.6	Na
Holes	18.9	1.2

The µXRF maps (Figure 4.15a) of the Peruvian pozzolane were obtained from a saw-cut, flat-ground solid sample from which a phase map was derived (Figure 4.15b). Idiomorphic crystals of Plagioclase and Amphibole, to a minor amount also Clino-pyroxene are dispersed with a non-resolvable, porous matrix. The matrix contains angular lithoclasts composed of a different matrix composition, high crystal density (idiomorphic Plagioclase, Ortho- and Clinopyroxene as well as Amphibole) showing a cummulate texture.

The matrix contains also a homogeneously dispersed Fe-bearing phase for which Amphibole has been assigned. Due to the small size (often smaller 25 pixel), the phase identification cannot be easily verified, however, based on a comparison of the grain shapes with the shapes of larger Amphibole crystals, a mis-identification might be quite likely. This is a good example that even at a very similar count rate for most major elements, a phase cannot be unequivocally identified if a) it is too small to obtain an XRF spectrum of the pure phase and b) if the contributing elements are in part too light and do not provide element maps with a sufficiently low noise. Here, the Fe-Xray signal of small particles may be attenuated by the surrounding and overlying matrix, rendering an assignment based on the elements (Al, Si, K,Ca, Fe, Ti) chosen for the segmentation very difficult. The problem of information depth becomes more obvious when an even heavier element is considered (Figure 4.16b). While the total SrO content of the plagioclase is only 0.3 wt% (or for the entire map 0.04 wt%), it can be seen that the plagioclase grain outlines obtained from Ca, roughly depict the grain intersection with the sample surface, while the Sr signal depicts a 3D grain shape as Sr-Kα Xrays have an energy sufficiently high to escape from larger depth of the sample. This restricts the usage of heavier elements (even if they are well detectable in low concentrations) for phase segmentation in combination with a lighter matrix. Usually Na and Mg do also not provide usable maps due to low count rates, resulting in a very high noise level which is unsuitable for image segmentation.

4 Industrial artificial pozzolanes (slags and ashes)

4 a1 Bottom ash from brown coal power plant

Typical industrial residues were included in this study as [1. 4 a1–12. 4 a1]. Their chemical and mineralogical compositions are given in Figure 4.17a–d and Tables 4.11 and 4.12. Brown coal fly ashes are discussed in chapter 2.1 of this book.

Larger ash particles were embedded in epoxy resin, cut and ground flat. The µXRF map was obtained on a large particle showing its heterogeneous composition with a porous core and a heterogeneous shell composed of several, well sorted but distinct components. While all phases were not identified, some phases are well compatible with those identified by XRD (an SiO_2 phase, cpx, cc and others). Figure 4.18b

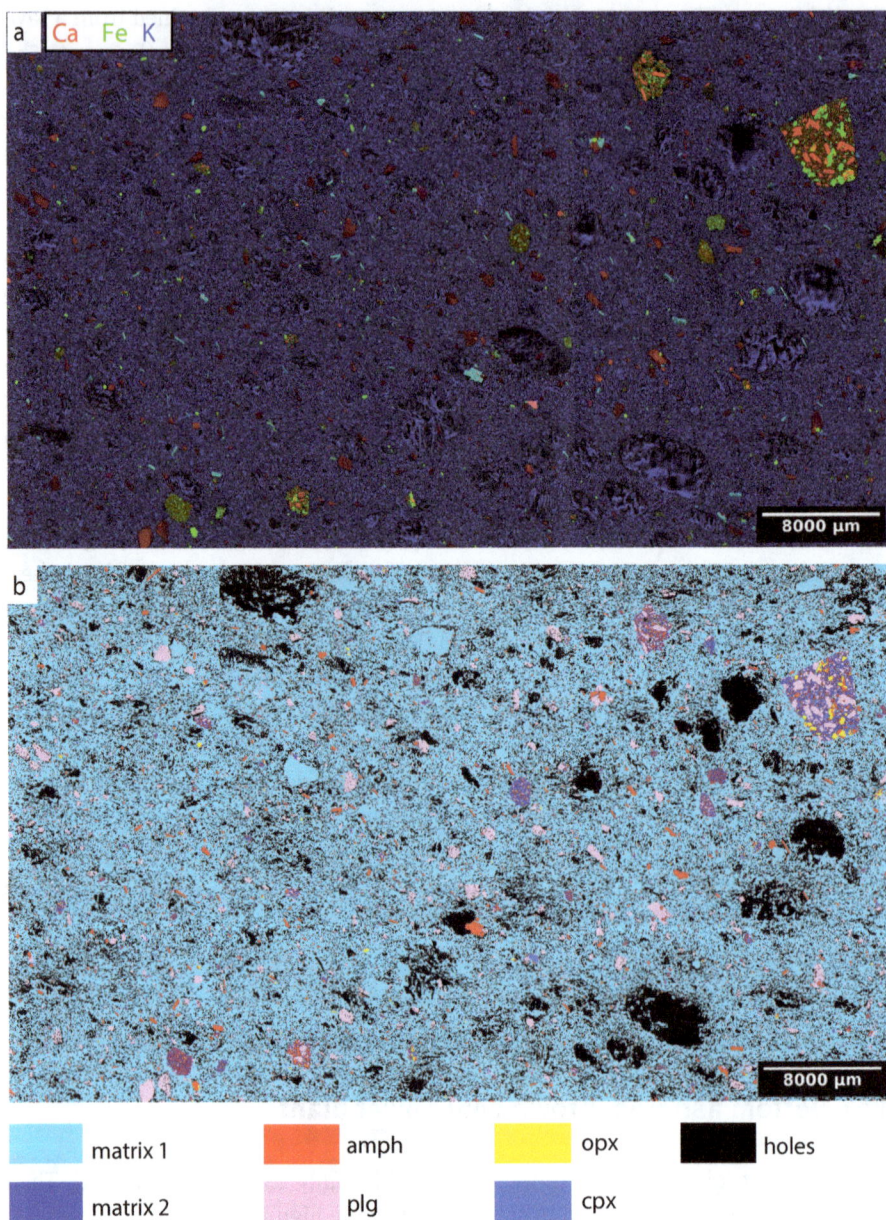

Figure 4.15: µXRF maps of Peruvian pozzolane. a) Ca-Fe-K composite map and b) kmeans-based segmentation for mayor constituent phases (minor amounts of zircon, apatite and sulfide phases are not considered). Given the low density of the matrix, the porosity is most likely underestimated. Pixel size 20 µm.

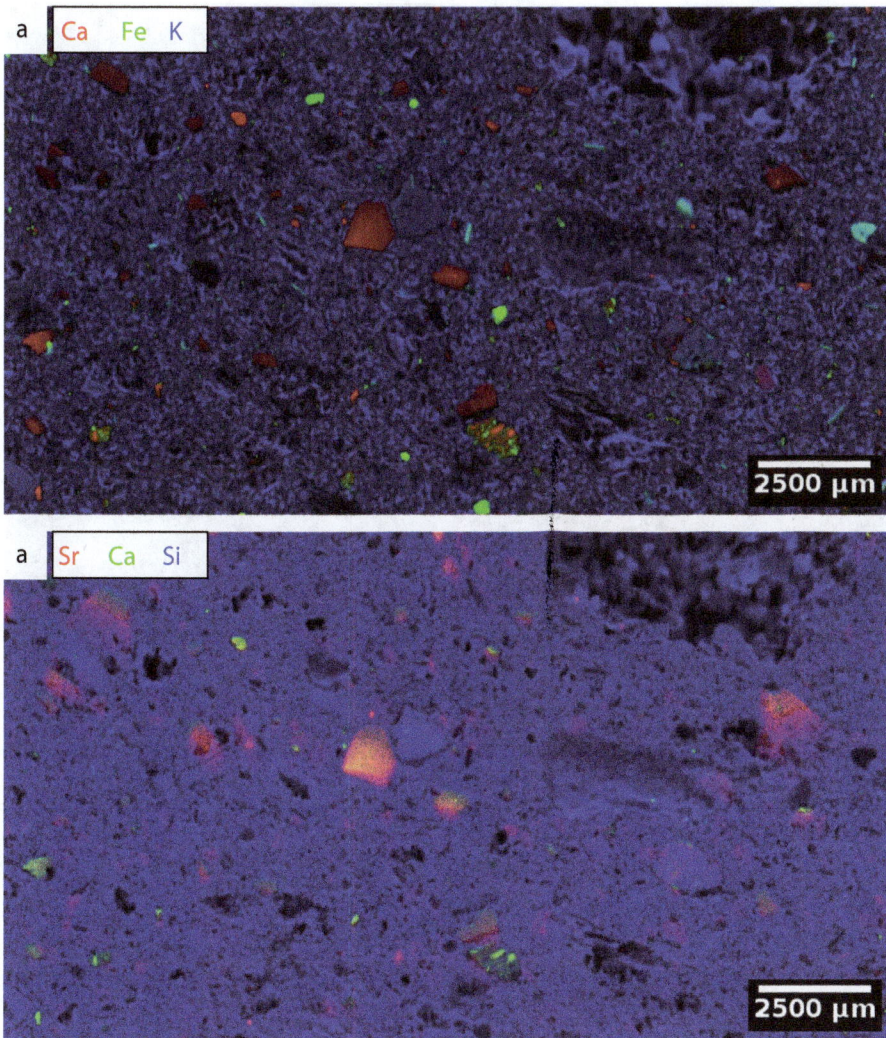

Figure 4.16: Close-up of µXRF maps of Peruvian pozzolane. a) Ca-Fe-K composite map, plagioclase appears as red grains. b) Sr-Ca-Si map of the same area show in (a), demonstrating the effect of a low density matrix where the x-ray signal of heavy elements can be followed to a much larger sample depth. While in this example Sr was not used for phase segmentation, in other systems this effect can result in a substantial overestimation of high density phases embedded in a low density matrix. Pixel size 20 µm.

shows apparently two distinct sulfur-bearing phases; one compatible with gypsum (turquoise = Ca and S counts) and one Ca-poor, sulfur-bearing phase (mottled green in Figure 4.18b). While gypsum was identified by XRD, the difference could arise due to sampling or can relate to the presence of a phase mixture of gypsum with at least one other phase which are smaller than the spatial resolution of the µXRF.

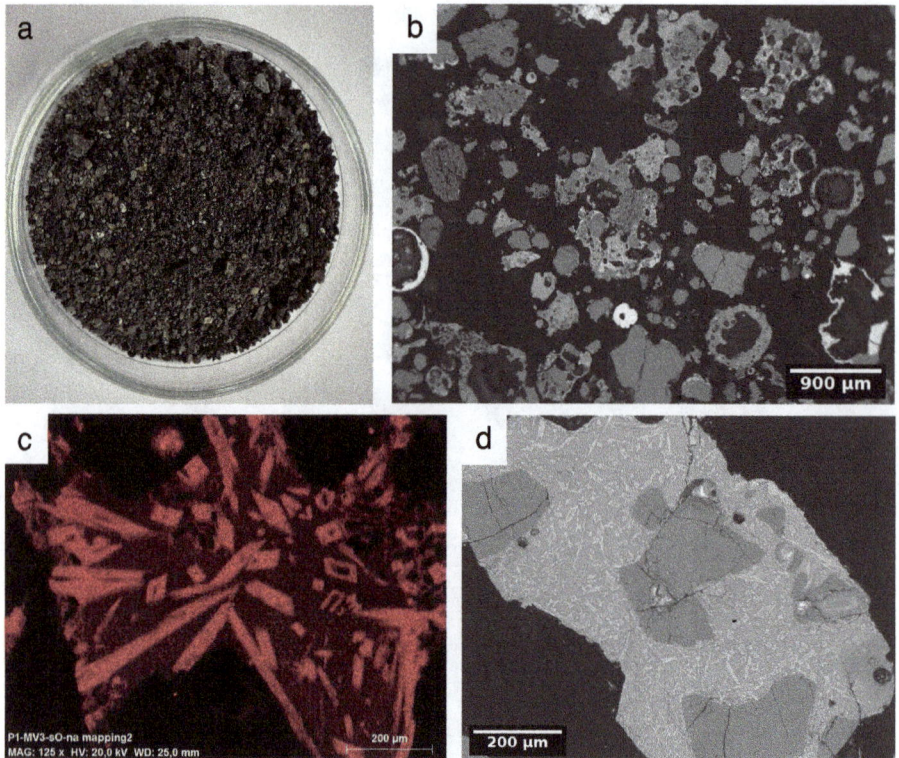

Figure 4.17: a):Brown coal bottom ash. b) SEM/BSE image of bottom ash, embedded in epoxy resin c) SEM/EDX derived Ca-distribution of silicate in amorphous matrix d) SEM/BSE-image of bottom ash.

Table 4.11: Chemical variance of bottom ash from brown coal plant, integrated composition obtained from µXRF map.

Chemistry	Chemical variance slag drilling	mass% (av. µXRF)
dry subst. DEV S3	72.2–96.2	
Al_2O_3	5.1–13.0	10.2
BaO	0.6–1.9	–
CaO	7.2–14.2	8.0
Cl	0.2–1.3	na
CoO	0.02–0.1	–
Cr_2O_3	0.4–2.4	0.0
CuO	0.6–1.3	–
FeO	12.6–24.4	14.1

Table 4.11 (continued)

Chemistry	Chemical variance slag drilling	mass% (av. μXRF)
K_2O	0.5–1.2	1.6
MgO	1.1–3.7	1.7
Na_2O	1.1–3.6	0.0
NiO	0.1–0.8	–
P_2O_5	0.6–2.5	0.0
PbO	0.1–0.9	0.05
SO_3	1.7–6.2	1.0
SiO_2	18.5–38.7	60.8
SrO	<0.01–0.08	0.1
TiO_2	2.0–3.5	1.0
ZnO	0.8–2.3	0.1

Table 4.12: Mineralogical composition of bottom ash from brown coal power plant by XRD.

primary phases	chemical formula
Gypsum	$CaSO_4.2H_2O$
Quartz	SiO_2
Diopsidic Augite	$Ca(Mg,Al)[(AlSi)_2O_6]$
Magnesioferrite	$MgFe_2O_4$
Hydrogarnet	$Ca_3(Al,Fe)_2[((OH)_4,SiO_4)_3]$
Anhydrite	$CaSO_4$
Lime	CaO
Periclase	MgO
Hematite	Fe_2O_3
Magnetite	Fe_3O_4
Perovskite	$CaTiO_3$
Brownmillerite	$Ca_4Al_2Fe_2O_{10}$
Larnite	Ca_2SiO_4
Gehlenite/Melilithe	$Ca_2Al[(Al,Si)_2O_7]$

Table 4.12 (continued)

primary phases	chemical formula
Calcite	$CaCO_3$
Merwinite	$Ca_3Mg[(SiO_4)_2]$
Mullite	$Al_6Si_2O_{11}$

Literature

[1. 4 a1]   Bambauer, H.U., Gebhard, G.,Holtzapfel, T., Krause, Ch., Willner, G.: (1988): Schadstoffimmobilisierung in Stabilisaten aus Braunkohleaschen und REA-Produkten, Fortschr.Min. 66, 253.

[2. 4 a1]   Brett B., Schrader D., Räuchle K., Heide G. & Bertau M. (2015 a): Wertstoffgewinnung aus Kraftwerksaschen – Teil 1: Charakterisierung von Braunkohlenkraftwerksaschen zur Gewinnung strategischer Metalle. Chem. Ing. Tech. 87, Nr. 10, S 1383–1391.

[3. 4 a1]   Feuerborn H.-J. (2007): Mittel- und ostdeutsche Braunkohlenflugaschen in hydraulischen Bindemitte ln. Dissertation Technische Hochschule Aachen, Verlagshaus Mainz GmbH, Aachen.

[4. 4 a1]   Gumz W., Kirsch H. & Mackowsky M.-T. (1958): Schlackenkunde – Untersuchungen über die Minerale im Brennstoff und ihre Auswirkungen auf den Kesselbetrieb. Springer-Verlag, Berlin/Göttingen/Heidelberg.

[5. 4 a1]   Keyn J., Schreiter P., Sansoni G. & Werner M. (1985): Zum Phasen- und Gefügeaufbau von Braunkohlenfilteraschen im Feinstkornbereich. Silikattechnik 36, Heft 11, S. 341–343

[6. 4 a1]   Münch U. (1995): Zu Konstitution, Elutionsverhalten und Kathodolumineszenz von Braunkohlenaschen. Dissertation Technische Universität Freiberg.

[7. 4 a1]   Ostrowski C. (1976): Einfluss des Kalziumhydroxids und des Gipses auf die puzzolanischen Eigenschaften von Flugaschen (Teil 1). Baustoffindustrie A6, S. 13–17

[8. 4 a1]   Ottemann J. (1951a): Über die Mineralbestandteile von Braunkohlenaschen und ihre Bedeutung für die Beurteilung von Aschenbindern. Mitteilungen aus den Laboratorien des Geologischen Dienstes Berlin, Akademie-Verlag Berlin, H. 1, S. 1–18

[9. 4 a1]   Ranneberg, M.: Quantifizierung und Optimierung von Deponiestabilisaten aus calciumreichen Braunkohlefilteraschen unter Verwendung statistischer Versuchsplanung – Dissertation Uni Halle, (2018)

[10. 4 a1] Vassilev S. V. & Vassileva C. G. (2005): Methods for Characterization of Composition of Fly Ashes from Coal-Fired Power Stations: A Critical Overview. Energy & Fuels 19, S. 1084–1098

[11. 4 a1] VGB POWERTECH (2014): Produktion und Verwendung von Kraftwerksnebenprodukten aus Kohlekraftwerken in Deutschland im Jahr 2011. www.vgb.org

[12. 4 a1] Walter G. & Gallenkemper B. (1996): Verwertung von Steinkohlen- und Braunkohlenaschen. In: Produktions- und produktintegrierter Umweltschutz, Handbuch des Umweltschutzes und der Umweltschutztechnik Bd. 2, Springer-Verlag Berlin/Heidelberg,S. 1037–1058

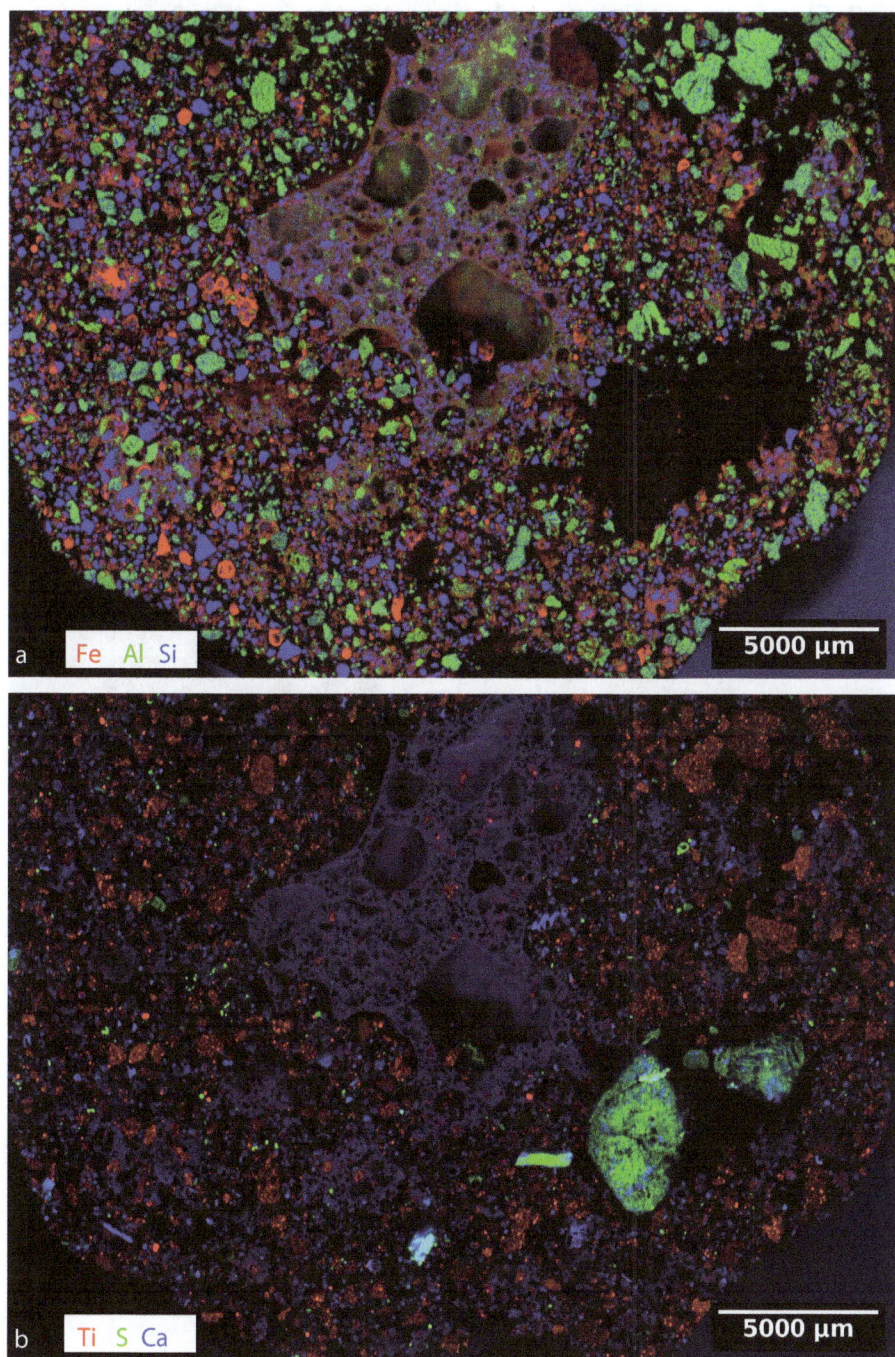

Figure 4.18: µXRF composite maps of bottom ash from brown coal power plant. a) Fe-Al-Si composite map, b) Ti-S-Ca composite map. Pixel size 15 µm.

4 a2 Slag from special waste incineration

A typical industrial residue comes from the incineration of different wastes – therefore a wide varying composition may occur. Very different wastes are included and only partially the mineralogical and chemical data can give all these details. The character of these materials depends on their origin.The here investigated material is described in Tables 4.13 and 4.14 and Figure 4.19a, b. Some overview of literature is given in [1. 4 a2–4. 4 a2]

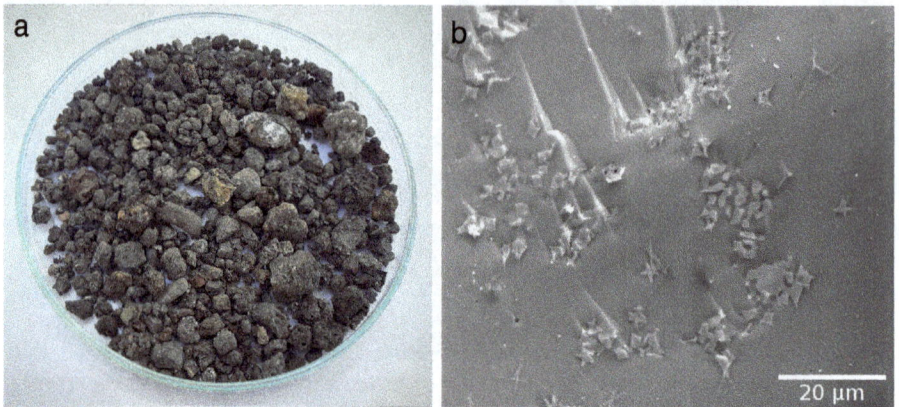

Figure 4.19: a) special waste incineration slag. b) SEM/SE image of dendritic recrystallization and surface efflorescence.

Table 4.13: Chemical variance of special waste incineration slag.

Chemistry	Chemical variance slag compositions
dry subst. DEV S3	72.2–96.2
Al_2O_3 (%)	5.1–13.0
BaO (%)	0.6–1.9
CaO (%)	7.2–14.2
Cl (%)	0.2–1.3
CoO (%)	0.02–0.1
Cr_2O_3 (%)	0.4–2.4
CuO (%)	0.6–1.3
FeO (%)	12.6–24.4

Table 4.13 (continued)

Chemistry	Chemical variance slag compositions
K_2O (%)	0.5–1.2
MgO (%)	1.1–3.7
Na_2O (%)	1.1–3.6
NiO (%)	0.1–0.8
P_2O_5 (%)	0.6–2.5
PbO_2 (%)	0.1–0.9
SO_3 (%)	1.7–6.2
SiO_2 (%)	18.5–38.7
SrO (%)	<0.01–0.08
TiO_2 (%)	2.0–3.5
ZnO (%)	0.8–2.3

Table 4.14: Mineralogical composition of waste incineration slag.

Minerals	Formula	Minerals	Formula
Lepidocrocite	$FeOOH$	Cerussite	$PbCO_3$
Quartz	SiO_2	Aragonite	$CaCO_3$
Gypsum	$CaSO_4.2H_2O$	Calcite	$CaCO_3$
Plagioclase	$(Na,Ca)((Al,Si)O_4)$	Malachite	$Cu_2(CO_3)(OH)_2$
Amorphous	Glass	Azurite	$Cu_3(CO_3)_2(OH)_2$
Ettringite	$3CaO.Al_2O_3.3CaSO_4.32H_2O$	Hydrocerussite	$Pb_3(CO_3)_2(OH)_2$
Fluorite	CaF_2	Chrysokoll	$H_4(Cu,Al)_4(OH)_8Si_4O_{10} \cdot 2H_2O$
Syngenite	$K_2Ca(SO_4)_2$	Palmierite	$(K,Na)_2Pb(SO_4)_2$
Aphtitalite	$K_3Na(SO_4)_2$	Anglesite	$Pb(SO_4)$
Pseudomalachite	$Cu_5(PO_4)_2(OH)_4$	Spertinite	$Cu(OH)_2$
Ramsbeckite	$Cu_{15}(SO_4)(OH)_{22}.6H_2O$	Linarite	$CuPb(SO_4)(OH)_2$
Thenardite	$Na(SO_4)$	Spangolithe	$Cu^{2+}Al(OH)_{12}(SO_4)Cl \cdot 3H_2O$
Goslarite	$Zn(SO_4) H_2O$	Elyite	$CuPb_4(SO_4)(OH)_8$
Zaratite	$Ni(CO_3)(OH)_4.4H_2O$	Spencerite	$Zn_4(PO_4)_2(OH)_2 \cdot 3H_2O$

Table 4.14 (continued)

Minerals	Formula	Minerals	Formula
Tolbachite	$CuCl_2$	Wroewolfeite	$Cu_4(SO_4)(OH)_6.2H_2O$
Ashoverite	$Zn(OH)_2$	Brochantite	$Cu_4(SO_4)(OH)_6$
Zincite	ZnO	Gordaite	$NaZn_4(OH)_6(SO_4)Cl \cdot 6H_2O$
Brochantite	$Cu_4(SO_4)4(OH)_6$	Jarosite	$KFe_3(SO_4)_2(OH)_6$
Rutile	TiO_2	Lanarkite	$Pb_2O(SO_4)$
Corundum	Al_2O_3	Trolleite	$Al_4(PO_4)(OH)_3$
Gibbsite	$Al(OH)_3$	Posnjakite	$Cu_4(SO_4)(OH)_6 \cdot H_2O$
Hematite	Fe_2O_3	Langite	$Cu_4(SO_4)(OH)_6 \cdot 2H_2O$
Magnetite	$FeFe^{3+}_2O_4$	Ktenasite	$(Cu,Zn)_5(SO_4)_2(OH)_6 \cdot 6H_2O$
Rockbridgeite	$Fe_5(PO_4)_3(OH)_5$	Namuwite	$(Zn,Cu)_4(SO_4)(OH)_6 \cdot 4H_2O$
Vivianite	$Fe_3(PO_4)_2 \cdot 8H_2O$	Serpierite	$Ca(Cu,Zn)_4(SO_4)_2(OH)_6 \cdot 3H_2O$
Graftonite	$Fe_3(PO_4)_2$	Chalcomenite	$Cu(SeO_3) \cdot 2H_2O$
Hopeite	$Zn_3(PO_4)_2 \cdot 2H_2O$	Iron phosphate	$Fe_3(PO_4)_2 2H_2O$
Erythrite	$Co_3(AsO_4)_2 \cdot 8H_2O$	Sarcopsid	$Fe_3(PO_4)_2$
Pyromorphite	$Pb_5Cl(AsO_4)_3$	and many others	

The slag from the special waste incineration is composed of 1–10 mm large granular aggregates, which were embedded in epoxy resin, saw cut and ground. μXRF composite maps immediately reveal the heterogeneous nature of the slag particles (Figure 4.20). Many show a fine grained, polyphase shells around a central particle. These cores of the aggregates are either mono- or polyphase with distinct compositions.

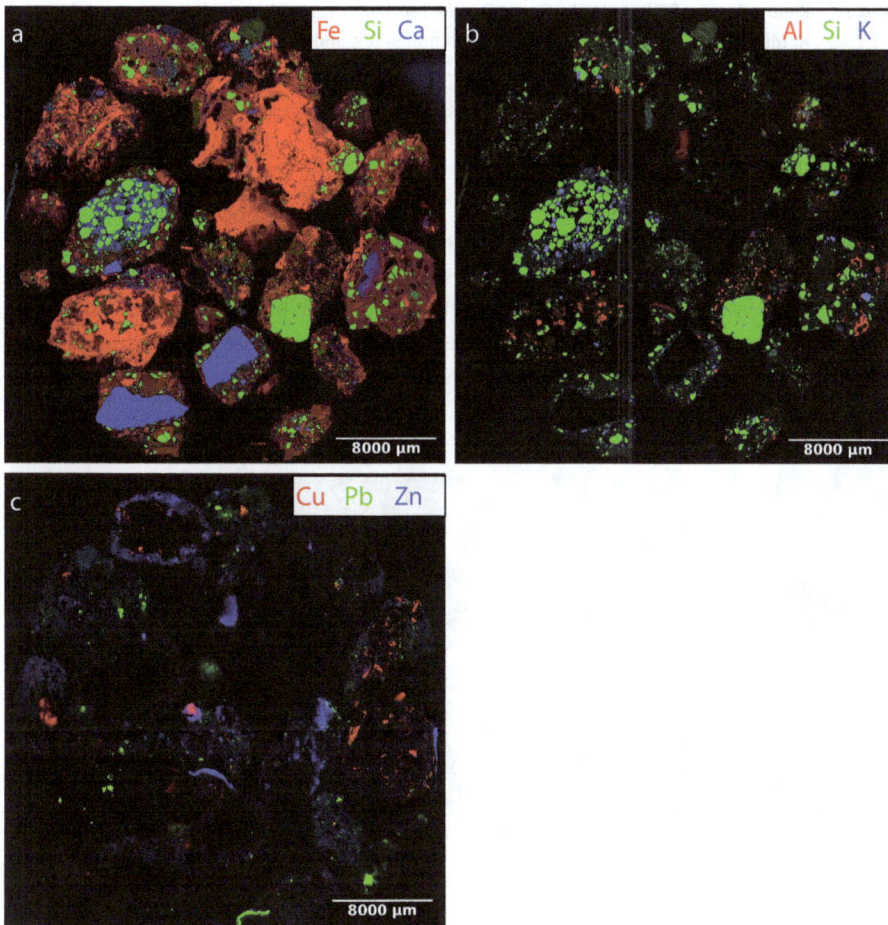

Figure 4.20: µXRF map showing the complex nature of the special waste incineration slag. Individual particles were selected at random and embedded in epoxy resin. a) Fe-Si-Ca composite map. Note the depth of the Fe-signal through the resin. b) Al-Si-K composite map with light elements showing in contrast to (a) the true cross sectional areas of particles. c) Cu-Pb-Zn composite map highlighting presence of minor heavy metal bearing phases.

Literature

[1. 4 a2] Gade, B., Pöllmann, H., Heindl, A., Westermann, H.: Long-term behaviour and mineralogical reactions in hazardous waste landfills: a comparison of observation and geochemical modelling, Environ. Geology 40(3),248-Abb.6, (2001)

[2. 4 a2] Pöllmann, H.: Mineralisation of industrial wastes, Shaker Vlg. (2010)

[3. 4 a2] Pöllmann, H.: Immobilisierung von Schadstoffen durch Speichermineralbildung, Shaker-Vlg., (2007)

[4. 4 a2] Riedmiller, A., Pöllmann, H., Gade, B., Westermann, H. and Amsoneit,
N: Mineralphasenbestand, Reaktionsmechanismen und Sickerwasserzusammensetzung
in einer Sondermülldeponie, Ber.Dt.Min.Ges. 6 (1),229, (1994)

4 a3 Hard coal fly ash

A typical residue from hard coal combustion is the hard coal fly ash which is quite
high in CaO-content and therefore already widely used in cement and concrete industry. The suppliers of these materials normally can provide hard coal fly ash according
to existing norms (Figure 4.21). These ashes and their usage are well documented in
literature for their composition (Tables 4.15 and 4.16), usage, properties and determination methods [1. 4 a3–9. 4 a3].

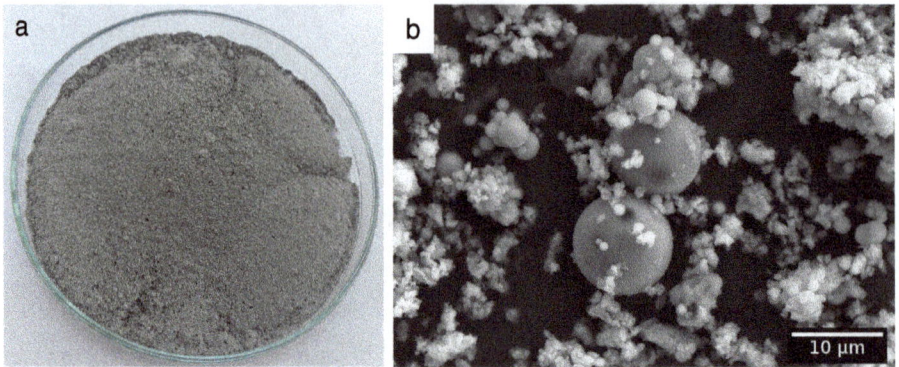

Figure 4.21: a) Fine grained hard coal fly ash b) SEM/SE image of hard coal fly ash.

Table 4.15: Chemical composition of hard coal fly ash. µXRF bulk chemical
composition is gives a total of 43% due to the high dilution by the resin
matrix. The analyzed area is 440 mm^2.

Chemistry	mass%	µXRF max%
LOI%	18,1	–
Na$_2$O (%)	0,7	0.5
MgO (%)	0,8	0.5
Al$_2$O$_3$ (%)	11,3	9.0
SiO$_2$ (%)	15,3	13.6
P$_2$O$_5$ (%)	1,1	0.9

Table 4.15 (continued)

Chemistry	mass%	µXRF max%
SO$_3$ (%)	14,2	17.8
K$_2$O (%)	0,6	0.6
CaO (%)	32,5	46.8
TiO$_2$ (%)	0,5	1.1
V$_2$O$_5$ (%)	0,0	0.1
Cr$_2$O$_3$ (%)	0,0	0.0
Mn$_2$O$_3$ (%)	0,1	–
MnO (%)	–	0.1
Fe$_2$O$_3$ (%)	3,8	–
FeO (%)	–	6.8

Table 4.16: Mineralogical composition of hard coal fly ash from XRD.

primary phases	chemical formula
Gypsum	CaSO$_4$.2H$_2$O
Quartz	SiO$_2$
Anhydrite	CaSO$_4$
Lime	CaO
Calcite	CaCO$_3$
Mullite	Al$_6$Si$_2$O$_{11}$
Portlandite	Ca(OH)$_2$

In this example, we find that segmentation of larger particles is straight forward (Figures 4.22 and 4.23). The spheroidal particles however are difficult to characterize since total composition is variable and also varies for different concentric shells. The circular intersection with the sample surface may be surrounded by a halo with highly variable x-ray spectrum which itself depends on the position of the intersecting surface relative to the particle.

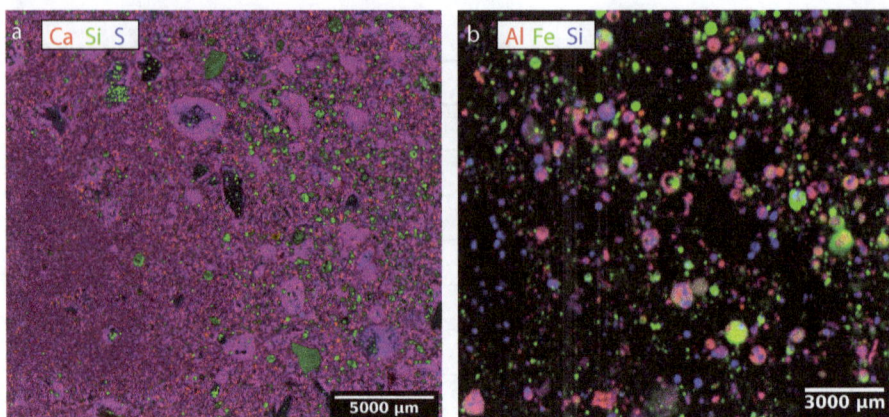

Figure 4.22: μXRF maps of coal fly ash. a) Ca-Si-S composite map of dispersed in epoxy resin. b) Detail of the same map showing the Al-Fe-Si composite. Note the complex particles, some showing shells of different compositions. Note that different intensities may arise from the element concentration or from the position with respect to the surface of the sample.

Literature

[1. 4 a3] Hower JC, Suarez-Ruiz I, Mastalerz M. An approach toward acombined scheme for the petrographic classification of fly ash: revision and clarification. Energy Fuels 2005;19: 653–5

[2. 4 a3] McCarthy, G. and Thedchanamoorthy: Semi-quantitative X-ray diffraction analysis of fly ash by the reference intensity ratio method, *MRS Online Proceedings Library* **136**, 67–76 (1988)

[3. 4 a3] G.J. McCarthy, in Fly Ash and Coal Conversion By-Products: Characterization. Utilization and Disposal IV, edited by G.J. McCarthy, F.P. Glasser, D.M. Roy and R.T. Hemmings, Mat. Res. Soc. Symp. Proc. Vol. 113 (Materials Research Society, Pittsburgh, 1988) pp. 75–86

[4. 4 a3] McCarthy, G., Manz, O., Stevenson, R.J.,Hassett, D.J. and Groenewold, G.J.-: Western Fly Ash Research, Development and Data Center, in Flv Ash and Coal Conversion Bv-Products: Characterization. Utilization and Disposal II, edited by G.J. McCarthy, F.P. Glasser and D.M. Roy, Mat. Res. Soc. Symp. Proc. Vol. 65 (Materials Research Society, Pittsburgh, 1986) pp. 165–16

[5. 4 a3] McCarthy, G., Manz, O., Johanson, D.,Steinwand, R., Stevenson, R. and Hassett, D.J.: in Fly Ash and Coal Conversion By-Products: Characterization. Utilization and Disposal III. edited by G.J. McCarthy, F.P. Glasser, D.M. Roy and S. Diamond, Mat. Res. Soc. Proc. Vol. 86 Materials Research Society, Pittsburgh, (1987) pp. 109–112.

[6. 4 a3] Manz OE. Coal fly ash: a retrospective and future look. Fuel1999;78:133–6.

[7. 4 a3] Pöllmann, H.: Mineralization of wastes and industrial residues, Shaker, (2010), 478 pages.

[8. 4 a3] Roy WR, Griffin RA. A proposed classification system for coalfly ash in multidisciplinary research. J Environ Qual 1982;11:563–568

[9. 4 a3] Vassilev, S. and Vassileva, C.: A new approach for the classification of coal fly ashes based on their origin, comcpoition, properties and behaviour, Fuel 86, (2007), 1490–1512

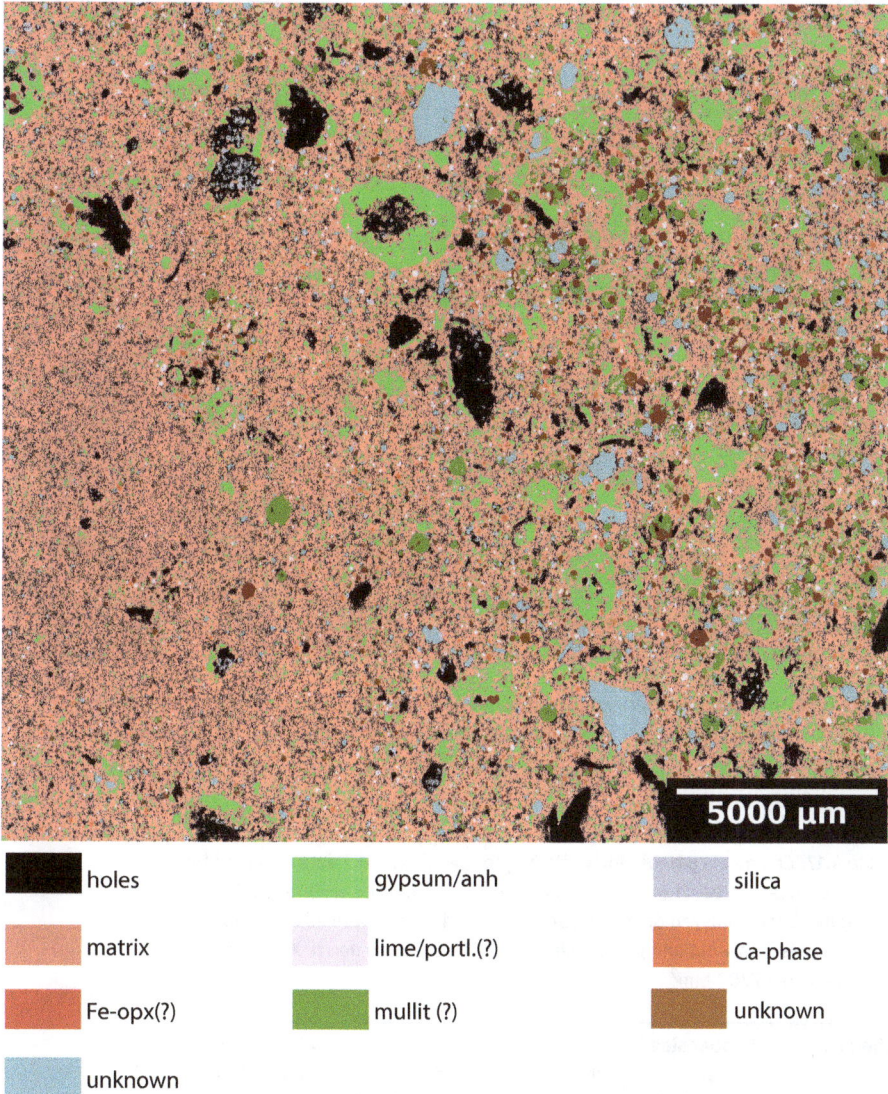

Figure 4.23: Attempt of segmentation of μXRF-derived element maps into a phase map. While some phases can be readily identified (Gypsum/Anhydrite), most complex phases are poorly determined which is most likely related to the large variability in composition (Figure 22 b above).

4 a4 Copper slag from copper mountains Helbra/Mansfeld

The famous residues from copper mining and the produced slags are described in Figure 4.24a–d and Tables 4.17–4.19. Literature is given by [1. 4 a4–43. 4 a4]. For longer exposed slags many newly formed minerals are described.

Figure 4.24: a) Copper slag stockpile Helbra/Mansfeld. b) Radial Pyroxene crystals (x = 1 cm). c) Radial pyroxene crystals (x = 1 cm). d) Cross-polarized light micrograph of Pyroxene crystals (Ranneberg).

Table 4.17: Chemical composition of Helbra copper slag. µXRF data obtained from two different samples. T1 (as shown in Figure 4.25 with three texturally different areas) has been analyzed on a saw-cut and ground slab as well as in a ground thin section. T2 is a saw-cut and ground slab which was used to derive XRF data (analyzed area 7700 mm^2).

Chemistry	Copper slag glassy	copper slag cristalline	T2 µXRF	T1 µXRF	T1 µXRF (thin section)
LOI%	0,0	0,0	–	–	–
Na$_2$O (%)	1,0	0,9	1.0	0.8	0.9
MgO (%)	5,7	5,4	5.9	7.4	8.0
Al$_2$O$_3$ (%)	13,9	14,0	14.1	13.8	14.8
SiO$_2$ (%)	46,6	47,7	46.8	46.1	50.0
P$_2$O$_5$ (%)	0,2	0,2	0.0	0.0	0.0
SO$_3$ (%)	0,0	0,0	0.7	0.7	0.7

Table 4.17 (continued)

Chemistry	Copper slag glassy	copper slag cristalline	T2 μXRF	T1 μXRF	T1 μXRF (thin section)
K$_2$O (%)	3,2	3,3	3.5	3.3	1.0
CaO (%)	19,6	19,4	19.0	18.8	18.2
TiO$_2$ (%)	0,7	0,8	0.8	1.1	1.0
V$_2$O$_5$ (%)	0,1	0,1	0.1	0.1	0.1
Cr$_2$O$_3$ (%)	0,1	0,1	0.0	0.1	0.1
Mn$_2$O$_3$ (%)	0,6	0,5	–	–	–
MnO (%)	–	–	0.5	0.6	0.4
Fe$_2$O$_3$ (%)	6,7	5,6	–	–	–
FeO (%)	–	–	6.1	6.6	3.8
NiO (%)	0,0	0,0	–	–	–
CuO (%)	0,3	0,2	0.2	0.2	0.1
ZnO (%)	0,6	0,5	0.6	0.2	0.1
SrO (%)	0,1	0,1	0.1	0.0	0.0
ZrO$_2$ (%)	0,0	0,0	0.0	0.1	0.0
BaO (%)	0,3	0,4	0.3	–	–

Table 4.18: Minerals in Helbra copper slag by XRD.

primary phases	chemical formula
copper	Cu
brass	(Cu,Zn)
lead	Pb
Chalkopyrite	CuFeS$_2$
no name	Cu-Sn-Phase
Djurleite	Cu$_{31}$S$_{16}$
Digenite	Cu$_9$S$_5$
Bornite	Cu$_5$FeS$_4$
Sphalerite	ZnS
Galenite	PbS

Table 4.18 (continued)

primary phases	chemical formula
Melilithe	$(Ca,Na)_2(Al,Mg,Fe)[(Si,Al)_2O_7]$
Forsterite	$Mg_2[SiO_4]$
Fayalite	$Fe_2[SiO_4]$
Leucite	$K[AlSi_2O_6]$
Willemite	$Zn_2[SiO_4]$
Hardystonite	$Ca_2Zn[Si_2O_7]$
Åkermanite	$Ca_2Mg[Si_2O_7]$

Some other secondary copper and lead minerals were described (Devilline, Brochantite, Antlerite, Malachite, Spertiniite, Copper, Atacamite, Clinoatacamite, Tenorite). Also Gypsum, Calcite, Jarosite, Anglesite and Lead oxide could be proved. But many more minerals are described in literature (Table 4.19).

Table 4.19: Table of known secondary minerals of Helbra copper slag heap (acc. WITZKE & PÖLLMANN 1996, WITZKE 1997, SIEMROTH & WITZKE 1999, KNOLL 2008, RANNEBERG 2009).

Secondary minerals	Formula	Secondary minerals	Formula
copper	Cu	Cerussite	$PbCO_3$
Silver	Ag	Aragonite	$CaCO_3$
Lead	Pb	Calcite	$CaCO_3$
Sulfurl	S	Malachite	$Cu_2(CO_3)(OH)_2$
Aluminium	Al	Azurite	$Cu_3(CO_3)_2(OH)_2$
Molybdenite	MoS_2	Callaghanite	$Cu_2Mg_2(CO_3)(OH)_6 \cdot 2H_2O$
Akanthite	AgS_2	Hydrocerussite	$Pb_3(CO_3)_2(OH)_2$
Covellite	CuS	Brianyoungite	$Zn_{12}(CO_3)_3(SO_4)(OH)_{16}$
Chalkosite	Cu_2S	Monohydrocalcite	$CaCO_3 \cdot H_2O$
Cotunnite	$PbCl_2$	Lansfordite	$Mg(CO_3)*5H_2O$
Nantokit	CuCl	Dypingite	$Mg_5(OH)_2(CO_3)_2 \cdot 5H_2O$
Laurionite	Pb(OH)Cl	Chrysokoll	$H_4(Cu,Al)_4(OH)_8Si_4O_{10} \cdot 2H_2O$
Atacamite	$Cu_2(OH)_3Cl$	Palmierite	$(K,Na)_2Pb(SO_4)_2$

Table 4.19 (continued)

Secondary minerals	Formula	Secondary minerals	Formula
Clinoatacamite	$Cu_2(OH)_3Cl$	Anglesite	$Pb(SO_4)$
Botallakite	$Cu_2(OH)_3Cl$	Spertinite	$Cu(OH)_2$
Penfildite	$Pb_4(OH)_2Cl_6$	Linarite	$CuPb(SO_4)(OH)_2$
Chlorixiphite	$Pb_3Cu^{2+}O_2(OH)_2Cl_2$	Spangolithe	$Cu^{2+}Al(OH)_{12}(SO_4)Cl \cdot 3H_2O$
Connelite	$Cu_{19}Cl_4SO_4(OH)_{32}$	Elyite	$CuPb_4(SO_4)(OH)_8$
Calumenite	$Cu(OH,Cl)_2 \cdot 2H_2O$	Chalkanthite	$Cu(SO_4) \cdot 5H_2O$
Diaboleite	$CuPb_2Cl(OH)_4$	Antlerite	$Cu_3(SO_4)(OH)_4$
Boleite	$Ag_9Cu_{24}Pb_{26}Cl_{62}(OH)_{48}$	Brochantite	$Cu_4(SO_4)(OH)_6$
No name	$Zn_4Na(OH)_6Cl(SO_4) \cdot 6H_2O$	Gordaite	$NaZn_4(OH)_6(SO_4)Cl \cdot 6H_2O$
No name	Pb-Cu-Oxichloride	Jarosite	$KFe_3(SO_4)_2(OH)_6$
No name	Pb-Oxichloride	Lanarkite	$Pb_2O(SO_4)$
Cuprite	Cu_2O	Gypsum	$CaSO_4*H_2O$
Tenorite	CuO	Posjakite	$Cu_4(SO_4)(OH)_6 \cdot H_2O$
Zincite	ZnO	Langite	$Cu_4(SO_4)(OH)_6 \cdot 2H_2O$
Lithargite	PbO	Ktenasite	$(Cu,Zn)_5(SO_4)_2(OH)_6 \cdot 6H_2O$
Massicotite	PbO	Namuwite	$(Zn,Cu)_4(SO_4)(OH)_6 \cdot 4H_2O$
Bunsenite	NiO	Serpierite	$Ca(Cu,Zn)_4(SO_4)(OH)_6 \cdot 3H_2O$
Minium	$Pb^{2+}_2Pb^{4+}O_4$	Chalcomenite	$Cu(SeO_3) \cdot 2H_2O$
Cassiterite	SnO_2	No name	$Pb_4O_3(SO_4) \cdot H_2O$
Corundum	Al_2O_3	No name	Cu-K-Sulfat
Delafossite	$CuFeO_2$	Mimetesite	$Pb_5Cl(AsO_4)_3$
Spinel	$MgAl_2O_4$	Erythrite	$Co_3(As_4)_2 \cdot 8H_2O$
Wulfenite	$PbMoO_4$	Ludlockite	$(Fe,Pb)As^{5+}_2O_6$
Trevoite	$NiFe^{3+}_2O_4$	Vivianite	$Fe_3(PO_4)_2 \cdot 8H_2O$
Spangolite	$Cu_6Al(OH)_{12}Cl(SO_4) \cdot 3H_2O$	No name	$CuHAsO_4?$
No name	"Grüner Rost"	No name	Cu-Cl-Arsenate
Chalkophyllite	$Cu_9Al(OH)_{12}(SO_4)_{1,5}(AsO_4)_2 \cdot 18H_2O$		

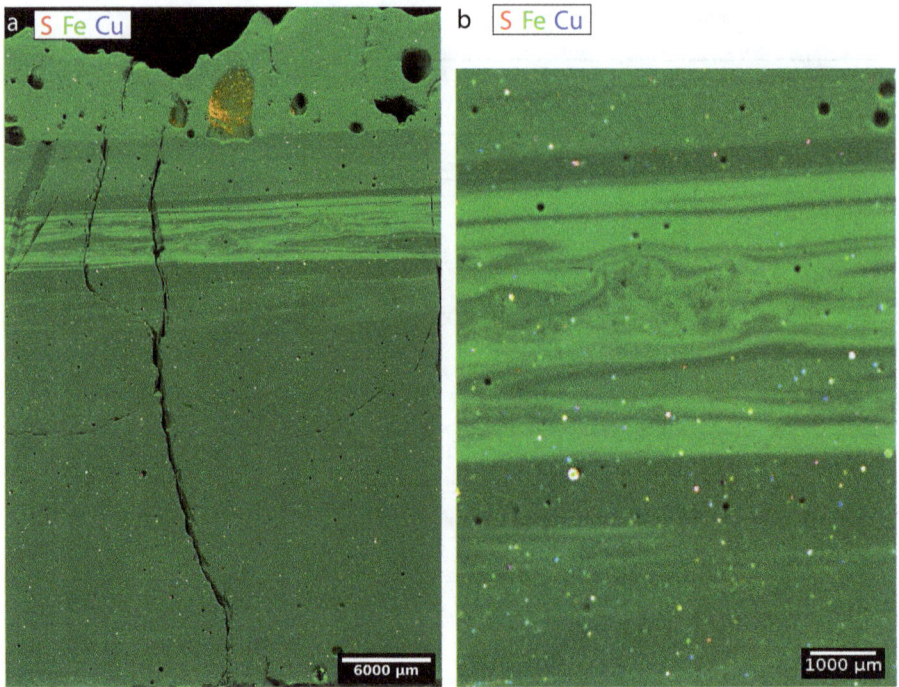

Figure 4.25: µXRF derived S-Fe-Cu composite map (a). b) shows a close-up of (a) showing a complex flow zone and dispersed particles of variable composition ranging from Fe to Fe-C to sulfidic composition. Sample T1.

The samples of the copper slag from Helbra consist of a very homogeneous silicate matrix and contain numerous inclusions (Figures 4.25a/b, 4.7). The inclusions show frequently a concentric structure with a compositional variation from core to rim. The exclusive occurrence of circular cross sections suggests that the particles in the slag have a spherical shape. Some of these particles seem to possess an Fe- rich core and a Cu-S rich rim (Figure 4.7). Given the relatively large excitation volume of Fe and Cu as well as the problem that for each circular section, the depth of the particle remains unknown, a more precise characterization is not straight forward. Apart from Cu-Fe-S phases as inclusions also Cu-rich or Fe-rich ones are observed. The low energy of related to S-Kα may actually cause the appearance of apparently Cu-rich but S-poor particles once those are submerged slightly below the surface of the sample. Based on Fe-Kα intensity, we may assume that at least one second type of particle, consisting of a rather Fe-rich phase, is present in the sample shown in Figure 4.25.

Furthermore, 3 different zones of the slag can be identified; a homogeneous internal part (Figure 4.25a bottom), a zone showing flow structures (Figure 4.25a center, 4.25b) and a porous outer (Figure 4.25a top) zone yielding large vugs. A notable result obtained from a different, slightly large sample is the fairly good agreement between the bulk XRF composition and the composition obtained from the integral spectrum of the µXRF map (Table 4.17).

Literature

[1. 4 a4]   Al-Jabri, K.S.; Taha, R.A.; Al-Hashmi, A.; Al-Harthy, A. S.: "Effect of copper slag and cement by-pass dust addition on mechanical properties of concrete." Construction and Building Materials 20. S. 322–331. 2005 / 2006.

[2. 4 a4]   Bharati; S.K. Dissertation: „Treatment of marine clay by using cement copper slag grout in low pressure grouting." 2012.

[3. 4 a4]   Bipra Gorai, R.K.; Premchand, J.: "Characteristics and utilisation of copper slag – /a review" Abstract; Resources, Conservation and Recycling 39; S. 299–313. 2002 / 2003.

[4. 4 a4]   Bilâl, S. (1983): Oxid-, Silikat- und Metallphasen in Schachtofen – Kupferschlacken. Diss. Universität Hamburg, Fachbereich Geowissenschaften, S. 127.

[5. 4 a4]   Shi, C.; Meyer, Ch.; Behnood, A.: "Utilization of copper slag in cement and concrete" Resources, Conservation and Recycling 52 (2008) S. 1115–1120. 2008.

[6. 4 a4]   Eisenächer W. & Jäger D. (1997): Die Verhüttung des Mansfelder Kupferschiefers unter besonderer Berücksichtigung der Verarbeitung von Rohhüttenschlacke. In: Reststoffe der Kupferschieferverhüttung. Teil 1: Mansfelder Kupferschlacken, Beiträge zum Workshop am 4. und 5. Dezember 1996 in Bad Lauchstädt, UFZ-Bericht Nr. 23/1997, 3–7.

[7. 4 a4]   Eisenächer W., Klette W. & Prohl H. (1999): Vom Kupferschiefer zum Metall – Die Verhüttung. In: Mansfeld – Die Geschichte des Berg- und Hüttenwesens. Hrsg.: Verein Mansfelder Berg und Hüttenleute e. V., Lutherstadt Eisleben und vom Deutschen Bergbaumuseum, Bochum, 206–360.

[8. 4 a4]   Eisenhut, K.-H. & Kautzsch, E. (1954): Handbuch für den Kupferschieferbergbau. Fachbuchverlag, Leipzig, 335 S.

[9. 4 a4]   Endell, K., Müllensiefen, W. & Wagenmann K. (1932): Über die Viskosität von Mansfelder Kupferhochofenschlacken in Abhängigkeit von Temperatur, chemischer Zusammensetzung und Kristallisation. Metall und Erz. Zeitschrift für Metallhüttenwesen und Erzbergbau einschließlich Aufbereitung, Heft 17, 368–375.

[10. 4 a4]   Faber, W. (1954): Mikroskopie der Metallhütten-Schlacken. In: Handbuch der Mikroskopie in der Technik. Bd. II, Teil 2, Hrsg.: H. Freund, Umschauverlag, Frankfurt/M., 519–593.

[11. 4 a4]   Galonska, K. (1997): Phasenbestand, Gefüge und Spurenelementgehalt sowie Elution von Kupferrohhüttenschlacken (Helbra/Eisleben). In: Reststoffe der Kupferschieferverhüttung. Teil 1: Mansfelder Kupferschlacken, Beiträge zum Workshop am 4. und 5. Dezember 1996 in Bad Lauchstädt, UFZ-Bericht Nr. 23/1997, 31–36.

[12. 4 a4]   Gerlach, R. (1995): Kluftgebundene Mineralisation im subsalinaren Tafeldeckgebirge des Harzvorlandes – Lagerstättentyp Mansfelder Rücken. In: Zur Geschichte des Mansfelder Kupferschieferbergbaus, Hrsg.: G. Jankowski, GDMB, Clausthal- Zellerfeld, 29–33.

[13. 4 a4]   Hammer, J., Rösler, H. J. & Niese S. (1988). Besonderheiten der Spurenelementführung des Kupferschiefers der Sangerhäuser Mulde und Versuche ihrer Deutung. Zeitschrift für angewandte Geologie, Vol. 34, Nr. 11, 339–343.

[14. 4 a4]   Hamroll, K. & Pöllmann H. (1997): Untersuchungen zur Speichermineralbildung aus Mansfelder Kupferschlacken. In: Reststoffe der Kupferschieferverhüttung. Teil 1: Mansfelder Kupferschlacken, Beiträge zum Workshop am 4. und 5. Dezember 1996 in Bad Lauchstädt, UFZ-Bericht Nr. 23/1997, 47–50.

[15. 4 a4]   Ihl, R. (1971): Kristallisationsverhalten von Mansfelder Rohhüttenschlacke. Diss. Karl –
Marx – Universität Leipzig, Sektion Chemie, 144 S. [unveröffentlicht]

[16. 4 a4]   Jahn, S., Hörold, H. & Friedrich, B. (2000): Der Kupferschieferbergbau bei Mansfeld,
Eisleben und Sangerhausen – Geschichte, Geologie und Mineralien (I). Mineralien –
Welt: Magazin für das Sammeln schöner Steine. Bd.11, Nr. 3, 17–32.

[17. 4 a4]   Jahn, S., Hörold, H. & Friedrich, B. (2000): Der Kupferschieferbergbau bei Mansfeld,
Eisleben und Sangerhausen – Geschichte, Geologie und Mineralien (II). Mineralien –
Welt: Magazin für das Sammeln schöner Steine. Bd.11, Nr. 4, 32–56.

[18. 4 a4]   Knitzschke, G. (1966) Zur Erzmineralisation, Petrographie, Haupt- und
Spurenelementführung des Kupferschiefers im SE – Harzvorland. Freiberger
Forschungshefte C207, Leipzig,147 S.

[19. 4 a4]   Knitzschke, G. (1999): Geologischer Überblick zur Kupferschieferlagerstätte. In:
Mansfeld – Die Geschichte des Berg- und Hüttenwesens. Hrsg.: Verein Mansfelder
Berg- und Hüttenleute e.V., Lutherstadt Eisleben und vom Deutschen Bergbaumuseum,
Bochum, 11–40.

[20. 4 a4]   Knitzschke, G. & Jankowski, J. (1995): Die Geologischen Verhältnisse. In: Zur Geschichte
des Mansfelder Kupferschieferbergbaus. Hrsg.: G. Jankowski, GDMB, Clausthal-
Zellerfeld, 3–29.

[21. 4 a4]   Köpernik, H. (1968): Beitrag zur Phasenanalyse der Mansfelder Rohhüttenschlacke.
Diplomarbeit, Sektion Chemie, Karl-Marx-Universität Leipzig.

[22. 4 a4]   Knoll, H. (2008): Interessante Mineralien aus den Schlacken der Kupferschiefer –
Verhüttung. Mineralienwelt, Heft 1, S. 36–47.

[23. 4 a4]   Murari, K.; Siddique, R.; Jain, K. K.: "Use of waste copper slag, a sustainable material."
J Mater Cycles Waste Manag. DOI 10.1007/s10163-014-0254-x. 2014

[24. 4 a4]   Lange, A. & Lindenlaub, W. (1959): Beitrag zur Abscheidung von suspendierten Metallen
und Metallverbindungen aus Schlacken, dargestellt an Mansfelder Rohhüttenschlacke.
Bergakademie Nr. 7, Freiberg, 399–407.

[25. 4 a4]   Moura; Washington Almeida Jardel Pereira Goncalves; Monica Batista Leite Lima:
copper slag waste as a supplementary cementing material to concrete"; Article; 4TH
BRAZILIAN MRS MEETING; J Mater Sci; S. 2226–2230; DOI 10.1007/s10853-006-0997-4
2007

[26. 4 a4]   Najimi, M., Sobhani, J.; Pourkhorshidi, A. R.: „Durability of copper slag contained
concrete exposed to sulfate attack" Construction and Building Materials 25.
S. 1895–1905. 2011.

[27. 4 a4]   Osborne, E.F., De Vries, R.C., Gee, K.H. & Kraner, H.- M. (1954): Optimum of blast
furnace slag as deduced from liquidus data for quarternary system CaO – MgO – Al_2O_3 -
SiO_2. Journal of Metalls Nr. 6, 33–45.

[28. 4 a4]   Pentinghaus, H. J., Istrate, G. & Schreck (1997): Mansfelder Kupferschlackenpflaster –
Phasenbestand, Gefüge und Verwitterung. In: Reststoffe der Kupferschieferverhüttung.
Teil 1: Mansfelder Kupferschlacken, Beiträge zum Workshop am 4. und 5. Dezember
1996 in Bad Lauchstädt, UFZ-Bericht Nr. 23/1997, 37–45.

[29. 4 a4]   Pompe, E. (1965): Orientierende Untersuchungen im Gebiet der Mansfelder
Kupferschlacke. Zum System $CaO-MgO-Al_2O_3-SiO_2$. Diplomarbeit, Institut für
anorganische und anorganisch-technische Chemie, TU Dresden[unveröffentlicht]

[30. 4 a4]   Ranneberg, M. Diplomarbeit Institut für Geowissenschaften der Martin-Luther-
Universität Halle-Wittenberg: „Charakterisierung und Untersuchung der primären und
sekundären Minerale der Kupferschlackehalde bei Hebra/Mansfeld-Südharz." 2009.

[31. 4 a4]   Rentzsch, J. & Knitzschke, G. (1968): Die Erzmineralparagenesen des Kupferschiefers
und ihre regionale Verbreitung. Freiberger Forschungshefte C 231, Leipzig, 189–211.

[32. 4 a4]   Sáncheza, M., Sudburyb, M.: „Physicochemical characterization of copper slag and alternatives of friendly environmental management" J. Min. Metall. Sect. B-Metall. 49 (2) B. S. 161–168. 2013.

[33. 4 a4]   Schreck P. (1997): Schadstoffausträge aus den Halden der Kupferschieferverhüttung. In: Reststoffe der Kupferschieferverhüttung. Teil 1: Mansfelder Kupferschlacken. Beiträge zum Workshop am 4. und 5. Dezember 1996 in Bad Lauchstädt, UFZ- Bericht Nr. 23/1997, 9–15.

[34. 4 a4]   Spangenberg, C. (1572): Mansfeldische Cronika. Auszugsweise in: Mansfeld – Die Geschichte des Berg- und Hüttenwesens. Hrsg.: Verein Mansfelder Berg- und Hüttenleute e.V., Lutherstadt Eisleben und vom Deutschen Bergbaumuseum, Bochum.

[35. 4 a4]   Schubert, E. (1965): Orientierende Untersuchungen im Gebiet der Mansfelder Kupferschlacke. Zum System CaO-FeO-Al$_2$O$_3$-SiO$_2$. Diplomarbeit, Institut für anorganische und anorganisch-technische Chemie, TU Dresden [unveröffentlicht].

[36. 4 a4]   Siemroth, J. & Witzke, T. (1999): Die Minerale des Mansfelder Kupferschiefers. Schriftenreihe des Mansfeld – Museums. Hrsg.: Förderverein Mansfeld – Museum e. V., gemeinsam mit dem Mansfeld – Museum Hettstedt, Neue Folge, Hettstedt, Nr. 4, 1–66.

[37. 4 a4]   Tewelde, M.: Dissertation Mathematisch-Naturwissenschaftlich-Technischen Fakultät der Martin-Luther-Universität Halle-Wittenberg: „Speichermineralbildung und Alinitherstellung aus MVA Flugasche, Mansfelder Kupferschlacke und Kalksteinmehl." 2004.

[38. 4 a4]   Tixier, R., Devaguptapu, R.; Mobasher, B.: „The effect of copper slag on the hydration and mechanical properties of cementitious mixtures." Cement and concrete research. Vol. 27. No. 10. S. 1569–1580. 1997.

[39. 4 a4]   Stedingk, K. (2008): Kupferschiefer. In: Geologie von Sachsen – Anhalt. Hrsg.: G.H. Bachmann, B.C. Ehling, R. Eichner & M. Schwab, E. Schweizerbart'sche Verlagsbuchhandlung (Nägele und Obermiller), Stuttgart, 524–534.

[40. 4 a4]   Viehl, W. (1997): Erfahrungen über die Einsatzmöglichkeiten des Mansfelder Kupferschlacke.Reststoffe der Kupferschieferverhüttung. Teil 1: Mansfelder Kupferschlacken, Beiträge zum Workshop am 4. und 5. Dezember 1996 in Bad Lauchstädt, UFZ-Bericht Nr. 23/1997, 51–54.

[41. 4 a4]   Wihsmann, F.G. (1966): Beiträge zur Gefügegenese von Formkörpern aus Kupferschlacke der Mansfelder Rohhütten. Diss. Bergakademie Freiberg, Fachbereich für Bergbau und Hüttenwesen.

[42. 4 a4]   Witzke, T. & Pöllmann, H. (1996): Mineralneubildungen in den Schlacken der Kupferschieferverhüttung des Mansfelder Reviers, Sachsen – Anhalt. Hallesches Jahrbuch für Geowissenschaften, Bd. 18, Reihe B, 109–118.

[43. 4 a4]   Witzke, T. (1997): Sekundärmineralbildung in Kupferschlacke als Indikator für Schwermetallmobilisierung und -fixierung. In. Reststoffe der Kupferschieferverhüttung. Teil 1: Mansfelder Kupferschlacken, Beiträge zum Workshop am 4. und 5. Dezember 1996 in Bad Lauchstädt, UFZ-Bericht Nr. 23/1997, 25–29.

4 a5 Grating slag from wood incineracion ash

As alternative energy production can include the incineration of wood, these incineration ashes can also be of some interest. Mineralogy and chemistry are given in Figure 4.26a, b and Tables 4.20 and 4.21. Some literature is given in [1. 4 a5–6. 4 a5].

The μXRF map obtained from one particle of the wood incineration ash reveals a complex structure of the ash, consisting of large pores a compositionally heterogeneous

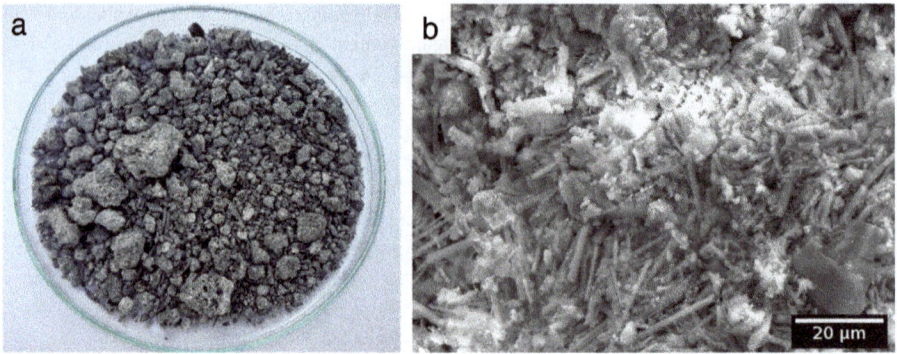

Figure 4.26: a) Bottom ash from wood incineration. b) Lathy crystals in wood bottom ash, SEM/SE image.

Table 4.20: Chemistry of bottom ash from wood incineration. µXRF data obtained from an area of approximately 120 mm² (excluding macroscopic porosity).

Chemistry	mass%	mass% (µXRF)
LOI%	4,5	–
Na_2O (%)	4,1	3.3
MgO (%)	2,5	2.5
Al_2O_3 (%)	6,2	6.9
SiO_2 (%)	50,6	59.8
P_2O_5 (%)	1,4	1.3
SO_3 (%)	0,0	0.4
K_2O (%)	2,1	1.2
CaO (%)	14,4	16.2
TiO_2 (%)	1,1	1.2
V_2O_5 (%)	0,0	0.0
Cr_2O_3 (%)	0,1	0.2
Mn_2O_3 (%)	0,2	–
MnO (%)	–	0.3
Fe_2O_3 (%)	11,9	–
FeO (%)	–	6.0

Table 4.21: Mineralogy of grating slag from wood incineration.

primary phases	chemical formula
Portlandite	$Ca(OH)_2$
Calcite	$CaCO_3$
Armacolite	$FeMgTi_4O_{10}$
Larnite	Ca_2SiO_4
Orthoclase	$K[AlSi_3O_8]$

matrix as well as angular clasts of at least 4 different phases. In contrast to the XRD results, also a potassium-bearing phase can be identified (Figure 4.27, green phase). In contrast to the XRD, we do not expect that Portlandite and Calcite can be easily distinguished.

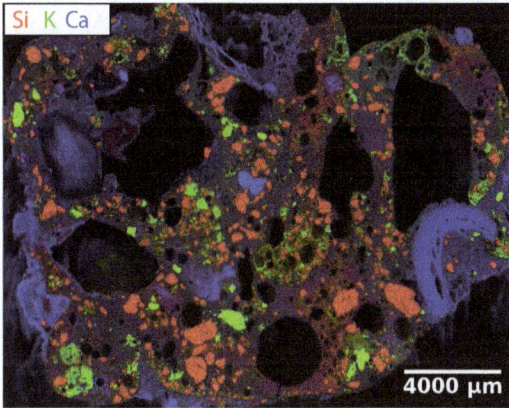

Figure 4.27: µXRF Si-K-Ca composite map of slag of ETU wood incineration. Single clast embedded in epoxy resin.

Literature

[1. 4 a5] Etiégni, L., Campbell, A.G., 1991: Physical and chemical characteristics of wood ash. Bioresource Technology 37: 173–178.

[2. 4 a] Kölling, C. und Stetter, U., 2008: Holzasche – Abfall oder Rohstoff? LWF aktuell Nr. 63, München: S. 54–56.

[3. 4 a] Noger, D., Felber, H., Pletscher, E., 1996: Verwertung und Beseitigung von Holzaschen.Schriftenreihe Umwelt Nr. 269, Holz/Boden. Bundesamt für Umwelt, Wald und Landschaft (BUWAL) Bern: 113 S.

[4. 4 a5] Obernberger, I., 1997: Aschen aus Biomassefeuerungen – Zusammensetzung und
          Verwertung. Institut für Verfahrenstechnik, Technische Universität Graz. In: VDI Bericht
          1319, pp. 199–222, 1997, „Thermische Biomassenutzung – Technik und Realisierung",
          ISBN 3-18-0913 19-3, VDI Verlag GmbH, Düsseldorf, Deutschland.
[5. 4 a5] Someshwar, A.V., 1996: Wood and combination wood-fired boiler ash characterization.
          J. Env. Qual. 25(5): 962–972.
[6. 4 a5] Zimmermann, S., Hässig, J., Landolt, W.: Literaturreview Holzasche-Wald, Studie im
          Auftrag des Bundesamtes für Umwelt BAFU, 2010.

4 a6 Chromium slag from RSA

Waste metal slags are probably not directly in the focus, due to their heavy metal contents, but are included here, because they also can be used after some special treatment in cement industry, for example of some specific leaching processes. Mineralogy and chemistry are given in Figure 4.28a, b and Tables 4.22 and 4.23.

Figure 4.28: a) Ferrochrome slag from RSA b) SEM/SE of ferrochrome slag powder.

Table 4.22: Chemistry of ferrochrome slag, µXRF data normalized from a total of 60.9%, analyzed area is 570 mm^2.

Chemistry	Mass%	mass% (µXRF)
LOI%	0,0	–
Na_2O (%)	0,1	0.2
MgO (%)	12,1	11.6
Al_2O_3 (%)	18,1	17.5

Table 4.22 (continued)

Chemistry	Mass%	mass% (µXRF)
SiO_2 (%)	41,9	44.4
P_2O_5 (%)	0,0	0.0
SO_3 (%)	0,0	0.7
K_2O (%)	0,2	0.3
CaO (%)	1,1	5.4
TiO_2 (%)	0,7	1.0
V_2O_5 (%)	0,2	0.0
Cr_2O_3 (%)	20,5	11.0
Mn_2O_3 (%)	0,1	–
MnO (%)		0.3
Fe_2O_3 (%)	10,4	–
FeO (%)	–	7.0

Table 4.23: Mineralogical composition of ferrochrome slag from XRD.

primary phases	chemical formula
Forsterite	$Mg_{1.814}Fe_{0.186}[SiO_4]$
Magnesiochromite	$MgCr_2O_4$
Chromite	$FeCr_2O_4$
Diopside	$Na_{0.25}Ca_{0.6}Mg_{0.7}Fe_{0.2}Al_{0.25}[Si_2O_6]$
Cristobalite	SiO_2

A sample of the chromium slag has been stabilized with epoxy resin, saw cut and ground flat. The slag is composed of roughly equi-sized, granular particles of different compositions. At least three different phases are easily identified (Figure 4.29). However, due to their fairly small size and the granular nature of particles – particles are cut at different diameters and some appear from below the sample surface through pores – a segmentation and characterization is not straight forward.

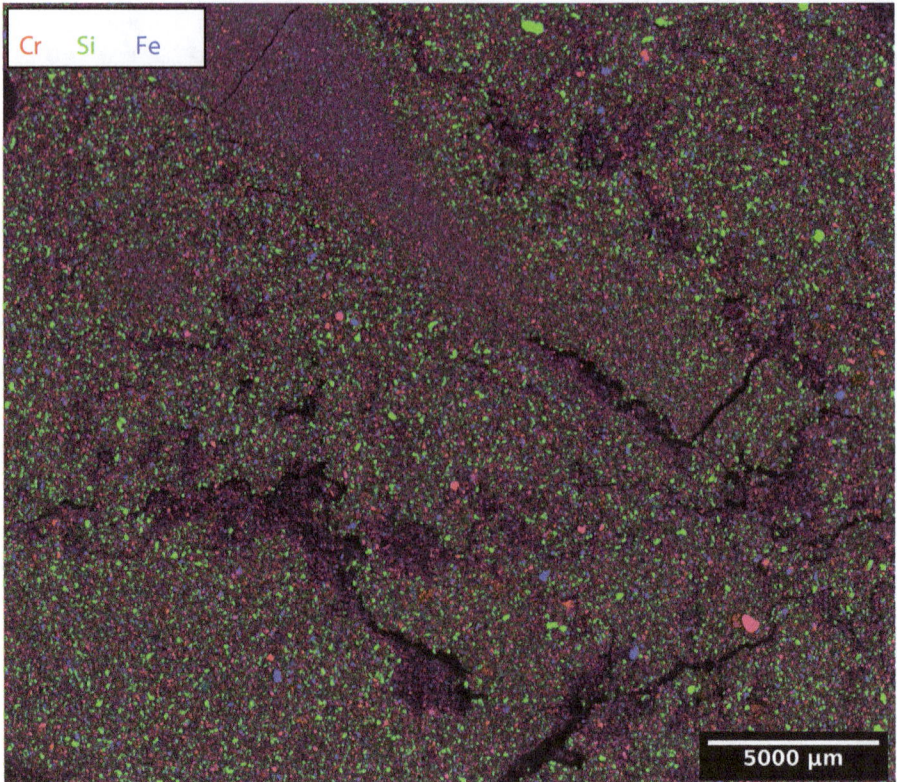

Figure 4.29: µXRF composite map of Cr-Si-Fe of a epoxy resin stabilized sample. Magenta colored areas relate to damage of the surface.

Literature

[1. 4 a6] Forsbacka, L., Holappa, L., Kondratiev, A.: Experimental staudy and modelling of viscosity of chromium containing slags, Stell research Int. (2016)

[2. 4 a6] Horckmann, L., Möckel, R., Nielsen, P., Kukurugya, F., Vanhooof, C., Morillon, A. and Algermissen, D.: Multi-analytical characterization of slags to determine the chromium concentration for a possible Re-Extraction, Minerals 9, 646, (2019)

[3. 4 a6] Niemalä, P. and Kauppi, M.: Production, characteristics and use of ferrochromium slags, INFACON XI, 171–179, (2007)

[4. 4 a6] Nkohla, M.A.: Characterization of ferrochrome smelter slag and ist implications in metal accounting, Dissertation, Univ. Stellenbosch, (2006)

[5. 4 a6] Sahu, N., Biswas, A. and Kapure, G.U.: A short review on utilization of ferrochromium slags, Mineral processing and extractive metallurgy review 37, 4, (2016)

4 a7 Titanium slag RSA

A typical titanium slag from RSA was included in this study due to the interesting composition. In fact probably these slags will be increasingly used for their vanadium contents primarily [1. 4 a7], [5. 4 a7]. This could be a similar leaching process as described for manganese slags [2. 4 a7 – 4. 4 a7]. Interestingly it is composed of low SiO_2-contents and still high TiO_2-contents. Mineralogy and chemistry are described in Figure 4.30a,b and Tables 4.24 and 4.25.

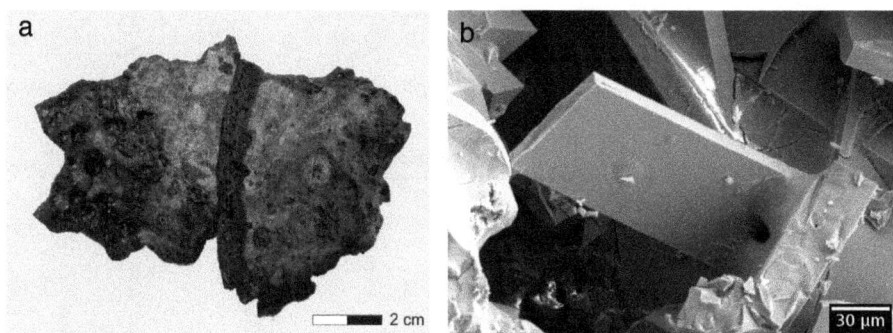

Figure 4.30: a) titanium slag from RSA b) SEM/SE of a Rutile crystal in small vug of slag.

Table 4.24: Chemical composition of titanium slag RSA, µXRF data normalized from a total of 63.6%, analyzed area is 572 mm^2.

Chemistry	mass%	mass% (µXRF)
LOI%	0,0	–
Na$_2$O (%)	0,0	0.0
MgO (%)	12,0	12.0
Al$_2$O$_3$ (%)	13,6	11.6
SiO$_2$ (%)	21,3	19.9
P$_2$O$_5$ (%)	0,0	0.0
SO$_3$ (%)	0,0	1.3
K$_2$O (%)	0,1	0.4
CaO (%)	15,2	15.2
TiO$_2$ (%)	29,6	33.4
V$_2$O$_5$ (%)	0,8	0.9

Table 4.24 (continued)

Chemistry	mass%	mass% (µXRF)
Cr_2O_3 (%)	0,2	0.2
Mn_2O_3 (%)	0,8	–
MnO (%)	–	0.9
Fe_2O_3 (%)	5,1	–
FeO (%)	–	3.3

Table 4.25: Mineralogy of Titanium slag/RSA from XRD.

primary phases	chemical formula
Pseudobrookite	$TiFe_2O_5$
Rutile	TiO_2
Diopsidic Augite	$Ca(Fe,Mg)[Si_2O_6]$
Spinel	$MgAl_2O_4$
Glass	

The titanium slag was saw-cut and flat ground for µXRF examination. It consists of variable porosity, a homogeneous, non-resolvable matrix, elongate needle-shaped grains of a Ti-enriched phase (e.g. Pseudobrookite), a V-rich phase as well as a droplet shaped Fe-phase (Figure 4.31). The size of the Ti- and the V-bearing phases is usually around several 100s of micrometers, which is sufficient to e.g. describe their shape, however, too small to provide an exact chemical characterization.

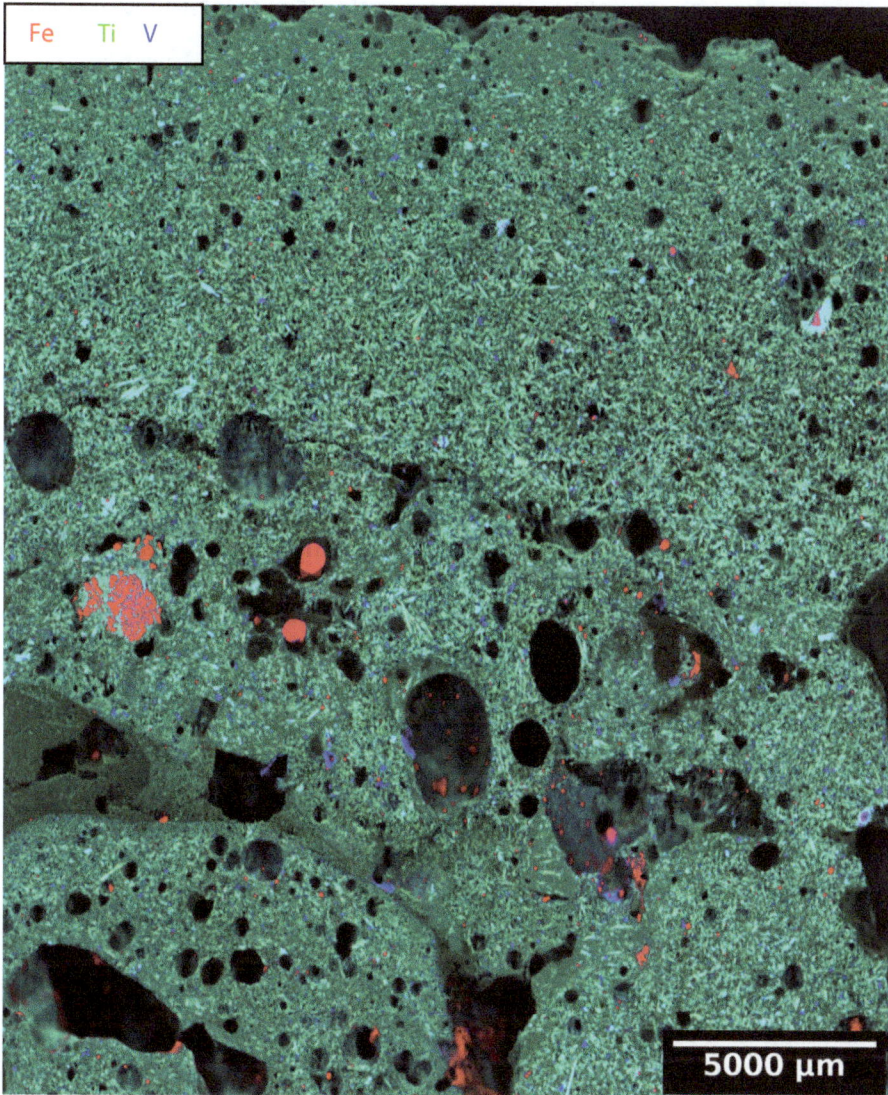

Figure 4.31: µXRF-derived composite map of Ti-slag. Saw-cut sample. Red grains are elementary Fe-droplets.

Literature

[1. 4 a7] Bunting, M.R.: Vanadium: How market developments affect the titanium industry. San Diego : International Titanium Association Conference, (2006)

[2. 4 a7] Groot, D.; Kazadi, D.; Pöllmann, H.; de Villiers, J.; Redtmann, T. and J. Steenkamp: Utilization of ferromanganese slags for manganese extraction and as a cement additive.

Advances in Cement and Concrete Technology in Africa, Emperor`s Palace, Johannesburg, pp 984–985, 28.-30 January (2013)

[3. 4 a7] Kazadi, D.M.; D.R. Groot; H. Pöllmann; J.P.R. de Villiers & T. Redtmann: Utilization of Ferromanganese Slags for Manganese Extraction as a Cement Additive. International Conference on Advances in Cement and Concrete Technology in Africa Proceedings. BMA Federal Institute for Materials Research and Testing (ISBN 978-3-9815360-3-4, (2013)

[4. 4 a7] Kazadi, D.M.; D.R. Groot; J.D. Steenkamp; H. Pöllmann: Control of silica polymerisation during ferromanganese slag sulphuric acid digestion and water leaching. HYDROM-04390; No of Pages 8, Hydrometallurgy 166, 214–221 (2016)

[5. 4 a7] Sadykhov, G.B. und Karyazin, I.A.: Titanium-Vanadium Slags upon the Direct Reduction of Iron from Titanomagnetite Concentrates. Russian Metallurgy (Metally). Bd. 6, (2007)

5 Advantages and disadvantages of characterization by µXRF

One strength of the µXRF is clearly the fast characterization and localization of phases in unknown samples. On average, the speed allows to analyze statistically representative sample sizes. The data permits to gain a fast overview of the number of different phases, their amounts, sizes, heterogeneity and distribution. For suitable samples, phase maps can be derived and from those maps additional microstructural parameters can be obtained such as volume percentages, grain size, porosity or the spatial interrelation of phases (e.g. intergrown crystals, distribution in a matrix, heterogeneity among others). At the other hand, minute phases with a distinct chemical signal can readily be identified and their volume contribution quantified. As a by-product, the chemical composition derived from the integral µXRF spectrum is often comparable to bulk XRF analysis obtained from glasses. For example, the bulk compositions of the glassy variety of the copper slag from Helbra are strikingly similar (Table 4.17). This applies even in cases where the analyzed total may be as low as 60%, e.g. due to high porosity. A clear limitation arises in the quantification of overall light samples which deviate most from the bulk XRF analyses. This is most likely a limitation of the used software, not particularly of the µXRF method itself.

The analysis of thick samples, especially porous samples with a light matrix (e.g. the puzzolane from Peru) yield the problem that heavier elements contribute from a much larger thickness of the sample. While a matrix correction in the case of bulk chemistry may compensate for that issue, care has to be taken when deriving phase maps from element count maps. Using matrix corrected (binned) maps of element concentrations may overcome this problem, however, at the moment there is no practical implementation available to compute concentration maps for the data sizes which can be acquired with the µXRF.

A way to overcome the problem of sampling depth is the analysis of thin sections. The disadvantage of thin sections is, that under certain conditions (very light samples, highly porous material) a characteristic contribution of Si and Ca from the glass may be present in the spectra. The other limitation of thin sections is their limited size and the difficulty to prepare large sections at a constant thickness. A way to minimize the chemical contamination which may arise from glass slides can be the preparation of sections on polycarbonate substrates.

A disadvantage of the method is the variable size of the excitation volume and the limited spatial resolution. While the spot can be considered 20 µm, a minimum size of a particle to be characterized depends on the chemical composition of the surrounding matrix and of the particle itself. As a rough number, particles should be at least 100 to 200 µm along their shortest dimension to be able to reasonably discriminate their composition from µXRF maps. For phase mixtures purely composed of light phases, as well mixtures of phases with a high difference in density, larger errors in composition have to be accepted.

6 Pozzolanity of different natural and artificial materials

The different natural and artificial materials were tested for their pozzolanic behaviour. The results are summarized in figure. For testing all materials were ground to specific surfaces according to Table 4.26. It is obvious that many slags and ashes, depending from their origin, are very different in their pozzolanic behaviour.

6 a Comparison of chemical compositions

The different chemical compositions of the used pozzolanes are summarized in ternary systems inclusing alkalies, CaO, aluminum oxide, iron oxide and SiO_2. In combination with the bonding of these oxides in mineral phases it gives some idea on their reactivity in a cementitious system. Also very important is their amorphous content and the content of the amorphous phase. In Figures 4.32a,b and 4.33 their main chemical compositions are summarized. Additionally it is highly important to characterize their specific surface for better reaction. In all cases at least some calorimetric, handling properties and strength investigations must accompany descriptions of usage of these materials.

In a typical binary system $(Na_2O + K_2O)$-SiO_2 the compositions of the materials are in regions of basalts, trachytes and also rhyolites. Two materials lie in regions with very low SiO_2-contents. Their potential reactivity of fine ground material was tested using DIN EN 196-5 method.

For pozzolanity test the used materials were ground intensively to specific surfaces acc. Table 4.26.

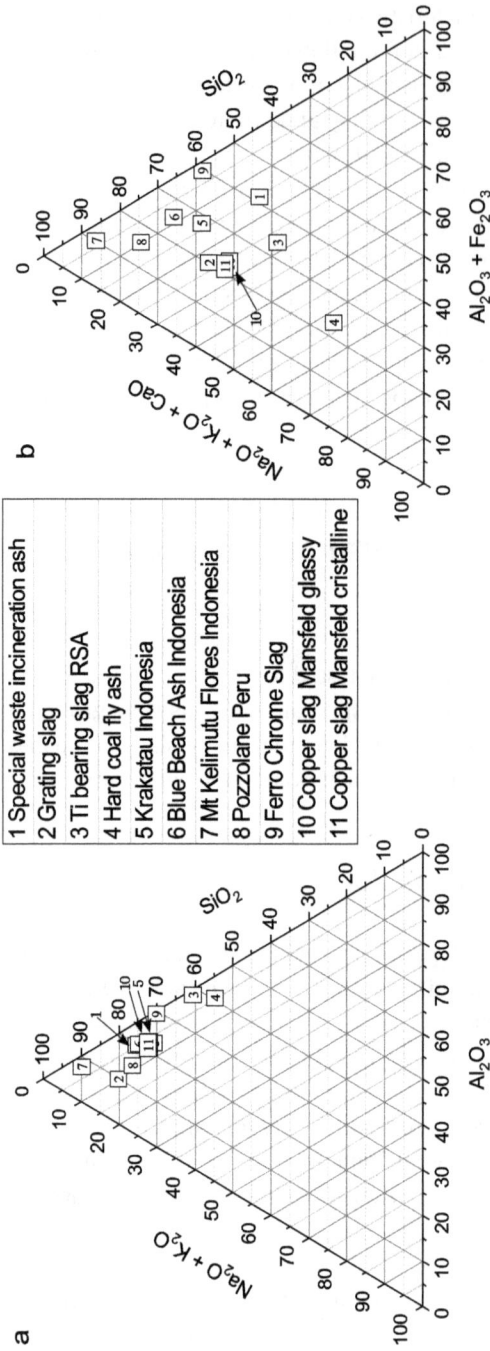

1 Special waste incineration ash
2 Grating slag
3 Ti bearing slag RSA
4 Hard coal fly ash
5 Krakatau Indonesia
6 Blue Beach Ash Indonesia
7 Mt Kelimutu Flores Indonesia
8 Pozzolane Peru
9 Ferro Chrome Slag
10 Copper slag Mansfeld glassy
11 Copper slag Mansfeld cristalline

Figure 4.32: a) Pozzolanes in the ternary system $Na_2O + K_2O - Al_2O_3 - SiO_2$ b) Pozzolanes in the ternary system $CaO + Na_2O + K_2O - Al_2O_3 + Fe_2O_3 - SiO_2$.

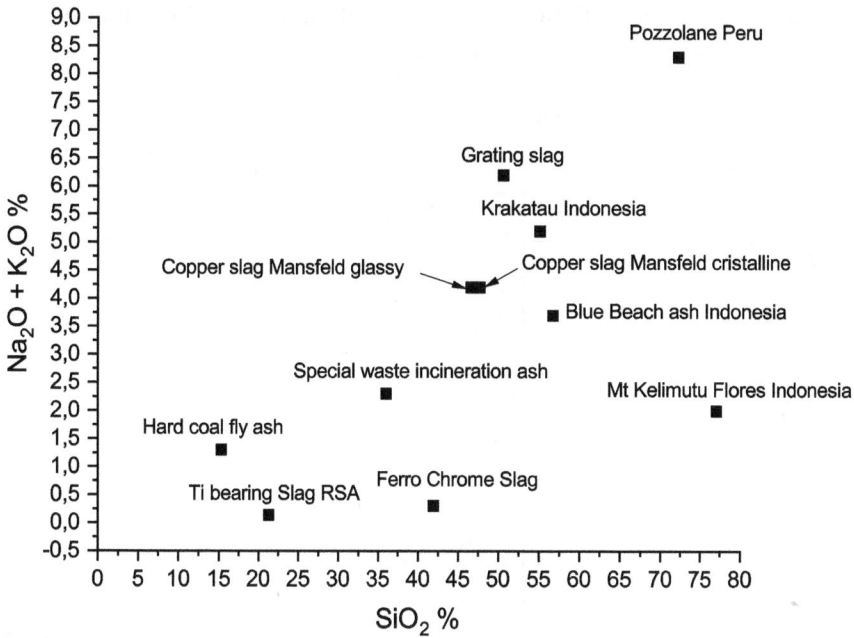

Figure 4.33: Pozzolanic materials in the system $(Na_2O + K_2O) - SiO_2$.

Table 4.26: Specific surfaces of pozzolanic materials.

Origin of material	Blaine S_V Average in cm²/g
Mt Kelimutu Flores Indonesia	7114,44
Pozzolane Peru	10230,96
Ferro Chrome Slag	5022,65
Ti bearing Slag RSA	5531,86
Hard coal fly ash	8096,17
Grating slag	7422,01
Special waste incineration ash	7900,21
Krakatau Indonesia	6390,83
Blue Beach Ash Indonesia	8572,14
Copper slag Mansfeld – cristalline	6377,57
Copper slag Mansfeld – glassy	3874,20

In Figure 4.34 the potential of the investigated natural and industrial pozzolanes is summarized as obtained from pozzolanity test.

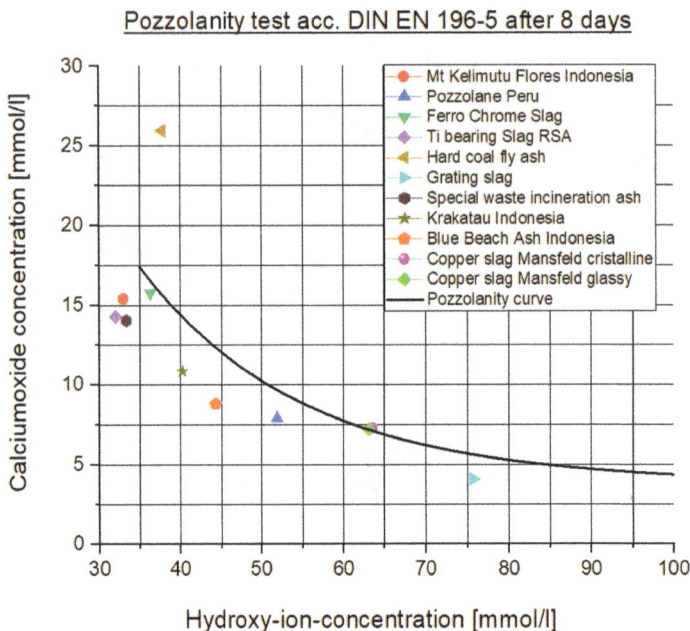

Figure 4.34: Pozzolanity data of all investigated pozzolanic materials.

As pozzolanes are very often used in mixtures with OPC's the determination of the relevant contents of pozzolanes in OPC mixtures the contents of these composition cement materials must be determined including the amorphous contents.

For the determination of these mixtures between OPC and pozzolanic materials the PLSR (Partial least squares refinement) method was successfully applied. For these investigations mixtures of OPC and pozzolanic material were used. For some applicable mixtures it could be shown that the different mixtures of cements and pozzolanes can be optimally characterized using PLSR and clustering procedure. Some examples of OPC composites (copper slag, natural pozzolane, hard coal fly ash) are given next in Figures 4.35a,b, 4.36a,b, 4.37a,b. Some more detailed examples and the methods applied are given in chapter 2. [1. 6 a–3.6a].

By using the PLSR quantification of cement pozzolane mixtures it becomes obvious that a clear prediction of the prognosed properties of the complex mixtures may be possible.

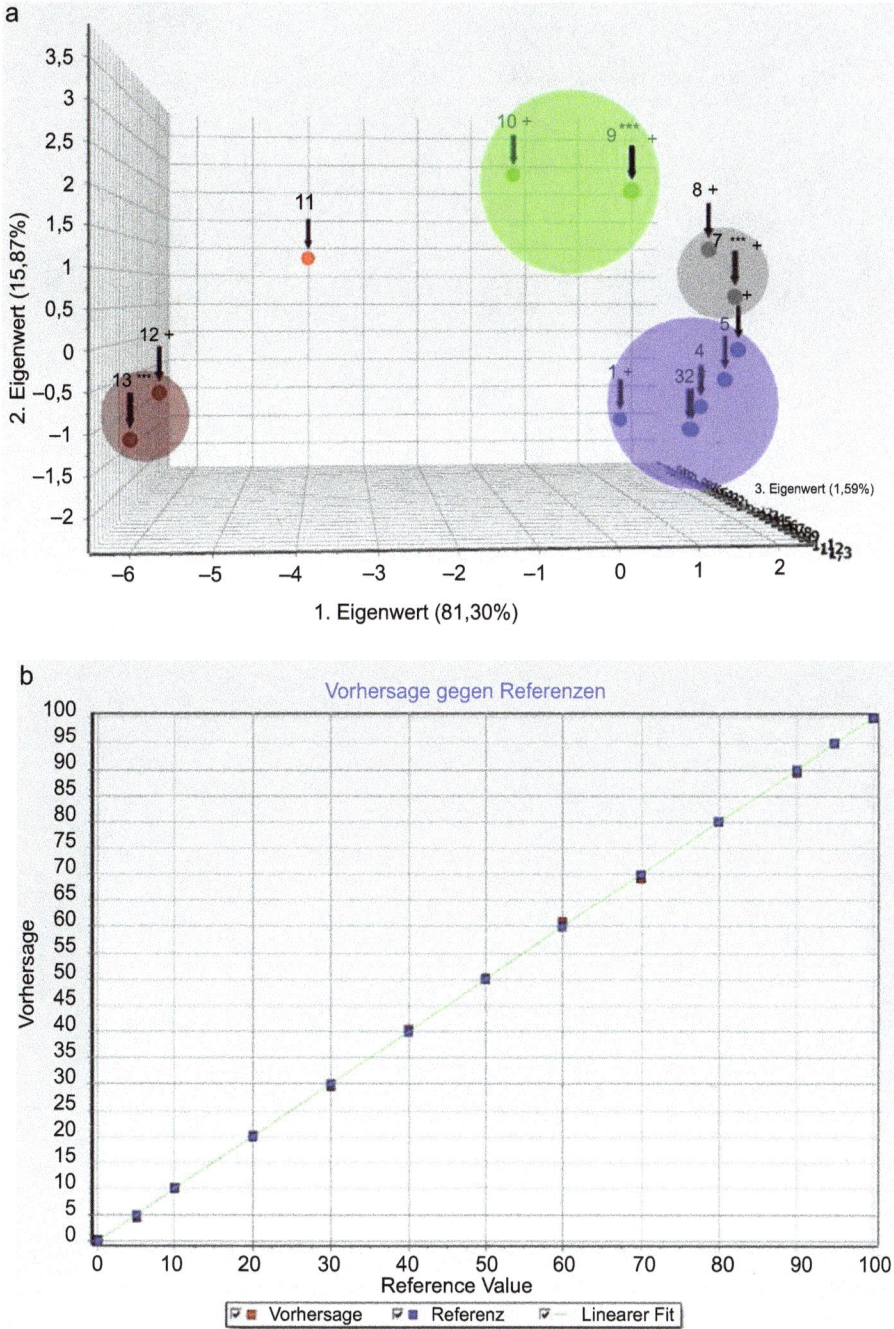

Figure 4.35: a) PCA - Principal component analysis, clustering of OPC/ copperslag b) Calibration curve of OPC/copper slag.

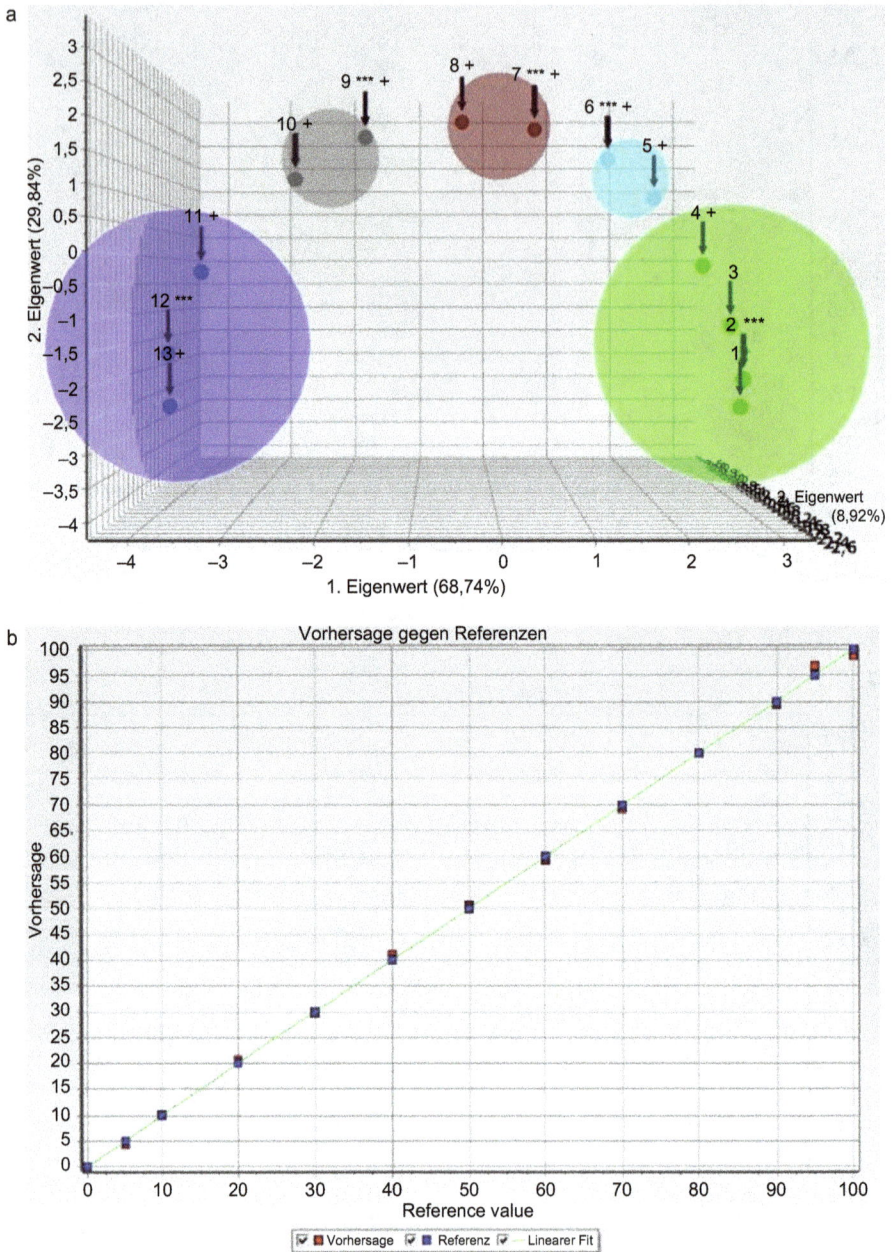

Figure 4.36: a) PCA - Principal component analysis, clustering of OPC / natural pozzolane mixtures b) Calibration curve of OPC / pozzolane mixtures.

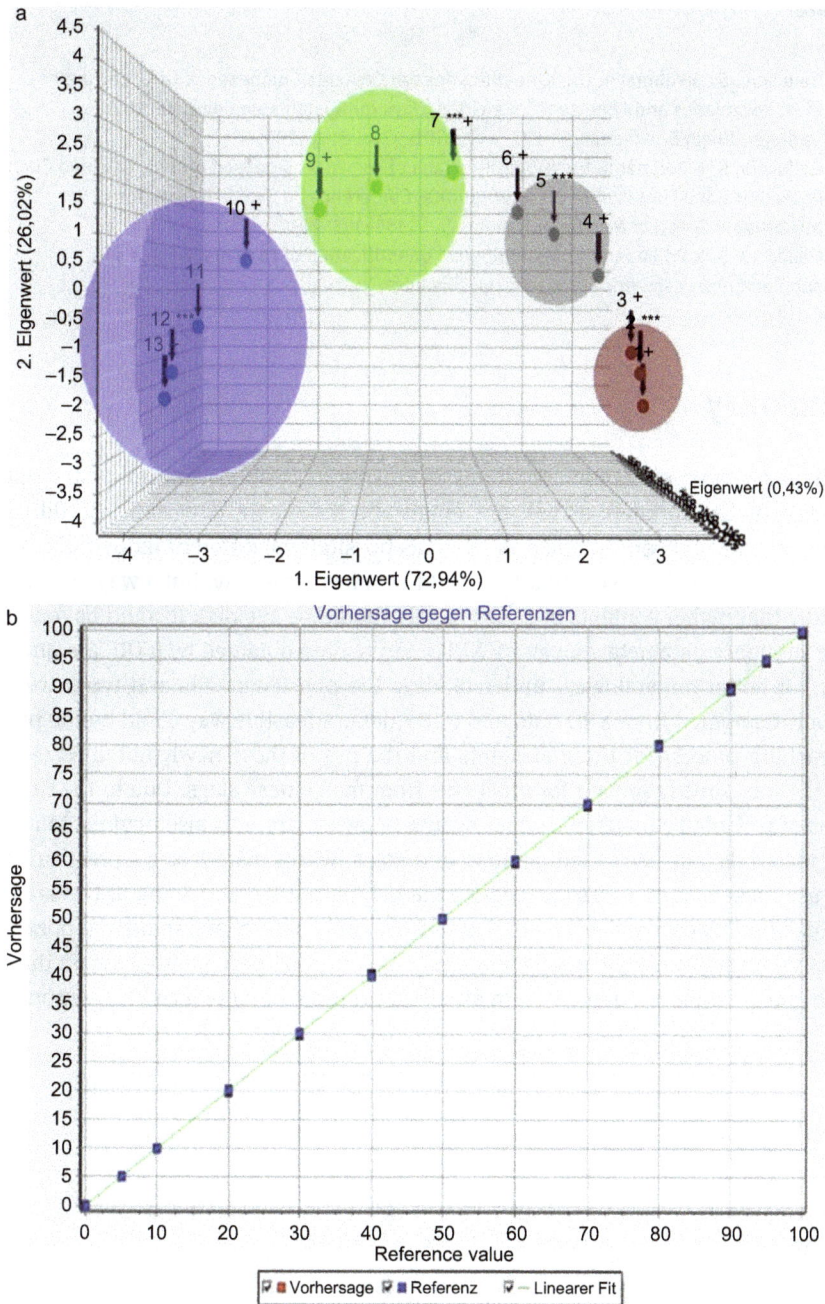

Figure 4.37: a) PCA - Principal component analysis, clustering of OPC / hard coal fly ash b) Calibration curve of OPC / hard coal fly ash.

Literature

[1. 6 a] Galluccio, S. & Pöllmann, H.: "Quantifications of Cements Composed of OPC, Calcined Clay, Pozzolanes and Limestone". Calcined Clays for Sustainable Concrete. 2020. Springer, Rilem Bookseries, S. 425–442, ISBN 978-981-15-2805-7

[2. 6 a] Galluccio, S. & Pöllmann, H.: "Quantifications of cements composed of OPC, Calcined Clay, Pozzolanes and Limestone." 3th International Conference on Calcined Clays for Sustainable Concrete New Delhi, Volume II. S. 135–153. 2019.

[3. 6 a] Galluccio, S. & Pöllmann, H.: Mineralogical quantification of cements, wastes and supplementary cementitious materials, this book.

7 Summary

Different pozzolanic materials are widely occurring as residues from industry but also in natural environment as volcanic ashes and sedimentary rocks. It should be an environmental concern to reuse as much as possible the different industrial residues and it seems highly likely that this could be a favorable way. But always it must be secured that higher contents of heavy metals should be avoided. It could be found that the summary parameters given by XRF and the data obtained by µXRF are quite similar. For better mineralogical understanding the phase contents and types were intensively compared from XRD data and µXRF data. A feasible way could be the primary leaching process for these elements and the use of those newly obtained residues. This was already proven for processes from manganese slags. Due to the high amounts of pozzolanic materials, which would be necessary, it is also obvious to use natural materials. The use of sedimentary and volcanic raw materials is a practicable way to use these natural rocks and reduce the CO_2-output by fabricating at the same time a high quality composite cement. Largely available natural and industrial pozzolanic materials are available in large amounts and can obviously be used for producing high quality standard cements with excellent properties and reduced CO_2-output.

Part 2: **Characterization of industrial residues**

Stefan Stöber, Herbert Pöllmann

Chapter 5
Characterization of supplementary cementitious materials: Brown coal fly ashes

Abstract: Lignite fly ash (LFA) is a waste product of the combustion of lignite, which is produced in large quantities (million tons) worldwide. Lignite fly ash is a heterogeneous mixture of partly round, partly irregularly round glass spheres, surrounding rock particles, the non-combustible, inorganic components of coal and unburned coal residues. Their chemical-mineralogical composition depends essentially on the initial composition of the raw lignite and the firing system used during combustion. On the basis of their chemism, LFAs rich in SiO_2, Al_2O_3, iron, lime and sulphate can be distinguished. The power plant ashes may have hydraulic or pozzolanic properties which can be used for building material applications. In order to guarantee the use of LFA e.g. in concrete, different characterization and testing methods have to be applied in order to evaluate the different properties of LFA, which are described in this article.

Keywords: Lignite fly ash, Properties, Characterization, Rietveld, Amorphous content

1 Introduction

Lignite is an energy source that is mainly used for the production of electricity for industrial purposes and also used for heating systems in family homes [1]. Lignite has a low price per unit of energy produced and therefore it is one of the main energy resources due to its large reserves, which are much higher than those of natural gas and crude oil [2]. Unfortunately, it has a low heating value (10–20 MJ), and a high moisture and ash content. According to the different mining areas, lignite contains different concentrations of ash and moisture. High-quality lignite with a high calorific value has low ash concentrations (e.g. Rhenish lignite < 7%) and low quality lignite fly ashes (LFAs) have low calorific values and high ash contents (20–30%) [3]. Lignite is one of the fossil energy sources. It evolved from the original plant material through biological and chemical conversion via various intermediate stages. A large part of the coal must have been deposited at the site of plant formation, so that the lignite

Stefan Stöber, Institute for Geological Sciences, Mineralogy/Geochemistry, Martin-Luther-University, Halle 06120 Halle (Saale), Germany, e-mail: stefan.stoeber@geo.uni-halle.de
Herbert Pöllmann, Institute of Geological sciences, Mineralogy/Geochemistry, Martin-Luther-University Halle, Halle (Saale), Germany

https://doi.org/10.1515/9783110674941-005

can be regarded as an autochthonous product of dead, rotten and decomposed plant parts [4]. Lignite has a low degree of carbonization, based on the young geological time of origin in the Tertiary, exactly between Eocene and Pliocene (45–12 million years), and a high proportion of volatile components. For example, Rhenish lignite has a high water content of 50–60% and a carbon content of 58–73% by mass of dry matter [5, 6]. The generation of energy from lignite produces considerable quantities of ashes. Power plant ashes occur in different processes or in different steps of combustion. A distinction is made between lignite fly ash (also called lignite filter ash or power plant ash (KWA) [7] and wet ash (also called boiler ash) [5]. LFAs account for 80–90% of power plant ashes, where lignite is used as fuel. The remaining 10–20% of the total ash quantity is present as boiler ash [8]. This consists of coarse particles, most of which come from the surrounding rock, and portions of unburnt coal or coke particles, which is very rich in quartz [9]. LFA is separated from the flue gas in the power plant by means of electrostatic precipitators or fabric filters. In Germany alone around 8.81 million tonnes of LFA were produced in 2011 according to the Association of Large Boiler Owners [10]. The International Energy Agency (IEA) estimates the current (2012) worldwide consumption of lignite at 1040 million tonnes. The chemical-mineralogical LFA composition depends mainly on the initial composition of the raw lignite and the combustion system used for combustion. On the basis of their chemistry, SiO_2– rich, Al_2O_3– emphasized, iron-, lime- and sulfate-rich LFA can be distinguished [11]. They are complex, heterogeneous, extremely fine-grained mixtures of substances consisting of a multitude of crystalline and amorphous components. They are generated during the combustion of lignite in pulverized coal firing systems with dry ash discharge. There is a large number of different ashes whose qualities depend on the raw lignite used and the combustion technology [12]. Certain factors are substantial for the constitution of LFA [13–15]. Due to the different physico-chemical properties of the LFAs regarding grain size, chemistry, mineralogy (active & inert phases) these ashes are used in very different technical applications:

- Utilization of fly ash as a soil amendment to improve soil quality thanks to its considerable K, Ca, Mg, S, and P contents [16]
- Utilization of fly ash to decrease the bulk density of soils, which, in turn, improved soil porosity and workability and enhanced water retention capacity [17]
- Wet and lignite fly ashes are dumped with the overburden in areas of opencast mines so that the mass deficit will be reduced and the resulting residual spaces become smaller [5].
- Utilization of LFA for the removal of Arsenic and Uranium from Mine Drainage [18]
- Application of fly ash for concrete production [19]

2 Chemical and physical properties of lignite fly ashes

2.1 Chemical composition

2.1.1 Main elements

In order to get a grip on the wide-ranging composition of lignite filter ash, numerous attempts have been made to clarify the differences in chemistry of the differently occurring LFAs. The chemical composition of LFA, like its physical properties, is subject to strong fluctuations. The main causes are the material variations of the lignites and the sometimes large differences between individual ash variants [20], but the boiler configuration, burning condition and temperature of the boiler, the particle size of the coal and the gas cleaning equipment play an important role, too. With regard to their particle content, LFAs are very heterogeneously composed mixtures of substances. Chemical analysis can only be used for orientation, but not for predicting material properties [11]. The chemistry of regionally very different lignite filter ashes was investigated [21–23], but in order to explain the different characterization models on the basis of the overall chemistry, the data of German lignite filter ashes are used in this chapter. The main chemical components of LFA in the traditional german areas are mainly SiO_2, CaO, Al_2O_3, Fe_2O_3, SO_3, MgO and C. As secondary components, Na_2O, K_2O, MnO, TiO_2, P_2O_5 and Cl are frequently found [11, 12, 14, 15, 20, 24–37]. Usually the sum of the main and minor components is between 97 and 99% [12]. The remaining part consists of already absorbed water and trace elements, to which heavy metals such as Co, Cr, Cu, Cd, Mo, Ni, Sn, Zn, Hg, Pb and rare earth metals such as e.g. Y, Ce and Nd [11, 12, 15, 34, 37–39]. Table 5.1 shows the fluctuation range of the main and secondary components and the loss on ignition of different LFA.

The chemical fluctuation ranges of various LFA for special power plants, discharged as mixed ashes (pre-cleaning, intermediate and post-cleaning) are summarized below in Table 5.2. A comparison of BFA from the Lusatian power plants (Boxberg, Jänschwalde & "Schwarze Pumpe") also shows that it is not possible to classify them into site-specific types. All ashes must be classified regardless of their geographical origin [11].

2.1.2 Trace elements

The informations about trace elements from LFA [42]. were summarized below (Table 5.3.) Data on binding to individual mineral phases and mobility are not available, but it is very likely that incorporation into hydrate phases occurs during the hydration process. The mineralogical composition of various LFAs is

Table 5.1: Fluctuation range of chemical major and minor components of different BFA from different coal fields [40].

Component	Rhineland area [Ma%]	Central German area [Ma%]	Lusatian area [Ma%]
SiO_2	20–80	18–36	32–68
Al_2O_3	1–15	7–19	5–14
Fe_2O_3	1,5–20	1–6	6–22
CaO	2–45	30–52	8–23
free Lime	2–25	9–25	0.1–4
MgO	0,5–11	2–6	2–8
SO_3	1,5–15	7–15	1–6
C	<2	<1	<2
K_2O	0,1–1,5	0,1–2	0,1–1
Na_2O	0,1–0,5	0,01–0,2	0,5–1,3
TiO_2	0,5–2	0,01–0,2	0,2–1
Cl	<0,2	<0,1	<0,02
LOI	<5	<5	<5

Table 5.2: Chemical composition of selected lignite ash [41].

Power station	Chemical composition [%]				
	MgO	CaO	Al_2O_3	Fe_2O_3	SiO_2
Boxberg***	4–9	20–40	5–17	18–38	9–43
Jänschwalde***	3–9	22–46	6–24	8–22	20–45
„Schwarze Pumpe"***	4–10	12–30	6–19	13–45	14–47
Thierbach**	1–4	19–51	10–32	1–15	20–52
Neurath*	4–6	20–44	5–7	5–19	22–50

summarized below with a description of the width of the quantitative fraction according to [11, 25, 31, 43–45].

These data are particularly important for the interpretation of the hydration behavior of different ashes, which includes not only the constitution of the LFAs but also the investigation of the reaction mechanism and the characterization of the hydration products [31]. Half of the Central German lignite ashes are used for above-

Table 5.3: Trace elements of lignite ashes in [41].

Element	Concentration [ppm] Pöhl [31]	Concentration [ppm] Wischnewski [46]
As	10–246	10–100
Be	1–3	-.-
Bi	1–12	-.-
Cd	0–1.5	<1 ppb
Cr	6–17	10–100
Cr	24–76	10–100
Se	1–8	<100 ppb
Sn	1–11	1–10
V	7–91	not determined
Zn	38–101	100–1 weight-%
Cu	17–55	10–100
Mo	1–30	not determined
Ni	7–104	10–100
Pb	7–100	10–100
Sb	1–12	<1
Au	-.-	<100 ppb

ground mine rehabilitation and another third for underground mine rehabilitation. For solidification of sludges and sediments approx. 10% are used and approx. 5% for other applications, for example in the building materials industry. ('Federal Environment Agency UB' 10.6.1997).

2.1.3 Classification of LFAs

The classification of LFAs is based on their average chemical composition. In the German-language literature, ashes are frequently classified into lime-rich, silica-rich and aluminosilicate LFAs (Table 5.4).The classification goes back to [26, 47–51]. The authors systematically investigated LFAs from lignite from the Lusatian and Central German lignite mining districts, which were produced in the large power plants of the former GDR.

FUNGK et al. (1969) [26] presented the mean values of the main chemical components from about 80 samples per power plant, in the ternary system (CaO + MgO –

Table 5.4: Classification by LFA according to chemical composition [52].

	High lime LFA [%]	Siliceous LFA [%]	Aluminosilicatic LFA [%]
CaO	>20	10–20	<10
SiO$_2$	–	>50, for high contents of free silica	>50, bei < 10 free silica
Al$_2$O$_3$	–	–	>20

SiO$_2$ – (Al$_2$O$_3$ – Fe$_2$O$_3$)) and supplemented the three-material diagram with the phase fields of the Portland cements, blast furnace slags and pozzolana (Table 5.5) (Figure 5.1).The position of the ashes in the ternary system illustrates the classification. It also allows comparisons with other hydraulic, latent hydraulic and pozzolanic binders.

Table 5.5: Average chemical composition of different LFA types from selected power plants [inMa%] [26].

	Calcareous LFA			LFA		Aluminosilicate LFA	
	Borna	Espenhain	Vockerode	Schwarze Pumpe	Lübbenau	Hagenwerder	Hirschfelde
SiO$_2$	34,5	31,1	32,5	46,2	54,3	42,7	51,5
Al$_2$O$_3$	6,2	14,2	19	12,8	9,9	26,6	32,0
Fe$_2$O$_3$	9,6	8,9	10,0	14,5	12,1	10,5	8,3
CaO	33,2	31,0	29,0	12,0	15,0	13,8	2,0
Freelime	6,6	3,2	4,0	-.-	0,6	-.-	-.-
MgO	2,5	4,0	2,0	2,9	2,5	2,4	1,1
SO$_3$	8,5	6,8	4,6	4,4	4,6	1,6	1,1

A more complex classification system for LFA was presented by ROY et al (1981) [53]. It is based on 479 published chemical analyses and includes the sulphate content, the most important secondary components, the pH value in aqueous solution and individual characteristics in the LFA chemistry. The classification system distinguishes seven LFA groups according to their chemical composition (Table 5.6) They are formed by the intersections of the end members in the ternary system (Figure 5.2). The end members include the sialic group (SiO$_2$ + Al$_2$O$_3$ + TiO$_2$), the calcic group (CaO + MgO + Na$_2$O + K$_2$O) and the ferric group (Fe$_2$O$_3$ + MnO + SO$_3$ + P$_2$O$_5$). The combination of the Sialic, Calcic and Ferric components results in the Ferrosialic, Ferrocalsialic, Calsialic and Ferrocalcic group. Figure 5.2 shows the position of the different LFA

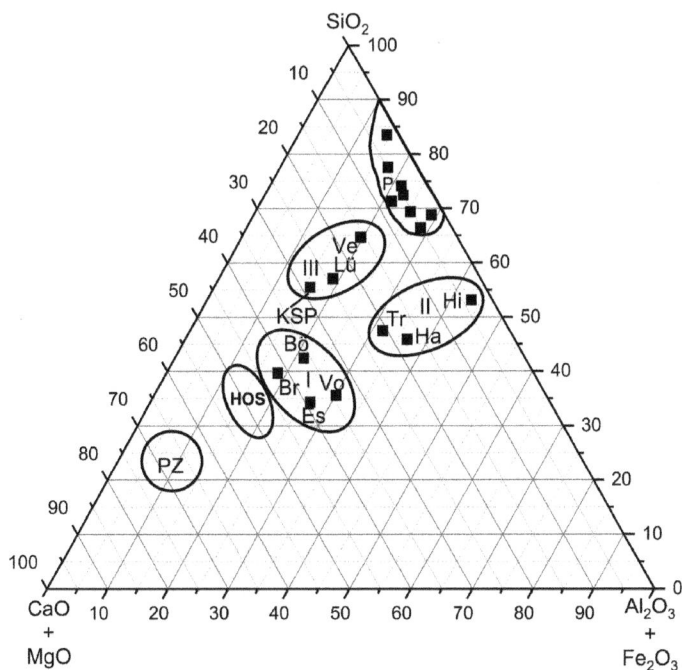

Figure 5.1: Ternary diagram (CaO + MgO) – SiO$_2$ – (Al$_2$O$_3$ – Fe$_2$O$_3$) with the stability field of compositional different lignite ashes from different mining areas drawn according to [26].

Table 5.6: Classification of LFA according to chemical composition [in %] [53].

Group	Sialic component	Calcic component	Ferric component
	(SiO$_2$ + Al$_2$O$_3$ + TiO$_2$)	(CaO + MgO + Na$_2$O+K$_2$O)	(Fe$_2$O$_3$ + MnO + SO$_3$ + P$_2$O$_5$)
Sialic	>88	0–12	0–12
Ferrosialic	48–88	0–29	23–52
Calsialic	<29–48	>29–5	0–23
Ferrocalsialic	>48–88	0–29	0–23
Ferric	<48	0–29	>23
Calcic	<48	>29	0–23
Ferrocalcic	<48	29–77	>23–71

groups in the ternary system (SiO$_2$ + Al$_2$O$_3$ + TiO$_2$) – (CaO + MgO + Na$_2$O + K$_2$O) – (Fe$_2$O$_3$ + MnO + SO$_3$ + P$_2$O$_5$).

In this classification system [53], the pH value, measured on a 1:1 mixture of LFA and distilled water, is prefixed to the respective LFA group. LFA with a pH value < 5

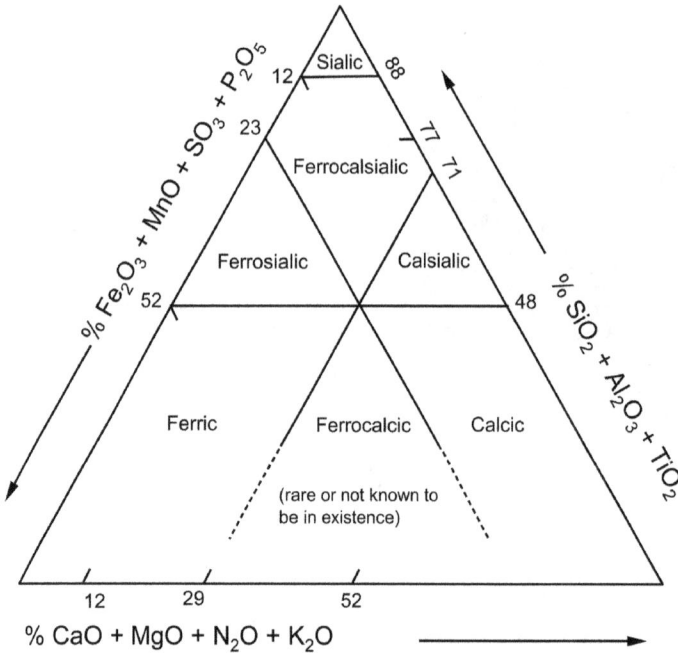

Figure 5.2: Location of different LFA groups in the ternary system (SiO_2 + Al_2O_3 + TiO_2) – (CaO + MgO + Na_2O + K_2O) – (Fe_2O_3 + MnO + SO_3 + P_2O_5) drawn according to [53].

have the prefix „Ad" (according to acidic), ashes with a pH value between 5–9 have the prefix „Nu" (according to neutral) and LFA with a pH value > 9 have the prefix „Al" (according to alcaline). Individual characteristics in the chemistry of LFA, e.g. a manganese content of 1500 mg/kg, are expressed with the corresponding element symbol before the prefix. According to the nomenclature shown, a LFA of the Calcic group, for example, which has a pH value in the basic range in aqueous solution and a relatively high manganese content of 1500 mg/kg, would bear the designation Mn-Alcalic.

2.2 General standardization of fly ashes

2.2.1 Chemical requirements

Fly ash is defined as a fine-grained dust, mainly consisting of spherical, vitreous particles, produced during the combustion of finely ground coal with or without co-combustion material(s), having pozzolanic properties and consisting essentially of SiO_2 and Al_2O_3, the content of reactive SiO_2 defined and described in DIN EN 197-1:2011-11 [54] being at least 25% by mass [55].

The loss on ignition is determined according to DIN EN 196-2:2013-10 [56]. Categories A, B & C are determined according to the level of volatile components A: ≤ 5%, B: 2,0–7,0% & C: 4,0–9,0%.

According to the DIN EN 450-1:2005 standard [55], the requirements and conformity criteria for the suitability of the fly ashes for use in concrete are tested by adding the material to a test cement (CEM I 42.5). Characteristic values for the suitability of fly ash in concrete are determined by chemical and physical requirements. The use of fly ash for concrete is determined by chemical analysis. A distinction is made between free [55] and reactive calcium oxide [54] and between reactive SiO_2 [57] or the sum of the contents of silicon dioxide (SiO_2), aluminium oxide (Al_2O_3) and iron oxide (Fe_2O_3). The free calcium oxide content shall not exceed 2,5% by mass. In the case of concentrations greater than 1,0% by mass, the volume stability shall be tested in accordance with DIN EN 450-1:2012-10 5.3.3 [55]. As in DIN EN 197-1:2000, 3.1 [54], reactive calcium oxide shall not exceed 10.0% by mass.

The content of reactive silica as described in DIN EN 197-1:2011, 3.2 [54] shall not be less than 25% by mass. The sum of the contents of SiO_2, Al_2O_3 and Fe_2O_3 is determined according to DIN EN 196-2 and shall be greater than or equal to 70% by mass. In addition, there are also limit values for Alkalis (Na_2O less than or equal to 5.0% by mass), MgO (less than or equal to 4% by mass), soluble phosphate (less than or equal to 100 mg/kg) (DIN EN 450-1:2005 (C), SO_3). (less than or equal to 3.0%) and the concentration of chloride ions less than or equal to 3.0%.

2.2.2 Chemical analysis of LFAs

2.2.2.1 Loss on Ignition (LOI)

The loss on ignition is determined for fly ashes in two different ways. To analyze the water content (ISO 11465) [58], the samples were dried at 105 °C. The loss in mass corresponds to the water content of the sample. In this way, the adhesive and crystal water is removed from the sample. The sample was then annealed at 815 °C. The loss on ignition corresponds to volatile phases (organic substances, CO_3^{2-}, etc.) without the water content (DIN 51719) [59].

2.2.2.2 Chemical analysis of major and minor element oxides

Major and minor element oxides are investigated by means of wavelength dispersive X-ray fluorescence analysis on a Siemens SRS3000 equipped with an end window X-ray source (AG66) with Rh anode and variable analyzer crystals (LiF 200, LiF 220, Ge, PET, OVO – 55, OVO – 160). Wax and fused tablets were produced for the analyses. The loss on ignition necessary for computer-aided evaluation of the results was determined in accordance with DIN 12879 [60], at 815 °C. Furthermore, trace elements of the samples were quantitatively determined by optical emission spectroscopy with inductively coupled plasma (ICP-OES).

2.2.2.3 Free lime determination

According to Franke (1941) [61], a wet-chemical quantitative analysis of the free lime and $Ca(OH)_2$ contents is carried out by extraction using ethyl acetate and isopropanol.

2.2.2.4 Testing the pozzolanicity of pozzolanic cements

To assess the pozzolanicity, the calcium hydroxide content of an aqueous slurry of pozzolanic cement after a certain time is compared with the calcium hydroxide content of a saturated solution with the same alkalinity. The test is considered as passed if the concentration of dissolved calcium hydroxide is lower than the saturation concentration [57].

2.2.3 Physical requirements

The fineness of fly ash is determined according to DIN EN 451-2 [62]. There are 2 categories N and S. Category N: The fineness must not exceed 40% mass fraction and must not deviate from the declared value by more than ± 10 percentage points. Category S: The fineness must not exceed 12% mass fraction. The limits for the deviation from the declared value of ± 10 percentage points do not apply. The activity index [63] (DIN EN 196-1) must be at least 75% after 28 days and at least 85% after 90 days. The room resistance (expansion) is determined according to [64] and must not exceed 10 mm. If the fly ash has a free calcium oxide content of less than 1% by mass, the requirement of stability in space is fulfilled. The grain density is determined according to [65]. It must not deviate by more than ± 200 kg/m^3 from the manufacturer's value. The start of solidification is determined according to DIN EN 196-3. It must not be later than 120 min later than the zero sample (test cement + mixing water). If fly ash is used, which is produced exclusively from finely ground coal, the requirements are considered to be fulfilled. The water content of fly ash of the fineness category S is tested according to DIN EN 450-1:2005 Methods Annex B [55]. It shall not be higher than 95% compared to the zero sample. In Europe exits the following standards (related to Germany):
- Fly ash for concrete – Part 1: Definition, specifications and conformity criteria; German version EN 450-1:2012 [55]
- Fly ash for concrete – Part 2: Conformity evaluation; German version EN 450-2:2005 [62]

- Method of testing fly ash – Part 1: Determination of free calcium oxide content; German version EN 451-1:201 [66]
- Method of testing fly ash – Part 2: Determination of fineness by wet sieving; German version EN 451-2:2017 [67]

The largest US-American rule setter and its internationally important, globally applied standards American Society for Testing and Materials ASTM International (ASTM)
- ASTM D5759-12, Standard Guide for Characterization of Coal Fly Ash and Clean Coal Combustion Fly Ash for Potential Uses [68]
- ASTM C618-19, Standard Specification for Coal Fly Ash and Raw or Calcined Natural Pozzolan for Use in Concrete [69]
- ASTM C311 / C311M-18, Standard Test Methods for Sampling and Testing Fly Ash or Natural Pozzolans for Use in Portland-Cement Concrete [70]

Coal Fly ashes are classified according to ASTM C618-19 in three different groups. Class N includes uncalcined or calcined some types of diatomaceous earth, opaline diatomaceous earth and slate, tuffs and volcanic ash or pumice, or materials (such as some clays and slates) that must be calcined to meet the requirements. Group F fly ash has pozzolanic properties. Class F is typically made from the combustion of anthracite or bituminous coal, but can also be made from sub-bituminous coal and lignite. In addition to pozzolanic properties, this Class C of fly ash also show some cement-like properties.The SiO_2, Al_2O_3 and Fe_2O_3 contents are at least 70%. Class C fly ash is typically produced from the combustion of brown coal or sub-bituminous coal, but can also be made from anthracite or bituminous coal. Class C fly ash typically has a total lime content, expressed as calcium oxide (CaO), that is higher than that of class F fly ash. Further more, SiO_2, Al_2O_3 and Fe_2O_3 contents are at least 50% but less than 70%.

2.3 Morphology

A large proportion of the BFA particles are, due to their formation, vitrified [12] have a spherical shape, which also dominates in grain size ranges below the resolving power of a light microscope [14]. From pre-cleaning to post-cleaning, the spherical habit of the particles increases due to stronger thermal influence of the smaller ash particles [28]. The spherical particles can be designed as solid or hollow spheres. Empty hollow spheres are called „cenospheres", hollow spheres filled with finest grain are called „plerospheres" [71–73] (Figure 5.3). The formation of plerospheres or „intraparticles" is based on the immiscibility of the glass melt [74]. The formation of cenospheres is influenced by melting of mineral inclusions in coal on a non-wetting surface, namely carbon. Furthermore, the optimal temperature for

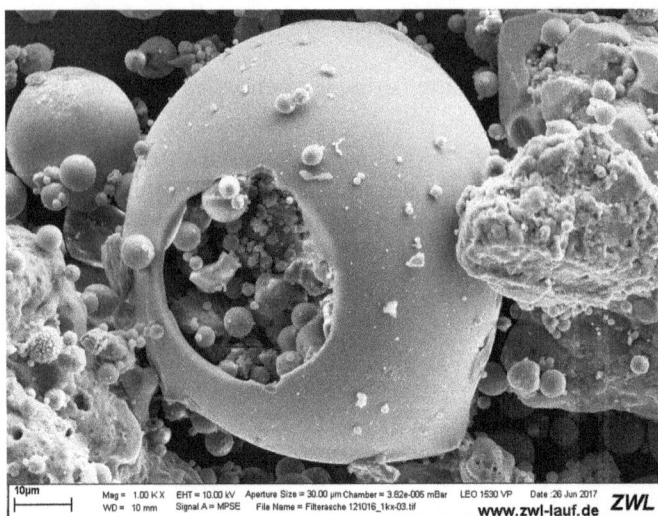

Figure 5.3: Plerosphere filled with much smaller spheres in a LFA powder.

the formation of cenospheres based on the ash density is at 1230 °C. At higher temperatures the gas evolution is too fast, so that gas escapes from the molten ash [75, 76]. In addition to the vitrified, spherical particles, there are also subordinate, compact, irregular coke particles with a porous structure [77] (Figure 5.4), irregularly shaped quartz grains, which originate from silt deposits of coal, asymmetric, inflated, highly porous particles, opaque iron oxide spheres with partially formed

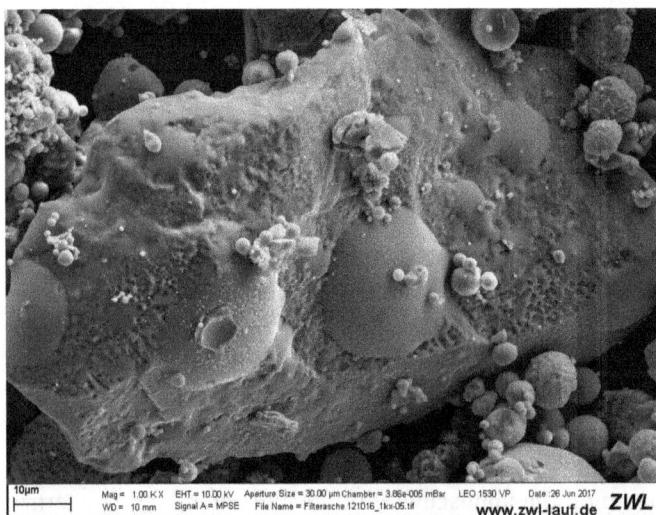

Figure 5.4: Irregular coke particles with a porous structure in a LFA powder.

crystal surfaces, heterogeneous sintering aggregates as well as dust particles < 1μm, which are predominantly adhesively bound as adhesive grains to coarser particles. Idiomorphous crystals are rare [11].

By means of element mapping the distribution of individual elements can be shown (Figures 5.8 and 5.9). The LFA of the Boxberg power plant shows a composition of irregularly shaped particles and a spectrum of roundish spheres, which rarely exceed 100 μm. The spheres are mainly composed of Ca-, Mg-, Fe- containing amorphous silicates. Associated with them are irregularly round Ca- and Al-containing iron oxides and titanium oxides (rutile).

Occasionally there are spheres which predominantly contain the elements Fe and O. They consist of magnetite (Fe_3O_4) and hematite (Fe_2O_3), respectively, which can be easily recognized by their surface relief. Magnetite melt spheres often show small octahedra on their surface (Figure 5.5). The surface of hematite spheres, on the other hand, consists of small hexagonal platelets (Figure 5.6). With the help of a magnet the magnetic part of the ashes, metallically shiny, roundish particles composed of magnetite, can be separated. In some cases, adhesions of magnetite (Fe_3O_4) and hematite (Fe_2O_3) can also be detected (Figure 5.7).

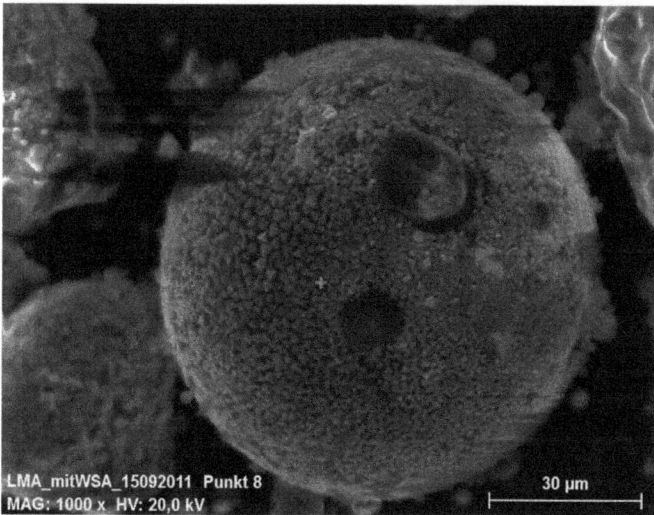

LMA_mitWSA_15092011 Punkt 8
MAG: 1000 x HV: 20,0 kV
30 μm

Figure 5.5: SE image: On the surface of the melting ball small octahedrons are visible, which are characteristic for magnetite (Fe_3O_4).

The silicate-rich aggregates contain Ca-sulphates, which were identified as anhydrite by X-ray diffraction. In combination with powder diffraction, Ca – & S- distribution patterns provide information about the distribution of anhydrite, bassanite or gypsum (Figure 5.8). Several mineral phases are characterized by a strong Ca

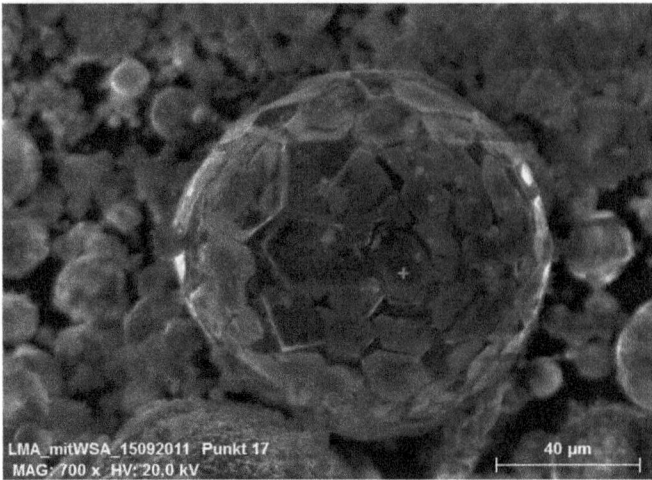

Figure 5.6: SE image: The surface of the sphere shows small platelets of hematite. They are formed by oxidation of the magnetite (martitization).

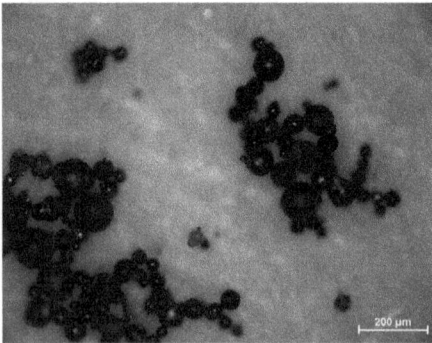

Figure 5.7: Microscopic image of the magnetic portion of the filter ash. Shiny metallic spheres can be seen, which consist mainly of magnetite.

coloring. In correlation with the S- distribution map particles of anhydrite ($CaSO_4$) can be identified. Occasionally, aluminosilicates containing K occur.

Figure 5.9 shows the Si – Al – K elemental distribution maps of the LFA. Besides quartz aggregates, an irregularly shaped particle with a correlatable Si-, Al- and K-concentration can be seen in the upper part. This is a K- containing aluminosilicate, probably orthoclase (P2) (Figure 5.10). The large sphere on the right edge is a Ca-containing silicate sphere (P1) (Figure 5.10), which has grown fine sulphate particles (anhydrite).

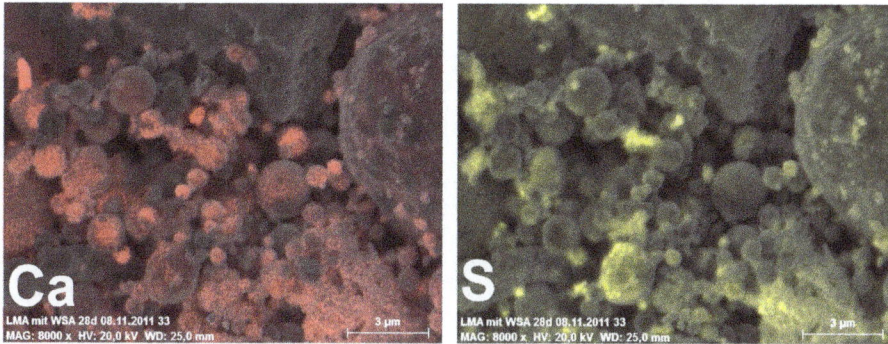

Figure 5.8: Ca, S – Mapping.

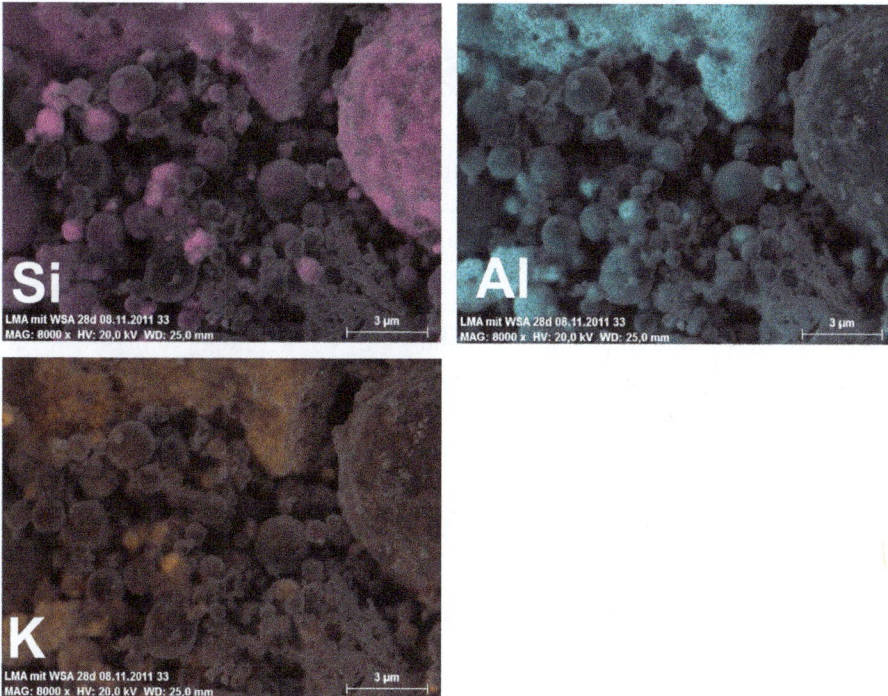

Figure 5.9: Si, Al, K – Mapping.

Rarely are trace elements combined with main elements. Cobalt, for example, shows an affinity to iron-containing mineral phases. Antimony is associated with Ca-containing mineral phases. The trace elements Cd, Cr, Hg, Cu, Pb, As, Mn, Zn and Sn are homogeneously distributed on the surfaces of the larger aggregates. Correlations with main mineral phases could not be observed so far.

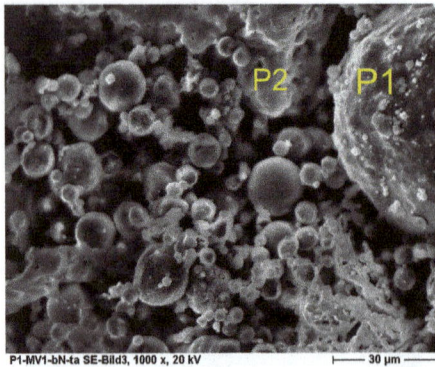

P1-MV1-bN-ta SE-Bild3, 1000 x, 20 kV ├── 30 µm ──┤

Figure 5.10: Secondary electron (SE) image of the LFA.

2.4 Grain size, density & specific surface

For rough estimates of the technical properties of a material such as permeability, strength and expansivity as well as its applicability in the cement industry, grain size distribution or particle size distribution are very helpful [78]. To describe the particle size distribution, the coefficient of uniformity C_u and the coefficient of curvature or gradation C_c are appropriate values. According to ASTM D-2487 gravel is classified as good if $C_u \geq 4$ and $1 < C_c < 3$; sand is classified as good (particle sizes are distributed over a wide range) if $C_u \geq 6$ and $1 \leq C_c \leq 3$. According to these criteria, Class C fly ashes were partly classified as well-graded and others as poorly graded in the investigations [78].

The grain size spectrum extends over the silt and sand range and is thus widely diversified. The distribution of the particles to individual fractions depends on the degree of grinding of the raw lignite, the dust filter system [71] used as well as the LFA variant (pre-, medium-, post-cleaning, mixed ash) [12]. In general, the average grain size decreases from pre-cleaning to post-cleaning. The diameter of individual particles can vary between < 1 µm (Figure 5.10) [12, 14, 15, 20, 26, 30–33, 35–37, 39, 73, 79].

Normally, the D_{50} values of the grain sum curve for LFA from pre-cleaning are in the range of 60 – 100µm, for LFA from middle/post-cleaning between about 10 – 40µm and for mixed ashes about 20 – 70µm. Exceptions are, for example, the average grain sizes of LFA from the pre-cleaning of the Jänschwalde power plant (Lusatian mining district), which are unusually fine-grained with about 35 – 70µm, or of the Frimmersdorf power plant (Rhenish mining district), which can be described as extraordinarily coarse-grained with about 165µm [11, 12, 15, 32, 35, 39] (Table 5.7).

Numerous studies have shown that there is a relationship between grain size and chemical composition. CaO, MgO and SO_3 increase with the grain fineness, SiO_2 and unburned components, however, decrease. The increase of SiO_2 with increasing grain

Table 5.7: Average particle sizes of BFA from different cleaning stages and districts [11, 12, 35, 39].

Power plant	VR [μm]	MR [μm]	NR [μm]	MA [μm]
Boxberg – Lusatian Mining District	75–100	20–40	15–30	25–60
Jänschwalde – Lusatian Mining District	35–70	10–20	10–20	30–60
Thierbach – Central German Mining District	86	38		62
– Central German Mining District	35, 85		22, 28	23–40, 60
Neurath – Rhenish mining district	83	11		35
Frimmersdorf – Rhenish mining district	165		15	

VR = pre-cleaning, MR = intermediate cleaning, NR = final cleaning, MA = mixed ash

size is due to an enrichment of quartz in the coarsest fractions. In contrast, free lime, periclase and anhydrite are concentrated in the finer fractions. Al_2O_3 and Fe_2O_3 behave rather unspecific. Their proportions can appear evenly distributed as well as decrease or increase with increasing grain size [11, 14, 15, 25–28, 31, 39, 51, 73]. The mean grain raw densities for the individual LFA variants are between 2.1–3.3 g/cm^3 [11, 12, 14, 15, 26, 35, 77]. The average bulk density is between 0.7–1.5 g/cm^3, while values in the range of 1.1 and 2.0 g/cm^3 have been measured for the average Proctor density at an optimal water content of 17–30% [36]. The average specific surface area according to BLAINE increases due to the decrease in particle size from pre-cleaning to post-cleaning. It varies, depending on the cleaning stage, between 1200–6800 cm^2/g [14, 15, 26, 35].

3 Mineralogy of LFA

The mineralogical composition of lignite fly ashes is very different, depending on the type of coal and the mining area. Furthermore, the size and distribution of minerals in the coal varies. The mineralogical composition of the coal influences the combustion process and the disposal of slag and ash. The mineralogical composition of LFA for different localities were reported. LFAs from Boxberg and Schwarze Pumpe power plants were described in [42] by [37], from power plants in Yugoslavia [80], from Megalopolis lignite fields, Peloponnese, Southern Greece [81], from Agios Dimitrios, Kardia, and Ptolemais power plants [82], from greek or Poland [21], fly ash from each of the lignite mining areas of North America [83], from China especially in the Yunnan province [84] The mineralogy of LFAs depend on the initial composition of the raw lignite and the combustion conditions. In general, the ash components can be divided into crystalline and amorphous components. They occur side by side and intergrown [12].

3.1 Selective dissolution techniques

For the identification of accessory phase fractions, which could not be reliably identified due to reflection overlaps or low peak intensities in the X-ray powder diffractograms, certain separation methods were applied. Selected filter ashes were subjected
to leaching with concentrated HCl, concentrated H_2SO_4, HCl + HNO_3 + H_2O (1:3:6) and
0.2 M HNO_3, in accordance with DIN EN 196-2 [56]. After leaching, the samples were
dried in a drying cabinet at 40 °C until their weight remained constant [42].

To enrich containing silica-phases in Iron- rich LFA of the Lower Lusatia the KOH-
Sucrose extraction method [85] has been used. KOH sugar extraction dissolves the aluminate phases contained in the mineral composite to enrich the silicate phases.

Furthermore the selective solution treatment, using Methanol and salicylic acid,
have been used to enrich calcium-aluminates, ferrite, periclase and sulphates. In this
solution process, the silicate phases and any free lime present are completely dissolved, leaving behind the enriched calcium aluminates, ferrites, periclase and sulphates [86–88].

3.2 Qualitative phase analysis of crystalline compounds

The crystalline components, determined by X-Ray diffraction, include mainly different oxides, silicates, aluminates, sulfates, carbonates and halides (Table 5.8). Depending on which type of LFA is present, the following mineral phases can occur in
varying proportions: Quartz, magnetite, hematite, anhydrite, lime, C_3S (Ca_3SiO_5),
C_2S (Ca_2SiO_4), C_3A ($Ca_3Al_2O_6$), C_2F ($Ca_2Fe_2O_5$), brownmillerite, Mg-Ca-aluminates,
hercynite, wollastonite, pseudowollastonite, mullite, gehlenite, calcite, MgO, BaO,
"metakaolin", rutile, halite, sylvine [41]. Furthermore, pseudobrookite, feldspars
[31], zircon, ilmenite [11] ye'elimite, CaS [32], merwinite [39], diopside and magnesite
[37]. Quartz, clay mineral relics, feldspars and some accessories (e.g. rutile, zircon,
Th-bearing ilmenite) belong to the old stock of coal. All other components are new
formations formed during combustion [11].

Figure 5.11 shows the complex phase composition of crystalline phases typical
for LFA from the Boxberg power plant. The amorphous content, which make up a
large part of the LFA components, consists of differently colored, transparent to
translucent particles, which are inhomogeneous, i.e. interspersed with secondary
crystallites, have a very wide range of chemical variations and can be interpreted as
glass [11]. In numerous single grain analyses it could be shown that the particles
mostly have nonstoichiometric compositions, so that the mineral or phase content
cannot be applied to these components [31].

Table 5.8: Classification of LFA components according to their reaction behavior [11, 32, 39, 41, 52] in [42].

Phase		Chemistry	Reaction behavior
Glas	Inert glass	SiO_2, Al_2O_3	weak reactive
Glas	Active Glass	CaO, Al_2O_3, SiO_2, Fe_2O_3, MgO	reactive
Quartz		SiO_2	inert
Magnetite		Fe_2O_3	inert
Hematite		Fe_2O_3	inert
Anhydrite		$CaSO_4$	reactive
Free Lime		CaO	reactive
Residual coal		C	inert
C_3S		$3\ CaO \cdot SiO_2$	reactive
C_2S		$2\ CaO \cdot SiO_2$	weak reactive
C_3A		$3\ CaO \cdot Al_2O_3$	reactive
C_2F		$2\ CaO \cdot Fe_2O_3$	weak reactive
Brownmillerite		$Ca_2(Al,Fe)_2O_5$	reactive
Mg-Ca-Aluminate		$Mg_xCa_yAl_2O_5$	reactive
Hercynite		$FeAl_2O_4$	inert
Wollastonite		$CaSiO_3$	inert
Pseudowollastonite		$CaSiO_3$	inert
Mullite		$Al_{4+2x}Si_{2-2x}O_{10-x}$	inert
Gehlenite		$Ca_2Al[AlSiO_7]$	inert
Calcite		$CaCO_3$	inert
Magnesite		$MgCO_3$	inert
Diopside		$CaMg[Si_2O_6]$	inert
Periklase		MgO	reactive
BaO		BaO	reactive
Pseudobrookite		Fe_2TiO_5	inert
Yeelimite		$Ca_4Al_6O_{12}(SO_4)$	reactive
Feldspars		e.g. Anorthite $(Ca,Na)x[(Al,Si)_3xO_8]$	inert
Merwinite		$Ca_3Mg[SiO_4]_2$	inert

Table 5.8 (continued)

Phase	Chemistry	Reaction behavior
Oldhamite	CaS	reactive
Metakaolin	$Al_2O_3 \cdot 2SiO_2$	weak reactive
Zircon	$ZrSiO_4$	inert
Ilmenite	$FeTiO_3$	inert
Rutile	TiO_2	inert
Halite	NaCl	soluble
Sylvite	KCl	soluble

3.3 Determination of quantitative phase assemblages in FA and FA – cement formulations

3.3.1 Strategies and quantification techniques

Quantitative phase analysis were performed mainly by using the Rietveld method together with different complementary methods such as NMR ^{29}Si and ^{27}Al MAS NMR) [89], or applying the Reference intensity method (RIM) (better known as RIR – or as the „Chung method") invented by [90, 91] in combination with the standard additions method (also known as spiking method) [92].

In the following, different approaches for the quantification of fly ashes and pretreatments will be introduced.

The authors [92] report that the vitreous fraction of fly ash can be determined indirectly by the difference of the total phase fraction minus the fraction of crystalline phases. For the application the Reference intensity method (RIM) was used [90, 91]. For the analysis a Siemens D5001 diffractometer with graphite monochromator and scintillation counter was applied. Measurements were performed in the range of 20 and 40° 2Θ, with a step size of 0.02°/s. For the quantification of the main phases anhydrite, quartz, mullite, K-feldspar, plagioclase, hematite and magnetite, different mixtures of ash/fluorite, quartz/fluorite, mullite/fluorite were prepared in a 50:50 ratio. Based on the results of the X-ray analysis, the constant K0 according to equation $I_0/I_f = K_0 \cdot X_0/X_f$ [93] (I_0 = quartz or mullite XRD intensity, I_f = fluorite XRD intensity, X_0 = concentration of quartz or mullite, X_f = concentration of fluorite, and K_0 = quartz or mullite constant) was determined. The concentrations of the minor constituents (illite, anhydrite, microcline-orthoclase, albite-anorthite, hematite, magnetite, calcite, and lime) were calculated based on the K-values of quartz. In order to verify the quantitative concentrations determined by the RIM

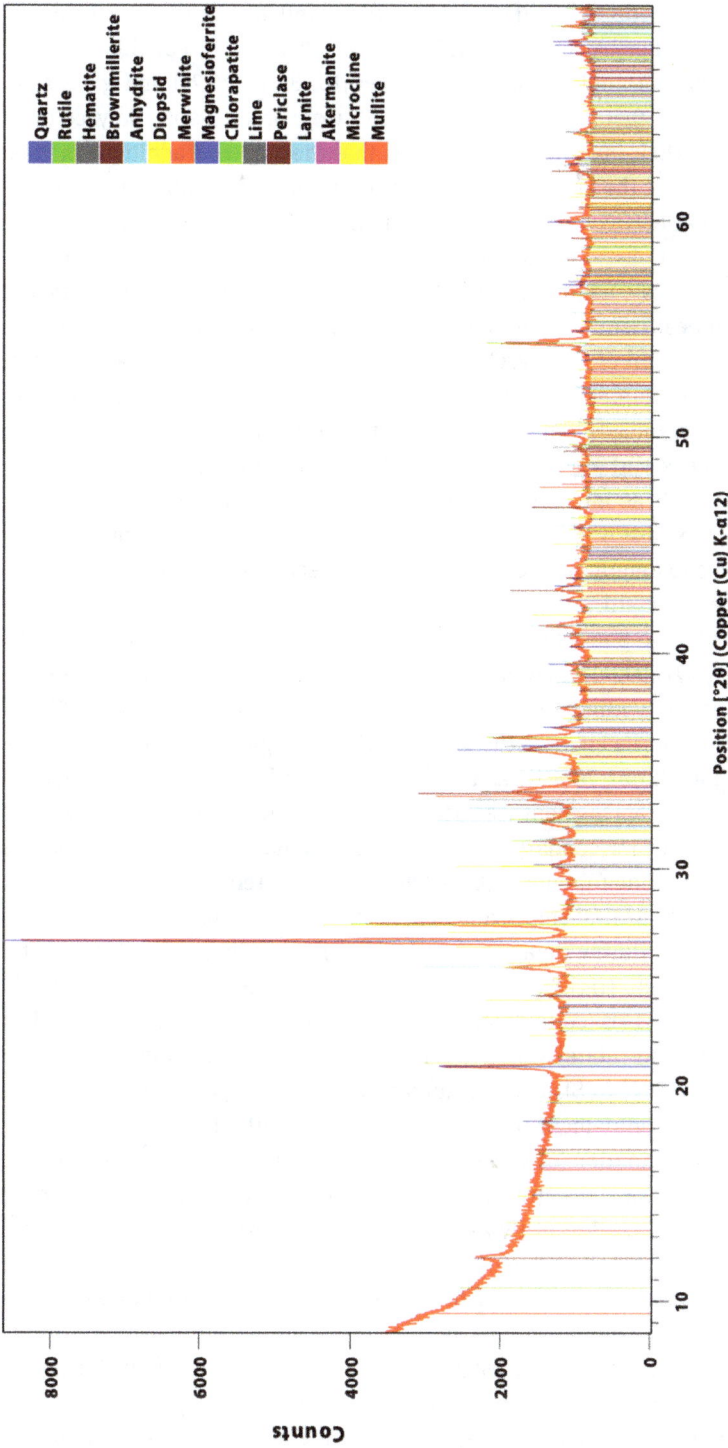

Figure 5.11: Qualitative phase content of a LFA sample determined by XRD.

method, the concentrations of quartz and mullite and the proportion of amorphous phases were checked by the standard addition method. For this purpose, different ash/quartz ash/mullite mixtures (95:5, 90:10, 85:15, 75:25) and glass standard/fly ash mixtures (50:50, 75:25, 65:35 and 90:10) were ordered and analyzed by X-Ray diffractometry. Their investigations revealed that the standard additions method yielded closer results of the amorphous fraction of fly ashes. The difference in amorphous phase content between standard addition and RIM processes is < 3%. According to their investigations, the amorphous phase inventory is determined in three steps, which means „calculation of the full area", „determination of the net area of amorphous broad peak" and determination of the amorphous phase fraction from the point of intersection of the correlation line with the X-axis in the diagram Standard (Y-axis) against flue ash concentration (X-axis). However, both methods also have different disadvantages [94]. The RIR and standard addition methods require a large number of steps and are limited to a small number of crystalline phases. Furthermore, effects such as microabsorption and preferred orientation are not taken into account. In the case of the RIR method, calibration experiments must be performed in advance to determine precise RIR ratios for each mineral phase relative to the added internal standard.

For the quantitative phase analysis of the type F filter ashes [89], used for alkaline cements, γ-Al_2O_3 (99.997% from Alfa) was used as internal standard material, which was pretreated by grinding (200 rpm/ 20min.) and by tempering (1200 °C/5h). For the quantitative phase analysis the grain fraction < 35 µm was used. The sample then consisted of a mixture 70% fly ash and 30% standard. For the measurements a D5000 diffractometer from Siemens was used with the following settings: Cu K$\alpha_{1,2}$ radiation (1.5418 Å) and a secondary curved graphite monochromator. The readings were taken in vertical Bragg-Brentano ($\Theta/2\Theta$) geometry (flat reflection mode) between 18 and 70° (2Θ) at 0.038° (2Θ) step size and 5 s counting time step. The flat samples were rotated at 15 rpm during data acquisition. Identical settings were also used for samples treated with 1% HF [95] to dissolve the amorphous part of the filter ash. Quantification of the phases (corundum, α-quartz, mullite, maghemite, calcium oxide, calcite aphthitalite and albite) was performed using the GSAS software package using the Inorganic Crystal Structure Database (ICSD). The parameters used and optimized during the refinement process are shown in [89]. The authors [89] concluded that the combination of chemical analysis, powder diffraction plus the Rietveld method and NMR spectroscopy yielded fairly accurate results regarding the amorphous phase fraction of fly ash. In addition, information on the thermal history of the ashes and the binding properties of SiO_2 and Al_2O_3 can be derived. This aspect is particularly important when filter ashes are activated alkaline as a reactive component of the binder.

Ward and French (2006) [96] investigated a series of 9 fly ashes from Australian power plants and a NIST fly ash 1633b with known qualitative and quantitative phase assemblages as reference sample. They applied two different methods to quantify the phase assemblages in the fly ashes. 15% ZnO or corundum used as internal standards were added to all samples (duplicate samples). Instead of adding a standard material, a powder diffractogram of a poorly crystalline silicate phase (metakaolin or tridymite) from the software database was used. During the quantification process this diffractogram was refined with the X-ray data of the fly ashes. The results of the investigations were mutually consistent and the results from the refinement of the NIST standard material are comparable to results from other published data [97, 98]. In addition, it was found that the use of tridymite or metakaolin as reference pattern in the refinement process seemed to be more appropriate for ashes from Australian and North American sources.

The analysis of cement mixtures with BFS and PFA additives by quantitative X-ray powder diffraction was already reported some years ago [99, 100]. The applicability of the Rietveld method for the routine quantitative phase analysis of materials with amorphous fractions was investigated and a systematic analysis was carried out of what this method is theoretically capable of [101]. For this purpose, a total of 149 fly ashes and four certified reference materials of NIST were investigated, either pure or in model mixtures with Portland cement. The model systems served as a reference to determine the accuracies. Theoretical considerations suggest the use of the Rietveld method to quantify amorphous fractions in the range of 20–100% amorphous portion seems reasonable. The non-linear function for the calculation of the amorphous fraction leads to the fact that the inaccuracy in the determination of the amorphous fraction decreases with increasing amorphous fraction. Below the 20% mark, the analytical inaccuracy takes on unacceptably high values. The applicability of the method to complex material systems could be demonstrated on the basis of fly ashes. In (simulated) routine use, analytical inaccuracies in the range of $\pm 1 \ldots \pm 2\%$ were found.

For the quantification of phase contents in fly ashes from Spanish and Dutch power plants, the Rietveld method was applied with phases pre-calibrated using a standard [94]. This method is called „Quantification of phases with partial or no known crystal structures" or „PONKCS" method [102]. For all phases using empirically derived structure factors ZMV „calibration constants" must be derived, e.g. via an internal standard. For the measurements a Bruker-AXS

(Siemens) D5005 powder diffractometer, Cu X-ray tube, a graphite monochromator and a scintillation detector were applied. Data collection were performed in the range of 4° and 60° 2Θ with a 0.05° 2Θ step size and a counting time of 3s. For the quantification procedure the software suite Topas 4.2 (Bruker AXS, 2003–2009) was used. Further details are available in [94]. In this publication the authors compare their results obtained using the PONKCS method with the results published in Font & Moreno 2010 [92], whereby the differences are negligible. Applying and fitting the amorphous phase contents from the RIR – and PONKCS – methods in a scatter diagram gives an increase of the quality factor to 1.013, which indicates good consistency of the results.

The PONKCS method was used for the precise and accurate quantification of amorphous supplementary cementitious materials here fly ash in Portland cement systems [103]. Further investigations applying the PONKCS method were carried out for the precise and accurate quantification of amorphous supplementary cementitious materials with fly ash in Portland cement systems for the quantification of Portland cement FA mixtures [104]. Singh & Subramaniam (2016) [105] analyzed binary and ternary mixtures of SCMs without OPC and binary mixtures of OPC and SCM, or binary mixtures of OPC and fly ash at 30–70%. Both studies reported the accuracy of this method to be approximately 2%. Recent investigations concerning the quantification of OPC – fly ash formulations were performed by [106]. They present the quantification of amorphous siliceous fly ash in hydrated cement systems using XRPD based on the Rietveld- PONKCS method (partial or no known crystal structure). To validate the results obtained by the Rietveld method, additional measurements, such as the quantification of the $Ca(OH)_2$ content, were performed by thermogravimetry and isothermal calorimetry to assess the degree of reaction of the fly ash. The results show a good accuracy in terms of quantification when the initial fly ash content is more than 10 wt.-%.

Investigations using in-situ XRD and heat flow calorimetry to detect the first 44 hours of hydration of ordinary Portland cement (OPC) mixed with silica fly ash (FA) [107]. The direct influence of FA on the silicate and aluminate reaction of OPC hydration could be determined by recalculating the heat flow using XRD data.

3.3.2 Quantification of brown coal fly ash

The software Highscore Plus (v.3.x) (Malvern – Panalytical) was used for refinement and quantification. The underground, used for the refinements, was determined manually. Before a quantitative analysis can be carried out a qualitative phase analysis must be performed. Therefore, the X-ray data was evaluated using the software package Highscore Plus (v.3.x) in combination with the ICDD database (International Centre for Diffraction Data). In complex multi-phase systems, such as in filter ashes, the peaks partially overlap and cannot be clearly verified in comparison with the database. In order for the mineral phases to be unambiguously determined, the mineral phases in the source material must be separated from each other to such an extent that no overlaps that interfere with the analysis are present. For this purpose, the starting material was selectively dissolved by KOH sugar extraction or methanol – salicylic acid. The structural data of the detected mineral phases were taken from the Inorganic Crystal Structure Database Leibniz Institute in Karlsruhe (ICSD). The peak shapes are described in the model purely geometrically, using mathematical profile shape functions (PSF). The internal standard method was used to determine the amorphous fraction. 10 Ma-% rutile was added to the filter ash sample and the mixture was homogenized. Data collection was performed applying a $\Theta - \Theta$

diffractometer X'Pert Pro (Malvern – Panalytical) with Bragg – Brentano geometry. Flat samples were prepared by back loading technique. The following instrument parameters for qualitative and quantitative analysis were selected (Table 5.9).

Table 5.9: Instrument parameters for XRD analysis.

	LFF -Tube	Cu LFF -Tube: Co
Parameters	Values	Values
Measuring range [°2 0]	5–70	5–90
Step [°2 0]		0,02
Counting time [s]		10
Divergence – & Anti – scatter slit H		1/8
Soller slits [®]		2.3
Voltage [kV]	45	40
Tension [mA]	30	35
β – filter	Ni	Fe

The investigated LFAs, independent of co-combustion, are dominated by amorphous glass phase (38–39%) and crystalline quartz (20–26%). Brownmillerite also shows high contents with 7–9%. The contents of the remaining mineral phases like anhydrite are moderate (≤5%). In order to verify the results of the rietveld refinement, model mixtures were prepared. Model mixtures of quartz and gypsum were prepared by adding known, increasing contents of the crystalline mineral phases, to the LFA, then the model mixtures were quantified as a first step applying the previously described refinement strategy and as a second step determining the net intensity of one not overlapped representative Peak (single peak method). The known contents of the admixed phase are plotted against the quantified contents in the model mixtures for both methods. The results are regression lines. If the straight line intersects the axis with the quantifications at 0% admixture, the phase concentration that was quantified for the LFA, must not be corrected. To verify the concentrations of quartz and anhydrite certain concentrations were added to the LFA.

The quartz content quantified initially by the Rietveld method is 24.8%. The regression line resulting from the removal of the added quartz against the quantified quartz content using the Rietveld method intersects the y-axis at a content of 23.2% (Figure 5.12). The deviation is 1.8%. The regression line, which is obtained by removing the added quartz against the quartz peak at 20.8°2Θ, intersects the x-axis at a quartz content of 25.0% (Figure 5.13). This value is slightly above the content quantified by Rietveld. The deviation with this method is 0.2%. Using the Rietveld

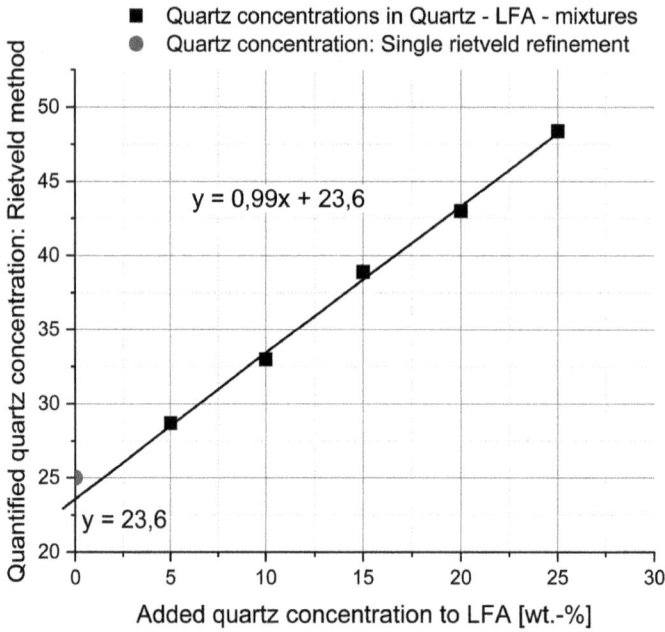

Figure 5.12: Quantification of quartz – LFA mixtures: Rietveld method.

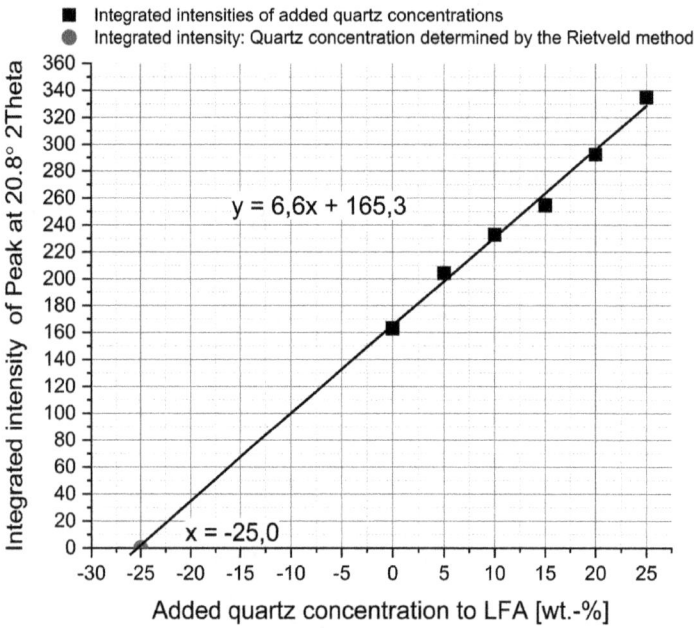

Figure 5.13: Quantification of quartz – LFA mixtures: Single peak method.

method quantified quartz contents of the filter ashes can therefore differ by 0.2 to 1.8% from the real quartz content in the ashes.

The anhydrite content quantified by the Rietveld method is 2.9%. The regression line resulting from the plotting of the added anhydrite against the quantified anhydrite content by the Rietveld method intersects the y-axis at a content of 6.75% (Figure 5.14). This significant deviation is due to the superposition of several anhydrite peaks with peaks of other mineral phases and also due to texture correction errors. The regression line resulting from the ablation of the admixed anhydrite components against the anhydrite peak at 25.5°2Θ intersects the x-axis at a content of anhydrite of 1.8% (Figure 5.15). This value is slightly below the content quantified by Rietveld. The deviation with this method is at least 1.1%. The anhydrite content of the filter ashes quantified by the Rietveld method can therefore deviate by at least 1.1% of the real anhydrite content in the ashes.

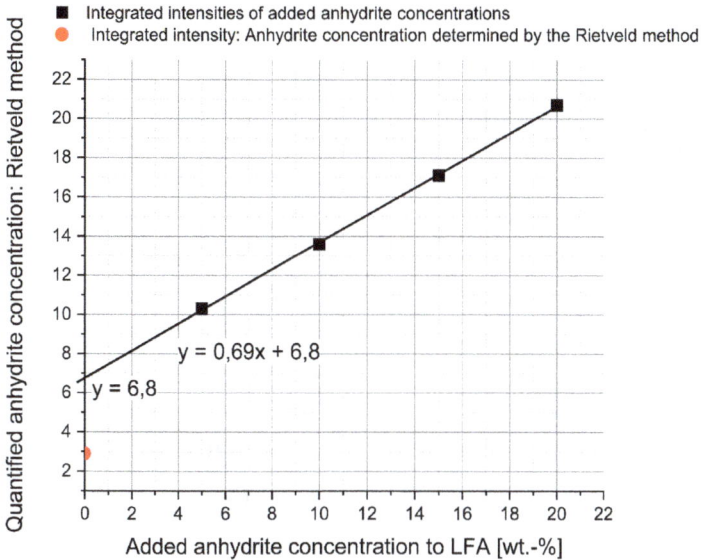

Figure 5.14: Quantification of anhydrite – LFA mixtures: Rietveld method.

To verify the content of the amorphous glass phase, amorphous „Poraver" glass powder was added. The increasing content of the added glass powder is plotted against the quantified amorphous portion. If the determined linear regression line intersects the quantification axis in the content of the quantified amorphous portion, the quantification strategy for determining the phase contents is considered verified (Figure 5.16). For the filter ash sample used for verification, an amorphous fraction of 40.3% was determined using the Rietveld method with an internal standard. The regression line resulting from the removal of the added glass components (Poraver glass powder) against the quantified amorphous portion using the Rietveld

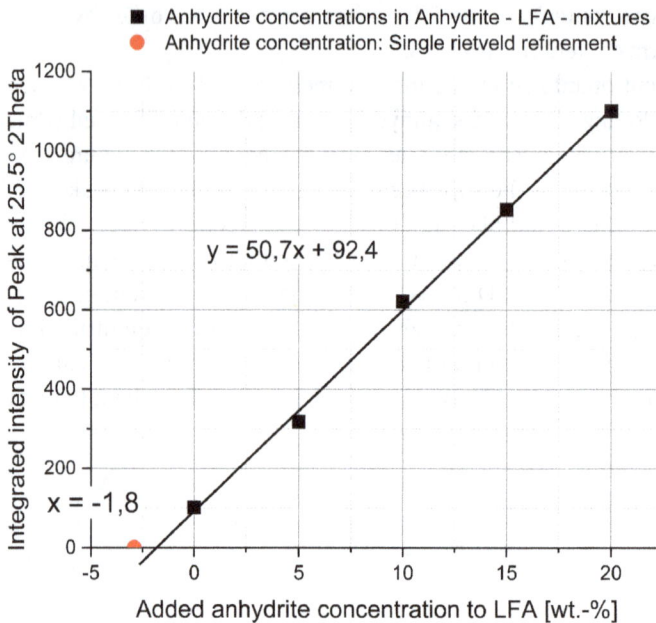

Figure 5.15: Quantification of anhydrite – LFA mixtures: Single peak method.

method intersects the y-axis slightly above the content quantified for the filter ash at 41% (Figure 5.16). The amorphous phase content quantified by the Rietveld method can therefore deviate by 1% from the real glass content in the filter ash.

4 Properties of the amorphous phase content in LFAs regarding its reactivity with H_2O

According to Pöllmann (2007) [41], the hydration behavior of LFAs can be summarized as follows. Reactive aluminates and ferrates take part in the reaction. The free lime content ensures that the latent hydraulic phases are activated. The dissolution rate of the active glass phase is essential for the hydration behavior of the LFA, in addition to the reactive phases. Ettringite is formed by the sulfate content and provides the initial strength. Final strength is also influenced by CSH phases. Glasses with high Fe content have a low dissolution rate. By adding CaO, a sulphate carrier mineral or Portland cement, solid setting formulations can be developed. These formulations are different for the individual deposits, purification stages and combustion technologies used. A literature review/data collection on the reactivity of LFA was published recently [42]. Due to the strong variation within the phase chemistry, there are also large differences in the reactivity properties of present phases. The knowledge

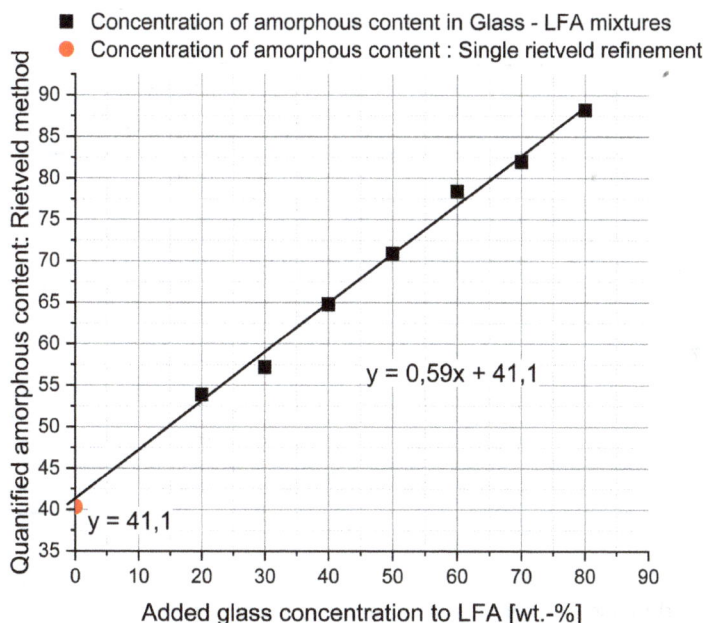

Figure 5.16: Quantification of the amorphous content: LFA – glass mixtures: Rietveld method.

of the mineral content is important to estimate the reactivity of the reactive components of the LFA in contact with aqueous solutions. The reactivity is different in varying milieus. In addition to the quantitative mineral content, the size and particle size distribution as well as the formation, adhesion and surface of the particles are important for the reactivity [43]. To assess the material properties, it is therefore necessary to differentiate between reactive and inert substances [11]. However, this classification is associated with greater difficulties, especially with regard to the amorphous components. In numerous older publications „active glasses" were distinguished from „inert glasses" [14, 25, 26, 29, 73]. The terms were introduced by [25] for HCl-soluble and HCl-insoluble glass types after [79] concluded that the hydraulically active, amorphous components consist of „basic, alumina-rich slag glasses", whereas „silica-rich slags" are inactive and only serve as fillers. According to Diamond (1983) [108], the glass phase can also be characterized by the CaO content and a distinction can be made between Ca-rich (>25% Ca weight-%), CaO-poor (<25% Ca weight-%) and CaO-free glasses. The higher the CaO content in the glass phase, the higher the reactivity.

The „inert glasses" are mainly composed of SiO2 and Al2O3, and to a lesser extent (< 10%) also of CaO, MgO, Fe2O3, TiO2, SO3; K2O and Na2O. Their density is on average greater than 2.5 g/cm3, the refraction of light is on average less than 1.55. In contrast, „active glasses" consist mainly of SiO_2, CaO, Al_2O_3, in addition to Fe_2O_3 and MgO (Table 5.10). Average densities of more than 2.65 g/cm^3 and refractions of more than 1.55 g/cm^3 were determined [11, 14, 25, 26, 29, 39, 52].

Table 5.10: Chemical composition of „active glasses" and „inert glasses" of various power stations [7].

Kraftwerk	Active glass					Inert glass			
	SiO_2	Al_2O_3	Fe_2O_3	CaO	MgO	SiO_2	Al_2O_3	Fe_2O_3	MgO
Espenhain	24,4	23,3	4,6	40,0	7,7	82,7	16,6	0,4	-.-
Vockerode	25,1	23,0	4,1	43,5	4,4	65,8	27,2	4,4	2,4
Borna	15,9	19,9	3,9	53,8	6,6	83,9	2,2	14,4	-.-
Böhlen	12,4	18,0	5,6	55,0	8,9	93,9	6,1	-.-	-.-
Lübbenau	27,8	28,1	1,9	37,8	4,2	87,9	5,8	5,3	-.-
Schwarze Pumpe	31,4	24,3	16,3	20,3	7,5	84,1	10,4	5,3	-.-
Hagenwerder	37,4	31,2	6,3	20,4	4,6	63,6	33,4	3,0	-.-
Hirschfelde	20,8	71,7	0,8	2,7	4,0	59,2	40,8	-.-	-.-

According to the chemical composition of the „active glasses", mineralogically both C_3S or C_3A related phases as well as the common compound gehlenite are possible. The probability of formation is determined by the ratio CaO to SiO_2 + Al_2O_3, whereby high lime excesses favor the formation of C_3S and C_3A. These components are detectable in lime-rich as well as in lime- and sulfate-rich LFAs. Silica-rich ashes have a lower proportion of „active" glasses due to lower lime contents. The „active" glasses have almost no significance in aluminosilicate LFA, since their chemical composition largely corresponds to that of Gehlenite [52].

Recently, investigations were carried out on LFA from lignite from the Central German, Rhenish and Lusatian coalfields [31]. „Inert glass" plays only a subordinate role, as it occurs only in small quantities in the HCl dissolving residue [11]. Pöhl (1994) [31] could show that the HCl-insoluble consists mainly of quartz and metakaolinitic particles or fritted clay minerals. She concluded that the „inert glass" is more closely related to metakaolin. The author redefined the HCl-soluble components. She distinguished between silicate-aluminate particles with an SiO_2 content > 10% and ferritic particles with a relatively high iron content and an SiO_2 content < 10%. Pöhl (1994) [31] regards both types of particles as agglomerates of glass matrix and crypto- or micro-crystalline phases (quartz, CaO, calcium ferrites, iron oxides), which are hydrolyzable in a basic aqueous environment.

Accordingly, the decisive difference between „active glasses" and „inert glasses" is that the „active glasses" react largely hydraulically, whereas the „inert glasses" are by no means inactive, but have a pozzolanic character. They do not react independently, but only in the presence of substances which release calcium hydroxide in the mixing water [52]. This assessment coincides with the information provided by [26], who established that the HCl dissolving residue in the ternary system SiO_2 – (CaO + MgO) – (Al_2O_3 + Fe_2O_3) is located in the field of the pozzolana or in its immediate vicinity, which in the opinion of the authors contradicts the view that the hydrochloric acid insoluble is to be regarded as a pure inert component.

According to current knowledge, all LFA glass particles exhibit hydrolysis in basic aqueous solution and participate in mineral regeneration reactions. Therefore, it has proven to be useful to calculate the average glass composition, even if the standard deviations are comparatively high. The results show that the differences between the average LFA glass composition of the different power plants are relatively small. This becomes particularly clear when the oxides Al_2O_3 and Fe_2O_3, which have the same effect in hydration reactions, are combined (Table 5.11).

Table 5.11: Average glass composition of BFA from various power plants [in Ma%] [11].

BFA	SiO_2	Al_2O_3	Fe_2O_3	CaO	SO_3
Jänschwalde	33 ± 12	14 ± 6	15 ± 8	32 ± 12	4 ± 2
Boxberg	25 ± 10	12 ± 6	26 ± 10	29 ± 7	6 ± 2
Thierbach	38 ± 14	21 ± 10	2 ± 3	32 ± 13	3 ± 1
Neurath	36 ± 14	11 ± 6	12 ± 7	32 ± 12	6 ± 2

References

[1] Ioakimidis C, Koukouzas N, Chatzimichali A, Casimiro S, and Itskos G. Assessment for Carbon Capture and Storage Opportunities: Greek Case Study. In: *Computer Aided Chemical Engineering*. Ed. by Pistikopoulos EN, Georgiadis MC, and Kokossis AC. Vol. 29. Elsevier, 2011,1939–43.

[2] Pawelec S and Bielowicz B. Petrographic Composition of Lignite from the Lake Somerville Spillway (East-central Texas). IOP Conference Series: Earth and Environmental Science 2017,95,022028.

[3] Papanicolaou C, Galetakis M, and Foscolos AE. Quality Characteristics of Greek Brown Coals and Their Relation to the Applied Exploitation and Utilization Methods. Energy & Fuels 2005,19,230–9.

[4] Lang R. Die Entstehung von Braunkohle und Kaolin im Tertiär Mitteldeutschlands Ein geologisch-bodenkundliches Problem. Jahrbuch des Halleschen Verbandes für die Erforschung der Mitteldeutschen Bodenschätze und Ihrer Verwertung 1920,2,65–92.

[5] Frank S, Eilert J, Acar T, Cremer N, Wohnlich S, and Wisotzky F. Elutionsverhalten von Braunkohleaschen des Rheinischen Braunkohlereviers. Grundwasser 2019,24,13–26.

[6] Schiffer HW. Der deutsche Braunkohlenbergbau im Jahre 1992; German brown coal mining in 1992. Glückauf (Essen) 1993,129,303–7.

[7] Kringel R. Untersuchungen zur Verminderung von Auswirkungen der Pyritoxidation in Abraumsedimenten des Rheinischen Braunkohlereviers auf die Chemie des Grundwassers. PhD thesis. Ruhr-Universität, Bochum, 1996.

[8] Drebenstedt C and Schollenbach G. Verwertung von Kraftwerksrückständen im Lausitzer Braunkohlenrevier. In: *Handbuch der Verwendung von Braunkohlenfilteraschen in Deutschland*. RWE AG Essen, Zentralbereich F&E, 1995,555–62.

[9] Cremer S Mobilisierung von Schwermetallen in Porenwässern von Deponien und belasteten Böden – Ermittlung relevanter Parame- ter des Hydrogeochemischen Milieus für die Entwicklung eines aussagekräftigen Elutionsverfahrens. PhD thesis. Ruhr-Universität, Bochum, 1992.

[10] Powertech V. Produktion und Verwendung von Kraftwerksnebenprodukten aus Kohlekraftwerken in Deutschland im Jahr 2016.

[11] Schreiter P, Bambauer HU, and Werner M. Erfahrungen mit BFA – Bindebaustoffen. In: *Handbuch der Verwendung von Braunkohlenfilteraschen in Deutschland*. RWE AG Essen, Zentralbereich F&E, 1995.

[12] Piekos S and Lemke D. Physikalische Eigenschaften und chemische Zusammensetzung von Braunkohlenflugaschen. In: *Handbuch der Verwertung von Braunkohlenfilteraschen in Deutschland*. RWE AG Essen, 1995.

[13] Meisel A Röntgenographische Untersuchungen von Braunkohlenfilteraschen. PhD thesis. Universität Leipzig, 1956.

[14] Zschach S Mineralogische Eigenschaften von Braunkohlenfilteraschen. Chemie der Erde 1978,37,330–56.

[15] Münch U Zu Konstitution, Elutionsverhalten und Kathodolumineszenz von Braunkohle-naschen. PhD thesis. Dissertation Technische Universität Freiberg, 1995.

[16] Masto RE, Ansari MA, J. G, Selvi VA, and Ram LC. Co-application of biochar and lignite fly ash on soil nutrients and biological parameters at different crop growth stages of Zea mays. Ecological Engineering 2013,58,314–22.

[17] Page AL, Elseewi AA, and Straughan IR. Physical and chemical properties of fly ash from coal-fired power plants with reference to environmental impacts. In: *Residue Reviews*. Ed. by Gunther FA and Gunther JD. Springer New York, 1979,83–120.

[18] Eberhard J, Burghardt D, Martin M, et al. From Waste to Valuable Substance: Utilization of Schwertmannite and Lignite Filter Ash for Removal of Arsenic and Uranium from Mine Drainage. In: *Mine Water – Managing the Challenges*. Ed. by Rüde RT, Freund A, and Wolkersdorfer C. 2011,359–63.

[19] Barthel M, Rübner K, Kühne HC, Rogge A, and Dehn F. From waste materials to products for use in the cement industry. Advances in Cement Research 2016,28,458–68.

[20] Bambauer HU, Gebhard G, Holzapfel T, Krause C, and Willner G. Schadstoff-Immobilisierung in Stabilisaten aus Braunkohlenaschen und REA-Produkten – Teil 1: Mineralreaktionen und Gefügeentwicklung; Chlorid-Fixierung. Fortschritte der Mineralogie 1988,66,253–79.

[21] Koukouzas N, Hämäläinen J, Papanikolaou D, Tourunen A, and Jäntti T. Mineralogical and elemental composition of fly ash from pilot scale fluidised bed combustion of lignite, bituminous coal, wood chips and their blends. Fuel 2007,86,2186–93.

[22] Hart BR, Powell MA, Fyfe WS, and Ratanasthien B. Geochemistry and Mineralogy of Fly-Ash from the Mae Moh Lignite Deposit, Thailand. Energy Sources 1995,17,23–40.

[23] Wu P, Li J, Zhuang X, et al. Mineralogical and Environmental Geochemistry of Coal Combustion Products from Shenhuo and Yihua Power Plants in Xinjiang Autonomous Region, Northwest China. Minerals 2019,9.

[24] Henning O and Danowski O. Untersuchungen an einigen Braunkohlenfilteraschen und deren Hydratationsprodukten. Wiss. Zeitschr. der Hochschule für Architektur und Bauwesen Weimar 1966,13,317–20.

[25] Schreiter P. Zum Phasenaufbau von Braunkohlenfilteraschen. Silikattechnik 1968,19,358–36.

[26] Fungk E, Ilgner R, and Lang E. Braunkohlenfilteraschen der DDR als Zumahlstoffe in der Zementindustrie. Silikattechnik 1969,20,302–7.
[27] Ostrowski C. Einfluss des Kalziumhydroxids und des Gipses auf die puzzolanischen Eigenschaften von Flugaschen (Teil 1). Baustoffindustrie 1976,A6,13–7.
[28] Schreiter P and Werner M. Gefügekatalog. Silikattechnik 1984,9.
[29] Werner M, Adam K, and Schreiter P. Die chemische Zusammensetzung der Glasphase und quantitative Phasenbestände von Flugaschen braunkohlebefeuerter Großkessel. Silikattechnik 1988,39,263–6.
[30] Auferheide C. Untersuchungen zum Abbindeverhalten von Flugaschen als Mittel zur Darstellung des Mischungszustandes der Asche beim Abzug eines Silos des Kraftwerks Jänschwalde. PhD thesis. Diplomarbeit Universität Münster, FB Chemie, 1994.
[31] Pöhl K. Zur Konstitution und Hydratation deutscher Braunkohlenfilteraschen. PhD thesis. Universität Leipzig, 1994.
[32] Seidel S. Korrosion und Stoffaustausch an Stabilisaten aus freikalkreicher Braunkohlenfilterasche aus dem Mitteldeutschen Revier. PhD thesis. Dissertation Universität Münster, FB Chemie, 1996.
[33] Mallmann R. Entwicklung hydraulischer Bindemittel mit rheinischen Braunkohlenfilteraschen. Dissertation Universität Siegen. PhD thesis. Dissertation Universität Siegen, Fachbereich 8, 2002.
[34] V. VS and G. VC. Methods for Characterisation of Composition of Fly Ashes from Coal-Fired Power Stations: A Critical Review. Energy & Fuels 2005,19,1084–98.
[35] Feuerborn HJ. Mittel- und ostdeutsche Braunkohlenflugaschen in hydraulischen Bindemitteln. PhD thesis. Technische Hochschule Aachen – Verlagshaus Mainz GmbH, 2007.
[36] VERKEHRSWESEN FFFSU. Merkblatt über die Verwendung von Kraftwerksnebenpro- dukten im Straßenbau. Forschungsgesellschaft für Straßen und Verkehrswesen, M KNP 624. Tech. rep. FGSV – Forschungsgesellschaft für Straßen und Verkehrswesen, 2009.
[37] Brett B, Schrader D, Räuchle K, Heide G, and Bertau M. Wertstoffgewinnung aus Kraftwerksaschen Teil I: Charakterisierung von Braunkohlenkraftwerksaschen zur Gewinnung strategischer Metalle. Chemie Ingenieur Technik 2015,87,1383–91.
[38] Blankenburg HJ and Sommer A. Freisetzung, Migration und Fixierung der anorganischen Komponenten der Braunkohle in Staubfeuerungsanlagen. Silikattechnik 0036,1.
[39] Heinrich-Bisping A. Mineralogisch-chemische Untersuchungen von Braunkohlenfilteraschen aus dem rheinischen Revier (KW Frimmersdorf). PhD thesis. Universität Münster, FB Chemie, 1995.
[40] Feuerborn HJ, Müller B, and Walter E. Use of Calcareous Fly Ash in Germany. In: *Proceedings of the EUROCOALASH 2012 Conference*, 2012.
[41] Pöllmann H. Immobilisierung von Schadstoffen durch Speichermineralbildung. Berichte aus der Geowissenschaft. Shaker Verlag Aachen, 2007.
[42] Ranneberg M. Quantifizierung und Optimierung von Deponiestabilisaten aus calciumreichen Braunkohlenfilteraschen unter Verwendung statistischer Versuchsplanung. Thesis. 2018.
[43] Bambauer HU. Mineralogische Schadstoffimmobilisierung in Deponaten Beispiel: Rückstände aus Braunkohlekraftwerken. BWK/TÜ/Umwelt-Special 1992,35–9.
[44] Schreiter P, Bambauer HU, Werner M, and Pöhl K. Puzzolanische und hydraulische Eigenschaften von Braunkohlenflugasche. In: *Handbuch der Verwendung von Braunkohlenfilteraschen in Deutschland*. RWE AG Essen, Zentralbereich F&E, 1995.
[45] Werner M. Hydratationsverhalten kalkreicher Kraftwerstrockenaschen. PhD thesis. Universität Leipzig, 1989.

[46]   Wischnewski C. Untersuchungen zu den Bindungsormen der anorganischen Komponenten in der Braunkohle. PhD thesis. Universität Leipzig, 1987.

[47]   Friedrich W. Chemische Schnellanalyse von Verbrennungsrückständen (Teil 1). Inst. f. Energetik 1965,71,1–10.

[48]   Friedrich W. Chemische Schnellanalyse von Verbrennungsrückständen (Teil 2). Inst. f. Energetik 1965,72,69–171.

[49]   Friedrich W. Chemische Schnellanalyse von Verbrennungsrückständen (Teil 3). Inst. f. Energetik 1965,71,137–71.

[50]   Langner K and Friedrich W. Physikalisch-chemische Schnellanalyse von Verbrennungsrückständen (Teil 1). Inst. f. Energetik Leipzig 1965,75,300–21.

[51]   Friedrich W. Aschenkatalog I Kontrollsystem für Aschen im Industriezweig Chemische Industrie. Inst. f. Energetik Leipzig 1967.

[52]   Ilgner R. BFA als Zumahlstoff für Zement; Verwertung von BFA in der Zementindustrie der ehemaligen DDR. In: *Handbuch der Verwertung von Braunkohlenfilteraschen in Deutschland*. RWE AG Essen, Zentralbereich F&E, 1995.

[53]   Roy WR, Thiery RG, Schuller RM, and Suloway JJ. Coal fly ash: a review of the literature and proposed classification system with emphasis on environmental impacts. Environmental Geology Notes 1981,96.

[54]   DIN EN 197-1:2011-11.Zement – Teil 1: Zusammensetzung, Anforderungen und Konformitätskriterien von Normalzement; Deutsche Fassung EN 197-1:2011. Beuth-Verlag, 2011.

[55]   DIN EN 450-1:2012-10.Flugasche für Beton – Teil 1: Definition, Anforderungen und Konformitätskriterien; Deutsche Fassung EN 450-1:2012. Beuth-Verlag, 2012.

[56]   DIN EN 196-2:2013-10. Prüfverfahren für Zement – Teil 2: Chemische Analyse von Zement; Deutsche Fassung EN 196-2:2013. Beuth Verlag, 2013.

[57]   DIN EN 196-5:2011-06Prüfverfahren für Zement – Teil 5: Prüfung der Puzzolanität von Puzzolanzementen; Deutsche Fassung EN 196-5:2011. Beuth-Verlag, 2011.

[58]   DIN ISO 11465:1996-12Bodenbeschaffenheit – Bestimmung des Trockenrückstandes und des Wassergehalts auf Grundlage der Masse – Gravimetrisches Verfahren (ISO 11465:1993). Beuth-Verlag, 2012.

[59]   DIN 51729-10: 2011-04.Prüfung fester Brennstoffe – Bestimmung der chemischen Zusammensetzung von Brennstoffasche – Teil 10: Röntgenfluoreszenz-Analyse (RFA). Beuth-Verlag, 2011.

[60]   DIN EN 12879:2001-02Charakterisierung von Schlämmen – Bestimmung des Glühverlustes der Trockenmasse; Deutsche Fassung EN 12879:2000. Beuth-Verlag.

[61]   Franke B. Bestimmung von Calciumoxyd und Calciumhydroxyd neben wasserfreiem und wasserhaltigem Calciumsilikat. Zeitschrift für anorganische und allgemeine Chemie 1941,247,180–4.

[62]   DIN EN 450-2:2005-05.Flugasche für Beton – Teil 2: Konformitätsbewertung; Deutsche Fassung EN 450-2:2005. Beuth-Verlag, 2005.

[63]   DIN EN 196-1:2016-11.Prüfverfahren für Zement – Teil 1: Bestimmung der Festigkeit; Deutsche Fassung EN 196-1:2016. Beuth-Verlag, 2016.

[64]   DIN EN 196-3:2017-03.Prüfverfahren für Zement – Teil 3: Bestimmung der Erstarrungszeiten und der Raumbeständigkeit; Deutsche Fassung EN 196-3:2016. Beuth-Verlag. 2017.

[65]   DIN EN 196-6:2019-03.Prüfverfahren für Zement – Teil 6: Bestimmung der Mahlfeinheit; Deutsche Fassung EN 196-6:2018. Beuth-Verlag, 2019.

[66]   DIN EN 451-1:2017-08.Prüfverfahren für Flugasche – Teil 1: Bestimmung des freien Calciumoxidgehalts; Deutsche Fassung EN 451-1:2017. Beuth-Verlag, 2017.

[67] DIN EN 451-2:2017-08.Prüfverfahren für Flugasche – Teil 2: Bestimmung der Feinheit durch Nasssieben; Deutsche Fassung EN 451-2:2017. Beuth-Verlag, 2017.

[68] ASTM D5759-12, Standard Guide for Characterization of Coal Fly Ash and Clean Coal Combustion Fly Ash for Potential Uses. West Conshohocken, PA: ASTM International, 2012.

[69] ASTM C618-19, Standard Specification for Coal Fly Ash and Raw or Calcined Natural Pozzolan for Use in Concrete. West Conshohocken, PA: ASTM International, 2019.

[70] ASTM C311/C311M-18, Standard Test Methods for Sampling and Testing Fly Ash or Natural Pozzolans for Use in Portland-Cement Concrete. West Conshohocken, PA: ASTM International, 2018.

[71] Siddique R and Khan MI. Fly Ash. In: *Supplementary Cementing Materials*. Berlin, Heidelberg: Springer Berlin Heidelberg, 2011,1–66.

[72] Mörtel A. Einsatzmöglichkeiten für Flugaschen. Berichte a. Keram. Ges. 1983,4,136–43.

[73] Keyn J, Schreiter P, Sansoni G, and Werner M. Zum Phasen- und Gefügeaufbau von Braunkohlenfilteraschen im Feinstkornbereich. Silikattechnik 1985,36,341–3.

[74] Hemmings RT and Berry EE. On the Glass in Coal Fly Ashes: Recent Advances. MRS Proceedings 1987,113,3.

[75] Bayat O. Characterisation of Turkish fly ashes. Fuel 1998,77,1059–66.

[76] Raask E. Slag-Coal Interface Phenomena. Journal of Engineering for Power 1966,88,40–4.

[77] Walter G and Gallenkemper B. Verwertung von Steinkohlen- und Braunkohlenaschen. In: Handbuch des Umweltschutzes und der Umweltschutztechnik: Band 2: Produktions- und produktintegrierter Umweltschutz. Ed. by Brauer H. Berlin, Heidelberg: Springer Berlin Heidelberg, 1996,1037–58.

[78] Bhatt A, Priyadarshini S, Mohanakrishnan AA, Abri A, Sattler M, and Techapaphawit S. Physical, chemical, and geotechnical properties of coal fly ash: A global review. Case Studies in Construction Materials 2019,11,e00263.

[79] Ottemann J. Über die Mineralbestandteile von Braunkohlenaschen und ihre Bedeutung für die Beurteilung von Aschenbindern. Mitteilungen aus den Laboratorien des Geologischen Dienstes Berlin. Akademie-Verlag, 1951.

[80] Ilic M, Cheeseman C, Sollars C, and Knight J. Mineralogy and microstructure of sintered lignite coal fly ash. Fuel 2003,82,331–6.

[81] Sakorafa V, Michailidis K, and Burragato F. Mineralogy, geochemistry and physical properties of fly ash from the Megalopolis lignite fields, Peloponnese, Southern Greece. Fuel 1996, 75,419–23.

[82] Kostakis G. Characterization of the fly ashes from the lignite burning power plants of northern Greece based on their quantitative mineralogical composition. Journal of Hazardous Materials 2009,166,972–7.

[83] McCarthy GJ, Johansen DM, Thedchanamoorthy A, Steinwand SJ, and Swanson KD. Characterization of North American Lignite Fly Ashes II. XRD Mineralogy. MRS Pro- ceedings 1987,113,99.

[84] Zhao Y, Zhang J, Tian C, Li H, Shao X, and Zheng C. Mineralogy and Chemical Composition of High-Calcium Fly Ashes and Density Fractions from a Coal-Fired Power Plant in China. Energy & Fuels 2010,24,834–43.

[85] Gutteridge WA. On the dissolution of the interstitial phases in Portland cement. Cement and Concrete Research 1979,9,319–24.

[86] Struble L. The effect of water on maleic acid and salicylic acid extractions. Cement and Concrete Research 1985x,9,319–24.

[87] Horn A and Pöllmann H. Quantification and verification of mineral phase content of iron – rich brown coal fly ash by Rietveld-method. 2014.

[88] Takashima S. Systematic Dissolution of Calcium Silicate in Commercial Portland Cement by Organic Acid Solution. In: *12th General Meeting, Tokyo, Japan*, 1958,12–13.

[89] Fernández-Jimenez A, Torre AG de la, Palomo A, López-Olmo G, Alonso MM, and Aranda MAG. Quantitative determination of phases in the alkali activation of fly ash. Part I. Potential ash reactivity. Fuel 2006,85,625–34.

[90] Chung FH. Quantitative interpretation of X-ray diffraction patterns of mixtures. I. Matrix-flushing method for quantitative multicomponent analysis. Journal of Applied Crystallography 1974,7,519–25.

[91] Chung FH. Quantitative interpretation of X-ray diffraction patterns of mixtures. II. Adia- batic principle of X-ray diffraction analysis of mixtures. Journal of Applied Crystallography 1974, 7,526–31.

[92] Font O, Moreno N, Querol X, et al. X-ray powder diffraction -based method for the determination of the glass content and mineralogy of coal (co)-combustion fly ashes. Fuel 2010,89,2971–6.

[93] Klug H and Alexander LE. X-ray diffraction procedures for Polycrystalline and Amorphous Materials. 2nd ed. New York: John Wiley and Sons, 1974.

[94] Ibáñez J, Font O, Moreno N, Elvira JJ, Alvarez S, and Querol X. Quantitative Rietveld analysis of the crystalline and amorphous phases in coal fly ashes. Fuel 2013,105, 314–7.

[95] Arjuan P, Silbee MR, and Roy DM. Quantitative determination of the crystalline and amorphous phases in low calcium fly ashes". In: *Proceeding of the 10th international congress of the chemistry of cement.* Vol. 3. 1997,2–6.

[96] Ward CR and French D. Determination of glass content and estimation of glass composition in fly ash using quantitative X-ray diffractometry. Fuel 2006,85,2268–77.

[97] McCarthy GJ and Johansen DM. X-Ray Powder Diffraction Study of NBS Fly Ash Standard Reference Materials. Powder Diffraction 1988,3,156–61.

[98] Winburn RS, Grier DG, McCarthy GJ, and Peterson RB. Rietveld quantitative X-ray diffraction analysis of NIST fly ash standard reference materials. Powder Diffraction 2000,15,163–72.

[99] Westphal T, Füllmann T, and Pöllmann H. Rietveld quantification of amorphous portions with an internal standard – Mathematical consequences of the experimental approach. Powder Diffraction 2009,24,239–43.

[100] Westphal T, Walenta G, Fullmann T, et al. Characterization of cementitious materials – Part III. Int Cem Rev July:47–51 (2002).

[101] Westphal T. Quantitative Rietveld-Analyse von amorphen Materialien am Beispiel von Hochofenschlacken und Flugaschen. Thesis. 2007.

[102] Scarlett NVY and Madsen IC. Quantification of phases with partial or no known crystal structures. Powder Diffraction 2006,21,278–84.

[103] Stetsko YP, Shanahan N, Deford H, and Zayed A. Quantification of supplementary cementitious content in blended Portland cement using an iterative Rietveld-PONKCS technique. Journal of Applied Crystallography 2017,50,498–507.

[104] Snellings R, Salze A, and Scrivener KL. Use of X-ray diffraction to quantify amorphous supplementary cementitious materials in anhydrous and hydrated blended cements. Cement and Concrete Research 2014,64,89–98.

[105] Singh GVPB and Kolluru VLS. Quantitative XRD Analysis of Binary Blends of Siliceous Fly Ash and Hydrated Cement. Journal of Materials in Civil Engineering 2016,28,04016042.

[106] Li X, Snellings R, and Scrivener KL. Quantification of amorphous siliceous fly ash in hydrated blended cement pastes by X-ray powder diffraction. Journal of Applied Crystallography 2019,52,1358–70.

[107] Dittrich S, Neubauer J, and Goetz-Neunhoeffer F. The influence of fly ash on the hydration of OPC within the first 44h – A quantitative in situ XRD and heat flow calorimetry study. Cement and Concrete Research 2014,56,129–38.

[108] Diamond S. On the glass present in low-calcium and in high-calcium fly ashes. Cement and Concrete Research 1983,13,459–64.

[109] Openshaw SC. Utilization of coal fly ash. Master's thesis. Report. 1992.

[110] Chou MIM. Fly Ash. In: *Encyclopedia of Sustainability Science and Technology*. Ed. by Meyers RA. New York: Springer New York, 2012,3820–43.

Andreas Ehrenberg

Chapter 6
Iron and steel slags: from wastes to by-products of high technical, economical and ecological advantages

Abstract: Worldwide about 620 mio. tons of different slags are produced yearly as by-products of iron and steel production. Their utilisation for different purposes, but mainly in the construction sector, is a significant contribution to resource conservation and CO_2 emissions reduction. This chapter covers the historical background and numerical data as well as an introduction of the different today's applications of iron and steel slags

Keywords: Blast furnace slag, Basic Oxygen Furnace (BOF) slag, Electric Arc Furnace (EAF) slag, slag granulation, slag utilization, slag cement, concrete, aggregate, road and waterway contruction, fertilizer, circular economy, CO_2 emissions, resource conservation, sustainability

1 Introduction

Worldwide the steel production of about 1.9 bn. tons yearly (2019) is inevitably linked to the production of different types of iron and steel slags which amount to about 620 mio. tons yearly (2019: 410 mio. t blast furnace slag, 210 mio. t steel slags) [1, 2]. This extensive amount of material offers a lot of technical, economical and ecological possibilities if the different specific material properties are evaluated and used in a proper way. For example in Germany the utilisation rate for iron and steel slags is about 95% since many years [3]. Most, but not all of the applications are related to building materials.

2 The different kind of iron and steel slags

The standard manual on all properties of iron and steel slags is the "Slag Atlas" [4]. It describes their composition, structure, phase diagrams, physical and thermal properties etc.

Andreas Ehrenberg, Department Manager Building Materials, FEhS- Institut für Baustoff-Forschung e.V., Bliersheimer Strasse 62, 47229 Duisburg, Germany, a.ehrenberg@fehs.de

https://doi.org/10.1515/9783110674941-006

2.1 Steel production routes and metallurgical tasks of slags

Today worldwide two steel production routes are the most relevant, the Blast Furnace + BOF (Basic Oxygen Furnace) route and the EAF (Electric Arc Furnace) route [5]. In some countries also Direct Reduced Iron (DRI) is re-melted in electric arc furnaces.

The two-stage Blast Furnace + BOF route is mainly applied on natural iron ores. In combination with coke or other reduction agents and additions like limestone to achieve lower melting points the iron ores are reduced and two immiscible melts are created: pig iron and blast furnace slag. The pig iron includes carbon and sulphur from the coke as well as other impurities like silicon, manganese or phosphorus added with gangue minerals. Thus, after tapping hot metal and slag and separating them by using the very different densities of both melts the pig iron is refined by oxygen injection in the BOF. As a result from additions like limestone and the refractory materials the BOF slag is created. The EAF route is mainly based on scrap as raw material input. The EAF slag again results from additives like limestone and sand. To further enhance processes in the BOF or EAF route some metallurgical challenges have been overcome in the secondary metallurgy, e.g. de-sulphurisation, de-oxydising or alloying. This was necessary to meet the very specific steel quality requirements of the products. In these metallurgical steps further secondary metallurgical slags are generated.

Iron and steel production slags are metallurgical tools and despite their often-times negative image they are an integral part of any steel production. The metallurgical tasks of blast furnace slags can be summarized as follows:
- decreasing the melting point by achieving a eutectic composition
- absorbing the non-metallic gangue minerals of ore, mineral additives (e.g. limestone) and the remaining coke/coal ashes
- releasing sulphur from the pig iron (coming from the reduction agents coke and coal)
- releasing Alkalis from the furnace in order to prevent the building of occur in the furnace
- achieving low viscosity to easily separate pig iron and molten slag and to be able to refine the metal effectively
- protecting the pig iron from re-oxidation by the hot blast

In earlier times the optical appearance of the hardened slag (see Figure 6.18) informed the blast furnace operator on the formerly unknown furnace processes itself. In 1897 it was written [6]: "If the slag before the tuyère was yellow and dark, then the furnace was too cold, if it was intensively white, then the furnace was too hot; a clear homogenous blue colour indicated the correct furnace process".

The metallurgical tasks of steel slags can be summarized as follows:
- absorbing undesirable elements like silicon, phosphorus and manganese
- protecting the liquid steel from hydrogen and nitrogen from the converter or furnace atmosphere

–   protecting the liquid steel from re-oxidation by the injected oxygen
–   avoiding a rapid heating of the melt
–   forming a splash guard during the oxygen injection process in the BOF

It is important to stress that "slag" is a metallurgical wording. In general materials like fly ash from coal burning or municipal waste incineration ash did not run through a melting process. In particular the latter one is often very inhomogeneous.

2.2 Slag handling processes

2.2.1 State of the art slag handling processes

The liquid slags are cooled down in different ways which strongly influence the technical and environmental properties of the hardened slags. After tapping the liquid slag either it runs into a slag ladle which transports it to a slag pit (Figure 6.1) or

Figure 6.1: BOF slag tapping into a slag pit.

it runs directly into slag pits nearby the blast furnace or in some EAF plants it runs in the cellar below the furnace. The liquid slag is solidifying into a compact and crystalline material. If liquid blast furnace slag is quenched with water ("granulation") it forms a granular and more or less glassy, amorphous material (Figure 6.2). The wet granulated blast furnace slag is dewatered in sieve drums, sieve wheels, silos, basins or other equipment and it is stored in interim storages having a residual moisture of about 10 wt.-% (Figure 6.3).

Figures 6.4–6.8 show typical examples of air-cooled and granulated iron and steel slags.

Figure 6.2: Water granulation of blast furnace slag [7].

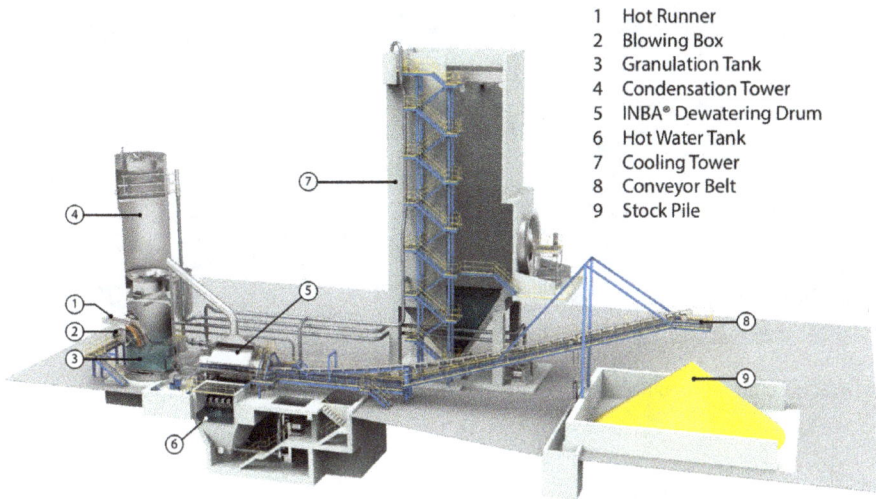

1 Hot Runner
2 Blowing Box
3 Granulation Tank
4 Condensation Tower
5 INBA® Dewatering Drum
6 Hot Water Tank
7 Cooling Tower
8 Conveyor Belt
9 Stock Pile

Figure 6.3: System concept of a modern water granulation plant for blast furnace slag [8].

Figure 6.4: Air-cooled blast furnace slag.

2.2.2 Other slag handling processes

In the past several other slag handling systems were developed, in particular for blast furnace slags. One challenge was avoiding water granulation in order to achieve a glassy, but dry material opposite to a wet slag which had to be dried before using it as a cement constituent (see 5.2.1). Some of the attempts were abandoned before entering the test phase. In several steel plants a rotating drum was used to "pelletize" the liquid blast furnace slag [9, 10]. However, without using a certain volume of water the glass content might be insufficient and crystalline blast furnace slag phases

Figure 6.5: Examples for different European water granulated blast furnace slags.

Figure 6.6: Air-cooled BOF slag.

Figure 6.7: Air-cooled EAF slag.

Figure 6.8: Secondary metallurgical slag (with C_2S disintegration).

like Merwinite, Gehlenite or Åkermanite (see 2.3.1) do not have hydraulic properties. Thus, a utilization as a cement constituent (see 5.2.1.1) is hindered. In a German steel plant a rotating drum was used [11]. Both technologies were discontinued. One main reason is that big blast furnaces with average pig iron capacities of 5,000 tons per day and a maximum of up to 12,000 tons per day produce 1,500–3,500 tons of slag daily. Thus, the slag flow during tapping might be up to 10 tons per minute requiring stable and sufficient slag handling facilities. The technologies mentioned above would need several installations to be operated in parallel resulting in a high area needed for slag handling. A technology being used e.g. in China is the "HK process" using a horizontal granulation wheel together with water [12].

For steel slags space-saving alternative cooling processes were also developed instead of slag pits, for example by Baosteel (China) [13]. The Baosteel's Slag Short Flow (BSSF) process is established in some steel plants worldwide. Liquid slag and water are introduced into a drum with steel balls. However, due to the fine grain size of the hardened steel slag being around 5 mm, a technical utilization (aggregate) is problematic. Therefore only the disposing process on dumps is eased. In China the "HK process" is used also for steel slags [12]. Another method is air-granulation of steels slags resulting in spherical particles [14]. The disadvantages of this system are that a utilization of the crystalline steel slag beads is difficult and that the chromium-III content of the slag might be oxidized to toxic chromium-VI. The advantage is that the free lime content of the BOF slag being problematic regarding its volume stability is minimized. Figure 6.9 shows the BOF slags cooled by the discussed alternative processes.

Today dry cooling processes for blast furnace slag, but also for steel slags are under development again [15–20]. However, the main interest is to use the thermal energy content of the liquid slag which is about 1.6–1.8 GJ/t. To recover this heat would be a contribution in the reduction of the CO_2 footprint of steel production. Water granulation results in warm water and steam without any technological

Figure 6.9: BOF slags after BSSF process (left) and after air-granulation (right).

benefit and in addition the wet slag has to be dried. After a dry cooling process as well the dry, but still glassy material as the hot exhaust air can be used e.g. in heat exchangers. The main challenges are safe handling of the liquid slag, production of a glassy material (blast furnace slag) and achieving a high temperature level to be used. Several technologies were developed. One of the interesting approaches is the "rotating cup" system [21] being developed in principle in the 1980/1990ies in UK [22, 23]. In that process the liquid slag flows vertically on a rotating disc or cup. At the edge of the disc the film-forming slag is divided into ligaments and then into spherical particles. Based on lab-scale (Figure 6.10) and semi-technical-scale tests [24] two research projects done in 2013–2019 in a pilot facility at the Voestalpine steel plant in Linz (Austria) demonstrated that the production of a glassy blast furnace slag solely by air-cooling is possible [25]. Unfortunately, besides several unsolved technical problems it was found that the reactivity of the air-cooled glassy slag in cementitious systems is lower compared to the same slag being water granulated [26, 27]. The reason is that the cooling velocity is lower resulting in a lower real glass transformation temperature $T_{fictive}$ and thus a lower enthalpy content of the slag glass. Figure 6.11 shows the glassy spherical particles produced in the rotating cup facility in Linz.

A very similar approach was investigated in Australia since 2002 [28, 29] and in China [30]. Also other ideas for dry slag granulation, like casting the liquid slag in a thin layer on a cooling conveyer, casting the slag in thin layers into moving moulds [31], spraying the liquid slag into a dry "powder" or adding big volumes of steel balls in troughs with liquid slag [17, 32] have been tested, some of them in pilot scale. However, no technology has been introduced into practice due to several reasons, e.g. low glass content, large space requirement, high energy demand, limited slag applications.

Figure 6.10: FEhS lab-scale tests for dry blast furnace slag quenching ("rotating disc" method); the slag particles agglomerates at the vertical wall of the un-cooled containment.

Figure 6.11: Blast furnace slag quenched with air ("rotating cup" method).

2.3 Slag properties

In 2010 iron and steel slags were registered under the European REACH (Registration, Evaluation, Authorisation and Restriction of Chemicals) regime [33]. Today this registration is the legal basis for the utilisation of slags as "products". "Wastes" have not to be registered. Table 6.1 shows the different entries. Together with the registration very detailed informations on chemical and mineralogical compositions, environmental behaviour etc. were given to the ECHA, the European Chemicals Agency [34].

Table 6.1: REACH registration numbers for the different types of iron and steel slags.

Family-No.	Slag type		CAS-No.	EINECS-No.
1	GBS	Slag, blast furnace (granulated and ground granulated)	65996-69-2	266-002-0
	ABS	Slag, blast furnace (air cooled)	65996-69-2	266-002-0
2	BOS	Slag, converter	91722-09-7	294-409-3
3a	EAF C	Slag, electric arc furnace (carbon steel production)	91722-10-0	294-410-9
3b	EAF S	Slag, electric arc furnace (stainless/high alloy steel production)	91722-10-0	294-410-9
4	SMS	Slag, steelmaking	65996-71-6	266-004-1

2.3.1 Chemical and mineralogical composition

In general iron and steel slags are described by the phase diagram $CaO - SiO_2 - Al_2O_3 - MgO$ for blast furnace slag and by the phase diagram $FeO_x - CaO - SiO_2 - (MnO_x, MgO)$ for BOF and EAF slags [35].

The chemical composition is mainly confined by the raw materials being used (raw ore, pellets, sinter, limestone, ilmenite etc.). In some cases modifications in basicity (adding limestone) or alumina (adding bauxite) were performed only to optimize the cementitious properties of granulated blast furnace slag (see section 5.2.1.1).

In the early times of iron and steel production the locally available raw ore was used. Thus, the chemical composition was eventually different for blast furnaces at different locations. However, the blast furnace operator aims to produce and adjust a stable composition in order to guarantee a stable iron making process. Today there are only few iron ore suppliers worldwide. Nevertheless, there is still a wide range in the chemical composition of the slags. In Table 6.2 the content of several main and minor constituents of blast furnace slags generated worldwide are listed. The data results from the database of FEhS institute [36]. As mentioned above in most cases the granulated blast furnace slag is more or less glassy (>95 vol.-%). Air-cooled blast

Table 6.2: Chemical composition of blast furnace slags worldwide [36].

CaO	MgO	SiO$_2$	Al$_2$O$_3$	TiO$_2$	S$_{total}$	CaO/ SiO$_2$	(CaO +MgO)/ SiO$_2$	F value*
		wt.-%					–	
29–49	1–17	23–42	5–23	0–4	0.1–2.2	0.7–1.7	0.9–2.1	1.0–2.6

* F = (CaO + 0.5 x S^{2-} + 0.5 x MgO + Al$_2$O$_3$) / (SiO$_2$ + MnO)

furnace slag mainly consists of Merwinite (3 CaO · MgO · 2 SiO$_2$) and Melilite, a solid solution of Åkermanite (2 CaO · MgO · SiO$_2$) and Gehlenite (2 CaO · Al$_2$O$_3$ · SiO$_2$). Dicalcium silicate (2 CaO · SiO$_2$) and Monticellite (CaO · MgO · SiO$_2$) are rarely found. In the first half of the 20th century the average basicity (CaO/SiO$_2$) of blast furnace slags was higher compared to the current practice. Thus, C$_2$S disintegration (α'- or β_H-C$_2$S $\rightarrow$ γ-C$_2$S) was possible ("limy slag"). Another problem could be the hydrolysis of FeS/MnS and the formation of Fe(OH)$_2$. Due to the change in chemical composition over the years (lower basicity, lower S content) today the typical mineralogical composition does not cause any risk regarding the volume stability.

Table 6.3 shows the chemical composition of German BOF slags [36]. The high content of iron oxides results in a very high density of air-cooled steel slags compared to most natural aggregates. The basicity of steel slags is much higher compared to blast furnace slags. Without a modification of the slag chemistry it is not possible to achieve a glassy BOF or EAF slag with common granulation facilities due to their high crystallization velocity. Moreover, due to some metal content in the liquid steel slags it is essential to adjust the slag and water flows precisely to avoid explosions. The free lime content might be a risk for the volume stability being necessary for technical applications because forming Ca(OH)$_2$ increases the slag grain volume. In order to avoid this problem several technologies have been developed, e.g. the injection of oxygen and sand [37], long-time storing, air-granulation, steam curing or the BSSF process mentioned in section 2.2.2. With respect to environmental properties it is important to stress that despite the oxidizing BOF process no chromium-VI occurs due to the thermodynamic preference of iron oxidation.

Table 6.3: Chemical composition of German BOF slags [36].

CaO	CaO$_{free}$	MgO	SiO$_2$	Al$_2$O$_3$	TiO$_2$	P$_2$O$_5$	Cr$_2$O$_3$	Fe$_{total}$	CaO/ SiO$_2$
			wt.-%						–
31–52	0.4–13.0	2–15	8–24	1–7	0.4–1.0	0.7–1.8	0.1–0.6	14–19	1.9–4.8

Table 6.4 shows the chemical composition of German EAF slags [36]. Due to the scrap basis the iron oxide as well as the chromium contents are higher. But no chromium-VI occurs. The basicity is lower compared to BOF slags. The free lime content is negligible. However, MgO might be formed as periclase which can react to $Mg(OH)_2$ being problematic regarding volume stability, too.

Table 6.4: Chemical composition of German EAF slags [36].

CaO	CaO$_{free}$	MgO	SiO$_2$	Al$_2$O$_3$	TiO$_2$	P$_2$O$_5$	Cr$_2$O$_3$	Fe$_{total}$	CaO/ SiO$_2$
				wt.-%					–
20–29	0.0–0.8	4–11	8–26	6–11	0.3–0.9	0.3–0.4	1.5–3.1	18–35	1.0–3.6

The dominant mineral phases of steel slags are dicalcium silicate (larnite), bredigite, calcio- or magnesiowuestite, magnetite, free lime and periclase.

2.3.2 Physical properties

An overview on some physical properties being relevant for iron and steel slags to be used as aggregates in different applications is given in Table 6.5, taken from [38]. In particular for steel slags the higher resistance to fragmentation as the higher apparent density compared to most of natural aggregates has to be stressed.

Table 6.5: Physical properties of iron and steel slags [38].

Slag type	Class	Water suction	Bulk density		Apparent density	Resistance to fragmentation	
			Grain class			Ballast	Chippings
		wt.-%	mm	g/cm^3	g/cm^3	wt.-%	
Air-cooled blast furnace slag	A	≤4	32/45	≥1.2	≥2.4	15–24	18–25
	B	≤6		≥0.9	≥2.1	20–33	22–34
	C	≤8		≥0.9	≥2.1	–	–
	D	–		≥0.8	≥2.0	–	–
Slag pumice	–	–	5/16	≤0.9	–	–	–
Granulated blast furnace slag	–	–	0/2	≥0.8	–	–	–
Steel slags	1	≤4	32/45	≥1.5	≥2.8	–	≤26
	2	≤6				–	–

2.3.3 Properties of the liquid slags

Iron and steel slags can be named "anthropogenic lava". All attempts to handle and to modify the liquid slags have to consider the dynamic viscosity of the slags. Usually a blast furnace slag at 1400–1550 °C has a viscosity of 0.1–0.8 Pa·s. Natural lava at 1500 °C ranges from 1–10^4 Pa. s, typical lime-soda-glass at 1400 °C has about 10 Pa·s. Pig iron at 1400–1550 °C has a low value of 0.001 Pa·s (like water at 20 °C). Figure 6.12 shows for some blast furnace slags measured viscosity-temperature curves as well as theoretical curves calculated by FEhS institute [39]. It is obvious that the real behaviour of slags during the cooling process deviates from theoretical expectations due to crystallisation. Few changes in chemistry can significantly influence the viscosity. Figure 6.13 shows the impact of an increase in TiO$_2$ content from 0.5 to 1.9 wt.-% which is a high, but not a typical value. The absolute viscosity value increases only from 0.47 Pa·s to 0.51 Pa·s although the slag behaviour was comparable to a slag being 100 °C colder. The slag viscosity strongly influences the metallurgical tasks (metal cleaning), an effective separation of metal and slag (metal yield), the slag handling (runner, ladle, granulation), the degassing (porosity of hardened slag) and the cooling rate during the quenching process in a granulation facility (glass structure).

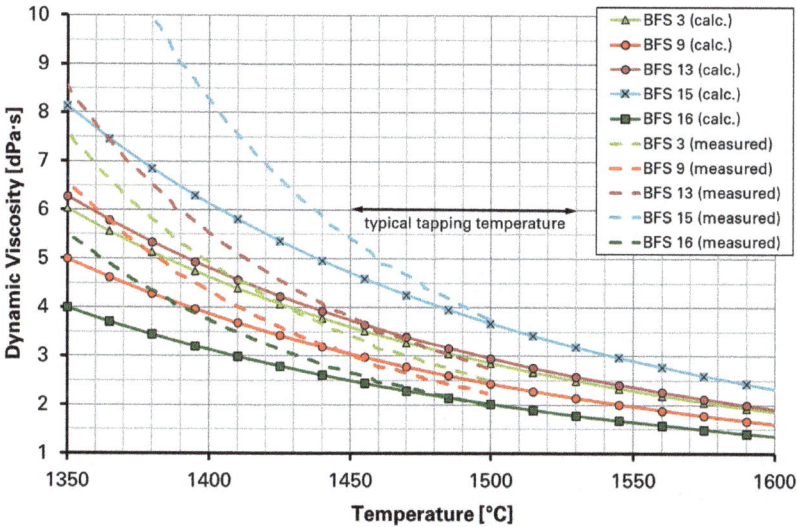

Figure 6.12: Dynamic viscosity of different liquid blast furnace slags (calculated and measured).

Slags have a low heat conductivity (≤1 W m^{-1} K^{-1}) as well as a low heat capacity. Therefore, liquid slags "freeze" very fast (see Figure 6.13, middle and right) which hinders slag handling and slag modification (e.g. injection of correction or reduction materials). In addition it decreases the yield of the slag granulation process if the liquid slag has to be transported in ladles to central granulation facilities instead of running directly into a granulation plant nearby to the blast furnace (see section 2.2).

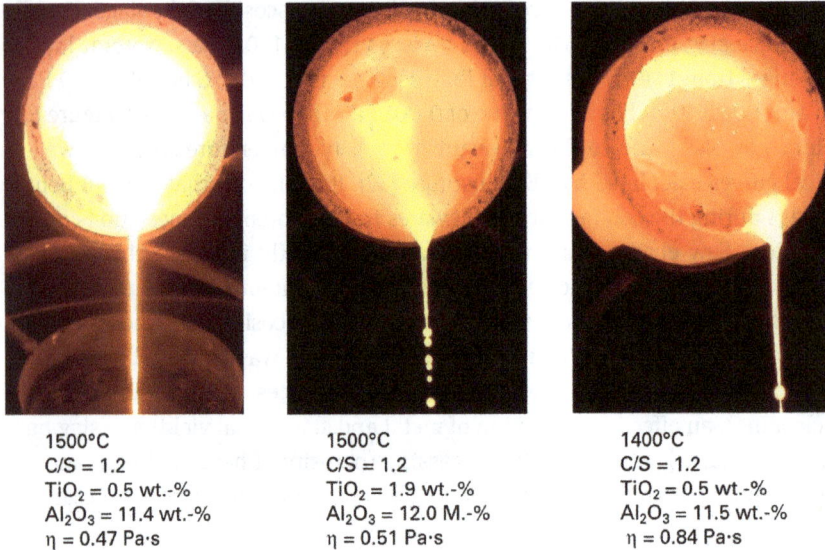

1500°C	1500°C	1400°C
C/S = 1.2	C/S = 1.2	C/S = 1.2
TiO_2 = 0.5 wt.-%	TiO_2 = 1.9 wt.-%	TiO_2 = 0.5 wt.-%
Al_2O_3 = 11.4 wt.-%	Al_2O_3 = 12.0 M.-%	Al_2O_3 = 11.5 wt.-%
η = 0.47 Pa·s	η = 0.51 Pa·s	η = 0.84 Pa·s

Figure 6.13: Influence of chemical variation on blast furnace slag viscosity.

3 Slag production data

Figure 6.14 shows the worldwide production of pig iron in blast furnaces and of crude steel in different process routes since 1990. Each process route results in different types of slags regarding chemical, mineralogical and technical properties. Their production is shown in Figure 6.15 [2]. On average the share of the process route blast furnace / BOF is 72% and the share of the EAF route is 28% (2019) [1]. The Open Hearth Furnace route is negligible. Compared to the pig iron production in blast furnaces (1281 mio. tons in 2019) the production of DRI without direct slag generation is very small (108 mio. tons in 2019). The slag/metal ratio is very different for the individual processes and vary in different countries. While in Germany the blast furnace slag/pig iron ratio and the BOF slag/crude steel ratio were 286 kg/t and 109 kg/t (2019), respectively, ratios of 346 kg/t and 137 kg/t (2002) were reported for China [2, 12]. In total it can be calculated that about 620 mio. tons of iron and steel slags are produced yearly.

The figures show that the huge increase in steel and slag production started after the year 2000, driven by China. It is obvious that China as the biggest pig iron and steel producer worldwide (63% and 53%, respectively, in 2019) also produces the highest slag volume. Another fast growing steel producing country is India. In

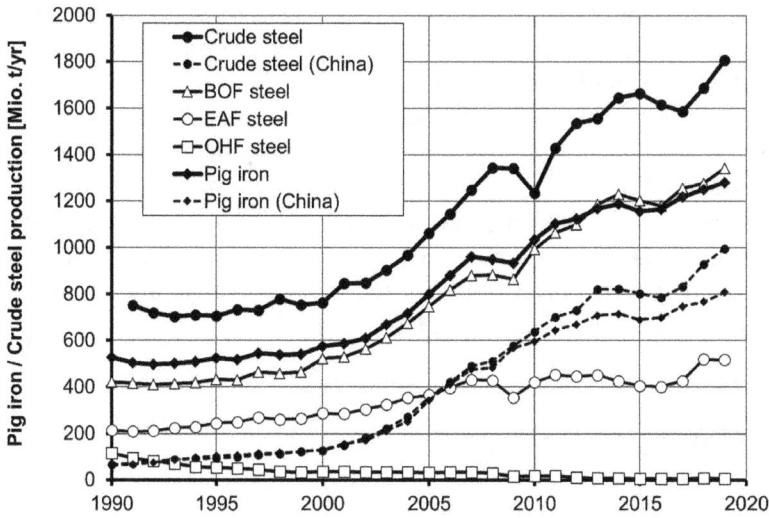

Figure 6.14: Production of pig iron (blast furnace) and crude steel in different process routes [1].

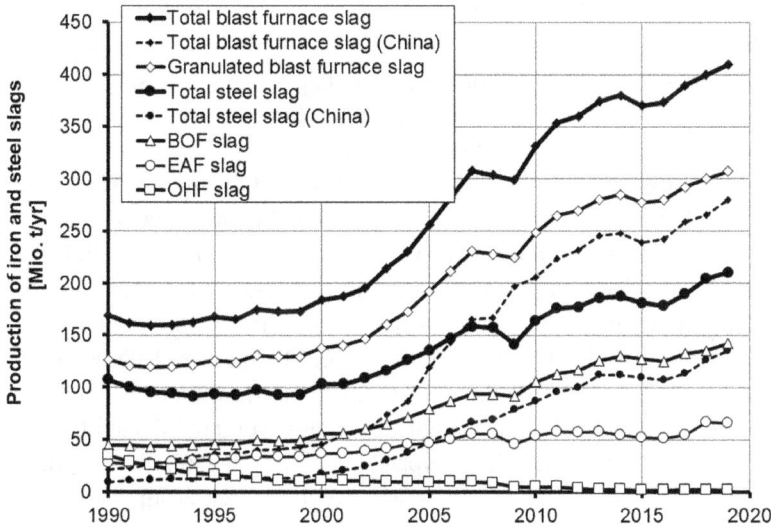

Figure 6.15: Production of blast furnace and steel slags in different process routes [2].

2019 the pig iron and crude steel production was 74 mio. tons and 111 mio. tons, respectively. However, according to the "steel policy 2017" India plans to increase its crude steel capacity up to 300 mio. tons per year by 2030/2031. This high growth is allotted to the blast furnace / BOF converter route (60–65% instead of 44% in 2018) [40]. The maximum of worldwide steel production is expected for 2070 with

2,800 mio. t [41]. Thus, the availability of iron and steel slags is secured for decades, also if there might be a transformation of the steel making processes resulting in a change of the slag properties (see section 7).

The share of the different steel production routes and respective types of iron and steel slags in the different countries varies considerably. In countries like Germany and the whole EU-28 the share of blast furnace / BOF route is near to the average (70%). In China the blast furnace / BOF route dominates (90%), whereas in the United States or in Turkey the EAF route prevails (70% and 68%, respectively) [1]. Based on that in different countries the interest in the different application fields for slags varies depending on the availability of the slag types.

4 Motivation to use slags

4.1 Motivation of the steel industry

Although in the very early beginning of metal and slag production different slag utilization was documented, the major motivation arose with the industrial revolution and the enormous increase of steel and slag production. Figure 6.16 shows typical capacities of blast furnaces and the associated slag/metal ratios. It has to be noted that due to the different densities of hot metal and slag a mass ratio slag/metal of 1:1 means a volume ratio of 3:1. Steel producers were forced to acquire expensive farmland only to dispose of unavoidable slag. For a steel plant in Dortmund-Hörde (Germany) it was written in 1853: "Beside this area reserved for the real steel plant the association additionally bought a place being nearly similar in size near to that in direction of the valley, preferentially to be used for slag dumping . . ." [42]. Therefore slag utilization avoided non-productive investment costs and in 1884 the steel plant Ilseder Hütte (Germany) vaunted that a slag dump was not necessary because "the composition of the slag being generated there (60,000 t per year) allows its utilization as road building material" [43]. However, in a compendium on steelmaking published in 1896 it was written, that "by far the biggest volume [of slag] forms a ballast for the steel plants and has to be dumped" [44]. For a steel plant in Oberhausen (Germany) in 1911 it was written that its slag dumping area is supposed to be bigger than the German island Helgoland [45]. Figure 6.17 shows this huge area in 1957. It is worth to note that during the following decades approximately all slag was quarried and in 2003 the whole area was used as an industrial park.

Driven by increasing slag volumes and accompanied by an intensive research work since the beginning of the 20th century slag utilization was more and more realized. Since many years stockpiling is not the method of choice. In many countries dumping areas are restricted and expensive. The different successful slag applications (see section 5) clearly show that slag utilization offers a positive contribution profit

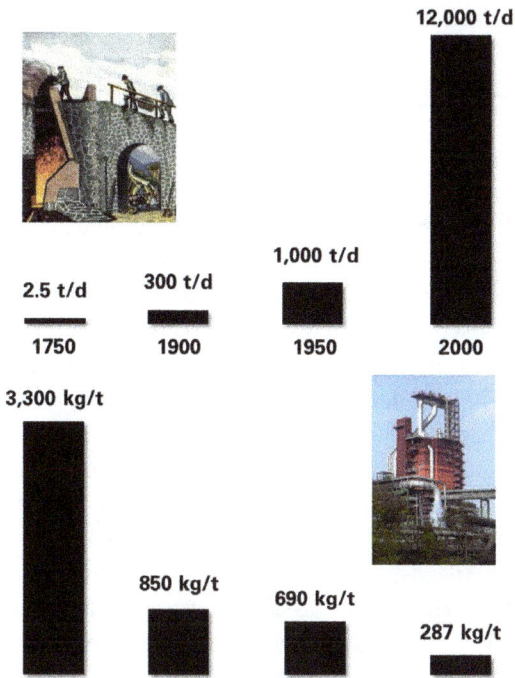

12,000 t/d

1,000 t/d

2.5 t/d **300 t/d**

1750   1900   1950   2000

3,300 kg/t

850 kg/t

690 kg/t

287 kg/t

Figure 6.16: Production capacity and slag/pig iron ratios of blast furnaces.

Figure 6.17: Stockpiling area of the former steel plant Gutehoffnungshütte in Oberhausen (Germany) in 1957 [46] (from West to East) and in 2003 (from East to West).

margin. The BAT reference document for iron and steel production includes slag utilization as state of the art [47]. For example, regarding blast furnace operation the document recommends "slag treatment, preferably by means of granulation (. . .), for the external use of slag (e.g. in the cement industry or for road construction)". And regarding BOF steelmaking the document recommends "slag treatment where market conditions allow for the external use of slag (e.g. as an aggregate in materials or for construction)".

4.2 Environmental motivation

Slag utilization offers a big potential to preserve natural resources, to reduce CO_2 emissions and to realize a circular economy. The advantages regarding CO_2 emissions are most important for the use of glassy granulated blast furnace slag in cement and concrete. Thus, this topic is discussed more in detail in section 5.2.1.1. The use of crystalline air-cooled blast furnace slag and steel slags is important regarding the substitution of natural aggregates for concrete production, road construction and fertilizer.

Aggregates are mainly extracted from natural deposits. In 2018 259 mio. tons of gravel and sand and 226 mio. tons of crushed natural rocks were mined in Germany alone. 95% are used for construction activities [48]. For gravel and sand open pits the yearly surface area equivalent is 9,6 km^2 (deposit thickness 15 m). For crushed natural rocks it is 3,5 km^2 (deposit thickness 25 m)! For the next decades a decrease of the enormous material demand is not expected. For Germany different scenarios anticipate a gravel/sand demand of 228–280 mio. t/yr [49]. This will be closely linked with a continuous and significant impact on nature accompanied by noise, dust and sometimes long transport distances by truck to the building materials (e.g. concrete) producers. In several regions of the world there are more and more critical discussions on these environmental impacts. Moreover, there are competing uses (development areas, nature and water protection areas) as well as economical reservations of landowners [50].

When looking at the worldwide demand for gravel and sand the situation is very critical regarding ecological, economical and therefore social and political aspects. According to UNEP [51] the annual demand for gravel, sand and crushed rock is about 40–50 bn tons. 26–30 bn tons are used for building purposes. By 2030 a rise up to 60 bn tons/yr is expected. According to OECD [52] for the years 2011–2060 an increase of the demand for non-metallic raw materials from 35 bn tons/yr up to 82 bn tons/yr (± 20%) is expected. In order to cover the increasing material demand "using alternative materials" and "by-products" is recommended inclusive "an integrated view on the governance" [51]. The OECD also recommends "the evolution of recycling and secondary materials use"[52].

The high availability, the advantageous technical and environmental properties of iron and steel slags and the long-term experiences advise to use these materials more intensive. Thus, against the background of the expected development of the future their use can be a contribution to attenuate the consequences for the nature and the mankind. In particular slags offer a high potential regarding resource efficiency, area consumption, lower energy demand and lower CO_2 emissions. Only for Germany it can be calculated that since 1948 the use of iron and steel slags saved about 603 mio. tons natural raw materials for road making and concrete aggregates, 85 mio. tons for fertilizer and 365 mio. tons natural raw materials for Portland cement clinker production and it prevented about 209 mio. tons of CO_2 emissions from clinker burning process.

5 Slag applications

Despite their early publication in 1934, 1949 and 1963, respectively, a good overview on blast furnace slag properties and utilisation is given in [53–55]. For steel slags a comparable overview is still missing. Informations on both, blast furnace and steel slags are given e.g. in [56–60]. In the proceedings of different slag related conferences many papers discuss all aspects of steel slag production and utilization [61, 62]. Also a chapter in the "slag atlas" [4] is dedicated to slag utilization [35]. Current informations are given on the websites of different organizations being confessed with slag topics [63–68].

5.1 Historic slag utilizations

From the early beginning of hot metal and steel production different types of slag utilization are known. The first target was to recover the remaining iron oxide or metal contents. Due to the production methods (low temperature, low reduction degree) slags from Roman period were so high in remaining iron oxides (60–70 wt.-%) that it later was worthwhile to use thousands of tons as anthropogenic raw materials for metal production, as it is described for the Forrest of Dean in Wales (UK) for the 17th century [69]. The liquid slags from charcoal blast furnaces in the 18th/19th century were so high in viscosity that they were handled manually with long rods. These slags often contained unburnt charcoal as well as metal droplets (about 6 wt.-% of the produced metal) caught within the viscous slag melt. Figure 6.18 shows such a slag sample being produced in a German charcoal blast furnace before 1843 [70]. The slag was crushed in a water-driven stamp mill to recover the metal. Liberated by the crushing process the high density metallic droplets could subsequently be collected and the crushed slag was removed with the river water [71]. Today blast furnace slags do not contain significant metallic exsolutions but for steel slags metal recovery is still an urgent topic for many plants.

For using slags as building materials it was obvious to cast the liquid slag into moulds. In 1728 a patent was granted in UK for casting bricks [72]. In Germany and in Sweden many houses were also built with casted slag stones [73]. However, these building stones were very dense and from the today's view of home environment and climate etc. they would not be acceptable. In 1766 a Swedish royal order even obligated

Figure 6.18: Air-cooled slag from a German charcoal blast furnace, before 1843.

blast furnace operators to cast slag bricks if the slag was suitable [74]. Despite the fact that the motivation for this decision was not an ecological but an economical as wood was saved for the charcoal production this order is a very early example for preferring a building material made from an industrial by-product instead using natural resources! The current procedure to crush the hardened slag for aggregate production had a predecessor as early as in the 19th century. In literature the slag aggregates are described as hard and durable material substituting crushed Basalt [75, 76].

The water granulation process was originally developed for an easier handling while disposing of the slag on dumps similar to the processing of e.g. Fe Ni slag today. However, in 1861 Emil Langen in Germany recognized the hydraulic property of ground granulated blast furnace slag. After activation and in contact with water it forms durable calcium-silicate-hydrates being comparable to those which are formed during Portland cement hydration. Despite the fact, in 1865 the first industrial application for granulated blast furnace slag was the production of bricks. Unground slag was mixed with hydrated lime and the bricks hardened through contact with air [77]. Millions of these bricks were produced for decades. Starting in the 1870ies the production of "slag cement" started by mixing 70–85 wt.-% ground granulated blast furnace slag with hydrated lime [78]. Since 1879 blast furnace cements were produced in Germany by mixing ground slag with Portland cement clinker as a more efficient activator and with own strength contribution which results from the hydration of the clinker phases, like

Tricalciumsilicate (Alite) or Dicalciumsilicate (Belite) [79]. That is the major application of granulated blast furnace slag till today (see section 5.2.1.1).

Based on the experiences with casted bricks in the 18th century and the utilization of copper slags in the 19th century [80] the casting of pavement stones was performed in several steel plants. Figure 6.19 shows the casting of liquid blast furnace slag into moulds and the utilization of the stones. Manually removing the hardened, but hot stones was hard physical labour. Moreover, the cooling process required about one week in order to avoid cracks. In addition, the moulds could only be used a limited number of times. Remaining pavement stones today exhibit a very good durability and resistivity to weathering.

Figure 6.19: Casting [81] and use of pavement stones made from blast furnace slag around 1930 and 1955, respectively.

When liquid blast furnace slag was atomised with high pressure steam slag wool was created. Comparable to rock or glass wool today it was used for insulation purposes [82]. Due to sulphur elution water contact had to be strictly avoided limiting the technical applications. Another logistical problem was the extensive volume of wool per ton of slag to be handled.

If blast furnace slag comes into contact with a limited water volume it starts to foam. This behaviour was used to produce slag pumice (Figure 6.20). Due to the low specific gravity the material was used for lightweight aggregates in concrete. Figure 6.21 shows an advertising brochure from the 1930ies and the same building today.

Figure 6.20: Blast furnace slag pumice.

Figure 6.21: Advertising brochure for slag pumice from 1935/36 and the same building today (2019).

The main utilization of steel slags from the Thomas process (predecessor of BOF process) in the late 19th century and in the first half of the 20th century was fertilizer production (see section 5.2.3). Before the steel slag was mainly dumped (about 230–240 kg per ton of steel).

The utilization of iron steel slags changed during the last 100 years. As an example Figure 6.22 shows the different applications for blast furnace slag in Germany in the year 1936 [83]. Most of them do not have any relevance today. Besides the high slag/metal ratio it is noteworthy that about one third of the blast furnace slag was dumped. For the USA it is written that in 1929 the utilisation rate for blast furnace slag was < 50%, too.

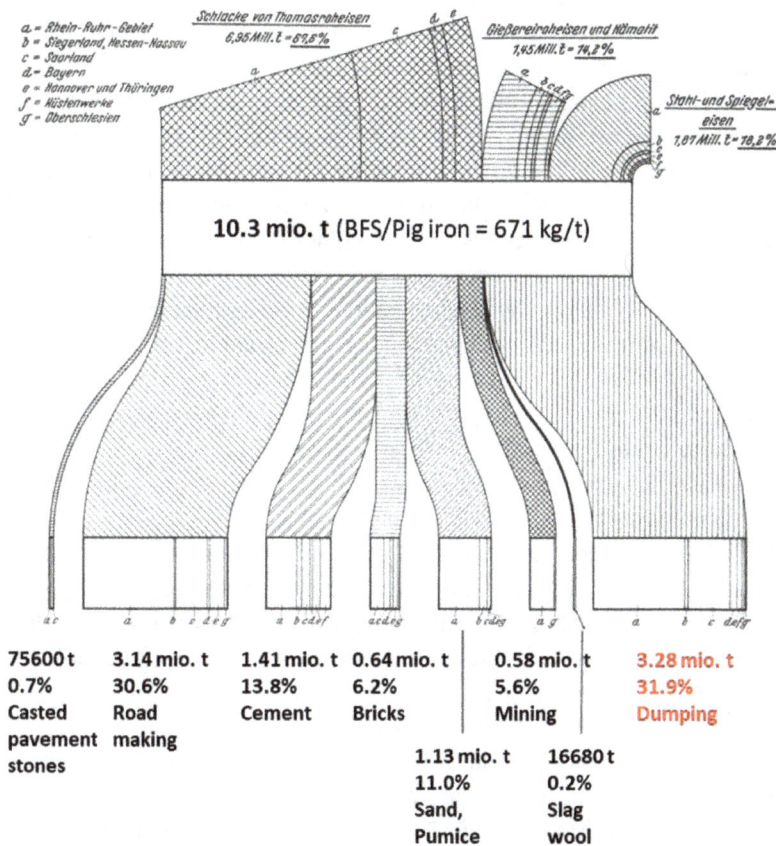

Figure 6.22: Share of different blast furnace slag applications in Germany in 1936 [83].

5.2 Slag utilization today

Figures 6.23 and 6.24 show the today's share of the different applications in Germany and in Europe, respectively. In Germany and in Europe the glassy granulated blast furnace slag is mainly used as a cement constituent substituting Portland cement clinker or as a concrete addition substituting cement. Crystalline air-cooled blast furnace slag and steel slags from BOF and EAF processes are used for different applications in road and waterway construction. A smaller part of BOF slags is used in agriculture. The remaining applications cover metallurgical recycling or interim stockpiles. In 2018 only 12% of the steel slags (5% of all iron and steel slags) were disposed on dumps in Germany. In most cases it is fine-grained material without technological benefits. The average figures for Europe are similar (13% and 6%, respectively). However, in some countries the share of dumped slag is much higher: E.g. for Sweden a value of 35% was reported for 2006 [84].

Comparing Figures 6.22 and 6.23 illustrate that several applications for blast furnace slags that had been established in the past (see section 5.1) were abandoned or replaced over time. Some of the production processes were highly dependent on tiresome and today expensive manual labour (casting pavement stones), produced a lot of emissions (pumice stones) or very high volumes of materials which cannot be handled for modern blast furnace capacities (slag wool). Today the use of granulated blast furnace slag as a latent hydraulic material in cement and concrete is the most promising one regarding technical, economical and ecological aspects.

5.2.1 Cement and concrete

5.2.1.1 Cement constituent
Worldwide the use as a cement constituent is the dominant application of granulated blast furnace slag (GBS). An overview on the major important aspects of granulated blast furnace slags and their use in cement and concrete is given e.g. in [86–93]. In some countries also the use of ground granulated blast furnace slag (GGBS) as a concrete addition is established [94, 95].

In contrary to Portland cement clinker, GBS is a glassy material. Figure 6.25 shows the fraction 40–63 µm of a ground GBS presented in Figure 6.5. This fraction is used for the optical evaluation of the glass content by transition light microscopy [96]. Today also XRD in combination with Rietveld analysis utilising an inner standard is used which allows to quantify the mineral phases. The difference to 100 wt.-% represents the glass content [97]. In order to be able to form CSH-phases the glass structure has to be solved at high pH values. In most cases this is done by alkaline activators, like Portlandite being formed during Portland cement clinker reaction. But any other alkaline materials might also be possible resulting in alkali activated slag binders [98–101].

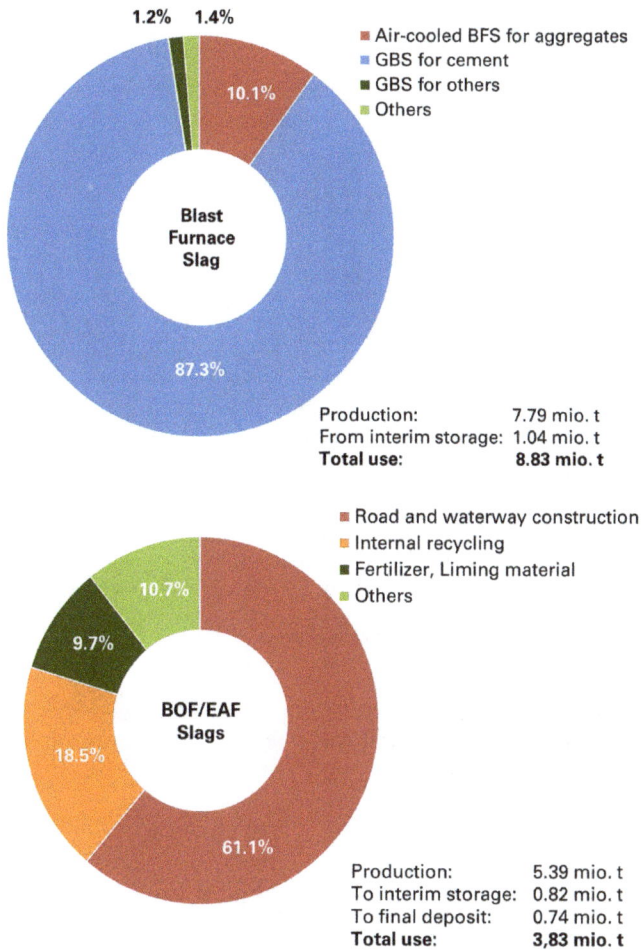

Figure 6.23: Share of the different slag applications in Germany in 2018 [3].

Unfortunately, there is no simple correlation between chemical composition and glass content of any GBS and its latent hydraulic reactivity or strength contribution in cementitious systems, also if frame conditions, like fineness, clinker type and content etc., have been kept constant [88–90, 102]. Many attempts tried to find a sufficient chemical sum parameter. Therefore, since 1909 the different German slag cement standards included different basicities (Table 6.6). Today some of them are still used in relevant standards of different countries worldwide. The reason for the insufficient correlation is the non-consideration of the thermal history of the slag and its influence on the slag glass properties [26].

The first German cement standard for a GBS containing cement was established in 1909 (Iron Portland Cement). In 1917 the first German standard for blast furnace

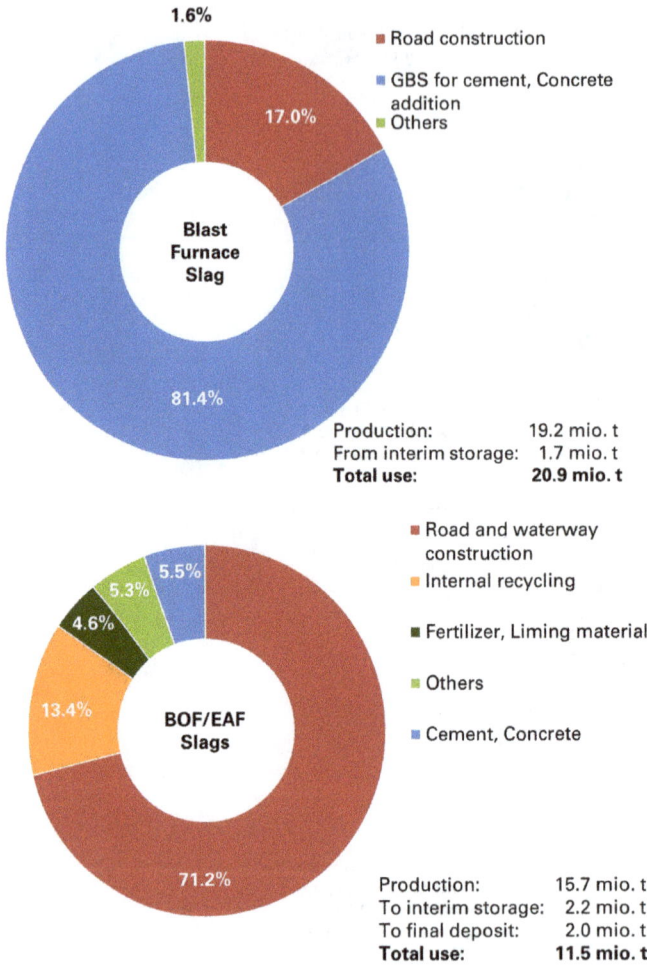

Figure 6.24: Share of the different slag applications in Europe in 2018 [85].

cement was published. The current harmonized European cement standard for "Common cements" (EN 197-1 [103]) contains 27 cement types. 9 of them are made with GBS from 6 to 95 wt.-%. Driven by the intention of the cement industry to reduce CO_2 emissions a new, non-harmonized standard EN 197-5 [104] will contain new cements with lower clinker contents and the possibility to use GBS together with other cement constituents, like pozzolana or limestone. Some standards were developed for special slag containing cements, e.g. "Supersulfated cement" (EN 15743), "Very low heat special cements" (EN 14216) or "Hydraulic road binders" (EN 13282). In European standards only granulated blast furnace slag is defined as a cement constituent. Only in one type of hydraulic road binder also BOF slag is allowed. However, in China a standard (GB/T 20491) exists for steel slag powder to be used in cement and concrete [12].

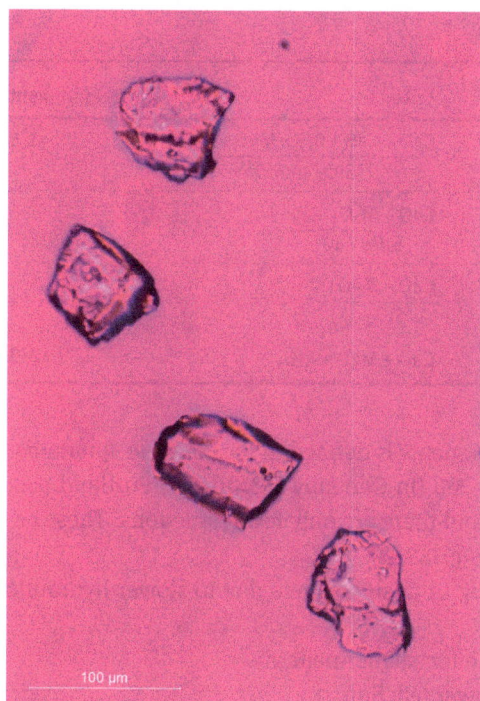

Figure 6.25: Glassy structure of the first granulated blast furnace slag in Figure 6.5 (fraction 40–63 µm of the ground sample).

Table 6.6: Chemical requirements for granulated blast furnace slags [102].

Standard/proposal	Year	Criteria	Requirement
"Simple" basicity	1886	CaO/SiO_2	>1
1st standard for iron Portland cement	1909	$\dfrac{CaO + MgO}{SiO_2 + Al_2O_3}$	≥1
1st standard for blast furnace cement	1917	$\dfrac{CaO + MgO + 1/3\ Al_2O_3}{SiO_2 + 2/3\ Al_2O_3}$	>1
DIN 1164	1932	$\dfrac{CaO + MgO + 1/3\ Al_2O_3}{SiO_2 + 2/3\ Al_2O_3}$	≥1
DIN 1164	1942	$\dfrac{CaO + MgO + Al_2O_3}{SiO_2}$	≥1
F value (Keil)	1942	$\dfrac{CaO + CaS + 0,5 \times MgO + Al_2O_3}{SiO_2 + MnO}$	>1,5

Table 6.6 (continued)

Standard/proposal	Year	Criteria	Requirement
F value (Sopora)	1959	$\dfrac{CaO + CaS + 0,5 \times MgO + Al_2O_3}{SiO_2 + MnO^2}$	>1,5
Basicity (Schwiete)	1963	$\dfrac{CaO + Al_2O_3 - 10}{SiO_2 + 10}$	–
DIN 1164-1	1994	$\dfrac{CaO + MgO}{SiO_2}$	>1
EN 197-1	2001		
EN 15167	2006	$CaO + MgO + SiO_2$	>2/3

The application of slag containing cements is defined in the concrete standards. For most of the exposure classes of EN 206 (in Germany DIN 1045-2) Portland slag and blast furnace cements are allowed and there are only few restrictions. These cements offer several technical advantages [91, 92]:

– good workability and no early setting of fresh concrete due to slower hydraulic reaction
– no segregation (bleeding) due to higher cement fineness
– low heat during hydration due to slower reaction
– high chemical resistance (chloride penetration, sulphate attack, acid attack) due to lower capillary porosity, very dense structure, high binding capacity and lower Portlandite content
– higher resistance against alkali silica reaction due to non-soluble alkalis
– higher flexural strength/compressive strength ratio (lower crack formation tendency)
– high sensitivity for thermal curing (pre-casted concretes)
– bright color

The disadvantages are limited to
– higher grinding energy demand due to the glassy structure of GBS
– lower early strength due to the time-consuming glass corrosion process
– higher sensitivity against insufficient curing conditions for the fresh concrete

Many concretes with GBS containing cements or with ground GBS as additions are used for regular building structures, in particular Portland slag cements with a GBS content of 6 to 35 wt.-% and blast furnace cements CEM III/A with a GBS content of 36 to 65 wt.-%. Special properties like low heat of hydration, very dense pore structure or high chemical resistance are important for buildings with special requirements, e.g. bridges, barrages, locks, TV towers, chimneys. To illustrate the durability of concrete structures made with GBS containing cements Figures 6.26–6.29 illustrate some examples.

Figure 6.26: Office building in Essen (Germany), erected 1928/29 [53]; situation 2003.

Figure 6.27: TV Tower in Dortmund, Germany, erected 1958/59 [105]; situation 2017.

Figure 6.28: Church in Duisburg (Germany), erected 1996 [106]; situation 2007.

Figure 6.29: Barrage for Saale river in Thuringia (Germany), erected 1936–1941 [107]; situation 2003 [108].

Figure 6.30 shows the dispatch development for different cement types in Germany since 1990 [109]. In the year of the German re-unification 1990 the market share of Portland cement was 72%. This share has decreased significantly to 28% in 2018 starting around 1999. At the same time the market shares of Portland slag cement and other Portland composite cements increased enormously from 5% and 8%, to 19% and 29%, respectively. The share of blast furnace cement increased from 15% to 24%.

The background of the market development is the discussion on the ecological consequences of cement production which causes 5–8% of all man-made CO_2 emissions [110, 111]. Cement is the binder for concrete, the most important building material worldwide which is designated to be responsible for 9% of all anthropogenic CO_2

emissions [52]. Besides burning fossil fuels and using electricity the dominant CO_2 source is the unavoidable decomposition of limestone during the clinker sintering process. Thus, in 2012 the German cement industry summarized its self-commitment efforts to reduce CO_2 emissions in the sentence: "the achieved decrease for cement is attributed mainly to the intensified production of cements with several main constituents" [112]. Granulated blast furnace slag is the most effective clinker alternative. It substitutes clinker 1:1 and with increasing slag content the CO_2 footprint of the cement decreases significantly. Figure 6.31 compares the CO_2 emission of Portland cement CEM I, a statistical average cement, Portland slag cement CEM II/B-S and blast furnace cement CEM III/A. A decisive question regarding the environmental benefit of slag cements is how much load is transferred from the main product pig iron (or steel made from it) to the by-product GBS [113]. To allocate the CO_2 load from the blast furnace process the economical allocation was used (EN 15804). Depending on the market price ratio hot-rolled steel/GBS this load is subject to minor fluctuations, thus, two values are given for the slag cements. The other important ecological advantages of slag cements are saving natural raw materials (about 1.6 ton per ton of clinker) and fossil fuels for clinker burning.

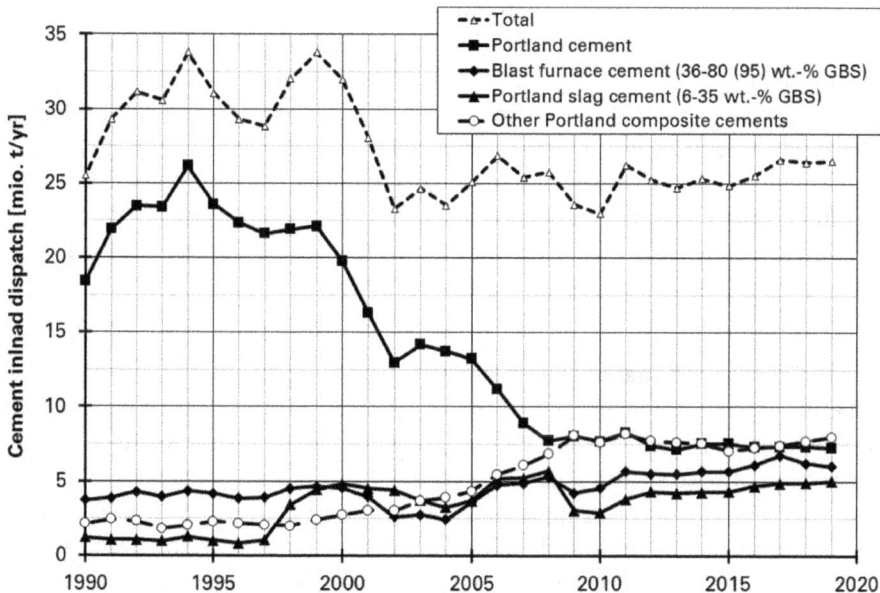

Figure 6.30: Inland dispatch of different cement types in Germany [109].

It can be expected that within the next years more and more the ecological aspects of buildings and thus of the used building materials will be a crucial factor for the public and other awarding authorities. By promoting and using "Green building" and "Green public procurement", authorities can provide industry with real incentives for

| GBS: | 0 | 14 | 30 | 30 | 60 | 60 wt.-% |

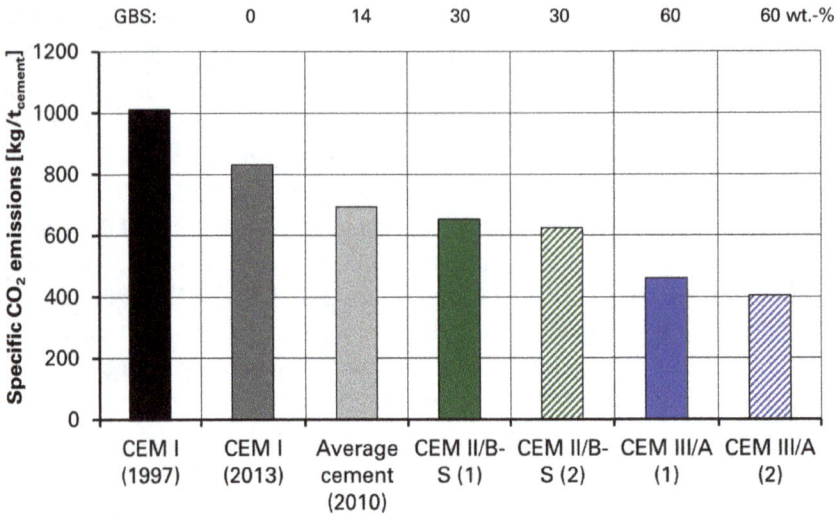

Figure 6.31: CO_2 footprint of different cement types [113].

using and developing green products. E.g. in the construction sector public purchasers command a significant share of the market and so their decisions have considerable impact [114]. Therefore, the producers of building materials which consume a relevant amount of natural raw materials and energy and which are responsible for significant anthropogenic CO_2 emissions are forced to reduce the ecological footprint of their products. For the cement and concrete industry the intensified reduction of the clinker share is an integral part of their CO_2 reduction roadmaps [110, 115, 116]. Thus, an intensified use of GBS, but also of (modified) steel slags (see section 5.3) and new slag types (see section 7) is an obvious contribution.

5.2.1.2 Substitute for natural raw materials

A specific aspect regarding cement is the use of iron and steel slags as raw materials for the Portland cement clinker burning process. As early as in 1883 the use of air-cooled blast furnace slag as a partial substitute for clay and limestone was realized. Until the 1970ies in Europe many steel plants operated also own cement plants in competition with the Portland cement producers as it is still the situation today in some countries like e.g. India. The steel plants had two main advantages: They could use as well own raw material (limestone being necessary for blast furnace and steel plant operation) as well own by-product (air-cooled blast furnace slag) for clinker burning. And they could use another by-product, granulated blast furnace slag, to produce slag cements. These advantages vanished when the steel industry closed or sold its own cement plants. With the upcoming discussion on the environmental impact of Portland cement clinker production, in particular the CO_2 emissions, the idea to use air-cooled blast furnace slag as a CO_2-free raw material was renewed. Calculations for a raw meal mixture containing 37 wt.-% air-cooled blast furnace slag resulted in an energy demand decrease of 27% and a CO_2 emission reduction of 33% [117].

However, in the meantime more and more blast furnace slag is granulated (Figure 6.15) and the availability of air-cooled slag as a raw material is very limited. Therefore, its use for clinker production, but also for coloured glass or rockwool production, is very limited.

Steel slags can further be used as raw materials for clinker production. From the chemical point of view in particular BOF slag might be suitable due to its high, CO_2 free lime content (see Table 6.3). In the United States the so-called CemStar^{SM} process was developed by the steel company TXI [118]. It can enable the use of an own by-product for a higher clinker production capacity. However, the process ignores the fact that during the oxidizing clinker sintering process the chromium-III content of the steel slags (see section 2.3.1) is converted into toxic chromium-VI. In Europe a 2 ppm-limit (0.0002 wt.-%) exists for cement and cement containing preparations since 2005. By only utilizing natural raw materials for clinker production up to 100 ppm chromium-VI are generated. It was stated that about 20% are water soluble [119–122]. Thus, approximately every cement producer already has to add reduction agents in order to fulfil the European 2 ppm requirement and a relevant increase in chromium-VI might not be controllable.

5.2.1.3 Concrete aggregate

The use of air-cooled blast furnace slag as concrete aggregate was established in the beginning of the 20th century and it is still common [53, 54, 123–126]. Figure 6.32 shows such concrete buildings in Youngstown/Ohio (USA) [53]. For example, building No. 5 and No. 7 still exist today.

In Germany the first guidelines how to use air-cooled blast furnace slag as a concrete aggregate were already published in 1917. Today EN 12620 covers air-cooled and granulated blast furnace slags as concrete aggregates. Due to the very good experiences since decades these slags only have to fulfil the technical requirements of this standard. There are no additional environmental requirements. Regarding the technical properties it has to be considered that air-cooled blast furnace slag as a crushed aggregate with an irregular shape (Figure 6.4) may increase the water demand of fresh concrete in order to achieve the same workability as concrete based on natural gravel. The porosity of the slag grains may cause a higher water demand. A simple way to avoid an impairment of the workability is pre-wetting of the aggregate [127]. The irregular grain shape results in a very good interlocking of grain and cement stone and thus in a higher concrete strength. The thermal properties of the blast furnace slag are also an advantage if concrete for high temperature applications is made, e.g. for coking plants or exhaust shafts [128, 129].

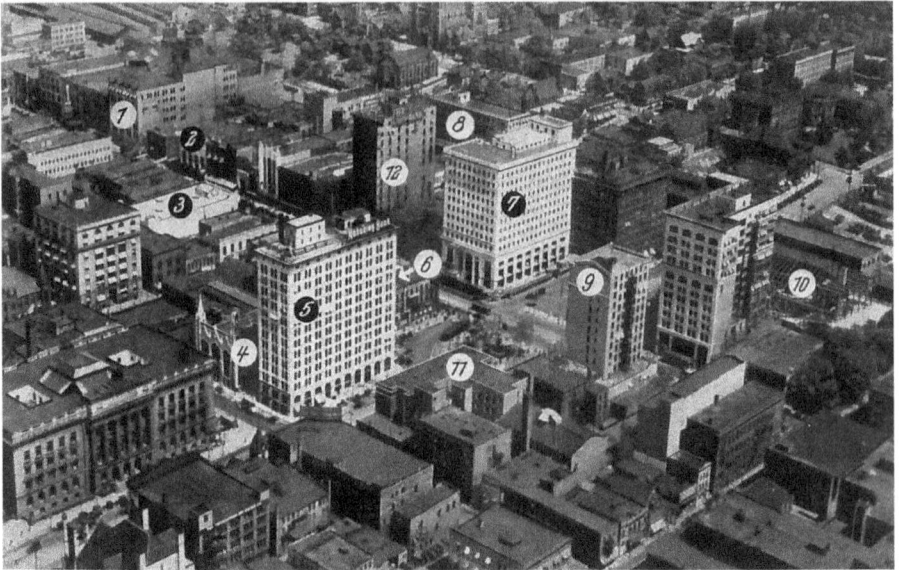

Figure 6.32: Concrete buildings in Youngstown/Ohio (USA) made with air-cooled blast furnace slag as aggregate, around 1930 [53].

Steel slags may also be used as concrete aggregates if their volume stability is guaranteed (see section 2.3.1). Regarding the concrete workability the same aspects have to be considered as described above. In addition, the higher density of the steel slags compared to blast furnace slag can be used to produce heavy weight concretes. Of course, the metal content of the slags has to be minimised in order to avoid rust formation resulting in discoloration and concrete spalling. In the framework of the worldwide increasing demand for aggregates and sand (see section 4.2) as well as the decreasing availability of air-cooled blast furnace slag instead of granulated slag (see section 3) it is worthwhile to further develop the application of steels slags. Many investigations have already been performed [130–132]. However, in many countries there are restrictions due to the heavy metal content of steel slags despite the fact that the leaching behaviour of heavy metals from concrete made with steel slags is unproblematic [133].

5.2.2 Road and waterway construction

Today the main application of steel slags is road construction [59, 61, 134–138]. Air-cooled blast furnace slag and less so granulated blast furnace slag is also used for that purpose. After excavating the cooled slags from the slag pits the material is crushed and screened into the different fractions being necessary for unbound and bound applications. The slags are used in the same way as natural aggregates (Figures 6.33 and 6.34).

Figure 6.33: Application of BOF slag for a bound bearing layer of a road.

Figure 6.34: Application of EAF slag for an unbound bearing layer of a road.

High hardness, low abrasion and a good grip are special features of steel slag aggregates. Another special positive feature is a consequence from their thermal properties. Due to the lower heat capacity and heat conductivity it was found that during road making the temperature of asphalt mixtures made with steel slags can be lower compared to natural aggregates. That reduces the thermal energy demand of the process. In addition, the temperature of the finished asphalt layer and its temperature variations, too, are lower resulting in less lane grooves.

Another positive effect is the (limited) self-hardening of mixtures from steel slag and secondary metallurgical slags (Figure 6.35, left). It reduces the dust formation on the road surface. Figure 6.35 (right) shows a small test road which was accompanied by FEhS institute for several years looking into its positive environmental behaviour regarding the leaching potential of heavy metals [139].

Figure 6.35: Application of EAF + secundary metallurgical slag (left) and BOF slag (right) for small unbound roads.

Besides the general technical requirements the relevant standards (e.g. EN 13242, EN 13043 [140, 141]) include some special requirements for iron and steel slags, e.g. regarding their volume stability. To evaluate the volume stability the so-called steam test was developed (EN 1744-1). For 24 hours (BOF slag) or 168 hours (EAF slag) saturated steam flows through a slag sample bed of defined grain sizes and compaction degree. The increase in sample height is measured. According to technical delivery standards the slag is designated volume stable if the increase is < 5 vol.-% (unbound application) or < 3.5 vol.-% (bound application), respectively.

Besides the technical properties the environmental properties of the slags have to be also considered. The requirements for products are defined by the individual countries, as there is no joint regulation up to date. The most relevant topic is the leaching behaviour of heavy metals. Since many years in Germany a 24 hours single batch leaching test is applied for assessing the materials leaching behaviour (EN 12457-4, liquid/solid = 10/1). Requirements have been established based on detailed investigations regarding pH value (all), sulphate (blast furnace slag), fluoride, total chromium and vanadium content (steel slags). In future another test method (DIN 19529, liquid/solid = 2/1) must be used resulting from the introduction of a new regulation for so-called substituting building materials. In other countries other test procedures and limiting values exist. For example, in The Netherlands an up-flow percolation test (NEN 7373) is used.

The higher density of steel slags is a good precondition to use them for waterway construction. Thus, e.g. the protection of bank stabilisations or pothole filling are typical applications (Figure 6.36). Of course, the application rules have to assure that there is no harmful increase in pH value, in particular in stagnant waters, like channels. Moreover, there are several investigations and pilot applications in marine environments [142].

Figure 6.36: Application of BOF slag for waterway bank stabilisation.

5.2.3 Fertilizer

For agricultural purposes two properties are used [143, 144]. Foremost, air-cooled or granulated blast furnace slag, BOF slags and some secondary metallurgical slags are used as liming agents. In Germany the use of blast furnace slag as soil conditioner is allowed since 1937. Compared to ground limestone the slags release Ca more slowly and therefore the slags are able to optimize soils more efficient. In addition, some minor or trace elements, like Mg, Si or Mo, have certain positive effects on the plants. Figure 6.37 shows corn field trials of FEhS institute performed for 19 years.

Figure 6.37: Corn field trials of FEhS institute in Germany (2011).

Further, steel slag can be applied as an actual fertilizer. The steel slag from the former Thomas process was enriched in phosphorus (16–20 wt.-% P_2O_5), which has been modified to a more or less soluble and plant available form (14–18 wt.-%) [145]. Thus, already in 1884 Sidney Gilchrist Thomas wrote in a letter to his cousin [146]: "However laughable you may consider the notion, I am convinced that eventually, taking cost of production into consideration, the steel will be the by-product, and phosphorus the main product". Figure 6.38 shows a French advertising brochure for the so-called "Thomas phosphate" which constituted of ground Thomas steel slag. When the Thomas process was substituted by the BOF process in the 1960s, the resulting slags have had very low in phosphorus content (0.7–1.8 wt.-%) and therefore the use of steel slag as phosphate fertilizer ended.

Today fine-grained iron and steel slags are mainly used as liming materials. However, the additional benefit of some trace elements or silica content on plant development is well-known [147].

Within a German pilot project the transformation of liquid BOF slag by solving P_2O_5 bearing sewage sludge incineration ash was successfully investigated [148]. This technology is comparable to the injection of sand and oxygen being used for free lime reduction (see section 2.3.1). While the phosphorus content of the ash itself

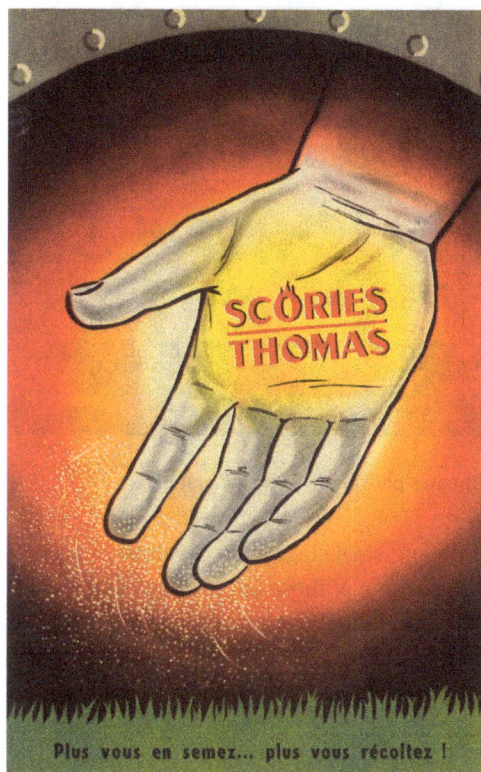

Figure 6.38: Old French advertising brochure for fertilizer made from Thomas steel slag.

is not plant available, the "Thomas phosphate – 2nd generation" has a comparable positive fertilizing effect as the original "Thomas phosphate" had due to the formation of Calcium-Phosphate-Silicates. A production on a larger scale might be realised if either the market price for natural phosphate fertilizers increase or the restrictions for established steel slag utilisations are surging. Figure 6.39 shows the positive influence of the modified BOF slag on plant growth being comparable to the effect of expensive (300–400 €/t) natural Triple Superphosphate.

5.3 Potential applications

The most relevant applications are discussed in section 5.2.1–5.2.3. Most of the historic slag applications being discussed in section 5.1 vanished due to several reasons: Technical limitation to small slag volumes, higher metallurgical efficiency, high costs due to extensive manual labour, changes in raw materials or process technology or other restrictions.

| Without Phosphate | Triple Super-phosphate | Thomas-phosphate | Sewage sludge ash | BOF slag + Sewage sludge ash | BOF slag + Sewage sludge ash | BOF slag + Sewage sludge ash | Rock phosphate |

Figure 6.39: Plant trials of FEhS institute with "Thomas phosphate – 2nd generation".

Several alternative applications for iron and steel slags were developed and presented on conferences and in papers [61, 62]. None of these applications reached an applicable state in terms of widely-used industrial realisation. Today, the potential use of steel slags for cement and concrete application retains again a lot of attention. This kind of potential use is not really a new idea. But against the background of the discussions on the worldwide increasing demand for building materials and in combination with the efforts to reduce greenhouse gas emissions and to save natural resources (see section 4.2) it has very high topicality.

The simplest approach to use steel slag as a cement constituent is to reduce its grain size to cement fineness ($d_{50} \approx 10$ μm). In China standards for cements containing steel slag were published [12]. However, in the European cement standard EN 197-1 steel slag is not mentioned as a possible constituent. The main reason is that like air-cooled blast furnace slag ground BOF or EAF slags of normal fineness do not show (latent) hydraulic properties being sufficient for binder production [149–153]. However, BOF slag can be hardened by carbonation. This process was used for example in Japan to form big slag blocks (0.5 m^3) for coastal protection purposes [154, 155].

Transformation of BOF slag into clinker like materials is apparent due to the high CaO content. As early as in the 1980s a pilot plant was successfully tested in Belgium [156]. The idea was rediscovered about 10 years ago [157]. However, it needs a significant energy demand to reduce the iron oxide content of the liquid slag, to achieve a sufficient workability of the high viscous reduced slag and to adjust as well the correct chemistry as the correct mineralogy.

The transformation of EAF slag into clinker-like or granulated blast furnace slag-like materials has been tested successfully on lab-scale, too [158]. Due to the differences in the chemical composition of BOF and EAF slags the transformation into a glassy material seems to be more promising. Overall, the energy demand might still remain a critical issue. Moreover, the steel plants have to invest into slag treatment and granulation facilities.

Summarising the topics being discussed above it can be stated that modified steel slags offer a high technical potential to be used as an additional Portland cement clinker substitute. The main task is to find a compromise between a good reactivity after grinding the slag to an acceptable fineness and the transformation and processing efforts/costs.

6 Legal status: Waste or by-product?

Despite the fact that this book is entitled "Industrial wastes" it is important to stress that iron and steel slags are not "wastes" but industrial "by-products". For technical applications it should be indifferent whether a material is considered to be a "waste" or a "by-product" if the technical and environmental properties are sufficient. However, practice shows that the acceptance in the markets might be very different [159]. Moreover, for "wastes" there are additional restrictions regarding handling, transport and documentation if a material is named to be a waste.

There is no unique legal situation in Europe regarding the classification of slags as "wastes" or "(by-)products" [160], and even within the one country different regulations may exist. Looking into the European Waste Catalogue [161] processed (e.g. granulated, air-cooled, crushed) slags are not classified as hazardous waste. Mentioned materials are only No. 10 02 01 "Waste from processing of slag" and No. 10 02 02 "Unprocessed slag". Clear parameters for the distinction of "waste" and "by-product" are given in the Waste Framework Directive, article 5 "by-products" [162]. If a further use is certain, if the material can be used directly without any further processing other than normal industrial practice, if it is produced as an integral part of a production process and if the further use is lawful, the material may be regarded as not being waste, but as being a by-product.

According to a communication of the European Commission from 2007 [163] it was stated that "Blast furnace slag can be used directly at the end of the production process, without further processing that is not an integral part of this production process (such as crushing to get the appropriate particle size). This material can therefore be considered to fall outside of the definition of waste".

The legal status of iron and steel slags is fiercely discussed in two position papers of the European Slag Association EUROSLAG [164, 165] as well as in two German legal opinions [166, 167].

In 2010 iron and steel slags have been registered under the REACH (Registration, Evaluation, Authorisation and Restriction of Chemicals) scheme [168] which is not necessary for wastes. For the registration (Table 6.1) a lot of chemical, mineralogical and technical data, examples for utilization and in particular of (eco-)toxicological tests and studies had to be prepared. All studies show that iron and steel slags are not hazardous [169].

7 Future steel and slag production

The main challenge for the future is related to the steel production transformation process starting in Europe. The blast furnace / BOF route causes about 1888 kg CO_2 per ton of crude steel. The EAF process, based on scrap, causes only about 455 kg/t [170]. However, the scrap availability is limited and a further switchover cannot be expected.

An alternative process is the smelting reduction (about 1349 kg CO_2 per ton crude steel [171]). Some "COREX" and "FINEX" industrials plants are already in operation since years [172, 173]. They produce a slag being comparable to blast furnace slag [174], but the slag/metal ratio is lower. In contrary, the new "HIsarna" process is still realized only in pilot plant scale with a yearly capacity of 60,000 tons hot metal [175]. The slag composition might be different compared to blast furnace slag due to the varying raw material input. Smelting reduction processes offer about 20% reduction of the CO_2 emissions.

The "Green deal" strategy of the European Commission from 2019 [176] aims to reduce the net CO_2 emissions to zero until 2050. In order to reach this task so-called breakthrough technologies have to be implemented, as well for the steel as other industries, like cement production.

Several steel producers defined strategies to substitute the blast furnace / BOF route by a combination of Direct Reduction Iron (based on "green" hydrogen reduction) with EAF, heated with renewable energy [171, 177]. Based on natural gas there is already a limited DRI production in the world (in 2019 108 mio. tons DRI compared to 1.3 bn. tons pig iron). It causes about 1487 kg CO_2 per ton crude steel [171]. The dry DRI process does not generate any slag, but the EAF will do. This slag will be different compared to today's EAF slag based on scrap re-melting.

All new steel production routes will produce new types of slags, too. However, the volume, the chemical and mineralogical composition and the physical properties of these new liquid and hardened slags are yet unknown. Thus, also their technical properties are still unknown.

8 Outlook

Since decades the successful utilization of iron and steel slags coming from the established steel making routes is a good example for saving natural resources. It is a worthwhile and environmental friendly contribution to a circular economy being realized many years before "sustainability" became part of the daily discussion. While the use of granulated blast furnace slag for cement and concrete is established worldwide, the use of steel slags as aggregates for road and waterway construction or concrete is often restricted. The increasing demand for natural resources, like gravel and sand, may convince the authorities and standardization committees to introduce these anthropogenic resources into the relevant system of rules as good experiences with the application have been made in different countries for several years.

In the future it will be a task to increase the steel slag utilization as a cement constituent. The motivation is the increasing demand on cement and concrete worldwide, but also upcoming restrictions in some countries regarding the classical utilization of steel slags in road making or agriculture and the higher profit margin for binder constituents compared to aggregates.

Another important task will be the utilization of the slags resulting from new steel production routes. Without slag utilization it seems to be very improbably that the new steel making processes, promising very high CO_2 reduction, would be established.

References

[1] Data of World Steel Association, Brussels.
[2] Calculations of FEhS institute, Duisburg.
[3] Merkel, T.: Data on production and utilization of iron and steel slags 2019 (in German). Report of FEhS Institute 27 (2020) 1, 29–30.
[4] Verein Deutscher Eisenhüttenleute: Slag Atlas, 2nd ed., Düsseldorf, 1995.
[5] Steel Institute VDEh: Steelmanual. Düsseldorf, 2015.
[6] Beck, L.: Geschichte des Eisens. Band 3, Das XVIII. Jahrhundert. Braunschweig, 1897.
[7] AJO-Anlagentechnik GmbH & Co. KG: Granulation von Hochofenschlacke, Freudenberg, 1995.
[8] Paul Wurth, Luxembourg, 2020.
[9] Kunicki, M.: La production des laitiers GALEX bouleté ou granulé et leurs applications. Silicates Industriel 42 (1977) 3, 91–98.
[10] Pereme, J.: Vitrified blast furnace slag: Granulated or pelletised. Proceedings of the 3rd European Slag Conference in Keyworth, 2002, EUROSLAG-Publication No. 2 (2003) 31–35.
[11] Jantzen, G.: Einrichtung zur Luftgranulation flüssiger Schlacken auf den Buderus'schen Eisenwerken, Stahl+Eisen 30 (1910) 20, 824–827.
[12] Li, G.: Slag valorisation in China – an overview. Proceedings of the 1st International Slag Valorisation Symposium, Leuven, 2009, 165–176.
[13] Huang, J.: The development of BSSF slag treatment technology. 8th European Slag Conference, Linz, 2015.
[14] Oh, S., Cha, S.: Diversified applications of PS Ball (Precious Slag Ball). Ecomaister Co. Ltd., 2008.
[15] Ehrenberg, A.: Dry granulation of liquid blast furnace slag – 2 tasks and 2 methods. Proceedings 14th International Congress on the Chemistry of Cements, Beijing, 2015.
[16] Algermissen, D. et al.: Heat recovery of EAF slag in consideration of further application. 9th European Slag Conference, Metz, 2017.
[17] Kappes, H., Michels, D.: Dry granulation and heat recovery. Proceedings of the 4th International Slag Valorisation Symposium, Leuven, 2015, 39–52.

[18] Moon, J.-W., Kim, H.-S., Sasaki, Y.: Energy recuperation from slags. Proceedings of the 1st-6th International Slag Valorisation Symposium, Leuven, 2009, 143–150.

[19] Hsieh, W., Tseng, Y.-H.: Dry slag granulation of modified BOF slag. Proceedings of the 4th International Slag Valorisation Symposium, Leuven, 2015, 101–105.

[20] Barati, M., Jahanshahi, S.: Granulation and heat recovery from metallurgical slags. Journal for Sustainable Metallurgy 6 (2020) 191–206 https://doi.org/10.1007/s40831-019-00256-4.

[21] Siemens AG: Method and device for using waste heat released in granulating a liquid slag. World patent WO 2011/036180 A1.

[22] Pickering, S. J. et al.: New process for dry granulation and heat recovery from molten blast-furnace slag. Ironmaking and Steelmaking 12 (1985) 1, 14–21.

[23] Davy McKee (Stockton) Ltd.: Slag granulation. World patent WO 95/05485.

[24] McDonald, I.: Dry slag granulation with heat recovery. 7th European Slag Conference, Ijmuiden, 09.-11.10.2013.

[25] Fenzl, T.: Dry granulated BF sand: A groundbreaking and sustainable innovation – production process, product grinding and building materials investigations. 14th Global Slag Conference, Aachen, 2019.

[26] Ehrenberg, A. et al.: Influence of the thermal history of granulated blast furnace slags on their latent-hydraulic reactivity in cementitious systems. Journal of Sustainable Metallurgy 6 (2020) 207–215 https://doi.org/10.1007/s40831-020-00269-4.

[27] Ehrenberg, A. et al.: Dry and wet granulated blast furnace slag – Comparison of their cementitious properties. 7th International Slag Valorisation Symposium, Leuven, 2021.

[28] Jahanshahi, S., Pan, Y., Xie, D.: Some fundamental aspects of the slag dry granulation process. 9th International Conference on Molten Slags, Fluxes and Salts, Beijing, 2012.

[29] Norgate, T. E., Xie, D., Jahanshahi, S.: Technical and economic evaluation of slag dry granulation. AISTech 2012 – The Iron & Steel Technology Conference and Exposition, Atlanta, 2012.

[30] Qin, R. et al.: Dry granulation of molten slag using a rotating multi-nozzle cup atomizer and characterization of slag particles. Steel Research 84 (2013) 9, 852–862 https://doi.org/10.1002/srin.201200325.

[31] Yasutaka, T., Tobo, H., Watanabe, K.: Development of manufacturing process for blast furnace slag coarse aggregate with low water absorption. JFE Technical Report (2018) 23.

[32] Kappes H., Michels D.: Dry slag granulation with energy recovery: From inception to pilot plant. European Steel Environment & Energy Congress (ESEC), Teesside, 2014.

[33] Regulation (EC) No 1907/2006 of the European Parliament and of the Council of 18 December 2006 concerning the Registration, Evaluation, Authorisation and Restriction of Chemicals (REACH), establishing a European Chemicals Agency. Official Journal L 396, 30.12.2006.

[34] https://echa.europa.eu/home (24/11/2020).

[35] Geiseler, J.: Composition and structure of slags. In: Verein Deutscher Eisenhüttenleute: Slag Atlas. 2nd edition, Düsseldorf, 1995, 215–224.

[36] FEhS – Institut für Baustoff-Forschung: Database on iron and steel slags.

[37] Drissen, P., Kühn, M., Schrey, H.: Successful treatment of liquid steel slag at Thyssen Krupp steel works to solve thepProblem of volume stability. 3rd European Oxygen Steelmaking Conference, Birmingham, 2000.

[38] DIN 4301:2009-06: Ferrous and non-ferrous metallurgical slag for civil engineering and building construction use.

[39] Mudersbach, D. et al.: Optimization of mathematical models to calculate the viscosity of slags (in German), Report des Forschungsinstituts 6 (1999) 2, 9–12.

[40] Indian Ministery of Steel: National steel policy 2017. New Delhi, 2017.

[41] European Steel Platform (ESTEP): Green Steel by EAF-Workshop report. Bergamo, 2019.

[42] Hörder Bergwerks- und Hüttenverein. Dortmund, 1853.

[43] Stahl+Eisen 4 (1884) 8, 500.

[44] Verein Deutscher Eisenhüttenleute: Gemeinfaßliche Darstellung des Eisenhüttenwesens. 3rd ed., Düsseldorf, 1896.

[45] Fleißner, H.: Eisenhochofenschlacken und ihre Verwendung, Halle a.S., 1911.

[46] thyssenkrupp Konzernarchiv, Duisburg.

[47] European Commission – Joint Research Centre: Best Available Techniques (BAT) Reference Document for Iron and Steel Production. Seville, 2013.

[48] Bundesanstalt für Geowissenschaften und Rohstoffe BGR: Deutschland – Rohstoffsituation 2018. Hannover, 2019.

[49] Bundesverband Baustoffe – Steine und Erden e.V.: The demand for primary and secondary raw materials in the mineral and building materials industry in Germany up to 2035. Berlin, 2016.

[50] Elsner, H.: Sand – Auch in Deutschland bald knapp? BGR/DRA Commodity TopNews 56 vom 23.02. 2018.

[51] United Nations Environmental Programme UNEP: 2019 – Sand and sustainability: Finding new solutions for environmental governance of global sand resources. Genf, 2019.

[52] Organisation for Economic Co-operation and Development OECD: Global material resourses outlook to 2060. Paris, 2018.

[53] Guttmann, A.: Die Verwendung der Hochofenschlacke im Bauwesen. 2nd ed., Düsseldorf, 1934.

[54] Josephson, G. W., Sillers, F., Runner, D. G.: Iron blast-furnace slag – Production, processing, properties, and uses. Washington, 1949.

[55] Keil, F.: Hochofenschlacke. 2nd ed., Düsseldorf, 1963.

[56] Lee, A. R.: Blastfurnace and steel slag. London, 1974.

[57] Commission of the European Communities: Proceedings of the Information day on utilization of blast furnace and steelmaking slags, Liège, 27.01.1988.

[58] Nippon Slag Association: Properties and effective uses of iron and steel slag. Tokyo, 1996.

[59] National Slag Association: Symposia 1928 ff.

[60] Forschungsgemeinschaft Eisenhüttenschlacken: Iron and steel slags – Properties and utilisation. Schriftenreihe der Forschungsgemeinschaft Eisenhüttenschlacken, No. 8, 2000.

[61] EUROSLAG: Proceedings of the 2nd-5th European Slag Conferences in Düsseldorf, 2000, Keyworth, 2002, Oulu, 2005, Luxembourg, 2007, Madrid, 2010.

[62] KU Leuven: Proceedings of the 1st-6th International Slag Valorisation Symposia in Leuven, 2009, 2011, 2013, 2015, 2017, Mechelen, 2019.

[63] FEhS – Building Materials Institute www.fehs.de.

[64] EUROSLAG – The European Slag Association www.euroslag.org.

[65] NSA – National Slag Association www.nationalslag.org.

[66] SCA – Slag Cement Association www.slagcement.org.

[67] Nippon Slag Association www.slg.jp.

[68] ASA – Australasian (iron and steel) Slag Association www.asa-inc.org.au.

[69] Beck, L.: Geschichte des Eisens. Band 2, Das XVI. und XVII. Jahrhundert. Braunschweig, 1895.

[70] Ehrenberg, A.: Schlacken auf der Eisenhütte St. Antony. Industrie-Kultur (2004) 3, 32–33.

[71] Tiemann, W.: Ueber Schlackentransport und Schlackengranulation. Stahl und Eisen 3 (1883) 10, 547–551.

[72] Abridgments of the specifications relating to the manufacturing of iron and steel. London, 1858.

[73] Gunnarsson, A., Nyblom, P.: Slaggsten och slagghus. Stockholm, 2016.

[74] Kongl. Maj: tsFörnyade Masmästare-Ordning. 26.06.1766.

[75] Solger, H.: Der Kreis Beuthen in Oberschlesien. Breslau, 1860.

[76] Bönisch: Vortrag über die Verwendung der bei der Eisen- und Zinkfabrikation gewonnenen Nebenprodukte zu baulichen Zwecken auf der Versammlung des Architekten-Vereins zu Berlin am 18.02.1865. Zeitschrift für Bauwesen 15 (1865) 383–388.

[77] Lürmann, F. W.: Mauersteine aus granulierten Schlacken. Stahl und Eisen 17 (1897) 23, 991–999.

[78] Bosse, R.: Ueber Cementfabrication aus Hochofenschlacke und deren neueste Vervollkommung. Stahl und Eisen 5 (1885) 9, 497–501.

[79] Locher, F. W.: Cement. Erkrath, 2013.

[80] 75 Jahre Mansfelder Pflastersteine 1863-1938. Eisleben, 1939.

[81] Lüer, H.: Teerstraßenbau unter besonderer Berücksichtigung der Hochofenschlacke, Essen, 1931, 131.

[82] Guttmann, A.: Verwendbarkeit und Eigenschaften von Schlackenwolle. Stahl und Eisen 49 (1929) 4, 97–101.

[83] Association for iron and steel slags, Düsseldorf, 1937.

[84] Haase, B.: Overview of residue utilisation in Sweden – Focus on by-products from the iron and steel industry. Proceedings of the 1st International Slag Valorisation Symposium, Leuven, 2009, 185–194.

[85] www.euroslag.org (07.06.2020).

[86] Keil, F.: Slag cements. Proceedings of the 3rd International Congress on the Chemistry of Cement. London, 1952, 530–580.

[87] Kramer, W.: Blast-furnace slags and slag cements. Proceedings of the 4th International Congress on the Chemistry of Cement, Washington, 1960, Vol. 2, 957–981.

[88] Schröder, F.: Slags and slag cement. Proceedings of the 5th International Congress on the Chemistry of Cement, Tokyo, 1968, Vol. 4, 149–199.

[89] Smolczyk, H. G.: Slag structure and identification of slags. Proceedings of the 7th International Congress on the Chemistry of Cement, Paris, 1980, Vol III, III-1/3–III-1/17.

[90] Ehrenberg, A.: Granulated blast furnace slag – From laboratory into practice. Proceedings of the 14th International Congress on the Chemistry of Cements, Beijing, 2015.

[91] Bijen, J.: Blast furnace slag cement for durable marine structures. 's-Hertogenbosch, 1996.

[92] Weber, R. et al.: Hochofenzement, 2. ed., Düsseldorf, 1998.

[93] Ehrenberg, A.: Granulated blast furnace slag – An efficient building material with tradition and future (in German). Beton-Informationen 46 (2006) 4, 35–63; 5, 67–95.

[94] American Concrete Institute: Ground granulated blast-furnace slag as a cementitious constituent in concrete. ACI 233-R-95, 1996.

[95] Reeves, C. M.: The use of ground granulated blastfurnace slag for within-mixer production of Portland-blastfurnace cement concretes: Background and development of the system in the U.K. Silicates Industriels (1982) 4/5, 109–113.

[96] Drissen, P.: Determination of the glass content in granulated blastfurnace slag. Zement-Kalk-Gips 48 (1995) 1, 59–62.

[97] Spieß, L. et al.: Moderne Röntgenbeugung – Röntgendiffraktometrie für Materialwissenschaftler, Physiker und Chemiker. 3rd edition, Berlin, 2019.

[98] Glukhovky, V. D.: Soil silicates. Kiev, 1959.

[99] Krivenko, P. V. (editor): 1st International Conference on Alkaline Cements and Concretes. Kiev, 1994.

[100] Shi, C., Krivenko, P. V., Roy, D.: Alkali-activated cements and concretes. New York, 2006.

[101] Provis, J. L., van Deventer, J.: Alkali Activated Materials. State-of-the-Art Report Volume 13, RILEM TC 224-AAM, Heidelberg, 2014.

[102] Ehrenberg, A. et al.: Granulated blastfurnace slag: reaction potential and production of optimized cements. Cement International 6 (2008) 2, 90–96, 3, 82–92.

[103] EN 197-1: 2011, Cement – Part 1: Composition, specifications and conformity criteria for common cements.

[104] prEN 197-5: 2020, Cement – Part 5: Portland-composite cement CEM II/C-M and Composite cement CEM VI.

[105] Der Fernmeldeturm in Dortmund. Beton-Informationen 1 (1961) 4, 41–42.

[106] Stratmann, E.: Der Kirchturm der katholischen Gemeinde "Christus – Unser Friede" in Duisburg-Meiderich. Beton-Informationen 36 (1996) 6, 87–94.

[107] Zementfabrik Thuringia: Thurament. 2nd ed., Unterwellenborn, 1938.

[108] BU Hydro Power Plants Germany.

[109] Data of Verein Deutscher Zementwerke, Düsseldorf.

[110] CEMBUREAU: The role of cement in the 2050 low carbon economy. Brussels, 2019.

[111] Lehne, J., Preston, F.: Making concrete change – Innovation in low-carbon cement and concrete. Chatham House Report, London, 2018.

[112] Verein Deutscher Zementwerke: Verminderung der CO_2-Emissionen. Monitoring-Abschlussbericht 1990–2012. Düsseldorf, 2013.

[113] Ehrenberg, A.: Ferrous slags – Really a contribution to low-carbon binders and concretes? 8th European Slag Conference, Linz, 2015.

[114] European Commission: Communication from the Commission to the European Parliament, the Council, the European Economic and Social Committee and the Committee of the Regions – Public procurement for a better environment. COM/2008/0400.

[115] CEMBUREAU: Cementing the European Green Deal. Brussels, 2020.

[116] Verein Deutscher Zementwerke: Dekarbonisierung von Zement und Beton – Minderungspfade und Handlungsstrategien. Düsseldorf, 2020. https://www.vdz-online.de/en/cement-industry /climate-protection (03.12.2020).

[117] Wolter, A., Locher, G., Geiseler, J.: Blast furnace slag (BFS) as a raw materials substitute for clinker burning. Proceedings of the VDZ Congress, Düsseldorf, 2002, 361–367.

[118] National Slag Association: CemStar[sm] process: Slag usage raises productivity, operational efficiency, lowers emissions. Technical Bulletin 201–1.

[119] Kersting, K., Adelmann, M., Breuer, D.: Bestimmung des Chrom(VI)-Gehaltes in Zementen. Staub – Reinhaltung der Luft 54 (1994) 11, 409–413.

[120] Pisters, H.: Chrom im Zement und Chromatekzem. Zement-Kalk-Gips 19 (1966) 10, 467–472.

[121] Frias, M. et al.: Contribution of toxic elements: Hexavalent chromium in materials used in the manufacture of cement. Cement and Concrete Research 24 (1994) 3, 533–541.

[122] Frias, M., Sanchez Rojas, M. I.: Determination and quantification of total Chromium and water soluble chromium contents in commercial cements. Cement and Concrete Research 25 (1995) 2, 433–449.

[123] American Concrete Institute: Blast furnace slag as concrete aggregate. Report of committee 201, 1931.

[124] Gutt, W., Kinniburgh, W., Newman, A.J.: Blastfurnace slag as aggregate for concrete. Magazine of Concrete Research 19 (1967) 59, 71–82.

[125] Morian, D. A., Van Dam, T., Perera, R.: Use of air-cooled blast furnace slag as coarse aggregate in concrete pavements. U.S. Dpt. of Transportation, Federal Highway Administration, FHWA-HIF-12-008. Washington, 2012.

[126] Verian, K. P., Panchmatia, P., Olek, J.: Investigation of use of slag aggregates and slag cements in concrete pavements to reduce the maintenance cost. Indiana Department of Transportation, FHWA/IN/JTRP2017/17, West Lafayette, 2017.

[127] Lang, E., Tabani, H.: Stahlwerksschlacken als Gesteinskörnung für Mörtel und Beton. Report des FEhS-Instituts 10 (2003) 2, 1–3.

[128] Gelfand, J.; Vinkeloe, R.; Witte, G.: Die neue Kokerei in Duisburg-Huckingen. Beton-Informationen 13 (1983) 5, 55–63.

[129] Knaack, A. M.; Kurama, Y. C.; Kirkner, D. J.: Stress-strain properties of concrete at elevated temperatures. Structural Engineering Research Report, Department of Civil Engineering and Geological Science at University of Notre Dame, Notre Dame, 2009.

[130] Papayianni, I., Anastasiou, E.: Utilization of electric arc furnace steel slags in concrete products. 6th European Slag Conference, Madrid, 2010.

[131] Faleschini, F., M.A. Zanini, M. A., Pellegrino, C.: New perspectives in the use of electric arc furnace slag as coarse aggregate for structural concrete. 8th European Slag Conference, Linz, 2015.

[132] Anastasiou, E.: Heavyweight concrete with ferronickel and steel slag aggregates. 10th European Slag Conference, Thessaloniki, 2019.

[133] Vollpracht A., Brameshuber, W.: Binding and leaching of trace elements in Portland cement pastes. Cement and Concrete Research 79 (2015) 76–92.

[134] Tikkakoski, A., Kujala, K., Mäkikyrö, M.: Geotechnical properties of blast furnace slag and LD-steel slag mixtures for road construction. 4th European Slag Conference, Oulu, 2005.

[135] Shiramata, E.K., Teixeira Moreira, R.F., Bernabe, V. L.: 10 years of "NOVOS CAMINHOS" program: Use of steel slag to promote mobility with sustainability. 9th European Slag Conference, Metz, 2017.

[136] Emery, J.: Steel slag utilization in asphalt mixes. National Slag Association Technical Bulletin 186–1.

[137] Australasian Slag Association: A guide to the use of iron and steel slag in roads. Wollongong, 2002.

[138] Nippon Slag Association: Iron and steel slag products – products for roads. http://www.slg. jp/e/slag/product/road.html (17.04.2020).

[139] Bialucha, R. Wetzel, T., Merkel, Th.: Verwendung von Stahlwerksschlacken in Landschaftsbaumaßnahmen und Lärmschutzparks. In: Schlacken aus der Metallurgie, Vol. 3. Neuruppin, 2014, 79–97.

[140] EN 13242:2008, Aggregates for unbound and hydraulically bound materials for use in civil engineering work and road construction.

[141] EN 13043:2002, Aggregates for bituminous mixtures and surface treatments for roads, airfields and other trafficked areas.

[142] Nakagawa, M. et al.: Technology of constructing seaweed beds by steel-making slag. 6th European Slag Conference, Madrid, 2010.

[143] Rex, M.: Blastfurnace and steel slags as liming materials for sustainable agricultural production. 2nd European Slag Conference, 2000, Düsseldorf.

[144] Pihl, U.: Slag based fertilizers – Best practice for circular economy. 10th European Slag Conference, Thessaloniki, 2019.

[145] Zanen, J. P.: Thomasphosphat in der europäischen Landwirtschaft. Essen, 2nd ed., 1954.

[146] Burnie, R. W.: Memoir and letters of Sidney Gilchrist Thomas, Inventor. London, 1891, 291–292.

[147] Fixariss, M.: Steel slag used as liming agent for agriculture. 14th Global Slag Conference, Aachen, 2014.

[148] Drissen, P. et al.: Enrichment of phosphorus in BOF-slag for improved application in agriculture. 8th European Slag Conference, Linz, 2015.

[149] Kollo, H.: Untersuchungen zur Frage der zementtechnologischen Eignung einer LD-Konverterschlacke als latent-hydraulischer Zumahlstoff. Doctorate thesis, Munich, 1985.

[150] Dongxue, L., et al.: Durability study of steel slag cement. Cement and Concrete Research 27 (1997) 7, 983–987.

[151] Sersale, R. et al.: Characterization and potential uses of a steel slag. Silicates Industriels 51 (1986) 11/12, 163–170.

[152] Adolfsson, D. et al.: Cementitious phases in ladle slags. Steel Research 82 (2011) 4, 398–403.

[153] Duda, A.: Hydraulic reactions of LD steelwork slags. Cement and Concrete Research 19 (1989) 5, 793–801.

[154] Isoo, T. et al.: Development of large steelmaking slag blocks using a new carbonation process. Advances in Cement Research 12 (2000) 3, 97–101.

[155] Nippon Kokan Kaisha: CO_2 reduction technology by slag & marine blocks. Tokyo, 2000.

[156] Piret, J., Dralants, A.: Verwertung von LD-Schlacke zur Erzeugung von Portlandzementklinker und Roheisen. Stahl und Eisen 104 (1984) 16, 42–46.

[157] Ludwig, H.-M.: Nutzungspotentiale für Stahlwerksschlacken in der Baustoffindustrie. 20. Internationale Baustofftagung ibausil, Weimar, 2018.

[158] Ehrenberg, A., Algermissen, D.: Transformation of electric arc furnace slag into a (latent) hydraulic material. 9th European Slag Conference, Metz, 2017.

[159] Bialucha, R., Merkel, T., Motz. H.: European environmental policy and its influence on the use of slag products. Proceedings of the 1st International Slag Valorisation Symposium, Leuven, 2009, 121–131.

[160] Kobesen, H.: Legal status of slag valorisation. Proceedings of the 2nd International Slag Valorisation Symposium, Leuven, 2011, 201–213.

[161] European Commission: European Waste Catalogue. 2014/955/EU.

[162] European Commission: Waste Framework Directive. 2008/98/EC.

[163] European Commission: Communication from the Commission to the Council and the European Parliament on the Interpretative Communication on waste and by-products. COM/2007/0059.

[164] EUROSLAG: Position paper the legal status of slags. Duisburg, 2006.

[165] EUROSLAG: Position paper on the legal status of ferrous slag complying with the Waste Framework Directive (Articles 5 / 6) and the REACH regulation. Duisburg, 2011.

[166] Versteyl, L.-A.: Eisenhüttenschlacken, Abfall oder Produkt? Schriftenreihe der Forschungsgemeinschaft Eisenhüttenschlacken, Heft 5, 1998.

[167] Versteyl, L.-A., Jacobj, H.: Gutachten über den rechtlichen Status von Schlacken aus der Eisen- und Stahlherstellung, Schriftenreihe des FEhS – Instituts für Baustoff-Forschung, Heft 12, 2005.

[168] Regulation (EC) No 1907/2006 of the European Parliament and of the Council of 18 December 2006 concerning the Registration, Evaluation, Authorisation and Restriction of Chemicals (REACH)

[169] Jochims, K., Bialucha, R.: Toxicological investigations of ferrous slag in the context of the REACH registration. Report - Research projects of the FEhS institute 26 (2019) 2, 33–34.

[170] Boston Consulting Group: Steel's contribution to a low-carbon Europe 2050. Boston, 2013.

[171] Arens, M. et al.: Pathways to a low-carbon iron and steel industry in the medium-term – the case of Germany. Journal of Cleaner Production 163 (2017) 84–98.

[172] Lüngen, H. B., Knop, K., Steffen, R.: State of the art of the direct reduction and smelting reduction processes. Stahl und Eisen 126 (2006) 7, 25–40.

[173] Primetals Technologies: COREX – Efficient and environmentally friendly smelting reduction. Linz, 2015.

[174] Alexander, M. G. et al.: Corex slag – properties and use of a new cementitious material. ZKG International 57 (2004) 10, 68–75.

[175] Santos, S. et al.: CO_2 capture in combination with the HIsarna process. METEC, Düsseldorf, 2019.

[176] European Commission: The Green Deal. Communication from the Commission to the European Parliament, the European Council, the Council, The European Economic and Social Committee and the Committee of the Regions, COM(2019) 640, 11.12.2019.

[177] Stockholm Environment Institute, Lund University: Hydrogen steelmaking for a low-carbon economy. EESS report No. 109, 2018.

Thomas A. Bier, Eva Kränzlein, Elsa Qoku, Sandra Waida

Chapter 7
Utilization of Supplementary cementitious materials (SCM) in Portland cement, alkali activated and ternary binders

Abstract: Amongst industrial wastes used in the construction business, so-called secondary cementitious materials (SCM) represent an important group. SCM are non-metallic fine powders, which are added to cement directly or replace cement partially in concrete and mortars. Their origin is either natural or artificial, where industrial wastes and industrial main products can be grouped.

The first part of the article describes the mostly used SCM by their origin, process of manufacture, chemical and physical properties and eventually their reactivity with water and usually lime rich materials (binders). Some industrial waste products, which do not fall under the notion of SCM are included as well.

One chapter is dedicated to the usage of SCM in OPC based systems, in alkali activated binders and ternary binders.

Keywords: origin, reactivity, rheological properties, mechanical properties, long-term behaviour

1 Introduction

1.1 General description

Most industrial wastes represent a side product or residue from a high temperature process such as coal power plants, iron, aluminium or other metallurgical processes. A larger number of these waste products are used in cement and concrete technology as dry, fine powders with a fineness comparable to flour (<100 µm). They are considered as supplementary cementitious materials (SCM). A comprehensive description of SCMs is given in the state-of-the-art report of the RILEM Technical Committee 238-SCM entitled "Properties of Fresh and Hardened Concrete Containing Supplementary Cementitious Materials" [1]. However, lime stone powder, metakaolin, calcined clay and ground rocks constitute as well widely

Thomas A. Bier, Eva Kränzlein, Elsa Qoku, Sandra Waida, TU Bergakademie Freiberg, IKFVW, Leipziger Straße 28, 09599 Freiberg

https://doi.org/10.1515/9783110674941-007

used SCMs in concrete technology. Due to their importance specifically when used in combination with waste materials, they are considered as well in this chapter.

The European Cement Standard EN 197 [2] specifies to which extent SCMs can be added to cement clinker for the five different cement types.

1.2 Powders involved in concrete and mortar technology

Modern concretes and mortars such as Self Compacting Concrete (SCC) are often considered as pentanary materials where aggregates, water, admixtures, cement and additives are combined. By this approach, admixtures are considered as chemicals, which are dosed in very small amounts in order to improve fresh properties such as retarders, plasticizers, accelerators etc. Additives are finely dispersed powders with a usual dosage above 5 percent of mass of the amount of cement used. Additives and/or cement make up the binder part of concrete and mortar. A non-metallic, inorganic binder is defined as a finely ground powder which forms a paste when mixed with water and hardens rock like as a function of time. Hardening happens as a hydraulic, hydratic, latent hydraulic or pozzolanic reaction with water to transform anhydrous phases to hydrates. In most recent publications the notion quaternary binders is used for OPC combined with three mineral additives such as Lime stone powder, fly ash, slag, metakaolin etc [2, 3]. or in few cases where pozzolanic materials are added to a ternary binder as defined above.

Three scenarios will be considered:
- The classical use of SCM in Portland cement, concrete and mortar. The binding part consists always of cement and SCM
- The use of SCM or waste materials without cement where the hardening is achieved by an alkali activation [3]
- The use of SCM with several cementitious or hydratable powders such as Portland cement, alumina cement and calcium sulphate. These binders are referred to as ternary binders [4]

2 Description and classification of major industrial wastes

2.1 General

Mineral or inorganic, non-metallic waste materials as well as classical raw materials (industrial, natural or chemical products) are listed – together with the abbreviations used throughout the text – in Table 7.1.

Table 7.1: Classification for inorganic waste and raw materials used in cement and concrete technology.

Inorganic waste materials Group I	Organic wastes or residues from the agro industry Group II	Inorganic raw materials Group III
GGBFS-Ground granulated blast furnace slag	RHA-Rice Husk Ash	LSP Limestone Powder
Metallurgical slag	BA Bagasse Ash	Metakaolin
Pulverized Fly Ash (pfa)	Palm oil fuel ash	Calcined Clays
Red Mud	Ashes from various plants such as cotton etc.	Natural Pozzolans
Silica Fume	Saw Dust	PC Portland Cement
Waste Glass		CAC Calcium Aluminate Cement
Recycled concrete		
Ground clay bricks		
Gypsum from flue gas desulfurization (fgd)		
Anhydrite as by-product from acid production		

Part of the materials are powder products (SCM and others) and a small part is used as crushed material. The powdery materials are typically used as part of a binder (cf. CEMII through CEMV according to EN 197) or added to concrete and high performance mortars. The materials are organised in three different groups.

Most of the waste products in group I and II represent inorganic, non-metallic or mineral powders from high temperature processes. Fly ash, other ashes and silica fume are already available as a fine powder where slags, waste glass, bricks and concrete need to be ground to cement fineness. Chemically the major components important for the utilization in cementitious materials are Al_2O_3, SiO_2, CaO and MgO.

2.2 Slags

Ground granulated blast furnace slag (GGBFS) represents a traditional waste product used in cement and concrete technology as a finely ground powder. GGBFS is a latent hydraulic material. Ehrenberg gives a comprehensive description of the state of art of GGBFS and metallurgical slags e.g. steel slag in chapter 2.3 of this book [5].

2.3 Silica Fume (SF) and Pulverized Fly Ash (PFA)

SF and PFA originate from combustion processes as fine rounded particles, which are recovered from the flue gases. SF comes with a high content of SiO_2 while PFA can be positioned in the so-called diagram Rankin (ternary phase diagram SiO_2 – Al_2O_3 – CaO) close to the binary line SiO_2 -Al_2O_3 for ashes with low CaO content. Ca rich PFA are roughly in the middle of the diagram close latent hydraulic materials. Silica Fume and Fly Ash are both used in cement technology since the 1980 and there are numerous publications and conferences dedicated to the subject. [cf [2]. and [6] in this book].

2.4 Metakaolin and limestone powder

Metakaolin, Calcined Clays, Natural Pozzolans and Limestone powder are no waste materials, but nowadays-important additives (SCM) to cement and concrete to reduce their CO_2 footprint. Furthermore, they serve as part of the composition of mineral additives, which consist of several powders. They are as well described in [2].

2.5 Red Mud

Red Mud originates from the leaching of bauxite ores (Bayer process) with NaOH. This waste is mostly deposited in large ponds, which represent an environmental hazard. Nevertheless, no large-scale use is being reported in literature. Some literature reports – similar to metallurgical slag – sinter and melting processes to extract metals from the waste material [7, 8]. Through a sintering process, Al_2O_3 can also be recovered and used as a raw material or partly substitution of raw materials in Portland cement or alumina cement production [9]. Due to its alkalinity, the use as a component for geopolymers has been reported in conjunction with slag, fly ash or rice husk ash [10, 11].

2.6 Ground clay bricks, waste glass and marble dust

While waste glass is used as raw material in glass smelters, publications during the last two decades report as well applications where glass powder replaces part of the binder in mortars and concrete [12, 13]. Similar investigations have been reported for the use of ground clay bricks [14, 15] and marble dust.

2.7 Rice husk ash

2.7.1 Introduction

Rice is considered a staple food throughout the world and is cultivated in many countries including China, India, Thailand, Brazil, the United States and Italy. The rice production worldwide reaches almost 700 million tons. During the process, the grains are ground in order to remove the outer shells, which are then called rice husks and are then regarded as agricultural waste material and only partly being used as fuel for rice drying or as a local fuel.

Rice husks represent 20% of the weight of rice and consist of 50% cellulose, 25%–30% lignin, 15%–20% silica and 10%–15% water [1]. When rice husks are burned rice, husk ash is generated. During the combustion process the evaporable components of rice husk escape and mostly silica remains. The properties of the rice husk ash depend on the combustion process. The outcome the of combustion process, namely the risk husk ash, is influenced by the following factors: composition of the rice husk, burning temperature, burning time, cooling rate and grinding.

The cultivation of rice will continue to increase over the next decades resulting in an increased amount of ash. Starting in the 1980s, research in some countries concerning the use of rice husk ash, such as India, Pakistan and the United States intensified. Rice husk ash has a high silica content of around 85%, more than other agricultural residues. With its high porosity, low weight and high surface, it is a valuable material for industrial applications.

2.7.2 The use of rice husk

Rice husks are already used in some areas, for example as a clarifying agent for beer, filler ingredient in cheap pet food, insulating material, fertilizer in agriculture, and as a fuel. The burning of the rice husks creates an ash with a wide range of applications. In Figure 7.1 is shown an overview of utilization. Every application requires an ash with specific properties. The burning temperature, the heating rate and particle size control these properties.

If the fired temperature above 800 °C, the ashes contain a high percentage of crystalline silica and a low percentage of carbon, and have a low thermal conductivity [16]. Thereby they are suitable as an insulating material. In the refractory areas, the ashes are used as a cover for the molten metal in tundish or as lining [17]. Another application of the ash is as a filler in various types of rubber. Tensile strength, tear strength and hardness were improved in epoxidized natural rubber [18].

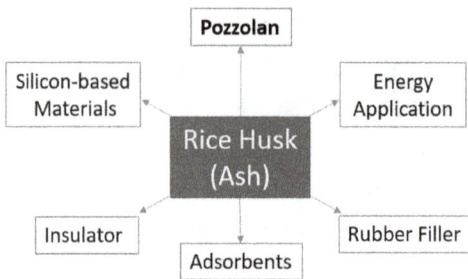

Figure 7.1: Schematic of applications of rice husk.

Silicon-based materials can also be manufactured using rice husk ash as a starting material, e.g. oxide ceramics (mullite, zeolite), non-oxide ceramics (silicon carbide, silicon nitride) and high-purified silicon [19–22].

Furthermore, adsorbents, which require a high specific surface, can be obtained from rice hush ash. A low firing temperature below 700 °C leads to an amorphous ash with a high surface area due to the structure of the shell. The carbon content can be influenced by the firing duration. The most common adsorbent is activated carbon and it is utilized in chemistry, medicine and wastewater treatment. Research has shown that activated carbon and highly microporous silica can be produced based on rice husk ash [23]. In Figure 7.2 a schematic for different ways to produce adsorbents is shown [24–27].

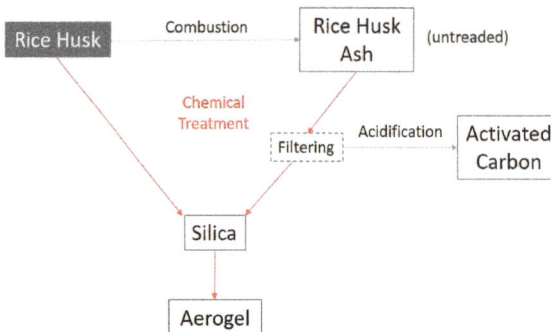

Figure 7.2: Schematic of treatments for adsorbents.

Amorphous rice husk ash has pozzolanic properties. For this reason, it is interesting to use rice husk ash in cementitious systems, e.g. in geopolymers, as an admixture and as a substitute for silica fume. Many researchers have investigated the use of rice husk ash as replacement for cement in different types of concrete.

2.7.3 Rice husk ash in cementitious materials

Combustion

The burning temperature and the duration have an influence on the composition of the rice husk ash. For cementitious materials, a controlled combustion under 700 ° C is useful, as these ashes have an amorphous phase, a high surface area and pozzolanic reactivity. Figure 7.3 shows an overview of the kilns used to produce these kinds of ashes. The furnaces can be divided into controlled and uncontrolled combustion. Because of the uncontrolled incineration, commonly ashes of poor quality are produced. Despite this problem, high quality ashes can be produced in rural areas showing pozzolanic properties and which are suitable as supplementary material [28].

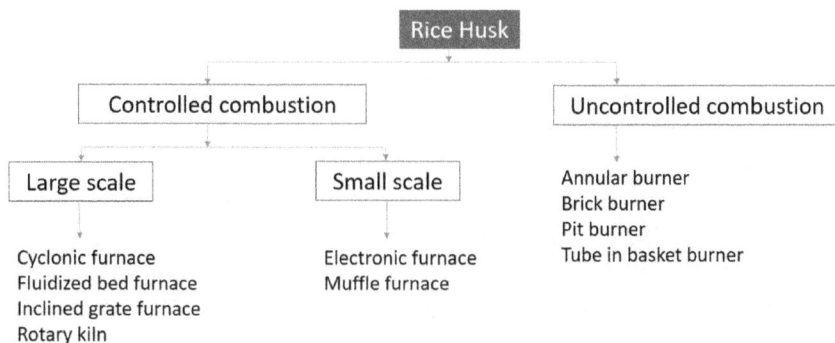

Figure 7.3: Production of rice husk ash by thermal treatment [29].

Moreover, the type of used furnace [30] influences the quality of the ashes. The fluidized bed reactor is suitable for the production of ashes on a larger scale. The controlled temperature and residence time in the firing chamber have an influence on the carbon level of ashes. The turbulences and the higher heat transfer allow a short residence time (<2 min) of the feed material in the combustion chamber [31, 32].

During the combustion, the process undergoes different stages, starting with the extraction of the residual moisture (drying). Then, the decompositions of lignin and cellulose begins (pyrolysis, degasification). The decomposition reactions increase with increasing temperature. Carbon-rich residues are formed during the process and burn at around 500 °C. Starting at around 700 °C, the amorphous silicon compounds convert into crystalline silica. A summary is shown in Table 7.2.

Table 7.2: Combustion of Rice Husk [29].

section	combustion stage			
	I	II	III	IV
temperature range [°C]	<100	100–350	350–700	>700
physico-chemical change	release of physically bound water	lignin, cellulose decomposition and volatiles liberation	carbon char burning and amorphous silica formation	amorphous silica conversion to crystalline form
surface area	–	increases	increases up to 500 °C, later decreases	decreases
grindability	–	–	increases up to 500 °C, later decreases	decreases

Characteristics

During the combustion, organic components evaporate, leaving an ash that is amorphous or crystalline, depending on the firing temperature, duration and cooling rate. If the firing temperature is below 700 °C, the ash is amorphous with a microporous cellular structure. This structure imaged by scanning electron micrograph is shown in Figure 7.4. The ash was produced under controlled combustion conditions in a muffle furnace at 580 °C and a holding time of two hours. Furthermore, a specific surface area of 150 m²/g was measured with nitrogen adsorption. Figure 7.5 shows the measurement of rice husk ash using X-ray diffraction. The amorphous phase is characterized by a broad hump between 15–30 2θ.

Figure 7.4: Scanning electron micrograph of rice husk ash.

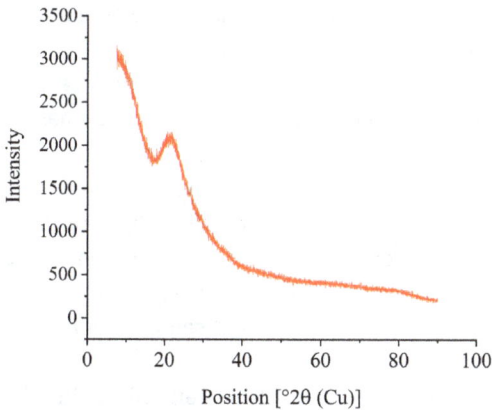

Figure 7.5: X-ray diffractogram of rice husk ash.

Rice husk ash is a fine-grained material and easily pulverized. The particles are angular and porous even after grinding. Even if the combustion process was not completely controlled and small residues of crystalline material and some carbon content remain, the two properties (water demand and pozzolanic activity) can be improved by finer grinding [33].

The microstructure of the rice husk ash is influenced by various parameters. These include temperature, time, type of furnace, chemical treatment and grinding. If the ash is treated, using an acidic leaching process, specific surfaces of 284 m^2/g can be achieved [34].

The reactivity of rice husk ash is influenced by two factors: the amorphous silica content and the specific surface. The chemical analyses of the ashes, which were produced by various researchers, indicate a silica content between 86–97% [35, 36]. If part of the cement is replaced by rice husk ash, the system shows an early hydration of C$_3$S [37]. The effect is attributed to the high surface area of the ashy particles. The fine particles act as crystallization nuclei for the formation of the hydrate phases in the cement paste. During the pozzolanic reaction, the amorphous silica reacts with the calcium hydroxide to form C-S-H phases [38]. This creates a denser paste with improved properties.

Addition to cement and concrete

Many researchers work on the use of rice husk ash in cement and concrete and investigate the fresh and hardened concrete properties. When using rice husk ash, the high specific surface area requires a higher water content in order to obtain the same workability compared to a reference sample without rice husk ash [39]. A possible solution, is the use or the higher dosage of superplasticizer (high-range water reducer), especially for high-performance concretes with a low water-cement ratio [40]. In high-performance concrete, mainly silica fume is used which has a similar composition to rice husk ash. Since silica fume is not available everywhere and is also expensive, rice husk ash is a suitable replacement material [41, 42].

Some studies have shown that rice husk ash as a partial replacement of cement enhances the compressive strength of concrete. The addition of rice husk ash has an optimal range of 10% to 30% [39]. The increase of compressive strength is achieved on the one hand by the filler effect and on the other hand by the pozzolanic reaction, which forms secondary C-S-H phases. The pozzolanic reaction modifies the pore structure, which thus densifies. The porosity of the cement paste with the addition of rice husk ash is reduced. This results in a higher compressive strength after a long-term storage (>90 d) compared to samples without rice husk ash and samples with silica fume as replacement material [35, 36, 40, 43].

The fine rice husk ash particles have a large water adsorption ability due the high specific surface and can reduce bleeding water on concrete (54). In addition, rice husk ash has the property of an internal curing agent. This effect has an influence on autogenous shrinkage. The replacement of cement by rice husk ash decreases the autogenous shrinkage as well as produces an autogeneous relative humidity change [44, 45].

The permeability of concrete plays an imported role in durability and is closely related to its pore size. The addition of rice husk ash refines the pore structure. This effect reduces the chloride penetration, decreases permeability and therefore enhances corrosion resistance properties [46].

Rice husk ash is a burnt agricultural by-product with a wide range of applications. Depending on the combustion process conditions, an ash with the required properties can be produced. If the rice husks are burned between 600 °C and 700 °C, the ash contains a high proportion of amorphous silica. This amorphous silica has pozzolanic properties, a high specific surface area and is considered a renewable raw material, which makes it highly interesting as a supplementary cementitious material. Rice husk ash offers the possibility to replace silica fume in high-performance concrete in rice-growing countries. The partial cement replacement in concrete enhances compressive strength in early age and after long-term storage. Furthermore, the rice husk ash in concrete shows good durability due the denser structure.

2.8 Bagasse ash and ashes from bio mass combustion

Besides rice husk ash, literature reports as well the use of ashes from Biomass such as Bagasse ash, Palm oil fuel ash, cotton stalks etc [47]. or even the addition of saw dust [48].

2.9 Recycled concrete

Recycled concrete is used as a crushed aggregate where maximum amounts of 20–30% by mass are allowed [49, 50].

2.10 Gypsum and anhydrite

Gypsum from flue gas desulfurization (fgd) in power plants and anhydrite from hydrofluoric and phosphoric acid production are widely used waste materials, but not considered as SCM. They are rather used as sole binders in the production of plasterboard (fgd) where Gypsum serves as a raw material to produce hemihydrate (plaster). Anhydrite from hydrofluoric acid production serves as a binder in flowable screeds as such and needs only to be activated. Additionally they are used as setting time regulator for Portland cement as well as a component in ternary binders (cf. chapter 5).

3 Use of SCM with Portland cement

3.1 General – classical concrete

As already mentioned in chapters 7.1 and 7.2.7, the use of waste materials as fine powders together with Portland cement can be considered a traditional or classical use. The addition of a number of waste materials is regulated by standards e.g. the European Standard EN 197.

When used in cement technology, part of the cement clinker is replaced or diluted by those fine powders and a composite binder or cement is achieved. In concrete technology, additional SCMs can be added when designing a mix. For both scenarios, which can be combined, the added waste material influences fresh and hardened properties as well. Sakir et al. [51] published a recent review, which focusses on the influence of various SCMs on fresh and hardened properties as well as on durability.

3.1.1 Fresh properties

Fresh properties such as workability and setting times are influenced by the SCM where their particle size distribution, particle shape, specific surface area and physico-chemical interaction with water influence the particle interactions on a microscopic level. Specifically

- Particle size distribution, influence together with the cement and aggregates used the particle packing in the fresh state (rheological properties such as workability, viscosity, yield stress etc.) as well as density and strength
- Particle shape, spherical particles such as silica fume and fly ash tend to improve flow properties as compared to the addition of angular particles such as ground slag, ground glass or limestone powder which tend to decrease flow values
- Specific surface area for fine powders increases with decreasing particle size. Moreover, the inner surface or tortuous surfaces contribute largely to an increase in surface area, which has a significant impact on water demand. Consequently, workability decreases significantly at constant water addition when waste powders with a high specific surface area are employed.
- Physico-chemical interaction with water during the first hours after water contact depends on the adsorption of water molecules on surfaces and/or the formation of interlayer water in narrow slits or pores. Additionally, modern concretes such as self-compacting concrete (SCC) or ultra-high performance concrete (UHPC) use admixtures such as plasticizers and viscosity modifying (VMA) or stabilizers which are adsorbed on surfaces as well and lead to a more complex influence of the powders nature to the technological properties of concrete. A chemical interaction in terms of hydration is less pronounced during the early times when fresh properties are the issue.

However, the microstructure of cement-based materials is characterized by a transition from a plastic or viscoelastic material to a solid-state material. After initial set of Portland cement based materials, the chemical mechanisms of hydration become much more important and lead to strength development. During the transition period or structuring period plastic shrinkage and depending on curing conditions autogenous shrinkage is observed. This early shrinkage depends largely on the nature of the waste powders added [52].

3.1.2 Hardened properties

The type of (waste) powder used will influence hardened properties such as the development of strength and porosity as well as durability to mention the most important properties more or less pronouncedly. A simplified concept considers two major influences:

- The particle size distribution together with the particle shape influence the packing density. There is for example the Andreasen [53] or Dinger–Funk [54] models, which optimize the particle packing based on the particle size distribution of the individual components of paste, mortar or concrete design. In addition, from the

Andreasen model an estimation of flow behavior (workability) can also be deduced. The packing density will influence strength, pore size distributions and
through permeability and diffusivity the durability of the materials.

- The chemical hydration reactions in combination with Portland cement occur as
latent hydraulic or pozzolanic reactions where the Ca^{2+} ions from the formed
calcium hydroxide can react over rather long periods to form additional C-S-H
phases. These chemical interactions play probably the most important role in
strength contribution and decrease of porosity. There is also inert waste materials or even natural SCM where no reaction occurs. Therefore, the reactivity of
SCM needs to be characterized by a test procedure involving soluble calcium,
as it is the case for mixtures with Portland cement [55]. In the end, the reactivity
depends on the solubility of Ca^{2+}, $Al(OH)_4^-$ and SiO_4^{4-} ions from the waste material. The solubility is generally associated with high surface areas and a poor
crystallinity of the waste material.

3.2 High Performance Concrete (HPC)

3.2.1 High strength concrete

High strength and ultra-high strength concretes have been developed in using particle packing especially with the addition of very fine (nano) particles such as silica
fume. Hans Hendrik Bache introduced this principle during the 1970's [56, 57] and
named the material DSP (densified by small particles). There have been numerous
application of ultra-high strength concrete [58] even for special, non-structural applications [59].

The principle of particle packing is also being used in refractory concretes through
the so-called low cement and ultra-low cement castable (LCC and ULCC) [60].

The use of the nano sized silica fume however, is only effective when the nanoparticles can be well dispersed. This is only possible with the use of deflocculants
or super-plasticizers.

3.2.2 Self compacting concrete

Self-compacting concretes (SCC) is a material which 'poured' into the formwork
compacts 'without any further compaction' and exhibits excellent flow behavior.
It's mix design calls for a very high amount of powder type components (approx.
600 kg/m^3) such as cement and fine inorganic, non-metallic powders, sand and relatively small sized aggregates < 16 mm. For the fine powders normally considerable
amounts of pozzolanic and inert minerals (SCM) are used to achieve the desired,

excellent rheological properties. The SCM's together with chemical admixtures are therefore the salient constituents of such systems. SCMs are used to modify the water demand of the system, to control early heat evolution, to reduce early shrinkage and to optimize the flow and place-ability in addition to improving the strength and microstructure [61].

4 Use of SCM in alkali activated binders

4.1 History and structure

The search for alternatives to OPC (ordinary Portland cement) started as soon as the 1930th with Kuhl experimenting using KOH and slag. In 1940, Glukhovsky started investigating alkali activated binding systems and introduced two different classes: $Me_2-Al_2O_3-SiO_2-H_2O$ and $Me-MO-Al_2O_3-SiO_2-H_2O$ (with Me = Na, K, . . . ; M = Ca, Mg, ..) [62]. Most research is focused on the first class, which includes five categories of alkali-activated cements:
- alkali activated slag-based cements
- alkali activated pozzolan cements
- alkali activated lime-pozzolan/slag cements
- alkali activated calcium aluminate blended cements
- alkali activated Portland blended cements (hybrid cements)

The second category of binders is called by different terms: "soil cements" Glukhovsky [63], "geocements" Krivenko [64], "inorganic polymers" van Deventer [65], "zeocements" Palomo [62] and geopolymers Davidovits [66].

The structure of alkali-activated materials can be best described as amorphous polymers composed of linked tetrahedrons with silicon or aluminum ions as centers. The 4 x -coordinated Al^{3+}-ions show a negative charge in the tetrahedrons which is balanced by the introduced alkali or earth alkali ions (Na^+-ions, Ca^{2+}-ions) [65–67]. A schematic description is given in Figure 7.6.

Geopolymers, or alkali activated cementitious systems can be synthesized using a variety of aluminum and silicate sources, e.g. metakaolinite, kaolinite, slag and fly ash (brown coal fly ash and black coal fly ash) [69]. The alumino silicates, either natural occurring, like kaolinite or industrial wastes like fly ash and slag, are activated by an alkaline solution, sometimes called geopolymer liquor [70], consisting of an alkali hydroxide solution and a sodium (or potassium) silicate solution.

This category includes the following systems, which are differentiated by their used raw materials [71]:
1. alkali activated fly ash cements
2. alkali activated natural pozzolan cement

3. alkali activated metakaolin cement
4. alkali activated soda lime glass cement

Of these cementitious systems, metakaolin and fly ash systems were investigated extensively over the last couple of years. In the following, the usage of secondary raw materials will be described in more detail.

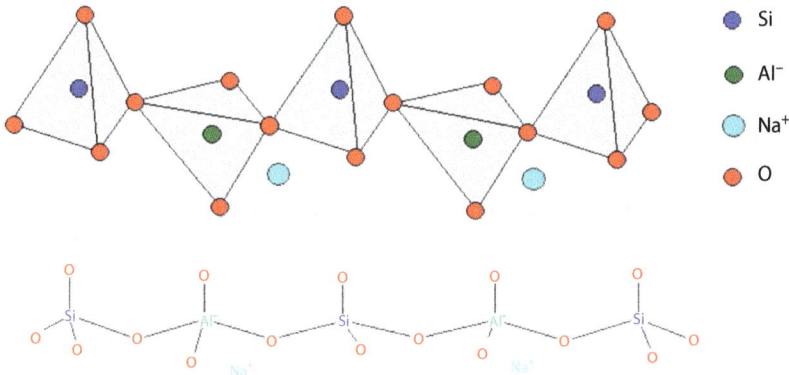

Figure 7.6: Theoretical schematic structure of linked tetrahedrons in the geopolymer structure. The negative charges are not located at the aluminum ion; however, they are because aluminum only has three valence electrons but has a four-fold coordination in the geopolymer structure [68].

4.2 Secondary raw materials in alkali activated binder systems

4.2.1 Fly ash

Fly ash, black coal fly ash as well as brown coal fly ash, is very well documented in the alkali-activated research. Especially class F fly ash, which has a lower Ca content than class C fly ash (ASTM C618), is widely used in the investigation of alkali-activated materials. However, recent research shows that the combination of Al content, Si content and Ca content in fly ash as starting material for AAMs is crucial [3]. Duxson and Provis designed a concept to implement the important characteristics of different fly ashes in correlation with the resulting compressive strength of the synthesized AAMs [3]. In their analysis, they conclude that a necessary amount of reactive aluminum and network-modifying ions, alkali -and alkaline earth metals, is needed however not sufficient to produce high strength AAMs. The underlying mechanisms and properties of the different fly ashes are too variable in their influence and outcome [72].

Nevertheless, a common distinction is made between high calcium containing fly ashes, such as brown coal fly ash and low calcium containing fly ashes, such as black coal fly ash.

Low Ca containing fly ash

Although a first description of these low Ca alkali activated binders was relatively early mentioned in the 1970s done by Davidovits [67], academic interest spiked in the early 2000 and is still apparent today [3]. Researchers, including J. Provis and A. Palomo, investigated the system thoroughly and published a highly cited review in 2007 [72].

The mechanism leading to the formation of the tetrahedral alumosilicate network in AAMs are also not completely understood, however, there has been research investigating these mechanisms. Fernández-Jimenez and Palomo et al. describe the mechanism in a graphic model based on MAS-NMR and FTIR [71]. Glukhovsky [63, 71] postulates on zeolite synthesis and in accordance with a previous described chemical model it. In this proposed mechanism, six steps are mentioned [71, 73]:

1. Chemical attack
2. Dissolution process
3. N-A-S-H precipitation of Gel 1
4. N-A-S-H precipitation of Gel 2
5. Polymerization
6. Crystal Growth

In the first step, the chemical attack of the highly alkaline activator onto the solid component, e.g. fly ash, the bonds between Si-O-Si, Al-O-Al, Al-O-Si and Me-O are severed. This leads directly to the second step, the dissolution. During these steps, a change in the electronic densities around the silicon atoms occurs and therefore the Si-O-Si bonds are attacked. The alkaline metal cations neutralize these anions, building $Si\text{-}O^-Na^+$, thereby preventing the reverse reaction. The attacked and severed Al-O-Si bonds form aluminates, which are stabilized by the hydroxyl groups, and form complexes like $Al(OH)_4^-$ or $Al(OH)_6^{3-}$. Which complexes are formed, is dependent on the pH of the solution [71, 74].

These monomers agglomerate to form an aluminum rich gel 1, which consists of alternating silicon and aluminum tetrahedral units. This structure was confirmed by NMR and FTIR data and is a metastable form during the reaction. One suggested reason why an aluminum rich gel is formed is that the Al-O bonds, which are weaker than the Si-O bonds, are severed firstly and to a higher amount. This leads to a higher concentration of Al^{3+} in the medium at the beginning of the reaction [75, 76].

This metastable gel 1 transforms into a gel 2, which has a Si/Al ratio of about two. This change in ratio comes from the fact that during the reaction, more Si-O groups dissolve. The change into a Si rich gel was proven by NMR study which provided the evidence of the formation of a Q^4 (3Al) and Q^4 (2Al) signal [77, 78].

The exact mechanisms of the formation of the N-A-S-H gels 1 and 2 were studied firstly by Iler [79] and were later on thoroughly investigated by Ikeda [70] Hereby they synthesized gels similar to these observed in alkali activated metakaolin and fly ash at different pH levels. At a pH ≥ 7 they found that polymerization is the main mechanism and which reaction rate increases with an increasing pH. Due to Ostwald ripening, colloids are generated which form cross-links and thereby generate 3D structures [79, 80].

Polymerization is the fifth step during the alkali activation of aluminosilicates at a pH of about 12.5 which also leads to growth of the colloid particles and therefore to a three-dimensional network [81].

Low calcium fly ash based geopolymers are also investigated in heavy metal immobilization as well as waste management capabilities [68, 82]. Research has shown that the ability of immobilization is dependent on the Si/Al and Na/Al ratio in AAMs [83].

High calcium containing fly ash

Fly ashes with a calcium content of >20% are considered class C fly ashes and are readily available as industrial waste, therefore research increases in using these materials as precursors for AAMs as well. Some literature proposes a higher compressive strength in AAMs from class C fly ashes, however a lesser stability against chemical attack [3]. This higher strength development might be due to the denser structure of class C based AAMs. In a study it was found, that class C fly ash based AAMs show higher compressive and flexural strength values than their class F counterparts [84]. However, not all authors have the same relationships, therefore it is concluded that many factors, including the SO_3 content and the Al_2O_3 content vary greatly between different countries and sources. Therefore, the class design of a high-quality class C fly ash AAM is more difficult than using a class F fly ash [3].

Environmental impact

Fly ash based AAMs are described as low CO_2 binder systems, which are beneficial for the environment. Considering the high amount of CO_2 emissions during cement production, this is obvious. The use of fly ash, which is readily available in many countries, such as China or India, is helpful in the concept of minimizing the need of landfill spaces and waste materials [3, 85]. However, countries like Germany and other European countries, which are planning the coal phase-out, do not have fly ash as precursor available. Another drawback might be the upkeep of coal mining and the use coal power plants in the future, as it is necessary to provide the building industry with needed raw materials. Therefore, some researchers, especially J. Davidovits, is arguing against fly ash based AAMs [86].

4.2.2 High calcium AAM systems (blast furnace slag)

Kühl [87] has investigated blast furnace slag as a starting material for the synthesis of AAM as early as 1908. In the last decades, same as with low calcium containing fly ashes, the interest in the research of these materials has increased drastically as the interest in secondary raw materials for cement production as well as for cement substitution increased. Blast furnace slag is a high in calcium industrial by product and therefore readily available in many countries. The chemical composition of high calcium AAMs in comparison to low calcium AAMs is quite different. In the following, the mechanism will be roughly shown and the differences will be laid out. Nevertheless, is the reader encouraged to read the original literature discussing the involved mechanisms.

It is commonly accepted that the structural development is governed by four mechanisms [3, 88]:
1. Dissolution of the glassy precursor articles
2. Nucleation and growth
3. Interaction and mechanical binding at the boundaries of the formed phases
4. Diffusion of reactive species

These mechanisms form a C-A-S-H type gel, which has a disordered tobermorite-like structure. This structure is also considered to play a part in hydrated Portland cement structures, however in AAMs; the Ca/Si ratio is lower than in hydrated Portland cement systems [89]. The exact type of gel and the Ca/Si (+Al) ratio is dependent on the activator used. NaOH leads to a more ordered system with a higher Ca/Si ratio whereas silicate activated binders show a lesser ordered structure [3, 90]. The binder provides an alkaline milieu, which leads to an increase in reaction speed and therefore to an acceptable timeframe. Both, alkali hydroxides and alkali silicates do this to an acceptable rate. Alkali carbonates and sulfates result in a comparably lower pH and hence to a slower hydration time [3].

Other important variables include the fineness of particles and the chemical and mineralogical composition of blast furnace slag. Due to the increasing reactivity of a finely ground material, blast furnace slag exhibits better reactivity when the fineness increases. An optimal fineness was found to be between 400 and 550 m^2 /kg [91]; however, fineness exceeding 450 m^2 /kg resulted in setting times between 1–3 min [92]. The slag's cooling conditions relate directly to the phase composition in the material, where slow cooling leads to an unreactive, crystalline material whereas rapid cooling leads to a highly amorphous [93], reactive material with only small crystalline amounts [3].

4.2.3 Municipal solid waste materials

Municipal solid waste is estimated to reach 2.2 thousand million tons in 2025 world-wide and 25% of this material is collected as municipal solid waste incinerator bottom ash (MIBA) [94]. Therefore, researchers started to use ground MIBA as precursor for AAMs starting in the early 2000 as MIBA consists mostly of aluminosilicates [91, 94]. Several studies found that it is possible to use MIBA at least as additional precursor to AAMs designed using blast furnace slag [95], metakaolin [96] or fly ash [97].

4.3 Applications of AAM

The use of alkali activated binder systems on an industrial scale is still very rare, due to the missing standard regarding the AAMs composition. To fix these missing standards, a working group was established, creating the last Rilem Report in 2014 [3]. In this report, the authors lay out a concept of how a standard for the construction industry could be structured and suggest different methods. In short, the authors suggest the development of two standards, one for an alkali activated binder system, and one for an alkali activated concrete. Both standards will be on performance of the AAMs and not on the composition of AAMs. This distinction is necessary, as it is possible to develop an alkali-activated concrete by mixing all components directly without premixing a binder system. Standards regarding cement are usually based on both with chemical composition and hydration products being an important component. In respect to AAMs, this approach is not suitable as most AAMs compositions, raw materials and activation systems vary widely. Hence, usually AAMs are designed in a specific region with specific raw materials and precursors, fitting the available components and purposes. Therefore, a performance-based standardization allows for these variables, nevertheless ensuring the safety and security needed for construction materials. It is suggested to use mostly well-established performance tests, such as workability tests, rheological behavior (including flow tests) and mechanical properties (strength tests) at different stages of the material.

Over the last decades, several buildings were built and infrastructure projects conducted using AAMs in comparison with normal cement-based concrete. These projects are scattered over the world, however mostly done not in European countries but in Russia, China and Australia. Especially the former USSR states used alkali activated slag-based systems far earlier than other countries and still; Russia and Ukraine have some projects going on. For more information about applications in the former USSR, the reader is advised to read book by Shi et al. describing construction in the former USSR [98].

Here, some European and Australian projects are shortly mentioned. In Australia, E-Crete™ developed by the Zeobond Group, is used in many projects, for example in the Melbourne Library a pre-cast E-Crete was used for the exterior of the building, Footpaths in Port Melbourne and Railway Sleepers [3]. In Finland, F-concrete based on blast furnace slag for roof tiling was developed and successfully tested [3]. In the Netherlands, ASCEM cement was industrialized producing concrete pavement units and sewer pipes [3, 99].

5 Use of SCM in ternary binders

5.1 Introduction to ternary binders

Ternary binder systems are composed of three mineral materials, which are Portland cement (PC), Calcium aluminate cement (CAC) and Calcium Sulphate ($C\bar{S}H_x$). These compositions are extensively used for dry mortars in the so–called "Building Chemistry Industry" as technical mortars, tile adhesives, waterproofing slurries, grouting mortars, screeds, self–levelling underlayments (SLU), repair or fast repair materials etc. A classification of the applied formulations in industrial mixes is given in Figure 7.7. Five main regions can be distinguished in the ternary diagram.

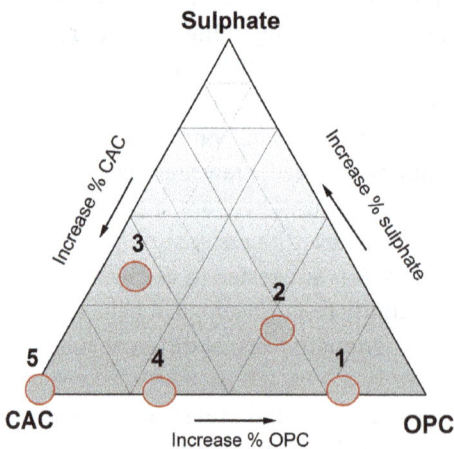

Figure 7.7: Distinct composition in ternary binder systems used for industrial application. Reproduced with permission from [4].

Two of these regions (1 & 2) are part of the so–called *PC – rich area*. They are characterized by high amounts of PC and moderate amounts of CAC (region 1) or moderate addition of CAC/$C\bar{S}H_x$ (region 2). The addition of CAC in region 1 provides early strength and quick setting. Improved hardening is achieved in region 2 by the addition of sulphates. However, the strength values for the mixes are below the strength values of the pure PC. Region 3 and 4 are part of the so-called *CAC rich area*. The compositions located in this zone of the ternary diagram exhibit fast hardening kinetics, fast setting, internal-drying capacity and shrinkage compensation due to the formation of ettringite. Region 5 corresponds to the plain CAC, which has been developed as cement with high sulphate resistance, due to the absence of portlandite [4].

One of the main characteristics of these compositions is the fast setting and the high strength development within hours. Although not really proven, it is believed [100–102] that ettringite is the prime and sole cause of the fast setting in ternary binders.

The hydration of ternary binders and the formation of hydrates is known to be very complex and strongly differs from the hydration of pure PC and CAC [102–105]. When mixtures of PC/CAC/ $C\bar{S}H_x$ encounter water some of the following reaction take place:

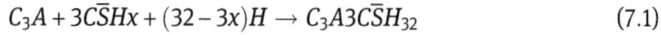

$$C_3A + 3C\bar{S}Hx + (32 - 3x)H \rightarrow C_3A3C\bar{S}H_{32} \tag{7.1}$$

$$3CA + 3C\bar{S}Hx + (38 - 3x)H \rightarrow C_3A3C\bar{S}H_{32} + 2AH_3 \tag{7.2}$$

where x = 0 for anhydrite, x = 0.5 for hemihydrate and x = 2 for gypsum.

When the calcium sulphate is depleted, ettringite reacts with remaining anhydrous CA or C_3A to form calcium monosulphoaluminate (AFm phase).

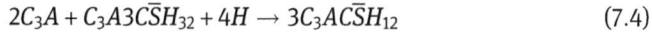

$$6CA + C_3A3C\bar{S}H_{32} + 16H \rightarrow 3C_3AC\bar{S}H_{12} + 4AH_3 \tag{7.3}$$

$$2C_3A + C_3A3C\bar{S}H_{32} + 4H \rightarrow 3C_3AC\bar{S}H_{12} \tag{7.4}$$

Hydration is strongly modified by the PC/CAC and CAC/ $C\bar{S}H_x$ ratios. In PC rich combinations, one of the main characteristic observed when the PC/CAC ratio falls into a certain range is the delay in the hydration of silicates [106–108]. The sulphate source is another factor that plays a key role in the hydration kinetics and phase formation of ternary binder systems. It has been shown that a certain form of calcium sulphate and the dosage influences the rate of hydrate phase formation, their morphology and spatial distribution due to their particular solubility and rate of dissolution [105, 109–112].

A visualization of the quantities of the main hydrate phases overall the PC-CAC-$C\bar{S}H_x$ ternary diagram as predicted from thermodynamic modelling is shown in Figure 7.8 C–S–H together with portlandite (Figure 7.8 a, b) are mostly located in the region close to the pure Portland cement. The content of both phases decreases towards lower Portland cement contents. The phases are absent in the region with PC < 30 percent by mass. Ettringite and monosulphate are present almost in the entire range of the ternary diagram with the exception of the regions where the plain CAC and sulphate are located.

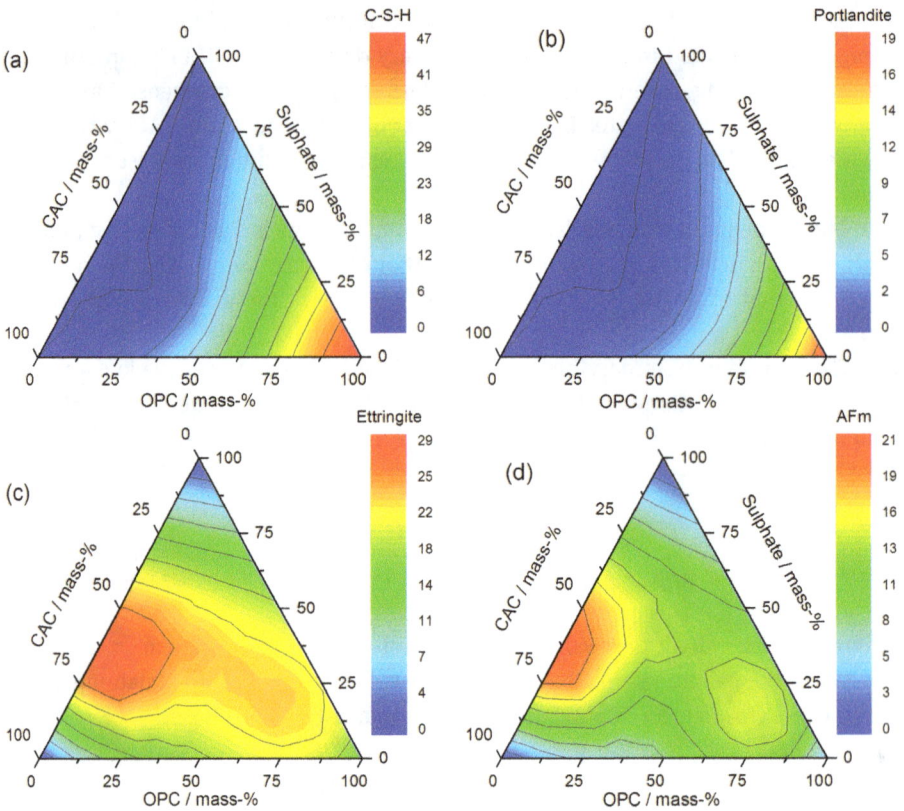

Figure 7.8: Distribution of hydrate quantities in the ternary diagram PC–CAC– CSHx after 90 days of hydration. (a) C–S–H, (b) portlandite, (c) ettringite and (d) monosulphate in g/100 g paste [105].

5.2 Role of slag, fly ash and limestone in ternary binders

5.2.1 General

Depending on the composition, mortars based on ternary binders exhibit specific characteristics, such as volumetric stability [113], fast strength development, rapid hardening and rapid drying [100], good fluidity [114] and good durability [107]. These properties might be improved or a degradation be compensated by the addition of secondary cementitious materials. Effects of SCM for system S1 (Portland cement cf. Figure 7.9a) are described in chapter 3. Most other studies in this direction are denoted to CAC.The main degradation process suffered by CAC is the conversion of hexagonal calcium aluminate hydrates to a cubic form. Mixes of CAC with silica fume, blast furnace slag, fly ash or metakaolin are considered an interesting alternative for the stabilization of the CAC hydrates. It has been reported by several authors [115–118] that replacement of CAC by blast furnace slag, or pozzolan such as microsilica, and fly ash could hinder the conversion process and lead to an increase in strength. The proposed mechanism hindering the conversion process in CAC follows: The silica present in the SCMs reacts with the calcium aluminates by avoiding the formation of C_2AH_8 and subsequently the conversion to C_3AH_6. Instead of C_3AH_6, the formation of strätlingite (C_2ASH_8) it is suggested to take place.

In spite of the studies denoted to understand the hydration of PC-SCMs or CAC-SCM mixtures, the role of SCMs in the hydrate phase assemblage of PC-CAC-$C\bar{S}H_x$ combinations has been hardly reported in the literature [119–122]. This chapter illustrates some effects of slag (S), fly ash (FA) and limestone powder (LSP) on PC-rich and CAC-rich ternary binders. Calorimetry, X-ray diffraction, thermal analysis, MIP and strength measurements are used to visualize their role.

5.2.2 Composition

Most of the research work in ternary systems is carried out using model formulations S1 to S7 as depicted in Figure 7.9a [105, 107, 123, 124]. In the ternary diagram, S1 and S7 refer to the pure PC and CAC respectively. Formulations S2 to S6 are typical combinations of PC–CAC–$C\bar{S}H_x$ reflecting binders of special products of the European market as depicted by Bier and Amathieu [125]. In Figure 7.9b a schematic visualization of two quaternary binders with 27 wt.% addition of SCM (i.e., Slag, Fly Ash and LSP) has been depicted. Table 7.3 gives the accurate composition of the quaternary binders. One of the quaternary binders is based on the S2 PC-rich

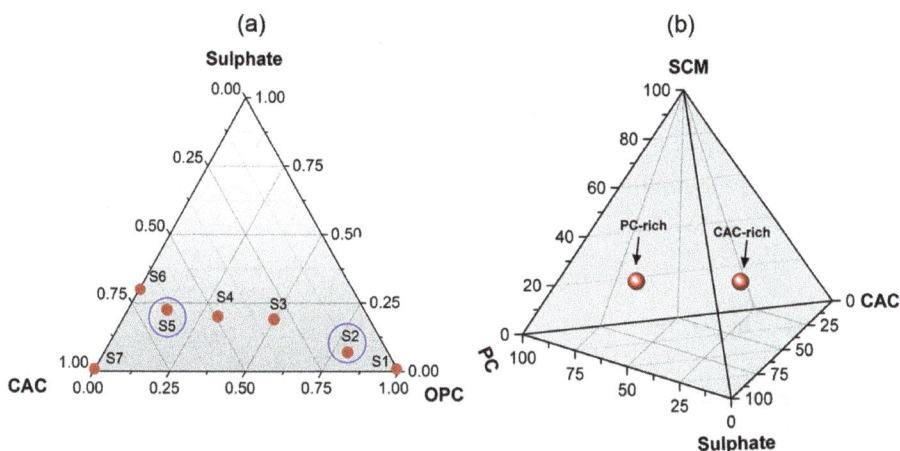

Figure 7.9: (a) Composition and classification of ternary binders according to Bier and Amathieu [125].(b) Quaternary binder compositions with 27 wt.% SCM addition (Slag/ Fly Ash/ LSP) based on the S2 and S5 formulations of the ternary diagram.

formulation, whereas the CAC-rich quaternary binders is based on the S5 combination depicted in the ternary diagram.

Table 7.3: Reference formulations of quaternary binders as depicted in Figure 7.9b.

Formulation	PC (wt.%)	CAC (wt.%)	C$\bar{\text{S}}$H$_x$ (wt.%)	SCM (wt.%)
PC-rich (S2)	58.4	9.6	5.2	27
CAC-rich (S5)	9.5	47.1	16.4	27

Table 7.4 summarizes the chemical composition of each of the used materials whereas; the mineralogical compositions are summarized in Table 7.5.

Table 7.4: Chemical composition of the used raw materials.

wt.%	CaO	SiO$_2$	Fe$_2$O$_3$	MgO	Al$_2$O$_3$	K$_2$O	P$_2$O$_5$	Na$_2$O	TiO$_2$	SO$_3$
PC	62.2	20	3.8	1.4	4.8	1	–	0.1	0.2	2.7
CAC	37.7	4.2	17.1	0.7	39.2	0.1	0.1	0.1	1.8	0.1
Anh	37.9	0.8	0.2	0.1	0.2	0.1	0.1	0.1	01.	58.2
HH	37.9	0.1	0.1	0.1	0.1	0.1	0.1	0.1	01.	51.3
Fly Ash	2.5	56.4	4.8	2.3	26.8	2.6	0.7	1.5	0.8	1.5
Slag	44.9	34.3	0.6	6.1	9.1	0.7	–	–	0.8	3.6
LSP	80.2	10.6	1.9	1.4	4.4	1.1	–	–	–	0.4
QP	–	98.6	0.9	–	0.4	–	–	–	–	–

Fly ash is of type C according to ASTM C618. Anhydrite was used as sulphate source for the PC-rich formulation, whereas in the CAC-rich combinations α-hemihydrate was applied.

5.2.3 Heat of hydration

Figure 7.10 depicts the heat flow curves for the PC-rich (Figures 7.10a, 7.10c) and CAC-rich (Figures 7.10b, 7.10d) quaternary systems.

Independently on the used SCM, the heat flow curves for the PC-rich quaternary formulations are relatively close to that of the S2 ternary binder [105]. There is hardly a difference between the heat of hydration evolved during the first 24 hours for the slag and fly ash S2 based pastes. However, the addition of fly ash in the PC-

Table 7.5: Mineralogical composition of the used raw materials. The highest uncertainty in the phase quantification is ± 2 wt%.

wt.%	PC	CAC	Anh	HH	Fly Ash	Slag	LSP	QP
C_3S	55.2	–	–	–	–	–	–	–
C_2S	16.4	2.6	–	–	–	–	–	–
C_4AF	14.8	14.7	–	–	–	–		–
C_3A_{cubic}	5.3	–	–	–	–	–		–
C_3A_{orth}	1.3	–	–	–	–	–		–
CA	–	61.3	–	–	–	–		–
$C_{12}A_7$	–	2.3	–	–	–	–		–
CA_2	–	–	–	–	–	–		–
C_2AS	–	4.4	–	–	–	–		–
$C\bar{S}$	0.5	–	98.3	–	2.0	–		–
$C\bar{S}H_{0.5}$	1.6	–	–	100	–	–		–
$C\bar{S}H_2$	–	–	–	–	–	–		–
$CaCO_3$	1.4	–	1.7	–	–	–	96.2	–
Arcanite	0.6	–	–	–	–	–		–
Perovskite	–	5.8	–	–	–	–		–
Magnetite	–	1.5	–	–	1.4	–		–
Mullite					11.3	–		–
Quartz					5.3	–	1.5	100
Ankerite							2.4	–
Amorphous	2.9	7.4	–	–	80.0	100	–	–

rich ternary binder slightly delays the main maxim of the hydration curve relative to that of slag and limestone containing pastes. Such an effect has been reported in previous investigations of Portland cement-fly ash-limestone ternary composites [126, 127]. The reference paste develops the highest heat release.

The CAC-rich pastes exhibit a much complex behavior compared to the PC-rich combinations. The three cement pastes (S5-S, S5-FA and S5-LSP) are characterized by a strong heat development during the initial period. Apart from the initial peak, there are two distinct exothermic peaks in the heat flow curves of the slag and fly ash-based formulations. The time interval between the two peaks is around 9 hours. In the S5-LSP paste, the third maxima, is not as pronounced as for the two other pastes.

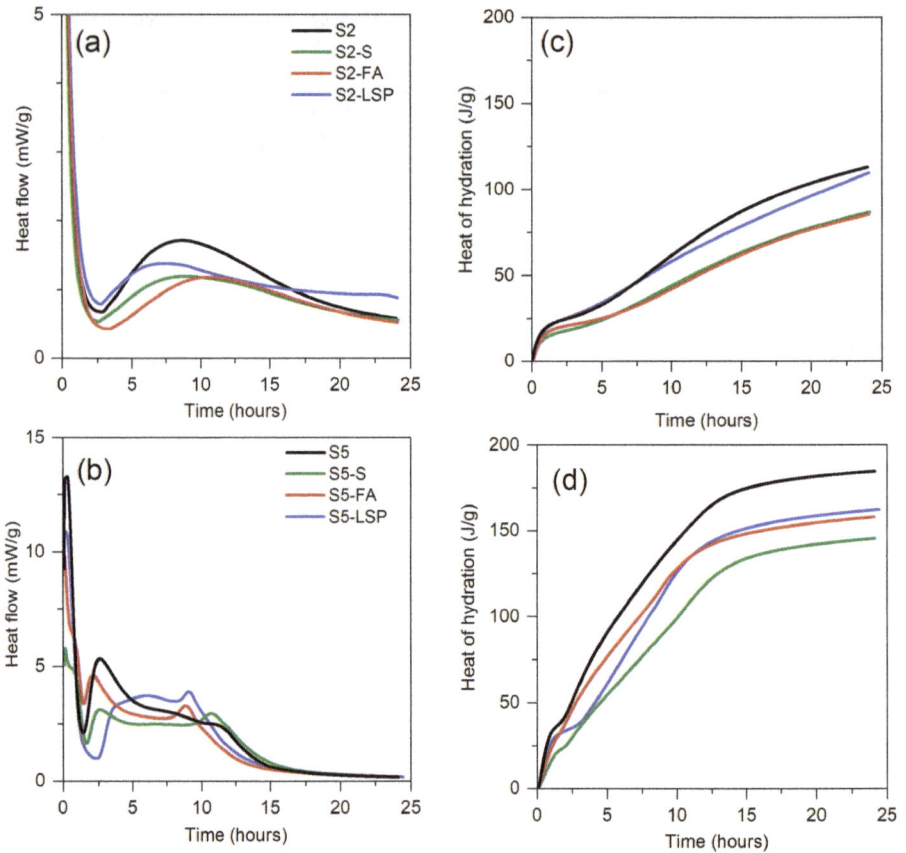

Figure 7.10: Heat of hydration for the S2 and S5 ternary and quaternary pastes as a function of SCM type: (a, b) heat flow curve over 24hrs, (c, d) cumulative heat flow curve over 24hrs.

5.2.4 Phase assemblage

Qualitative phase analysis

Figure 7.11 shows the XRD patterns for the PC-rich (left hand graph) and CAC-rich (right hand graph) of the quaternary formulations at 1 and 28 days of hydration. The XRD analysis reveals the presence of C_3S, C_2S, CA, C_4AF and calcite as main anhydrous phases in all PC rich combinations. Ettringite and portlandite stand as main crystalline hydration products. C–S–H as another major hydrate phase is X–ray amorphous, thus not detected in any of the X–ray patterns. In presence of slag and fly ash, the hemicarbonate phase is detected at 28 days of hydration, whereas in presence of limestone both hemicarbonate and monocarbonate are identified, hence indicating a conversion of hemicarbonate to the latter phase [128].

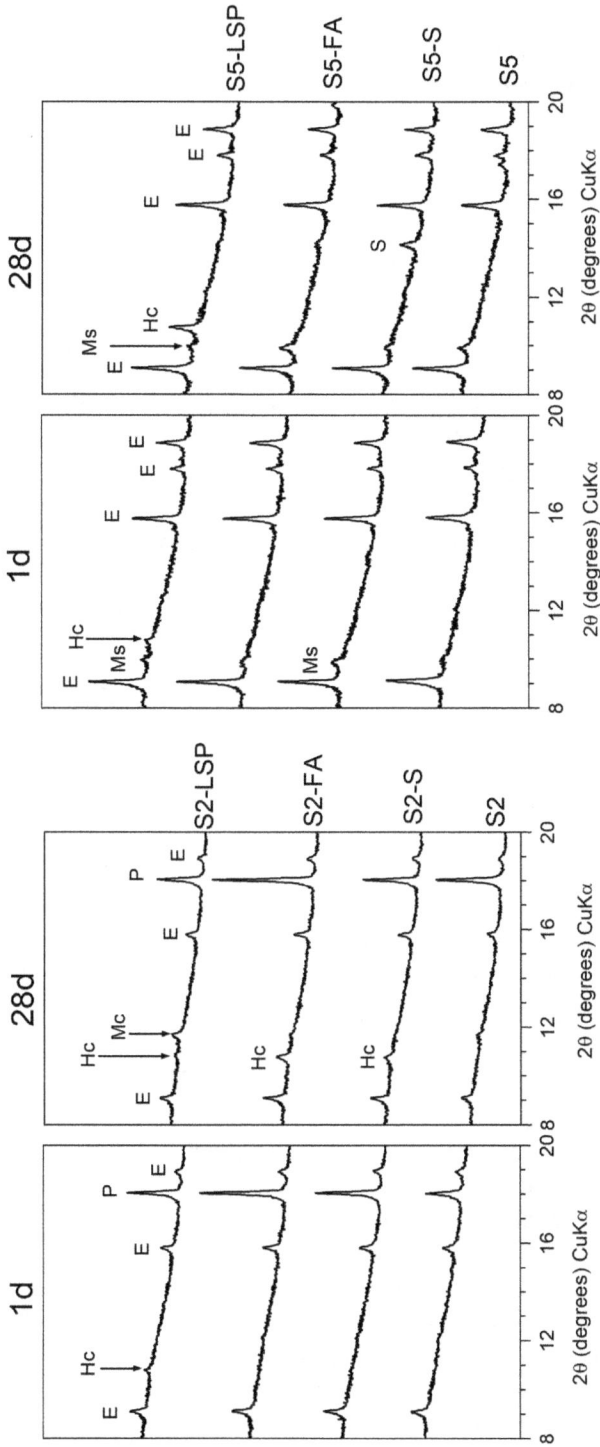

Figure 7.11: XRD diffractograms of the S2 (left hand graph) and S5 (right hand graph) ternary and quaternary pastes at 1 and 28 days of hydration. Phase abbreviation: (E) for ettringite, (P) for portlandite, (Ms) for monosulphate (Hc) for hemicarboaluminate (Mc) for monocarboaluminate and (S) for strätlingite. The samples were prepared with a water to binder ratio w/b = 0.55.

Unreacted CA, C_2S, C_3S, C_4AF and calcite are identified in the CAC-rich formulations. Ettringite stands as main hydrated crystalline phase in all samples. Traces of monosulphate are also identified. Additionally, in presence of limestone, the formation of hemicarbonate takes place. From the other hand, in the slag-based paste, a peak centered around $2\theta = 14.25°$ at 28 days from hydration has been related to the strätlingite phase. Traces of the latter compound are also observed in the fly ash-based sample at 28 days from hydration.

Figures 7.12 and 7.13 show the TGA/DTG data analysis after 1 and 28 days of hydration. The data are in good agreement with the XRD results, hence confirming the precipitation of ettringite, C-S-H and portlandite for the OPC-rich combination (Figure 7.12). Portlandite content exhibits an increase over time in all samples. The AFm phases, including monosulphate, monocarbonate and hemicarbonate strongly overlap to each other and a broad peak in the range ~ 250–350 °C [105] is attributed to the aforementioned phases.

Ettringite, AH_3, monosulphate, AFm-carbonate equivalents and calcite are indicated in the DTG curves of the CAC-rich blends. An additional exothermic peak correlated to strätlingite in the range between 160–175 °C [105] is evident at 28 days from hydration. The peak is relatively pronounced in the slag-based paste at 28 days. Additionally, the presence of AH_3 is no longer detected in this composition, thus suggesting that the crystallization of strätlingite occurs through the reaction of silicate bearing phases with AH_3. The occurrence has been previously reported in CSA based compositions [129] as well as in ternary binders [105].

Quantitative phase analysis

PC-rich quaternary binders

The influence of slag, fly ash and limestone on the content of portlandite, ettringite and X-ray amorphous hydrates for the PC based formulations is displayed in Figure 7.14. The amount of portlandite slightly increases over time (Figure 7.14a). When comparing the respective amounts of portlandite formed in the reference sample and in the pastes with slag/fly ash/LSP addition, it can be observed that the highest amount of the phase precipitates in the control ternary binder and in the LSP based composition. Whereas the slag and fly ash samples deliver lower contents of portlandite. The increase of portlandite content in the LSP based sample indicates that the former phase does not contribute to the formation of AFm-carbonate equivalents.

Most importantly, in agreement with the data from the DTG curves (see Figure 7.12), no consumption of CH in the slag and fly ash-based samples as function of time is quantified. The occurrence suggests that in both quaternary blends, the pozzolanic reaction has not yet commenced. Depending on the reactivity of the fly ash used, different results on the onset of the pozzolanic reaction have been reported in mixtures with

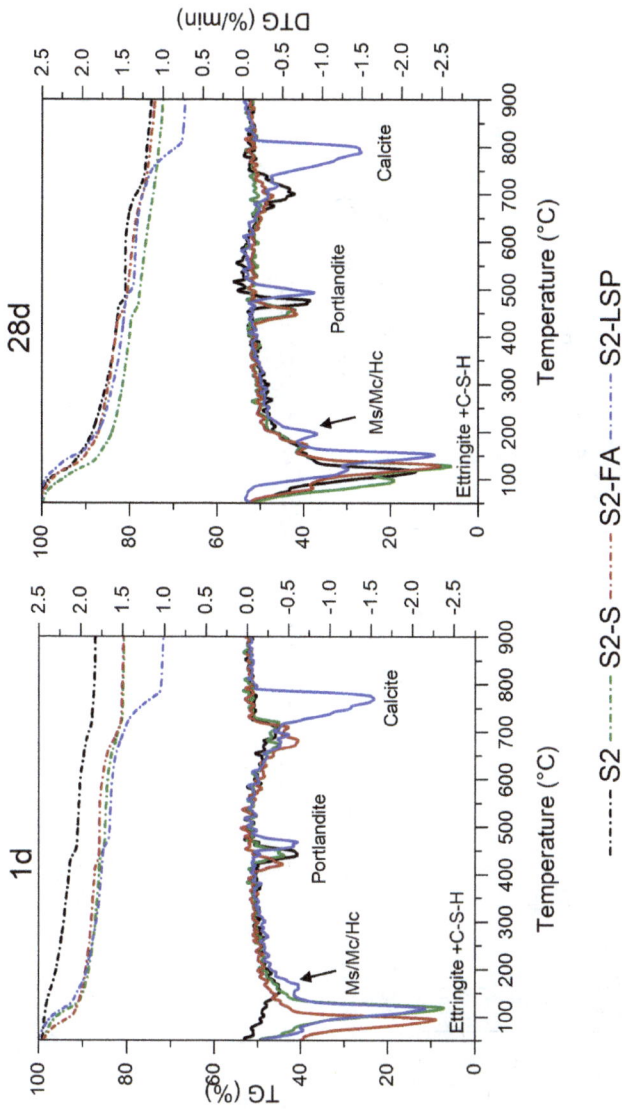

Figure 7.12: TGA and DTG curves at 1 and 28 days of hydration for the PC- rich ternary and quaternary pastes. Phase abbreviation: (Ms) for monosulphate, (Hc) for hemicarboaluminate and (Mc) for monocarboaluminate. The samples were prepared with a water to binder ratio w/b = 0.55.

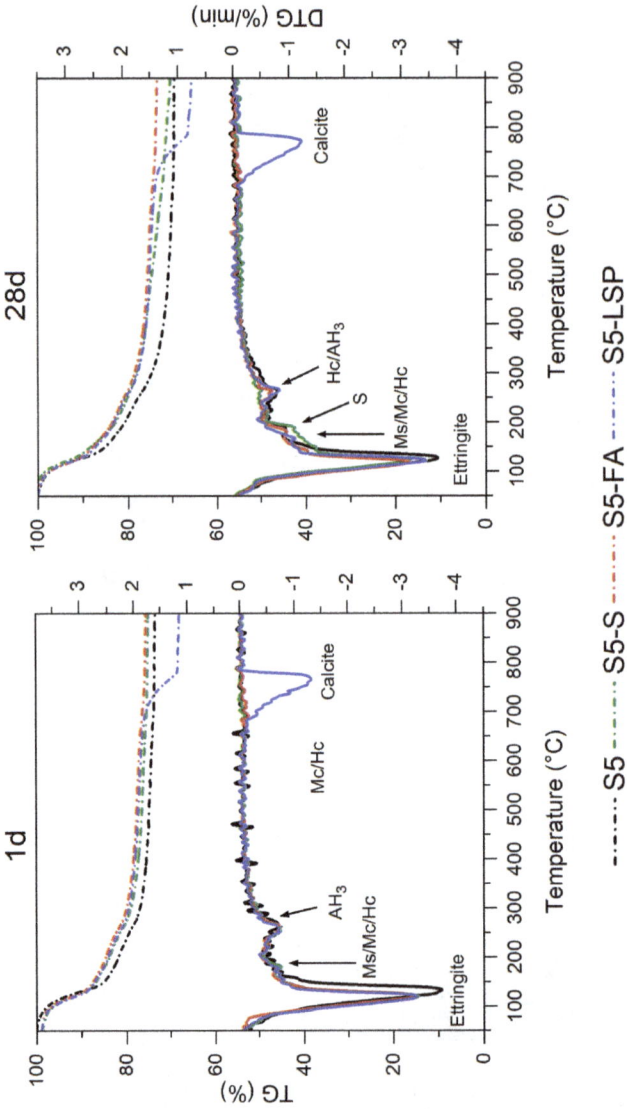

Figure 7.13: TGA and DTG curves at 1 and 28 days of hydration for the CAC- rich ternary and quaternary pastes. Phase abbreviation: (Ms) for monosulphate, (Hc) for hemicarboaluminate, (Mc) for monocarboaluminate and (S) for strätlingite. The samples were prepared with a water to binder ratio w/b = 0.55.

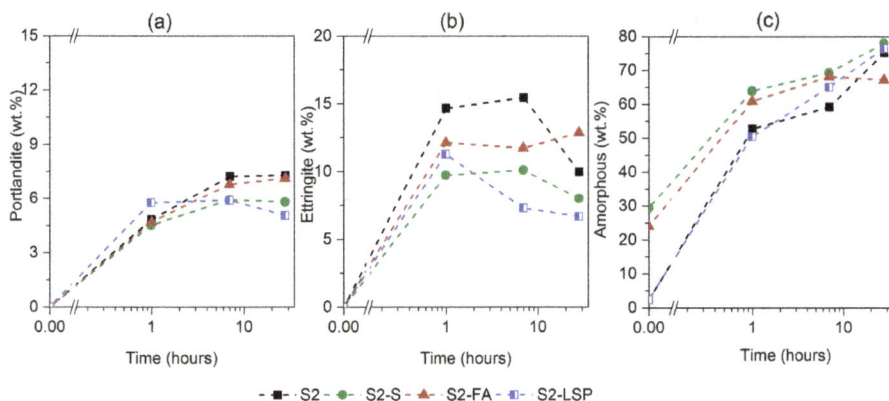

Figure 7.14: Solid phase composition as a function of time from Rietveld refinement for PC- rich ternary and quaternary pastes up to 28 days from hydration: (a) portlandite content, (b) ettringite content and (c) amorphous content. The samples were prepared with a water to binder ratio w/b = 0.55.

Portland cement. Usually, the decrease of portlandite has been measured after 7 days of hydration [130–132] and in particular cases after 28 days [133]. It has also been shown that at early ages fly ash may acts as a filler [126, 130], providing additional nucleation sites (seeding effect) on the surface of the grains for the hydration of PC. Our results are in agreement with the aforementioned studies, indicating that during the first 28 days of hydration, fly ash acts as a filler in the quaternary system. An initiation of the pozzolanic activity is expected at later ages. Similar conclusions can be drawn for the slag-blended cements. Thermodynamic modelling conducted in PC-Slag blends has shown that the destabilization of the portlandite phase takes place at high replacement levels of PC (> 30% substitution level) [134].

As the content of ettringite is concerned (Figure 7.14b), the highest amount is quantified in the LSP based paste, followed by the slag and fly ash bearing blends. In the control ternary binder, the ettringite content decreases starting from 7 days of hydration, indicating its conversion to monosulphate according to eq. (7.4). As expected, the addition of limestone stabilizes ettringite, by hampering its conversion to monosulphate and leading to the formation of AFm-carbonate equivalents. The inhibition of ettringite to monosulphate conversion appears also to take place in presence of slag and fly ash, as up to 28 days no decrease of the ettringite content is registered.

Finally, in terms of X-ray amorphous content (Figure 14c), from the Rietveld refinement, the highest fractions are calculated in the slag and fly ash cement blends as compared to the control ternary pastes. The lowest content is found in the limestone powder-based sample.

CAC-rich quaternary binders

The effect of slag, fly ash and limestone on the content of ettringite, monosulphate and X-ray amorphous hydrates for the CAC based blends is displayed in Figure 7.15.

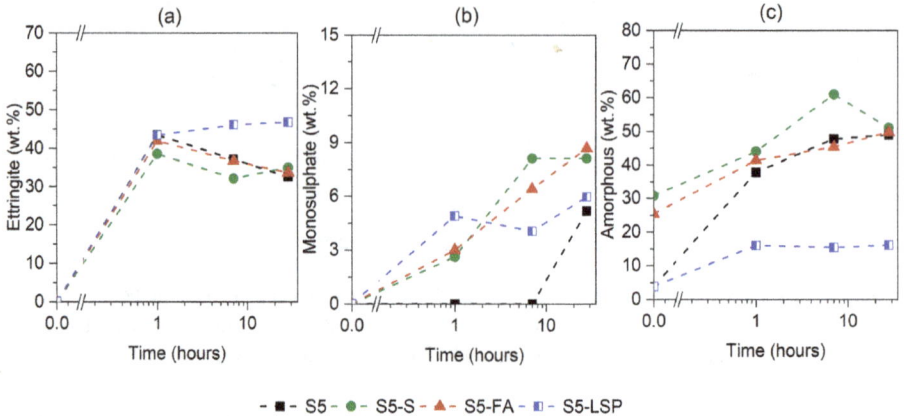

Figure 7.15: Solid phase composition as a function of time from Rietveld refinement for CAC-rich ternary and quaternary pastes up to 28 days from hydration: (a) ettringite content, (b) monosulphate content and (c) amorphous content. The samples were prepared with a water to binder ratio w/b = 0.55.

The content of ettringite is shown in Figure 7.15a. The limestone powder-based blend develops the highest amount of ettringite and the lowest content of monosulphate (Figure 7.15b) as compared to all other samples. The occurrence is due to the formation of AFm-carbonate equivalents as suggested from the qualitative XRD analysis (see Figure 7.11). The content of ettringite in the control ternary binder is relatively close to the amounts of the phase quantified in the fly ash and slag bearing pastes. In these three cases, the decrease of ettringite begins after one day. Most importantly, both fly ash and slag-based pastes develop the highest content of monosulphate. This means that the use of fly ash and slag at this specific addition promotes the destabilization of ettringite and its conversion to AFm. The occurrence might be closely related to the level of SCM addition. For instance, in ternary blends of CAC-CŠH$_x$ -Fly Ash, an increase of ettringite content with the increase of fly ash addition has been reported [119].

In terms of X-ray amorphous phase content, similarly to the PC rich formulation, slag and fly ash-based samples develop the highest amorphous fractions as compared to the reference, whereas the LSP based paste the lowest.

Reactivity – seeding effect or chemical reaction?

In order to observe whether there is a promotion of the PC and CAC hydration by the addition of the former SCMs, the hydration degree of the major anhydrous phases (i.e. C_3S for PC, CA for CAC) was calculated for the first 28 days of hydration and the data are shown in Figure 7.16.

From the graph, it can be observed that for the C_3S phase a higher degree of hydration is achieved in presence of limestone and fly ash in comparison with the control paste. More than 75% of C_3S has reacted at 1 day of hydration and more than 90% after 28 days for the tested combinations. This confirms the seeding effect of fly ash, as indicated in the previous section. From the other hand, limestone appears to induce two main effects on the system: (i) that of a nucleus by accelerating the hydration of C_3S and (ii) the chemical effect by reacting with the aluminates thus leading to the formation of AFm-carbonate equivalents as shown by the XRD data. The hydration degree of C_3S in presence of slag is relatively close to that of the reference paste, indicating that the material does not significantly promote its hydration.

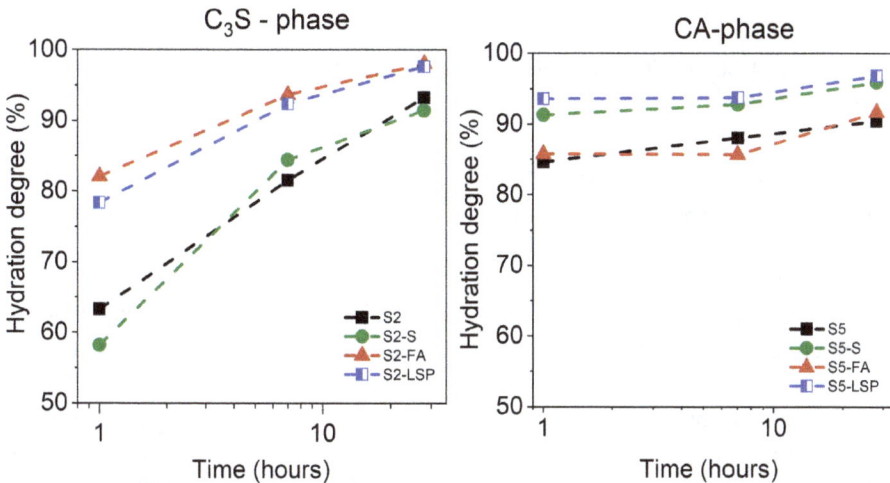

Figure 7.16: Hydration degree of C3S and CA as a function of the SCM type for the control ternary binders (S2 & S5) and the corresponding quaternary binders during the first 28 days of hydration.

A different behavior is observed for the CA phase. In the reference sample, already at 1 day of hydration almost 85% of the phase has reacted. The phase appears to react faster in presence of limestone and slag, whereas no major enhancement in presence of fly ash is observed. Here also limestone seems to induce two effects, that of a filler and that of reactant, although with a slower pace as compared to the PC rich quaternary blends.

The changes in the content of calcite as a function of time in both blends are depicted in Figure 7.17.

The initial content of calcite in both mixtures is 27 wt.%. At 28 days, ~21 wt.% and 24 wt.% are quantified in the S2-LSP and S5-LSP blends respectively. This corresponds to a hydration degree of ~22% and 10% of calcite. Although the quantification

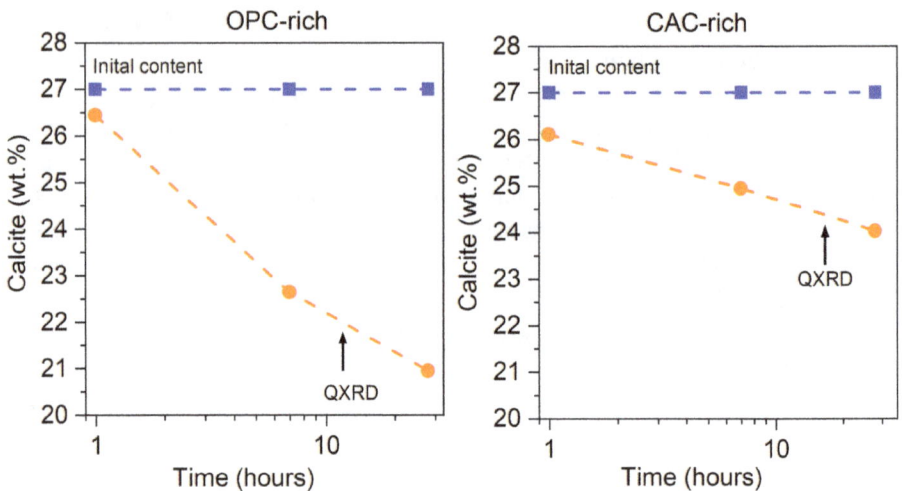

Figure 7.17: Content of calcite as a function time for the S2-LSP and S5- LSP corresponding quaternary binders.

of calcite in the matrix is experimentally difficult as both methods (XRD and TGA) are not sufficiently sensitive to determine the small changes, a general decrease of the calcite content can be postulated. The occurrence seems to be more enhanced in the PC rich formulation.

This role of calcite as reactive component in the cement matrix, by enabling the formation of hemi and monocarboaluminate, thus stabilizing ettringite has been already reported in C_3A-CaO-gypsum-calcite systems [128], LSP-PC systems [135], LSP-PC-sulphate systems [136], PC-CAC-LSP [122] and CAC- $C\bar{S}H_x$-LSP [120] formulations as well. In agreement with our observations, it has been noted [122] that in presence of CAC the reactivity of limestone is limited, in spite of the provision of sufficient water, aluminum and calcium. This suggests the significance of the type aluminum source independently of the content of alumina in the system. Additionally, the reaction of limestone in CAC– $C\bar{S}H_x$ rich binders depends also on the balance between aluminate and sulphates [120]. It has been shown [120] that for systems close to S6 (see Figure 7.9a) with an excessive use of sulphates (>40 mol%), all calcium aluminate reacts with sulphate ions to form ettringite and aluminium hydroxide, whereas limestone most likely acts as a filler.

5.2.5 Strength development and porosity

Figure 7.18 shows the compressive strength of the quaternary systems as a function of the open porosity at 7 days hydration.

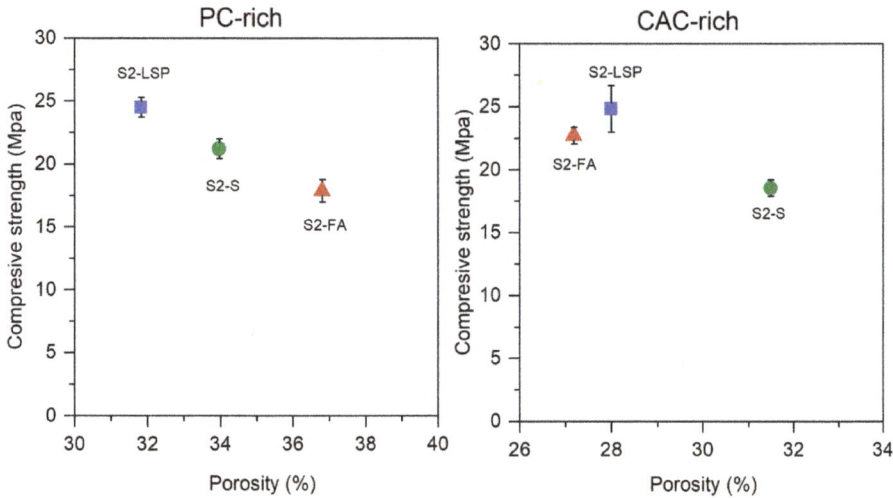

Figure 7.18: Compressive strength as a function of the open porosity at 7 days from hydration measured with MIP.

In the PC rich formulation, the limestone-based paste develops the highest compressive strength and lowest porosity. Whereas the slag and fly ash-based samples show a reduction in the compressive strength and increase in porosity respectively.

In the CAC-rich quaternary binder, the limestone-bearing blend develops the highest compressive strength whereas its open porosity is somehow higher in the fly ash-based sample. This highlights the fact that there is not a unique relation between porosity and compressive strength, especially in two chemically different system. The lowest compressive strength in the CAC-rich sample is measured in the slag-based composition.

Overall, the results suggest a potential increase of the compressive strength in presence of limestone powder, which can be directly linked, with formation of AFm-carbonate equivalents, thus increased amount of ettringite and increased volume of hydrates.

6 Summary

Besides recycled concrete and crushed slag used as aggregates, many industrial waste materials can be used as fillers in cement and concrete technology. This holds specifically true for their application as secondary cementitious materials (SCM) as an additive to Portland cement and concrete where international standards specify their acceptable properties and the amount of additions allowed for. Their contribution to physical and chemical properties and the development of microstructure and the resulting engineering properties is well known and described in literature, regulations and standards.

SCMs originating from the combustion of agricultural waste products show as well a positive contribution to Portland cement hydration and strength development, when the combustion process is optimized. However, they are not included in standards and regulations. Furthermore, more studies are necessary to evaluate their full potential.

Besides their use in Portland cement based construction materials, waste materials can also be successfully used in Alkali Activated Materials (AAM). Although these materials are as of today used as niche products for special applications, many researchers and developers believe that their use should be largely extended in the future, specifically as a standard construction material, since the environmental impact is less pronounced as for Portland cement. AAMs fit better to a circular economy.

The use of SCM in ternary binders or new cement types such as CSA cement is very little investigated. This might be because these special binders are not used for structural applications in most parts of the world. First results suggest, that amongst SCM, lime stone powder, which is not a waste material, plays an important role in terms of reactivity in lime and sulphate rich binders whereas for alumina and sulphate rich binders a stabilization of ettringite is achieved. Puzzolanic SCM from waste such as fly ash and slag seem to act rather a nucleus than to undergo a pozzolanic reaction with eventually available calcium.

References

[1]   De Belie, N., Soutsos, M. & Gruyaert, E.: Properties of Fresh and Hardened Concrete Containing Supplementary Cementitious Materials, State-of-the-Art Report of the RILEM Technical Committee 238-SCM, Working Group 4, Springer 2018, ISBN 978-3-319-70606-1.
[2]   DIN EN 197.
[3]   Provis J.L., van Deventer J.S.J.: Alkali Activated Materials, State-of-the-Art Report of the RILEM Technical Committee 224-AAM, Springer 2014, ISBN 978-94-007-7671-5.
[4]   Bier Th. A.: Composition and properties of ternary binders; in Cementitous Materials – Composition, Properties, Application, Ed. Herbert Pöllmann, De Gruyter, 2017.
[5]   Ehrenberg, A.: Iron and steel slags –from wastes to by–products of high technical, economical and technological advantages; in Industrial wastes – Characterization, Modification and Application of industrial residues, Ed. Herbert Pöllmann, De Gruyter, 2021.
[6]   Heuss-Abichler S.: Fly ash from municipal solid waste incineration – a potential resource for zinc?; in Industrial wastes – Characterization, Modification and Application of industrial residues, Ed. Herbert Pöllmann, De Gruyter, 2021.
[7]   Kaußen F., Friedrich B.: Reductive Smelting of Red Mud for Iron Recovery; Chemie Ingenieur Technik, DOI: 10.1002/cite.201500067.
[8]   Kaußen F., Friedrich B.: Soda Sintering Process for the Mobilisation of Aluminium and Gallium in Red Mud; Bauxite Residue Valorisation and Best Practices, Leuven, 2015.

[9]   R. Feige, G. Merker,: SEROX – ein Tonerderohstoff aus der Aufarbeitung von Aluminium-Salzschlacke; in "Einsatz von Abfallstoffen im Bereich Keramik, Glas, Baustoffe" Freiberger Forschungshefte, A 864 Verfahrenstechnik/Umwelttechnik 2001, S. 107–117.

[10]   Jian He,Yuxin Jie, Jianhong Zhang, Yuzhen Yu, Guoping Zhang: Synthesis and characterization of red mud and rice husk ash-based geopolymer composites; Cement and Concrete Composites, Volume 37, March 2013, Pages108–118.

[11]   Smita Singh, Rahul Das Biswas, Aswath M.U.: Experimental Study on Redmud Based Geopolymer Concrete With Fly Ash & GGBS in Ambient Temperature Curing; International Journal of Advances in Mechanical and Civil Engineering, ISSN: 2394-2827 Special Issue, Sep.- 2016.

[12]   Carsana M., Frassoni M., Bertolini L.: Comparison of ground waste glass with other supplementary cementitious materials, Cement and Concrete Composites, 2014, 45, pp. 39–45.

[13]   Federico L.: Waste Glass – A Supplementary Cementitious Material, PhD Thesis, McMaster University, Hamilton, Ontario, Canada, 2013.

[14]   Kartini, K., Rohaidah, M.N., and Zuraini, ZA.: Performanceof Ground Clay Bricks as Partial Cement Replacement in Grade 30 Concrete, International Journal of Civil and Environmental Engineering, Vol 6, No 8, 2012.

[15]   Olofinnade O.M., Ede1 A.N., Ndambuki J.M., Bamigboye G.O.: StructuralProperties of Concrete Containing Ground Waste Clay Brick Powder as Partial Substitute for Cement, Materials Science Forum, ISSN: 1662-9752, Vol. 866, pp 63–67.

[16]   Goncalves, M. R. & Bergmann, C. P., 2007. Thermal insulators made with rice husk ashes: Production and correlations between properties and microstructure. Construction and Building Materials, Volume 21, pp. 2059–2065.

[17]   Ahmed, Y. M., Ewais, E. M. & Zaki, Z. I., 2008. Production of porous silica by the combustion of rice husk ash for tundish lining. *Journal of University of Science and Technology Beijing*, 3(15), pp. 307–313.

[18]   Ismail, H., Ishiaku, U. S., Arinab, A. R. & Mohd Ishak, Z. A., 1997. The Effect of Rice Husk Ash as a Filler for Epoxidized Natural Rubber Compounds. International Journal of Polymeric Materials and Polymeric Biomaterials, Issue 36, pp. 39–51.

[19]   Chen, J.-P. et al., 2020. Preparation of SiC whiskers using graphene and rice husk ash and its photocatalytic property. Journal of Alloy and Compounds, Volume 833.

[20]   Hossain, S. K., Mathur, L. & Roy, P. K., 2018. Rice husk/rice husk ash as an alternative source of silica in ceramics: A review. Journal of Asian Ceramic Societies, 6(4).

[21]   Hossain, S. K., Pyare, R. & Roy, P. K., 2020. Synthesis of in-situ mullite foam using waste rice husk ash derived sol by slip-casting route. Ceramics International, 8(46), pp. 10871–10878.

[22]   Amick, J. A., 1982. Purification of Rice Hull as a Source of Solar Grade Silicon for Solar Cells. *Journal of The Electrochemical Society*, Volume 129, pp. 864–866.

[23]   Liu, Y. et al., 2012. Simultaneous preparation of silica and activated carbon from rice husk ash. Journal of Cleaner Production, Volume 32, pp. 204–209.

[24]   Soltani, N., Bahrami, A., Pech-Canul, M. & Gonzales, L., 2015. Review on the physiochemical treatsments of rice husk for production of advanced materials. Chemical Engineering Journal 264, pp. 899–935.

[25]   Feng, Q. et al., 2018. Synthesis of high specific surface areo silica aerogel from rice husk ash via ambient pressure drying. Colloids and Surfaces A, Volume 539, pp. 399–406.

[26]   Moayedi, H. et al., 2019. Applications of rice husk ash as green and susutainable biomass. Journal of Cleaner Production, 237(117851), pp. 1–10.

[27]   Chandrasekhar, S. et al., 2003. Review Processing, properties and applications of reative silica from rice husk – an overview. Journal of Materials Science, Volume 38, pp. 31159–3168.

[28] Nair, D. G., Jagadish, K. S. & Fraaij, A., 2006. Reative pozzolanas from rice husk ash: An alternative to cement for rural housing. Cement and Concrete Research, Volume 36, pp. 1062–1071.

[29] Bapat, J. D., 2013. Rice Husk Ash. In: T. &. F. Group, ed. Mineral Admixtures in Cement and Concrete. s.l.:CRC Press.

[30] Fernandes, I. J. et al., 2016. Characterization of rice husk ash produced using different biomass combustion techniques for energy. *Fuel*, Volume 165, pp. 351–359.

[31] Rozainee, M., Ngo, S. P., Salema, A. A. & Tan, K. G., 2008. Fluidized bed combustion of rice husk to produce amorphous siliceous ash. Energy for Sustainable Development, 12(1), pp. 33–42.

[32] Natarajan, E., Nordin, A. & Rao, A. N., 1998. Overview of combustion and gasification of rice husk in fluidizied bed reactors. Biomass and Bioenergy, 14(5/6), pp. 533–546.

[33] Agarwal, S. K., 2006. Pozzolanic activity of various siliceous materials. Cement and Concrete Research, Volume 36, pp. 1735–1739.

[34] Gholizadeh Vayghan, A., Khaloo, A. R. & Rajabipour, F., 2013. The effects of a hydrochloric acid pre-treatment on the physiochemical properties and pozzolanic performance of rice husk ash. Cement & Concrete Composites, Volume 39, pp. 131–140.

[35] Bui, D. D., Hu, J. & Stroeven, P., 2005. Particle size effect on the strength of rice husk ash blended gap-graded portland cement concrete. *Cement & Concrete Composites*, Volume 27, pp. 357–366.

[36] da Silva, F. G., Liborio, J. B. & Helene, P., 2008. Improvement of physical and chemical properties of concrete with brazilian silica rice husk (SRH). Revista Ingenieria de Construccion, Volume 23, pp. 18–25.

[37] Vieira, A. P., Filho, R. D. T., Tavares, L. M. & Cordeiro, G. C., 2020. Effect of particle size, porous structure and content of rice husk ash on the hydration process and compressive strength evolution of concrete. Construction and Building Materials, Volume 236, pp. 1–9.

[38] Isaia, G. C., Gastaldini, A. L. & Moraes, R., 2003. Physical and pozzolanic action of mineral additions on the mechanical strength of high-performance concrete. Cement & Concrete Composites, Volume 25, pp. 69–76.

[39] Ganesan, K., Rajagopal, K. & Thangaval, K., 2008. Rice husk ash blenden cement: Assessment of optimal level of replacement for strength and permeability properties of concrete. Construction and Building Materials, Volume 22, pp. 1675–1683.

[40] Cordeiro, G. C., Filho, R. D. T. & Fairbairn, E. d. M. R., 2009. Use of ultrafine rice husk ash with high-carbon content as pozzolan in high-performance concrete. *Materials and Structure*, Volume 42, pp. 983–992.

[41] Zhang, M.-H. & Malhotra, V. M., 1996. High-Performance Concrete Incorporating Rice Husk Ash as a Supplementary Cementing Material. ACI Materials Journal, 93(6), pp. 629–636.

[42] Salas, A., Delvasto, S., de Gutierrez, R. M. & Lange, D., 2009. Comparison of two processes for treating rice husk ash for use in high performance concrete. Cement and Concrete Research, Volume 39, pp. 773–778.

[43] van Nguyen, T., Ye, G., Breugel, K. v. & Copuroglu, O., 2011. Hydration and mircostructure of ultra high performance concrete incorporating rice husk ash. Cement and Concrete Research, Volume 41, pp. 1104–1111.

[44] Ye, G. & van Nguyen, T., 2012. Mitigation of Autogenous Shrinkage of Ultra-High Performance Concrete by Rice Husk Ash. Journal of the Chinese Ceramic Society, 40(2), pp. 212–216.

[45] de Senale, G. R., Ribeiro, A. B. & Goncalves, A., 2008. Effects of RHA on autogenous shrinkage of Portland cement cement" pastes. *Cement & Concrete Composites*, Volume 30, pp. 892–897.

[46] Saraswathy, V. & Song, H.-W., 2007. Corrosion performance of rice husk ash blended concrete. Consrtuction and Building Materials, Volume 21, pp. 1779–1784.

[47] Nicoara A.I., Stoica A.E., Vrabec M., Rogan N.S., Sturm S., Ow-Yang C., Gulgun M.A., Bundur Z.B., Ciuca I and Vasile B.S.: End-of-Life Materials Used as Supplementary Cementitious Materials in the Concrete Industry, Materials 2020, 13, 1954; doi:10.3390/ma13081954.

[48] Usman M., Khan A.Y., Farooq S.H., Hanif A., Tang S., Khushnood R.A., Rizwan S.A.: Eco-friendly self-compacting cement pastes incorporating wood waste as cement replacement: A feasibility study, Journal of Cleaner Production 190 (2018) 679e688.

[49] Jan Pizon J, Gołaszewski J., Alwaeli M. and Szwan P.: Properties of Concrete with Recycled Concrete Aggregate Containing Metallurgical Sludge Waste, Materials 2020, 13, 1448; doi:10.3390/ma13061448.

[50] Oikonomou N.D.: Recycled concrete aggregates, Cement & Concrete Composites 27 (2005) 315–318.

[51] Sakir S., Raman S.N., Safiuddin Md., Amrul Kaish A. B. M. and Mutalib A.A.: Utilization of By-Products and Wastes as Supplementary Cementitious Materials in Structural Mortar for Sustainable Construction, Sustainability 2020, 12, 3888; doi:10.3390/su12093888.

[52] Bier T.A., Rizwan S. A.: Early Shrinkage In Self-Compacting Mortars Using Secondary Raw Materials. In:: 9th Symposium on High Performance Concrete, Rotorua, NZ, August 9–12, 2011.

[53] Andreasen, A H M. „Über die Beziehung zwischen Kornabstufung und Zwischenraum in Produkten aus losen Körnern (mit einigen Experimenten)". In: Colloid & Polymer Science 50. 3 (1930), S. 217–228.

[54] Funk, J.E. and Dinger, D.R. (1994) Predictive Process Control of Crowded Particulate Suspensions – Applied to Ceramic Manufacturing. Boston: Kluwer Academic Publishers.

[55] DIN EN 450.

[56] Bache H.H.: The processing of fresh concrete, in Fresh Concrete, ed. A.M. Neville and D. Slater, Proc. RILEM seminar, Leeds 1973.

[57] Bache H.H.: Densified cement/ultra-fine particle-based materials, Second International Conference on Superplasticizers in Concrete, Ottawa, 1981.

[58] Designing and building with UHPFRC / edited by Jacques Resplendino, François Toutlemonde, Wiley, ISBN 978-1-84821-271-8 (hardback).

[59] Wise, S., Satkowski, J.A., Scheetz, B., Rizer J.M., Mackenzie, M.L., Double, D.D.: Development Of A High Strength Cementitious Tooling/Molding Material, Materials Research Society Symposia Proceedings, 42, pp. 253–263, 1985.

[60] Clavaud B., Kiehl J.P. and Radal J.P.: Anew generation of Low Cement Castables, New Developments in Monolithic Refractories, Advances in Ceramics, Am. Ceram.Soc., 1984.

[61] Rizwan, S.A and Bier, Th. A.: Self-Consolidating Mortars Using Various Secondary Raw Materials", ACI Materials Journal, 106, pp. 25–32, 2009.

[62] Palomo, A.; Glasser, F. P. (1992): Chemically-bonded cementitious materials based on metakaolin. In: British ceramic. Transactions and journal 91 (4), S. 107–112.

[63] Glukhovsky, V. D. (1967): Soil Silicate Articles and Structures. In: Budivelnyk Publisher, 156 pp.

[64] Krivenko, P. V. (1997): Alkaline cements. Terminology, classification, aspects of durability. Goeteborg, Sweden (Proceedings of the 10th International Congress on the Chemistry of Cements).

[65] van Deventer, J.S.J.; Provis, J. L.; Duxson, P.; Lukey, G. C. (2007): Reaction mechanisms in the geopolymeric conversion of inorganic waste to useful products. In: Journal of Hazardous Materials 139 (3), S. 506–513

[66] Davidovits, Joseph (2015): Geopolymer Chemistry and Applications. 4. Aufl.: Institut Géopolymère.

[67] Davidovits, Jospeh (1991): Geopolymers. Inorganic Polymeric New Materials. In: Journal of Thermal Analysis 37, S. 1633–1656

[68] Kränzlein, Eva (2019): Immobilization mechanisms of lead(II) and zinc(II) ions in fly ash based geopolymers depending on the Na/Al and Si/AL ratio. Martin-Luther-Universität Halle-Wittenberg. Online verfügbar unter http://dx.doi.org/10.25673/13503.

[69] Komnitsas, Kostas; Zaharaki, Dimitra (2007): Geopolymerisation. A review and prospects for the minerals industry. In: Minerals Engineering 20 (14), S. 1261–1277. DOI: 10.1016/j.mineng.2007.07.011.

[70] Ikeda, K.; Nunohiro, T.; Iizuka, N. (1998): Consolidation of silica sand slime with a geopolymer binder at room temperature and the strength of the monoliths. In: Chem Pap 52 (4), S. 214–217.

[71] Shi, Caijun; Jiménez, A. Fernández; Palomo, Angel (2011): New cements for the 21st century. The pursuit of an alternative to Portland cement. In: Cement and Concrete Research 41 (7), S. 750–763. DOI: 10.1016/j.cemconres.2011.03.016.

[72] Duxson, Peter; Provis, John L. (2008): Designing Precursors for Geopolymer Cements. In: Journal of the American Ceramic Society 91 (12), S. 3864–3869. DOI: 10.1111/j.1551-2916.2008.02787.x.

[73] Duxson, P.; Fernández-Jiménez, A.; Provis, J. L.; Lukey, G. C.; Palomo, A.; van Deventer, J. S. J. (2007): Geopolymer technology. The current state of the art. In: J Mater Sci 42 (9), S. 2917–2933. DOI: 10.1007/s10853-006-0637-z.

[74] Swaddle, Thomas W.; Salerno, Julian; Tregloan, Peter A. (1994): Aqueous aluminates, silicates, and aluminosilicates. In: Chem. Soc. Rev. 23 (5), S. 319–325. DOI: 10.1039/CS9942300319.

[75] Provis, John L.; Lukey, Grant C.; van Deventer, Jannie S. J. (2005): Do Geopolymers Actually Contain Nanocrystalline Zeolites? A Reexamination of Existing Results. In: Chemistry of Materials 17 (12), S. 3075–3085. DOI: 10.1021/cm050230i.

[76] A. Fernández-Jiménez, A. Palomo M.M. Alonso (2005): Alkali activation of fly ashes. Mechanisms of reaction. Breno, Czech Republic (Proceeding non traditional cement and concrete).

[77] Fernández-Jiménez, Ana; Palomo, Angel; Sobrados, Isabel; Sanz, Jesús (2006): The Role Played by The Reactive Alumina Content in The Alkaline Activation of Fly Ashes 91, S. 111–119.

[78] Palomo, Angel; Alonso, Santiago; Fernandez-Jimenez, Ana; Sobrados, Isabel; Sanz, Jesús (2004): Alkaline activation of fly ashes. NMR study of the reaction products. In: Journal of the American Ceramic Society 87 (6), S. 1141–1145.

[79] Iler, R. K. (1955): The colloid chemistry of silica and silicates: Cornell University Press % This file was created with Citavi 5.5.0.1 (George Fisher Baker non-resident lectureship in chemistry at Cornell University). Online verfügbar unter https://books.google.de/books?id=BHwGAQAAIAAJ.

[80] Madras, Giridhar; McCoy, Benjamin J. (2001): Distribution kinetics theory of Ostwald ripening. In: The Journal of Chemical Physics 115 (14), S. 6699–6706.

[81] Garcia, I. (2008): Compatiblidad de geles cementantes CSH y NASH. Estudios en muestras reales y en polvos sintéticos. Thesis.

[82] Khale, Divya; Chaudhary, Rubina (2007): Mechanism of geopolymerization and factors influencing its development. A review. In: J Mater Sci 42 (3), S. 729–746. DOI: 10.1007/s10853-006-0401-4.

[83] Kränzlein, E.; Harmel, J.; Pöllmann, H.; Krcmar, W. (2019): Influence of the Si/Al ratio in geopolymers on the stability against acidic attack and the immobilization of Pb2+ and Zn2+. In: Construction and Building Materials 227, S. 116634. DOI: 10.1016/j.conbuildmat.2019.08.015.

[84] Dlaz-Loya, E., Allouche, E.N., Vaidya, S. (2011): Mechanical Properties of Fly-Ash-Based Geopolymer Concrete. In: ACI Materials Journal 108 (3), S. 300–306.

[85] Bajpai, Rishabh; Choudhary, Kailash; Srivastava, Anshuman; Sangwan, Kuldip Singh; Singh, Manpreet (2020): Environmental impact assessment of fly ash and silica fume based geopolymer concrete. In: Journal of Cleaner Production 254, S. 120147. DOI: 10.1016/j.jclepro.2020.120147.

[86] Davidovits, J. (2020): A continent is on fire. STOP promoting fly ash-based cements. Online verfügbar unter https://www.geopolymer.org/news/a-continent-in-on-fire-stop-promoting-fly-ash-based-cements/zuletztaktualisiertam 15.05.2020.

[87] Kühl, H.: Slag cement and process of making the same. Veröffentlichungsnr: 900,989.

[88] Bernal, Susan A.; Provis, John L.; Rose, Volker; Mejía de Gutierrez, Ruby (2011): Evolution of binder structure in sodium silicate-activated slag-metakaolin blends. In: Cement and Concrete Composites 33 (1), S. 46–54. DOI: 10.1016/j.cemconcomp.2010.09.004.

[89] Provis & Bernal: Geopolymer and related alkali activated materials 2014.

[90] Escalante-García, Jose I.; Fuentes, Antonio F.; Gorokhovsky, Alexander; Fraire-Luna, Pedro E.; Mendoza-Suarez, Guillermo (2003): Hydration Products and Reactivity of Blast-Furnace Slag Activated by Various Alkalis. In: Journal of the American Ceramic Society 86 (12), S. 2148–2153. DOI: 10.1111/j.1151-2916.2003.tb03623.x.

[91] Wang, Shao-Dong; Scrivener, Karen L.; Pratt, P. L. (1994): Factors affecting the strength of alkali-activated slag. In: Cement and Concrete Research 24 (6), S. 1033–1043. DOI: 10.1016/0008-8846(94)90026-4.

[92] Brandstetr, B. and TallingJ.: Present State and Future of Alkali-Activated Slag Concretes. In: ACI Symposium Publication 114. DOI: 10.14359/1873.

[93] Silva, R. V.; Brito, J. de; Lynn, C. J.; Dhir, R. K. (2017): Use of municipal solid waste incineration bottom ashes in alkali-activated materials, ceramics and granular applications. A review. In: Waste management (New York, N.Y.) 68, S. 207–220. DOI: 10.1016/j.wasman.2017.06.043.

[94] Zhang, Jian; Shi, Caijun; Zhang, Zuhua; Ou, Zhihua (2017): Durability of alkali-activated materials in aggressive environments. A review on recent studies. In: Construction and Building Materials 152, S. 598–613. DOI: 10.1016/j.conbuildmat.2017.07.027.

[95] Lee, W.K., Son, Y.G (2012): Effect on the compressive strength of pastes made from MSWI bottom ash with mixing ratio of sodium silicate and potassium silicate. In: Waste Manage (29), S. 93–97.

[96] Lancellotti, Isabella; Ponzoni, Chiara; Barbieri, Luisa; Leonelli, Cristina (2013): Alkali activation processes for incinerator residues management. In: Waste management (New York, N.Y.) 33 (8), S. 1740–1749. DOI: 10.1016/j.wasman.2013.04.013.

[97] Jo, B.W., Kim, K.I., Park, J.C., Park, S.K. (2006): Properties of chemically activated MSWI (Municipal Solid Waste Incinerator) mortar. In: J. Korea Concr. Inst. (18), S. 589–594.

[98] Shi, C., Krivenko, P.V., Roy, D.M. (2006): Alkali-Activated Cements and Concretes.

[99] Buchwald, A. (2012): ASCEM\textregistered cement-a contribution towards conserving primary resources and reducing the output of CO 2. In: Cem. Int 10 (5), S. 86–97.

[100] Amathieu L., Bier Th. A. and Scrivener K. (2001) Mechanisms of set acceleration of portland cement through CAC addition. Int.Conf. on Calcium Aluminate Cement pp 303–17.

[101] Justness H. Acceleration by retardation SINTEF Building and Infrastructure (Trondheim, Norway).

[102] Kighelman J (2007) Hydratation and structure development of ternary binder system as used in self-levelling compounds. PhD Thesis, EPFL.

[103] Onishi K. and Bier T. A. (2010) Investigation into relations among technological properties, hydration kinetics and early age hydration of self-leveling underlayments Cem. Concr. Res. 40 1034–40.

[104] Torres D. (2013) Mezclas ternarias de Cemento Portland cemento de aluminato de Calcio y sulfato cálcico: Mecanismos de expansión. PhD Thesis.

[105] Qoku E. (2019) Characterization and quantification of crystalline and amorphous phase assemblage in ternary binders during hydration. PhD Thesis, TU Bergakademie Freiberg.

[106] Gu P., and Beaudoin J. J. (1997) A conduction calorimetric study of early hydration of Ordinary Portland cement /high alumina cement pastes J. Mater. Sci. 32 3875–81.

[107] Lamberet S (2005) Durability of ternary binders based on portland cement, calcium aluminate cement and calcium sulfate. PhD thesis, EPFL.

[108] Qoku E., Bier T. A. and Westphal T. (2017) Phase assemblage in ettringite -forming cement pastes: A X-ray diffraction and thermal analysis characterization J. Build. Eng. 12.

[109] Bayoux J.P., Bonin A., Marcdargent S., Mathieu A. M. V. (1990) Study of the hydration properties of aluminous cement and calcium sulfate mixes in calcium aluminates cement. Proceedings of the International Symposium on Calcium Aluminate Cements, pp 220–49.

[110] Evju C. and Hansen S. (2005) The kinetics of ettringite formation and dilatation in a blended cement with β-hemihydrate and anhydrite as calcium sulfate Cem. Concr. Res. 35 2310–21.

[111] Puri A., Voicu G. and Badanoiu A. (2010) Expansive binders in the Portland cement – Calcium aluminate cement – calcium sulfate system Rev Chim 61 740–4.

[112] Xu L., Wang P. and Zhang G. (2012) Calorimetric study on the influence of calcium sulfate on the hydration of Portland cement -calcium aluminate cement mixtures J. Therm. Anal. Calorim. 110 725–31.

[113] Mori H., Maruya E., Atarashi D. and Etsua S. (2014) Effect of temperature on length change of cementitious materials using Portland cement –calcium aluminate cement – anhydrite blast furnace slag system Calcium Aluminate Cements, Proceedings of the international conference pp 407–21.

[114] Emoto T. and Bier T. A. (2007) Rheological behavior as influenced by plasticizers and hydration kinetics Cem. Concr. Res. 37 647–54.

[115] Majumdar A. J. and Singh B. (1992) Properties of some blended high-alumina cements Cem. Concr. Res. 22 1101–14.

[116] Hidalgo A., García Calvo J. L., Alonso M. C., Fernández L. and Andrade C. (2009) Microstructure development in mixes of calcium aluminate cement with silica fume or fly ash. J. Therm. Anal. Calorim. 96 335–45.

[117] López A. H., Garcia Calvo J. L., Olmo J. G., Petit S. and Alonso M. C. (2008) Microstructural evolution of calcium aluminate cements hydration with silica fume and fly ash additions by scanning electron microscopy, and mid and near-infrared spectroscopy J. Am. Ceram. Soc. 91 1258–65.

[118] Ding J., Fu Y. and Beaudoin J. J. (1995) Strätlingite formation in high alumina cement – silica fume systems: Significance of sodium ions. Cem. Concr. Res. 25 1311–9.

[119] Fernández-Carrasco L. and Vázquez E. (2009) Reactions of fly ash with calcium aluminate cement and calcium sulphate Fuel 88 1533–8.

[120] Bizzozero J. and Scrivener K. L. (2015) Limestone reaction in calcium aluminate cement-calcium sulfate systems Cem. Concr. Res. 76 159–69.

[121] Westphal T., Bier T.A and Bajrami B. (2012) Einfluss von Fließmittel und Kalksteinmehl auf die frühe Phasen- und Strukturentwicklung in Systemen mit ternären Bindemitteln. Tagung Bauchemie.

[122] Puerta-Falla G., Balonis M., Le Saout G., Falzone G., Zhang C., Neithalath N. and Sant G. (2015) Elucidating the role of the aluminous source on limestone reactivity in cementitious materials J. Am. Ceram. Soc. 98 4076–89.

[123] Westphal T., Bier T. A and Dlugosch F. (2011) Influence of admixtures on phase development in ternary binder systems 13th Int. Congr. Chem. Cem. 1–7.

[124] Bier T. A., Bajrami A., Westphal T., Qoku E. and Qorllari A. (2015) Influence of re-dispersible powders on very early shrinkage in functional mortars Adv. Mater. Res. 1129 77–85.

[125] Bier T. A. and Amathieu L. (1997) Calcium aluminate cement (CAC) in building chemistry formulations CONCHEM congress.

[126] De Weerdt K., Sellevold E., Kjellsen K. O. and Justnes H. (2011) Fly ash-limestone ternary cements: Effect of component fineness Adv. Cem. Res. 23 203–14.

[127] De Weerdt K., Haha M. Ben., Le Saout G., Kjellsen K. O., Justnes H. and Lothenbach B. (2011) Hydration mechanisms of ternary Portland cements containing limestone powder and fly ash Cem. Concr. Res. 41 279–91.

[128] Kuzel H.J. and Poellman H. (1991) Hydration of C_3A in the presence of $Ca(OH)_2$, $CaSO_4$ $2H_2O$ and $CaCO_3$. *Cem. Concr. Res.* **21** 885–95.

[129] Winnefeld F. and Lothenbach B. (2016) Phase equilibria in the system $Ca_4Al_6O_{12}SO_4$ – Ca_2SiO_4 – $CaSO_4$ – H_2O referring to the hydration of calcium sulfoaluminate cements *RILEM Tech. Lett.* **1** 10.

[130] Deschner F., Winnefeld F., Lothenbach B., Seufert S., Schwesig P., Dittrich S., Goetz-Neunhoeffer F. and Neubauer J. (2012) Hydration of Portland cement with high replacement by siliceous fly ash Cem. Concr. Res. 42 1389–400.

[131] Martin L. H. J., Winnefeld F., Tschopp E., Müller C. J. and Lothenbach B. (2017) Influence of fly ash on the hydration of calcium sulfoaluminate cement Cem. Concr. Res. 95 152–63.

[132] Rahhal V. and Talero R. (2004) Influence of two different fly ashes on the hydration of portland cements J. Therm. Anal. Calorim. 78 191–205.

[133] Jun-Yuan H., Scheetz B. E. and Roy D. M. (1984) Hydration of fly ash-portland cements Cem. Concr. Res. 14 505–12.

[134] Lothenbach B. Scrivener K. and Hooton R. D. (2011) Supplementary cementitious materials Cem. Concr. Res. 41 1244–56.

[135] Lothenbach B., Le Saout G., Gallucci E. and Scrivener K. (2008) Influence of limestone on the hydration of Portland cements Cem. Concr. Res. 38 848–60.

[136] Zajac M., Rossberg A., Le Saout G. and Lothenbach B. (2014) Influence of limestone and anhydrite on the hydration of Portland cements Cem. Concr. Compos. 46 99–108.

Chubaakum Pongener, Koweteu Sekhamo,
Rajib Lochan Goswamee

Chapter 8
Study of some physico chemical properties of plastic clays belonging to Girujan deposits from Chumoukedima Nagaland, India and their prospective industrial applications

Abstract: Highly clayey soil samples collected from its native locations in Chumouke-dima area of south western part of Nagaland, India, on detailed mineralogical and physico-chemical characterisation showed that they possess different important properties of good quality plastic clay having immense industrial prospects. The argilla-ceous samples after air drying, ball milling and sieving were subjected to various classical as well as high end instrumental analysis. FT-IR spectroscopy was used to provide information of various functional groups present. Oriented powder-XRD combined with ethylene glycol intercalation showed that it mainly consisted of kaolin, illite, traces of quartz, muscovite, rutile and absence of any swelling clay. SEM-EDXA showed presence of stacked layers having nano pores in the basal planes, important from the catalyst perspectives. Solid state NMR supported by XPS analysis data gave the finer structural details of important constituent ions like Si, Al and Fe. Isomor-phous substitution of Si^{4+} by Al^{3+} in the clay tetrahedral layers results in generation of a negative surface charge. This is supported by Zeta potential measurements as well as high hydrational nature of causing higher Atterberg plasticity index. The re-sults of particle size distribution combined with Winkler diagram analysis showed the highly plastic clay samples suitable for thin walled hollow brick production.

Keywords: Chumoukedima, Kaolinite, Girujan Clay, ^{27}Al and ^{29}Si-NMR, Winkler's Diagram

Acknowledgement: One of the authors Dr. Chubaakum Pongener is grateful to CSIR-New Delhi for fund-ing the research work (under RA sanction No 31/025(0145)/2019 EMR-5 dated 31-03-2019, institutional manuscript no NMN 20201). Director CSIR-NEIST Jorhat for the permission to carry out the work and Directorate of Geology and Mining Government of Nagaland, Dimapur for help in samples collection.

Chubaakum Pongener, Advanced Materials Group, Materials Science and Technology Division, CSIR-NEIST, Jorhat, Assam, 785001, India, e-mail: chubaakum@gmail.com
Koweteu Sekhamo, Department of Geology, Patkai Christian College, Dimapur, Nagaland, India -797115
Rajib Lochan Goswamee, Advanced Materials Group, Materials Science and Technology Division, CSIR-NEIST, Jorhat, Assam, 785001, India

https://doi.org/10.1515/9783110674941-008

1 Introduction

Since ancient times, due to the abundances of different clays in earth's crust, depending on their local availability and various physico chemical properties different clays are occupying a very significant position in various day to day applications of the human society such as pottery, building materials, paints, paper, cosmetics and medicine, polymers and catalysis [1, 2]. Therefore, in every continent physico-chemical characterisation of clay bearing rocks and soils, identification and quantification of the primary clay minerals present in them has became an important aspect of satisfying the desires of answering the inquests on geochemical information as well as their prospective economic and industrial applications.

South Asia, especially the North Eastern Region of India including its neighbouring areas, is the home of large typical clay bearing geological formation named as Girujan clay stage. It is extended over an area of several hundred kilometres in length. Girujan clay stage belong to the middle Tipam series deposits of Miocene geological age [3] and occupies various depths from low lying flat ground with the clay thickness varying from 500–850 m to around 2300 m in Kumchai–Manabhum area near the Mishimi uplift [3, 4].In its continued extensive locations near Naharkatiya in the famous upper Assam petroleum basin the Girujan formation is a proven oil and gas producing zone also [5].

In the state of Nagaland in the North Eastern Region of India despite having this huge clay deposit, to the best of our knowledge, so far no report has been recorded on the quality assessment and gradation of the deposit with a view of prospecting some industrial use of it. However, historically the nearby famous terracotta artefacts and sandstone pillars of aboriginal Kacharis of Dimapur boldly declares the intimate relationship of clay and clay products with the society for around thousands of years [6]. In this backdrop in the present work a mineralogical and physico-chemical characterization of the said plastic clay deposits was carried out. Accordingly, high clay bearing soil samples were collected from its native locations in Chumoukedima area of south western part of Nagaland, India. The samples were located at a GPS coordinate of 25°76′34″ N, 93°82′53″ E at an elevation of 350 meters from the mean sea level on the banks of a river locally known as Chathe.

Apart from classical evaluation methods like fractionation using Stokes sedimentation [7], wet and dry sieving, Atterberg plasticity measurement, cation exchange capacity determination and oxidic composition analysis of the collected samples were subjected to a thorough physico-chemical evaluation by the use of different high precision analytical techniques like BET surface area pore volume analysis, particle size analysis, zeta potential measurement, XRD, SEM, EDX, FT-IR, Solid State-NMR, XPS to know the nature and structure of the constituent clay and other minerals.

2 Materials and methods

2.1 Study area and sampling

The clayey soil sample was collected from exposed surface of the Girujan deposit as starting raw material. The deposit is located (Figure 8.1) about 16 km south-east from the Chumoukedima, Dimapur district, Nagaland [8] and as per the geological report, it is composed of paleogene sediments [3, 9]. Due care was taken during sampling to have representative and distributed samples from within a limited collection area of 3 km². The collected dried samples after coning and quartering were sieved through different mesh size sieves ranging from +212 to −53 ≤µm. In the next step −53 ≤µm clay fractions was ball milled by grinding 1:2 weight ratio of clay and ceramic ball (500 g clay and 1000 g ceramic ball) for 1 hour at 120 RPM to obtain maximum uniform clay size materials which was then stored in an air tight container for further analysis.

2.2 Analytical methods

Soil texture analysis was done using conventional textural triangle diagram for clay, silt and sand [10, 11]. Atterberg limits experiment were used to understand the consistency of the soil which measures the physical transformations when wet soil hydrates with expansion of volume and gives the limit of water at which the behaviour of a clayey soil changes from plastic to liquid. Thus, water limit, plasticity and plasticity index were done in accordance with standard procedures [12].

FT-IR spectroscopy was carried out in the region from 4000 to 200 cm^{-1} in an IR spectrometer using KBr pellet to identify the presence of different surface functional groups in the clay.

The ion exchange capacity of the −53 µm fraction clay samples were determined after exchanging all the exchangeable cations on the clay surface with Ca^{2+} ion by treating it in aqueous 1M $CaCl_2$ solution. This was followed by thorough washing and replacing the exchanged Ca^{2+} ions with K^+ ions [13] and the replaced Ca^{2+} obtained in the filtrate fraction was estimated titrimetricaly to calculate the ion exchange capacity.

To study the particle size distribution of the sample, the clay fractions were further refined using Stoke's law based clay sedimentation technique [14]. The particle size distribution provides vital information on the suitability of the clay for different purposes such as building materials, paints or ceramic [15]. The surface charge density of the clay defines the nature of electrokinetic potential of the clay and also has important significances on ions migration in soil [16, 17] and aqueous swelling ability. Thus, accordingly the particle size analysis and the surface charge density of the clay sample were measured in a particle size analyser, Master sizer 2000 (Malvern) and Zeta potential analyser (Zeta Asizer-Nano ZS). In both the cases,

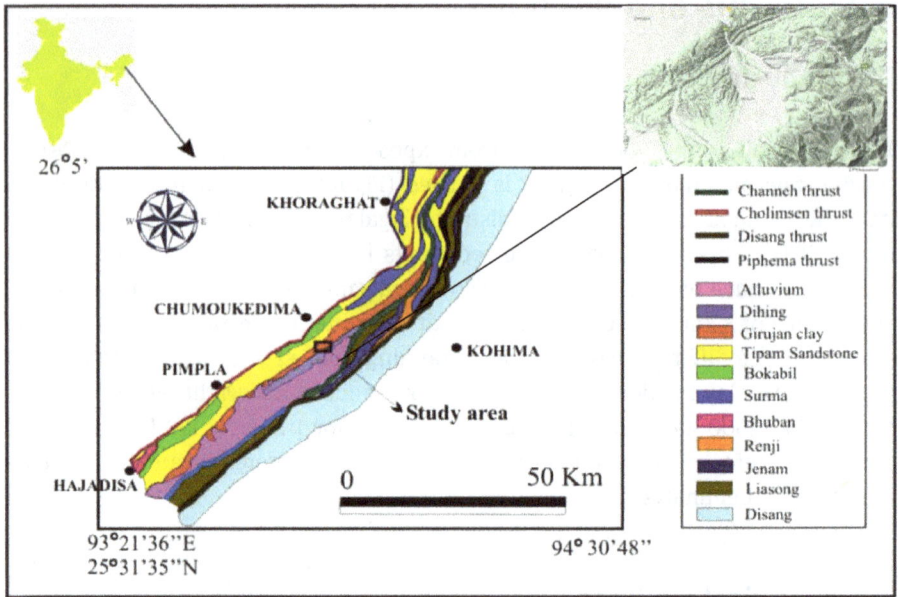

A

B

Figure 8.1: a. Geological map of the study area. b. Photograph of clay deposit site.

2% weight/volume dispersion of −53 µm clay fraction were used after dispersing it in double distilled water and sonicating for 2 hours. Winkler's and Shepard's diagram were constructed from the particle size analysis data to infer the probable industrial applicability of the clay under investigation.

Thermal analysis mainly TGA was done in an inert nitrogen atmosphere by starting from room temperature to 1000 °C at a heating rate of 10 °C/min. The different weight loss regions were recorded having different DTG peak temperatures viz. 103, 360, 600 and 1000 °C. The BET surface area, pore volume and pore radius were obtained from BET surface analyser (Quantra chrom, USA-Autosorb IQ), using nitrogen as purge gas. Powder XRD data were recorded in a Rigaku diffractometer (model Ultimate IV) using CuK$_\alpha$ (λ = 1.5405 Å) radiation at a scan rate of 0.04° 2θ/second where X-Ray generator had a current setting of 40KV and 40milli ampere. In order to identify the presence of any swelling clays like montmorillonite or chlorite, the oriented thin films of both the raw clay and −53 µm clay samples were examined after air drying as well as after their saturation with ethylene glycol vapour (EG). The SEM and EDX patterns were taken to understand the surface morphology and the elemental composition of the clay particles. Single pulse Magic Angle Spinning (MAS) NMR was done for Al and Si nuclei using Jeol ECZ400 magnet at the spin rate of 10 MHz. The XPS analysis of clay was performed on a spectrometer (Thermo-Scientific ESCALAB Xi+) having a monochromatic Al Kα X-ray source (1486.6 eV). XPS analyses for Al, Si and O were carried out to understand the bond parameters from the positions of the energy peaks.

3 Results and discussion

As the deposit is located by the side of proposed Asian Highway-1 which would connect Turkey from Tokyo and lies specifically in the historic Burma Road section, linking Dimapur in India and Tamu in Myanmar, hence it bears good prospect for exploring establishment of some industrial ventures for high valued products. On traversing across the length and breadth of accessible and exposed sections of the virgin deposit an initial visual understanding of its size is obtained which could be not less than approximately 50,000 tonnes of clayey soil. This figure however can be thoroughly estimated once the whole deposit site is surveyed by standard geological means and methods, which is beyond the scope of the present initial work.

Initial estimation of different clay fractions obtained by using sieves of 212 µm and 53 µm were evaluated, and accordingly the percentage compositions of different particle size were calculated; the percentage of +212 µm size was 0.053%, +53 µm size 3.105% and −53 µm size 96.84%. Thus, the soil sample with more than 96% of −53 µm clay fraction is an indicative of high clay bearing soil material. The first step in the chemical characterisation of the clay as carried out by following standard classical methods prescribed in books and international standards [18, 19] showed that the Iron oxide content in the clay is low with 0.4% w/w level, while Magnesium and Calcium oxide were found to be 3.5% and 2.8% w/w respectively; mixed oxides accounted for 35% and SiO_2 content was found to be 44% w/w respectively. Because of the low iron content in the clay sample, the colour of the clay appears whitish and the clay becomes suitable for refractory purposes. The loss on ignition after firing at 1000 °C for 1 h was found to be 7.57% which suggest that the clayey sample is high in inorganic components (Table 8.1).

Table 8.1: Chemical analysis of clayey soil.

Oxides/LOI	% Weight/Weight Content
Fe content (as oxides)	0.4%
Mixed oxides	35%
Calcium oxide	2.8%
Magnesium oxide	3.5%
Silicon dioxide	44%
K_2O	2.6%
Na_2O	3.8%
LOI	7.57%

Accordingly, the Winkler's and Shepard's diagram were constructed by analysing the particle size data to target some probable high temperature industrial product from of the samples of clayey soil under investigation. The representative raw sample was initially investigated to determine its soil type by using texture triangle diagram [10, 11] of clay, silt and sand content (Figure 8.2a) from size fraction data. As shown in the Figure 8.2a the soil type belongs to the clay type as the coarse fraction, silt fraction and clay fraction were found 11.49%, 29.88% and 58.62%w/w respectively. Further refining the same data by plotting in Shepards diagram [20], shows it to be a silty clay type (Figure 8.2a–b).

Further analysing the results of Laser Doppler particle size analysis of the Girujan stage clay of Chumoukedima it was found that it has a rather wide range of particle size distribution (Figure 8.3d) from less than 2 µm sizes to 100 µm sizes with a large volume percent of the clay belonging to less than 20 µm fractions. So far as application prospects are concerned following Shepard's classification scheme [20] of soil, the silty clay and clayey silts are the suitable raw materials for fabricating building material products [2].

As said above the particle size analysis of clayey sample shows wide range of particle size distribution with coarse fraction >20 µm, silt fraction in the range of 2–20 µm and clay fraction <2 µm contents as 25.31%, 45.37% and 27.31% V/V respectively

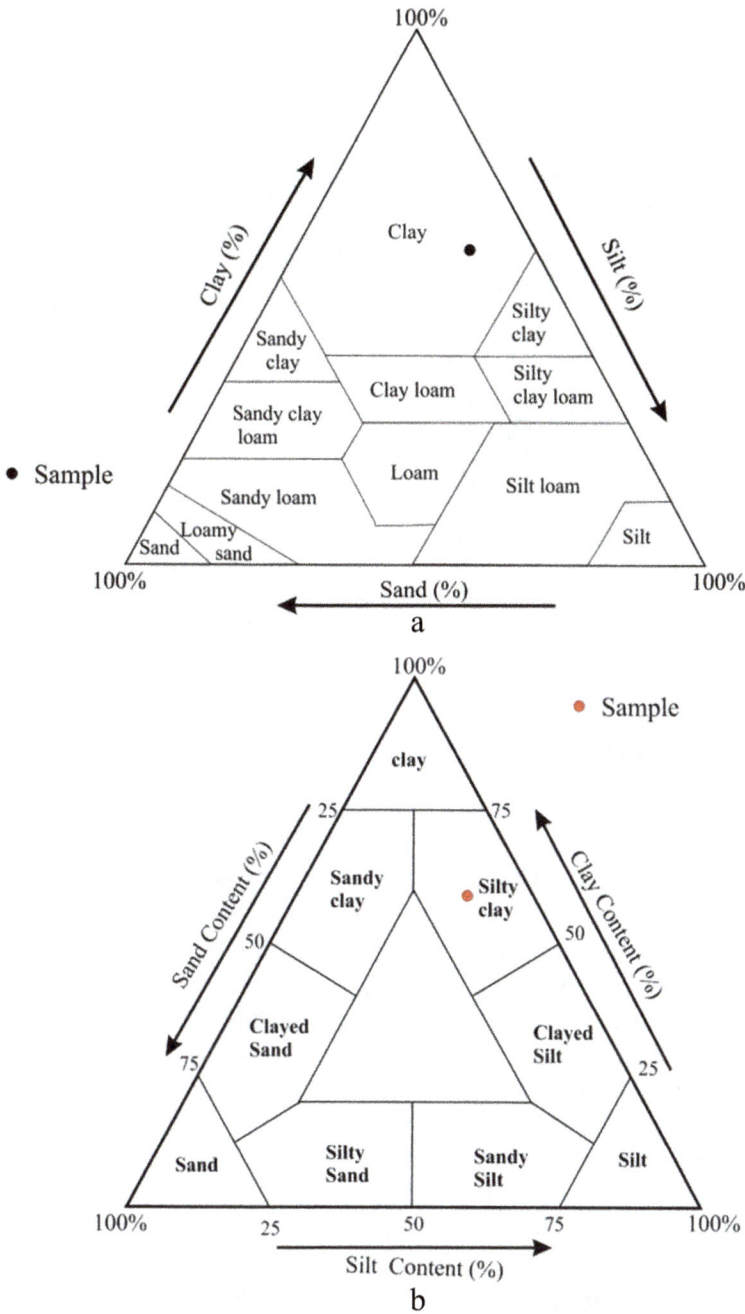

Figure 8.2: a. Textural triangle diagram of clay, silt and clay. b. Shepards diagram.

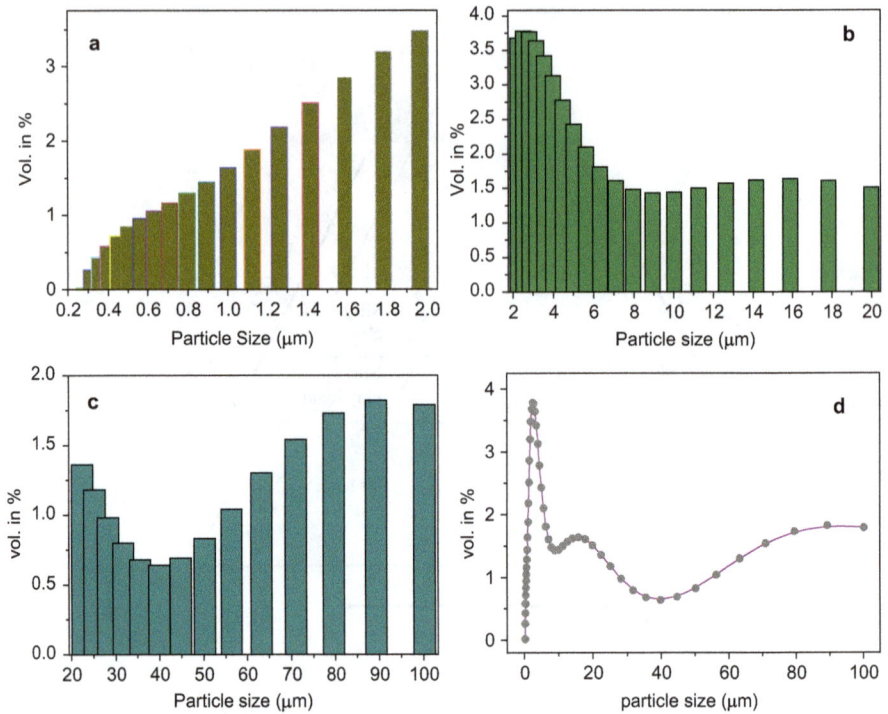

Figure 8.3: Particle size of: a. <2 µm, b. 2–20 µm, c. >20 µm clay fraction, d. Overall particle size distributions.

(Figure 8.3a–c). In fact, Figure 8.4d shows that the particle size distribution is more distinct within the three narrow regions, i.e. 0.17–8.95 µm, 8.95–39.75 µm and 39.75–101.1 µm respectively and from the figure it is evident that the clayey sample possesses higher percentage of finer particles than coarse particles.

The data of different particle size based trilinear plots shown in above Figure 8.2 were used to refine the classification parameters of the clays as per Shepard's diagram [21]. Subsequently, the refined data from the size-fraction analysis were further considered to analyse the product prospects as per Winkler diagram (Figure 8.4), i.e. plotting percentage volume fraction <2 µm against >20 µm for the clayey sample [22, 2] and accordingly from the plot, the clayey sample was found to be suitable for thin walled hollow bricks [21]. So far as the Nagaland or North Eastern Region of India is concerned the positive aspect of introducing such products could be that it will be suitable for the region as it has geographically a hilly topography and geophysically a very high seismicity.

Apart from the particle size distribution another important property of the particle is their morphology. The SEM morphology study of the samples reveals they have a layered structure with the presence of fine nano pores in the basal surfaces.

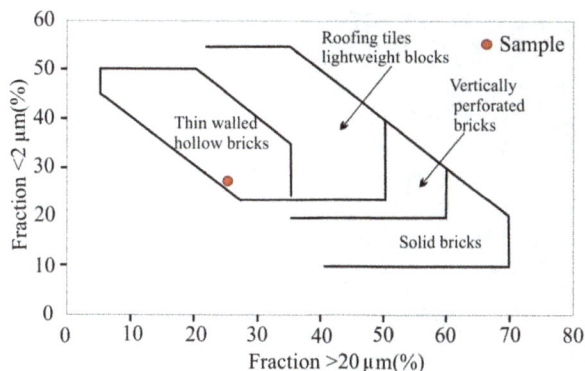

Figure 8.4: Winkler's diagram of the clay sample.

The layered morphology (Figure 8.5a–b) of the crystals grouped in a stacked bundled aggregation is a characteristic feature of the occurrence of the clay materials in nature [23]. The corresponding EDX study of the particles indicated the presence of different major, minor and trace elements like Al, Si, Fe, O, Na, K, Mg, C in the layered crystals (Figure 8.5c) which are common in kaolin and illite clay minerals.

Finer atom coordination level structure was elucidated with equipment like solid state MAS NMR. Which gave the information about the tetrahedral or octahedral coordination structure of Si and Al in the hydroxidic framework that generally constitute the structure of 2:1 or 1:1 phyllosilicate [24]. The MAS NMR spectrums of the ^{29}Si and ^{27}Al nuclei are given below in Figure 8.6a–b. The ^{27}Al MAS NMR spectra showed that at field strength of 9.4 T (Tesla) and spin rate of 10 MHz, the signal splits into two peaks one at around 77.1ppm and another at 1.1 ppm indicating the presence of both tetrahedral and octahedral Al in the structure [25], which suggest that isomorphic substitution of Si^{4+} by Al^{3+} in the tetrahedral sheet of clay structure. However, in ^{29}Si MAS NMR spectra a broad single peak at around –90 ppm was observed, indicating the existence of Si atom only in tetrahedral environment [26, 27]. Isomorphous replacement of tetravalent Si^{4+} by trivalent Al^{3+} in the tetrahedral sheet causes charge imbalance in the sheet which in this case is counter balanced by exchangeable cations in the inter layer. It is the reason for the high negative zeta potential over the clay surface and consequent hydration of the clay by osmotic swelling [28, 29] and high plasticity of the Chumoukedima clay.

Presence of exchangeable ions like Na and K in the surfaces is shown by surface survey scan of clay by XPS also (Figure 8.7a), apart from these elements surface survey also shows the presence of other important atoms. The chemical composition of surface as obtained from XPS scan in atom percentage of the elements O, Si, Al, Fe, K, Na and Mg were 57.89%, 16.18%, 10.43%, 1.57%, 1.02%, 0.9% and 0.83% respectively. The higher resolution spectrum (Figure 8.7b) of Si 2p show binding energy of 102.60 eV; however, as per the literature [30] the binding energy of Si in quartz

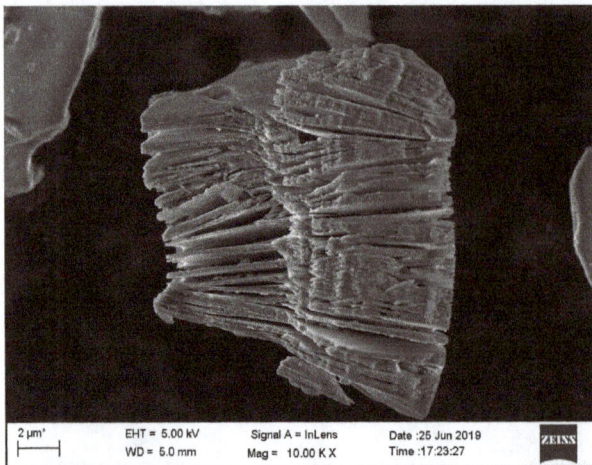

a

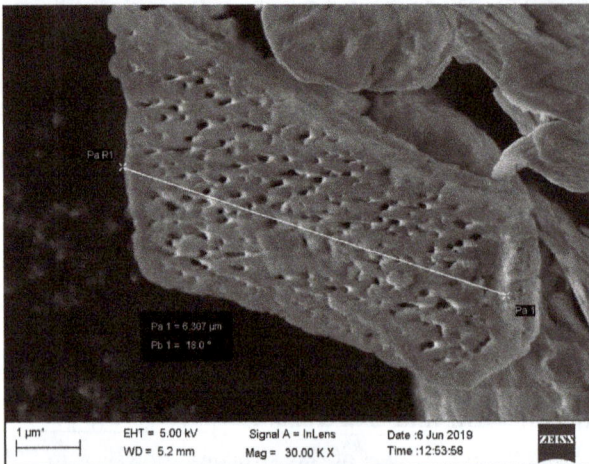

b

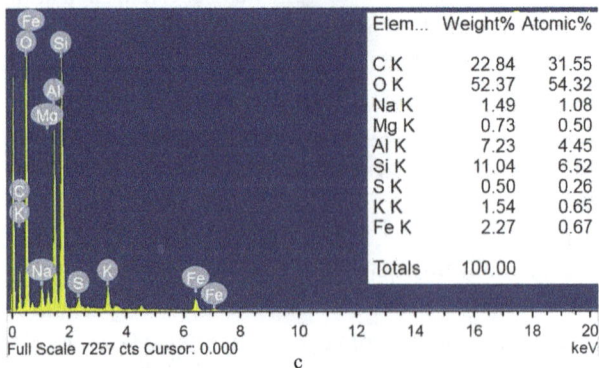

Elem...	Weight%	Atomic%
C K	22.84	31.55
O K	52.37	54.32
Na K	1.49	1.08
Mg K	0.73	0.50
Al K	7.23	4.45
Si K	11.04	6.52
S K	0.50	0.26
K K	1.54	0.65
Fe K	2.27	0.67
Totals	100.00	

Full Scale 7257 cts Cursor: 0.000

c

Figure 8.5: a and b, SEM micrograph, and c. EDX of clay sample.

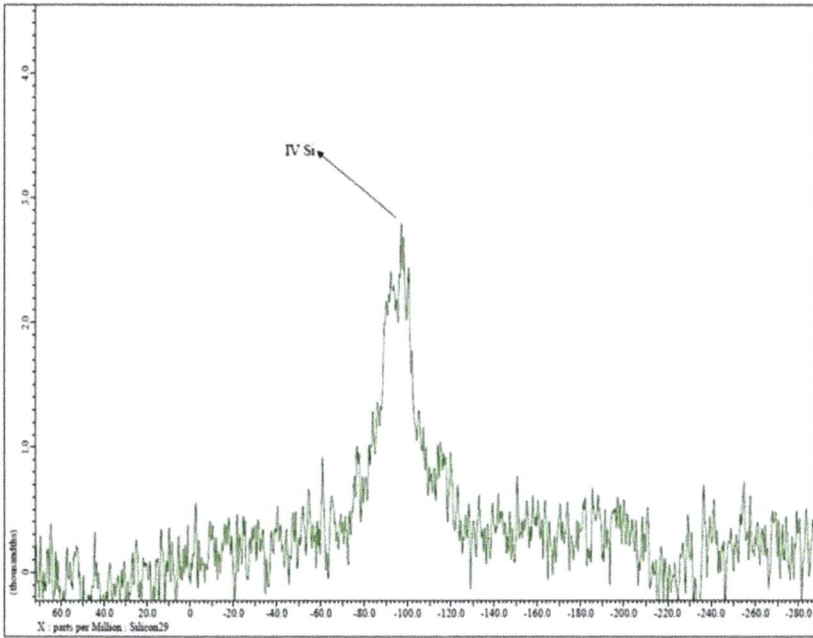

a

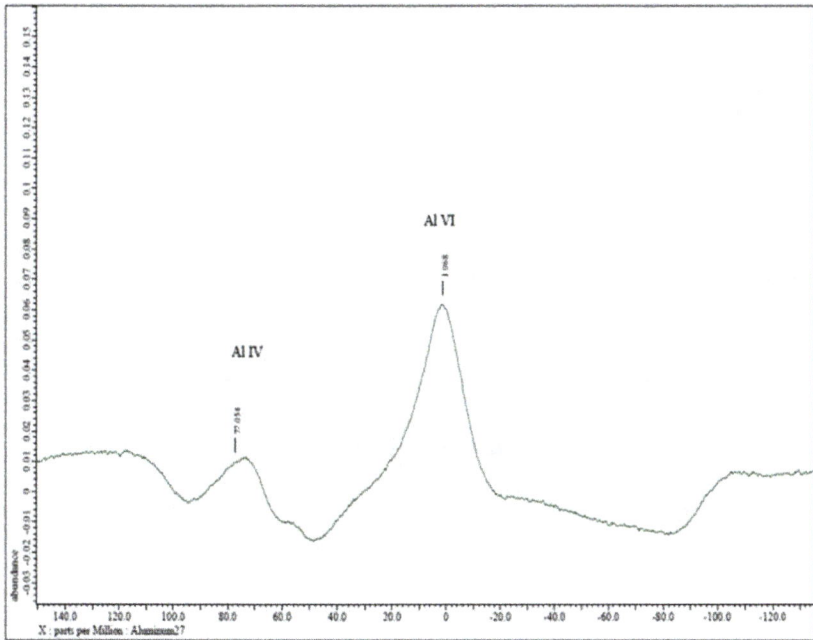

b

Figure 8.6: Solid State NMR spectra of: a. NMR of Si^{29} spectra. b. NMR of Al^{27} spectra.

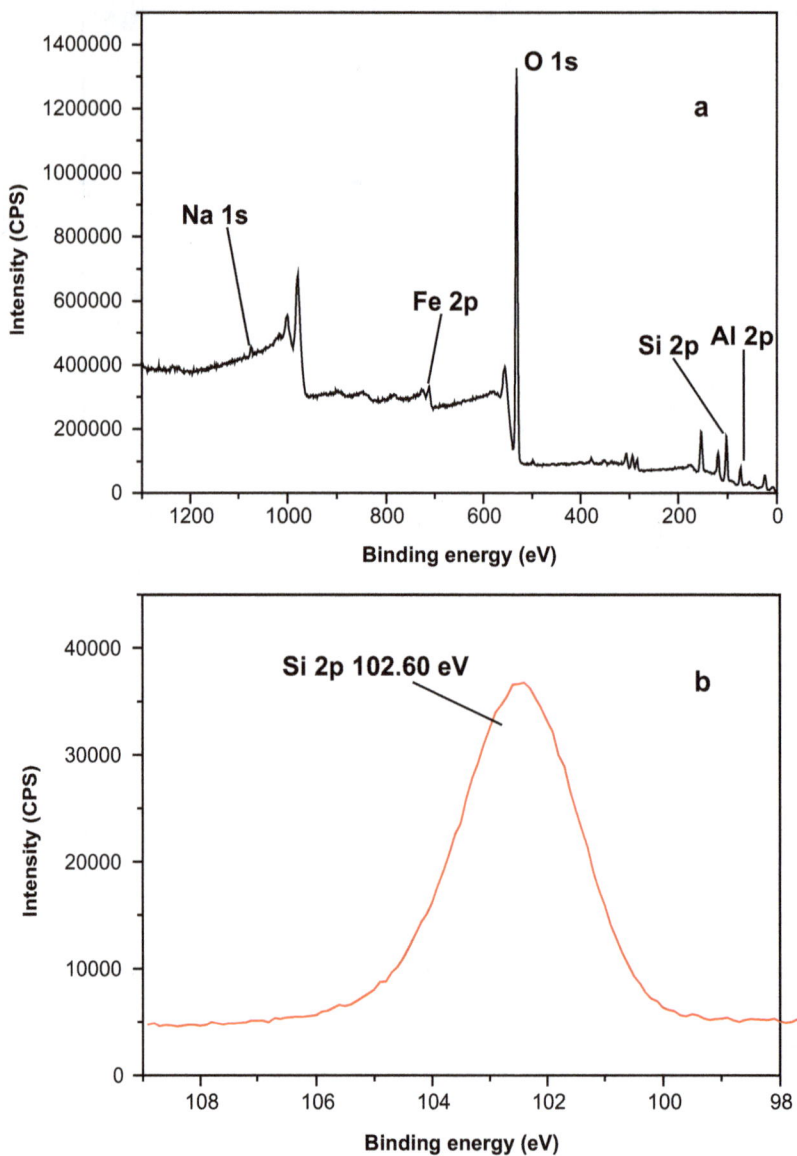

Figure 8.7: a. XPS survey scan. b, c and d. High resolution spectrum Si 2p, Al 2p and O 1s.

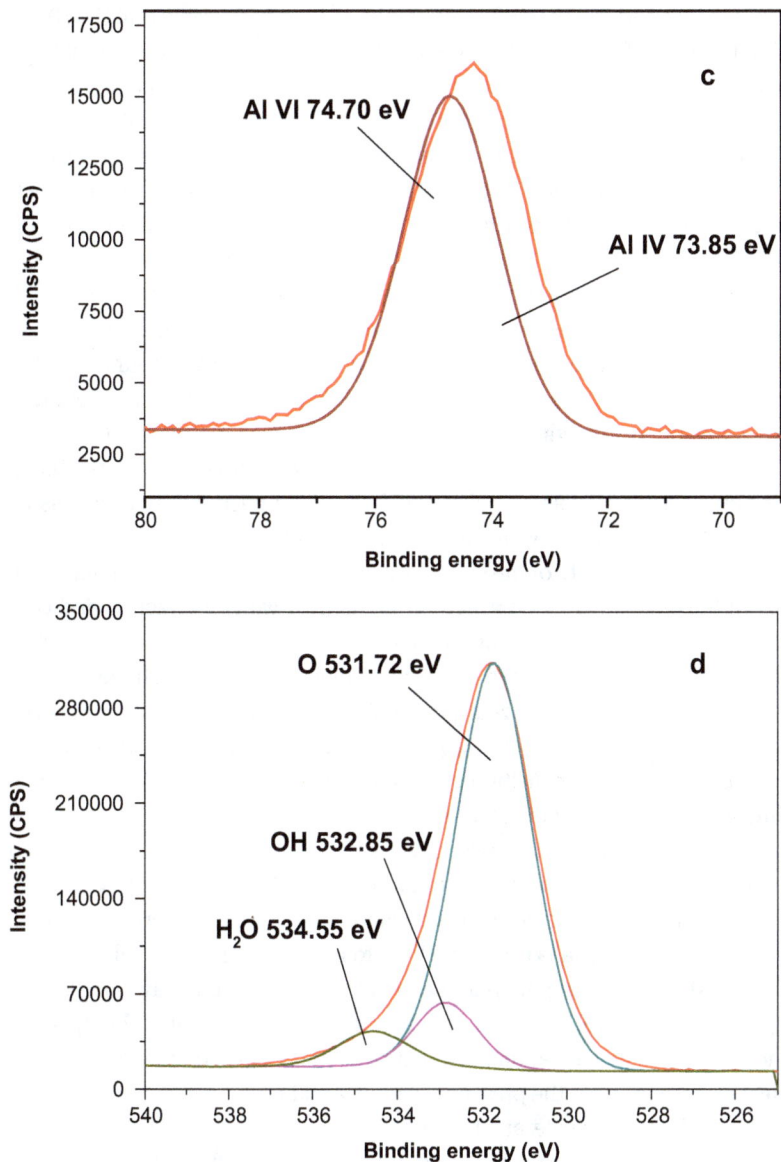

Figure 8.7 (continued)

(α-SiO$_2$) is around 104.17 eV, this further suggests that some Si atoms in tetrahedral sheet has been substituted by Al atoms which affects the binding energy of Si 2p. Since, there is isomorphous substitution in the tetrahedral sheet of Si^{4+} by Al^{3+} and the electronegativity of Si center is higher than Al^{3+} center. Therefore, O^{2-}centerswill be more closely bonded to Al centers in Al-O-Si linkage than in a pure Si-O-Si linkage, resulting in the decrease of binding energy [29, 31]. The existence of Al in both octahedral (74.70 eV) and tetrahedral (73.85 eV) form further corroborates the isomorphic substitution of Si by Al atom (Figure 8.7c) in the clay structure. Higher resolution scan of O 1s indicate the presence of three transition peaks associated with O atom around 531.72 eV, OH group around 532.85 eV and small amount of water in the clay system around 534.55 eV. Table 8.2 present the major elements of clay tetrahedral and octahedral sheet and their binding energies. The result obtained from XPS analysis is in good agreement with solid state NMR both indicating the presence of tetrahedral Si sheet as well as octahedral Al sheet with some substitution of Si by Al atoms in tetrahedral sheet.

The physical properties of the clay sample were evaluated from the data obtained from different experiments. The Atterberg liquid limit of the clay was around 29.63%, while the value of plastic limit was 20% and accordingly plasticity index was found to be 9.63 which correspond to medium plastic nature of the clay with a good cohesive behaviour [32]. The high-water absorbance capacity of the clay also could be attributed to the porous nature of the clay as was observed in the SEM micrograph as well as by the hydration of interlayer cations by means of osmotic swelling.

As the surface charge density of the clay plays a very important role in understanding the nature of the clay minerals especially so far as their water interaction and consequent crystalline and osmotic swelling is concerned [28]. Swelling is an important property either to disperse the clay or to make some high solid content shape castable green body from it. Accordingly, the zeta potential of the clay sample was measured and was found to be −34.7 mV (Figure 8.8). The −ve zeta value can be attributed to isomorphous substitutions either in the octahedral aluminium hydroxide sheet or in the tetrahedral silicate sheet by some central metal ions of lower positive valency. The same has already been reported in the present work ^{29}Si and ^{27}Al NMR properly backed up by XPS analysis in a section above. The total CEC of the clay after $CaCl_2$ saturation was found to be 26.048 meq/g, which corresponds to illite group of minerals [33] again indicating that the clay sample falls under medium to high plasticity clay [32].

The powder XRD crystallographic parameters obtained from the clay samples are shown in Figure 8.8. The P-XRD spectra of untreated oriented samples clearly show peaks presence of various layered alumino-silicate minerals corresponding to kaolinite, illite, and muscovites apart from the presence of quartz and rutile minerals. The d spacings of the identified phyllosilicate are 14.1, 9.9,7.1, 3.3, 3.2, 2.48, 1.9 Å respectively. Most of these are *00l* basal plans of the corresponding clay minerals as their intensities are enhanced in the oriented pattern than in non-oriented pattern due to better exposition effect. The study of these basal peaks on Ethylene Glycol exposition shows only traces of montmorillonite type smectitic clays in the samples.

The identification of surface functionality in the samples were carried out by FT-IR spectroscopy (Figure 8.9a–b) which shows presence of IR active groups in the clays corresponding to quartz and traces of carbonate minerals. IR bands at 3621, 3433, 1632, 1031, 921, 1440, 1010, 796, 775, 694, 618, 536 and 471 cm^{-1} are indicative of presence of kaolinite, montmorillonite, muscovite and illite minerals [34–36]. The peak at

Table 8.2: Binding energies of certain major element of Chumoukedima clay as obtained from XPS analysis.

Sl. No.	Element	Binding energy (eV)
1	O 1s	531.72 (O 82.49%) 532.85 (OH 10.81%) 534.55 (H_2O 6.70%)
2	Si 2p	102.64
3	Al 2p	74.70 (Al VI 72.26%) 73.85 (Al IV 27.74%)
4	Fe 2p (3/2)	712.8

1031 cm^{-1} (Si-O stretching), 921 cm^{-1} (Al-OH bending), deformations peaks of Si-O-Al at 536 cm^{-1} and Si-O-Si at 471 cm^{-1} correspondsto kaolinite and 1440 cm^{-1} indicate traces of carbonate in kaolinite mineral. The bands at 3621 and 3433 cm^{-1} correspond to stretching band of structural hydroxyl group and broad stretching band of water respectively which are characteristics of montmorillonite and illite type of minerals, the band at 1632 cm^{-1} also represent water in the system [34, 35]. The peaks at 1032 and 618 cm^{-1} correspond to Si-O stretching, deformation band of Al-Al-OH vibration and Si-O respectively which are characteristics of montmorillonite and illite minerals. The band at 775 cm^{-1} is in plane vibration of Al-O-Si of illite. The FT-IR spectrum also recorded non-clay mineral, bands at 777 and 694 cm^{-1} correspond to quartz [34]. Thus, the FT-IR data are also in accordance with solid state NMR, XPS and powder XRD data.

Another important property studied in characterisation of clay materials is their thermal stability and structural breakdown temperatures. Accordingly, the thermal behaviour of the present clay sample was measured (Figure 8.10). From the thermo-gram, it can be observed that the initial weight loss occurs around 50–103 °C due to loss of pore condensed and surface adsorbed water, while the next minor onset of weight loss started around 360 °C and continues up to 400 °C which indicates pres-ence of minor iron hydroxide minerals [15], third major weight loss occurs at around 400–620 °C which indicate structural degradation of kaolinitic clays. The final minor weight loss events are distributed between 850–1000 °C corresponds to break down of any trace carbonate minerals and gradual decomposition of meta-kaolin to free silica and alumina [15, 37].

The overall weight loss till 1000 was found to be around 7.61% which combines both the structural and interlayer water loss, this indicate that the clayey sample show excellent thermal stability which is suitable for high temperature materials. The total TGA weight loss corresponds well with Loss on Ignition data of the clay.

Another aspect of surface characterisation of clay mineral is to know their specific surface area. Clays possessing higher specific surface area have better porosity, pore

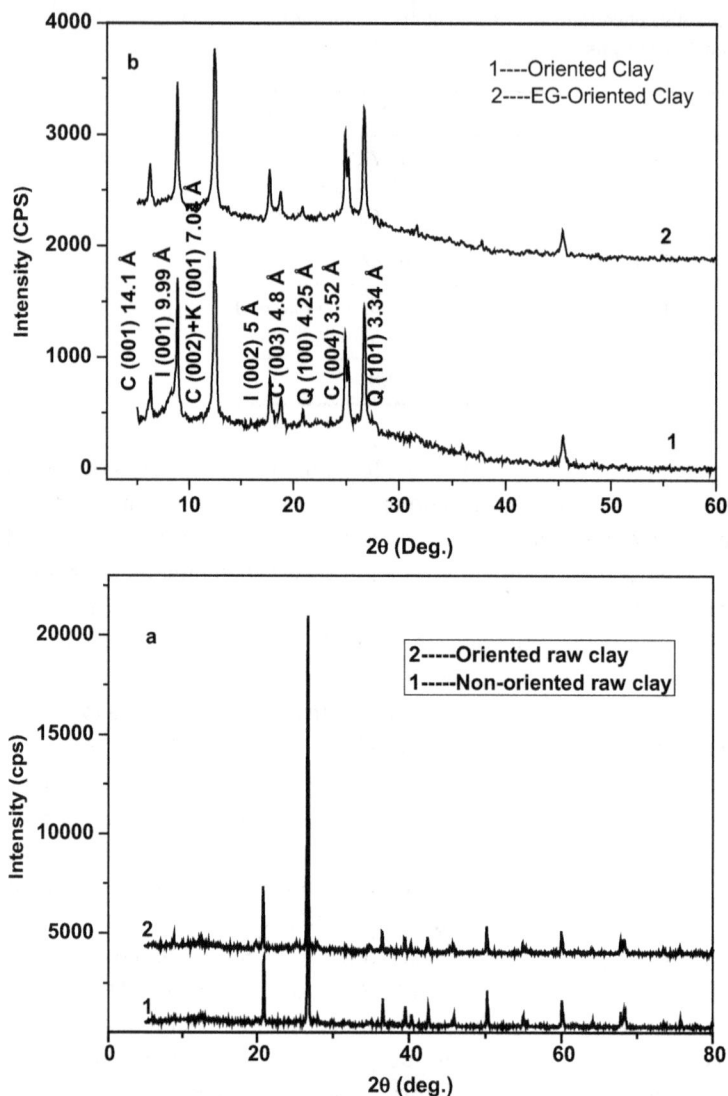

Figure 8.8: P-XRD spectra of clay sample. a. Raw Clay. b. Clay fraction. (C-Chlorite = PDF no. 1–83-1381, Q-Quartz = PDF no. 46–1045, K-Kaolinite = 3–0058, I–Illite = PDF no. 2–0050, M-Muscovite = 01-089-5401, R-Rutile = 01-086-0147).

size distribution, shape and size of the constituting clay nano sheets. High specific surface area is important for the preparation of adsorbents and catalysts. To know this property an adsorption isotherm (Figure 8.11) of the clay sample was obtained by measuring the amount of gas adsorbed by the clayey sample across a wide range of relative pressures at a low temperature like boiling point of nitrogen (77.15 K). Based on the nitrogen adsorption and desorption capacity, different parameters like surface area,

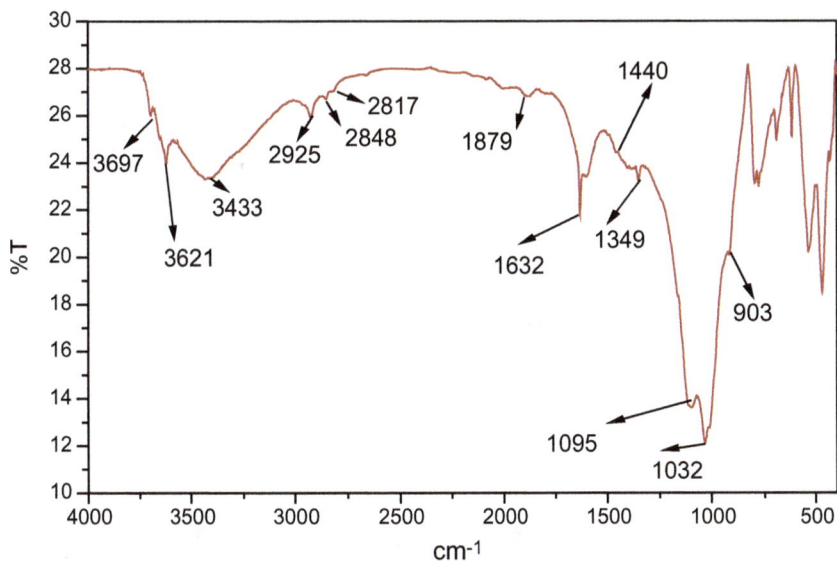

Figure 8.9: FT-IR spectra of clay sample.

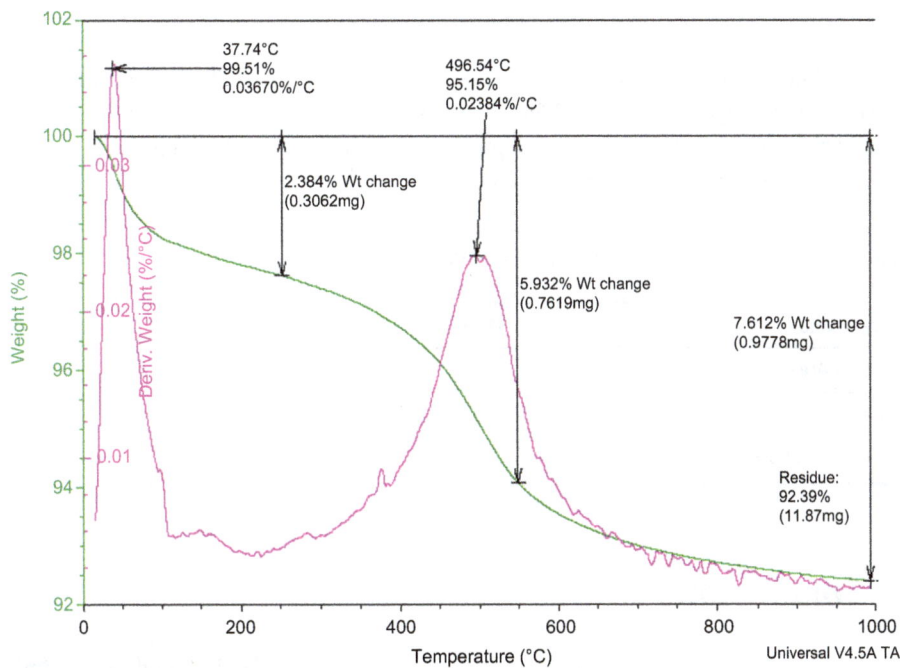

Figure 8.10: Thermogram of clay sample.

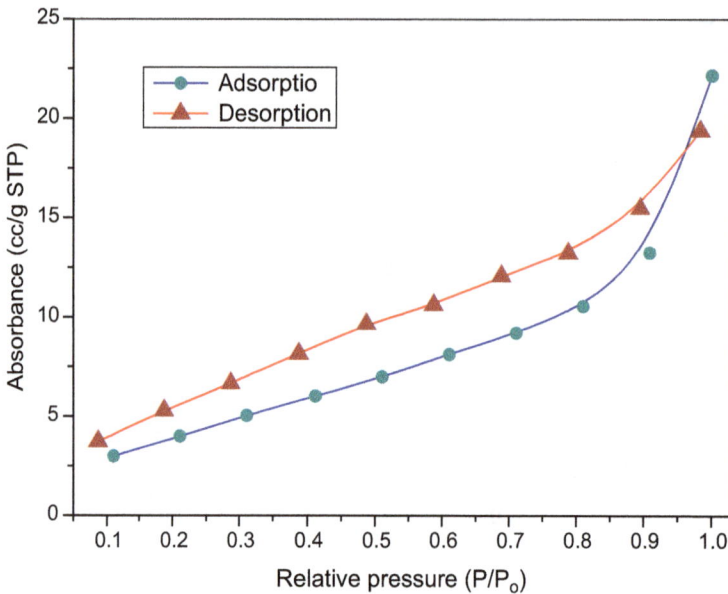

Figure 8.11: Adsorption isotherm of Nitrogen by clay sample.

pore volume, pore radius were determined for the present clay (Table 8.3). The presence of an average porosity of 17.68 Å, very near to 2 nm bench mark, in the sample emphasises its mesoporous nature a property suitable for development high surface area catalyst or adsorbent material from clay.

Table 8.3: Surface properties of −53 um clay fraction.

Parameters	−53 um clay
Surface Area	11.64 m^2/g
Pore Volume	0.02 cc/g
Pore Radius Dv(r)	17.68 Å

4 Conclusions

The collected sample show high clay content, and also exhibits good water limit and cohesive plastic nature. From the results of Winkler's diagram, it is found that the clay can be used for thin walled hollow bricks manufacturing, which is certainly a suitable product for a place like Nagaland. Because of the porous nature and wide range of particle size distribution it can also be used for fabrication of ceramic membranes. Major clay mineral present is kaolin, while other minor minerals found in the sample were muscovite, illite, rutile and carbonate. The presence of tetrahedral and octahedral Al center was also confirmed from XPS and solid state NMR. Suggesting isomorphic substitution in the tetrahedral Si sheet of clay structure which renders charge imbalance with generation of negative the surface charge and duly balanced by exchangeable cations. The clay sample also exhibited high thermal tolerance capacity. Thus, Chumoukeidima deposit is a clay rich soil of a deposit size of approximately 50,000 tons, which can be explored and used for setting up potential clay based industries for bricks or ceramic production.

References

[1]   Galindo Alberto L, Viseras C. Pharmaceutical and cosmetic applications of clays, clay surfaces: fundamentals and applications. Interface Science and Technology, 2004, 1, 267–289

[2]   Milošević M, Logar, M, Kaluđerović, L, Jelić, I. Characterization of clays from Slatina (Ub, Serbia) for potential uses in the ceramic industry, E EnergyProcedia, 2017, 215, 650–655.

[3]   Geological Survey of India (GSI). Geology and Mineral resources of Manipur, Mizoram, Nagaland and Tripura. 2011. Miscellaneous Publication, No. 30, Part IV Vol-1 (Part-2).

[4]   Bhattacharya N. Girujan clays of upper Assam-Marine or Continental? Geological Society of India,1980, 21, 154–157.

[5]   Raju SV, Mathur N. Petroleum geochemistry of a part of Upper Assam Basin, India: a brief overview. Organic Geochemistry, 1995, 23, 55–70.

[6]   Gait, E. A. A history of Assam. 1906. Calcutta: Thacker, Spink & Co.

[7]   Stokes, G.G. (1850). On the effect of the internal friction of fluids on the motion of pendulums. Transactions of the Cambridge Philosophical Society,9, 8–106.

[8]   Ghosh T, BasuS, Hazra S. Geological mapping of the Schuppen belt of north-east India using geospatial technology. Journal of Asian Earth Sciences, 2014, 79, 97–111.

[9]   Odyuo M, Ramasamy S, Parthasarathy P, Stephen A. Clay mineralogy, palynology and geochemistry of the paleogene sediments in inner fold belt of Nagaland, Northeast India. International Research Journal of earth sciences, 2016, 4, 1–18.

[10] Davis ROE, Bennett HH. Grouping of soils on the basis of mechanical analysis. Department circular (United States. Department of Agriculture), 1927, 419.

[11] Groenendyk DG, Ferré Tpa, Thorp KR, Rice AK. Hydrologic-process-based soil texture classifications for improved visualization of landscape function. Public Library of Science one, 2015, 10, 1–17.

[12] ASTM Standard D4318. Standard Test Methods for Liquid Limit Plastic Limit,and Plasticity Index of Soils, ASTM International, 2005.

[13] Baruah B, Mishra M, Bhattacharjee CR, Nihalani MC, Mishra SK, Baruah SD, Phukan P, Goswamee RL. The effect of particle size of clay on the viscosity build up property of mixed metal hydroxides (MMH) in the low solid-drilling mud compositions. Applied Clay Science, 2013, 80, 169–175.

[14] Bergaya F, Lagaly G. Handbook of clay science vol. 5, 2nd Edition, Part A. Fundamentals. Elsevier publication. 2013, 218–351.

[15] Milošević M, Logar M. Properties and characterization of a clay raw material from Miličinica (Serbia) for use in the ceramic industry. Clay Minerals,2017, 52,2, 329–340.

[16] Delgado A, Gonzalez-Caballero F, Bruque JM. On the zeta potential and surface charge density of montmorillonite in aqueous electrolyte solutions. Journal of Colloid and Interface Science. 1986, 113, 203–211.

[17] Moghimi AH, Hamdan J, Shamshuddin J, Samsuri AW, Abtahi A. Physicochemical properties and surface charge characteristics of arid soils in south eastern Iran. Applied and Environmental Soil Science, 2013, 1–11.

[18] ASTM C323 – 56. Standard test methods for chemical analysis of ceramic whiteware clays. 2016.

[19] Grim, R.E. Clay Mineralogy. McGraw-Hill Inc., US, ISBN-10: 0070248362. 1968.

[20] Shepard FP, Nomenclature based on sand-silt-clay ratios: Jour. Sed. Petrology, 1954, 24, 151–158.

[21] Dondi M, Fabbri B, Guarini G. Grain-size distribution of Italian raw materials for building clay products: a reappraisal of the Winkler diagram. Clay Minerals, 1998, 33,435–442.

[22] Winkler HGF. Importance of the grain size distribution and the mineral content of clays for the production of heavy clay products. Reports of German Ceramic Society, 1954, 31, 337–343.

[23] Jasmund K, Lagaly G. Tonminerale und Tone. Steinkopf Verlag, Darmstad (ISBN No. 3-79885-0923-9). 1992.

[24] Gieseking JE. Soil Components: Vol. 2: Inorganic Components, Springer Books. 1975.

[25] Stebbins Jf. Quadrupolar NMR inEarth Sciences. 2011, DOI:10.1002/ 9780470034590. emrstm1217.

[26] Fafard J, Terskikh V, Detellier C. Solid-state 1H and27Al NMR studies of DMSO-Kaolinite intercalates. Clays and Clay Minerals, 2017, 3, 206–219.

[27] Thompson JG, Barron PF. Further consideration of the 29si nuclear magnetic resonance spectrum of kaolinite. Clays and Clay Minerals, 1987, 35, 38–42.

[28] Olphen HV. An Introduction to Clay Colloid Chemistry: For Clay Technologists, Geologists, and Soil Scientists, Interscience Publishers. 1963.

[29] Stucki JW, Banwart WL. Advanced chemical methods for soil and clay mineral research. D. Reidel Publishing Co: Dordrecht, The Netherlands. 1980.

[30] Kloprogge JT, Wood BJ. Baseline studies of the clay minerals society source by X-Ray photoelectron spectroscopy. Clay Science.2018, 22, 85–94.

[31] Meunier A. Clays. 2005, Springer, Berlin, ISBN 3-540-21667-7.

[32] Andrade FA, Al-Qureshi HA, Hotza D. Measuring the plasticity of clays: A review. Applied Clay Science, 2011, 51,1–7.

[33] Grim Re. Clay Mineralogy. McGraw-Hill, New York, 384. 1953.

[34] Farias TMB, Gennari RF, Chubaci JFD, Watanabe S. FTIR spectra and TL properties of quartz annealed at high temperatures. Physics Procedia, 2009, 2, 493–496.

[35] Ritz M, Kova LV, Plevova E. Identification of clay minerals by infrared spectroscopy and discriminant analysis. Applied Spectroscopy, 2010, 64, 1379–1387.

[36] Madejová J, Gates WP, Petit S. (2017). IR Spectra of Clay Minerals. Developments in Clay Science. 2017, 8, 107–149.

[37] Fouzia C, Hamidouche M, Belhouchet H, Jorand Y, Doufnoune R, Fantozzi G. Mullite fabrication from natural kaolin and aluminium slag. Boletín de la Sociedad Española de Cerámica y Vidrio, 2018, 57, 169–177.

Part 3: **Use and application of industrial residues**

N. B. Singh
Chapter 9
Conversion of CO$_2$ into useful products

Abstract: Extensive use of fossil fuels started due to Industrial revolution. Because of this CO$_2$ gas is released into the atmosphere. Over-emission of CO$_2$ has never been mentioned in the History of civilization. Svante Arrhenius predicted for the first time in 1880s that global warming may occur due to emissions of large amount CO$_2$. It may be due to ability of CO$_2$ to absorb and emit infrared radiation. Number of options for reducing CO$_2$ emission used is: reduction of fossil fuels, energy consumption reduction and capture and storage of CO$_2$. However, better option for fossil fuels has not been economically found. One of the best ways for minimizing CO$_2$ may be conversion into useful products. CO$_2$ can be a source of carbon in the synthesis of chemicals and fuels. It has been always a debate whether carbon dioxide is a waste or wealth and this debate is still going on. In this article methods have been discussed for conversion of CO$_2$ into useful products.

Keywords: Carbon dioxide, Products, Reduction, Photocatalysts, Methanol, Carbon Nanotube

1 Introduction

CO$_2$ in the atmosphere increased from 325 ppm in 1970 to 414.40 ppm in Feb. 2020. Because of this change people are worried for climate change. Our today's Society depends on fossil fuels for different products and services, which emit CO$_2$ gas (Figure 9.1) [1].

In the atmosphere gases that trap heat such as carbon dioxide (CO$_2$), water vapor (H$_2$O), nitrous oxide (N$_2$O), methane (CH$_4$), ozone, and chlorofluorocarbons (CFCs), are called greenhouse gases (GHGs). CO$_2$ is the most important GHG CO$_2$ in the atmosphere is obtained by burning fossil fuels (oil, coal and natural gas), trees, solid waste, other biological materials, and also by anthropogenic activities such as certain chemical reactions (e.g., manufacture of cement) (Table 9.1) [2].

Percent emission of CO$_2$ from different sectors and country is given in Figure 9.2 [3].

N. B. Singh, Department of Chemistry and Biochemistry, SBSR, Sharda University, Greater Noida, India, e-mail: nbsingh43@gmail.com

https://doi.org/10.1515/9783110674941-009

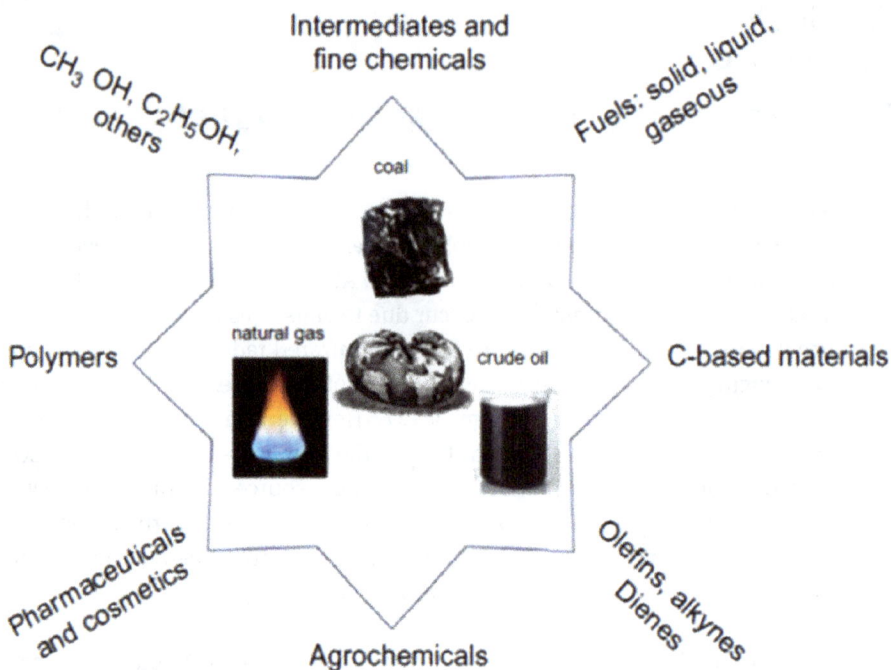

Figure 9.1: CO_2 Emission and utilizations [1].

Table 9.1: CO_2 emission sources [2].

Natural Sources	Mobile Sources	Stationary Sources
Animals	Trucks and Buses	Independent Power Producers
Humans	Cars, and Sports Utility Vehicles	Fossil Fuel-based Electric Power Plants
Land Emission/Leakage	Trains & Ships	Commercial & Residential Buildings
Plant & Animal Decay	Aircrafts	Manufacturing Plants in Industry
Earthquake	Military Vehicles & Devices	Military & Government Facilities
Volcano	Construction Vehicles	Flares of Gas at Fields

a) A lot of CO_2 gas emits in the manufacture of cement.

In CO_2 cycle, CO_2 evolution and absorption are in equilibrium. Anthropogenic activities disturb this equilibrium. If the global mean temperature increases by more than 2 °C – the warming up level corresponds to a CO_2 concentration level of around 450 ppm in the atmosphere. Approximately 2–3 ppm CO_2 increases each year, so that the critical value may be reached in 25–30 years from now [4].

In January 2020, CO_2 levels in atmosphere reached to 412 ppm, which is approximately 48% higher than that of before Industrial Revolution. Such levels are much higher than that found at least 5 million years before. Variation of CO$_2$ emission in the atmosphere is continuously increasing and given in Figure 9.3 [4].

■ Residential, 6%   ■ Road transport, 16%
■ Energy, 41%   ■ Industries, 20%
■ Other transport, 6% ■ Other sectors, 10%

■ China, 30%   ■ EU, 16%
■ Other HICs, 8% ■ India, 7%
■ USA, 195   ■ Other MICs, 11
■ Russia, 7%   ■ Japan, 4%

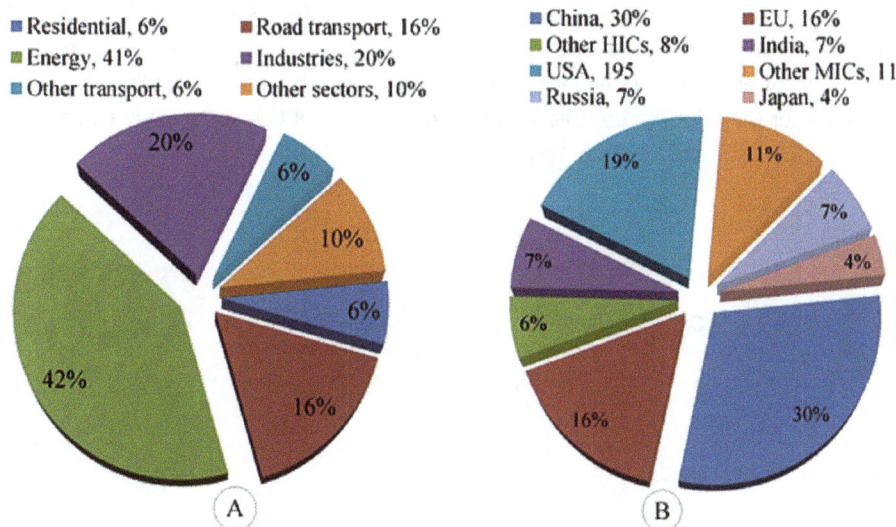

Figure 9.2: CO$_2$ emissions from different sector and country [3].

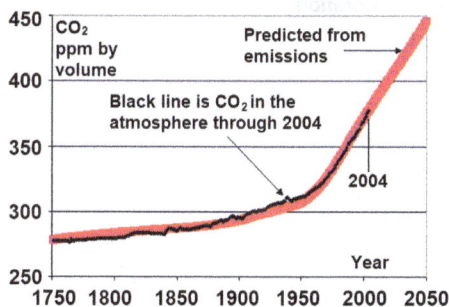

Figure 9.3: Variation of CO$_2$ emission in the atmosphere [4].

Thus there is a need to minimize CO$_2$ emissions in the atmosphere or convert CO$_2$ into useful products so that amount of CO$_2$ in the atmosphere can be reduced. Researches in the carbon management field are now concern to CO$_2$ utilization in energy and chemical industries, directly or indirectly. In this chapter conversion of CO$_2$ into useful products has been discussed. However, before conversion, some properties of CO$_2$ are also given in this chapter.

2 Structure and vibrational modes of CO_2

Structure of CO_2 is given in Figure 9.4.

A linear triatomic molecule like CO_2 should show $3n - 5 = 4$ normal vibrational frequencies (IR bands) as given in Figure 9.5. Since the normal vibrations v_{2a} and v_{2b} should occur at the same frequency, one would expect three different vibrational frequencies for CO_2. But the infrared absorption spectrum of CO_2 shows only two strong bands at 557 cm^{-1} and 2350 cm^{-1} [5].

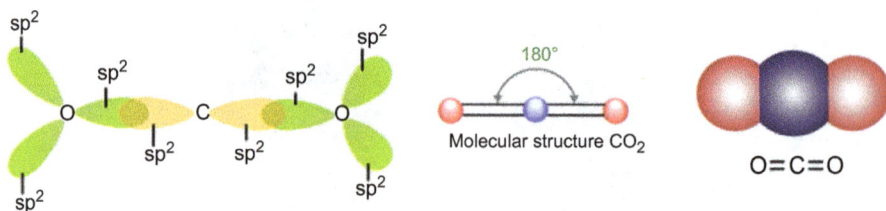

Figure 9.4: Structure of CO_2.

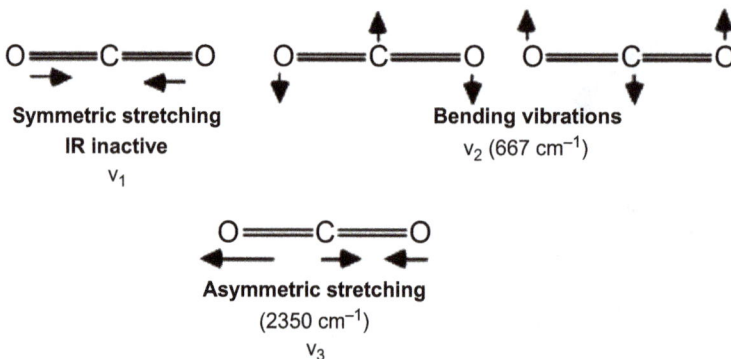

Figure 9.5: Vibrational frequencies of CO_2 molecule.

The infrared (IR) active vibrations of CO_2 molecule are responsible for its role as a GHG. The Earth's atmosphere is transparent to visible light coming from the sun and striking the Earth's surface and IR radiation is reemitted. As atmospheric CO_2 concentration increases, the amount of IR radiation trapped increases. Number of gases (from methane to water vapor) traps heat, but CO_2 is of greatest risk of irreversible changes if it continues to increase and accumulate. GHGs trap heat radiating from the Earth. This heat, in the form of IR radiation, gets absorbed and then emitted in the atmosphere. As a result the lower atmosphere and the surface get warmed.

3 Phase diagram of CO$_2$

The phase diagram of carbon dioxide is given in Figure 9.6 [5]. The triple point comes at 5.1 bar at −57 °C. This is the minimum pressure for existence of liquid CO$_2$. The critical temperature and pressure for carbon dioxide are 31 °C and 73.8 bar respectively, above which carbon dioxide forms a supercritical fluid.

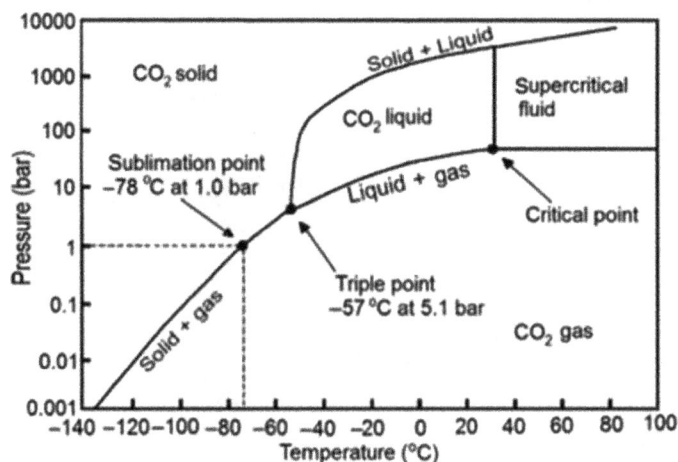

Figure 9.6: Phase diagram of CO$_2$ [5].

4 CO$_2$ emissions, uses and conversion into useful products

With Industrial Revolution and coming up of number of chemical industries, CO$_2$ has been found useful in various industries like beverage industries for carbonated drinks. Further it is used as dry ice, fire extinguishers, welding medium, etc. In addition CO$_2$ is converted into different useful products in different ways. Figure 9.7 gives a schematic representation of CO$_2$ conversion and utilization strategies and Figure 9.8 gives physical use of CO$_2$ [6].

Utilization of CO$_2$

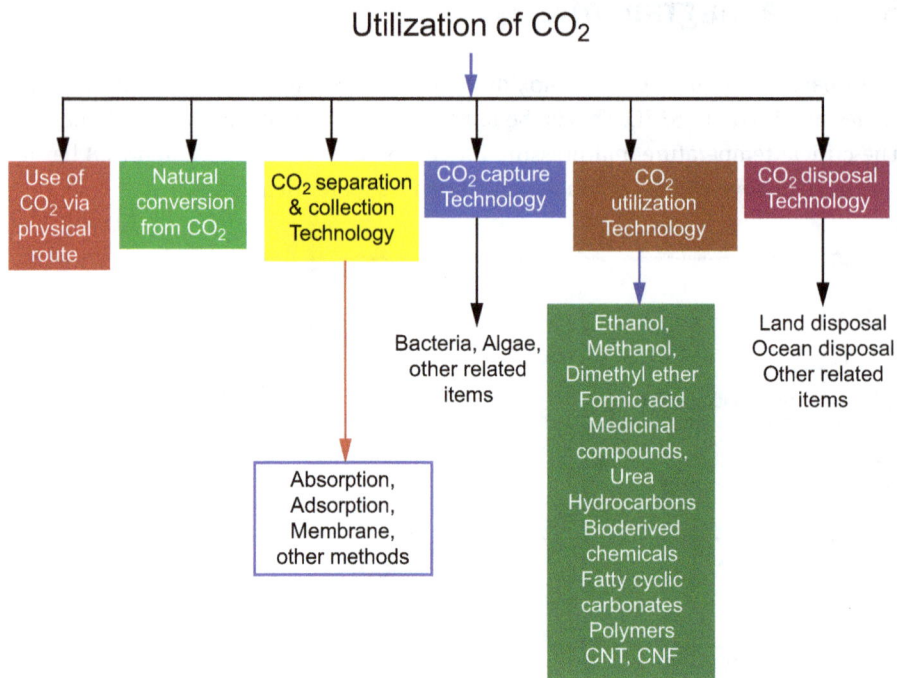

| Use of CO$_2$ via physical route | Natural conversion from CO$_2$ | CO$_2$ separation & collection Technology | CO$_2$ capture Technology | CO$_2$ utilization Technology | CO$_2$ disposal Technology |

Bacteria, Algae, other related items

Ethanol, Methanol, Dimethyl ether Formic acid Medicinal compounds, Urea Hydrocarbons Bioderived chemicals Fatty cyclic carbonates Polymers CNT, CNF

Land disposal Ocean disposal Other related items

Absorption, Adsorption, Membrane, other methods

Figure 9.7: CO$_2$ utilization strategies.

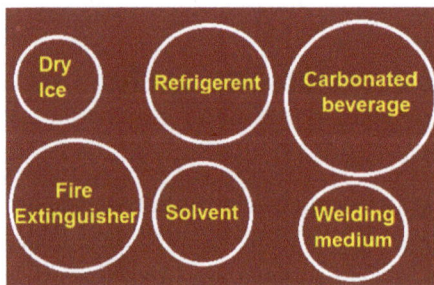

Dry Ice Refrigerent Carbonated beverage

Fire Extinguisher Solvent Welding medium

Figure 9.8: Physical uses of CO$_2$.

5 CO$_2$ conversion into products strategies

Conversion of CO$_2$ can be classified into two ways (Scheme 9.1) [7]: non-reductive (A and B) and reductive (C + D). Reductive process uses less energy and non-reductive process uses higher energy to reduce CO$_2$.

Basically CO$_2$ can be converted into useful products by using number of methods such as Chemical, Photochemical, Biological, Electrochemical, and Thermal etc. (Figure 9.9) [8].

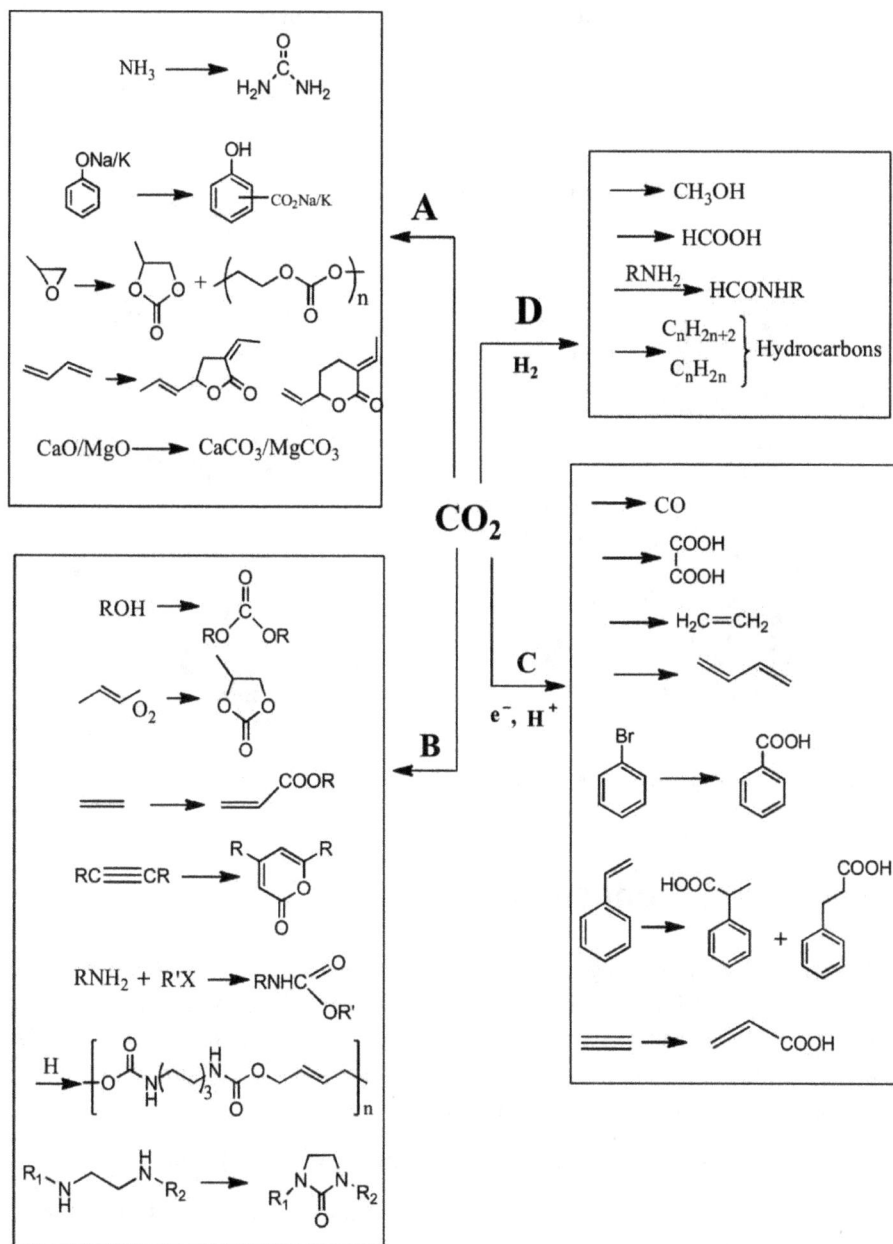

Scheme 9.1: Chemicals obtained from CO$_2$ via non-reductive routes (A + B); and reductive routes (electroreduction (C) and hydrogenation (D)) [7].

CO$_2$ is used as a raw material for synthesis of number of synthetic materials (Scheme 9.2) [4].

Advantages and disadvantages of different conversion methods are given in Table 9.2 [9]

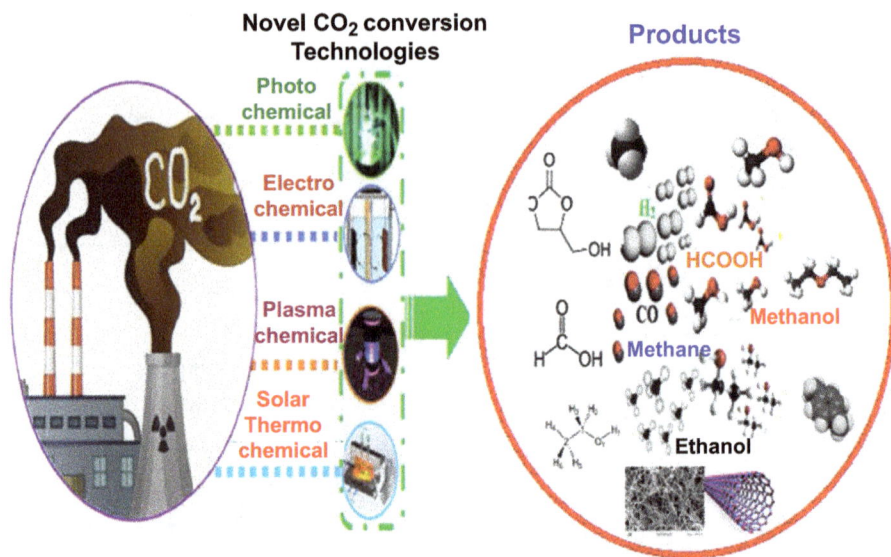

Figure 9.9: CO$_2$ conversion processes and chemicals produced [8].

6 Methanol production from CO$_2$

Methanol (CH$_3$OH) has a large application in number of chemical industries and in recent times as a fuel in fuel-cell cars. CH$_3$OH can be obtained by chemical recycling of CO$_2$. Currently conversion of CO$_2$ to methanol with various reforming technologies is one of the major activities in research and industrial sphere (Figure 9.10) [6]. CO$_2$ conversion to methanol occurs as given below.

$$CO_2 + 3H_2 \leftrightarrow CH_3OH + H_2O; n\Delta H298 = -49.5\,kJ/mol$$

Fanavaran Petrochemical plant in Iran produces one million tonnes of methanol per year through CO$_2$ utilization.

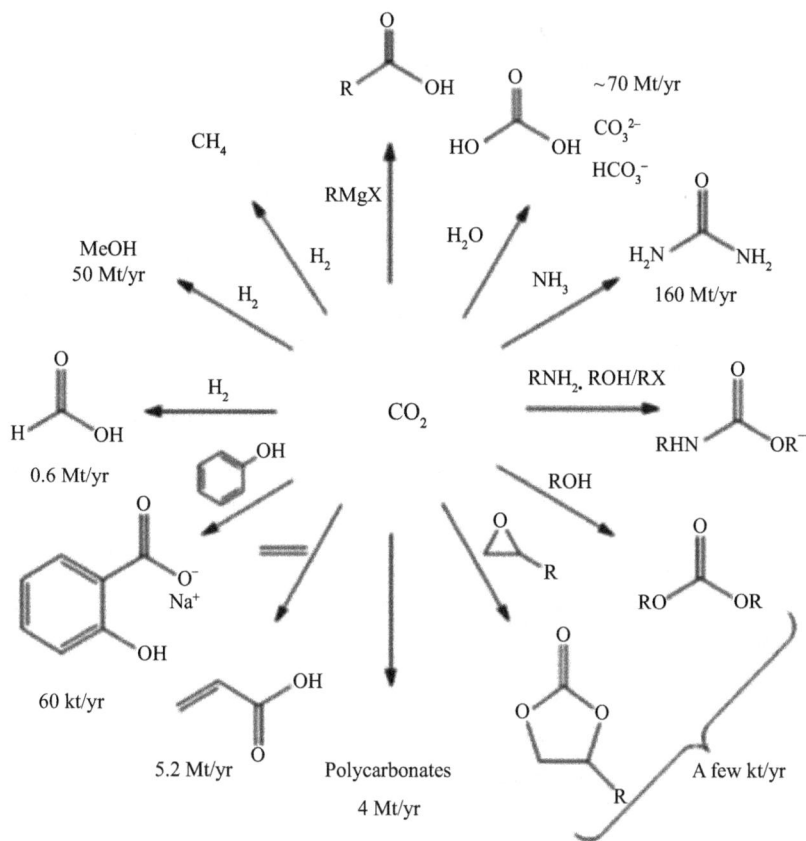

Scheme 9.2: Various CO_2 conversion routes for chemicals [4].

Table 9.2: Advantages and disadvantages of different methods for conversion of CO_2 [9].

Technology	Advantages	Disadvantages
Electrochemical	Room temperature is enough for the process; products obtained may be used as source for electricity generation	Catalytic life is short; operating cost is high; requires electrical energy; slow kinetics.
Photochemical	High yield; mild conditions; safe, cost-effective and ecofriendly.	Product and catalyst separation is difficult; product selectivity is a challenge; photoreactor designing is complex; yield is low.

Table 9.2 (continued)

Technology	Advantages	Disadvantages
Thermal	Syngas is produced which can be converted into potent liquid fuel.	High energy and temperature are required; instability of the catalyst; fuel formation is risky.
Biochemical	Energy required for carbon capture is less; utilization of algae is cost-effective	Climatic condition is major drawback; low energy harvesting system and adsorption capacity and efficiency is low.
Chemical	CO_2 is used as a raw material for chemical feedstock	CO_2 utilization is less; high cost compared to other techniques; process development must be high.

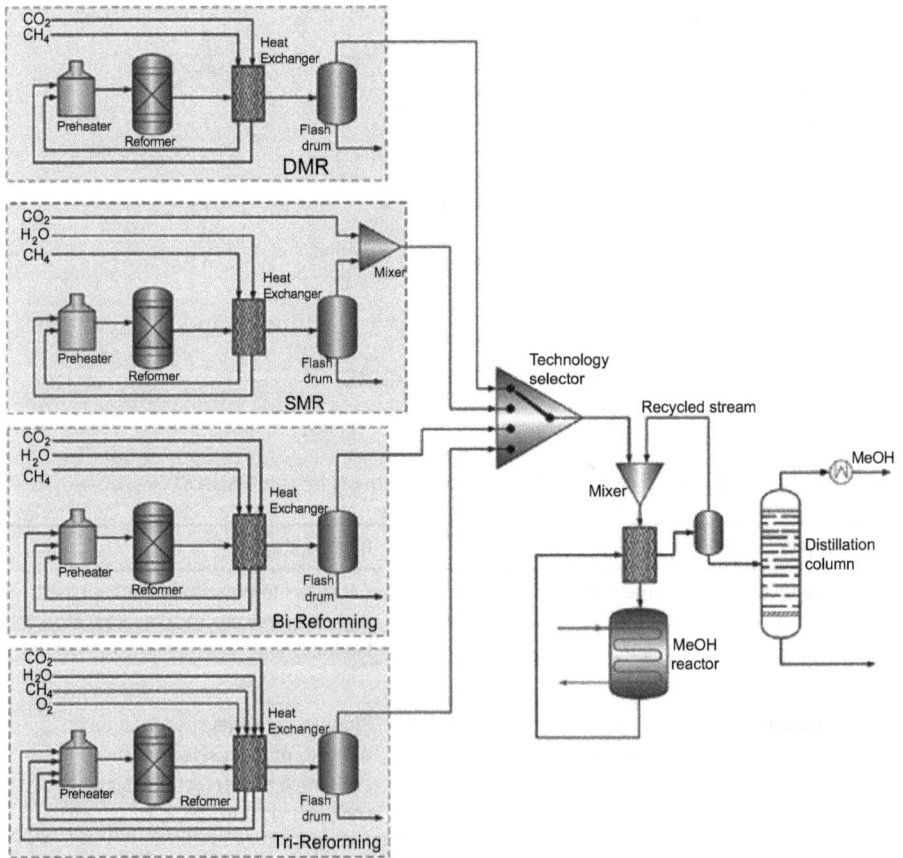

Figure 9.10: Methanol production with various reforming technologies [6].

7 CO$_2$ hydrogenation to CH$_3$OH

CO$_2$ conversion to CH$_3$OH has been studied extensively. CH$_3$OH synthesis via CO$_2$ hydrogenation in presence of catalysts has been studied. Cu is found to have superior catalytic performances. It is normally used with various modifiers such as Ti, Zn, Ce, Al, Si and so on as shown in Table 9.3 [10]. A typical catalyst is Cu/ZnO/Al$_2$O$_3$ for industrial CH$_3$OH synthesis from CO$_2$ and H$_2$ at conditions (200–300 °C, 5–10 MPa).

Table 9.3: Catalysts for synthesis of CH$_3$OH through CO$_2$ hydrogenation [10].

Catalyst systems	CO$_2$ Conversion rate (%)	CH$_3$OH selectivity (%)	T. (°C)
Cu/ZrO$_2$	6.3	48.8	240
Cu/B/ZrO$_2$	15.8	67.2	250
Cu/Ga/ZnO	6.0	88.0	270
Cu/Zn/Ga/SiO$_2$	5.6	99.5	270
Cu/Zn/ZrO$_2$	12.0	71.1	220
Cu/Zn/Ga/ZrO$_2$	–	75.0	250
Cu/Zn/ZrO$_2$	21.0	68.0	220
Au/Zn/ZrO$_2$	1.5	100	220
Cu/Zn/Al/ZrO$_2$	18.7	47.2	240
Ag/Zn/ZrO$_2$	2.0	97.0	220
Cu/Ga/ZrO$_2$	13.7	75.5	250
Pd/Zn/CNTs	6.3	99.6	250

8 Thermodynamic parameters for CO$_2$ conversion to CH$_3$OH

CO$_2$ is an important product in any combustion process. From a thermodynamic standpoint, CO$_2$ conversion is a difficult process. For conversion reactions (Table 9.4) [10], Gibbs free energy change is positive indicating non-spontaneity of process. Further the ΔH values are positive, showing that CO$_2$ conversion processes are endothermic under a standard state.

Table 9.4: Calculated ΔG, ΔH and ΔS) of some CO_2 conversion reactions [10].

Product	Net reaction	ΔG (kJ mol^{-1})	ΔH (kJ mol^{-1})	ΔS (J mol^{-1} K^1)
Hydrogen	$H_2O \rightleftharpoons H_2\uparrow + 0.5O_2\uparrow$	237.17	285.83	163.30
Carbon monoxide	$CO_2 + H_2O \rightleftharpoons CO\uparrow + H_2O + 0.5O_2\uparrow$	257.38	283.01	86.55
Formic acid	$CO_2 + H_2O \rightleftharpoons HCOOH + 0.5O_2\uparrow$	269.86	254.34	−52.15
Formaldehyde	$CO_2 + H_2O \rightleftharpoons HCHO + O_2\uparrow$	528.94	570.74	140.25
Methanol	$CO_2 + 2H_2O \rightleftharpoons CH_3OH + 1.5O_2\uparrow$	701.87	725.97	80.85
Ethanol	$2CO_2 + 3H_2O \rightleftharpoons C_2H_5OH + 3O_2\uparrow$	1325.56	1366.90	138.75
Propanol	$3CO_2 + 4H_2O \rightleftharpoons C_3H_7OH + 4.5O_2\uparrow$	1962.94	2021.24	195.65
Methane	$CO_2 + 2H_2O \rightleftharpoons CH_4\uparrow + 2O_2\uparrow$	818.18	890.57	242.90
Ethane	$2CO_2 + 3H_2O \rightleftharpoons C_2H_6\uparrow + 3.5O_2\uparrow$	1468.18	1560.51	309.80
Ethylene	$2CO_2 + 2H_2O \rightleftharpoons C_2H_4\uparrow + 3O_2\uparrow$	1331.42	1411.08	267.30

From a thermodynamic point of view conversion reaction is spontaneous if ΔG is negative. A catalyst if used, lowers the activation energy and makes the conversion spontaneous (Figure 9.11) [10].

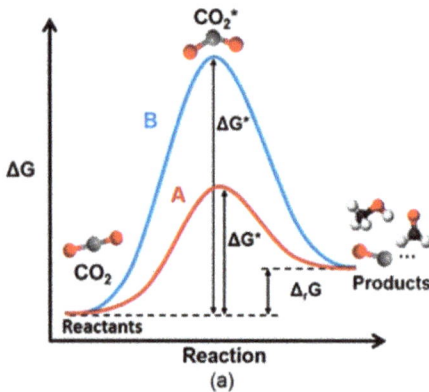

Figure 9.11: CO_2 conversion in presence of catalysts [10].

9 Conversion of CO_2 to Ethanol

In waste-to-fuel technology, electrochemical process using tiny spikes of carbon and copper to turn carbon dioxide into ethanol have been developed [11]. The major reduction product of CO_2 is ethanol, which corresponds to a 12 e⁻ reduction with H_2O as the H^+ source.

$$2\,CO_2 + 9\,H_2O + 12\,e^- \rightarrow C_2H_5OH + 12\,OH^-\ E^0 = 0.084\ V$$

where E^0 is the equilibrium potential.

10 Conversion of CO_2 to urea

Urea ($(NH_2)_2CO$) is an important fertilizer and plays an important role in mammals' metabolism. CO_2 is converted to Urea as shown in Figure 9.12 [6]. Following reactions are involved in the conversion process.

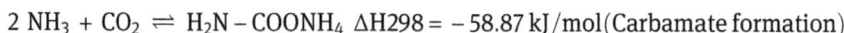

$$2\,NH_3 + CO_2 \rightleftharpoons H_2N-COONH_4\ \Delta H298 = -58.87\ kJ/mol\,(\text{Carbamate formation})$$

$$H_2N-COONH_4 \rightleftharpoons (NH_2)_2CO + H_2O\ \Delta H298 = -56.46\ kJ/mol\ (\text{Urea conversion})$$

The first reaction is highly exothermic and occurs at high temperature and high pressure to produce ammonium carbamate (H_2N-$COONH_4$), which further decomposed via a slow endothermic reaction in the second stage to form urea.

11 Dimethyl carbonate from CO_2 and methanol

Dimethyl carbonate (DMC) is a linear carbonate and it is used extensively in industry. When CO_2 is converted, following reaction occurs [12]. Because of thermodynamic limitations, yield is very low but can be increased in presence of catalysts.

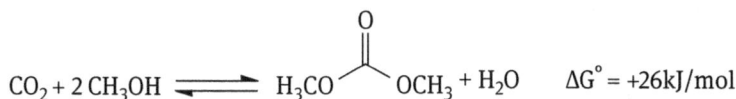

$$CO_2 + 2\,CH_3OH \rightleftharpoons H_3CO-\overset{\displaystyle O}{\overset{\|}{C}}-OCH_3 + H_2O \qquad \Delta G^\circ = +26kJ/mol$$

Yield of DMC very much depends on temperature and pressure. It has lot of industrial applications.

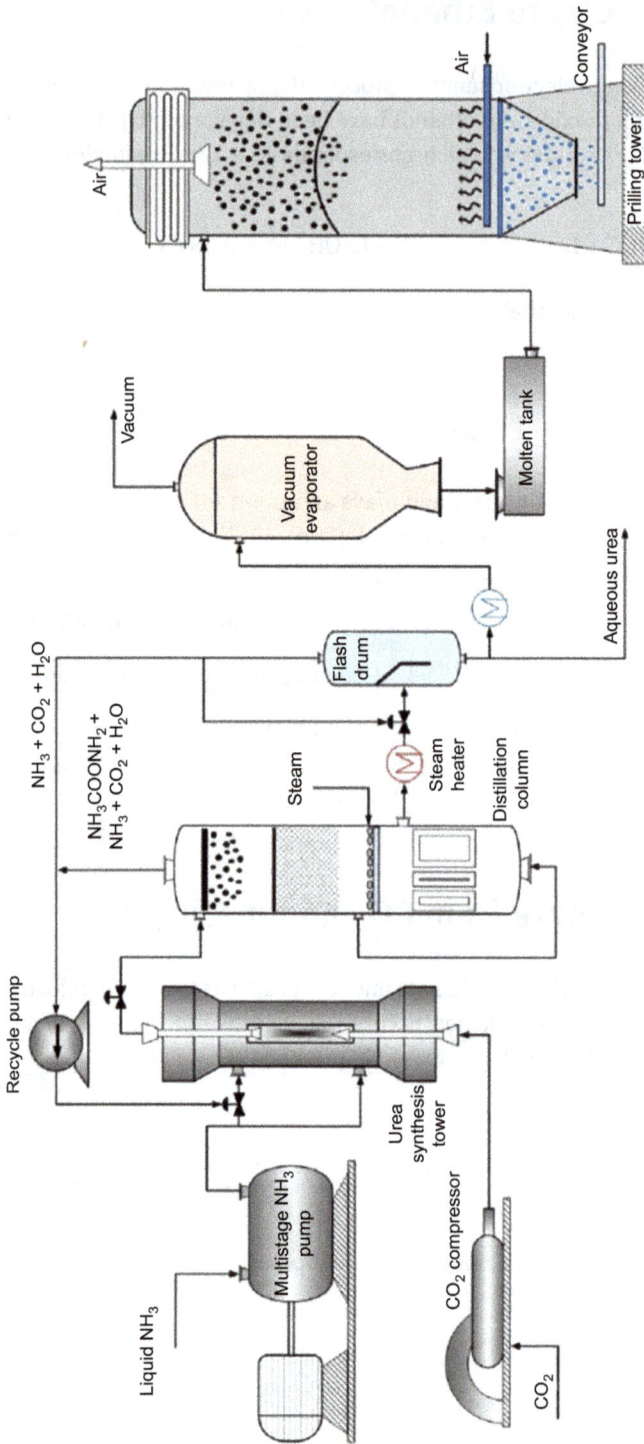

Figure 9.12: Urea manufacture from CO_2 [6].

12 Conversion of CO_2 to formic acid

Hydrogenation of CO_2 to liquid formic acid is an endoergonic process because of the strong entropic contribution and occurs in presence of catalysts as given below [1].

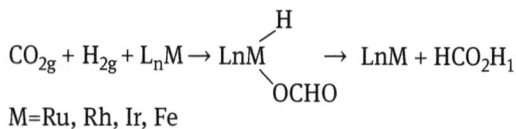

$$CO_{2g} + H_{2g} + L_nM \rightarrow LnM\overset{H}{\underset{OCHO}{<}} \rightarrow LnM + HCO_2H_1$$

M=Ru, Rh, Ir, Fe

The separation of catalyst from liquid formic acid is very slow process because of kinetic and thermodynamic reasons. Therefore industrial production is not feasible.

13 Polymers from carbon dioxide

CO_2 is first converted to urea, which then combines in the following way with form-aldehyde giving polymer [13, 14].

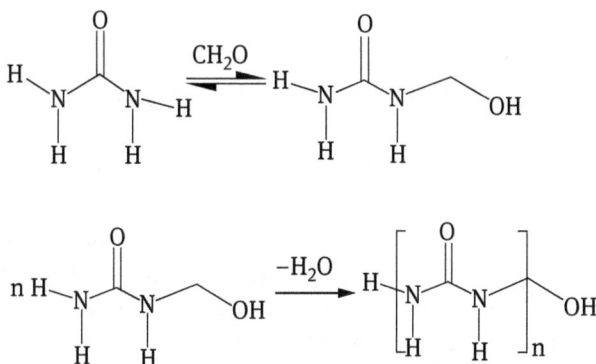

Melamine is produced from urea according to:

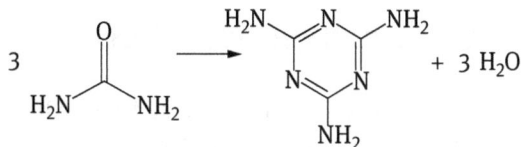

Similarly, thermosetting Melamine Formaldehyde resins are obtained [13].

14 Production of inorganics

Number of inorganic chemicals as given below are produced from CO_2 [13].

$$Metal\ oxide + CO_2 \rightarrow Metal\ carbonate + heat$$

The amount of heat released is significant and it depends on the specific metal, such as calcium. Natural silicates take part in the following exothermic carbonation reactions:

$$Mg_2SiO_4 + 2CO_2 \rightarrow 2MgCO_3 + SiO_2 + 89\ kJ/molCO_2$$
(Olivine)

$$Mg_3Si_2O_5(OH)_4 + 3CO_2 \rightarrow 3MgCO_3 + 2SiO_2 + 2H_2O64\ kJ/molCO_2$$
(Serpentine)

$$CaSiO_3 + CO_2 \rightarrow CaCO_3 + SiO_2 + 90kJ/molCO_2$$
(Wollastonite)

Formations of carbonates are exothermic and hence thermodynamically feasible at lower temperature.

15 Electrochemical conversion of CO_2

Number of pathways for the electrochemical conversion of CO_2 has been proposed (Figure 9.13) [10] . The pathways are of two types: high-temperature ($\geq$800 °C) and low-temperature ($\leq$200 °C) conversions. In high temperature conversion, gaseous techniques like solid oxide cells and solid oxide electrolysis cells are used whereas at low temperature aqueous and non-aqueous techniques like transition metal electrodes in liquid electrolytes (eg. methanol, acetonitrile, etc.) are used. Different products obtained with high-temperature conversions is much low than that of low-temperature conversion technique.

16 Photocatalytic conversion of CO_2

CO_2 is very stable linear symmetrical molecule and the reduction of CO_2 is highly thermodynamically unfavorable [14]. Photocatalytic CO_2 reduction is extremely complicated, of low efficiency and selectivity of different products. Photocatalytic conversion of CO_2 into various products is given in Figure 9.14 [15].

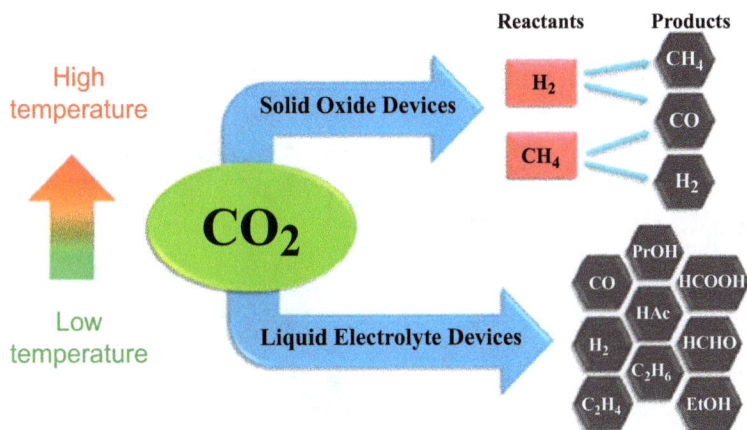

Figure 9.13: Pathways for electrochemical conversion of CO$_2$ [10].

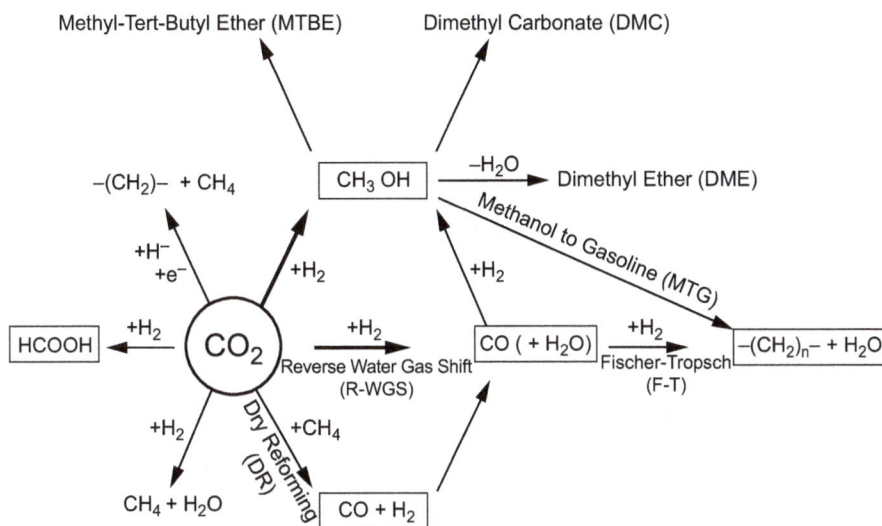

Figure 9.14: Photoreduction CO$_2$ for large-scale[15].

CO$_2$ can be converted into methanol by multi-functionalized TiO$_2$ photocathodes (Figure 9.15A) [13] and copper-based semiconductors convert CO$_2$ into number of products in presence of light (Figure 9.15B) [15]

The photocatalytic conversion of CO$_2$ using TiO$_2$/ rGO/ CeO$_2$ catalyst by UV radiation in the presence of water to methanol and ethanol is given in Figure 9.16 [16].

Metal-organic frameworks (MOFs) are well known materials used for different applications using photocatalytic conversion of CO2 into number of useful products (Figure 9.17) [17].

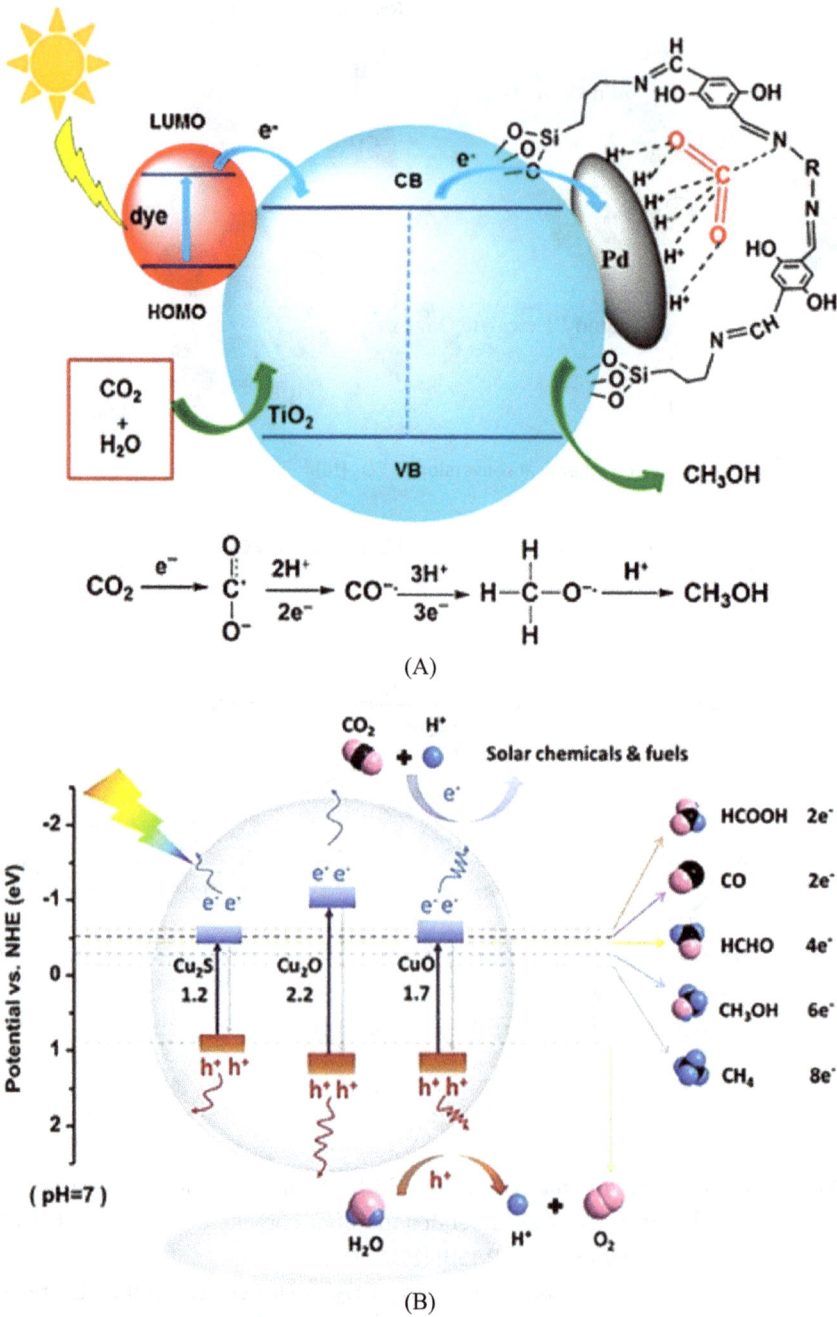

Figure 9.15: Photocatalytic conversion of CO_2 into products: A- in presence of multi-functionalized TiO_2 photocathodes, B- in presence of Copper based catalyst [13, 15].

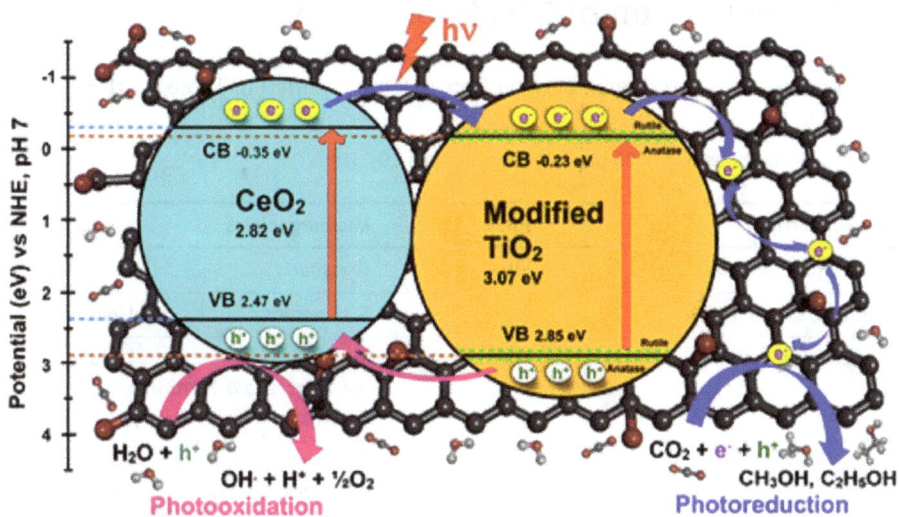

Figure 9.16: Conversion of CO_2 with H_2O to methanol and ethanol in presence of TiO_2/rGO/CeO_2 photocatalysts [16].

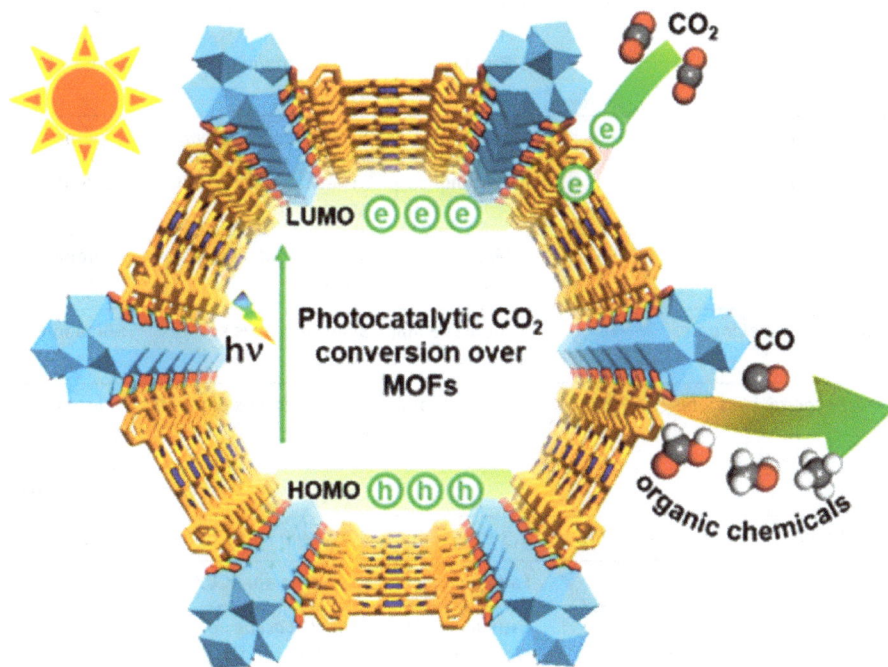

Figure 9.17: CO_2 reduction into CO and organic chemicals by MOFs photocatalysts [17].

17 Bioconversion of CO₂ to products

Different bio-based products obtained from CO_2 are reported in Table 9.5 [18]

Table 9.5: Some value-added bioproducts derived from CO_2. [18].

Microorganisms	Product	Microorganisms	Product
Scenedesmus obliquus	Ethanol	*C. vulgaris*	Fatty acids
Synechocystis sp. PCC6803	Ethanol	*S. dimorphus*	Lipid
S. cereivisiae	Ethanol	*Moorella thermoacetica & Yarrowia lipolytica*	Lipid
Clostridium ljungdahlii	Ethanol	*Synechococcus elongates* PCC7942	Biodiesel (FAEE)
Clostridium ljungdahlii	Ethanol	*Chlorella vulgaris*	Biodiesel (FAME)
Synechococcus elongates PCC7942	1,3-Propanediol	*S. obliquus*	Biodiesel (FMAE)
Synechococcus elongates PCC7942	1,3-Propanediol	*Bacillus* sp. SS105	Biodiesel (FAME)
Synechococcus PCC 7942	1-Butanol	*Monoraphidium dybowskii*	Biodiesel
*Synechococcus elongates*PCC7942	Isobutanol	*Synechococcus elongates*	Isoprene
Synechocystis sp. PCC6803	Isobutanol	*Synechocystis* sp. PCC 6803	Isoprene
Synechococcus elongates PCC7942	2-Methyl butanol	*Synechocystis* sp. PCC 6803	Isoprene
Synechococcus elongates	2,3-Butanediol	*Synechococcus* sp. PCC7002	α-Bisabolene
Synechococcus elongates PCC7942	2,3-Butanediol	*Synechococcus elongates* PCC 7942	Amorpha-4, 11-diene (Farnesyl di-phosphate)
Synechococcus elongatus PCC 7942	2,3-Butanediol	*Anabaena* sp. PCC7120	Farnesene
Synechocystis sp. PCC6803	Fatty alcohol	*Anabaena* sp. PCC7120	Limonene
Synechocystis sp. PC6803	Fatty alcohol (1-Octanol)	*Synechococcus* sp. PCC7002	Limonene

Table 9.5 (continued)

Microorganisms	Product	Microorganisms	Product
Synechococcus sp. 7002	Fatty acids	*Synechocystis* sp. PCC6803	Ethylene
Synechocystis sp. PCC6803	Fatty acids	*Synechocystis* sp. PCC6803	Ethylene
Chlorococcum littorale	Fatty acids	*Synechococcus elongates* PCC7942	Ethylene
Synechocystis sp.	Ethylene	*Synechocystis* sp. PCC 6803	Lactate
Synechocystis sp. PCC6803	Heptadecane	*Synechococcus elongates* PCC7942	Lactate
Synechococcus sp. NKBG15041	Heptadecane, Nonadecene, Nonadecadiene	*Synechococcus elongates* PCC7942	Lactate
Synechococcus sp. PCC7002	1-Nonadecene, 1,14-Nonadecadiene	*Synechocystis* sp. PCC 6803	Lactate
Biofilm(*Burkholderiaes, Clostridiales, Natranaerobiales* & *Metanobacteriales*)	Acetate	*Chlorella* sp.	Lactate
Acetobacterium woodi	Acetate	*Synechococcus elongates* PCC7942	Sucrose
Clostridium sp.	Butyrate	*Synechococcus elongates* PCC 7942	Sucrose
Synechococcus elongates PCC7942	Succinate	*Synechococcus elongates* PCC 7942	Sucrose
Citrobacter amalonaticus	Succinate	*Synechococcus elongates* PCC 7942	Sucrose
Synechocystis sp. PCC 6803	Succinate & Lactate	*Chlorella zofingiensis*	Starch
Chlorella sp. AE-10	Starch	*Cupriavidus necator*	PHB
Cupriavidus eutrophus B-10646	PHA	*Cupriavidus nectar*	PHB
S.elongatus & *P. putida*	PHA	*Cupriavidus nectar*	PHA
Ideonella dechloratanus	PHB		

18 Photosynthesis

Natural conversion of CO_2 is photosynthesis (Figure 9.18)

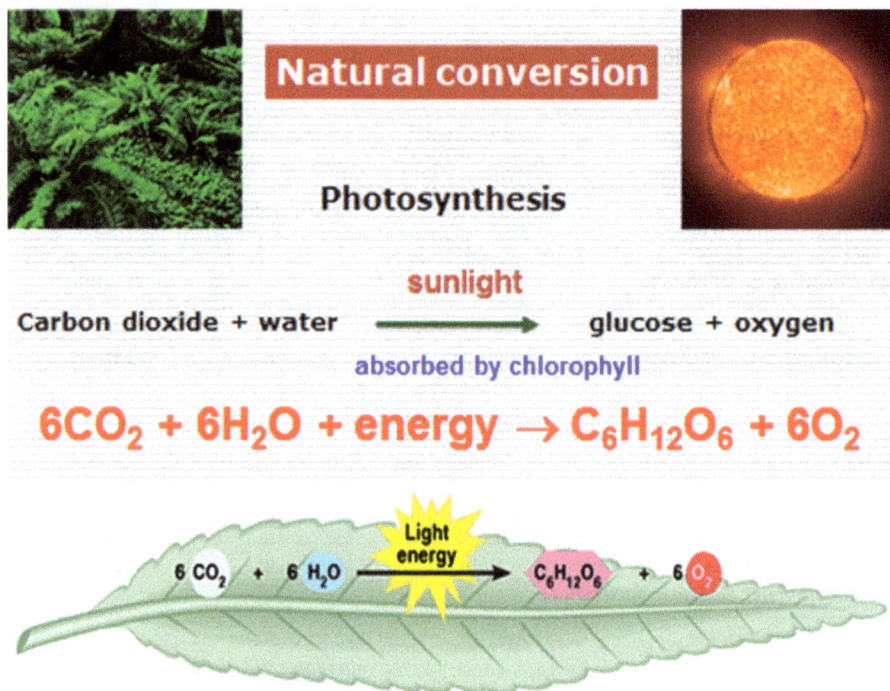

Figure 9.18: Photosynthesis from CO_2.

19 CO_2 conversion to CNT

One person uses 1 ton of concrete per year. Portland cement is the main binding material of concrete and 10 kg of cement produces 9 kg of CO_2. In general cement industry emit about 8% CO_2 gas all over the globe. CO_2 emitted from different processes in cement industry is shown in Figure 9.19.

Cement plant can be joined with the inlet of a CNT chamber where CO_2 is transformed by electrolysis of Li_2CO_3 in nickel crucible at a nickel anode and a steel cathode (Figure 9.20) [19]. In this CO_2 is converted to CNT at the cathode, and O_2 at the anode. Following reactions take place.

$$Li_2CO_3(l) \rightarrow CNT + Li_2O(dissolved) + O_2(g) \tag{9.1}$$

As long cement manufacture will depend on following reaction, it is difficult to avoid emission

CaCO$_3$ ➔ CaO + CO$_2$

From burning of fuels

35%

50%

15%

Indirect emissions from electricity use. cement grinding, raw material grinding, etc.

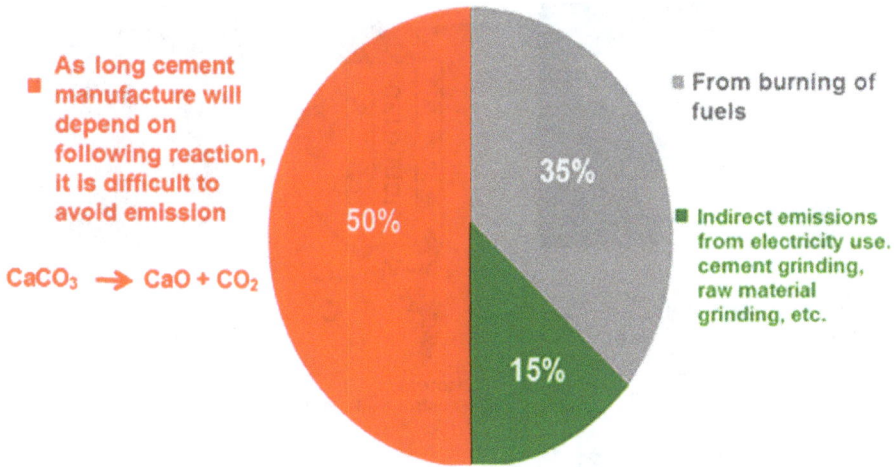

Figure 9.19: CO$_2$ emission in cement industry.

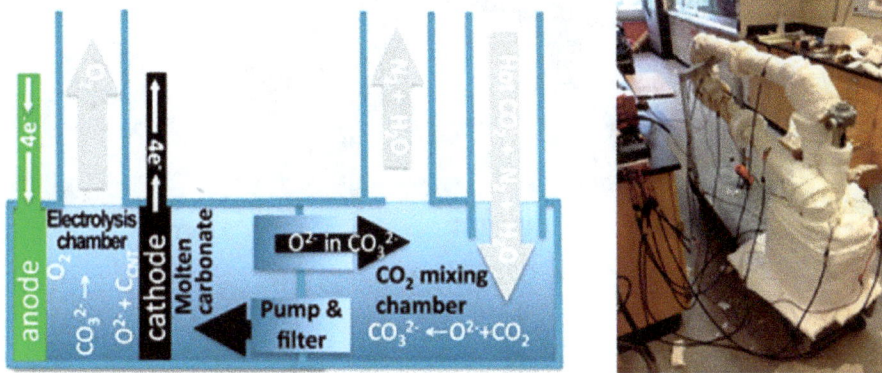

Figure 9.20: Electrolysis of LiCO$_3$ [19].

Li$_2$CO$_3$ used up by electrolysis is continuously regenerated by reaction of Li$_2$O with CO$_2$:

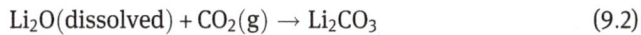

$$Li_2O(dissolved) + CO_2(g) \rightarrow Li_2CO_3 \tag{9.2}$$

Net reaction (combining eqs. (9.1) and (9.2)):

$$CO_2(g) \rightarrow CNT + O_2(g)$$

During this process carbon dioxide is converted to CNT (Figure 9.21) [19]

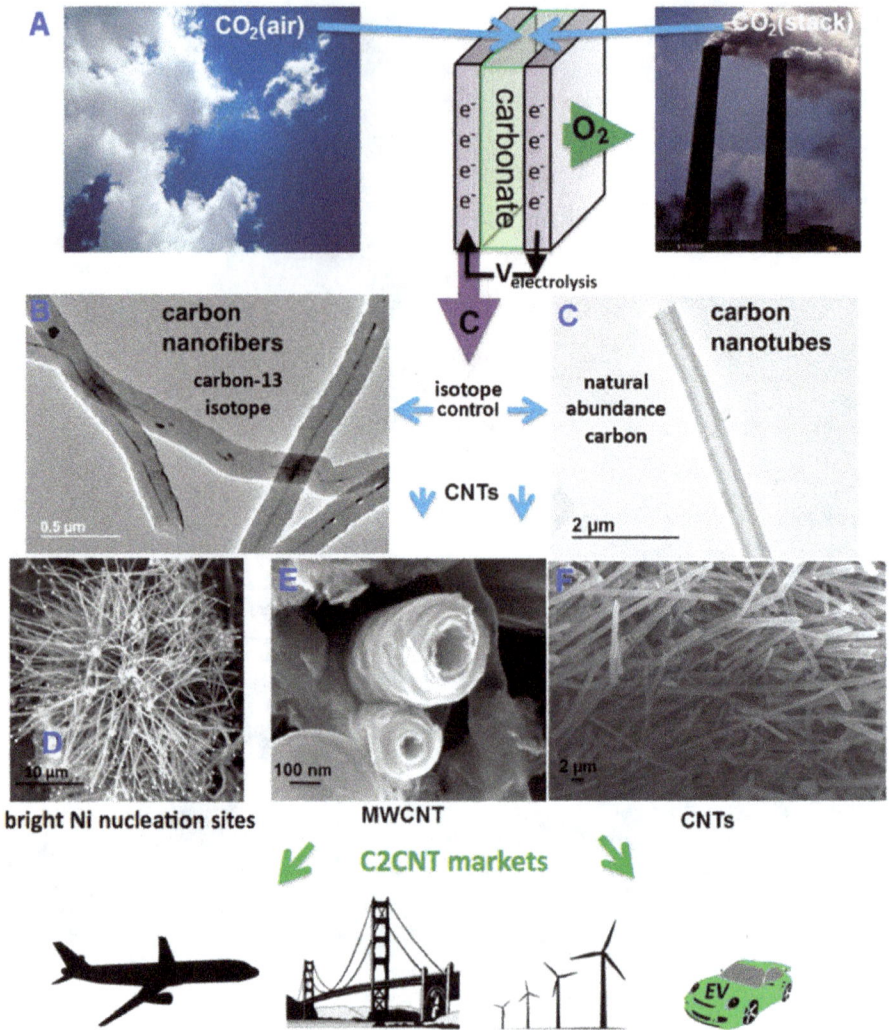

Figure 9.21: CO_2 to CNT during electrolysis [19].

20 Conclusions

CO_2 emissions from different sector are very harmful particularly for global warming. Therefore, its use and conversion to different products is essential. Attempts have been made to summarize CO_2 conversion into different useful products. However, this is a developing field and there are lot of challenges in terms of feasibility of the process and economics. Extensive research is needed.

References

[1]   Aresta Michele, Nocito Francesco, Dibenedetto Angela. What Catalysis Can Do for Boosting
        CO_2 Utilization. Advances in Catalysis 2018,62, 49–111
[2]   Chunshan Song. CO_2 Conversion and Utilization: An overview, Chapter 2002,1, 1–20.
[3]   Mugahed Amran YH, Alyousef Rayed, Alabduljabbar Hisham, El-Zeadani Mohamed. Clean
        production and properties of geopolymer concrete: A review. Journal of Cleaner Production
        2020,251,119679.
[4]   Aulice Scibioh M, Viswanathan B. CO_2 conversion – Relevance and Importance. Carbon
        Dioxide to Chemicals and Fuels 2018, 1–22.
[5]   Michael North. What is CO_2? Thermodynamics, Basic Reactions and Physical ChemistryChapt.
        1 in Carbon Dioxide Utilisation 2015, 3–17.
[6]   Rafiee Ahmad, Khalilpour Kaveh Rajab, Milani Dia, Panahi Mehdi. Trends in CO_2 conversion
        and utilization: A review from process systems perspective, Journal of Environmental
        Chemical Engineering 2018,6, 5771–94.
[7]   Olajire Abass A. Recent progress on the nanoparticles-assisted greenhouse carbon dioxide
        conversion processes. Journal of CO_2 Utilization 2018,24,522–47.
[8]   Mustafa Azeem, Lougoua Bachirou Guene, Shuai Yong, Wang Zhijiang, Tana Heping. Current
        technology development for CO_2 utilization into solar fuels and chemicals: A review. Journal
        of Energy Chemistry 2020,49, 96–123.
[9]   Yaashikaa PR, Kumar P Senthil, Varjani Sunita J, Saravanand A. A review on photochemical,
        biochemical and electrochemical transformation of CO_2 into value-added products. Journal of
        CO_2 Utilization 2019,33,131–47.
[10]   Zheng Yun, Zhang Wenqiang, Li Yifeng, Chena Jing, Yu Bo, Wang Jianchen, Zhang Lei, Zhang
        Jiujun. Energy related CO_2 conversion and utilization: Advanced materials/ nanomaterials,
        reaction mechanisms and technologies. Nano Energy 2017,40, 512–39.
[11]   Song Yang, Peng Rui, Hensley Dale K, Bonnesen Peter V, Liang Liangbo, Wu Zili, Meyer
        Harry M, Chi Miaofang, Ma Cheng, Sumpter Bobby G, Rondinone Adam J. High-Selectivity
        Electrochemical Conversion of CO_2 to Ethanol using a Copper Nanoparticle/N-Doped
        Graphene Electrode. Energy Technology & Environmental Science. Chemistry Select 2016,1,
        6055–61
[12]   Marciniak Aryane A, Alves Odivaldo C, Appel Lucia G, Mota Claudio JA. Synthesis of dimethyl
        carbonate from CO2 and methanol over CeO2: Role of copper as dopant and the use of methyl
        trichloroacetate as dehydrating agent. Journal of Catalysis 2019,3, 88–95.
[13]   Xu Yanjie, Jia Yongjian, Zhang Yuqian, Nie Rong, Zhu Zhenping, Wang Jianguo, Jing
        Huanwang. Photoelectrocatalytic reduction of CO_2 to methanol over the multi-functionalized
        TiO_2 photocathodes. Applied Catalysis B: Environmental 2017,205,254–61.
[14]   Alper Erdogan, Orhan Ozge Yuksel. CO_2 utilization: Developments in conversion processes.
        Petroleum 2017,3, 109–26.
[15]   Gao Yu, Qian Kun, Xu Baotong, Li Zheng, Zheng Jiaxin, Zhao Shan, Ding Fu, Sun Yaguang, Xu
        Zhenhe. Recent advances in visible-light-driven conversion of CO_2 by photocatalysts into
        fuels or value-added chemicals. Carbon Resources Conversion 2020,3, 46–59
[16]   Seeharaj Panpailin, Kongmun Panyata, Paiplod Piyalak, Prakobmit Saowanee, Sriwong
        Chaval, Kim-Lohsoontorn Pattaraporn, Vittayakorn Naratip. Ultrasonically-assisted surface
        modified TiO_2/rGO/CeO_2 heterojunction photocatalysts for conversion of CO_2 to methanol
        and ethanol. Ultrasonics – Sonochemistry 2019,58,104657.
[17]   Li Dandan, Kassymov Meruyert, Cai Xuechao, Zang Shuang-Quan, Jiang Hai-Long.
        Photocatalytic CO_2 reduction over metal-organic framework-based materials. Coordination
        Chemistry Reviews 2020,412,213262

[18] Salehizadeh Hossein, Yan Ning, Farnood Ramin. Recent advances in microbial CO_2 fixation and conversion to value-added products. Chemical Engineering Journal 2020,390, 124584.

[19] Licht Stuart, Co-production of cement and carbon nanotubes with a carbon negative footprint, Journal of CO_2 Utilization 201,18,378–89

Mirja Illikainen, Jenni Kiventerä

Chapter 10
Mine tailings as precursors for alkali-activated materials and ettringite binders

Abstract: Mine tailings are a major waste flow generated in the mining industry. Most of the tailings are discarded into the landfill area near mining sites with limited practical use. mine tailings have gone through several processing steps and the final material has fine particle size and high specific surface area. The tailings consist of equal chemical elements than cementitious materials. Because of the favorable physical and chemical properties, the use of tailings as precursors to alkali-activated materials and other cementitious binders is an increasing research area. Construction industry is looking for alternatives to traditional cement. In addition, there is a need to develop new safety methods to utilize and storage mine tailings instead of land filling in tailings dams.

Mineralogy is the most important factor defining the suitability of the recycling of the mine tailings in cementitious systems. Additionally, the geochemistry of the produced mine tailings is important to analyze prior utilizing the generated mine tailings in new cementitious materials. Usually mine tailings contain minerals with low reactivity under alkaline conditions. In some cases, the reactivity can be enhanced by pre-treatment methods such as thermal or mechanical treatment methods. The silicate mine tailings rich in phyllosilicates have the highest potential for alkali-activated materials. They also have the highest response to the mechano-chemical or thermal pretreatment before the alkali activation.

Sulfidic mine tailings can contain harmful elements and acid generating potential which hinder their utilization in alkali-activated materials. For sulfidic tailings, the binders based on ettringite formation are the most promising since sulfates can participate in the ettringite reaction. Ettringite can efficiently immobilize sulfides and oxyanions, which are difficult to immobilize with alkali-activated materials. The ettringite binding system is a novel method that requires more research to understand the long-term durability of the formed material. The most potential applications for cementitious binders based on mine tailings include mine backfill materials and cover layers on the surface deposits as well as safety storage of the tailings.

Keywords: mine tailings, ettringite, alkali-activated material, silicate mineral, sulfidic minerals

Mirja Illikainen, University of Oulu, Fibre and Particle Engineering Research Unit, P.O.Box 4300, 90014 University of Oulu, Finland, e-mail: mirja.illikainen@oulu.fi
Jenni Kiventerä, University of Oulu, Fibre and Particle Engineering Research Unit, P.O.Box 4300, 90014 University of Oulu, Finland

https://doi.org/10.1515/9783110674941-010

1 Introduction

Different industrial fields generate substantial amounts of inorganic residues or waste materials that could be considered as secondary resources instead of waste. At European level, the amount of inorganic waste generated is more than 1.2 billion tonnes per year. The extractive industry including mining activities, mineral processing and metallurgical extraction is responsible for most of those (Table 10.1). Significant part of this is mine tailings (the global annual production of 7–14 billion tons [1]).

At the same time, construction industry is looking for alternatives to cement, gravel and sand. Cement production causes significant CO_2 emissions, and lack of high-quality sand and aggregates is an increasing global challenge. The use of industrial residues to replace virgin raw materials in construction products is an intensive research area.

Some industrial residues have established use as cementitious binders, such as blast furnace slag or coal combustion fly ash, both used as supplementary cementitious materials or precursors to alkali-activated materials. However, there is a wide variation of industrial residues that could be better utilized. This chapter focuses on mine tailings, the major waste flow from mining process, which currently has limited practical use. The properties of the tailings are discussed, as well as their potential for alkali-activated materials and cementitious systems based on ettringite formation.

Table 10.1: Annual waste generation from four aggregated industrial sectors in 2016 [2].

Industrial sector	waste generation (millions of tonnes)
Extractive industry	632.98
Manufacturing industry	250.28
Energy supply	97.19
Waste	253.61

2 The properties of mine tailings

In mining industry, valuable metals or minerals are extracted in the process that includes crushing, grinding, and different separation processes (Figure 10.1). The mine tailings form in the later stage of the process when finely ground valuable particles are separated from the gangue phases. After that, the tailings are typically stored as a slurry form to large tailing bonds mainly under water coverage to isolate them from surroundings. The properties of the tailings are thus determined not only by the origin of the material but also the processing steps and storage conditions it has passed.

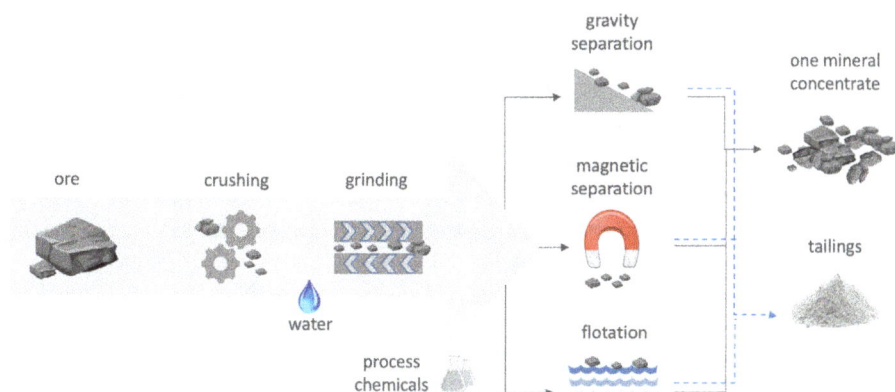

Figure 10.1: A simplified flow diagram of mineral processing and mine tailing production (modified from [5]).

The tailing's ponds need to be designed carefully to ensure the safety of the deposit. Tailings ponds can cause serious environmental concerns into the surroundings if the contaminants are released into the air or ground water and the environmental problems can occur long time after closure of mining site [3]. Even failure of dams can happen because of earthquakes, blasting vibration and extreme weather if the ponds are losing their engineering properties [4]. Tightening legislative norms by environmental authorities have caused a pressure to plan more safety management technologies for mining waste in the future. The recycling potential, new safety storage method using mine tailings in cementitious purposes or in alkali activation interest many researchers globally.

2.1 Physical properties

As mining process is based on the liberation of valuable minerals, the mine tailings are finely ground material with low grain size and high specific surface area, both beneficial for cementitious applications. The grain size of tailings ranges from 2 µm to 2 mm, and they consist of 70–80% of sand sized particles and 20–30% of finer clay-sized particles [1]. Fine particle size can cause air pollution without proper management, which is prevented by disposing the mine tailings under water coverage in the tailing ponds.

2.2 Chemical composition

During the beneficiation process, the tailings go through mainly mechanical processing. Thus, the composition of the tailings is close to that of the treated orebody although the processing chemicals can slightly affect the chemistry of the tailings. Consequently, the tailings consist mainly of chemical elements that are the most abundant in the earth crust (Figure 10.2). The same elements are the building blocks of different cementitious materials, which makes tailings interesting raw materials for the cementitious binders.

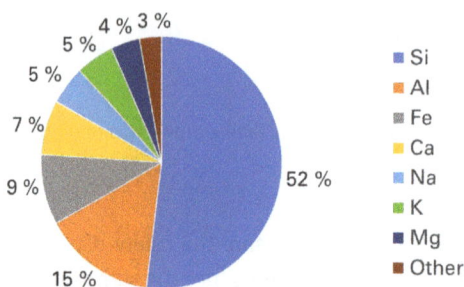

Figure 10.2: The elemental composition (excluding oxygen) of the earth crust.

In addition to the main chemical elements, the tailings can contain wide variety of minor chemical components, some of those harmful to environment. Thus, not only the total concentration of the elements is important but also the leachable, minor substances that significantly affect the usability of the tailings. Typical minor components in the tailings include metals like Cu, Zn, Pb, Ni, Cd, Co, Hg, Al, Mn, U, and metalloids As, Sb, And Se [6].

Investigations about the use of tailings in cementitious systems are typically discussed from the chemical composition point of view, which is an important aspect. However, the mineralogy and the geochemistry of the tailings have not always been considered even if they are critical factors affecting the reactivity of the tailings and thus the suitability of the minerals for cementitious systems [6].

2.3 Mineral composition

The mineralogical composition of the tailings is close to that of original ore and gangue minerals. Most of the mine tailings consist of silicate minerals, which are the most common minerals in the earth crust (Table 10.2) [7]. The other main groups include carbonates, oxides and sulfates. In addition to primary minerals, the tailings can contain secondary minerals formed due to weathering, as well as chemical precipitates formed during and after mineral processing and disposal. The potential behavior of different mine tailings (both fresh and land filled tailings) under different environmental conditions need to be understood prior planning to utilize as cementitious binders [6]. The geochemistry of the generated mine tailings is important to understand also the long-term stability of the mine tailing based concrete materials.

Table 10.2: The mineralogical composition of the earth crust [7].

Mineral group	Minerals	%
tectosilicates	feldspar group	57.9
inosilicates and nesosilicates	pyroxenes, amphiboles, olivine	16.4
oxide	quartz	12.6
oxide	Fe oxides	3.7
phyllosilicates	Mica	3.3
carbonate	Calcite	5.0
phyllosilicates	Clay minerals	1.0
other	Other	3.6

2.3.1 Silicate tailings

The silicates are a biggest group of minerals, so most tailings are rich in silicate minerals. The silicate minerals are classified according to their structure to 3D silicate framework, sheet silicates, chain silicates, ring silicates, and finally isolated silicon tetrahedra (Figure 10.3) [7]. The chemical composition of the silicate minerals can be complex, and silicate minerals are typically very stable having high resistance to thermal treatment.

The reactivity of the silicate tailings in cementitious systems depends on their crystalline structure. Ouffa et al. [8] studied the reactivity of different silicate minerals for alkali-activated systems and concluded that phyllosilicates (kaolinite, muscovite, chlorite and biotite) show potential for alkaline solubility while framework and chain silicates have very low reactivity. Also, Kinnunen et al. [9] concluded that most reactive minerals belong to the group of 1:1 layer lattice aluminosilicates (phyllosilicate group).

2.3.2 Sulfidic tailings

Another interesting group of tailings is sulfidic tailings. Many valuable metals are incorporated to sulfidic minerals, so the tailings from metal mines often contain abundant sulfides. Sulfides exist in minerals such as pyrite FeS_2, pyrrhotite FeS, arsenopyrite (FeAsS), chalcopyrite ($CuFeS_2$), galena (PbS), gersdorffite (NiAsS) and sphalerite (ZnS) [10].

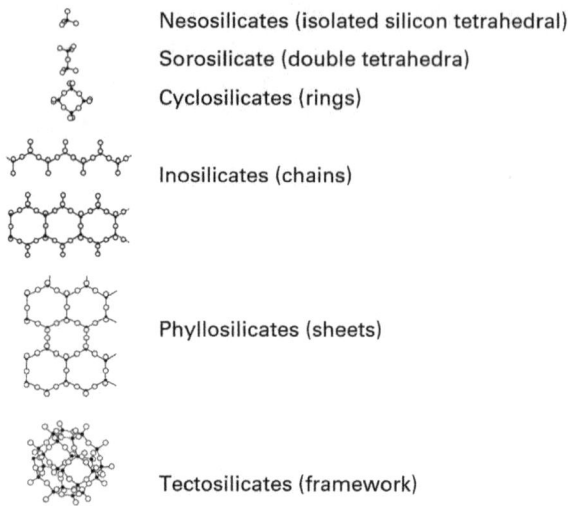

Nesosilicates (isolated silicon tetrahedral)

Sorosilicate (double tetrahedra)

Cyclosilicates (rings)

Inosilicates (chains)

Phyllosilicates (sheets)

Tectosilicates (framework)

Figure 10.3: Silicate minerals (modified from [7]).

The sulfidic tailings have a special feature to oxidize and to produce acid water containing sulfate, heavy metals and metalloids. For example, pyrite is the most common sulfide mineral found in earth crust. Pyrite is stable underground but after excavation and in the presence of oxygen, water it starts to be oxidized releasing ferrous ions, sulphates and H^+ into the environment [11]. This may cause acid mine drainage (AMD), which is the largest environmental problem in the mining industry. The oxidation can occur in short-term or long time after mining closure. Many abandoned mining sites contain still acid generating mine tailings which should be stabilized to prevent the further environmental problems. To face this problem, different strategies are applied, such as isolation of tailings, controlling pH or blending the tailings with other materials [1]. Stabilization of tailings by different cementitious systems is an active research area (e.g. [12, 13]).

2.4 Harmful substances

The tailings can contain components harmful for human health and environment, as summarized in [14]. The mining activities can release highly toxic components such as arsenic, or minerals causing irritation, lung diseases (e.g. asbestos or silicosis) or even cancer. Environmental legislation has been set the limits for the solubility of certain elements and according to the solubility of different elements the mine tailings can be classified either regular or hazardous waste.

The potential hazards of the tailings must be considered when producing cementitious binders. On the other hand, the leachability of the potential harmful components needs to be analyzed from the end product to ensure the safety of the materials. There are several leaching tests utilized for cementitious materials. The most used leaching tests to evaluate S/S of heavy metals in AAM are Toxicity Characteristic Leaching Procedure (TCLP) and NEN 7345 [15, 16], in both tests extractant is water at an acidic pH. In Finland EN 12457 is used to evaluate the material for landfill disposal and used also for waste based alkali-activated materials [17, 18].

3 Alkali-activated materials based on mine tailings

3.1 Alkali-activated materials

Alkali-activated materials have received a lot of attention during the last decades. With an alkaline activator, several industrial residues can react to form cementitious binders. In alkali activation, an alkaline solution is used to dissolve chemical species from the solid precursor, and in the liquid phase, the dissolved components react to form a hardening matrix. The suitable precursor for alkali-activated materials has reactive aluminate and silicate groups that dissolve and form an aluminosilicate network during the alkali activation process. Also, the materials containing calcium, iron and magnesium have successfully used as precursors for alkali-activated materials.

Alkali activation can be used for the immobilization of the harmful substances of solid waste, which is why it is attracting method also for using in mine tailings stabilization/solidification technology. The immobilization of heavy metals can occur by chemical bonding, physical encapsulation or precipitation due to high alkaline conditions and all these immobilization mechanisms can occur simultaneously [19]. The use of tailings in alkali-activated materials is an emerging research area (Figure 10.4). Different kinds of mine tailings (gold, vanadium, copper, nickel, lithium etc.) with different mineralogies have been studied as material for alkali activation. The aim of the studies is to reveal the potential of mine tailings for mine backfilling or construction materials, or to immobilize harmful substances to safe form.

3.2 Mine tailings as precursor for alkali-activated materials

Mine tailings as such typically have low reactivity for alkali-activated systems, resulting in low compressive strength especially when cured at ambient temperature. When the mine tailings have been used as a sole precursor, the proposed application is typically mine backfilling since the strength requirements for the backfilling materials are only 0.5 to 5 MPa [20]. Even if some tailings could achieve reasonable compressive strength for the backfilling applications at long term, co-binders are needed to achieve reasonable early age strength for practical applications [8]. The

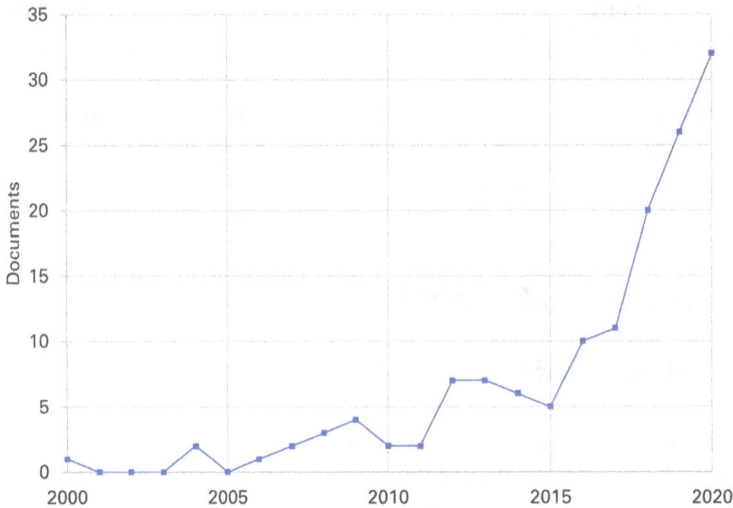

Figure 10.4: Number of documents by year. TITLE-ABS-KEY (("*mine waste*" OR "*mining waste*" OR "*mine tailings*" OR "*mine tailing*") AND (*alkali-activated* OR "*alkali activated*" OR *geopolymer* OR "*inorganic polymer*" OR "*activated cement*")) Scopus 4.12.2020.

addition sub-micron tailings [21] or small amount of co-binder [22] have been proposed to enhance the early age strength of the alkali activated tailings.

The mine tailings with low reactivity work more as fillers than reactive components [23]. In most of the studies related to alkali-activated mine tailings, co-binders have been used to adjust the properties of the material. Blast furnace slag [22], basic oxygen furnace slag [24], and fly ash [25], among others, have been used to improve the mechanical properties of the material.

Most of the studies about alkali-activated mine tailings have considered construction materials, such as bricks [26], concrete or aggregates [27] as target products. However, it was also proposed [28] that utilization of the alkali-activated materials should be used directly on the mine site. In addition to backfill materials, potential application includes cover materials for surface deposits of tailings in order to store them ecologically safe and sealed off from the environment.

3.3 Pre-treatment of mine tailings

In order to improve the reactivity of the tailings, different pre-treatment methods can be used (Figure 10.5). Thermal treatment is the most frequently studied pre-treatment method and phyllosilicate group minerals have shown good response to thermal treatment. The reactivity after thermal treatment is based on the transformation of the crystalline structure into amorphous and the removal of hydroxyl group from the starting material [29].

Thermally treated muscovite rich tailings have been extensively studied as precursors for alkali-activated materials [30]. Moreover, tailings rich in kaolinite [31], phlogopite [31] and montmorillonite [32] have shown potential for alkali-activated materials after thermal treatment. In addition to phyllosilicates, carbonates (calcite and dolomite) as well as albite (belongs group of framework silicates) have been affected by the thermal treatment.

The tailings have been treated also by thermal alkali activation process, where sodium hydroxide was mixed with the tailings before the thermal treatment. The treatment destroyed phlogopite structure while quartz and fluorapatite remained after the calcination [33]. However, in this study, the tailings alone did not result high mechanical properties for alkali-activated materials.

Another way to treat mine tailings is mechanochemical activation. In mechanochemical treatment, mineral structure is destroyed by mechanical forces, which is seen in increased amorphous content. Phyllosilicate group minerals seem to have highest response also to mechanochemical activation. Niu [34] treated phlogopite rich tailings with vibratory disc mill and found almost complete collapse of phlogopite mineral structure together with calcite and dolomite. Adesanya [35] treated waste mineral rich in muscovite and chamosite, and found disruption of minerals structure resulting increased reactivity for alkali-activation purposes.

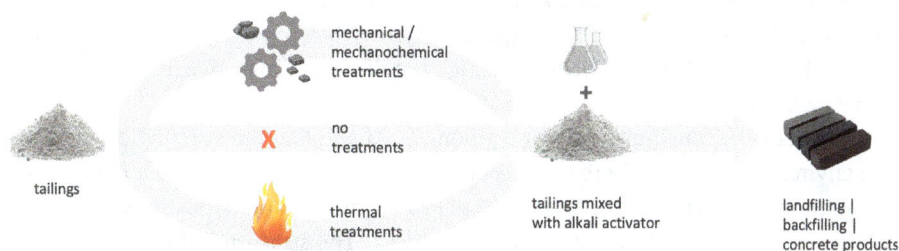

Figure 10.5: Different treatment methods used for mine tailings (modified from [23]).

3.4 Immobilization of heavy metals by alkali-activated materials

An important area of research is to use alkali activation for immobilizing the harmful substances of mine tailings. Especially, sulfidic tailings can cause environmental problems due to high contents of sulfides and heavy metal(loid)s such as arsenic.

Alkali-activated materials have ability to immobilize different substances in their structure. The immobilization mechanism can be either chemical or physical. In physical immobilization, the harmful components can be encapsulated within the alkali-activated material preventing their leachability into the surroundings.

When alkali-activated materials are formed, the negative charge occurs in the structure. To be balanced the negative charge, some cationic species need to be involved into the structure. The cationic elements can come from waste and in alkali-activated systems, the cationic species are often effectively immobilized [19, 36]. Anionic components, on the other hand, are more problematic since they form oxyanions at highly alkaline conditions. Especially arsenic has been problematic in alkali-activated systems [37].

The same applies in the studies related to the mine tailings. The solubility of As has shown to increase after alkali-activation compared with the original tailings if no co-binder is used [38, 39]. With fly ash [38], slags [24] or metakaolin [39] as co-binders the immobilization is more effective, especially with extremely long curing times. Even if other heavy metals, such as Cu, Cr, Pd an Cd, have been efficiently immobilized, alkali activation seems not to be the most effective way to immobilize heavy metal containing mine tailings.

4 Binders based on ettringite formation

4.1 Cementitious systems based on ettringite formation

Ettringite is the main reaction product in calcium sulfoaluminate belite (CSAB) cements [40], super sulfated cements [41] or other ettringite based binders [42]. Ettringite mineral has high capacity to uptake different heavy metals into its structure (Figure 10.6), which makes it interesting to sulfidic tailings rich in sulfides and heavy metals [43].

Theoretically, ettringite can form when Ca, Al and SO_4 are reacting under alkaline environment. When there are presence other elements such as Fe it can replace Al components (Figure 10.6). According to [44, 45] the ettringite has shown high capability to immobilize also oxyanions such as As. The substitution of different elements is based on the ion charge and size. When there are elements competing the same place of the ettringite structure, the most dominant ion will be substituted most easily.

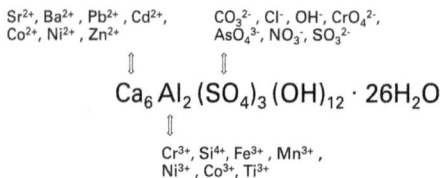

$$Sr^{2+}, Ba^{2+}, Pb^{2+}, Cd^{2+}, \quad CO_3^{2-}, Cl^-, OH^-, CrO_4^{2-},$$
$$Co^{2+}, Ni^{2+}, Zn^{2+} \quad\quad AsO_4^{3-}, NO_3^-, SO_3^{2-}$$
$$\Updownarrow \quad\quad\quad \Updownarrow$$
$$Ca_6 Al_2 (SO_4)_3 (OH)_{12} \cdot 26H_2O$$
$$\Updownarrow$$
$$Cr^{3+}, Si^{4+}, Fe^{3+}, Mn^{3+},$$
$$Ni^{3+}, Co^{3+}, Ti^{3+}$$

Figure 10.6: Ettringite capability to immobilize heavy metals [43].

Ettringite is known to be formed at pH from 10.5 to 13 and the stability of ettringite is highly dependent on the pH and temperature [46, 47]. If heavy metals are substituted into ettringite structure the stable pH can be changed [48]. In ettringite forming materials, the immobilization capacity is based on also the other hydration phases, which can support the immobilization of different elements.

4.2 Sulfidic tailings in ettringite binders

The research related to ettringite-based binders based on sulfidic tailings is limited. Sulfidic tailings have been studied as a precursor to materials based on ettringite formation in [27], where tailings were treated with lime, aluminum and fly ash to produce hardened samples with ettringite and gypsum as the main reaction products. The monolithic product had good resistance towards freeze and thaw cycles and low permeability. However, the long-term hydration resulted in lowering pH and disintegration of formed ettringite. The recipe was further developed with higher content of aluminum [49], which results stable product (measured up to 660 days).

Activation of sulfidic tailings with lime and 10–25% of GGBFS resulted effective immobilization of sulfates and arsenic. Over 90% of sulfates and more than 99% of arsenic were immobilized only after one week of curing. In this system, As was found in form of $Ca_3(AsO)_2$ but also ettringite was found in all samples. Sulfates were immobilized by forming gypsum and by attending in ettringite formation [50].

Mine tailings have been used as a sulfate source for CSAB cement hydration process [13]. By mixing CSAB cement with $CaSO_4$, the ettringite is the dominated phase formed in the resulting matrix and the ettringite formation is highly dependent on the sulfate source. If the sulfate source is depleted during the hydration, the monosulfate can be formed. Both phases are able to immobilize heavy metals including oxyanions. By mixing CSAB cement, pure gypsum and sulfidic gold tailings with water the hydration results ettringite and monosulfate with ability to stabilize sulfates and oxyanions into its structure. All harmful elements were effectively immobilized, and the material resulted good mechanical properties. The immobilization can occur by chemical immobilization and physical encapsulation simultaneously within the formed material.

5 Future perspectives

As the mines are often locating in remote areas, the most promising approach is to utilize mine tailings on-site. The use of alkali-activated tailings as backfilling material or as a covering layer for the surface deposits are interesting options [28]. The strength properties of the alkali-activated tailings seem to meet the requirements of backfilling. However, practical aspects (such as potentially long pumping distances) should be considered as they may set further requirements for e.g. setting behavior of the material. When utilizing alkali-activated tailings as the covering layers for the surface deposits, the environmental conditions of the mining site should be taken into account.

In order to utilize mine tailings in alkali-activated materials or in ettringite based materials in practical applications, more valuable research data is needed in both laboratory and pilot scale. Especially, a comprehensive understanding on the long-term mechanical and environmental performance of the materials is required.

6 Conclusions

Mine tailings are finely ground materials with high specific surface area and fine particle size. Additionally, the mine tailings consist of the most abundant elements of the earth crust, which are also the main building blocks of cementitious materials. Because lack of economic value, billions of tons of mine tailings are discarded into the nature in tailings dams under water coverage which can cause problems to the nature and human society. New sustainable and safe management methods are required. An interesting option is the use of mine tailings as raw materials in cementitious materials.

Mineralogy is the most important factor defining the suitability of mine tailings for cementitious systems. Minerals found in the mine tailings are usually unreactive in alkaline conditions being not suitable straight to be utilized in alkali-activated materials. Mine tailings can be utilized as an inert filler or the reactivity can be enhanced with suitable pre-treatment methods such as thermal or mechanical treatment. The silicate mine tailings rich in phyllosilicates have highest potential for alkali-activated materials. They also have the highest response to the mechano-chemical or thermal pretreatment before the alkali activation.

Many valuable metals are found in sulfidic minerals and the sulfidic rich mine tailings with heavy metals are one of the biggest group of tailings in the mining industry. Sulfidic mine tailings are not suitable raw materials for alkali activation technology. For sulfidic tailings, the binders based on ettringite formation are the most promising since sulfates can participate in the ettringite reaction being immobilized within this reaction. Simultaneously, ettringite can efficiently immobilize oxyanions, which are difficult to immobilize with alkali-activated materials. For ettringite formation, the co-binder material and chemical environment play a critical role. The use of tailings in ettringite binders is still at initial stage, so further studies are needed to understand the long-term performance of these cementitious binders.

The most potential applications for cementitious binders based on mine tailings include mine backfill materials and cover layers on the surface deposits as well as the safety storage of the tailings.

References

[1] B. B. Lottermoser, *Mine wastes : characterization, treatment and environmental impacts*. Springer Verlag.

[2] Eurostat, "Generation and treatment of waste. Report on environment and energy.," European statistical data support, 2009.

[3] F. Rao and Q. Liu, "Geopolymerization and Its Potential Application in Mine Tailings Consolidation: A Review," *Miner. Process. Extr. Metall. Rev.*, vol. 36, no. 6, pp. 399–409, Nov. 2015, doi: 10.1080/08827508.2015.1055625.

[4] M. Rico, G. Benito, and A. Díez-Herrero, "Floods from tailings dam failures," *J. Hazard. Mater.*, vol. 154, no. 1–3, pp. 79–87, Jun. 2008, doi: 10.1016/j.jhazmat.2007.09.110.

[5] J. Kiventerä, P. Perumal, J. Yliniemi, and M. Illikainen, "Mine tailings as a raw material in alkali activation: A review," *Int J Min. Met. Mater*, vol. 27, no. 8, p. 13, 2020.

[6] H. E. Jamieson, S. R. Walker, and M. B. Parsons, "Mineralogical characterization of mine waste," *Appl. Geochem.*, vol. 57, pp. 85–105, 2015, doi: 10.1016/j.apgeochem.2014.12.014.

[7] S. K. Haldar and J. Tišljar, "Chapter 2 - Basic Mineralogy," in *Introduction to Mineralogy and Petrology*, S. K. Haldar and J. Tišljar, Eds. Oxford: Elsevier, 2014, pp. 39–79.

[8] N. Ouffa, M. Benzaazoua, T. Belem, R. Trauchessec, and A. Lecomte, "Alkaline dissolution potential of aluminosilicate minerals for the geosynthesis of mine paste backfill," *Mater. Today Commun.*, vol. 24, 2020, doi: 10.1016/j.mtcomm.2020.101221.

[9] P. Kinnunen, *et al.*, "Recycling mine tailings in chemically bonded ceramics–a review," *J. Clean. Prod.*, vol. 174, pp. 634–649, 2018.

[10] I. Park, *et al.*, "A review of recent strategies for acid mine drainage prevention and mine tailings recycling," *Chemosphere*, vol. 219, pp. 588–606, 2019, doi: 10.1016/j.chemosphere.2018.11.053.

[11] K. K. Kefeni, T. A. M. Msagati, and B. B. Mamba, "Acid mine drainage: Prevention, treatment options, and resource recovery: A review," *J. Clean. Prod.*, vol. 151, pp. 475–493, 2017, doi: 10.1016/j.jclepro.2017.03.082.

[12] P. Desogus, P. P. Manca, G. Orrù, and A. Zucca, "Stabilization-solidification treatment of mine tailings using Portland cement, potassium dihydrogen phosphate and ferric chloride hexahydrate," *Miner. Eng.*, vol. 45, pp. 47–54, 2013, doi: 10.1016/j.mineng.2013.01.003.

[13] J. Kiventerä, K. Piekkari, V. Isteri, K. Ohenoja, P. Tanskanen, and M. Illikainen, "Solidification/ stabilization of gold mine tailings using calcium sulfoaluminate-belite cement," *J. Clean. Prod.*, vol. 239, p. 118008, Dec. 2019, doi: 10.1016/j.jclepro.2019.118008.

[14] S.K. Haldar, Josip Tišljar, "Hazards of Minerals - Rocks and Sustainable Development," in *Introduction to Mineralogy and Petrology*, Burlington: Elsevier, 2013, pp. 306–323.

[15] H. Xu and J. S. J. van Deventer, "Effect of Source Materials on Geopolymerization," *Ind. Eng. Chem. Res.*, vol. 42, no. 8, pp. 1698–1706, Apr. 2003, doi: 10.1021/ie0206958.

[16] "NEN 7341. Leaching Characteristics of Solid (Earthy and Stony) Building and Waste Materials. Determination of the Availability of Inorganic Components for Leaching.," vol. 1995.

[17] "SFS-EN 12457-3. Characterisation of waste. Leaching. 2002."

[18] "SFS-EN 12457-2. Characterisation of waste. Leaching." 2002.

[19] J. G. S. Van Jaarsveld, J. S. J. Van Deventer, and L. Lorenzen, "The potential use of geopolymeric materials to immobilise toxic metals: Part I. Theory and applications," *Miner. Eng.*, vol. 10, no. 7, pp. 659–669, 1997, doi: 10.1016/S0892-6875(97)00046-0.

[20] C. Chen, X. Li, X. Chen, J. Chai, and H. Tian, "Development of cemented paste backfill based on the addition of three mineral additions using the mixture design modeling approach," *Constr. Build. Mater.*, vol. 229, 2019, doi: 10.1016/j.conbuildmat.2019.116919.

[21] M. Falah, K. Ohenoja, R. Obenaus-Emler, P. Kinnunen, and M. Illikainen, "Improvement of mechanical strength of alkali-activated materials using micro low-alumina mine tailings," *Constr. Build. Mater.*, vol. 248, p. 118659, Jul. 2020, doi: 10.1016/j.conbuildmat.2020.118659.

[22] J. Kiventerä, L. Golek, J. Yliniemi, V. Ferreira, J. Deja, and M. Illikainen, "Utilization of sulphidic tailings from gold mine as a raw material in geopolymerization," *Int. J. Miner. Process.*, vol. 149, pp. 104–110, Apr. 2016, doi: 10.1016/j.minpro.2016.02.012.

[23] Jenni Kiventerä, Priyadarshini Perumal, Juho Yliniemi, Mirja Illikainen, "Mine tailings as a raw material in alkali-activation – a review."

[24] T. Falayi, "A comparison between fly ash- and basic oxygen furnace slag-modified gold mine tailings geopolymers," *Int. J. Energy Environ. Eng.*, vol. 11, no. 2, pp. 207–217, 2020, doi: 10.1007/s40095-019-00328-x.

[25] L. Zhang, S. Ahmari, and J. Zhang, "Synthesis and characterization of fly ash modified mine tailings-based geopolymers," *Constr. Build. Mater.*, vol. 25, no. 9, pp. 3773–3781, 2011.

[26] F. A. Kuranchie, S. K. Shukla, and D. Habibi, "Utilisation of iron ore mine tailings for the production of geopolymer bricks," *Int. J. Min. Reclam. Environ.*, vol. 30, no. 2, pp. 92–114, 2016, doi: 10.1080/17480930.2014.993834.

[27] Yliniemi, Paiva, Ferreira, Tiainen, and Illikainen, "Development and incorporation of lightweight waste-based geopolymer aggregates in mortar and concrete," *Constr. Build. Mater.*, Nov. 2016, doi: 10.1016/j.conbuildmat.2016.11.017.

[28] R. Obenaus-Emler, M. Falah, and M. Illikainen, "Assessment of mine tailings as precursors for alkali-activated materials for on-site applications," *Constr. Build. Mater.*, vol. 246, 2020, doi: 10.1016/j.conbuildmat.2020.118470.

[29] C. Ferone, *et al.*, "Thermally treated clay sediments as geopolymer source material," *Appl. Clay Sci.*, vol. 107, pp. 195–204, 2015.

[30] F. Pacheco-Torgal, J. Castro-Gomes, and S. Jalali, "Durability and environmental performance of alkali-activated tungsten mine waste mud mortars," *J. Mater. Civ. Eng.*, vol. 22, no. 9, pp. 897–904, 2010.

[31] P. Perumal, K. Piekkari, H. Sreenivasan, P. Kinnunen, and M. Illikainen, "One-part geopolymers from mining residues–Effect of thermal treatment on three different tailings," *Miner. Eng.*, vol. 144, p. 106026, 2019.

[32] S. Mabroum, A. Aboulayt, Y. Taha, M. Benzaazoua, N. Semlal, and R. Hakkou, "Elaboration of geopolymers based on clays by-products from phosphate mines for construction applications," *J. Clean. Prod.*, vol. 261, p. 121317, Jul. 2020, doi: 10.1016/j.jclepro.2020.121317.

[33] J. Wu, J. Li, F. Rao, and W. Yin, "Mechanical property and structural evolution of alkali-activated slag-phosphate mine tailings mortars," *Chemosphere*, vol. 251, 2020, doi: 10.1016/j.chemosphere.2020.126367.

[34] H. Niu, P. Kinnunen, H. Sreenivasan, E. Adesanya, and M. Illikainen, "Structural collapse in phlogopite mica-rich mine tailings induced by mechanochemical treatment and implications to alkali activation potential," *Miner. Eng.*, vol. 151, 2020, doi: 10.1016/j.mineng.2020.106331.

[35] E. Adesanya, K. Ohenoja, J. Yliniemi, and M. Illikainen, "Mechanical transformation of phyllite mineralogy toward its use as alkali-activated binder precursor," *Miner. Eng.*, vol. 145, p. 106093, Jan. 2020, doi: 10.1016/j.mineng.2019.106093.

[36] J. G. S. Van Jaarsveld, J. S. J. Van Deventer, and A. Schwartzman, "The potential use of geopolymeric materials to immobilise toxic metals: Part II. Material and leaching characteristics," *Miner. Eng.*, vol. 12, no. 1, pp. 75–91, Jan. 1999, doi: 10.1016/S0892-6875(98)00121-6.

[37] A. M. Fernandez Jiminez, E. E. Lachowski, A. Palomo, and D. E. Macphee, "Microstructural characterisation of alkali-activated PFA matrices for waste immobilisation," *Cem. Concr. Compos.*, vol. 26, no. 8, pp. 1001–1006, Nov. 2004, doi: 10.1016/j.cemconcomp.2004.02.034.

[38] N. Cristelo, J. Coelho, M. Oliveira, N. C. Consoli, A. Palomo, and A. Fernández-Jiménez, "Recycling and application of mine tailings in alkali-activated cements and mortars-strength development and environmental assessment," *Appl. Sci. Switz.*, vol. 10, no. 6, 2020, doi: 10.3390/app10062084.

[39] J. Kiventerä, I. Lancellotti, M. Catauro, F. D. Poggetto, C. Leonelli, and M. Illikainen, "Alkali activation as new option for gold mine tailings inertization," *J. Clean. Prod.*, vol. 187, no. 20, pp. 76–84, June, doi: https://doi.org/10.1016/j.jclepro.2018.03.182.

[40] E. Gartner and H. Hirao, "A review of alternative approaches to the reduction of CO2 emissions associated with the manufacture of the binder phase in concrete," *Cem. Concr. Res.*, vol. 78, Part A, pp. 126–142, Dec. 2015, doi: 10.1016/j.cemconres.2015.04.012.

[41] "Hydration mechanisms of super sulphated slag cement - ScienceDirect." https://www.science direct.com/science/article/pii/S0008884608000574?via%3Dihub (accessed Jun. 30, 2020).

[42] H. Nguyen, *et al.*, "Byproduct-based ettringite binder – A synergy between ladle slag and gypsum," *Constr. Build. Mater.*, vol. 197, pp. 143–151, Feb. 2019, doi: 10.1016/j. conbuildmat.2018.11.165.

[43] Q. Y. Chen, M. Tyrer, C. D. Hills, X. M. Yang, and P. Carey, "Immobilisation of heavy metal in cement-based solidification/ stabilisation: A review," *Waste Manag.*, vol. 29, no. 1, pp. 390–403, Jan. 2009, doi: 10.1016/j.wasman.2008.01.019.

[44] M. Chrysochoou and D. Dermatas, "Evaluation of ettringite and hydrocalumite formation for heavy metal immobilization: Literature review and experimental study," *J. Hazard. Mater.*, vol. 136, no. 1, pp. 20–33, Aug. 2006, doi: 10.1016/j.jhazmat.2005.11.008.

[45] P. Kumarathasan, G. J. McCarthy, D. J. Hassett, and D. F. Pflughoeft-Hassett, "Oxyanion Substituted Ettringites: Synthesis and Characterization; and their Potential Role In Immobilization of As, B, Cr, Se and V," *MRS Proc.*, vol. 178, Jan. 1989, doi: 10.1557/PROC-178-83.

[46] D. Damidot and F. P. Glasser, "Thermodynamic investigation of the CaO-Al2O3-CaSO4-H2O system at 50°C and 85°C," *Cem. Concr. Res.*, vol. 1992, no. 22, pp. 1179–1191.

[47] D. Damidot and F. P. Glasser, "Thermodynamic investigation of the CaO-Al2O3-CaSO4-H2O system at 25°C and the influence of Na2O," *Cem. Concr. Res.*, vol. 1993, no. 23, pp. 221–238, 1993.

[48] R. B. Perkins and C. D. Palmer, "Solubility of Ca6‰Al. . .OH †6Š2. . .CrO4†3 Á 26H2O, the chromate analog of ettringite; 5±758C," *Appl. Geochem.*, p. 16, 2000.

[49] A. M. O. Mohamed, M. Hossein, and F. P. Hassani, "Role of Fly Ash and Aluminum Addition on Ettringite Formation in Lime-Remediated Mine Tailings," *Cem. Concr. Aggreg.*, vol. 25, no. 2, pp. 49–58, 2003.

[50] J. Kiventerä, H. Sreenivasan, C. Cheeseman, P. Kinnunen, and M. Illikainen, "Immobilization of sulfates and heavy metals in gold mine tailings by sodium silicate and hydrated lime," *J. Environ. Chem. Eng.*, vol. 6, no. 5, pp. 6530–6536, Oct. 2018, doi: 10.1016/j.jece.2018.10.012.

Thomas Neumann

Chapter 11
Industrial waste as fuel and raw material in the cement industry

Abstract: There is an increase in the use of secondary or alternative fuels in cement plants worldwide. This can conserve fossil energy sources and raw materials and reduce CO_2 emissions. Of particular importance are plastic waste, waste tyres, sewage sludge and meat and bone meal. It must be taken into account that considerable amounts of ash are introduced into the clinker with these fuels. The composition of these fuel ashes differs significantly from a raw meal or clinker composition. This influencing factor must already be taken into account in the composition of the raw meal for clinker production. In addition, water and other volatile elements can enter the kiln via secondary fuels. To ensure trouble-free kiln operation in the cement plant, the process technology must be adapted to the respective fuel mix. Even with optimal operational management, qualitative influences on the clinker or cement produced must be expected if secondary fuels are used.

Keywords: secondary fuel, alternative fuel, refused derived fuels (RDF), waste tyres, meat and bone meal (MBM), municipal solid waste (MSW), wood, sewage sludge, used foundry sand, calciner, burner lance, pyrolysis, burn-out behaviour, fuel ash

1 Introduction

The production of cement or cement clinker is an energy-intensive process. Although the thermal efficiency of modern cement plants is over 70%, the specific energy requirement of cement clinker production is with about 3,300 kJ/kg very high. About 90% of this energy is supplied in the form of fuels as thermal energy [1].

In the past, mainly fossil fuels such as crude oil, natural gas and coal were used for cement production. Due to the current CO_2 problem and due to the increased fuel costs, alternative fuels are increasingly used for cement clinker production. These alternative fuels are also referred to as alternitive or secondary fuels. These materials are essentially industrial by-products or waste materials. The proportion of secondary fuels used in the cement industry for thermal purposes is increasing worldwide. In Austria the average secondary fuel rate is already over 80% and in Germany over 65% [2, 3].

Thomas Neumann, SCHWENK Zement GmbH & Co.KG, Laudenbacher Weg5, 97753 Karlstadt, Germany

https://doi.org/10.1515/9783110674941-011

The use of secondary fuels avoids the landfilling of residual or waste materials. The incineration in the cement plant is carried out without residues. Any pollutants that may be present can be rendered harmless by incineration. The fuel ashes are bound in the cement clinker. In addition to the use of energy of the residual and waste materials, there is also partial material substitution. Due to the high gas temperatures of up to 2,000 °C, a fuel-related increase in waste gas emissions, e.g. for dioxins or furans, is not to be expected.

As a matter of principle, the recycling of residual materials is always preferable to thermal recycling. However, if the recycling of materials requires extreme expenditures for auxiliary materials or energy, thermal recycling should be preferred. A further advantage of thermal use of residual or waste materials is that no harmful methane emissions are produced compared to landfilling or composting. In many cases, secondary fuels contain high biogenic contents. The CO_2 produced during combustion is climate neutral for these components [4, 5].

With all secondary fuels it must be taken into account that the range of variation in composition is always significantly higher than with fossil fuels. In addition to different moisture contents and calorific values, the ash composition in particular must be taken into account. After burning, the fuel ash combines with the raw material and reacts to cement clinker at the end of the kiln. In case of high ash inputs, the fuel ash content in the clinker may exceed 5% [6].

The heavy metals contained in cement and cement clinker are mainly introduced via the raw materials. When using secondary fuels, the proportion of some heavy metals or trace elements can be increased significantly. This applies especially to the elements antimony, cadmium and zinc. A small increase is recorded for the elements lead, cobalt and vanadium [7]. A determination of the maximum element concentrations and a careful monitoring of the input streams are the prerequisites for the use of alternative fuels in cement plants. Furthermore, it must be taken into account that heavy metals and trace elements influence the formation of the melting phase and mineralisation in the clinker.

2 Types of secondary fuels

An overview about the heating values, the ash contents and composition is given in Table 11.1. The content of heavy metalls and trace elements of different secondary fuels is given in Table 11.2.

2.1 Refused Derived Fuels (RDF)

The RDF is waste from industrial and commercial processes. RDF mainly consists of paper, plastic and other organic materials [31]. Normally waste is first collected by recycling companies and processed for the cement industry. After shredding with

Table 11.1: Heating values, ash contents and composition of fuel ashes; mean values or usual range of variation.

unit	Heating-value MJ/kg	Ash-content % of dry sample	Ash composition % of ash										Reference
			CaO	SiO_2	Al_2O_3	Fe_2O_3	SO_3	P_2O_5	MgO	K_2O	Na_2O	Cl	
natural gas	34	0.1											[8]
fuel oil	41.0	0.2					2.5 S*			$Na_2O + K_2O$: < 0,0153			[8, 9]
petrol coke	29–31	0.2–1					2–7.5 S*						[9–11]
lignite	22	2–20	2–52 ***	15–80 9	1–23	1.5–22	1–15		0.5–11	0.1–2	0.1–2	0.03*	[12–14]
hard coal	26–31	5–15	1–10	45–50	25–30	5–15	< 2		1.7–7	3–4 **		0.1–1*	[14, 15]
tyres	26–36	20–76				17–71	2.5–5.5						[16, 17]
shredded tyres	26–37	3.1–18.6	15–39	33–42	2.4–4.1	2.9–6.7	10–14	0.3	1.1–1.7	0.7–1.3	0.3–0.5	6.6	[17–20]
waste oil	33	1											[9, 10]
RDF	17–21	13.0	26.1	27.5	17.0	5.0	2.9	0.9	2.1	1.3	3.1	8.8	[9, 10, 21]
MBM	16–20	18.4–28.3	35–50	1.3–2.6	0.1–0.3	0.2–0.3	2.1	34–40	1.2–1.3	4.7	6.0	3.7	[17, 22, 23]
municipal solid waste	14–25.7	14–15.3 13.6–46.7	26–32.1	22–31	7.8–60	4.3–6.8	0.25–2.5					0.8–4.3	[13, 23, 24]

(continued)

Table 11.1 (continued)

	Heating-value	Ash-content	Ash composition										Reference
			CaO	SiO$_2$	Al$_2$O$_3$	Fe$_2$O$_3$	SO$_3$	P$_2$O$_5$	MgO	K$_2$O	Na$_2$O	Cl	
unit	MJ/kg	% of dry sample	% of ash										
waste solvent	24	4											[9, 10]
waste wood	13–19	0.3–6	0.6–25	8–79	0.9–28	0.9–14.2	0.5	0.5	0.7–9.3	0.4–10.4	0.1–6.5	<4	[9, 10, 23, 24]
sewage sludge	11 (at 7% moisture)	41 21–67	19.3	25.9	9.8	14.2	3.7	16.7	2.3	1.1	0.9	0.12	[9, 17, 25]

* . . . related to fuel ** . . . Na$_2$O-equivalent *** . . . 0.1–25% free lime.

Table 11.2: Average contents of trace elements in different fuels.

mg/kg*	fuel oil	petrol-coke	lignite (rheinisch)	hard coal	tyres	waste oil	RDF	MBM	municipal solid waste	waste wood	sewage sludge
As	<0.1	1.7	0.82	13.6	0.3–20	2.4	0.9	<7.4	0.3	1.2	7.2
Cd	<0.4	1.0	0.01	0.3	0.1–10	0.8	1.8	<0.7	0.8	1.0	1.4
Co		1.5	3.5	16.7	1–250	1	3.3	3.4	3.7	4.0	12.0
Cr	2–4	3.0	5.1	26.5	2.1–97	12	27	14–32	40	13	109
Cu		1.2	1.2	33	1.5–450	51	123	30–70	230	15	375
Hg		0.0	0.05	0.2	<0.43		0.2	<0.1	0.2	0.24	0.3
Mn			116	125	2.8–750		30	42.0	39	143	784
Ni	5–43	350	9.3	45	1.2–77	20	16	9–18	10	7	43
Pb	1–34	1.5	0.81	68	3–760	151	62	<5.0	50	303	62
Sb		0.6	0.24	2.5	0.7	1	28	<9.4	12	1.4	9.4
Sn		0.3	1.2	5	10	6	11	0.9	8.3	1.8	80.0
Ti					450**		2.000**	<100			1.639
Tl	<0.12	<3.1	0.09	0.3	0.2–5		0.3	<5.7	0.1	<0.2	
V	2–117	1500	2	75	5–5.3	2	3.6	3.8–6.3	3.1	23	56.0
Zn	Mai 85	16.0	3.9–22	10–300	3.000–20.500	700	331	39–144	331	440	1039
Reference	[26]	[7, 13, 27]	[28, 29]	[13]	[16, 17, 19]	[7]	[7, 13, 19]	[22, 30]	[7, 13]	[7, 21]	[25]

* . . . related to dry material ** . . . single value.

granulators, hammer mills or similar equipment, the waste is classified into different particle sizes. Dense and compact particles are separated from light film-like particles by ballistic sorting. Metals can be sorted out via magnetic separators and eddy current separators (non-ferrous metals). PVC can be detected by NIR sensors and discharged via compressed air nozzles. For cement production, the input of chlorine via fuels must be limited. The content of chlorine in cement is limited to 0.10%. This prevents corrosion of the reinforcing steel in hardened concrete. In many cases an upper limit of 1% chlorine has been established for the fuel [5]. When using fuels with a high chloride content, a furnace with gas bypass is necessary to reduce the absolute chlorine content in the cement and to remove alkali salts from the process. The latter form deposits in the preheater and disturb the production process. In some cases, the ashes of secondary fuels cause an increased input of aluminium or aluminium oxide into the cement clinker. In material terms, this proportion must be taken into account in the raw meal composition [32, 33].

The maximum particle size of RDF usually used in cement plants is 25 mm. Cement plants with calciners can also burn particles with larger dimensions. In some cases RDF is dried before burning in the cement plant. This is usually done by using the waste heat from the klin. RDF is characterised by very high specific calorific values. The material also has very good ignition properties. Due to the low ash content, RDF burns almost residue-free in the rotary kiln or calciner of the cement plant [34].

2.2 Tyres

Tyres are used car or truck tyres. Tyres are characterised by a very high calorific value of. They can be delivered to the rotary kiln both as complete tyres or in shredded form. It must be noted that tyres always contain iron (tyre carcass). There is a very large difference in iron content between car and truck tyres. For example, the carcass content in truck tyres at $\leq24\%$ is significantly higher than in car tyres at $\leq12\%$. This iron content in particular must be taken into account in the raw meal composition. The proportion of Fe_2O_3 can easily amount to 70% of the tyre mass [16]. Furthermore, the carcass material introduced into the cement clinker can lead to locally reducing conditions, e.g. within the clinker granules. Ideally, the steel carcass is completely oxidized during the burning process, so that integration into the clinker phases is possible without any problems. Apart from iron, tyres contain a significant amount of silicon oxide and sulphur. Sulphur binds alkalis in the kiln system. The resulting alkali sulphates or alkali-earth-alkali double salts (e.g. Langbeinite) leave the kiln via the bypass or are discharged with the clinker. Another typical accompanying element of tyres is zinc [35]. Zinc has a mineralizing effect during clinker burning [36].

2.3 Meat and bone meal (MBM)

Every year, the U.S.A. alone produces 25Miot of animal waste. A large part of it is processed into MBM. MBM and bone meal are processed slaughterhouse waste and animal carcasses. Whereas MBM used to be mainly processed into animal feed and fertilizer, since the TSB crisis there has been increased use as a fuel, particularly in the cement industry. MBM is partly divided into animal meal, bone meal and blood meal.

During processing, the shredded animal components (<50 mm) are sterilized at 3 bar and a temperature of 133 °C for at least 20 min. The intermediate product is then mechanically dehydrated to a maximum moisture content of 5% and reduced to a maximum fat content of 12%. A grinding process produces animal meal with a maximum particle size of 2 mm. In Germany alone, over 1 million tons of meat and bone meal are produced annually. The phosphorus content in MBM correlates directly with the bone content and amounts to 1–3%. When burning MBM, the high alkali, chlorine and fluorine loads can cause increased build-up and blockages in the preheater. MBM can also accelerate the aging of catalysts for nitrogen oxide reduction due to increased arsenic content [14].

2.4 Municipal solid waste

Waste separation systems have been established in many countries. Usually, separation into paper/, cardboard, plastic, metal, glass, compostable materials, and residual waste is carried out. In some cases, these groups are combined or subdivided in more detail. This measure enables the targeted reuse or recycling of waste materials and the reduction of non-recyclable waste. Due to the high consumption of land, municipal solid waste is incinerated rather than landfilled in many countries. Attempts to use the ash and slag produced in this process as a substitute for gravel and sand in concretes have so far been unsuccessful. The glass particles contained in the ash are problematic, as they can trigger a damaging alkali-sicate reaction in the alkaline environment of the concrete. Furthermore the aluminum components lead to hydrogen formation in the plastic state of concrete, resulting in an increased porosity [37].

More and more often municipal solid waste is used as fuel in cement plants. Especially the high-calorific fractions are interesting in this context.The high temperatures enable complete oxidation of metallic components and complete integration of the mineral components into the clinker phases. The raw material composition must be adapted to the ash input. This essentially means an increase in the lime standard of the raw meal, since the calcium content of the ash is very low [38]. With the increase of the amount of packaging materials in municipal solid wastes an increase of the aluminium (foil) content in clinker can be expected. This can cause a change in clinker reactivity [39].

The composition of municipal waste differs greatly by region. In addition, a change in the composition of waste over time can be recorded. If municipal waste is incinerated in a cement plant, this material must be carefully processed. It must also be noted that the use of municipal solid waste increases the input of trace elements and heavy metals [40].

2.5 Waste wood

Waste wood is often coated with "anti fouling paints". This applies to wood used for boat hulls as well as wood that has been used outdoors, e.g. in the ground or as sleepers. Besides copper and zinc, these coatings often contain higher contents of silver, cadmium, chrome, nickel, lead and tin [41].

2.6 Sewage sluge

Depending on the respective region and treatment in the sewage treatment plant, sewage sludge of different composition and with different degrees of dewatering is produced. Usually the digested sewage sludge is dewatered by chamber or membrane filter presses, sometimes also by centrifuging. In addition, the sewage sludge can be conditioned with polymers or lime. The water content of the mechanically dewatered sewage sludge is between 25% and 80%. Thermally dried sewage sludge has a moisture content of about 10%. It must be noted that sewage sludge contains high concentrations of phosphorus, trace elements and heavy metals. Due to cosmetics and pharmaceuticals, sewage sludge often contains high concentrations of mercury. During combustion, this mercury is emitted via the gas phase. A careful incoming inspection must rule out the possibility of mercury emissions being exceeded. Mercury emissions in the furnace system can be further reduced by activated carbon filter systems [14].

In relation to the dried sewage sludge, the ash content is approx. 41% [25]. The composition of these ashes differs significantly from the composition of raw meal. When using sewage sludge as fuel, the raw material composition must be corrected accordingly. The thermal drying of sewage sludge improves its calorific value and increases the efficiency of the kiln system in the cement plant. The waste heat available in cement plants is particularly suitable for drying mechanically dewatered sewage sludge [34].

2.7 Used foundry sand

In foundries, moulding materials are used for lost-foam-process or common casting processes. These moulding materials consist of a quartz sand of a defined grain size and a reaction resin (furan resin + hardener). After casting, the sands are recovered so that the resin is enriched to a certain degree. Due to metallic residues, foundry sands often contain high concentrations of heavy metals and especially of chrome. In the first instance foundry sands used in cement industry must considered as a silicon-rich raw material and less as a fuel. This raw material is fed directly into the kiln for correction e.g. in the case of high free lime values. Foundry sands are not suitable as raw meal constituents, as they would cause high CO or volatile organic carbon emissions in the preheating process. The direct introduction into the rotary kiln allows complete combustion of the organic components.

2.8 Other fuels

In addition to the secondary fuels mentioned above, there are a number of other fuels. In some cement plants, for example, **waste oils** or **solvent-containing residues** are burned. **Bleaching earth**, a waste product from the refining of mineral or vegetable oil or from the production of stearin, is also partly used as a fuel. In some countries, such as Brazil, **charcoal** is used to fire cement kilns [18].

During the refining of crude oil, petroleum coke is produced as a distillation residue. This fuel is usually dried and ground prior to thermal utilization in cement plants [11]. With petroleum coke, vanadium can be introduced into the clinker [7]. Like petroleum, petroleum coke is characterized by a very low ash content.

3 Combustion of secondary fuels in the kiln system of a cement plant

A prerequisite for the burning of secondary fuels in the kiln system of a cement plant is compliance with the respective prescribed emission limit values. For this purpose, the firing should guarantee a minimum temperature of 850 °C. For hazardous waste with >1% chlorine, this minimum temperature must even be 1,000 °C. This minimum temperature must be reached for at least 2 seconds [42].

Even though most heavy metals are introduced into the clinker via the raw materials (limestone, sand, iron ore, . . .), some fuels lead to a significant increase in certain heavy metal or trace element concentrations. Mercury is considered to be particularly critical. As most kiln systems do not have activated carbon filter systems, this element is almost completely emitted.

Compared to fossil fuels such as coal, secondary fuels often contain more water or moisture. When burning, this water evaporates and thus increases the gas volume in the kiln system. It must be taken into account that as the water content of

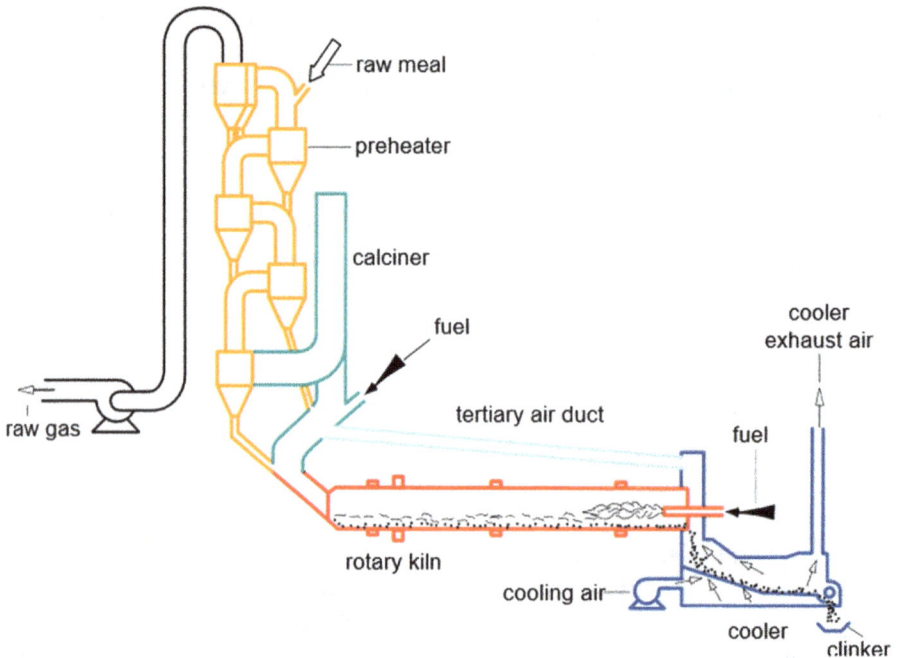

Figure 11.1: Rotary kiln with preheater, precalciner and clinker cooler [kindly supportet by VDZ].

the fuels increases, both the thermal energy demand increases (evaporation of the water) and the electrical energy demand (fan power for conveying the furnace exhaust gas). In the hot furnace atmosphere, water is also available as a binding partner for alkalis. Internal alkaline cycles are changed by the input of water [43].

If sulphur is introduced into the furnace system via the fuel, alkali sulphates or sulphides are formed. Due to their low evaporation temperature, these salts change into the gas phase. In the countercurrent process of the rotary kiln, they are then transported with the exhaust gas flow in the direction of the preheater and condense on the cold raw meal. (The basic design of a cement kiln is shown in Figure 11.1) In this way, the compounds are transported into the rotary kiln again and can evaporate into the gas phase. Similar to sulphur, chlorine also binds alkalis and leads to the formation of cycles. The vapour pressure of the alkali chlorides is even lower than that of the alkali sulphates. Beside the alkali cycles the alkali chlorides and sulphates introduced can condense in the refractory lining of the furnace and cause damage there [44].

A so-called gas bypass is often used to relieve alkali sulphate or chloride circuits. Here, part of the hot gas is removed from the process at the transition between the rotary kiln and the preheater and quenched. During this cooling process, these alkaline sulphur and chlorine compounds condense and can be separated by conventional filter technology. The cleaned exhaust gas is then returned to the heat exchanger.

4 Combustion in the precalciner

The most energy-intensive partial step during clinker burning is the decomposition of the calcium carbonate. This partial step takes place almost completely in the preheater (usually a cyclone preheater). Modern cement kiln systems are equipped with so-called precalciners. These allow thermal energy to be fed directly into the preheater. In systems with precalciners, up to 60% of the thermal energy can be introduced into the preheater via the precalciner.

Precalciners allow the combustion of large quantities of fuel without loading the rotary kiln with additional steam [23]. Over the past decades a number of different designs of precalciners have been developed. What they have in common is that the combustion air in the precalciner is supplied via so-called tertiary air lines. For this purpose, a partial flow of the oxygen-rich and preheated secondary air is branched off from the cooler and conducted to the precalciner. Another advantage of the technology is the possibility to use of unprocessed fuel or larger fuel particles [28].

5 Combustion in the rotary kiln firing

In rotary kiln plants without a calciner, most of the thermal energy is provided by the primary firing at the main burner Until the 1980s, burners were used for this purpose which were developed exclusively for the use of pulverized coal or for oil or natural gas. With these burners, only the primary air component could be varied. From the 1980s onwards, the first burners for co-combustion of secondary fuels were developed. Today the plant manufacturers offer special multi-channel burners, low-NOx burners and secondary fuel burners for the combustion of secondary fuels. These rotary kiln burners allow the feeding of different types of fuel simultaneously. In addition to the adjustment of the primary air volume, the air flows (e.g. axial or swirl air) can also be varied with these burners. When burning secondary fuels the aim is to keep these fuels in the gas phase as long as possible due to the slower combustion process. An optimised air flow enables a good burnout behaviour of these fuels. For energy reasons, kiln burners should operate with a minimum of primary air [26]. In some cases, in addition to these main burners, auxiliary burner lances are also used to feed fuels into the rotary kiln.

Compared to fossil fuels such as coal, secondary fuels often contain a higher volatile content. This means that the combustion of the secondary fuel cannot be understood as a direct oxidation. Instead, combustion takes place in various stages. After a stage of heating and drying, pyrolysis and outgassing of the volatile components takes place. During this time the carbon-rich residue of the fuel particle cannot burn, as the pyrolysis gas produced initially prevents oxygen from reaching the surface. Only after pyrolysis is completed the carbon-rich residue can oxidize and the ash formation begins [5, 23].

Irrespective of the burner technology used, it must be avoided that incompletely burnt fuel particles enter the clinker and are capsulated. In this manner granulated fuel particles can lead to locally reducing conditions in the clinker. As a result, Fe^{3+} can be reduced to Fe^{2+} or in extreme cases to metallic iron [45]. If fewer trivalent iron ions are available, the ferrite content decreases. As a result, less Al_2O_3 is bound in this phase and consequently, more reactive aluminate phase (C_3A) is formed [41, 46]. For this reason, clinkers from reducing firing tend to set rapidly as cement [47]. In addition, brown discolouration can occur in the processed end product (concrete).

In addition to the local reducing conditions, the oxygen content of the kiln atmosphere must also be taken into account. A low-oxygen atmosphere can reduce the existing sulphates to sulphites or sulphides. The alkali sulphates consequently decompose at low oxygen partial pressures. If the alkalis of the binding partners are missing, they are absorbed by the melt phase and incorporated into the aluminate [48].

Due to the changed burn-out behaviour of the secondary fuels, the sintering zone in rotary kilns changes. While classical coal firing plants lead to short sintering zones which are located near the kiln outlet, secondary fuels lead to longer sintering zones. In addition, the sintering zone shifts towards the kiln inlet. When secondary fuels are used, the residence time of the fuel in the sintering zone is consequently extended. Furthermore, the pre-cooling zone in the rotary kiln is extended. The formation of the clinker phases is directly influenced by the temperature profile of the kiln. Thus, larger alites and belite crystallites are formed in a longer sintering zone. A longer pre-cooling zone leads to a changed interstital phase structure [26, 29, 39, 49]. The effects of the changed temperature profile on the strength development of the corresponding cements cannot be evaluated independently of the fuel ash input. In general, it can be stated that the use of secondary fuels leads to slower strength development [50].

6 Influence of secondary fuels on clinker burning

Compared to fossil fuels, such as coal, secondary fuels often have significantly higher moisture contents. The evaporation of this water requires a higher energy input. In addition, the volume of gas to be passed through the furnace system increases. This reduces the furnace capacity [51]. In order to maintain a high kiln capacity despite high secondary fuel rates, fuel dryers operating with kiln waste heat are increasingly used [34, 52].

Compared to fossil fuels, secondary fuels, especially used tyres or sewage sludge, lead to an increased input of corresponding ashes. In some cases clinker-related ash inputs of over 5% are achieved. The composition of these ashes differs very clearly from that of raw meal or clinker. The consequence is a heterogeneity within the combustible material caused by secondary fuels [6].

The heavy metals introduced by secondary fuels often have a mineralizing effect or reduce the melting temperature. Alite formation in the rotary kiln is improved. Some of these elements can change the viscosity and surface tension of the melt and influence crystal growth. A reduction in the melt phase viscosity and an increase in the surface tension of the melt accelerate the conversion of free lime and belite to alite [53]. Especially elements with a small atomic radius or a high electronegativity reduce the free lime content of clinkers. Zinc, as introduced in particular by tyres, is a typical example of this [35]. A decrease in the required sintering temperature in the rotary kiln has also been demonstrated for other secondary fuels [49].

7 Influence of alternative fuels on the clinker and cement produced

Some elements that are introduced into the clinker via secondary fuels have a direct influence on the clinker and cement quality. Besides the trace elements and heavy metals described above heavy metals described above, this applies in particular to the element phosphorus. Phosphorus from MBM or from sewage sludge is almost completely absorbed in Alit and Belit. At the same time, the proportion of Al_2O_3 and Fe_2O_3 is increased and the proportion of SiO_2 is reduced [54]. Phosphorus stabilizes in particular belite. With high phorphorus contents in the clinker, the alite content is reduced accordingly. This influence directly reduces the strength of the corresponding cements [39]. In addition to this direct influence of secondary fuels on clinker and cement quality, there are a number of indirect interactions, which are described as follows.

If reducing conditions occur in the furnace system as described above or by granulation of incompletely burnt residues, Fe^{3+} ions are reduced to Fe^{2+} ions. These Fe^{2+} ions substitute calcium ions in the alite structure. During subsequent oxidation in the pre-cooling zone of the kiln or in the clinker cooler, the iron is reoxidized and separated from the alites as calcium ferrate (CaF_2). Due to the lack of iron bonding in the calcium aluminate ferrite (C_4AF), significantly more calcium aluminate C_3A is formed in these clinkers, which increases the reactivity of corresponding clinkers and cements [45].

Chlorine introduced via fuels increases the volatility of the alkalis. The alkali bond in C_3A is thus reduced. This also changes the reactivity of the clinker. Conversely, secondary fuels usually reduce the sulphur input (e.g. through coal), which reduces the volatility of the alkalis. The pre-cooling zone influenced by the changed flame profile also has a direct effect on the formation of the intermediate phase.

Surprisingly, an increase in the ferrite content is often observed when using aluminium-rich fuel [39]. High-ash fuels, which are fed into the rotary kiln via main or satellite burner lances, encounter partially formed clinker granules in the sintering zone, consisting of alite, belite and liquid melt. In all cases, the fuel ash has a lower lime saturation factor than the melting phase of the clinker. If the ash residues diffuse into the granules, the calcium content of the melt phase is reduced locally. When the granules cool down, the calcium-poor state can be frozen. Ferrite is the phase with the lowest calcium content of the four main clinker phases. For this reason, the aluminate/ferrite ratio can shift towards ferrite when there is a high ash input into the rotary kiln. The excess Al_2O_3 is incorporated within the ferrite and the A/F ratio within this phase increases [6].

Calcium depletion of the melt can also occur in the presence of free lime. Fuel ashes are often not homogeneously distributed in the clinker. Instead, these ashes accumulate on the surfaces of already formed granules. Due to the high surface/volume ratio of small clinker granules, the ash influence is particularly noticeable in these fractions. While the centres of large granules often have a higher lime standard, small granules or the surfaces of large granules are characterised by a reduced lime standard. Similarly, high and low free radical concentrations are produced locally in the clinker. Especially the ash-rich areas are characterised by the above-mentioned shift towards higher ferrite contents and an increase in the A/F ratio within the ferrite.

These mineralogical changes have a direct effect on the hydration behaviour of corresponding cements. If the aluminate content of the intermediate phase decreases due to fuel ashes and, conversely, the ferrite content increases, the clinker reactivity to water is reduced. Less ettringite or monophases are formed at early times. The water binding capacity of the clinker is thus reduced. This relationship can be transferred to corresponding cements. The higher excess water reduces the early heat of hydration and early strength of corresponding cements. Due to the lack of heat of hydration from the aluminate reaction, the autocatalytic self-heating of the mortar or concrete is also reduced, which also slows down the strength development of corresponding concretes.

References

[1]   Hoenig, V.; Müller C., Palm, S.; Reiners, J.; Fleiger, P.; Koring, K.: Energy efficiency in cement production; part 1. Cement International. 3/2013. pp. 50–67
[2]   VDZ: Cement International. 2/2020. p.11
[3]   VÖZ, „Emissionen aus Anlagen der österreichischen Zementindustrie", Berichtsjahr 2017.
[4]   Kirsch, J.: Umweltentlastung durch Verwertung von Sekundärbrennstoffen. Zement-Kalk-Gips. 12/1991. pp. 605–610.
[5]   Martens, H.; Goldmann, D.: Energetische Verwertung von festen Abfällen und Einsatz von Ersatzbrennstoffen. Recyclingtechnik. 2. Auflage. 2016. pp. 501–531.
[6]   Neumann, T.: Einfluss veränderter Brennbedingungen auf die Klinkerzwischenphase und die daraus resultierende Beeinflussung der Hydratation. Diss. MLU-Halle. 2019.
[7]   Achternbosch, M.; Bräutigam, K.-R.; Hartlieb, N.; Kupsch, C.; Richers, U.; Stemmermann, P.: Impact of the use of waste on trace element concentrations in cement and concrete. Waste Management & Research. 2005. S. 328–337.

[8]   Philipp, O.; Voigt, H. Wächtler, H. J.: Einfluß der Brennstoffarten auf die Ausbildung der
      Klinkerphasen. TIZ-Fachberichte. Vol. 107. No. 11. 1983.
[9]   Cordes, W.; et al.: Einfluss von Sekundärbrennstoffen auf die Klinkereigenschaften.
      Fachtagung Zementchemie. Düsseldorf. VDZ. 2011.
[10] Tsakalakis, K. G.: Einsatz von Altreifen in der Zementindustrie der EU – eine Betrachtung aus
      ökonomischer und Umweltsicht. ZEMENT-KALK-GIPS-International. 2007.
[11] Turnell, V.J.: Brennstoffwechsel bei Zementdrehöfen-Grundlagen und
      technische Möglichkeiten. part 1. ZKG-International. No. 4/2001. pp. 174–179. /
      part 2. ZKG-International. No. 7/2001. pp. 372–378.
[12] FGSV: Merkblatt über die Verwendung von Kraftwerksnebenprodukten im Straßenbau.
      FGSV. 2009.
[13] NRW: Leitfaden zur energetischen Verwertung von Abfällen in Zement-, Kalk- und Kraftwerken
      in Nordrhein-Westalen. 2. Auflage. Ministerium für Umwelt Naturschutz und
      Verbraucherschutz. 2005.
[14] VDS: Einsatz von Ersatzbrennstoffen in kohlebefeuerten Kraftwerken. Merkblatt zur
      Schadenverhütung. Hrsg. Gesamtverband der Deutschen Versicherungswirtschaft
      e. V. (GDV). 200.
[15] vom Berg, W.; et al.: Handbuch Flugasche im Beton. Verlag Bau + Technik. 2004.
[16] Fehrenbach, H.; et al.: Ergebnisbericht zum Forschungsvorhaben "Ökologische Bilanzen in
      der Abfallwirtschaft" Fallbeispiel: Verwertung von Altreifen. Inst. für Energie- und
      Umweltforschung Heidelberg GmbH. Auftraggeber Umweltbundesamt. 1997.
[17] Schmidthals, H: Luftvergasung von Altreifen zur integrierten stofflichen Nutzung im
      Klinkerprozeß. Dissertation. 2001. Ruhr-Universität Bochum.
[18] Naredi, R.: Using solid fuels in supplementary firing of rotary cement kils. ZKG-International.
      No. 4/1983. pp. 185–189.
[19] VDZ: Tätigkeitsbericht. 2012 – 2015.
[20] Weislehner, G.: Verbrennung von Altreifen bei Rohrbach-Zement in Dotternhausen.
      ZEMENT-KALK-GIPS. 1983. S 454–457.
[21] Own data, Cement plant Karlstadt, Germany.
[22] Cyr M, Ludmann C. Low risk meat and bone meal (MBM) bottom ash in mortars as sand
      replacement. Cement and Concrete Research. 2006. pp. 469–80.
[23] Larsen, M. B.: Alternative fuels in cement production – a new opportunity. Cement
      International. 2/2008. pp. 58–65.
[24] Chowdhury, S.; Mishra, M., Suganya, O.: The incorporation of wood waste ash as a
      partialcement replacement material for making structuralgrade concrete: An overview. Ain
      Shams Engineering Journal. Article in press. Download from: https://www.researchgate.net/
      publication/270163491.
[25] Krüger, O.; Adam, C.: Monitoring von Klärschlammmonoverbrennungsaschen hinsichtlich
      ihrer Zusammensetzung zur Ermittlung ihrer Rohstoffrückgewinnungspotentiale und zur
      Erstellung von Referenzmaterial für die Überwachungsanalytik. Umweltbundesamt. 2014.
[26] VDZ 2012 VDZ-Merkblatt Vt 17. Drehofenfeuerung für den Sekundärbrennstoffeinsatz. VDZ
      gGmbH. 2012.
[27] Sprung, S.; Rechenberg, W.: Schwermetallgehalte im Klinker und im Zement. ZEMENT-KALK-
      GIPS. Nr.5/1994. S. 258–263.
[28] Baier, H.; Menzel, K.: Alternative fuels in the cement industry. Zement-Kalk-Gips. 10/2011.
      S. 50–58
[29] Ono, Y.: Microscopical observation of clinker, for estimation of burning condition, grindability
      and hydraulic activity. Proceedings of the 3rd International Conference on cement
      microscopy. Houston/ Texas. 1981. S. 198–210.

[30] VDZ-Tierm: Bewertung des Einsatzes von Tiermehlen und Tierfetten in Drehofenanlagen der Zementindustrie. Verein Deutscher Zementwerke. 2001.

[31] VDZ: Zahlen und Daten 2020.

[32] Baier, H.: Proven experiences with alternative fuels in cement kilns. Proceedings of 13th International Conference on the Chemistry of Cement. 2011. Madrid.

[33] Saint-Jean, S. J.; Jøns, E.; Lundgaard, N.; Hansen, S.: Chlorellestadite in preheater system of cement kilns as an indicator of HCl formation. Cement and Concrete Research 35. 2005. pp.431–437.

[34] Trenkwalder, J.; Schwörer, W.: BGS-Trocknung in den Zementwerken Bernburg und Karlstadt. VDZ-Fachtagung Zementverfahrenstechnik. Düsseldorf. 12. Februar 2020.

[35] Krivoborodov, Y.; Samchenko, S.; Burlov, I.; Kouznetsova T.: Influence of alternative fuel on cement qualitiy. Proceedings of 14th International Conference on the Chemistry of Cement. Beijing. 2015.

[36] Knöfel, D.: Beeinflussung einiger Eigenschaften des Portlandzementklinkers und des Portlandzementes durch ZnO und ZnS. ZEMENT-KALK-GIPS. 1978. S 157–161

[37] Müller, U.; Rübner K.: The microstructure of concrete made with municipal bottom ash as an aggregate component. Cement and Concrete Research. 36.2005. pp. 1434–1443.

[38] Shih, P.-H.; Chang, J.-E.; Chiang, L.-C.: Replacement of raw mix in cement production by municipal solid waste incineration ash. Cement and Concrete Research 33. 2003. pp. 1831–1836.

[39] Klaska, R.; Baetzner, S.; Möller, H.; Paul, M.; Roppelt, T.: Auswirkungen von Sekundärbrennstoffen auf die Klinkermineralogie. Cement International. 4/2003. S. 88–98.

[40] Aubert, J.E.; Husson, B.; Vaquier, A.: Use of municipal solid waste incineration fly ash in concrete. Cement and Concrete Research 34. 2004. pp. 957–963.

[41] Singh, N.; Turner, A.: Trace metals in antifouling paint particles and their heterogeneous contamination of coastel sediments. Marine Pollution Bulletin 58. 2009. pp. 559–554.

[42] Bundesimmisionsschutzverordnung – BimSchV.

[43] Enders, M.; Haeseli, U.: Reactions of alkalis, chlorine and sulfur during clinker production. Cement International. 3/2011. S. 38–53.

[44] Klischat, H.-J.; Liever, H.: Innovative refractory materials for the use of secondary fuels in the cement industry. Cement International. 4/2003. pp. 78–87.

[45] Klauß, J.: Reduzierendes Brennen von Zementklinker im Drehofen, ein Weg zur Senkung der NOx-Emissionen. ZEMENT-KALK-GIPS. Nr.3/2000. S. 132–144.

[46] Locher, F.W.; Richartz, W.; Sprung, S.; Sylla, H.-M.: Erstarren von Zement. Teil III: Einfluß der Klinkerherstellung. ZKG-International. 12/1982. pp. 669–676.

[47] Sylla, H.-M.: Einfluß reduzierenden Brennens auf die Eigenschaften des Zementklinkers. ZEMENT-KALK-GIPS. Nr. 12/ 1981. S.618–630.

[48] Long, G.R.: The effect of burning environment on the microscopic characteristics of Portland cement. Proceedings of th 4th International Conference on Cement Microscopy. Las Vegas. 1982. S. 128–140.

[49] Knöpfelmacher, A.: Beurteilung des Klinkerbrennprozesses mittels röntgenografischer und mikroskopischer Analysemethoden. Fachtagung „Zementchemie" Düsseldorf, 30. März 2017.

[50] Möller, H.: Control of product quality during the use of secondary materials. 6th International VDZ Congress. Düsseldorf. 2009.

[51] Krennbauer, F.: Secondary fuels and their influence on the cement burning prosess. ZKG International. No. 5/ 2006. pp. 63–71.

[52] Trenkwalder, H.: Waste heat recovery for drying of sewage sludge. 6th Int. VDZ-Congress. Düsseldorf. 2009. pp.176–180.

[53] Timashev, V.V.: The kinetics of clinker formation. The structure and composition of clinker and its phases. Proceedings of the 6th International Congress on the Chemistry of Cement. Moskau. 1974. I-3.

[54] Ifka, T.; Palou, M.; Baraček, J.; Šoukal, F.; Boháč, M.: Evaluation of P2O5 distribution inside the main clinker minerals by the application of EPMA method. Cement and Concrete Research. 2014. S. 147–154.

[55] Sieber, R.; Pöllmann, H.; Brunner, P.: Investigations on clinker, microstructure, hydration characteristics and quality coming from varying burning conditions in the kiln using waste fuels. Proceedings of the 18th International Conference on Cement Microscopy. Houston/Texas. 1996. S. 284–303.

Kai Tandon, Soraya Heuss-Aßbichler

Chapter 12
Fly ash from municipal solid waste Incineration: from industrial residue to resource for zinc

Abstract: Since the beginning of human existence, waste has been constantly produced. With industrialization, waste generation has been steadily increased worldwide, and at an alarming rate. By 2050, waste generation is estimated to increase by about 50%.

Originally, incineration of waste was used to tackle epidemics; today, it is a convenient strategy to manage the ever-growing amount of waste. Modern municipal waste is therefore classified as a substitute fuel due to its high calorific value (caused by the high proportion of plastics). The incineration process reduces mass and volume by 70 wt.-% and 90 vol.-%, respectively, and recovers the heat transported with the exhaust gases as energy. The main products of waste incineration are bottom ash (20-30 wt%), boiler ash (0,4 wt%) and residues of the flue gas treatment (FGT) including Fly ash (1-5 wt%) and air pollution control (APC) residues. FGT residues are considered hazardous waste due to the high content of (volatile) heavy metals, easily soluble salts and persistent organic pollutants and are disposed of in special landfills. Often, MSWI fly ash is treated and/or processed before disposal, e.g., through stabilization or solidification.

Various methods have been developed to utilize fly ash: some approaches aim to use fly ash as a raw material substitute (e.g., as an additive in cement) in order to save landfill space. However, these methods are not effective, because the ash negatively affects the properties of the respective products; moreover, the metals in the fly ash dissipate and are thus lost as a resource. Fly ash is mainly enriched in Zn (3,000 up to 70,000 mg/kg Zn) and can reach the content of natural ore deposits. The variation of the Zn content depends on the flue gas treatment (FGT) procedure. The addition of Ca as milk of lime during the flue gas treatment reduces the content of all heavy metals contained in the fly ash by dilution effect.

Acknowledgements: The authors want to thank MSWI plant Ingolstadt for providing the materials and their support, special thanks to Michael Funk. Furthermore, we want to thank Tracy Phillips for proof-reading.

Kai Tandon, Soraya Heuss-Aßbichler, Section Mineralogy, Petrology & Geochemistry, Department of Earth and Environmental Sciences, Ludwig-Maximilians-Universität München, Theresienstr. 41, 80333 Munich, Germany

https://doi.org/10.1515/9783110674941-012

Only a few processes focus on the recovery of metals, especially Zn, as a resource. As a first step, metals are leached out of the fly ash. In acidic fly ash washing (FLUWA), the acidic scrub water from a wet flue gas treatment is combined with the fly ash resulting in a metal-enriched Cl-rich solution. In a second step, the novel FLUREC process is used to recover zinc as a metal with high purity from MSWI fly ash. However, the process with low environmental impact is rather complex and only economically feasible if the concentration of Zn is above 40.000 mg/kg Zn in MSWI fly ash. An alternative process under development, called SPOP, is based on chemical precipitation. It aims to enrich zinc in the residue while keeping the Cl concentration in the solid low, which is a prerequisite for recycling the precipitation residues.

Keywords: Municipal solid waste incineration, fly ash, recycling, circular economy, heavy metals

1 Waste treatment throughout time – the example of municipal solid waste incineration

Municipal solid waste (MSW) and the way it is treated reflect the history of human development and their habits and customs [1, 2]. Waste is generated daily all over the world, and therefore its treatment was essential from the very beginning [1]. Archaeologists can draw conclusions about the culture and technology of a particular society from the composition of the waste or the absence of certain components in it [2, 3]. In the 14th and 15th centuries, the disposal of waste dumped in open passageways was an unresolved problem in rapidly growing cities. It became essential in the 19th century, when repeated epidemics and diseases of industrialised societies made the improvement of public health increasingly urgent. In addition to the construction of water treatment and sewage systems, the need for an appropriate handling of waste was seen as imperative [4–6].

A significant change occurred after the Second World War. With the economic boom and technological progress, the rapidly growing population consumed steadily more, and in consequence, the volume of waste increased enormously [1, 5]. On the one hand, the demand for new landfill sites raised. On the other hand, the resistance of the population against new landfills grew. For protection of the environment and the population, many legislations were introduced [1]. The waste hierarchy (3R: reduce, reuse, recycle) and the waste directive were adopted to reduce the waste volume and to ensure the safety of environment and health through a proper treatment of waste [7, 8]. In the last years, security of resource supply and thus recovery of valuable elements from waste has become increasingly important. In 2015, agenda 2030 and the 17 sustainable development goals (SDGs) were adopted as a transnational guideline to sustainability [9]. A key aspect of the SDGs is the realisation of sustainable development by taking into account economic, social and environmental aspects. Waste Management is no longer just a question of providing disposal security and pollution control. It is regarded as essential for fostering a circular economy, as the recovery of raw materials from waste should contribute to a sustainable future [6, 10]. However, this change is a long and complex process. In Germany, one of the pioneer countries in recycling [11], only 14% of the raw materials used originate from recycling processes [12]. In comparison, 9% of the global economy is circular [13].

Treatment of the residues after incineration reflects the changing needs of the modern society. Waste incineration was introduced at the end of the 1880s, primarily for hygienic reasons. The significant reduction in the volume of waste was also swiftly perceived as a major advantage of incineration [2, 5]. The stench and air pollution which were caused by the incineration at that time were extreme, resulting in a high resistance to incineration [1]. Another disadvantage was the low calorific value of the waste, because at the beginning of the 20th century, it consisted of 60% of ash, 20–25% organic waste and only small amounts of paper, paperboard, glass and metals [1]. Despite the poor calorific value, initial attempts were made to recover the heat released during combustion [3, 5]. Due to high operation costs, recurring technical issues and other reasons, waste incineration was not widespread in many European cities until the middle of 20th century [1]. The construction of MSWI plants was promoted to cope with the increased waste volume. Changing lifestyles led to a change in the composition of waste. The growing proportion of plastics in waste and the resulting increase in the calorific value made heat recovery more attractive [1, 3, 5], and various types of waste-to-energy power plants were constructed [14]. The acceptance of the population was achieved by the development of a sophisticated air pollution control system for all MSWI plants [2, 3, 5].

At the same time, more attention was paid to the residues after waste incineration – bottom ash, boiler ash, fly ash and the remains from air pollution control – since, against initial expectations, the residues are not inert materials. They all contain a considerable amount of heavy metals, salts and other pollutants, that are harmful to the human health and environment, e.g., [3, 15, 16]. Studies also showed that fresh bottom ash can produce heat [17] and release H_2 [18]. First concepts about the utilization of MSWI residues, mainly of the bottom ash, were developed. In general, after quenching the hot bottom ash with water, separation and recycling of the non-ferrous and ferrous metals occurs [19–21]. Depending on the region, the remains of this process are landfilled or used as backfill or road construction material. At the same time, alternatives to the use of bottom ash were sought. Because of its similarity to gravel, it is partly applied as a substitute for construction materials or as additive in the cement industry, e.g., [22], while only few applications are established for fly ash and air pollution control (APC) residues. Fly ashes are mainly disposed of in landfills for hazardous waste because of their toxicity and easily soluble components.

Following, we present an overview of the composition of municipal solid waste and how the living standards affect it (chapter 2). Next, we give a short introduction into the principles of a MSWI plant and the types of residues obtained after combustion of waste (chapter 3). Chapter 4 gives an overview of the methods introduced or proposed to treat the residues with focus on fly ash. Zinc is the most abundant metal in MSWI fly ash [23, 24]. How much zinc is concentrated in the ashes in various countries and how the flue-gas treatment affects the concentration will be discussed in chapter 5. Finally, we present new concepts developed to recover the metals in fly ashes with Zn as an example for closing the loop (chapters 6 & 7).

2 Composition of municipal solid waste (MSW)

[T]he world is on a trajectory where waste generation will drastically outpace population growth by more than double by 2050. Although we are seeing improvements and innovations in solid waste management globally, it is a complex issue and one that we need to take urgent action on [25].

It is expected that waste generation will increase from 2.02 billion in 2016 to 3.4 billion tons in 2050. Therefore, various studies highlight the importance of an environmentally sound waste treatment on global scale e.g., [25]. Low-income countries, where only 5% of the world's waste is currently produced [25], will contribute more and more to the fast-growing waste come up due to increasing wealth, urbanization and population growth [6, 25]. The German Advisory Council on Global Change defines the 21st century to be the century of cities and estimates that by 2050, around two thirds of the world-wide population will live in cities [26]. Especially in these urban areas, tremendous amounts of municipal solid waste are generated every year. Main producers are households, including commercial wastes.

In general, municipal solid waste is a very inhomogeneous material and can consist of paper, paperboard, plastics, glass, non-ferrous metals, iron, food, garden waste, rubber, leather, textiles, various composite materials, mineral wastes and others [20, 27, 28] changing from day to day and from season to season [29]. The composition of municipal solid waste also depends strongly on the income level, as can be seen in Figure 12.1.

In lower to middle income countries, the proportion of food and green waste with 50 vol.-% is by far the highest. Although the total amount of organic waste is more or less the same in all countries, the percentage in high income level countries is comparably lower, due to a higher proportion of plastics and other inorganic wastes [25]. The higher the income level, the more recyclable materials (paper, cardboard, plastic, glass, metal) are in the waste, with a maximum of 50% in high income countries.

During the last decades, waste prevention legislations (waste directive and waste hierarchy) were introduced, and a sophisticated waste collection system helped to decrease the amount of waste that finally has to be treated. In the last 20 years, incineration has been accepted in the industrial nations as an effective contribution to waste management, including energy recovery during the thermal treatment of waste. The composition of the waste as an extremely heterogeneous material influences the composition of the residues after incineration.

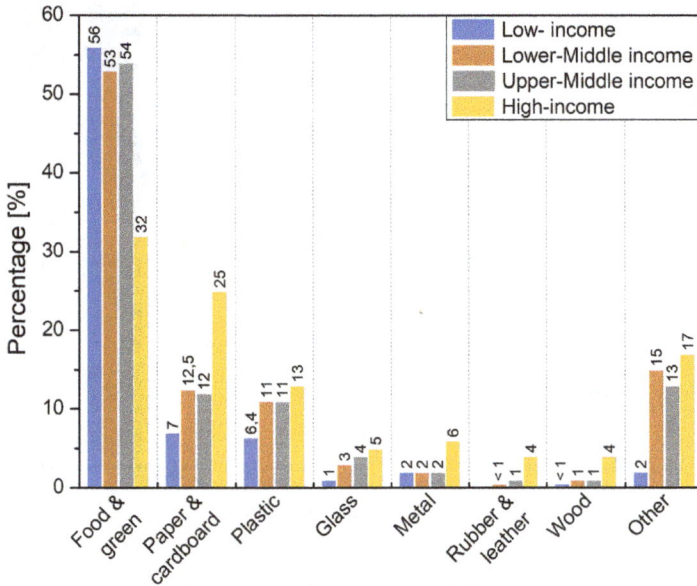

Figure 12.1: Global waste composition in dependence of income level of countries in 2016, after [25].

3 Principle of MSW incineration plant and types of MSWI residues

The incineration process reduces the mass and the volume by 70 wt.-% and 90 vol.-%, respectively [10, 27, 29, 30]. Figure 12.2. shows a general scheme of a municipal solid waste incineration (MSWI) plant. Waste is unloaded in the waste bunker, homogenized and stored several days to decrease the moisture which may vary between 20–30 vol.-%. Afterwards, it is fed into the combustion chamber by a crane. The incineration temperature must be above 850 °C in the combustion chamber for at least 2 seconds to completely destroy persistent organic pollutants (POP's), e.g., dioxins (polychlorinated dibenzo-p-dioxins, PCDDs) and furans (polychlorinated dibenzofurans, PDCFs) [31]. Main products of waste incineration are bottom ash, boiler ash and residues of the flue gas treatment. Before the flue gas is released to the atmosphere, hazardous components and pollutants such as ash and dust particles, acid gases (e.g., HCl, HF, SO_X, NO_X) as well as organic pollutants (dioxins, furans, CO, hydrocarbon) are removed. A sophisticated technology recovers the heat transported with the exhaust gases as energy. Fly ash and flue gas treatment residues are considered as pollutant sink. They are classified as hazardous waste due to their toxic components.

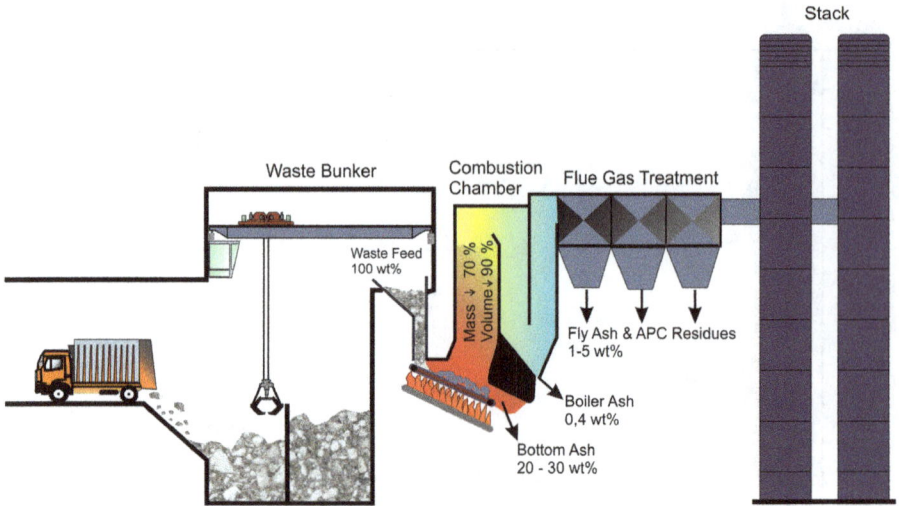

Figure 12.2: General scheme of a municipal solid waste incineration plant.

Bottom ash with 20–30 wt.-% is the main product of the MSW incineration process [15, 27, 31–33] and is collected in a quenching/cooling tank after the combustion chamber [15]. Main components of the bottom ash are vitrified particles as well as partly coarse-grained materials, which pass the combustion chamber without any changes, e.g., glass, ceramics, stones, metals as well as unburnt organic matter. Heavy metals are especially enriched in the finer fraction of the bottom ash [20, 31].

Boiler ash is only partially entrained with the flue gas and remains in the boiler. It represents 0.4 wt.-% of the original waste input and is usually treated together with the fly ash [27, 32].

Residues of the flue gas treatment (FGT) consist primarily of pollutants in the gaseous phase which are trapped in solid particles after the cleaning process, including fly ash and air pollution control (APC) residues. Accordingly, these residues are enriched in heavy metals, salts and organic micro-pollutants [15].

Fly ash makes up about 1–5 wt.-% of the initial waste input [27, 31, 32, 34] and is smaller in size than bottom ash. The dust and ash particles generally consist of amorphous phases (silicate glass), oxides and chlorides that are trapped by the flue gas and pass the heat recovery units [15], as shown in Figure 12.2. Volatile heavy metals are especially enriched in fly ash due to their low boiling point [35]. They either condense on the surface of the ash particles or form new, non-volatile species during cooling on the way from the combustion chamber to the flue gas treatment unit [15, 36–38]. In the first stage of the gas cleaning process, the fly ash is removed by bag filters or electrostatic precipitators.

APC residues are generated in the next stage during the flue gas treatment of the remaining acid gases and organic pollutants [27]. Basically, there are two different techniques: (1) dry and semi-dry or (2) wet FGT. (1) In **dry** and **semi-dry** FGT, sorbents are fed into the flue gas stream to compensate for the acidity introduced by the gases. During dry FGT, coke, hydrated lime or sodium bicarbonate with high-porosity are injected as powder. In case of semi-dry FGT, the sorbent is an aqueous solution, i.e. milk of lime [27]. By spraying it into the flue gas stream, the solution evaporates. Semi-dry methods are usually more efficient, because the addition of the solution is easier to control and therefore less excess of alkaline reagents is needed, which decreases the total volume of APC-residue. In both cases the resulting residues are dry. (2) The **wet** treatment process differs significantly. Flue gas is injected into a scrubber that usually contains water. There, the acid gases dissolve and a strongly acidic solution results (pH 0–1), especially enriched in Cl. For efficient flue gas cleaning, parts of the scrubber water must be constantly exchanged, and the very acidic wastewater has to be treated. Usual methods are either neutralizing the wastewater with an alkaline agent (i.e. milk of lime) and disposal of the solid remains or spraying the wastewater back into the flue gas for evaporation in order to avoid wastewater treatment [27].

The distribution of elements (fractionation), and in particular the heavy metals in the respective incineration products depends on various parameters, which are [27]:
- combustion conditions (temperature, oxygen supply, furnace type, residence time)
- Cl-concentration in the waste
- flue-gas treatment (FGT) system

One main factor for the element distribution is the furnace technology, as it affects temperature, oxygen supply, and residence time in the combustion chamber.

In general, the waste composition determines the total sum of elements that can be found in the respective residues after the incineration process. Material flow analyses of the individual MSWI components were performed, with focus on the heavy metals, e.g., [29, 31, 37]. Belevi & Mönch [32] conducted a detailed study on this speciation. Figure 12.3. shows that about 79 wt.-% of the initial waste input is transferred to the exhaust gas, 0.36 wt.-% to the boiler ash, 1.9 wt.-% to the FGT-residues and about 19 wt.-% to the bottom ash.

Looking at the fractionation of heavy metals between the different types of MSWI residues, the major part (>80 wt.-%) of Co, Cu, Cr, Fe, Mn, Mo and Ni, in addition to critical elements Ba, Li, P and Ti, is concentrated in the bottom ash. It contains also the main part (>50 wt.-%) of As, and only minor parts (<25 wt.-%) of Cd, Pb, Sb, Sn and Zn. The rest of these elements are distributed over the boiler ash and

FGT-residues. It is noteworthy that the main fractions of the heavy metals (Cd, Sb, Sn, Zn) and the halogens (Br, Cl, F) including S are concentrated in the FGT residues, whereas Pb and Cd are more evenly distributed over all types of MSWI residues. The highly toxic and volatile Hg is mainly found in the exhaust gas and the FGT residues. In case of wet FGT, it is treated separately due to its hazardousness.

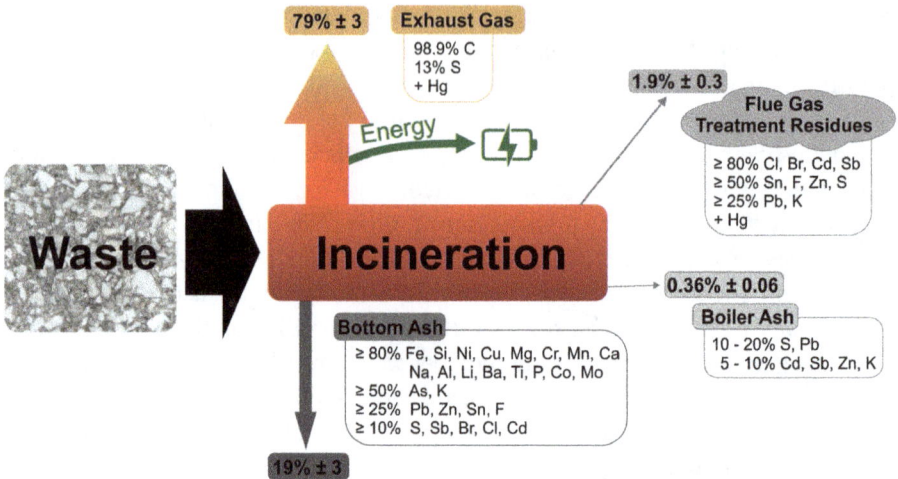

Figure 12.3: Mass transfer and fractionation of various elements in the exhaust gas, flue gas treatment residues, boiler and bottom ash after incineration of waste in wt.-%. The graph is based on numbers from [32].

As shown in Figure 12.3, highly volatile elements are enriched in the fly ash, while less volatile elements are accumulated in the bottom ash [31]: with increasing furnace temperature, the concentration of Zn, Cu, Cd, Pb and Sn in the flue gas treatment residues rises [31, 32], depending on the melting and boiling point of the respective metal (Me)-compounds. Decreasing or low oxygen supply during incineration leads to an enrichment of Sn and As and a depletion of Zn, Cu, Mo and Sb in the fly ash. The oxidation conditions have only little effect on the fractionation of Cd and Pb [32].

The concentration of Cl in the waste has an enormous impact on the fractionation of Zn, Cu, Cd, Pb, Sn, Mo, Sb, As; increasing Cl contents promote the accumulation of these metals in the fly ash [32]. This can be explained by the low boiling point of the respective Me-chlorides [37, 39] and hence, a higher vapor pressure compared to the Me-oxides [29]. The FGT system determines the distribution of the heavy metals in the flue gas cleaning residues.

4 Treatment and utilization of MSWI fly ash

MSWI fly ash is considered the most hazardous fraction of the residues after the combustion process, because of its high content of potentially toxic pollutants such as heavy metals, easily soluble salts and POPs [39–41]. According to the European Waste Framework Directive 2008/98/EC [7] (article 7, list of waste), it is classified as hazardous, marked by waste codes EWC 19 01 13* (fly ash containing hazardous substances) apart from EWC 19 01 14 (fly ash other than those mentioned in 19 01 13*).

In the course of time, various methods have been developed and proposed for processing and/or treatment of MSWI fly ash. In principle, these techniques can be distinguished according to their objective: 1) treatment and/or disposal, 2) reuse in new products.

1) Treatment and disposal

Safe disposal of fly ash and APC residues is a priority in order to prevent the dissemination of pollutants into the environment, as salts together with heavy metals, mainly Pb and Zn, can be leached upon contact with water [42]. The solubility of heavy metals is dependent on pH; due to the amphoteric properties of the metals, the concentrations may increase strongly at both high and low pH values [43]. Fly ash is mostly filled into "big bags" stored e.g., in parts of salt-mines where backfilling is required [44], as shown in Figure 12.4, or underground caverns to avoid contact with water and for stabilization [41].

Various treatment methods have been developed to fix the pollutants before landfilling with the aim to minimize the quantity of hazardous waste and thus to reduce the disposal costs [31, 40]. The most used method worldwide for the entrapment of heavy metals in MSWI fly ash is the stabilization/solidification (S/S) process [27, 31, 40, 43, 45]. This process utilizes chemical or solidifying agents such as cements, iron oxides, phosphates, silicates or organic acids to fix the pollutants in the matrix [31]. Portland cement is the most commonly used binder. In this case, the fly ash is mixed with cement and water. This increases the weight and volume of the residue [42] that finally has to be disposed of.

By washing the fly ash before S/S processes with neutral or basic water and thus removing soluble salts containing heavy metals, the long-time stability of the cement can be increased, and its toxicity reduced [42]. This effect is further enhanced by washing the fly ash with acids, as this hydrometallurgical process promotes the dissolution of most metals in the fly ash [46]. Another possibility for treating MSWI fly ashes are thermal methods such as sintering, melting or vitrification by using additives; all these methods have a high energy demand [41].

Figure 12.4: Utilization of FGT residues stored in big bags for stabilization in parts of salt-mines where backfilling is required.

2) Use as product phases

Fly ash from municipal solid waste can be exploited either by using it directly or by improving some material specifications to manufacture a product, e.g., [23]. Quina et al. [41] presented in their paper the following possible applications of MSWI fly ash: ceramic materials, glass-ceramics, cement, secondary building materials, epoxy composites, zeolite-like material, adsorbent and thermal energy storage materials.

Various methods have been proposed for the treatment of fly ash, such as separation techniques (based on magnetic properties, size or density), washing, carbonation or stabilization, to improve the technical properties of fly ash as a product or additive and to reduce the hazard potential. Washing, for example, removes highly soluble heavy metals with the salts. In most cases, however, fly ash is added to the actual raw materials only in small quantities in order to avoid a negative influence on the chemical or mechanical properties. For example, only 1–10 wt.-% fly ash can be added for the production of ceramics [47–50]. Even an addition of 5 wt.-% fly ash has a significant effect on the properties of the ceramics, among others the heavy metals contained in the ashes can be released to such an extent that local regulations cannot be met [50]. The situation is similar in cement production. Between 5–15 wt.-% [51] and 5–35% of dry mass of the fly ashes [52], washed and unwashed, were used for clinker production. With increasing addition of fly ash, the mechanical properties decrease dramatically. A substitution of 35 wt.-% clinker with fly ash lead to a negligible compressive strength of the cement after 28 days [52]. Moreover, it is not quite clear to what extent the heavy metals can be released through contact with water. A similar observation was made in relation to the use of fly ash in brick production, where only 2.5–7.5 wt.-% ash was added. With the increasing amount of added ash, the H_2O absorption in bricks increases, too, and the compressive strength decreases.

5 Fly ash – an anthropogenic resource for Zn?

As shown in the previous chapter, fly ashes are mainly disposed of in landfills for hazardous waste because of their toxicity and soluble components. In consequence, the metals they contain are lost as a resource. Recently, waste is increasingly discussed as a potential source for valuable compounds. Especially Zn is enriched in the fly ashes. Table 12.1. shows the Zn-concentration in mg/kg of MSWI fly ash from different countries. It shows, that the Zn-concentration in MSWI fly ash varies significantly, from 3.000 mg/kg up to 70.000 mg/kg Zn. The values differ even within a country. The Ca-content can be by a factor 10 higher than Zn, varying between 58.000 mg/kg and 410.000 mg/kg Ca. High Ca contents are caused by addition of milk of lime during wet or semi-dry/dry FGT.

Table 12.1: Zn- and Ca-concentrations of selected literature in mg/kg. It is marked, if Ca was added during the FGT (yes, no, no Info). The values were differentiated by country.

Country	Zn [mg/kg]	Ca [mg/kg]	Ca-Addition	Reference
Denmark	31.700	181.000	no	[53]
	9.100	331.000	yes	
Italy	13.293	58.409	yes	[43]
Portugal	5.826	243.300	yes	[54]
	4.308	287.300	yes	
	6.367	349.600	yes	
	4.939	361.600	yes	
Sweden	5.500	315.000	yes	[55]
	3.300	410.000	yes	
	5.800	360.000	yes	
	26.000	210.000	yes	
	55.000	120.000	no	

Table 12.1 (continued)

Country	Zn [mg/kg]	Ca [mg/kg]	Ca-Addition	Reference
Suisse	22.000	184.000	no	[32]
	26.770	191.500	no	[56]
	48.190	168.900	no	
	65.170	151.150	no	
	71.190	143.814	no	[23]
	69.345	140.266	no	
	51.940	176.698	no	
	61.348	170.870	no	
	62.182	164.105	no	
Taiwan	9.620	337.660	no info	[57]
UK	5.800	260.000	yes	[35]
	7.300	270.000	yes	
	2.600	224.000	yes	
	7.300	240.000	yes	
	7.100	270.000	yes	
	7.100	240.000	yes	
	6.600	230.000	yes	
	3.700	320.000	Yes	
Germany	49.100	124.100	no info	[58]
	39.341	164.602	no	[24]
	28.131	128.259	no	
	33.955	152.039	no	
	9.831	231.367	no	
	8.560	302.880	yes	
	13.126	110.138	no	

The correlation between Zn-concentrations in fly ashes and Ca-concentration shows Figure 12.5, based on the data listed in Table 12.1. Pure fly ash is generally enriched in Zn (>20.000 mg/kg Zn). In general, the Ca-concentration – except for one sample – is always <200.000 mg/kg Ca. The effect of adding Ca as milk of lime during semi-dry/dry treatment is evident as it leads to the opposite trend: in this case the Zn concentrations are generally low (<10.000 mg/kg Zn), while the samples are enriched in Ca (>200.000 mg/kg Ca). The dilution leads not only to a decrease of Zn content, but also of all other heavy metals contained in the MSWI fly ash, a trend which is also observed during a long-time study of Haberl et al. 2018 [24].

In two cases (red triangles (1) and (2) in Figure 12.5.), no information was given by the authors regarding the treatment process. However, based on the Ca concentration, it can be clearly assigned that sample (1) represents an example of dry/semi-dry treatment and sample (2) a pure fly ash. The most significant deviation from this trend shows sample (3), where Ca was added according to the authors [43]. The relatively low values of Zn and Ca is possibly the effect of incomplete digestion of the sample and/or choice of the analysis method. Insufficient combustion conditions during incineration can be excluded because of the very low Ca content of the sample. In total, these analyses show that the concentration of heavy metals is high enough to be regarded as a potential source for Zn.

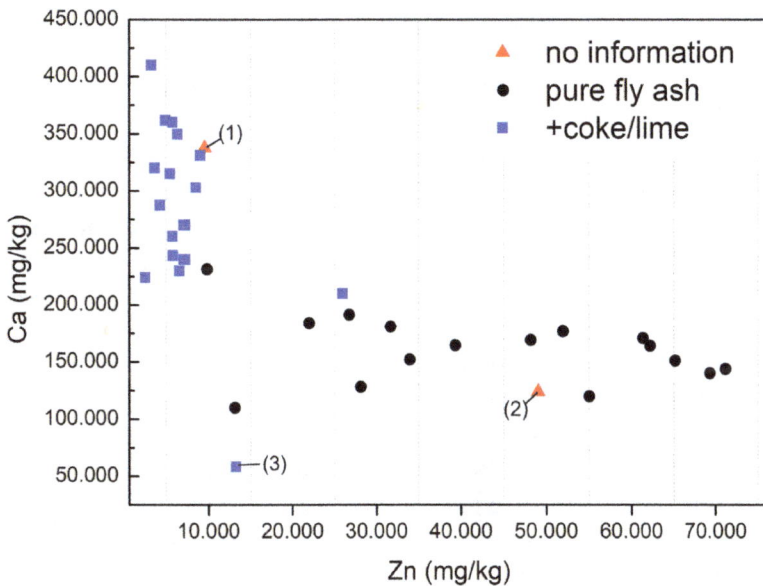

Figure 12.5: Zinc-concentration as a function of Calcium-concentration in mg/kg in MSWI fly ash in various countries. The symbols indicate whether milk of lime was added (blue square) during FGT or not (grey circle), or if no information was available (red triangle). Data is based on the literature listed in Table 12.1.

6 Recovery of metals from MSWI fly ash

Various concepts were developed with the aim to treat and utilize MSWI residues, in particular focused on bottom ash e.g., [59]. However, fly ashes from MSWI should also be taken into account because of its high zinc content, which may reach the grade of natural ore deposits [60]. In recent decades, increased research has been conducted into how elements from MSWI fly ash can be recovered. In a review article, Quina et al. [41] stated that not only the recovery of Zn, but also Cu, P, other (precious) metals and rare earth elements are in focus of various treatment processes.

As was shown before, MSWI fly ash is very inhomogeneous, and the composition varies significantly day by day and season by season (see Chapter 3). Therefore, processes for the potential recovery of zinc from fly ash have to be very robust to handle the aforementioned fluctuations in chemistry. In the following, two methods are presented, focusing 1) on the leaching of metals from MSWI fly ash (e.g., FLUWA), and 2) recovering Zn from MSWI fly ash (e.g., FLUREC and SPOP).

6.1 Leaching of metals from MSWI fly ash

There are two different types of studies that deal with the "accessibility" of heavy metals in fly ash: Most of the studies were carried out to test the long-term stability of the fly ashes in contact with water or acids, with the aim to predict the potential risk to the environment and thus making recommendations for avoiding environmental pollution, e.g., [46, 61].

Recent studies focus on optimizing the extraction of metals from fly ash with the aim to recover the metals contained by hydrometallurgical processes, e.g., [36, 62–65]. In general, highest leaching rates for heavy metals are obtained by using acids [62, 64]. These processes are strongly dependent on the extraction conditions, such as type of acid, addition of oxidizing agent(s), pH, L/S ratio, extraction temperature and time. For example, Tang et al. [64] achieved a yield of 70% Cu, 80% Zn, 93% Cd and 35% Pb under optimal conditions using HCl, pH 2, a L/S ratio of 20, a temperature of 20 °C and 20 h leaching time.

The acidic washing of MSWI fly ash is already in large-scale use in Switzerland since 1997 [65, 66]. This so-called **FLUWA process** has been further developed over the years and is now state of the art in more than 60% of waste incineration plants in Switzerland [41, 65]. Prerequisites for the FLUWA process are MSWI plants equipped with a wet FGT system.

In the first step of the process, the alkaline fly ash is mixed with the (Hg depleted) acidic flue-gas-scrub water produced in the wet FGT process and then remains in a three-stage tank cascade for 60 minutes. Table 12.2 shows a typical elute composition obtained after the FLUWA process [66]. One parameter to optimize the leaching conditions is the addition of H_2O_2 as an oxidizing agent to increase the yield of all metals, especially of the redox sensitive metal Cu. Weibel et al. [67] obtained an optimal yield of 55% Cu, 68% Zn, 97% Cd and 62% Pb by using HCl 5% at pH 2.5, 60 °C and addition of 30L H_2O_2/t fly ash. The extraction of metals can be further enhanced by a second leaching step using concentrated NaCl-solution (NaCl 300 g/l, pH 3.5, 85 °C) leading to a yield up to 81% Cu, 78% Zn, 100% Cd and 98% Pb [67].

Table 12.2: Typical composition of the elute after FLUWA process. Values after [66].

Element	Concentration	Element	Concentration
Cl [mg/l]	30.000–70.000	Zn [mg/l]	4.000–12.000
F [mg/l]	200–600	Pb [mg/l]	500–2.000
SO4 [mg/l]	1.000–5.000	Cu [mg/l]	20–300
Ca [mg/l]	3.000–12.000	Cd [mg/l]	40–200
Mg [mg/l]	500–2.000	Hg [mg/l]	0.003–0.01
		Fe [mg/l]	<5
		Ni [mg/l]	<1

The addition of H_2O_2 promotes the oxidation of metals, e.g., Fe^{2+} to Fe^{3+}, and promotes the precipitation of Fe-hydroxide which can be removed before further treatment [65]. The solids in the suspension are then separated from the liquid by using a vacuum-belt-filter, and the filter cake depleted in toxic heavy metals and salts can be treated together with the bottom ash [41]. After the FLUWA process, the metal(s) in the filtrate are usually neutralized with milk of lime, as a state-of-the-art treatment of heavy metals containing wastewater which is an easy-to-apply and cheap method (see also [68]).

6.2 Recovery of Zn from MSWI fly ash

A new process called **fly ash recycling (FLUREC)** is complimentary for FLUWA and was developed with the aim to recover Zn as a valuable resource. This process was first described by Schlumberger and Schuster in 2007 [69] and optimized over the last years [65]. It is now in use on an industrial scale in a waste incineration Plant in Zuchwil, Switzerland [69]. The overarching goal of the FLUREC process is the recovery of Zn as a high-grade metal (SHG, 99.995%) that can be used as a high-quality raw material. The hydrometallurgical process is divided into three steps: 1) cementation, 2) solvent-extraction and 3) electrolysis.

1) As the first step and in preparation for solvent extraction, all heavy metals except zinc are removed by cementation. Therefore, Zn-powder is added to the metals

enriched filtrate after the FLUWA process. Hereby, according to the electrochemical series, all metals which are nobler than zinc are reduced. The dissolution of metallic Zn^0-powder leads to the precipitation of Cu^0, Pb^0 and Cd^0 as cementate (see eq. (12.1)).

$$Cd^{2+}, Pb^{2+}, Cu^{2+} + 3\ Zn^0 \rightarrow Cd^0 \downarrow, Pb^0 \downarrow, Cu^0 \downarrow + 3\ Zn^{2+} \qquad (12.1)$$

The filtered cementate has a high dry substance content (>60%) and can be sent to the smelter. It is of interest due to its high content of Pb^0 (50–70 wt.-%) [66].

2) The selective extraction of Zn from the solution occurs by a solvent-solvent extraction process in a counter-current mixer-settler setup. A chelating agent forms an organic complex with Zn, which is insoluble in water and strongly pH dependent. Optimal condition for Zn complexation is pH 2.7–3.0 [69]. Another important parameter is the contact time of the immiscible phases to form these complexes. The smaller the droplets, and hence the higher the contact surface of the two phases, the faster the exchange. However, too small droplets cause a stable suspension which inhibits the next step: the separation of the two liquid phases [69]. Under optimal conditions, 99.5% of Zn can be stripped by this process. In the stripped solution, <25 mg/l Zn remains, which is then treated with milk of lime to meet the local regulations of discharge [66]. The remains of the organic phase in this solution need to be treated, too.

3) The next step is washing the loaded organic phase to remove impurities that would cause problems during electrolysis. Finally, sulfuric acid (at adjusted pH) is added to re-extract Zn from the organic phase to achieve a concentrated Zn-sulphate solution (>160 g/l Zn). The organic phase is regenerated for the complexing process (2) [66, 70]. The recovery of Zn is performed by electrolysis using an aluminum cathode and a titanium anode. By applying a DC voltage, SHG zinc is deposited on the aluminum cathode as plates (99.995% Zn) and H_2-, O_2-, and traces of Cl_2-gas are generated on the anode. These zinc plates can then be mechanically separated from the aluminum cathode and sold. The generated gases are treated in-house. The reaction tank is constantly filled with a Zn-rich solution to compensate for the loss of Zn during electrolysis. The potential of combining leaching and selective extraction of copper and zinc was also recognized by Tang and coworkers on laboratory [63] and pilot scale [71]. They have developed a method similar to the FLUREC process by first leaching the metals with HCl and then recovering first Cu and then zinc by solvent-extraction.

The **Specific Product Oriented Precipitation (SPOP)** is a new approach for the recovery of Zn from MSWI fly ash: The chemical precipitation takes place under consideration of the mineralogy of the residues. SPOP has two objectives: a) to recover metals from wastewater as a resource and b) to purify simultaneously the wastewater to fulfill local discharge regulations. Depending on the reaction conditions, heavy metals can be precipitated as zero valent metals, metal oxides or metal ferrites. Optimally, the precipitation products are free of hydroxides, which means a significant reduction of the precipitates volume. The solids have the potential to be used directly as a commercial product, or the material can be processed as an anthropogenic resource in recovery facilities. Generally, after the process, the treated wastewaters comply with the regulations for indirect or direct discharge. The SPOP process was first developed in model systems on basis of sulfuric acids and up to 25 g/l Zn, Cu, Ni, Mn, Pb, Sn, Pd, Ag, Au and/or Fe, with typical recovery rates > 99,9% [72–79]. The controlling parameters are the concentration of heavy metals in the wastewater, the reaction temperature, Fe^{2+}-addition, the alkalization conditions, pH-value, aging conditions and addition/exclusion of O_2 [80]. The addition of Fe^{2+} is optional. Various industrial wastewaters have already been successfully treated in laboratories [72–74, 77–80], and first positive tests were performed with a pilot plant [81].

SPOP is also a promising process for the recovery of Zn from the Cl-rich solutions after the extraction of heavy metals by the FLUWA-process [82], because the residues obtained meet three important criteria: a high Zn content, a low Cl-content and small quantities of hydroxides. On a laboratory scale, the process has been shown to be robust, easy to apply, and providing the preconditions for an economic implementation. At first, the experiments were performed with synthetic Zn-Cl solutions based on the eluate concentrations after FLUWA, provided by the MSWI plant in Ingolstadt. The only chemical required for the multi-stage process is NaOH. After the treatment, the Zn content in the filtrate was 0.46 mg/l, below the limit value of 1 mg/L Zn for indirect discharge. Experiments with real wastewater from MSWI plant Ingolstadt, the first plant in Germany using the FLUWA process, show comparable results.

Table 12.3 shows the composition of the residues after application of the adapted SPOP process (before washing) compared to a slurry after the conventional treatment with milk of lime. Even without washing out the soluble salts, the Zn concentration in the residue could be almost three times as high, while the Cl concentration is only half as high as in the conventional process. The washing and thus reduction of the Na, K and Cl content in the residue causes a further increase of the Zn content in the solid. The enrichment of Zn in the residue after SPOP is already high enough for resource recovery before washing, i.e. by smelting. Washing out the salts will further enrich the Zn-concentration in the residue.

Table 12.3: Comparison of the residues obtained after treatment with milk of lime with those obtained after applying the adapted SPOP process to synthetic wastewater (before washing to remove salts).

Parameter	Unit	Milk of Lime	SPOP
Dry mass substance	%	25.7	46
Na	mg/kg	n. A.	72.000
K	mg/kg	28.000	7.000
Ca	mg/kg	88.000	12.000
Cl	mg/kg	190.000	99.000
Zn	mg/kg	190.000	560.000

7 Closing the loop – MSWI fly ash as a resource for zinc

As shown in this article, the concept of waste has continuously changed over time in terms of material and composition: what has been considered as waste and thus worthless for years and decades is in a new context valuable, sometimes out of necessity, often for economic reasons and, recently, for reasons of sustainability [1]. This transition in just a few decades impressively reflects the history of waste incineration. Created to eliminate catastrophic hygienic conditions, it soon proved to be a useful tool for waste management as it eliminates the huge quantities of waste while reducing its toxicity. Pressure from the population enforced the development of effective flue gas cleaning systems. Its product, MSWI fly ash, is regarded as hazardous waste due to its high content of easily soluble salts, heavy metals and POPs, and therefore is disposed of in special landfills, either directly or after processing by S/S techniques. A number of approaches are aimed at using fly ash as a raw material substitute (e.g., as an additive in cement) and thus saving landfill space [41, 83]. However, only small quantities of ash can be added without negatively affecting the properties of the respective products. In addition, the toxic heavy metals are introduced into the material cycle, which can make subsequent recycling of the building materials problematic. It also leads to the loss of metals like Zn as a resource.

Therefore, practicable methods are required for an effective recovery of the resources contained in fly ashes.

The gas cleaning system has a massive influence on the efficiency of such a process. The addition of milk of lime to fly ash during dry/semi-dry FGT considerably dilutes the heavy metal content in the ash, making metal recovery redundant. Not so with wet flue gas scrubbing. The often-cited disadvantage of this process – the formation of an acidic solution – can be used to mobilize the metals from the fly ash, as demonstrated by the FLUWA process (see chapter 6). By optimizing the leaching conditions, Zn, Pb, Cd and Cu can be leached from the fly ash. However, the next step – the treatment of the metals-enriched solution with milk of lime – leads to a considerable increase in the residues volume and reduces the concentration of metals in the neutralization sludge. This makes the recovery of metals uneconomical, and the sludge represents a hazardous waste.

Only a few processes, such as FLUREC or SPOP, focus on the recovery of metals like Zn as a resource. FLUREC is a novel but also complex process leading to the recovery of zinc as a high-grade metal from MSWI fly ash after the leaching process. Huber et al. [44] evaluated the FLUREC process with various other treatment and disposal methods for MSWI fly ash, including the use of life cycle assessment (LCA). They concluded that the FLUREC-process has the lowest environmental impact. On the basis of the prevailing commodity prices for Zn, however, it is only economically feasible if the concentration of Zn in MSWI fly ash is above 40.000 mg/kg Zn [44]. The alternative SPOP process is still under development. It is based on chemical precipitation and enables the enrichment of zinc in the residue while the Cl concentration is kept low. The next step will be to scale up the process and to test whether the process is economically and environmentally feasible on industrial scale.

Such innovative concepts which meet the criteria of circular economy [84] in consideration of the sustainable development goals (SDG) contribute to save primary raw materials and avoid hazardous waste. The recovery of valuable materials from anthropogenic sources in order to reintroduce them into production helps to close the gap in the circular economy and is thus an important step towards an "industrial ecosystem" [83].

Literature

[1] R. Keller, *Müll- Die gesellschaftliche Konstruktion des Wertvollen*, 2nd ed. Wiesbaden: VS Verlag für Sozialwissenschaften, 2009.

[2] S. Heuss-Aßbichler and G. Rettenberger, "Geschichte der Deponie – ist Deponie Geschichte?" in *Inwastement – Abfall in Umwelt und Gesellschaft*, J. Kersten, Ed. Bielefeld: Transcript Verlag, 2016, pp. 109–130.

[3] J. A. Chandler *et al.*, "Municipal solid waste incinerator residues," Elsevier Ltd, Amsterdam, 1997.

[4] F. W. Geels and J. Schot, "Typology of sociotechnical transition pathways," *Res. Policy*, vol. 36, no. 3, pp. 399–417, 2007.

[5] L. Makarichi, W. Jutidamrongphan, and K. Techato, "The evolution of waste-to-energy incineration: A review," *Renew. Sustain. Energy Rev.*, vol. 91, no. November 2017, pp. 812–821, 2018.

[6] D. C. Wilson *et al.*, *Global Waste Management Outlook*. United Nations Environment Programme (UNEP) and International Solid Waste Association (ISWA), 2016.

[7] EU, "Directive 2008/98/EC of the European Parliament and of the Council of 19 November 2008 on waste and repealing certain Directives – Statement of the European Parliament and the Council of the European Union," 2008.

[8] S. Schulze, "Abfall- und Kreislaufwirtschaft im Kontext der Agenda 2030," in *Die Agenda 2030 als Magisches Vieleck der Nachhaltigkeit: Systemische Perspektiven*, E. Herlyn and M. Lévy-Tödter, Eds. Wiesbaden: Springer Fachmedien Wiesbaden, 2019, pp. 179–197.

[9] United Nations, "Resolution A/RES/70/1 adopted by the General Assembly on 25 September 2015 – Transforming our world: the 2030 Agenda for Sustainable Development. Outcome document of the united Nations summit for the adoption of the post-2015 development agenda," 2015.

[10] A. Pinasseau, B. Zerger, J. Roth, M. Canova, and S. Roudier, *Best Available Techniques (BAT) Reference Document for Waste treatment Industrial Emissions Directive 2010/75/EU (Integrated Pollution Prevention and Control); EUR 29362 EN*. Luxembourg: Publications Office of the European Union, 2018.

[11] X. Bing, J. M. Bloemhof, T. R. P. Ramos, A. P. Barbosa-Povoa, C. Y. Wong, and J. G. A. J. van der Vorst, "Research challenges in municipal solid waste logistics management," *Waste Manag.*, vol. 48, pp. 584–592, 2016.

[12] U. Schneidewind, *Die Große Transformation. Eine Einführung in die Kunst gesellschaftlichen Wandels*. Frankfurt am Main: Fischer Verlag, 2018.

[13] P. for A. the C. E. PACE, "The Circularity Gap Report 2019: Closing the Circularity Gap in a 9% World," 2019.

[14] C. Paraskevi, P. C. S., and T. N. J., "WTE plants installed in European cities: a review of success stories," *Manag. Environ. Qual. An Int. J.*, vol. 27, no. 5, pp. 606–620, Jan. 2016.

[15] T. Sabbas *et al.*, "Management of municipal solid waste incineration residues," *Waste Manag.*, vol. 23, no. 1, pp. 61–88, 2003.

[16] D. Sager, "Lösungsprozesse und Transport leichtlöslicher Salze in Monodeponien für Rückstände aus der Müllverbrennung," Ludwig-Maximilians-Universität München, 2007.

[17] C. Speiser, T. Baumann, and R. Niessner, "Characterization of municipal solid waste incineration (MSWI) bottom ash by scanning electron microscopy and quantitative energy dispersive X-ray microanalysis (SEM/EDX)," *Fresenius. J. Anal. Chem.*, vol. 370, no. 6, pp. 752–759, 2001.

[18] S. Heuss-Aßbichler, G. Magel, and K. T. Fehr, "Abiotic hydrogen production in fresh and altered MSWI-residues: Texture and microstructure investigation," *Waste Manag.*, vol. 30, no. 10, pp. 1871–1880, 2010.

[19] L. Muchová, "Wet physical separation of MSWI bottom ash," Ponsen & Looijen b.v., 2010.

[20] R. Bunge, "Wieviel Metall steckt im Abfall?" *Miner. Nebenprodukte und Abfälle – Aschen, Schlacken, Stäube und Baurestmassen*, pp. 91–104, 2014.

[21] P. Lechner, P. Mostbauer, and K. Böhm, "Grundlagen für die Verwertung von MV-Rostasche. Teil A: Entwicklung des Österreichischen Behandlungsgrundsatzes," Wien, 2010.

[22] U. Müller and K. Rübner, "The microstructure of concrete made with municipal waste incinerator bottom ash as an aggregate component," *Cem. Concr. Res.*, vol. 36, no. 8, pp. 1434–1443, 2006.

[23] A. Gianoncelli, A. Zacco, R. P. W. J. Struis, L. Borgese, L. E. Depero, and E. Bontempi, "Fly Ash Pollutants, Treatment and Recycling," in *Pollutant Diseases, Remediation and Recycling*, E. Lichtfouse, J. Schwarzbauer, and D. Robert, Eds. Cham: Springer International Publishing, 2013, pp. 103–213.

[24] J. Haberl, R. Koralewska, S. Schlumberger, and M. Schuster, "Quantification of main and trace metal components in the fly ash of waste-to-energy plants located in Germany and

Switzerland: An overview and comparison of concentration fluctuations within and between several plants with particular focus on valuable metals," *Waste Manag.*, vol. 75, pp. 361–371, 2018.

[25] K. Silpa, Y. Lisa C., B.-T. Perinaz, and V. W. Frank, *What a Waste 2.0 : A Global Snapshot of Solid Waste Management to 2050*, 1st ed. Washington: World Bank, 2018.

[26] WBGU – German Advisory Council on Global Change, "Humanity on the move : Unlocking the transformative power of cities," Berlin, 2016.

[27] F. Neuwahl, G. Cusano, J. Gómez Benavides, S. Holbrook, and S. Roudier, *Best Available Techniques (BAT) Reference Document for Waste Incineration; Industrial Emissions Directive 2010/75/EU (Integrated Pollution Prevention and Control); EUR 29971 EN*. Luxembourg: Publications Office of the European Union, 2019.

[28] Eurostat, "OECD/Eurostat joint questionnaire on municipal waste."

[29] S. Abanades, G. Flamant, B. Gagnepain, and D. Gauthier, "Fate of heavy metals during municipal solid waste incineration," *Waste Manag. Res.*, vol. 20, no. 1, pp. 55–68, Feb. 2002.

[30] O. Hjelmar, "Disposal strategies for municipal solid waste incineration residues," *J. Hazard. Mater.*, vol. 47, no. 1, pp. 345–368, 1996.

[31] Z. Youcai, *Pollution Control and Resource Recovery: Municipal Solid Wastes Incineration Bottom Ash and Fly Ash*. Oxford: Butterworth-Heinemann, 2017.

[32] H. Belevi and H. Moench, "Factors Determining the Element Behavior in Municipal Solid Waste Incinerators. 1. Field Studies," *Environ. Sci. Technol.*, vol. 34, no. 12, pp. 2501–2506, Jun. 2000.

[33] K. Yin, A. Ahamed, and G. Lisak, "Environmental perspectives of recycling various combustion ashes in cement production – A review," *Waste Manag.*, vol. 78, pp. 401–416, 2018.

[34] F.-Y. Chang and M.-Y. Wey, "Comparison of the characteristics of bottom and fly ashes generated from various incineration processes," *J. Hazard. Mater.*, vol. 138, no. 3, pp. 594–603, 2006.

[35] A. Bogush, J. A. Stegemann, I. Wood, and A. Roy, "Element composition and mineralogical characterisation of air pollution control residue from UK energy-from-waste facilities," *Waste Manag.*, vol. 36, pp. 119–129, 2015.

[36] H. Luo, Y. Cheng, D. He, and E.-H. Yang, "Review of leaching behavior of municipal solid waste incineration (MSWI) ash," *Sci. Total Environ.*, vol. 668, pp. 90–103, 2019.

[37] C. Ferreira, A. Ribeiro, and L. Ottosen, "49 Possible applications for municipal solid waste fly ash.," *J. Hazard. Mater.*, vol. 96, no. 2–3, pp. 201–216, 2003.

[38] J. Evans and P. T. Williams, "Heavy metal adsorption onto flyash in waste incineration flue gases," in *2nd International Symposium on Incineration and Flue Gas Treatment Technologies*, 2000, vol. 78, pp. 40–46.

[39] Z. Youcai, S. Lijie, and L. Guojian, "Chemical stabilization of MSW incinerator fly ashes," *J. Hazard. Mater.*, vol. 95, no. 1–2, pp. 47–63, 2002.

[40] A. Zacco, L. Borgese, A. Gianoncelli, R. P. W. J. Struis, L. E. Depero, and E. Bontempi, "Review of fly ash inertisation treatments and recycling," *Environ. Chem. Lett.*, vol. 12, no. 1, pp. 153–175, 2014.

[41] M. J. Quina *et al.*, "Technologies for the management of MSW incineration ashes from gas cleaning: New perspectives on recovery of secondary raw materials and circular economy," *Science of the Total Environment*. 2018.

[42] M. J. Quina, J. C. Bordado, and R. M. Quinta-Ferreira, "Treatment and use of air pollution control residues from MSW incineration: An overview," *Waste Manag.*, vol. 28, no. 11, pp. 2097–2121, 2008.

[43] L. Benassi *et al.*, "Chemical Stabilization of Municipal Solid Waste Incineration Fly Ash without Any Commercial Chemicals: First Pilot-Plant Scaling Up," *ACS Sustain. Chem. Eng.*, 2016.

[44] F. Huber, D. Laner, and J. Fellner, "Comparative life cycle assessment of MSWI fly ash treatment and disposal," *Waste Manag.*, vol. 73, pp. 392–403, 2018.

[45] E. Bontempi, A. Zacco, L. Borgese, A. Gianoncelli, R. Ardesi, and L. E. Depero, "A new method for municipal solid waste incinerator (MSWI) fly ash inertization, based on colloidal silica," *J. Environ. Monit.*, vol. 12, no. 11, pp. 2093–2099, 2010.

[46] M. J. Quina, J. C. M. Bordado, and R. M. Quinta-Ferreira, "The influence of pH on the leaching behaviour of inorganic components from municipal solid waste APC residues," *Waste Manag.*, vol. 29, no. 9, pp. 2483–2493, 2009.

[47] M. M. Jordan, J. M. Rincón, and B. Rincón-Mora, "Rustic ceramic covering tiles obtained by recycling of marble residues and MSW fly ash," *Fresenius Environ. Bull.*, vol. 24, pp. 533–538, 2015.

[48] M. J. Quina, J. M. Bordado, and R. M. Quinta-Ferreira, "Recycling of air pollution control residues from municipal solid waste incineration into lightweight aggregates," *Waste Manag.*, vol. 34, no. 2, pp. 430–438, 2014.

[49] M. J. Quina, M. A. Almeida, R. Santos, J. M. Bordado, and R. M. Quinta-Ferreira, "Compatibility analysis of municipal solid waste incineration residues and clay for producing lightweight aggregates," *Appl. Clay Sci.*, vol. 102, pp. 71–80, 2014.

[50] O. Kizinievich, V. Voishniene, V. Kizinievich, and E. Shkamat, "Effect of Municipal Solid Waste Incineration Fly Ash on the Properties, Durability, and Environmental Toxicity of Fired Aluminosilicates," *Glas. Ceram.*, vol. 76, no. 7, pp. 307–310, 2019.

[51] L. Wang, I. A. Jamro, Q. Chen, S. Li, J. Luan, and T. Yang, "Immobilization of trace elements in municipal solid waste incinerator (MSWI) fly ash by producing calcium sulphoaluminate cement after carbonation and washing," *Waste Manag. Res.*, vol. 34, no. 3, pp. 184–194, Dec. 2015.

[52] A. Bogush *et al.*, "Co-processing of raw and washed air pollution control residues from energy-from-waste facilities in the cement kiln," *J. Clean. Prod.*, vol. 254, Jan. 2020.

[53] J. Hyks, T. Astrup, and T. H. Christensen, "Long-term leaching from MSWI air-pollution-control residues: Leaching characterization and modeling," *J. Hazard. Mater.*, vol. 162, no. 1, pp. 80–91, 2009.

[54] M. J. Quina, R. C. Santos, J. C. Bordado, and R. M. Quinta-Ferreira, "Characterization of air pollution control residues produced in a municipal solid waste incinerator in Portugal," *J. Hazard. Mater.*, vol. 152, no. 2, pp. 853–869, 2008.

[55] K. Karlfeldt Fedje, C. Ekberg, G. Skarnemark, E. Pires, and B.-M. Steenari, "Initial studies of the recovery of Cu from MSWI fly ash leachates using solvent extraction," *Waste Manag. Res.*, vol. 30, no. 10, pp. 1072–1080, Apr. 2012.

[56] G. Weibel, U. Eggenberger, S. Schlumberger, and U. K. Mäder, "Chemical associations and mobilization of heavy metals in fly ash from municipal solid waste incineration," *Waste Manag.*, vol. 62, pp. 147–159, 2017.

[57] T. Y. Huang and P. T. Chuieh, "Life Cycle Assessment of Reusing Fly Ash from Municipal Solid Waste Incineration," *Procedia Eng.*, vol. 118, pp. 984–991, 2015.

[58] A. P. Bayuseno, W. W. Schmahl, and T. Müllejans, "Hydrothermal processing of MSWI Fly Ash-towards new stable minerals and fixation of heavy metals," *J. Hazard. Mater.*, vol. 167, no. 1–3, pp. 250–259, 2009.

[59] X. Dou *et al.*, "Review of MSWI bottom ash utilization from perspectives of collective characterization, treatment and existing application," *Renew. Sustain. Energy Rev.*, vol. 79, no. May 2016, pp. 24–38, 2017.

[60] A. W. Richards, "Zinc processing," *Encyclopædia Britannica, inc.*, 2019. [Online]. Available: https://www.britannica.com/technology/zinc-processing/Ores. [Accessed: 06-Aug-2020].

[61] P. Ni, H. Li, Y. Zhao, J. Zhang, and C. Zheng, "Relation between leaching characteristics of heavy metals and physical properties of fly ashes from typical municipal solid waste incinerators," *Environ. Technol.*, vol. 38, no. 17, pp. 2105–2118, Sep. 2017.

[62] K. Karlfeldt Fedje, C. Ekberg, G. Skarnemark, and B. M. Steenari, "Removal of hazardous metals from MSW fly ash-An evaluation of ash leaching methods," *J. Hazard. Mater.*, vol. 173, no. 1–3, pp. 310–317, 2010.

[63] J. Tang and B. M. Steenari, "Solvent extraction separation of copper and zinc from MSWI fly ash leachates," *Waste Manag.*, vol. 44, pp. 147–154, 2015.

[64] J. Tang and B. M. Steenari, "Leaching optimization of municipal solid waste incineration ash for resource recovery: A case study of Cu, Zn, Pb and Cd," *Waste Manag.*, vol. 48, pp. 315–322, 2016.

[65] G. Weibel, "Optimized Metal Recovery from Fly Ash from Municipal Solid Waste Incineration," Universität Bern, 2017.

[66] S. Schlumberger, "Neue Technologien und Möglichkeiten der Behandlung von Rauchgasreinigungsrückständen im Sinne eines nachhaltigen Ressourcenmanagements," 2010.

[67] G. Weibel *et al.*, "Extraction of heavy metals from MSWI fly ash using hydrochloric acid and sodium chloride solution," *Waste Manag.*, vol. 76, pp. 457–471, 2018.

[68] I. Anagnostopoulos and S. Heuss-Aßbichler, "Residues of Industrial Wastewater Treatment: Hazardous Waste or Anthropogenic Resource?" in *this book*, 2021.

[69] S. Schlumberger, M. Schuster, S. Ringmann, and R. Koralewska, "Recovery of high purity zinc from filter ash produced during the thermal treatment of waste and inerting of residual materials," *Waste Manag. Res.*, vol. 25, no. 6, pp. 547–555, Dec. 2007.

[70] S. Schlumberger and J. Bühler, "Metallrückgewinnung aus Filterstäuben der thermischen Abfallbehandlung nach dem FLUREC-Verfahren," pp. 377–397, 2013.

[71] J. Tang, M. Petranikova, C. Ekberg, and B. M. Steenari, "Mixer-settler system for the recovery of copper and zinc from MSWI fly ash leachates: An evaluation of a hydrometallurgical process," *J. Clean. Prod.*, vol. 148, pp. 595–605, 2017.

[72] S. Heuss-Aßbichler, M. John, and A. Huber, "A new procedure for recovering heavy metals in industrial wastewater," in *WIT Transactions on Ecology and the Environment*, 2016.

[73] S. Heuss-Aßbichler and M. John, "Gold, silver, and copper in the geosphere and anthroposphere: Can industrial wastewater act as an anthropogenic resource?" in *Highlights in Applied Mineralogy*, S. Heuss-Aßbichler, G. Amthauer, and M. John, Eds. De Gruyter, 2017, pp. 137–151.

[74] S. Heuss-Aßbichler, M. John, D. Klapper, U. W. Bläß, and G. Kochetov, "Recovery of copper as zero-valent phase and/or copper oxide nanoparticles from wastewater by ferritization," *J. Environ. Manage.*, vol. 181, pp. 1–7, 2016.

[75] M. John *et al.*, "Low-temperature synthesis of $CuFeO_2$ (delafossite) at 70°C: A new process solely by precipitation and ageing," *J. Solid State Chem.*, vol. 233, pp. 390–396, 2016.

[76] M. John, S. Heuss-Aßbichler, and A. Ullrich, "Conditions and mechanisms for the formation of nano-sized Delafossite ($CuFeO_2$) at temperatures ≤90 °C in aqueous solution," *J. Solid State Chem.*, vol. 234, pp. 55–62, 2016.

[77] M. John, S. Heuss-Aßbichler, A. Ullrich, and D. Rettenwander, "Purification of heavy metal loaded wastewater from electroplating industry under synthesis of delafossite (ABO_2) by 'Lt-delafossite process," *Water Res.*, vol. 100, pp. 98–104, 2016.

[78] M. John, S. Heuss-Aßbichler, K. Tandon, and A. Ullrich, "Recovery of Ag and Au from synthetic and industrial wastewater by 2-step ferritization and Lt-delafossite process via precipitation," *J. Water Process Eng.*, vol. 30, 2019.

[79] A. Huber, M. John, and S. Heuss-Aßbichler, "Is an effective recovery of heavy metals from industrial effluents feasible?" in *Recy&Depotech*, 2016.

[80] M. John, "Low temperature synthesis of nano crystalline zero-valent phases and (doped) metal oxides as $A_xB_{3-x}O_4$ (ferrite), ABO_2 (delafossite), A_2O and AO. A new process to treat industrial wastewaters?" Ludwig-Maximilians-Universität München, 2016.

[81] I. Anagnostopoulos, J. Knof, and S. Heuss-Aßbichler, "Industrieabwasser als Rohstoffquelle für Buntmetalle: Abwassereinigung und Rohstoffrückgewinnung mit einer portablen Technikumsanlage mit Hilfe des SPOP Verfahrens," in *Deutsche Gesellschaft für Abfallwirtschaft, 9. Wissenschaftskongress*, 2019.

[82] K. Tandon, M. John, S. Heuss-Aßbichler, and V. Schaller, "Influence of salinity and Pb on the precipitation of Zn in a model system," *Minerals*, vol. 8, no. 2, 2018.

[83] J. A. Stegemann, "The potential role of energy-from-waste air pollution control residues in the industrial ecology of cement," *J. Sustain. Cem. Mater.*, vol. 3, no. 2, pp. 111–127, 2014.

[84] C. Helbig, A. Thorenz, and A. Tuma, "Quantitative assessment of dissipative losses of 18 metals," *Resour. Conserv. Recycl.*, vol. 153, 2020.

Iphigenia Anagnostopoulos, Soraya Heuss-Aßbichler

Chapter 13
Residues of industrial wastewater treatment: Hazardous waste or anthropogenic resource?

Abstract: Water is an essential resource for metal processing. Therefore, huge amounts of heavy metal loaded wastewater is produced not only during manufacturing. The electroplating industry is one of the sectors producing large amounts of complex wastewaters enriched with metals like copper, nickel or chromium. These wastewaters are challenging to treat due to their diverse and complex properties and are generally considered as hazardous, as they are toxic to humans and the environment.

Serious pollution incidents in the past have led to the development of various water treatment technologies with the aim of removing metal load from the wastewater and thus enabling the water to be discharged safely. The most common conventional treatment method is the neutralization process. The precipitation of heavy metals as hydroxides, however, causes large amounts of voluminous sludge which, after a physical-chemical treatment, are usually disposed of in hazardous landfills. Another well-established purification method is ion-exchange, which uses exchange resins to concentrate heavy metals. The concentrates are hazardous waste and must be treated afterwards, which is in many cases done by neutralization precipitation. Ion exchange is not cost-effective for large quantities of wastewater. Other technologies to extract metals from wastewater are e.g., adsorption, membrane filtration, coagulation-flocculation, flotation, or electrochemical methods.

Recently, waste has received greater attention as a potential secondary metal resource for economic, but especially strategic reasons. In 2017 a total of 418.700 Mg of waste were treated in Germany according to EWC 11 01 09* "sludges and filter cakes containing heavy metals". Only 5.200 Mg were recycled, which corresponds to a recycling rate of 1%. The economic potential of this waste results in a loss of about 3.0 Mio € for Cu, 4.0 Mio € for Ni, 1.0 Mio € for Zn and 1.7 Mio € for Sn. Accordingly, research is done to improve wastewater treatment methods to extract

Acknowledgements: This article was written as part of the project BAF01SoFo-71263 funded by the Bavarian State Ministry of the Environment and Consumer Protection (StMUV). Thanks to Valerie Strassmann-Kayan for proof reading this article.

Iphigenia Anagnostopoulos, Soraya Heuss-Aßbichler, Section Mineralogy, Petrology & Geochemistry, Department of Earth and Environmental Sciences, Ludwig-Maximilians-Universität München, Theresienstr. 41, 80333 Munich, Germany

https://doi.org/10.1515/9783110674941-013

metals from wastewater with higher efficiency. An innovative concept, the Specific Product-Oriented Precipitation (SPOP), which is currently being developed, is based to the principles of circular economy, as it meets both high recovery rates (> 99 %) of metals as oxides and the purification of wastewater.

Keywords: wastewater, anthropogenic resource, recovery, urban mining, heavy metal

1 Introduction

The supply of metals and the access to clean water have been crucial to mankind since ancient times. The first settlements were trans-regionally close to rivers, as they provided drinking water and food. Water was also important for the production of goods, and it is mainly essential for metal processing. Since the beginning of metallurgical processes, elevated emission of Pb- and Cu-aerosols have been witnessed due to the usage of smelting furnaces [1]. Likewise, water enriched with heavy metals has led to the pollution of rivers and basins [2]. Already in the Third millennium B.C., (Cu) mining and smelting activities caused regional pollution of bodies of water, for example in the Odiel river in the Iberian Pyrite Belt, where increased amounts of Cu, Zn and As were discharged on a larger scale into the river [3]. During the First millennium B.C., the Etruscans developed advanced metallurgical techniques. Their products, like bronze vessels, were in great demand in the Mediterranean region. However, their intensive mining and refining caused heavy metal poisoning of inhabitants and the severe pollution of heavy metals, especially As, a side-product of Cu-smelting, was responsible for the abandonment of a number of Etruscan settlements [4].

In the last century, the industrial revolution caused an exponential increase in extraction of mineral raw materials, especially coal and iron. In consequence, industrial wastewater was generated in huge amounts. At that time, basically all branches of industry, not only mine operators, for example, but also chemical industries or dye factories discharged their wastewater untreated into rivers, taking advantage of the dissipative dynamics of water. Rivers were considered as "self-cleaning" and the strategy for wastewater management was in line with the principle "the solution to pollution is dilution" [5].

However, at the beginning of the 20th century, several examples of serious environmental pollution that poisoned the population made headlines. A remarkable example is the Kamioka mine (Honshu, Japan) which has been exploited for Zn extraction since the 16th century [6]. Since 1905, Cadmium (Cd), a by-product of the mining process, was discharged into the Jinzu River with the mine`s effluent, whose water was used in basins for rice cultivation. In sum, a total Cd discharge of 3000 t was estimated until 1977 [7]. During this time period, a painful disease affected the farmers in the vicinity which caused them to cry out loudly, and therefore called Itai-Itai disease (jap. "It hurts! It hurts"), as a result of softening of the bones. At the end it turned out that the residents were poisoned by eating rice contaminated with Cd growing on the fields irrigated with the mine water [7].

Apart from the mining sector the production sector also led to serious environmental pollution. A catastrophic event presents the Minamata disease caused by ingestion of toxic methylmercury (MeHg). It was caused by the prolonged discharge of wastewater from a large, then modern Chisso Corporation chemical plant into Minamata Bay (Kyushu, Japan) first in 1932 and a second time in 1968. At first, fish and shellfish died, birds suddenly fell off in flight and cats could no longer walk straight ahead and collapsed dead (this is why Minamata disease was firstly called "the dancing cat disease"). But soon afterwards the inhabitants developed chronic symptoms and infants were born with congenital Minamata disease. As part of a comprehensive investigation, 2010 ppm of Hg were detected in the sludge close to the drainage channel of the Chisso industrial plant [8]. The microbiological conversion of Hg^0 into methylmercury enabled the uptake of Hg by plankton and thus enriched Hg in the food chain [8]. The environmental damage was enormous and irreversible. Meanwhile, Minamata Bay has been covered with plastic and soil to seal millions of tons of mercury sludge. As a result, the Minamata Convention, which came into force in 2017, was created to protect human health and the environment from effects of Hg [9]. It includes the control of anthropogenic release of Hg by e.g. banning of new Hg mines and termination of the operation of existing Hg mines. It regulates also informal, artisanal and small-scale gold mining because gold refining consumes large quantities of Hg, which are emitted into the environment. The convention includes the implementation of actions to phase out Hg-containing products by 2020.

In Europe too, industrialization caused severe pollution of European rivers. Emissions of toxins to the environment reached such a life-threatening dimension that in the second half of the 20th century extensive efforts were made to revitalize rivers and in consequence, various wastewater purification techniques were developed. The first legislative actions were started in 1960. For example, in the Federal Republic of Germany rivers and lakes became legally protected by the "Wasserhaushaltsgesetz". In 1972, the EU began an Environment Action Programme, based on the idea that prevention is better than cure and the introduction of the "polluter pays" principle. The Water Framework of the European Commission (2000) addressed water protection to ensure clean waters and required the remediation of polluted waters. In 2008, the European Waste Framework Directive of 1975 (Directive 2000/60/EC) for the prevention and recycling of wastes was revised and basic concepts for waste management were introduced (Directive 2008/98/ EC on waste). The regulations provide limit values for pollutants in industrial wastewaters which have to be met before direct or indirect discharge into rivers.

Heavy metal bearing industrial wastewaters are challenging to treat due to their diverse and complex properties. A brief overview of the wastewater production in electroplating industry, as an example of heavy metal bearing industrial wastewater, is presented in Section 2. A number of treatment methods, which were developed in consequence, are presented in Section 3. Recently, the demand for recycling of heavy metals from wastes is increasing, as natural mineral deposits are widely consumed and their mining is getting more and more extensive. Therefore, in addition to the recovery potential of heavy metals from wastewater, the challenges connected with the recovery of heavy metal from wastewaters are discussed in Section 4.

2 Wastewater production in electroplating industry

A prominent example for heavy metal containing wastewater is the one that is generated in the electroplating industry. The finishing of metals and non-metals with metallic coatings was applied shortly after the principles of galvanic cells were found by Luis Galvani in the late 18th century. Therefore, electroplating is one of the first developed industrial technologies. By 1840 the electrochemical deposition of various metals on goods were ready [10]. Even though electroplating exists for 200 years, its scope of application is still growing. Branches like the semiconductor industry, who use latest technologies in metal deposition, like sputtering and evaporation, are lately considering electroplating as an optional process of production, as it provides a very pure material deposition accompanied by a high deposition rate of 4 µm/min [11]. Nowadays, there is hardly any industrial branch that does not use metal finished products, e.g., the automotive, aerospace, military, and medical industry, to name just a few. The functionality and technological progress of these sectors were not possible without electrodeposition and surface treatment.

Metal finishing is a key technology that requires many resources such as raw materials including metal salts and other chemicals and water. The trend moves towards the use of critical raw materials in addition to standard metals. The use of water, which is an essential medium to produce coated surfaces, has led to an enormous water consumption. Consequently, the advantages of this technology are associated with an enormous volume of waste, especially heavy metal bearing wastewater, which is regularly generated by the exchange of spent process baths and as rinsing water. The wastewater reflects the composition of the metals used for electroplating. Very common elements are Cu, Ni, Cr or Zn, but also noble metals such as Pd, Au or Ag may occur. Depending on the requirements, the volumes of the process and rinsing baths vary from a few liters, e.g. for noble metal coatings, up to 500 m^3 for components suppliers.

Over time, different types of coating techniques have been developed. Figure 13.1 shows a schematic process flow sheet of the electroplating process [12, 13]. In general, the whole process consists of three stages with sequences of different plating and rinsing baths. The respective waste streams must be collected, treated and disposed of in accordance with the Waste Directive, whereby hazardous materials are indicated with * in the waste code.

The first step is a pretreatment of the products to remove oily and fatty components, dirt, mechanical residuals and corrosion products on the surface. It is carried

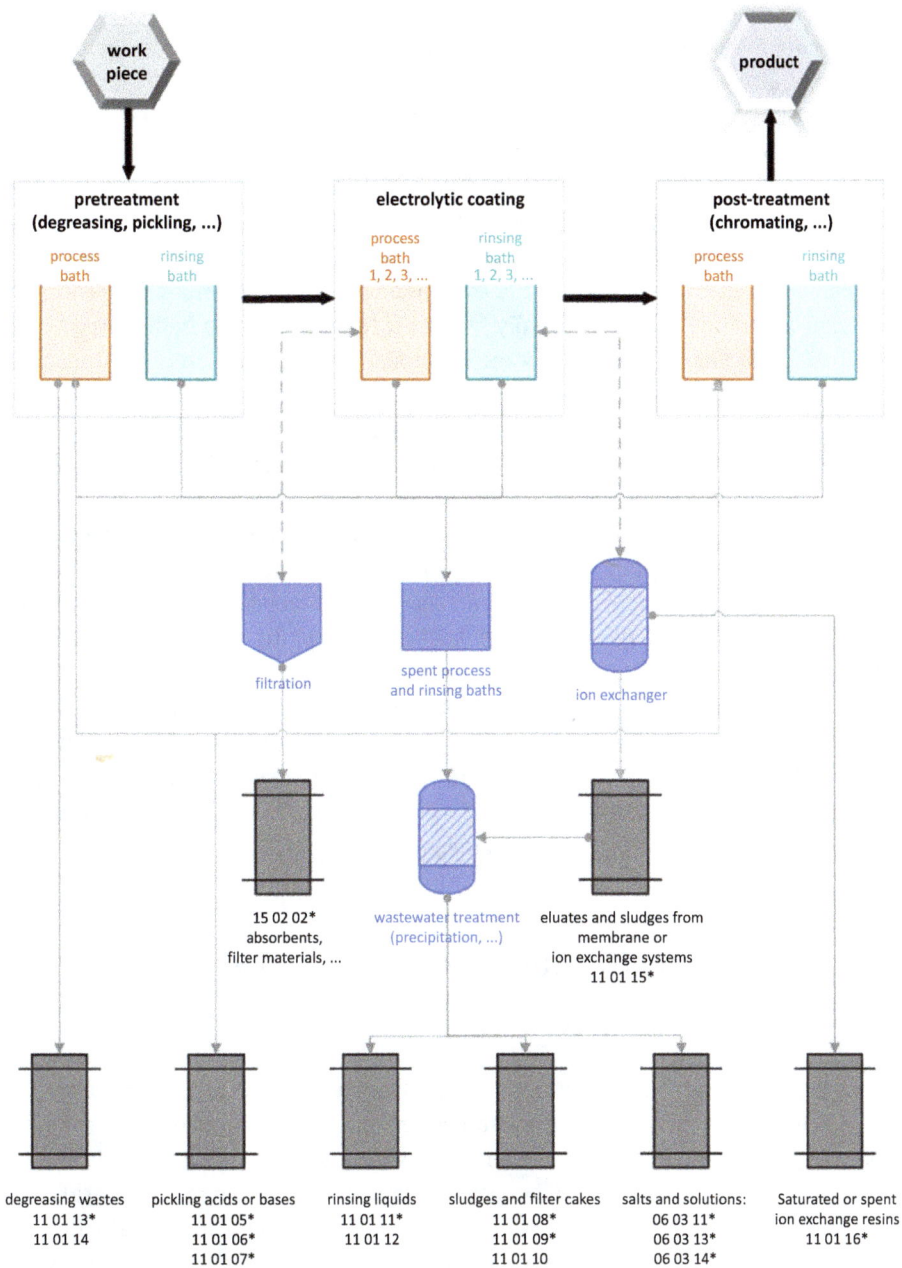

Figure 13.1: Schematic process flow sheet of the electroplating process with accompanied wastes. Waste codes marked with * indicate hazardous materials. Modified after [12, 13].

out e.g., with hot cleaning or halogenated hydrocarbons and the respective waste is listed under the waste code 11 01 13* and 11 01 14. The surface is activated by pickling with e.g., (noble metals containing) mineral acids (waste codes 11 01 05*, 11 01 06*, 11 01 07*). The main step is the coating process, which is applied to the cleaned and activated surfaces. The specific metals are electrodeposited either together as alloys or sequentially in the form of multiple layers. Depending on the intended purpose, both acidic or basic metal-containing electrolytes can be used. Basic electrolytes need additional chelating agents, like cyanide complexes, to keep the metals in solution. Typical feed materials are metal -sulfates, -fluoroborates, or -chlorides. Solid filtrable materials and adsorbents are partially separated from the process baths (waste code 15 02 02*) and the remaining process solution is reused. In a third step, a post-treatment can be applied to improve the corrosion protection or to achieve decorative effects. In most cases this is done by chromating using chromic acid. In addition, surface treatment processes such as acid dipping can be interposed, involving chemicals such as sulfuric acid, hydrochloric acid or hydrofluoric acid.

The used-up process and rinsing baths are collected and treated in the central wastewater treatment plant, which is commonly a chemical precipitation process initiated by adding a reaction agent. In this step voluminous hydroxide sludge containing heavy metals is generated and collected under the waste code 11 01 09*, which is partly disposed of in landfills. The rinsing baths can contain up to 20% of the initial metal salts present in the plating baths [14] and are often purified by ion exchange and reused before they are finally treated by chemical precipitation. The accruing heavy metal waste water and ion exchange eluates are collected and purified in the central wastewater treatment system (waste code 11 01 09*). The spent ion exchange resins are disposed or incinerated (waste code 11 01 16*).

3 Treatment techniques for heavy metal bearing wastewater

Heavy metals occur in acidic solutions as dissolved cations. The dissolved heavy metals are inorganic pollutants as they can contaminate the environment and drinking water when discharged to surface waters. Heavy metals are regarded as persistent environmental toxins because, unlike other contaminants such as pharmaceuticals and most dyes, they cannot be metabolized to carbon dioxide and water or detoxified by chemical or biological remediation processes [15]. For this reason, treated wastewater is required to meet the strict criteria of environmental regulations to prevent pollution of surface waters with heavy metals after direct or indirect discharge.

In practice, the various wastewater streams are collected and mixed to a final wastewater. Accordingly, the wastewater consists of a mixture of different types of metals with enormous fluctuations in concentration (from mg/l to several g/l) and pH value. A number of processes have been developed to treat these complex wastewaters. In addition to the most industrially established processes, various alternative and promising methods are presented in the following chapter, some of which have been intensively investigated, others which are still at an early stage of development.

3.1 Chemical precipitation

Wastewater treatment by chemical precipitation is a robust and technically highly mature process [16, 17] and with an implementation rate of 75% by far the most widely used treatment process in electroplating industry [18]. The principle is based on the pH adjustment of the wastewater to a range, in which the metal hydroxides (M-OH) show minimal solubilities and consequently precipitate. Afterwards the generated metal hydroxide sludge is separated from the liquid phase by sedimentation and/or filtration. Hydroxides or sulfides are conventionally used as precipitants.

3.1.1 Hydroxide precipitation

Various alkalizing agents are used for the hydroxide precipitation of metals from wastewater of the electroplating industry. Milk of lime ($Ca(OH)_2$) is mainly applied because it is low-priced, can be dosed easily and shows good coagulation of particles [19]. Caustic soda (NaOH) is less frequently used to form metal hydroxides due to the higher costs. In various studies soda (Na_2CO_3), sodium bicarbonate ($Na(HCO_3)_2$ and magnesia (MgO) have been investigated as alkalization agents [20, 21] and studies showed that soda ash (Na_2CO_3) is an effective alternative to hydroxide precipitation for Cu, Zn and Pb, as carbonate precipitation takes place at a comparatively lower pH of around 9 [22].

The chemical precipitation of metal cations to metal hydroxides runs by eq. (13.1):

$$M^{2+} + 2(OH)^- \leftrightarrow M(OH)_2 \downarrow \tag{13.1}$$

where M^{2+} represents the dissolved metal cation, OH^- the alkaline precipitant and $M(OH)_2$ the resulting metal hydroxide. Metal hydroxides have their specific minimum solubilities at different pH values; any deviation from the optimal pH for precipitation causes the metals to dissolve. Table 13.1 gives an overview of the optimum pH values for the precipitation of some metal hydroxides.

Due to the complex mixture of different metals in the final wastewater, it is difficult to determine the best precipitation condition. Therefore, the wastewater is alkalized in the pH range of the main component. As a result, metals with a different minimum solubility than the target metal remain partly in solution. Due to this limitation caused by the complex composition of the wastewater, further treatment techniques are often applied after the hydroxide precipitation.

Table 13.1: Overview of the final precipitation pH values for selected metal hydroxides and sulphides.

Metal	Hydroxide precipitation		Sulphide precipitation	
	Precipitation pH	Ref.	Precipitation pH	Ref.
Cd	10,4–10, 9	[20]	4	[23]
Co(II)	10	[24]	ca. 3	[25]
Cr(III)	8,7	[26]	ca. 6	[27]
Cu(II)	8,1–11,1	[28]	2,5	[23]
Fe(III)	6,2–7,1	[28]	3,6–5,7	[29]
Mo	5–7	[24]	–	–
Ni(II)	10,5–11	[20]	7,5	[23]
Pb(II)	7,8–8,8	[28, 30]	7,5–8,5	[31]
Zn(II)	8,7–9,6	[28]	5,5	[23]

A main problem is the insufficient sedimentation of the precipitates due to their colloidal properties which hinders the liquid/solid separation [32]. Techniques like ultracentrifugation or colloid filtration could be applied, but the recent industrial standard is the addition of flocculants to initiate coagulation. Using a synthetic wastewater consisting of Zn, Cd, Mn and Mg, it was shown that the addition of a coagulant after the precipitation with lime can significantly reduce the residual heavy metal concentration in the solution [33]. Another approach to improve the separation process is the addition of fly ash as a crystallization seed can enhance the precipitation of heavy metals together with milk of lime. It was found that the addition of coal fly ash in combination with exposure to CO_2 gas increases the particle size of the precipitate, resulting in better sedimentation of the sludge and increased efficiency of heavy metal removal [34]. Another problem is posed by strongly complexing organic agents like EDTA (ethylenediaminetetraacetic acid), which are often present in wastewater from electroplating industry and inhibit the removal of heavy metals with hydroxide precipitation [35]. The addition of Fe^{3+} was found to reduce the residual Ni^{2+} concentration in the solution as it is adsorbed onto precipitating Fe^{3+}-containing phases like ferrihydrite [35].

Hydroxide precipitation produces large volumes of sludges which are highly enriched in water and low-concentrated in metals [36]. Therefore, the sludges have to be dewatered and the remaining sludge are mostly disposed of in hazardous landfill sites [37]. Some research has been done on the use or detoxification of these sludges. The utilization of dried metal hydroxide sludge from electroplating industry was studied to solve reactive dye wastewater problems and it was shown that the sludge is an effective positively charged adsorbent for azo reactive dyes [38]. With an enhanced electrokinetic process it was possible to remove 34–69% of Cr, Zn, Ni, Cu and Pb from electroplating sludge by using citric acid as electrolytes in electrode chambers [39]. Experiments performed with Cr containing electroplating sludges by using HCl achieved an extraction rate of 97,6% [40]. With a combined process of acid-leaching, ammonium jarosite precipitation and electro-deposition more than 95% of Cu and Ni were recovered from electroplating sludge [41]. Electroplating sludge enriched with 50% Sn was treated to recover Sn by leaching with concentrated HCl at room temperature for one month or by boiling in 10% HCl solution for two hours. Subsequent electrowinning achieved a Sn recovery of 93% [42].

Recently, upcycling methods for electroplating sludges were investigated. Two specific electroplating sludges were tested as adsorbents for Zn, Cu, Ni, and Co containing electroplating wastewater [43]. The sludges contained greater amounts of Fe, Co and Cr, respectively, and after a hydrothermal treatment, particles containing erdite (a sulfide containing Fe, Na and H_2O) were formed, which were subsequently used as adsorbents.

Summary: Chemical precipitation with hydroxides is the most frequently applied method for heavy metal bearing wastewaters. The process is robust, technically easy, and used chemicals are cheap. However, the liquid/solid separation process is time consuming, it is strongly pH dependent and the removal rates are partly limited due to re-concentration of metals in wastewaters. Additionally, with respect to water purification hydroxide precipitation is ineffective for low concentrated wastewater [17]. The generated sludges contain hazardous materials and pose a burden for the environment. In practice a recovery of the heavy metals from hydroxide sludge is rarely applied, as it is sumptuous and the metal content is often too low for recovery. As a result, the sludges have to be transported and deposited safely at great expense.

3.1.2 Sulphide precipitation

Sulphide precipitation needs precipitation agents such as solid (FeS, CaS), aqueous (Na_2S, NaHS) or gaseous sulphides (H_2S) [44]. The reaction involved in sulphide precipitation shows eq. (13.2). [45].

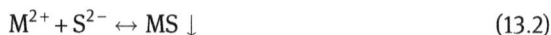

$$M^{2+} + S^{2-} \leftrightarrow MS \downarrow \tag{13.2}$$

M^{2+} represents the dissolved metal cation, S^{2-} the sulphide precipitant, and MS the resulting precipitates. Sulfide precipitation is often combined with metal hydroxide precipitation and/or coprecipitation at alkaline pH in order to reduce the theoretical sulphide dosage [45].

Sulphide precipitation is not a common method because of technical difficulties and criticalities. These involve an insufficiently controlled dosing of sulphide [46] and a potential for H_2S gas evolution [47]. However, it shows some conceptual advantages compared to hydroxide precipitation. Contrary to hydroxides, metal-sulphides show no amphoteric properties and the solubilities of metal-sulphides are lower than to metal-hydroxides [47, 48]. Therefore, precipitation with sulphides can be applied over a broader pH range and the acidic wastewaters do not need to be highly alkalized to achieve a high degree of metal removal [17] (see also Table 13.1).

Furthermore, treatment with sulphides has a potential for selective precipitation. For example, treatment of Zn +Cu solution with Na_2S leads to the precipitation of CuS (covellite) and ZnS (sphalerite) [49]. The reaction rate is fast, the settling properties are high due to the high density of the particles and afterwards the Me-sulphides can be roasted to oxides. The resulting Me-sulphide sludge is easier to concentrate than hydroxide sludge and needs no dewatering process [17]. However, the fine-grained Me-sulphide precipitates makes the separation process laborious and flocculants need to be added [50]. Combined processes of sulphide and hydroxide precipitation have also been applied for selective metal recovery. An example is the recovery of Cd from an electroplating wastewater, containing contaminants (Fe, Cu, Ni, Zn) and impurities (Co, Mn) [51]. At first, the contaminants and impurities were removed by acidifying the wastewater with HNO_3 to precipitate the metals as cyano-metal complexes. Hereby, it is critical to control the pH. Secondly, the wastewater was alkalized with NaOH to precipitate $Cd(OH)_2$. In a final polishing step, sulphide precipitation with Na_2S was conducted to precipitate the remaining Cd in the alkaline pH range.

Summary: The sulphide precipitation process has some advantages over the hydroxide precipitation process, e.g. higher degree of metal removal at lower pH values and selective metal precipitation, but dosing of sulphides and separation of the fine precipitate is technically difficult. Therefore, treatment of electroplating wastewater with sulphides is rarely applied compared to hydroxide precipitation. For highly effective metal removal, combinations of hydroxide and sulphide precipitation are applied. Metal sulphide sludges are more stable than hydroxide sludges with respect to leachability, but criticalities arising from metal sulphide sludges are, for example, the possibility of release of H_2S and redissolution of heavy metals due to oxidation of metal sulphides to sulfate [47].

3.1.3 Heavy metal chelating precipitation

The presence of complexing agents like EDTA hinders the removal of heavy metals from electroplating wastewater by conventional hydroxide or sulphide precipitation. For example, when Ni and Cr form a complex with cyanide, the removal efficiency is impaired [52]. Therefore, stronger chelating agents are used for precipitation to form complexes with both the metal cations and the metal chelate present. Many commercial chelating agents exist on the market, some for non-selective, some for selective metal removal and research for new materials goes on. For the removal of different metals from electroplating wastewater nonstoichiometric interbiopolyelectrolyte green complexes (NIBPEGCs) were found to be suitable [53], while chitosan and its derivate proved to be particularly effective for the selective precipitation of Pd from electroplating wastewater with up to 95% removal efficiency [54].

The resulting sludges contain hazardous substances, and the chemical extraction of metals from chelated sludges is very costly, as it requires strong chemicals and generates other wastes. For noble metals, such as Pd, a recovery process with aqua regia has proven to be possible [54]. Sludges resulting from chelate precipitation are often disposed of in hazardous waste landfills.

3.1.4 Specific Product Oriented Precipitation (SPOP)

The Specific Product Oriented Precipitation Process (SPOP) is a new precipitation method to effectively treat wastewater while efficiently recovering heavy metals in industrial wastewater without high energy consumption. The goals are (1) to achieve water quality that meets the limits for discharge with respect to heavy metal concentration, and (2) to recover heavy metals as an anthropogenic resource. The SPOP process was inspired by the ferrite process, which was developed to treat low-concentrated laboratory wastewaters by binding the heavy metals in ferrites and use their magnetic properties for separation [55]. With the development of a modified ferrite process, Cu could be precipitated not only as Cu-ferrite $((Cu,Fe)Fe_2O_4)$, but also as other valuable phases like cuprite (Cu_2O), tenorite (CuO) and zero-valent Cu. The controlling factors, apart from the Cu^{2+} concentration (up to 10 g/l), are the pH value, the reaction temperature (<70 °C), and the aging conditions [56]. Green rust, a $Fe^{2+} - Fe^{3+}$ containing double layered hydroxy-sulphate observed as the main phase in fresh samples, proved to be a precursor for the formation of magnetite (Fe_3O_4) [56].

With variation of molar ratios of Fe:Me, further stable metal containing phases were identified. At a Fe:Cu = 1:1, delafossite $(CuFeO_2)$ was precipitated at temperatures <80 °C both from Cu^{2+} containing aqueous solutions and from wastewater from electroplating industry; it demonstrated the possibilities of recovering valuable phases from wastewater [57–59]. Phases with delafossite structure are of technical importance due to their interesting conductivity properties. For example, Cu-Fe-containing delafossite is a promising material for gas sensing applications and has the potential for selective gas sensing of reducing and oxidizing gases [60]. The treatment of zinc electroplating wastewater with traces of Ni, Fe, Cu and Cr led to the precipitation of doped ZnO nanoparticles and sole traces of $ZnCO_3$ and $Zn(OH)_2$ [61]. Synthetic and industrial wastewater containing either ionic Ag or Au was treated with a 2-step process. Au was precipitated as zero-valent phase while Ag was recovered as either single phased delafossite $(AgFeO_2)$ or composite particles of Ag^0 together with delafossite or magnetite [62].

In general, the treatment of salt rich solutions is sophisticated and requires enhanced technics and materials. However, it was possible to extend the application field of the SPOP process to highly saline wastewater [63]. At low salt concentrations, Zn precipitate as ZnO, and with increasing salinity the stability of various Zn-sulphate-hydroxide phases containing Na^+, Cl^- and SO_4^{2-} is favored [63]. In one study, this process was used to recover Zn from fly ash from municipal solid waste incineration (see also Chapter 3.4). Fly ash is particularly enriched with Zn and Pb, and also contains other elements such as Cu, Cd and Cr. An acidic extraction of fly ash yields a solution highly enriched in Cl and especially Zn. The application of SPOP led to a concentration of Zn in the precipitate and at the same time a purification of the wastewater [64].

In all studies applying SPOP, the recovery rates are generally high (>99,9%) and in laboratory scale the limits for discharge are mostly achieved after the first treatment. To make the variety of product phases possible, the properties of the wastewater has to be analyzed with respect to its metal composition and concentration. Therefore, the SPOP process is no panacea and possible additives like chelating agents can influence the product synthesis and recovery rate.

In order to test the feasibility of the SPOP process in practice, a customizable pilot plant was constructed and operated in a quasi-batch process [65]. The principle process is shown in Figure 13.2. The wastewater is first tempered (T < 70 °C) and subsequently alkalized with NaOH (≤ pH 12). For product variability, it is optionally possible to add Fe^{2+} -containing solution to the wastewater. After the precipitation process the particles are separated and altered, if required.

In synthetic wastewaters containing Cu^{2+} or Zn^{2+}, Cu was recovered as cuprite (Cu_2O), while Zn was precipitated solely as zincite (ZnO). Figure 13.3 shows the precipitated zincite with a spherical morphology; the nanoparticles exhibit good particle agglomeration.

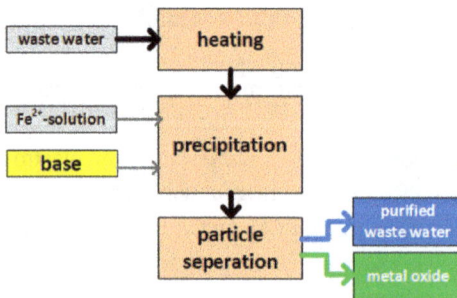

Figure 13.2: Process flow chart of the Specific Product Oriented Precipitation (SPOP) process. The heavy metal bearing waste water is tempered to reaction temperature (<70 °C), alkalized with NaOH (≤ pH 12) and optionally supplemented with Fe2+ solution. In a sedimentation vessel the particles are separated from the treated waste water.

Summary: with the SPOP process it is possible to precipitate different metal containing phases like oxides with ferrite or delafossite structure and doped metaloxides, zero-valent phases or hydroxides by adjusting the treatment parameters. The first results of the experiments with the pilot plant confirm that it is not only possible to purify heavy metal bearing wastewater but also to recover metals as a resource. The metals are highly enriched in the residues and the volume of the powdery precipitation products is very small compared to hydroxide sludge. The generated phases from wastewater have the potential to be reused in other industrial applications. In the worst-case the precipitation products can be subjected to hydrometallurgical processes for reuse.

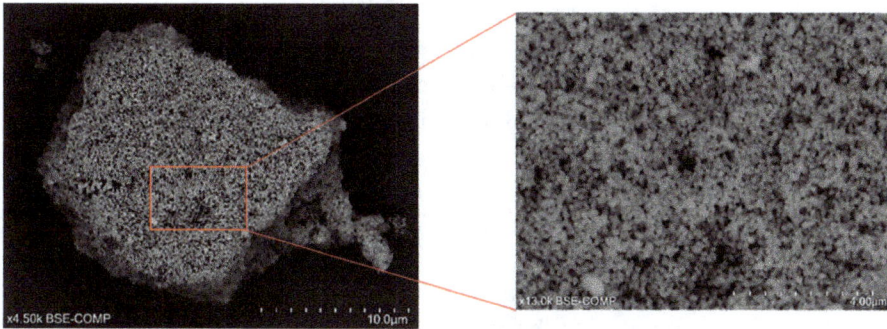

Figure 13.3: Scanning electron microscope picture of zincite precipitated from a model wastewater with SPOP in pilot scale. The nanosized particles show a spherical morphology with good agglomeration to micrometer sized compounds.

3.2 Ion exchange

Ion exchange has been intensively studied and is a well-established process to remove heavy metals from wastewater. Advantages are high treatment capacities, high removal efficiencies and fast kinetics [17, 66]. In general, both anions and cations can be exchanged, but in the case of heavy metals, cation exchangers must be considered. The technique is based on an ion exchange between a cation (B^+) dissolved in a liquid phase and a loaded resin ($M^- A^+$). The ion transfer between the cation (A^+) bound on the resin M^- and the cations of the wastewater (B^+) can be shown by following eq. (13.3) [67]

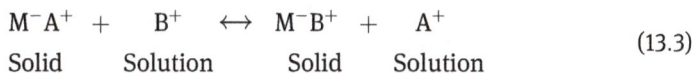

$$M^-A^+ + B^+ \longleftrightarrow M^-B^+ + A^+$$
$$\text{Solid} \quad \text{Solution} \quad \text{Solid} \quad \text{Solution} \tag{13.3}$$

According to eq. (13.3), the number of ions in the wastewater remains the same, since the adsorption of the toxic heavy metal cations is compensated by the desorption of harmless cations like Na^+, Ca^{2+} or Mg^{2+}.

The resin must meet high requirements to guarantee a good cation exchange efficiency, such as a hydrophilic structure with regular and reproducible shape, a controlled exchange capacity and chemical and thermal stability, to name just a few [67]. Natural materials like e.g. alumosilicates fulfill these high requirements only to a certain limit. Modern ion exchange materials are synthetically produced organic compounds like e.g. styrenic or acrylic resins. The exchange-active groups are e.g. sulpho groups ($SO_3^-H^+$) or carboxyl groups (COO^-H^+). The exchange of the dissolved cations with hydrogen ions means a conversion of the metal salt solution into free acids and thus an acidification of the waste water [68]. After consumption of the exchange active hydrogen ions of the resin, the ion exchanger can be treated with caustic soda and regenerated with e.g. hydrochloric acid. After that, the concentration of cations contained in the regenerate is higher than in the original wastewater [68]. Therefore, ion exchange is also a method for concentrating ions in a liquid phase. In the surface coating industry, ion exchangers are used for rinse water recirculation, recovery of materials from rinse water, regeneration of process solutions and subsequent purification of treated waste water [68].

To determine the kinetics and equilibrium for the removal of Zn and Ni, the removal efficiency of the ions as a function of pH, resin dosage and contact time were studied for a specific ion exchange resin (Dowex HCR S/S). It was concluded, that pH is crucial as it influences the ionization of surface functional groups and hydronium ions (H_3O^+) are competing adsorbates reducing the exchange rate of metal cations [69]. The optimum adsorption of Ni^{2+} and Zn^{2+} occurred at pH 4 and 6, respectively; at higher pH values $Ni(OH)_2$ and $Zn(OH)_2$ are formed. With a higher dosage of resin, the removal efficiency improved as the number of adsorption sites was increased. Another important parameter for the effectiveness of ion exchange is the contact time. For the tested resin, contact times of 90 min and 120 min were required for Ni and Zn, respectively [69]. The treatment of polymetallic wastewaters with an ion exchanger is difficult because of the competitive replacement of ions with the resin. For example, Cr^{3+} can re-mobilize Co^{2+} and Ni^{2+} that had already been removed from wastewater [66]. Therefore, it is necessary to consider the heavy metal composition in the wastewater before applying ion exchange. Combined processes are applied to selectively concentrate metals by ion exchange and subsequently, to precipitate with chemical treatment methods. For example, Cr^{6+} in electroplating wastewater was removed with Cr-selective resins and reduced to Cr^{3+} by hydroxide precipitation using NaOH [70].

Natural examples of ion exchange materials are zeolites, which have a good ion exchange capacity due to their crystallographic structure. Clinoptilolite, for example, is a zeolite of the Heulandite group, which is used to remove heavy metals from industrial effluents. Both, natural [71–73] and synthetic clinoptilolite [74, 75] were studied for their exchange capabilities. Clinoptilolite consists of SiO_2 and AlO_4 tetrahedra, with Al^{3+} positioned in the center of the tetrahedra. By replacement of Al^{3+} by Si^{4+}, one negative charge is created in the lattice. The cations can be replaced by other cations in solutions or wastewater. Investigations showed that clinoptilolite is particularly useful for the exchange with Pb^{2+} [75], and thus can selectively bind Pb^{2+} to the structure. In addition, other heavy metal ions (Zn^{2+}, Cd^{2+}, Cu^{2+}) can be removed by ion exchange depending on the pH, the solid/liquid ratio and the existence of competitive heavy metal cations [75]. Tests with natural and synthetic zeolites used in the metal finishing industry to remove heavy metal cations show that the sorption capacity of synthetic zeolites is about ten times higher than that of natural zeolites [76].

Summary: Ion exchange is an established purification method for heavy metal bearing wastewaters. The exchange resins are partly ion selective, but treatment of multi-element wastewaters is still challenging due to remobilization of metal cations. To ensure an effective purification many parameters like pH, contact time and dosage need to be adjusted. Long contact time slows down the process. Ion exchange enables the concentration of heavy metal-rich solutions, but the concentrate is often treated with chemical precipitation afterwards. Treatment of large volumes of low-concentration wastewater with ion exchange is not cost-effective and is therefore not applied on a large scale [17]. An additional problem is subsequent pollution caused by the regeneration of used ion exchange resins with chemical reagents. After the ion exchange resins have been used up, they are either disposed of or incinerated [68].

3.3 Adsorption

The principle of binding atoms or compounds to the surface of a solid phase by adhesion is commonly called adsorption. Adsorption processes are reversible because the adsorbents can be regenerated by desorption. Finding new efficient adsorption materials has recently been in the focus of research, as they offer an effective method for treatment of heavy metal bearing wastewater [17].

One standard adsorbent is activated carbon (AC) because of its large surface area created by large micro- and mesopores [17]. It is therefore often used in waste water treatment. However, commercial coal-based AC is generally expensive and prices are constantly rising [17, 77]. Several materials are proposed as alternatives for AC. Cheap and easily available agricultural wastes, like rice husk [77], coconut shell [78], honeydew melon peel [79], groundnut shell, Indian beech and onion skin [80] were investigated to remove Ni^{2+} [77], Cu^{2+} [78], Fe^{2+} [77, 78], Pb^{2+} [77, 78], Zn^{2+} [78, 79], Cr^{3+} [79], and Cd^{2+} [80] ions from electroplating wastewater. The activation of the alternative carbon sources is achieved either by pre-treatment with chemicals like zinc chloride [78] or acids like H_2SO_4 and H_3PO_4 in combination with high-temperature carbonization [79] and final grinding and sieving [80]. The adsorption capacity strongly depends on the adsorbent dosage, contact time, pH and stirring rate [78]. For example, coconut shell-based activated carbon has the following highest removal rates: 93.37% for Fe, 92.22% for Cu, 60.52% for Zn and 100% for Pb [78].

In another approach the reuse of electroplating sludge and calcinated electroplating sludge as adsorbents for Ni^{2+} from electroplating wastewater were tested [81]. The adsorption capacity achieved for the electroplating sludge and the calcinated sludge was 135,7 and 118,3 mg/g Ni^{2+} respectively, higher than that of some commercially available adsorbents [81].

Another example for the treatment of wastewater with materials to be disposed of is the usage of drinking water treatment residual (DWTR) for the adsorption of metals. To remove particles and dissolved materials from water supplies, coagulants like alum, iron (III) chloride or iron (III) sulfate are used to produce drinking water. The resulting amorphous sludge consist mainly of Al, Fe oxyhydroxides and organic matter with functional groups on the surface making them an adsorbent for a variety of metals including V, Ga, As, Se and B [82]. However, the application of DWTR has limitations, as industrial waste often contains toxins such as Hg and As, which can be released during the application [82, 83]. A recovery process for metals from DWTR has yet to be developed [82].

Summary: Adsorption processes for the removal of heavy metals from aqueous solutions are effective but limited to low concentrated wastewaters. The use of conventional activated carbon is limited due to its high cost [17]. Recently, alternative adsorbents have been studied and show promising results, but will still need to be optimized regarding adsorption efficiency. Furthermore, the activation of alternative carbon adsorbents requires treatment with acids or other chemicals and/ or carbonization at high temperatures. For more information on industrial waste as a low-cost potential adsorbent, see [84]. On the development and preparation of sludge-based adsorbents refer to [85].

3.4 Membrane filtration

Membrane filtration is the separation of species in a solution by selective transport of individual atoms or components through a membrane. Different procedures have been developed depending on the particle sizes to be separated: Microfiltration is used for the separation of particles in the range of micro range and larger, ultrafiltration for colloids, nanofiltration for macromolecules and reverse osmosis for molecules and ions.

Since different particle sizes occur in the liquids of the surface coating industry, different membrane separation techniques are applied. In a review paper on membrane technology for the removal of heavy metals from wastewater, the most important methods were presented [86]: (1) polymer-enhanced ultrafiltration (PEUF), (2) micellar-enhanced ultrafiltration (MEUF), (3) adsorptive ultrafiltration mixed matrix membranes (UF MMMs), (4) nanofiltration (NF), (5) reverse osmosis (RO), (6) forward osmosis (FO), (7) liquid membrane (LM), and (8) electrodialysis (ED). Usually MEUF, PEUF and UF MMMs are used for the removal of heavy metals from low concentration wastewater, while NF, RO and FO can be applied on high concentration wastewater.

The heavy metal rich concentrates generated with membrane technologies, like concentrates from ion exchange, are finally treated with the central wastewater treatment plant of the industrial facility. The spent membranes are disposed of in land-fills or incinerated. For 2015, the global mass of spent RO membranes from desalina-tion industry was estimated to 12.000 Mg, indicating the potential disposal problem [87].

3.4.1 Ultrafiltration

The removal of colloids and dissolved materials is carried out by ultrafiltration (UF). Separation is achieved by a membrane which filters solutes based on the membrane pore size and the molecular weight of the solutes [88]. Since the pore sizes of UF membranes (ca. 0,01 μm) are 400–500 times larger than dissolved metal ions in the form of hydrated ions, these can pass through the membrane [86]. Therefore, im-proved UF processes such as PEUF, MEUF and UF MMMs have been developed to remove heavy metal ions from wastewater. For this purpose, heavy metals are either bound to micelle compounds (MEUF) or charged groups to form large filterable complexes (PEUF, UF MMM).

3.4.1.1 Micellar-Enhanced Ultrafiltration (MEUF)

MEUF uses charged surfactants that form larger micelle agents by attachment and thus can be retained by the membrane. The micelles show strong electrostatic forces and bind oppositely charged metal ions or other compounds at the micellar surface [89]. Therefore, MEUF can be applied for the removal of anions or cations from wastewater. The removal efficiency depends on the used surfactant, operating pres-sure, pH, and temperature of the solution [88]. Micelle formation is only possible within a narrow concentration range of cationic or anionic surfactants, also called critical micelle concentration (CMC) [89].

One problem is the subsequent contamination by high residual concentrations of non-aggregated surfactant in the permeate [86]. Additionally, the surfactant itself contributes to a large extend to the process costs [90]. In order to decrease costs, methods for separating the metal ions from the micelles are being studied in regard to recovering and reusing the surfactants. Three methods were tested to separate Cu and Cd from the surfactant: (1) acidification to a strongly acidic pH and subsequent UF, (2) complexation with addition of the chelating agent EDTA followed by UF, and (3) precipitation of Cu and Cd by ferric- and ferrocyanide and surfactant recovery by subse-quent centrifugation [90]. Highest surfactant recovery with almost 100% was achieved with (2) centrifugation after complexation [90]. Good success has been achieved in the latest laboratory-scale research dealing with the problem of subsequent contamination by surfactants. During the treatment of a synthetic Hg containing wastewater with a combined process of MEUF and activated carbon fiber, >97% of surfactant was re-moved from the permeate [91].

3.4.1.2 Polymer-Enhanced Ultrafiltration (PEUF)

Polymer-enhanced ultrafiltration (PEUF) or also called CEUF (complexation-enhanced UF) is based on the addition of water-soluble polymers with chelating capability to the wastewater to form larger heavy metal complexes and thus prevent their passage through a microporous membrane [86]. The metal ions are electrostatically bound to the ligands of the polymers. The most frequently used complexing polymers are poly (acrylic) acids, carboxyl methyl cellulose, and polyvinylethyleneimine [86]. The polymer adsorption depends on surface wettability, surface charge, and surface roughness [92]. In practice, a suitable polymer and a corresponding membrane are selected depending on the wastewater properties.

Acids or alkalis are needed for the regeneration of the polymers, making the process less environmentally friendly. A modified process developed for electroplating wastewater provides a remedy [93]. In a shear-induced dissociation coupled with ultrafiltration the metals (Zn^{2+}, Cu^{2+}, Ni^{2+}, Cr^{3+}, and Fe^{3+}) are selectively bound by a complexing agent and then separated from the complexant by dissociation of metal-polymer complexes.

Even though PEUF has been in development since the 1960s, most studies are currently carried out on lab- or pilot-scale. The main challenges that restrict PEUF from large scale applications are (1) the regeneration of the polymers for reuse, (2) the low metal concentration recovered from the polymer, and (3) the decline in water flux caused by polymer deposition on the membrane surface or in the pores [92]. Another limitation is the narrow range of critical surfactant concentration for optimal treatment to avoid poor removal rates [86]. Additionally, waste ligands or micelles and other harmful products can accumulate [86].

3.4.1.3 UF adsorptive mixed matrix membranes (UF MMMs)

In recent years, adsorptive mixed metal membranes (MMM) have been studied especially for solutions containing low heavy metal concentrations [86]. The great advantage of MMMs is their independence from chemicals to remove heavy metal ions from solutions. This implies the addition of nanoparticles to the membrane matrix to improve the physicochemical properties of the membranes for better removal efficiency [86]. But membrane thickness, pressure and leaching of the nanomaterials in the treated water can limit the process [86]. To ensure effective adsorption, an optimal thickness of the membrane is required. A possible leaching of nanoparticles from MMMs carries the risk of toxic secondary pollutants. Therefore, the quantity of nanomaterials added to the MMM needs to be determined to keep leaching processes to a minimum [86].

3.4.2 Nanofiltration (NF)

Membranes for nanofiltration (NF) are thin-film composites made of synthetic poly-mers containing charged groups [94]. The pore sizes are small enough to separate metallic ions [86] and in order to achieve a significant removal of metal ions, modi-fied membranes are required [86]. The separation of metals by nanofiltration is com-plex and not only driven by the charged groups but also by solution diffusion, dielectric exclusion, electromigration, and the Donnan effect [94] which describes an unequal distribution of permeable ions between two sides of a membrane. The properties of the membranes, i.e. the surface charge and the pore size are pH depen-dent, making the pH of the wastewater a controlling parameter for the separation capacity of NF membranes [94].

For the treatment of electroplating wastewater two NF membranes were prepared by two different preparation techniques (cast and spray) and cast membranes were found to be less porous, making them more suitable for metal retention [95]. The max-imum retention capacity for cast membranes were 94% for Zn and 93% for Fe [95].

3.4.3 Reverse osmosis (RO)

In a normal osmosis process, a solvent flows from an area with a low solute concen-tration (high water potential) to an area with a high solute concentration (low water potential). Reverse osmosis (RO) is a diffusion-controlled process, which runs in op-posite direction by applying a pressure to force water molecules to move against the concentration gradient [86]. RO membranes are very dense and semi-permeable. Therefore, high retention rates for heavy metals can be achieved. Accordingly, RO can be applied to produce highly purified water.

Summary: In general, the use of energy-intensive membrane technologies in in-dustrial applications is restricted by high costs, process complexity, membrane fouling and low permeate flux [17].

3.5 Coagulation and flocculation

Colloidal particles carry surface charges that repel each other and hinder the par-ticles to settle down. The addition of chemicals destabilizes colloids in wastewater and promotes the coagulation of compounds to larger aggregates. The added coagu-lants, e.g. alum, polyaluminium chloride (PAC), ferric chloride, or hydrated lime [96] carry the opposite charge to the colloids to neutralize the repulsive effect.

Flocculation is a process which enables the contact and adhesion between col-loids to form larger aggregates. Flocculants enhance the sedimentation rate of par-ticles and thus accelerate the sedimentation process of the produced sludge. Many different flocculants exist on the market, e.g., PAC, polyferric sulfate (PFS), or poly-acrylamide (PAM) [17].

The removal efficiency of heavy metals by coagulation-flocculation depends on the type and dosage of used coagulant, pH, and concentration of metal ions [96]. For mixed metal containing wastewater the separation efficiency can be reduced by coagulation-flocculation in comparison to single metal wastewaters [96]. In general, heavy metals cannot be completely removed by coagulation-flocculation [17, 97]. To ensure the complete removal of heavy metals, various treatment techniques are combined with coagulation-flocculation. In addition to high input of chemicals, co-agulation-flocculation contributes to an increased sludge volume [17].

3.6 Flotation

Flotation processes separate particles by using differences in the surface properties of different particles [98]. The separation of heavy metals from liquid phase is achieved by the attachment of hydrophobic particles to air bubbles that rise to the surface [98]. The most effective flotation process for heavy metal bearing wastewater is dissolved air flotation (DAF) because it can separate light and small suspended particles [98]. The DAF process uses collectors to selectively adsorb onto the surface of specific minerals and form a monolayer of non-polar hydrophobic hydrocarbons [98, 99]. Flotation shows a high removal efficiency and is partly metal selective, but it is rarely applied in industrial applications, as it requires high initial capital costs and high maintenance [17].

3.7 Electrochemical treatment methods

All electrochemical wastewater treatment techniques are based on an electrical current which is set to the wastewater via electrodes. In principle, it is the same process used in electroplating facilities for surface plating with metal. Electrochemical techniques for the purification of heavy metal bearing wastewater are getting more and more attention because it does not need high temperatures and pressures, less usage of chemicals, and the process is robust and easily controlled [100]. Electrodeposition is the most frequently applied method among the electrochemical treatment techniques and enables the deposition of zero-valent metals onto the cathode [17]. For medium (100 mg/l) to highly concentrated wastewaters (g/l), metal deposition can be achieved on simple plate electrodes. However, removal efficiency is reduced with decreasing metal cation concentrations since the current efficiency of simple plate electrodes also decreases. Therefore, residual concentrations of heavy metal cations remain in the electrochemically treated wastewater and have to be removed otherwise, e.g. by ion exchange. Newly developed substrate materials try to solve this process limiting factor. A low cost and non-toxic material discussed as substrate are hydrotalcite-like compounds (layered double hydroxides) [101]. In a lab scale study, using $ZnAl\text{-}CO_3$ and MgAl-H hydrotalcite as substrate, 75% Cd^{2+} and almost 100% Pb^{2+} were removed from aqueous solution [101]. Pilot-scale experiments were conducted to remove Ni^{2+} from real electroplating wastewater by a combined process of adsorption and electro-deposition [102]. First the Ni^{2+} ions were concentrated by adsorption with a commercially available cation-exchange resin, and then the regenerate solution (>30 g/l) was reduced to Ni sheet metals by electrodeposition. Due to relatively large capital investment and expensive electricity supply, electrochemical treatment methods have not been widely installed [17].

4 Recovery potential of metals in residues from electroplating industry

In recent years, the recovery of heavy metals from wastes has gained increasing importance, particularly to extract metals like Co or V, which are classified as critical raw materials. Hydroxide precipitation is the most commonly applied treatment technique for heavy metal bearing wastewater and the resulting metal containing sludges are usually disposed of in hazardous landfills. Therefore, the question arises to what extent these sludges are a source for metal recovery.

The recovery potential of raw materials in wastes, and in this case from sludges from electroplating wastewater treatment, can be derived from the amount of waste produced per year and the concentration of specific elements. In Europe, all waste is classified according to the European Waste Catalogue (EWC), in which they are listed according to industrial sector and sector-typical processes (EWC codes). The recovery potential of raw materials from the waste cannot be calculated directly from the EWC database as this classification does not take into account the chemical composition of the waste [103].

In Germany, a total of 418.700 Mg of waste were treated in 2017 according to EWC 11 01 09* "sludges and filter cakes containing heavy metals" [104]. Figure 13.4 shows the quantities of deposited and treated waste: A large proportion of it, 360.400 Mg, underwent physico-chemical processes [104]. 22.200 Mg of waste went directly to landfills, 34.300 Mg of waste were treated with other methods and 1.800 Mg of waste were thermally treated. Of the total waste volume, only 5.200 Mg were recycled, which corresponds to a recycling rate of 1%. For the year 2013 a correspondingly low recycling rate of 3% was calculated [103].

Chemical composition of random samples were published by the German Federal Environmental Agency in 2016, based on the results from the Waste Analysis Data Base (ABANDA) and from random samples of industrial companies who participated voluntarily [105]. The average metal concentrations found in waste 11 01 09* for the years 2010 and 2011 were for Fe 58.823 mg/kg, Zn 53.773 mg/kg, Cu 40.458 mg/kg, Ni 35.868 mg/kg,

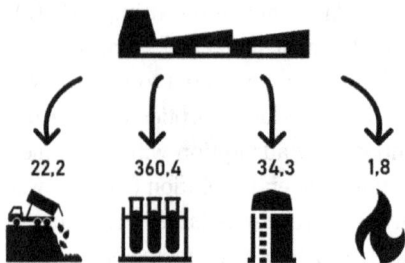

Figure 13.4: Proportion of sludges deposited and treated in Germany in 2017 according to EWC 11 01 09* in 1000 Mg, after [104]. From left to right: 22.200 Mg of waste were directly landfilled, 360.400 Mg were treated with physico-chemical processes, 34.300 Mg were treated with other process methods and 1.800 Mg were thermally treated.

Al 27, 855 mg/kg, Cr 27.355 mg/kg, Sn 19.677 mg/kg, Pb 10.000 mg/kg, Mn 1.430 mg/kg and for Pb, Sb, Co, Mo, V and W 1.000 mg/kg each [105]. The amount of Fe found in the sludges arise mostly from the Fe-containing substrate due to etching. Compared to all other listed wastes, waste 11 01 09* owns the highest concentration of Sn and the second highest concentration of Cr [105]. It should be emphasized that the values are based on random samples and are therefore not representative.

Figure 13.5 shows the amount of metals dissipated in 2011, calculated as the product of the mean value of metal concentrations and the landfilled waste (20.200 Mg) according to EWC 11 01 09* [106]. The values serve only for orientation because the real quantity behind the analysis is not known. Accordingly, the following quantities were removed from the material cycle: Fe (1.188,2 Mg/a), Zn (1.086,2 Mg/a), Cu (817,3 Mg/a), Ni (724,5 Mg/a), Al (562,7 Mg/a), Cr 552,6 Mg/a, Sn 397,5 Mg/a and Mn 28,9 Mg/a [106].

A rough estimate of the economic potential of waste 11 01 09*, based on a resale (scrap) value of metals [107], results in a loss of 3.02 Mio € for Cu, 3.98 Mio € for Ni, 1.03 Mio € for Zn and 1.7 Mio € for Sn. However, the metal concentrations and

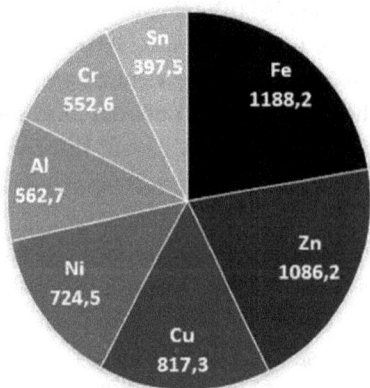

Figure 13.5: Quantity of metals [Mg/a] determined for waste EWC 11 01 09* based on the proportion of waste going to landfill (20.200 Mg), after [105, 106].

therefore the economic loss for individual companies might be much higher. A survey amongst 10 companies in the electroplating or smelting industry resulted in higher concentrations with 2.000 Mg/a for Cu and 5.300 Mg/a for Ni [103]. Wastes from the numerous smaller surface processing and coating companies is particularly suitable for recycling because the waste contains a higher proportion of metals [105]. However, the amount of generated waste per company is relatively small [105].

It should be noted that companies must pay for the transport of neutralization sludge from the emitter to the waste treatment facilities or landfills. Depending on parameters like total volume, metal content (recycling capability/toxicity), and transport distance, prices range from 100 to 200 €/t [103].

In general, the recycling of waste becomes interesting when the concentrations of the metals are in the range of concentrations found in ores. Table 13.2 summarizes the lower limit of commercial viability of the metals Cr, Cu, Fe, Ni, Pb, Sn and Zn in comparison to the concentrations in sludges and filter cakes of waste 11 01 09* [105]. Accordingly, the marketability of the waste is fulfilled for the following metals: Cr (265.469 ppm in sludge), Ni (36.922 ppm in sludge, 6.315 ppm in filter cake) and Zn (114.100 ppm in filter cake).

Table 13.2: Comparison of metal concentrations determined in waste EWC 11 01 09* and commercial viability [105].

	Cr	Cu	Fe	Ni	Pb	Sn	Zn
				ppm			
commercial viability	35.000	4.000	350.000	4.000	40.000	25.000	50.000
11 01 09* sludge	265.469	42	38.742	36.922	25	50	50
11 01 09* filtercake	11.353	87	132.257	6.315	1.646	50	114.100

5 Conclusion

For the last 70 years, the focus has been set on the treatment of heavy metal bearing wastewaters. Accordingly, the goal of the treatment processes was to bind or discharge the heavy metals and to safely dispose of the hazardous sludge. Worldwide increasing waste generation has become a global environmental problem [108]. In recent years, awareness of the loss of heavy metals as valuable materials with waste has increased, leading to the concept of urban mining and waste as an anthropogenic resource. The recovery of heavy metals from waste is politically demanded. Considering waste as a resource is a key element towards a circular economy addressed by the European Commission in its 7th Environment Action Programme elaborated in 2015. Reuse and recycling are of major importance to keep metals in the cycle of matter.

Landfilling is therefore the last favorable option in the waste hierarchy. It was also recognized that obligatory measures are needed to encourage innovation in recycling techniques and to enable efficient metal recovery. To this end, discussions will include whether the limit values for metals in waste going to landfill should be lowered and whether molybdenum (Mo) should be added to the list [105].

A sustainable handling of heavy metals in wastewater from the electroplating industry requires an improved waste management system. The analysis of waste 11 01 09* showed that many sludges with higher heavy metal content are landfilled [105]. The treatment methods presented here are designed for wastewater treatment only and are often unsuitable for recycling. Novel treatment methods like SPOP are promising and should be brought to industrial scale, with the aim of reusing the heavy metals contained in the wastewater at least as raw materials for the production of materials. Such a solution requires a change in conceptual rethink: treatment should no longer carried out centrally with multi-element wastewater but directly at the source of wastewater production in order to achieve the highest possible quality of product phases with high degree of purity.

Bibliography

[1]   S. Hong, J. P. Candelone, M. Soutif, and C. F. Boutron, "A reconstruction of changes in copper production and copper emissions to the atmosphere during the past 7000 years," *Sci. Total Environ.*, vol. 188, no. 2–3, pp. 183–193, 1996, doi: 10.1016/0048-9697(96)05171-6.

[2]   A. Martínez Cortizas, L. López-Merino, R. Bindler, T. Mighall, and M. E. Kylander, "Early atmospheric metal pollution provides evidence for Chalcolithic/Bronze Age mining and metallurgy in Southwestern Europe," *Sci. Total Environ.*, vol. 545–546, pp. 398–406, 2016, doi: 10.1016/j.scitotenv.2015.12.078.

[3]   F. Nocete, E. Álex, J. M. Nieto, R. Sáez, and M. R. Bayona, "An archaeological approach to regional environmental pollution in the south-western Iberian Peninsula related to Third millennium BC mining and metallurgy," *J. Archaeol. Sci.*, vol. 32, no. 10, pp. 1566–1576, 2005, doi: 10.1016/j.jas.2005.04.012.

[4]   A. P. Harrison, I. Cattani, and J. M. Turfa, "Metallurgy, environmental pollution and the decline of Etruscan civilisation," *Environ. Sci. Pollut. Res.*, vol. 17, no. 1, pp. 165–180, 2010, doi: 10.1007/s11356-009-0141-5.

[5]   G. Lofrano, G. Libralato, F. G. Acanfora, L. Pucci, and M. Carotenuto, "Which lesson can be learnt from a historical contamination analysis of the most polluted river in Europe?" *Sci. Total Environ.*, vol. 524–525, pp. 246–259, 2015, doi: 10.1016/j.scitotenv.2015.04.030.

[6]   M. Kasuya *et al.*, "Water pollution by cadmium and the onset of Itai-itai disease," *Water Sci. Technol.*, vol. 26, no. 11, pp. 149–156, 1992, doi: 10.2166/wst.1992.0286.

[7]   F. Yoshida, A. Hata, and H. Tonegawa, "Itai-Itai disease and the countermeasures against cadmium pollution by the Kamioka mine," *Environ. Econ. Policy Stud.*, vol. 2, no. 3, pp. 215–229, 1999, doi: 10.1007/BF03353912.

[8]   M. Harada, "Minamata Disease: Methylmercury Poisoning in Japan Caused by Environmental Pollution," *Crit. Rev. Toxicol.*, vol. 25, no. 1, pp. 1–24, 1995, doi: 10.3109/10408449509089885.

[9]   United Nations Environment Programme, "Minamata Convention On Mercury," Kumamoto, 2013.

[10]   F. Neukirchen, "Industrielle Revolution und Hightech," in *Von der Kupfersteinzeit zu den Seltenen Erden*, Springer-Verlag Berlin Heidelberg, 2016, pp. 113–152.

[11]   K. Witt, "Electroplating: An Old Technology For The Future," *Silicon Semiconductor*, 2020. [Online]. Available: https://siliconsemiconductor.net/article/97266/Electroplating_An_old_technology_for_the_future/feature. [Accessed: 09-Jul-2020].

[12]   Landesamt für Umwelt Baden-Württemberg, "Abfallsteckbrief 1101 Chemische Oberflächenbearbeitung – Galvanikschlamm," 2017. [Online]. Available: https://www.abfall bewertung.org/repgen.php?report=ipa&char_id=1101_GalS&lang_id=de&avv=&synon=&ka pitel=6&gtactive=no.

[13]   ABAG-itm GmbH, "Abfallsteckbrief 1101 Chemische Oberflächenbearbeitung – Galvanikschlamm," 2008. [Online]. Available: https://abag-itm.de/startseite/.

[14]   K. Upadhyay, "Solution for wastewater problem related to electroplating industry: An overview," *J. Ind. Pollut. Control*, vol. 22, no. 1, pp. 59–66, 2006.

[15]   A. S. Mohammed and E. Al., "Heavy Metal Pollution: Source, Impact, and Remedies," in *Biomanagement of Metal-Contaminated Soils*, vol. 20, no. March, M. S. Khan, A. Zaidi, R. Goel, and J. Musarrat, Eds. Springer Netherlands, 2011, pp. 1–28.

[16]   Y. Ku and I. L. Jung, "Photocatalytic reduction of Cr(VI) in aqueous solutions by UV irradiation with the presence of titanium dioxide," *Water Res.*, vol. 35, no. 1, pp. 135–142, 2001, doi: 10.1016/S0043-1354(00)00098-1.

[17]   F. Fu and Q. Wang, "Removal of heavy metal ions from wastewaters: A review," *J. Environ. Manage.*, vol. 92, no. 3, pp. 407–418, 2011, doi: 10.1016/j.jenvman.2010.11.011.

[18]   K. G. Karthikeyan, H. A. Elliott, and F. S. Cannon, "Enhanced metal removal from wastewater by coagulant addition," in *Proc. 50th Purdue Industrial Waste Conf*, 1996, vol. 50, pp. 259–267.

[19]   K. A. Baltpurvins, R. C. Burns, G. A. Lawrance, and A. D. Stuart, "Effect of electrolyte composition on zinc hydroxide precipitation by lime," *Water Res.*, vol. 31, no. 5, pp. 973–980, 1997, doi: 10.1016/S0043-1354(96)00327-2.

[20]   J. W. Patterson, H. E. Allen, and J. J. Scala, "Carbonate Precipitation for Heavy Metals Pollutants," *J. Water Pollut. Control Fed.*, vol. 49, no. 12, pp. 2397–2410, 1977.

[21]   X. Lin, R. C. Burns, and G. A. Lawrance, "Heavy metals in wastewater: The effect of electrolyte composition on the precipitation of cadmium(II) using lime and magnesia," *Water. Air. Soil Pollut.*, vol. 165, no. 1–4, pp. 131–152, 2005, doi: 10.1007/s11270-005-4640-9.

[22]   Q. Chen, Y. Yao, X. Li, J. Lu, J. Zhou, and Z. Huang, "Comparison of heavy metal removals from aqueous solutions by chemical precipitation and characteristics of precipitates," *J. Water Process Eng.*, vol. 26, no. 12, pp. 289–300, 2018, doi: 10.1016/j.jwpe.2018.11.003.

[23]   M. Gharabaghi, M. Irannajad, and A. R. Azadmehr, "Selective Sulphide Precipitation of Heavy Metals from Acidic Polymetallic Aqueous Solution by Thioacetamide," *Ind. Eng. Chem. Res.*, vol. 51, no. 2, pp. 954–963, 2012, doi: 10.1021/ie201832x.

[24]   J. H. Huang, C. Kargl-Simard, M. Oliazadeh, and A. M. Alfantazi, "pH-Controlled precipitation of cobalt and molybdenum from industrial waste effluents of a cobalt electrodeposition process," *Hydrometallurgy*, vol. 75, no. 1–4, pp. 77–90, 2004, doi: 10.1016/j.hydromet.2004.06.008.

[25]   J. Jandová, K. Lisá, H. Vu, and F. Vranka, "Separation of copper and cobalt-nickel sulphide concentrates during processing of manganese deep ocean nodules," *Hydrometallurgy*, vol. 77, no. 1–2, pp. 75–79, 2005, doi: 10.1016/j.hydromet.2004.10.011.

[26]   S. A. Mirbagheri and S. N. Hosseini, "Pilot plant investigation on petrochemical wastewater treatment for the removal of copper and chromium with the objective of reuse," *Desalination*, vol. 171, no. 1, pp. 85–93, 2004, doi: 10.1016/j.desal.2004.03.022.

[27]   J. R. Mudakavi, G. Venkateshwar, and M. Ravindram, "Removal of chromium from electroplating effluents by the sulphide process," *Indian J. Chem. Technol.*, vol. 2, no. 2, pp. 53–58, 1995.

[28]   F. M. Pang, S. P. Teng, T. T. Teng, and A. K. Mohd Omar, "Heavy Metals Removal by Hydroxide Precipitation and Coagulation-Flocculation Methods from Aqueous Solutions," *Water Qual. Res. J. Canada*, vol. 44, no. 2, pp. 174–182, 2009, doi: 10.2166/wqrj.2009.019.

[29]   D. Wei and K. Osseo-Asare, "Particulate pyrite formation by the Fe3+ /HS- reaction in aqueous solutions: effects of solution composition," *Colloids Surfaces A Physicochem. Eng. Asp.*, vol. 118, no. 1–2, pp. 51–61, 1996, doi: 10.1016/0927-7757(96)03568-6.

[30]   C. Sist and G. P. Demopoulos, "Nickel Hydroxide Precipitation from Aqueous Sulfate Media," *JOM*, vol. 55, no. 8, pp. 42–46, 2003, doi: 10.1007/s11837-003-0104-0.

[31]   T. T. Hien Hoa, W. Liamleam, and A. P. Annachhatre, "Lead removal through biological sulfate reduction process," *Bioresour. Technol.*, vol. 98, no. 13, pp. 2538–2548, 2007, doi: 10.1016/ j.biortech.2006.09.060.

[32]   A. v. Krusenstjern and L. Axmacher, "Zur Neutralisation von Galvanikabwasser mit Kalzium- und Natriumhydroxid," *Metalloberfläche*, vol. 18, no. 3, pp. 65–69, 1964.

[33]   L. Charerntanyarak, "Heavy Metals Removal By Chemical Coagulation And Precipitation," *Water Sci. Technol.*, vol. 39, no. 10–11, pp. 135–138, 1999.

[34]   Q. Chen, Z. Luo, C. Hills, G. Xue, and M. Tyrer, "Precipitation of heavy metals from wastewater using simulated flue gas: Sequent additions of fly ash, lime and carbon dioxide," *Water Res.*, vol. 43, no. 10, pp. 2605–2614, 2009, doi: 10.1016/j.watres.2009.03.007.

[35]   X. Lin, R. C. Burns, and G. A. Lawrance, "Effect of electrolyte composition, and of added iron(III) in the presence of selected organic complexing agents, on nickel(II) precipitation by lime," *Water Res.*, vol. 32, no. 12, pp. 3637–3645, 1998, doi: 10.1016/S0043-1354(98)00131-6.

[36]   B. R. Babu, S. U. Bhanu, and K. S. Meera, "Waste minimization in Electroplating Industries: A review," *J. Environ. Sci. Heal. – Part C*, vol. 27, no. 3, pp. 155–177, 2009, doi: 10.1080/ 10590500903124158.

[37]   L. G. Twidwell and D. R. Dahnke, "Treatment of metal finishing sludge for detoxification and metal value," *Eur. J. Miner. Process. Environ. Prot.*, vol. 1, no. 2, pp. 76–88, 2001.

[38]   S. Netpradit, P. Thiravetyan, and S. Towprayoon, "Application of 'waste' metal hydroxide sludge for adsorption of azo reactive dyes," *Water Res.*, vol. 37, no. 4, pp. 763–772, 2003, doi: 10.1016/S0043-1354(02)00375-5.

[39]   G. Peng and G. Tian, "Using electrode electrolytes to enhance electrokinetic removal of heavy metals from electroplating sludge," *Chem. Eng. J.*, vol. 165, no. 2, pp. 388–394, 2010, doi: 10.1016/j.cej.2010.10.006.

[40]   P. T. De Souza E Silva et al., "Extraction and recovery of chromium from electroplating sludge," *J. Hazard. Mater.*, vol. B128, no. 1, pp. 39–43, 2006, doi: 10.1016/ j.jhazmat.2005.07.026.

[41]   P. P. Li, C. S. Peng, F. M. Li, S. X. Song, and A. O. Juan, "Copper and nickel recovery from electroplating sludge by the process of acid-leaching and electro-depositing," *Int. J. Environ. Res.*, vol. 5, no. 3, pp. 797–804, 2011, doi: 10.22059/ijer.2011.386.

[42]   T. Stefanowicz, T. Golik, S. Napieralska-Zagozda, and M. Osińska, "Tin recovery from an electroplating sludge," *Resour. Conserv. Recycl.*, vol. 6, no. 1, pp. 61–69, 1991, doi: 10.1016/0921-3449(91)90006-A.

[43]   Y. Liu *et al.*, "Upcycling of Electroplating Sludge to Prepare Erdite-Bearing Nanorods for the Adsorption of Heavy Metals from Electroplating Wastewater Effluent," *Water*, vol. 12, no. 1027, pp. 1–16, 2020, doi: 10.3390/W12041027.

[44]   A. E. Lewis, "Review of metal sulphide precipitation," *Hydrometallurgy*, vol. 104, no. 2, pp. 222–234, 2010, doi: 10.1016/j.hydromet.2010.06.010.

[45] D. Bhattacharyya, A. B. Jumawan Jr., and R. B. Grieves, "Separation of Toxic Heavy Metals by Sulfide Precipitation," *Sep. Sci. Technol.*, vol. 14, no. 5, pp. 441–452, 1979, doi: 10.1080/01496397908058096.

[46] A. H. M. Veeken and W. H. Rulkens, "Innovative developments in the selective removal and reuse of heavy metals from wastewaters," *Water Sci. Technol.*, vol. 47, no. 10, pp. 9–16, 2003, doi: 10.2166/wst.2003.0525.

[47] R. W. Peters and J. Ferg, "The Dissolution/Leaching Behavior of Metal Hydroxide/Metal Sulfide Sludges from Plating Wastewaters," *Hazard. Waste Hazard. Mater.*, vol. 4, no. 4, pp. 325–355, 1987, doi: 10.1089/hwm.1987.4.325.

[48] B. M. Kim and P. A. Amodeo, "Calcium sulfide process for treatment of metal-containing wastes," *Environ. Prog.*, vol. 2, no. 3, pp. 175–180, 1983, doi: 10.1002/ep.670020309.

[49] R. M. M. Sampaio, R. A. Timmers, Y. Xu, K. J. Keesman, and P. N. L. Lens, "Selective precipitation of Cu from Zn in a pS controlled continuously stirred tank reactor," *J. Hazard. Mater.*, vol. 165, no. 1–3, pp. 256–265, 2009, doi: 10.1016/j.jhazmat.2008.09.117.

[50] J. S. Whang and D. Young, "Soluble-Sulfide Precipitation for Heavy Metals Removal from Wastewaters," *Environ. Prog.*, vol. 1, no. 2, pp. 110–113, 1982, doi: 10.1002/ep.670010306.

[51] S. Islamoglu, L. Yilmaz, and H. O. Ozbelge, "Development of a Precipitation Based Separation Scheme for Selective Removal and Recovery of Heavy Metals from Cadmium Rich Electroplating Industry Effluents," *Sep. Sci. Technol.*, vol. 41, no. 15, pp. 3367–3385, 2006, doi: 10.1080/01496390600851665.

[52] R. Naim et al., "Precipitation Chelation of Cyanide Complexes in Electroplating Industry Wastewater," *Int. J. Environ. Res.*, vol. 4, no. 4, pp. 735–740, 2010.

[53] E. A. López-Maldonado, O. G. Zavala García, K. C. Escobedo, and M. T. Oropeza-Guzman, "Evaluation of the chelating performance of biopolyelectrolyte green complexes (NIBPEGCs) for wastewater treatment from the metal finishing industry," *J. Hazard. Mater.*, vol. 335, pp. 18–27, 2017, doi: 10.1016/j.jhazmat.2017.04.020.

[54] Q. Xie, G. Liang, T. Lin, F. Chen, D. Wang, and B. Yang, "Selective chelating precipitation of palladium metal from electroplating wastewater using chitosan and its derivative," *Adsorpt. Sci. Technol.*, vol. 38, no. 3–4, pp. 113–126, 2020, doi: 10.1177/0263617420918729.

[55] Y. Tamaura, T. Katsura, S. Rojarayanont, T. Yoshida, and H. Abe, "Ferrite Process; heavy metal ions treatment system," *Water Sci. Technol.*, vol. 23, no. 10–12, pp. 1893–1900, 1991, doi: 10.2166/wst.1991.0645.

[56] S. Heuss-Aßbichler, M. John, D. Klapper, U. W. Bläß, and G. Kochetov, "Recovery of copper as zero-valent phase and/or copper oxide nanoparticles from wastewater by ferritization," *J. Environ. Manage.*, vol. 181, pp. 1–7, 2016, doi: 10.1016/j.jenvman.2016.05.053.

[57] M. John et al., "Low-temperature synthesis of CuFeO2 (delafossite) at 70 °c: A new process solely by precipitation and ageing," *J. Solid State Chem.*, vol. 233, pp. 390–396, 2016, doi: 10.1016/j.jssc.2015.11.011.

[58] M. John, S. Heuss-Aßbichler, and A. Ullrich, "Conditions and mechanisms for the formation of nano-sized Delafossite (CuFeO2) at temperatures ≤90 °c in aqueous solution," *J. Solid State Chem.*, vol. 234, pp. 55–62, 2016, doi: 10.1016/j.jssc.2015.11.033.

[59] M. John, S. Heuss-Aßbichler, A. Ullrich, and D. Rettenwander, "Purification of heavy metal loaded wastewater from electroplating industry under synthesis of delafossite (ABO 2) by 'Lt-delafossite process,'" *Water Res.*, vol. 100, pp. 98–104, 2016, doi: 10.1016/j.watres.2016.04.071.

[60] J. Patzsch, I. Balog, P. Krauß, C. W. Lehmann, and J. J. Schneider, "Synthesis, characterization and p-n type gas sensing behaviour of CuFeO2 delafossite type inorganic wires using Fe and Cu complexes as single source molecular precursors," *RSC Adv.*, vol. 4, no. 30, pp. 15348–15355, 2014, doi: 10.1039/c3ra47514j.

[61] M. John, S. Heuss-Aßbichler, and A. Ullrich, "Recovery of Zn from wastewater of zinc plating industry by precipitation of doped ZnO nanoparticles," *Int. J. Environ. Sci. Technol.*, vol. 13, no. 9, pp. 2127–2134, 2016, doi: 10.1007/s13762-016-1049-5.

[62] M. John, S. Heuss-Aßbichler, K. Tandon, and A. Ullrich, "Recovery of Ag and Au from synthetic and industrial wastewater by 2-step ferritization and Lt-delafossite process via precipitation," *J. Water Process Eng.*, vol. 30, no. December, pp. 0–1, 2019, doi: 10.1016/j.jwpe.2017.12.001.

[63] K. Tandon, M. John, S. Heuss-Aßbichler, and V. Schaller, "Influence of salinity and Pb on the precipitation of Zn in a model system," *Minerals*, vol. 8, no. 2, pp. 1–16, 2018, doi: 10.3390/min8020043.

[64] K. Tandon and S. Heuss-Aßbichler, "Fly Ash from Municipal solid waste Incineration – from industrial residue to resource for zinc," in *Industrial wastes – Characterization, Modification and Application of industrial residues*, Pöllmann, Ed. DeGruyter.

[65] I. Anagnostopoulos, J. Knof, and S. Heuss-Aßbichler, "Industrieabwasser als Rohstoffquelle für Buntmetalle: Abwassereinigung und Rohstoffrückgewinnung mit einer portablen Technikumsanlage mit Hilfe des SPOP Verfahrens," in *Deutsche Gesellschaft für Abfallwirtschaft, 9. Wissenschaftskongress*, 2019, pp. 195–199.

[66] S. Y. Kang, J. U. Lee, S. H. Moon, and K. W. Kim, "Competitive adsorption characteristics of Co2+, Ni2+, and Cr3+ by IRN-77 cation exchange resin in synthesized wastewater," *Chemosphere*, vol. 56, no. 2, pp. 141–147, 2004, doi: 10.1016/j.chemosphere.2004.02.004.

[67] C. E. Harland, "Discovery and Structure of Solid Inorganic Ion Exchange Materials," in *Ion Exchange: Theory and Practice*, Second Ed., Cambridge: The Royal Society of Chemistry, 1994, pp. 1–20.

[68] G. Dietrich, *Hartinger Handbuch Abwasser- und Recyclingtechnik*, 3rd ed. Hanser, 2017.

[69] B. Alyüz and S. Veli, "Kinetics and equilibrium studies for the removal of nickel and zinc from aqueous solutions by ion exchange resins," *J. Hazard. Mater.*, vol. 167, no. 1–3, pp. 482–488, 2009, doi: 10.1016/j.jhazmat.2009.01.006.

[70] Z. Ye *et al.*, "An integrated process for removal and recovery of Cr(VI) from electroplating wastewater by ion exchange and reduction–precipitation based on a silica-supported pyridine resin," *J. Clean. Prod.*, vol. 236, p. 117631, 2019, doi: 10.1016/j.jclepro.2019.117631.

[71] M. Vaca Mier, R. López Callejas, R. Gehr, B. E. Jiménez Cisneros, and P. J. J. Alvarez, "Heavy metal removal with mexican clinoptilolite: Multi-component ionic exchange," *Water Res.*, vol. 35, no. 2, pp. 373–378, 2001, doi: 10.1016/S0043-1354(00)00270-0.

[72] V. J. Inglezakis, M. D. Loizidou, and H. P. Grigoropoulou, "Ion exchange of Pb2+, Cu2+, Fe3+, and Cr3+ on natural clinoptilolite: Selectivity determination and influence of acidity on metal uptake," *J. Colloid Interface Sci.*, vol. 261, no. 1, pp. 49–54, 2003, doi: 10.1016/S0021-9797(02)00244-8.

[73] I. Rodríguez-Iznaga, A. Gómez, G. Rodríguez-Fuentes, A. Benítez-Aguilar, and J. Serrano-Ballan, "Natural clinoptilolite as an exchanger of Ni2+ and NH4+ ions under hydrothermal conditions and high ammonia concentration," *Microporous Mesoporous Mater.*, vol. 53, no. 1–3, pp. 71–80, 2002, doi: 10.1016/S1387-1811(02)00325-6.

[74] L. Ćurković, Š. Cerjan-Stefanović, and T. Filipan, "Metal ion exchange by natural and modified zeolites," *Water Res.*, vol. 31, no. 6, pp. 1379–1382, 1997, doi: 10.1016/S0043-1354(96)00411-3.

[75] Y. Li, P. Bai, Y. Yan, W. Yan, W. Shi, and R. Xu, "Removal of Zn2+, Pb2+, Cd2+, and Cu2+ from aqueous solution by synthetic clinoptilolite," *Microporous Mesoporous Mater.*, vol. 273, no. 5, pp. 203–211, 2019, doi: 10.1016/j.micromeso.2018.07.010.

[76] E. Álvarez-Ayuso, A. García-Sánchez, and X. Querol, "Purification of metal electroplating waste waters using zeolites," *Water Res.*, vol. 37, no. 20, pp. 4855–4862, 2003, doi: 10.1016/j.watres.2003.08.009.

[77] H. A. Hegazi, "Removal of heavy metals from wastewater using agricultural and industrial wastes as adsorbents," *HBRC J.*, vol. 9, no. 3, pp. 276–282, 2013, doi: 10.1016/j.hbrcj.2013.08.004.

[78] E. Bernard, A. Jimoh, and J. O. Odigure, "Heavy Metals Removal from Industrial Wastewater by Activated Carbon Prepared from Coconut Shell," *Res. J. Chem. Sci.*, vol. 3, no. 8, pp. 3–9, 2013.

[79] Z. M. Yunus, A. Al-Gheethi, N. Othman, R. Hamdan, and N. N. Ruslan, "Removal of heavy metals from mining effluents in tile and electroplating industries using honeydew peel activated carbon: A microstructure and techno-economic analysis," *J. Clean. Prod.*, vol. 251, no. 4, p. 119738, 2020, doi: 10.1016/j.jclepro.2019.119738.

[80] S. N. Vinaykumar, B. C. Reddy, and A. B. P. Sah, "Removal of Cadmium from Electroplating Industrial Waste Water using Natural Adsorbents," *Int. Res. J. Eng. Technol.*, vol. 6, no. 5, pp. 86–92, 2019.

[81] G. Peng, S. Deng, F. Liu, T. Li, and G. Yu, "Superhigh adsorption of nickel from electroplating wastewater by raw and calcined electroplating sludge waste," *J. Clean. Prod.*, vol. 246, no. 2, p. 118948, 2020, doi: 10.1016/j.jclepro.2019.118948.

[82] C. Shen, Y. Zhao, W. Li, Y. Yang, R. Liu, and D. Morgen, "Global profile of heavy metals and semimetals adsorption using drinking water treatment residual," *Chem. Eng. J.*, vol. 372, no. 2, pp. 1019–1027, 2019, doi: 10.1016/j.cej.2019.04.219.

[83] Y. Zhao, R. Liu, O. W. Awe, Y. Yang, and C. Shen, "Acceptability of land application of alum-based water treatment residuals – An explicit and comprehensive review," *Chem. Eng. J.*, vol. 353, no. 5, pp. 717–726, 2018, doi: 10.1016/j.cej.2018.07.143.

[84] M. Ahmaruzzaman, "Industrial wastes as low-cost potential adsorbents for the treatment of wastewater laden with heavy metals," *Adv. Colloid Interface Sci.*, vol. 166, no. 1–2, pp. 36–59, 2011, doi: 10.1016/j.cis.2011.04.005.

[85] G. Xu, X. Yang, and L. Spinosa, "Development of sludge-based adsorbents: Preparation, characterization, utilization and its feasibility assessment," *J. Environ. Manage.*, vol. 151, no. 5, pp. 221–232, 2015, doi: 10.1016/j.jenvman.2014.08.001.

[86] N. Abdullah, N. Yusof, W. J. Lau, J. Jaafar, and A. F. Ismail, "Recent trends of heavy metal removal from water/wastewater by membrane technologies," *J. Ind. Eng. Chem.*, vol. 76, no. 8, pp. 17–38, 2019, doi: 10.1016/j.jiec.2019.03.029.

[87] W. Lawler *et al.*, "Towards new opportunities for reuse, recycling and disposal of used reverse osmosis membranes," *Desalination*, vol. 299, no. 8, pp. 103–112, 2012, doi: 10.1016/j.desal.2012.05.030.

[88] I. G. Wenten, K. Khoiruddin, A. K. Wardani, and I. N. Widiasa, "Synthetic polymer-based membranes for heavy metal removal," in *Synthetic Polymeric Membranes for Advanced Water Treatment, Gas Separation, and Energy Sustainability*, 1st ed., A. F. Ismail, S. Wan Norharyati Wan, and Y. Norhaniza, Eds. Elsevier Inc., 2020, pp. 71–101.

[89] A. Yusaf, S. Adeel, M. Usman, A. Mansha, and M. Ahmad, "Removal of Heavy Metal Ions from Wastewater Using Micellar- Enhanced Ultrafiltration Technique (MEUF): A Brief Review," in *Textiles and Clothing: Environmental Concerns and Solutions*, M. Shabbir, Ed. New Jersey: Wiley and Scrievering, 2019, pp. 289–315.

[90] H. Kim, K. Baek, J. Lee, J. Iqbal, and J. W. Yang, "Comparison of separation methods of heavy metal from surfactant micellar solutions for the recovery of surfactant," *Desalination*, vol. 191, no. 1–3, pp. 186–192, 2006, doi: 10.1016/j.desal.2005.09.013.

[91]   M. Yaqub and S. H. Lee, "Micellar enhanced ultrafiltration (MEUF) of mercury-contaminated wastewater: Experimental and artificial neural network modeling," *J. Water Process Eng.*, vol. 33, no. 8, p. 101046, 2020, doi: 10.1016/j.jwpe.2019.101046.

[92]   Y. Huang and X. Feng, "Polymer-enhanced ultrafiltration: Fundamentals, applications and recent developments," *J. Memb. Sci.*, vol. 586, no. 5, pp. 53–83, 2019, doi: 10.1016/j.memsci.2019.05.037.

[93]   S. Y. Tang and Y. R. Qiu, "Selective separation and recovery of heavy metals from electroplating effluent using shear-induced dissociation coupling with ultrafiltration," *Chemosphere*, vol. 236, no. 12, p. 124330, 2019, doi: 10.1016/j.chemosphere.2019.07.061.

[94]   B. A. M. Al-Rashdi, D. J. Johnson, and N. Hilal, "Removal of heavy metal ions by nanofiltration," *Desalination*, vol. 315, no. 4, pp. 2–17, 2013, doi: 10.1016/j.desal.2012.05.022.

[95]   A. G. Boricha and Z. V. P. Murthy, "Preparation, characterization and performance of nanofiltration membranes for the treatment of electroplating industry effluent," *Sep. Purif. Technol.*, vol. 65, no. 3, pp. 282–289, 2009, doi: 10.1016/j.seppur.2008.10.047.

[96]   F. M. Pang, P. Kumar, T. T. Teng, A. K. Mohd Omar, and K. L. Wasewar, "Removal of lead, zinc and iron by coagulation-flocculation," *J. Taiwan Inst. Chem. Eng.*, vol. 42, no. 5, pp. 809–815, 2011, doi: 10.1016/j.jtice.2011.01.009.

[97]   Q. Chang and G. Wang, "Study on the macromolecular coagulant PEX which traps heavy metals," *Chem. Eng. Sci.*, vol. 62, no. 17, pp. 4636–4643, 2007, doi: 10.1016/j.ces.2007.05.002.

[98]   H. Al-Zoubi, K. A. Ibrahim, and K. A. Abu-Sbeih, "Removal of heavy metals from wastewater by economical polymeric collectors using dissolved air flotation process," *J. Water Process Eng.*, vol. 8, no. 12, pp. 19–27, 2015, doi: 10.1016/j.jwpe.2015.08.002.

[99]   G. Liu, X. Yang, and H. Zhong, "Molecular design of flotation collectors: A recent progress," *Adv. Colloid Interface Sci.*, vol. 246, no. 8, pp. 181–195, 2017, doi: 10.1016/j.cis.2017.05.008.

[100] T. K. Tran, K. F. Chiu, C. Y. Lin, and H. J. Leu, "Electrochemical treatment of wastewater: Selectivity of the heavy metals removal process," *Int. J. Hydrogen Energy*, vol. 42, no. 45, pp. 27741–27748, 2017, doi: 10.1016/j.ijhydene.2017.05.156.

[101] M. A. González, R. Trócoli, I. Pavlovic, C. Barriga, and F. La Mantia, "Capturing Cd(II) and Pb (II) from contaminated water sources by electro-deposition on Hydrotalcite-like Compounds," *Phys. Chem. Chem. Phys.*, vol. 18, no. 3, pp. 1838–1845, 2015, doi: 10.1039/c5cp05235a.

[102] T. Li *et al.*, "Recovery of Ni(II) from real electroplating wastewater using fixed-bed resin adsorption and subsequent electrodeposition," *Front. Environ. Sci. Eng.*, vol. 13, no. 6, p. 91, 2019, doi: 10.1007/s11783-019-1175-7.

[103] S. Heuss-Aßbichler, A. L. Huber, and M. John, "Recovery of Heavy Metals From Industrial Wastewater – Is It Worth It?" *5th Int. Conference "Industrial Hazard. Waste Manag.*, no. September 2016, 2016.

[104] Statistisches Bundesamt, "Umwelt Abfallentsorgung 2017," Wiesbaden, 2019.

[105] Umweltbundesamt *et al.*, "Überprüfung der Grenzwerte von Metallen in Abfällen, bei deren Überschreitung eine Verwertung mit Metallrückgewinnung der einfachen Abfallverwertung im Versatz oder auf Deponien vorgeht," Dessau-Roßlau, 2016.

[106] Statistisches Bundesamt, "Umwelt Abfallentsorgung 2011," Wiesbaden, 2013.

[107] Springer & Sohn GmbH, "Metallabrechnung," 2020. [Online]. Available: https://www.springer-und-sohn.de/uploads/springer-und-sohn_rv-preise.pdf.

[108] L. Liu, Y. Liang, Q. Song, and J. Li, "A review of waste prevention through 3R under the concept of circular economy in China," *J. Mater. Cycles Waste Manag.*, vol. 19, no. 4, pp. 1314–1323, 2017, doi: 10.1007/s10163-017-0606-4.

Susmita Sarmah, Jitu Saikia, Pinky Saikia, Champa Gogoi,
Rajib Lochan Goswamee

Chapter 14
Composites of some sustainable siliceous materials for the removal of fluoride from ground water and immobilization of the sludge generated

Abstract: Across the world, with an ever increasing population burden the pressures on meeting the demand supply position of quality drinking water is becoming day by day serious both from the quality and quantity aspects. In this context the fluoride contamination of ground water is a common public health problem and to control the same a large number of methods are evolving. Being an inorganic contaminant it is not possible to drive away fluoride from water by photo or bacterial degradation. Thus adsorption over cheaper surfaces is the most common economic approach for it. However, disposal of the sludge produced in the process is a major technical hurdle. In this work a common potter's clay has been surface modified by oligomeric hydroxyl-alumina to adsorb fluoride. It was observed that fluoride removal capacity is favoured by high temperature and low pH. Adsorption kinetics as well as adsorption isotherms have been investigated. The spent adsorbent from the hydroxyl-alumina activated potter's clay has been studied for the immobilization of the sludge. By utilizing the sludge as a raw mix substituent for cement preparation the sludge could be managed safely in the lab scale studies.

Keywords: Potter's clay, defluoridation, sludge management

Acknowledgements: The authors are grateful Director CSIR-NEIST for allowing to publish the work. The authors are grateful to CSIR India, Ministry of Environment, Forests and Climate Change, New Delhi, Govt of India for supporting the work at the initial stages. Also, authors are grateful to Water Technology Initiative (WTI) scheme of Department of Science and Technology, New Delhi, Govt of India for supporting the work both in the laboratory and up-coming scale up activities.

Susmita Sarmah, ICAR NBSS-LUP (NER CENTER) Jamuguri Road, Jorhat 785004, Assam, India
Jitu Saikia, Dept of Chem., Pandit Deendayal Upadhyaya Adarsha Mahavidyalaya, Ratowa, Biswanath, Assam, India, 784184
Pinky Saikia, Dept of Chem., Joya Gogoi College, Khumtai, Golaghat, Assam, India, 785619
Champa Gogoi, Dept of Chem., C N B College- Bokakhat 785612, Golaghat, Assam, India
Rajib Lochan Goswamee, Advanced Materials Group, Materials Science and Technology Division Council of Scientific and Industrial Research – North East Institute of Science and Technology (formerly RRL Jorhat) Jorhat, Assam, India 785006, e-mail: goswamirl@neist.res.in, rajibgoswamee@yahoo.com

https://doi.org/10.1515/9783110674941-014

1 Introduction

1.1 Ground water quality

Water is the most essential component to the living organisms. Without water no one can survive on the earth. Although, about 70% of the earth's surface is covered by water yet out of it only 2.80% is safe for use and the remaining 97.20% is saline and not of potable grade. Potable water is considered as fit for consumption by humans and other animals but all the potable water sources are not safe. A safe drinking water has no significant risk to health when consumed over a long time [1]. Water sources are contaminated through both natural and anthropogenic activities. According to World Health Organization (WHO) about 58% death of the global population are due to the water borne diseases.

The major chemical water contaminants include some heavy metals, non-metals, free radicals and chemical compounds ranging from iron, lead, mercury, chromium, cadmium, zinc, uranium, copper, arsenic, fluoride, chloride, nitrate, sulphate, different colouring substances thrown out from dye, textile and paper industries, oil and grease bearing effluents from refineries and petrochemical plants, pesticide and its residues from agricultural fields or large plantations, grain storage sites to name a few. Amongst all of them, fluoride and arsenic are very common harmful contaminants found in ground water across the globe.

Fluoride is one of the most important elements that serve to maintain healthy teeth and bones but when its concentration in water exceeds 1.5 mgL^{-1} (ppm) [1–4] the consumption of such water becomes threat to human health. The toxic effect of exposure to excessive fluoride intake is manifested in the form of skeletal and non skeletal disorders together called fluorosis.

1.2 Common techniques used to improve quality of fluoride bearing water

As globally, ground water use is increasing accordingly side by side there is an urgent increasing need of developing newer technology options to reduce the excess fluoride from water. Therefore, a large amount of research attentions have been focussed on the area with the offering of a number of technology solutions viz. precipitation [5, 6], electro-coagulation [7, 8], ion exchange [9, 10], nano filtration [11], adsorption over a large number of adsorbent surfaces [12–14]. Among all the removal techniques, adsorption is the most convenient method, mainly due to its simple design and low process cost [15]. Nalgonda technique [16] is a well known adsorptive fluoride removal process, where fluoride is basically arrested over alumina surface and used widely in the developing countries like India. Literature also says that activated alumina is the most efficient adsorbent for the removal of fluoride from water [17]. The high fluoride adsorption by aluminium oxide is due to favourable hard-hard interaction of Al^{3+} and F^- centers. In the present era some newer higher surface area bearing nano materials are also prescribed for defluoridation e.g. Ce-Ti@Fe_3O_4 and Ce-Ti oxide nano particles [18] Layered Double Hydroxide (LDH) based nano sheets [19, 20], nano oxide based composites [21, 22].

Apart from simplicity of design and cost, the success of a defluoridising adsorbent depends on the issue of post adsorption safe disposal of the sludge generated. Although, number of adsorbents like activated aluminium oxide, other alumina based compound, activated carbon, calcite, clay, tree bark, saw dust, rice husk, bone char are reported to be good adsorbent materials but most of them fail on the ground of safe disposal of the sludge generated.

In this present work, some easily available potter's clay was surface modified with nano aluminium oxide and studied as a defluoridising adsorbent material. Clay was selected because of its wide availability, relatively inert toxicity and its ease of interaction with an additional precursor compound called hydroxyl-alumina taken for this study. Hydroxyl-alumina (OH-Al) is also called as Poly-Basic Aluminium Chloride or partially hydrolysed Aluminium Chloride. The compound is basically a polycation with a Keggin type structure of 13 Aluminium centers having a tetrahedral aluminium ion surrounded by 12 numbers of hydroxidic octahedras of aluminium (Figure 14.1) with shared edges and faces and positive charge over each oligomeric centers [23]. Its chemical formula can be written as $AlO_4 Al_{12}(OH)_{24}(H_2O)_{12}^{7+}$ [24]. The complex has a high charge to radius ratio and gets easily converted to Al_2O_3 on application of temperature as low as 80 °C [25].

Such an oligomeric water soluble hydrolytic polycation can immediately react over any colloidal solid surface having a net negative surface charge. Clay surfaces are well known candidates for the same, which is the basis of a common prescription for formation stabilisation during water injection in secondary oil recovery wells [26].

In the present context of preparation of Al_2O_3 nano-particle loaded clay based defluoridation adsorbent preparation, the water soluble Al oligomers were first prepared by a reaction of metallic aluminium powder with aluminium chloride at a proper stoichiometric ratio [25]. This was followed by its coating it over clay surface by impregnation in aqueous solution and development of Al_2O_3 nano centers by heat treatment of the hydroxyl-alumina impregnated clay at around 80–100 °C. This way the Al_2O_3 nano-particles are supported over the clay surface which facilitates high fluoride adsorptivity. Adsorption of fluoride over such a composite surface is enhanced by hard-hard interaction of Al and fluoride atoms [27]. Although, number of researcher have reported the application of different types of clays for fluoride removal [28–30] but decorating the surface of the clay with hard acid type nano oxides to develop some superior adsorbents has not been reported. Another important aspect is the prospect of safe use of the spent adsorbent as cement raw mix component, which is the focus of the present report.

Structurally, clay contains both SiO_2 and Al_2O_3 in high amount and presence of high amount of Al_2O_3 is theoretically advantageous for better fluoride adsorption. Apart from that clay is a low cost material, having high water sorption capacity, high ion exchange capacity, high surface area, favourable molecular sieve like porous structure, good chemical stability, biological and environmental benignity which make it a good adsorbent surface. Many of these properties are further enhanced on heat treatment. To confirm enhanced adsorption various associated parametric evaluations like determination of the adsorption isotherm, kinetics of the rate of adsorption, the effects of adsorbent dose, pH, temperature and co-existing ion on defluoridation is carried out. In the present work plastic clay rich soil sample was collected from the agricultural fields of Dhodar Ali, Tanti Gaon from Titabor area of Jorhat District, Assam, India with a GPS location of N = 26°34/36.04// and E = 94°9/54.46//. A portion of soil weighing around 20 Kg from potter's own clay yard was collected. As the collected sample was rich in clay and was taken as such for experimentation, it was impregnated with a solution of hydroxyl-alumina.Here a reported method [27] is followed to synthesise the chemical activating agent, i.e. hydroxyl-alumina in a ratio of the clay and 0.5% hydroxyl-alumina aqueous solution at 1:20 ratio. The required time was 3 hrs and after reaching the required time the solution was filtered and the clay was dried at about 80 °C in an air oven. The dried mass is the required adsorbent for adsorption studies.

Theoretically, the adsorption of fluoride on to solid adsorbent generally takes place via three steps, first through the external mass transfer reaction i.e. diffusion or transport of fluoride ions to the external surface of the adsorbent, secondly fluoride adsorption on to the particle surface and finally by transfer of fluoride ions into the internal surface of the porous adsorbent which is known as intra particle diffusion [31]. To study the adsorption process, the experimental data of the kinetics study were fitted in different kinetics models such as pseudo-first order kinetic model, pseudo-second order kinetic model, Elovich model and intra-particle diffusion model (Table 14.1). Where q_t and q_e are the adsorption capacities at time t and at equilibrium, respectively (mg g^{-1}), k_1 (min^{-1}) and k_2 (g mg^{-1} min^{-1}) are the rate constants of the first- and second-order sorption, respectively, k_d is the intra-particle diffusion rate constant (mg g^{-1} min$^{-1/2}$), and C is a constant in the intra-particle diffusion equation, which corresponds to the thickness of the boundary layer, α is the initial adsorption rate defined in the Elovich equation (mg g^{-1} min^{-1}) and β is the Elovich constant (g mg^{-1}) [31, 32].

The experimental results from kinetics study were plotted in different models like pseudo first order model, pseudo second order model [33], Elovich model and intra particle diffusion models [34, 35]. The linear fitting results from fitting experimental data to pseudo-first-order, pseudo- second-order, Elovich, and intra-particle diffusion models are shown in Table 14.1. Comparing with the correlation coefficient value of the linear plot of each model, it can be concluded that the kinetics follow the pseudo second order with $R^2 = 0.99$, from which it can be said that the present adsorption reaction is a chemical sorption or chemisorptions, through sharing or exchange of electrons between adsorbent and adsorbate [36].

Table 14.1: Comparison of parameters of different kinetic-models of defluoridation using clay adsorbent.

Kinetic models	Parameters	Values
Pseudo First order $$\log(q_e - q_t) = \log q_e - \frac{k_1}{2.303} t$$	K_1 (min^{-1})	5.5272×10^{-3}
	q_e, cal (mg/g)	0.340
	R^2	0.307
Pseudo Second order $$\frac{t}{q_t} = \frac{1}{k_2 q_e^2} + \frac{t}{q_e}$$	K_2 (g/mg min)	0.130
	q_e, cal (mg/g)	1.275
	R^2	0.999
Intra particle diffusion $$q_t = k_d t^{1/2} + C$$	K_d $(mg/g \ min^{1/2})$	0.0059
	C	1.095
	R^2	0.772
Elovich $$q_t = \frac{1}{\beta} \ln(\alpha\beta) + \frac{1}{b} \ln t$$	β (g/mg)	24.925
	α [mg/(g min)]	2.3326×10^9
	R^2	0.943

2 Immobilization of the generated sludge

2.1 Immobilization of fluoride adsorbed over PHA and clay surface

As the fluoride toxicity is a major environmental problem and adsorption over suitable adsorbent surface is a most common solution, consequently the problem of disposal of the fluoride containing spent adsorbent sludge is posing a serious threat to the environment. Therefore, it is extremely important to develop some safe solutions for the permanent disposal of such harmful sludges. Additionally, if the sought for safe solution entails some valorization of the process by conversion of the sludges to any economically useful material it may even offer comparative cost advantages to the overall fluoride mitigation process. One such relevant method of stabilisation of the Fluoride emitted from ceramic tiles manufacturing industry has been reported by Ponsot et al where the Fluoride bearing flue gas is adsorbed in saturated lime sludge for paving the way to ultimately dispose them as calcium silicate [34]. In this method the acidic fumes are absorbed over lime and the exhausted saturated lime is treated with waste glasses like broken Cathode Ray Tubes (CRT) along with clay at around 900 °C giving a Cupsidine ($Ca_4Si_2O_7F_2$) like phase capable of stabilizing Fluoride with a reduced leaching rate.

While the above approach might be useful for acidic Fluoride fumes adsorption over lime surface in ceramic industry, an identical approach of sintering with waste glasses may not be appropriate for water treatment sludges particularly produced during decentralised small community based defluoridation facilities in rural areas. This is because the quantity of weekly production of such sludges would not only be small in amount to afford a complex post adsorption high temperature disposal facility within its immediate neighbourhood nor transport of toxic sludges to a centralised collection repository is allowed by the regulatory authorities. In such a context the present group of authors in an earlier report have proposed an idea of adsorption of Fluoride over commonly available and affordable media viz. Paddy Husk Ash (PHA) [27, 35] where surface is modified by nano-alumina centers derived from oligomeric hydroxy-polycation of alumina as shown in Figure 14.1. The surface modified PHA not only shows good defluoridation efficiency as well as facile routes for sludge handling. The process offered the opportunities of treating the dried sludge either in lime through the so-called 'lime – silica' hydration reaction or using it as a partial substituent of clay and lime in raw mixture for black meal cement manufacturing process. In practice the lime-silica reaction is a very old concept adopted by construction specialists from ancient times. Romans used the process even for construction of underwater structures. In India also various age old structures exist where this methodology was adopted by different rulers, for example the four hundred fifty years old *Rang Ghar* an amphitheatre like structure built by Ahom rulers of Thai origin in Sivasagar Assam or more than two hundred year old British colonial administrative buildings across various states in the sub-continent (Figure 14.2.) of pre-Portland cement era have been basically made by such construction practices. These structures are standing solid and strong till now indicating the prospect of exploring the suitability of the stabilization mechanism working within it for immobilization of toxic ions like Fluoride in its present day contexts and requirements.

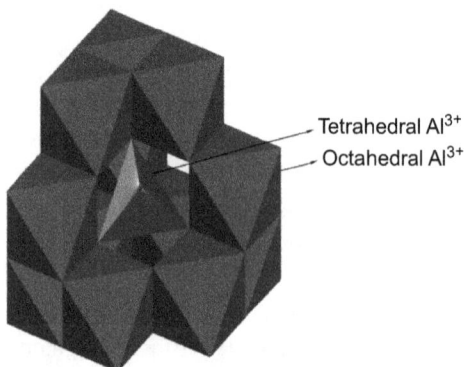

Tetrahedral Al^{3+}
Octahedral Al^{3+}

Figure 14.1: Structure of $AlO_4 Al_{12}(OH)_{24}(H_2O)_{12}^{7+}$ type complex (Redrawn from ref. [23] Copyright 1981 with the permission of Wiley).

Figure 14.2: Image of an old British period court building in India (in Jorhat, Assam) and the amphitheatre (*Rang Ghar*) made by Ahom king (in Sivasagar, Assam) which were constructed using 'Lime-Silica' cement.

However, since PHA is available mainly in paddy growing tropical countries only, therefore effort was made to substitute PHA in the presently described work by common potters' clay available in different forms around the world. The inherent aluminium oxide of the clay as well as surface situated nano aluminium oxide grains derived from hydroxyl alumina precursor together not only contributed towards the Fluoride adsorption also high surface area of clay contributes to faster 'lime-silica' reaction and source of additional alumina for fixing the raw material composition during eventual burning for stabilisation through clinkerisation. As well as it offers the ease of formation of high temperature phases through clinkerisation assistance observed due to the presence of Fluoride species during high temperature firing.

In this context it may be further added in the present day world different type of construction related industries have shown interest in the recycling of industrial and civic wastes generated by them. Partial or complete substitution of waste sludge materials into the admixture composite in concrete industry saves the disposal site as well as can reduces the cost and consumption of natural raw materials for construction activity [36]. Here the fluoride bearing waste clay sludge was used as a substituent for silica and alumina in the raw mix composite of cement clinker preparation. It is well known that lime-stone and clay are the main raw mix composite for preparation of commercial cement clinker. Keeping the same concept, it was tried to substitute the clay part of the cement admixture with the fluoride containing clay adsorbent for preparation of good quality cement clinkers. Various studies are there for gainful management of waste sludge into concrete [37–39].

2.1.1 Preparation of cement clinkers by utilizing the clay sludge

As stated above the study of disposal of sludge was carried out by utilizing the fluoride bearing waste to prepare cement clinker and studying the leaching of fluoride from the same. It is well known that lime and natural alumino-silicate clay are the two main raw mix components used to prepare clinker in cement industry. This basic principle was applied in this present approach also. Initially a 2–3% of the fluoride bearing clay was substituted in the raw mix composite. After different optimization experiments it was observed that up to 100% of the plain clay can be replaced with fluoride bearing clay sludge to obtain some good quality cement clinker.

For the preparation of initial admixture compositions needed for making the cement clinkers, the dried lime and the spent adsorbent or sludge was thoroughly ground into fine powder in an Agate mortar followed by its grinding in a planetary ball mill. The admixture was prepared as par black meal cement manufacturing process. The hand pressed clinker balls were burnt in digital muffle furnace at 1450 °C and 1350 °C for about 45 minutes in air atmosphere.

Generally, the progress of clinkerisation behaviour is described by the formation of four mineral phases by observing under powder XRD along with decrease of free lime content. These four phases are tricalcium silicate (C_3S), dicalcium silicate (C_2S), tricalcium aluminate (C_3A) and tetra-calcium aluminoferrite (C_4AF). From the XRD characterization (Figure 14.3) it was observed that clinker made at 1450 °C by using fluoride bearing clay sludge have the same phases with raw clinker which are prepared in controlled condition without the addition of fluoride sludge. Side by side it was observed from the XRD study that the raw clinkers prepared at 1350 °C in controlled experiments i.e. without the addition of fluoride sludge have only a part of the desired phases formed with reduced intensity and complete absence of C_3A phase but as we prepare the clinker by substituting the clay part with fluoride bearing sludge even after burning at 1350 °C the intensity of those phases were found sharp.

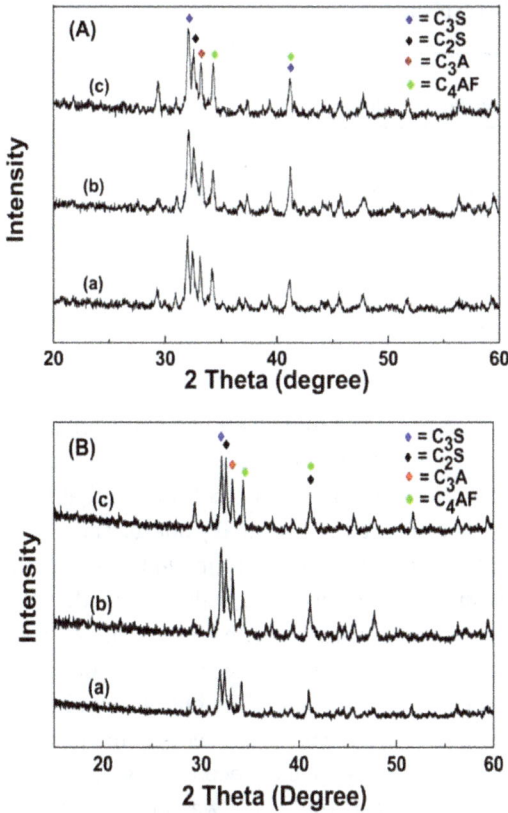

Figure 14.3: XRD study of the clinkers prepared at (A) 1450 °C and (B) 1350 °C, here (a) cement clinker prepared from clay and limestone with 0% replacement of clay by sludge, (b) cement clinker prepared from clay and limestone with replacement of clay by fluoride bearing sludge obtained after successful adsorption of fluoride from fluoride spiked water, and (c) cement clinker prepared from clay and limestone with replacement of clay by fluoride bearing sludge obtained after successful adsorption of fluoride from fluoride contaminated ground water.

Also, it was found that such clinkers had reduced amount of free lime. The free or uncombined lime (CaO) in the burnt product is an important factor for the study of clinkerisation process, which is determined by standard ethylene glycol method [40]. It involves extraction of free CaO with ethylene glycol at 65–70 °C in the form of calcium glycolate and titrating the same with hydrochloric acid to give back ethylene glycol and calcium chloride [41]. Bromocresol green is used as indicator for this titration. When in a clinker the percentage (wt.%) of free lime found to be only 0.2% clinker formed is considered as of good quality.

The composition of the raw mixtures used for clinker burning is given in Table 14.2. Odler and Maula [42] in a detailed similar study have reported the positive effect of CaF_2 addition in raw meal on clinker formation and its effect on lowering of

Table 14.2: Raw mix composite for cement clinker preparation.

Sample	% of chemical grade lime	% of Titabor clay	% of Sludge
Sample 1	74.40	25.54	0
Sample 2	74.40	22.986	2.554
Sample 3	74.40	20.432	5.108
Sample 4	74.40	12.77	12.77
Sample 5	74.40	0	25.54

clinkerization temperature. They have reported that during the process the formation temperature of C_3S phase may even go down as low as 1250 °C. Also, they reported that during the process only a small amount of Fluoride is lost as fumes and at high Fluoride level some of the C_3A phase gets converted to $C_{11}A_7.CaF_2$ phase. They have reported that hydration of such cement leads to accelerated formation of ettringite (AFt) phase also. Thus, it is known that in the substituted clinker fluoride sludge can not only acts as a fluxing agent for easier clinker burnability with consequent reduced thermal energy consumption [36] also the clinkers produced have better performance so far as the AFt phase formation is concerned. From these different observations it can be concluded that fluoride as a mineraliser can decrease clinker burning temperature around 100 °C which can be an energy saving step. If the economic gain of such energy savings can be transferred to rural water users the cost of water defluoridation can be expected to go down.

2.1.1.1 Fluoride leaching study

The study of leaching from ultimate disposable solids, bearing the toxic components is an important factor in developing the effective adsorbent materials. Here we have applied some previously reported method [35, 43] for the leaching study of fluoride from the prepared clinkers. For the determination of the leachate fluoride, about 1 g of the finely ground fluoride bearing clinker was shaken in 50 ml of double distilled water at room temperature and at a higher temperature of 90 °C for 2 hrs. After filtration, the volume was made up to 100 ml and leached out fluoride concentration was tested in the solution. The observed results are discussed in the result and discussion section.

Additionally, standard Toxicity Characteristics Leaching Protocol, TCLP [44] was also followed for checking of fluoride leachate from the clinkers in a more acidic condition. For it, 1 g finely powdered clinker was suspended in 20 ml TCLP fluid at pH 4.9 containing 1:1 distilled water and acetic acid. The suspension was stirred by a Teflon coated magnetic needle for 24 hours at 100 rpm over a magnetic stirrer. After the required time, the suspension was filtered and the pH of the filtrate further brought down to 2.5 by adding 6 M HNO$_3$. The leachate fluoride of the resulting solution was measured with using ion-sensitive electrode.

From the fluoride leaching tests, no significant fluoride leaching was observed for clinkers. From the absence of fluoride in the leachate it could be accepted that strong retention of fluoride occurs within the clinker structure. Therefore, it can be concluded that the clinker produced from fluoride bearing sludge are non-hazardous and will have no negative effect on the environment.

2.1.2 Permanent immobilization of Fluoride retained in ceramic burnt clay adsorbents

As adsorption is one of the most convenient users friendly, cost efficient, high throughput process for water treatment in recent days a large number of newer adsorbents are also developed. One of them is ceramic bodies of different forms like ceramic barriers of flat, nodular, tubular etc shapes. Many of them are used for various applications towards wastewater treatment in biotechnology, pharmaceutical and food industries depending upon different water quality parameters. Like in case of PHA or clay based adsorbents in case of ceramic adsorbents the problem of waste disposal can be solved by immobilization of the waste materials in different composites such as the preparation of paver block, production of Portland cement, glass ceramics, concrete preparation, as component in cementitious composites pastes for masonary work [36, 45–54].

Ceramic based adsorbent barriers have their own advantages over other adsorbent materials and polymeric membrane in terms of their high thermal, mechanical and chemical stability. Use of clay based ceramics on this context have some advantages like easy availability of the material, low cost and comparatively lower sintering temperature as compared to metal oxide membranes. A variety of local easily available raw materials including plastic clays, natural phosphate, stone quarry dust, volcanic pozzolana, fly ash and natural apatite are used for preparation suitable low cost ceramics of different shapes. Figure 14.4 shows the nodular and flat ceramic barriers prepared from a local potters' clay.

Different research workers are working on the development of cost efficient and environment friendly methods for the preparation and application of ceramics for the separation of toxic pollutants from water and air. Use of various ceramic adsorbents used for removal of different contaminants is presented in Table 14.3.

As said above, Goswamee and coworkers have prepared cost efficient ceramic barriers of different shapes i.e. circular dish shaped and nodules using locally available raw materials such as potters' clay, stone quarry dust and tea waste for efficient adsorption of toxic water pollutants such as fluoride and dyes after modification of the barriers with carbon layer [55, 56]. These reports show that the barriers could be used efficiently for the treatment of different water contaminants and they show very good adsorption capacity. The main advantage of this method is

Figure 14.4: Photographs of (A) raw clay based ceramic nodules and (B) burnt clay based ceramic nodules obtained on firing at 1000 °C (Reproduced from ref. [55] Copyright 2017 with the permission of Elsevier) (C) clay based flat ceramic membrane, (D) clay based flat ceramic membrane after sintered at 900 °C (Reproduced from ref. [64] Copyright 2019 with the permission of Springer).

the recyclable way of handling the toxic sludge generated during the adsorption process. Like in case of nano alumina modified clay as described in the section 2.1.1, a method was also developed for permanent immobilization of toxic fluoride adsorbed ceramics in cement clinkers by reacting it with lime and clay by black meal method of clinker preparation.

Apart from CaF_2 different workers across the globe have reported various methods of addition of mineralizers like AlF_3, ZnO, MnO_2, SnO_2, CuO to the raw mixture to form stable clinkers and many of them brings down the clinker formation temperature, which ultimately would result in decrease of the energy requirement and the

Table 14.3: Comparison between various clay and ceramic based adsorbents used for pollutant removal.

Name of the adsorbent	Contaminant type	Adsorption capacity (mg/g)	References
Potters' clay based flat ceramic	Dye	–	57
Granular ceramic adsorbent	Fluoride	0.93–0.99	58
carbon coated ceramic barriers	Fluoride	–	59
Porous granular ceramic containing dispersed Aluminium and Iron oxide	Fluoride	1.79	51
Fe-Al impregnated granular ceramic adsorbent	Fluoride	3.56	60
Ceramic adsorbent	Fluoride	2.16	61
Microporous hydroxyapetite ceramic bead	Fluoride	7–12	62
HAC modified ceramic nodule	Fluoride	6.32	56
Zirconium metal-organic frameworks modified alumina membrane	Fluoride	102.40	63

CO_2 emission to the environment [57–59]. Globally the reduction of CO_2 from the cement manufacturing to the atmosphere is taken as a very urgent programme.

It was observed by the present group of authors that all the necessary phases like tricalcium silicate ($3CaO.SiO_2$), dicalcium silicate ($2CaO.SiO_2$), tricalcium aluminate ($3CaO.Al_2O_3$) and tetra calcium aluminoferrite ($4CaO.Al_2O_3.Fe_2O_3$) found in a typical Portland cement clinker are observed in the prepared clinkers containing waste ceramic based fluoride adsorbents. The study also showed that conventional firing temperature of 1450 °C for clinker formation could be lowered to 1350 °C by the addition of fluoride adsorbed ceramic waste due to the said mineralization activity Al-F species present in adsorbent surface. In the reported work it is shown that lowering of burning temperature of the clinkers by 100 °C did not change the peak positions and the all the phases were found to form properly.

The XRD patterns of the clay based ceramics and the clinkers prepared from the fluoride adsorbed ceramics are shown in Figure 14.5. Quartz is the main crystalline phase of the clay based ceramics with other minor phases like trydimite, mullite and cristobalite. The major crystallographic phases formed during the cement clinker formation are Tricalcium silicate, C_3S ($3CaO.SiO_2$); dicalcium silicate, C_2S ($2CaO.SiO_2$); tricalcium aluminate, C_3A ($3CaO.Al_2O_3$) and tetracalcium aluminoferrite, C_4AF ($4CaO.Al_2O_3.Fe_2O_3$) [57].

FESEM monographs of clay based ceramics and the clinker obtained from fluoride adsorbed clinker is shown in Figure 14.6. A rough, discontinuous and porous surface can be observed in case of the ceramic barriers due to the melting of the

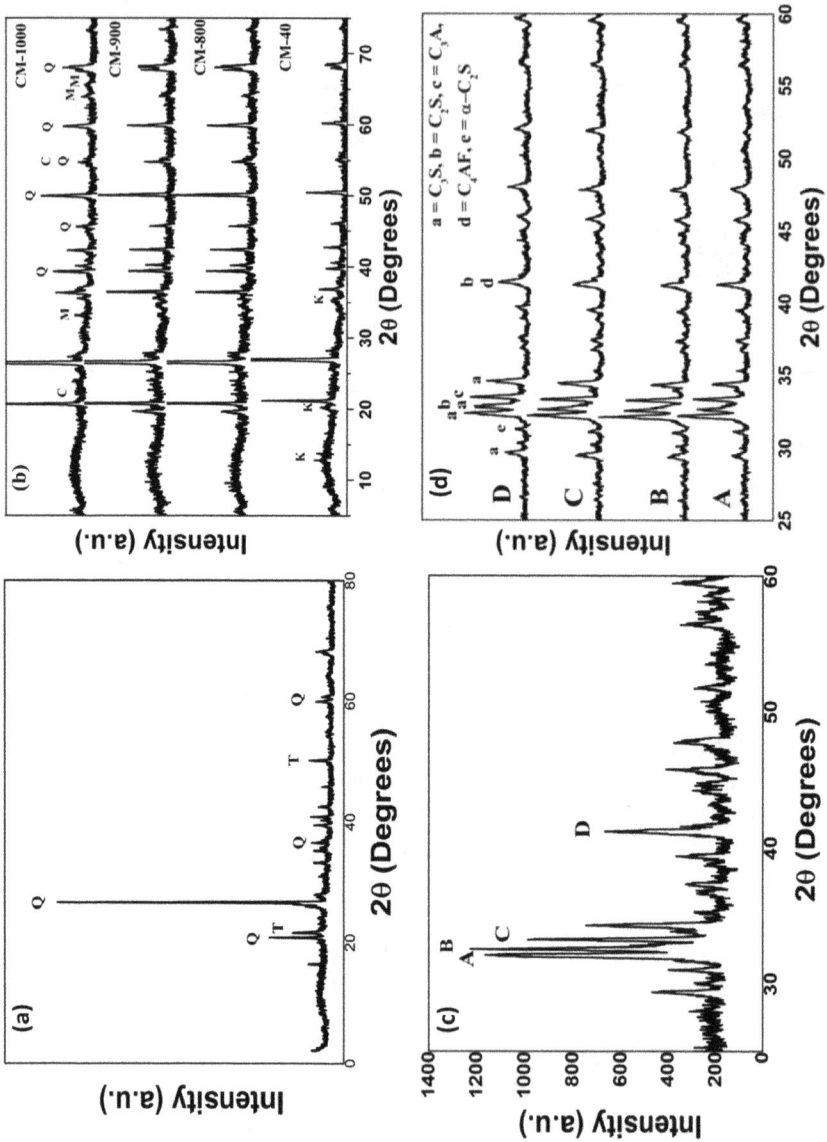

Figure 14.5: PXRD patterns of (a) clay based ceramic nodules obtained on firing at 1000 °C, (b) clay based flat ceramic membrane sintered at variable temperatures (Reproduced from ref. [64] Copyright 2019 with the permission of Springer), (c) Clinker prepared from lime, clay and fluoride adsorbed modified ceramic nodules obtained on firing at 1000 °C (A = C_3S, B = C_2S, C = C_3A and D = C_4AF) (d) Clinker prepared from lime, clay and fluoride adsorbed nano-aluminum oxyhydroxide deposited carbon (NAOC) coated clay based flat ceramic barrier, (A) Ceramic membrane fired at 1450 °C, (B) Fluoride adsorbed Ceramic membrane fired at 1450 °C, (C) Ceramic membrane fired at 1350 °C and (D) Fluoride adsorbed Ceramic membrane fired at 1350 °C (Reproduced from ref. [56] Copyright 2019 with the permission of Springer).

clay minerals at high firing temperature. From the FESEM monograph of the waste derived clinker, crystalline phases were obtained which is a characteristics of the high temperature fired materials and is mainly found in cement clinkers.

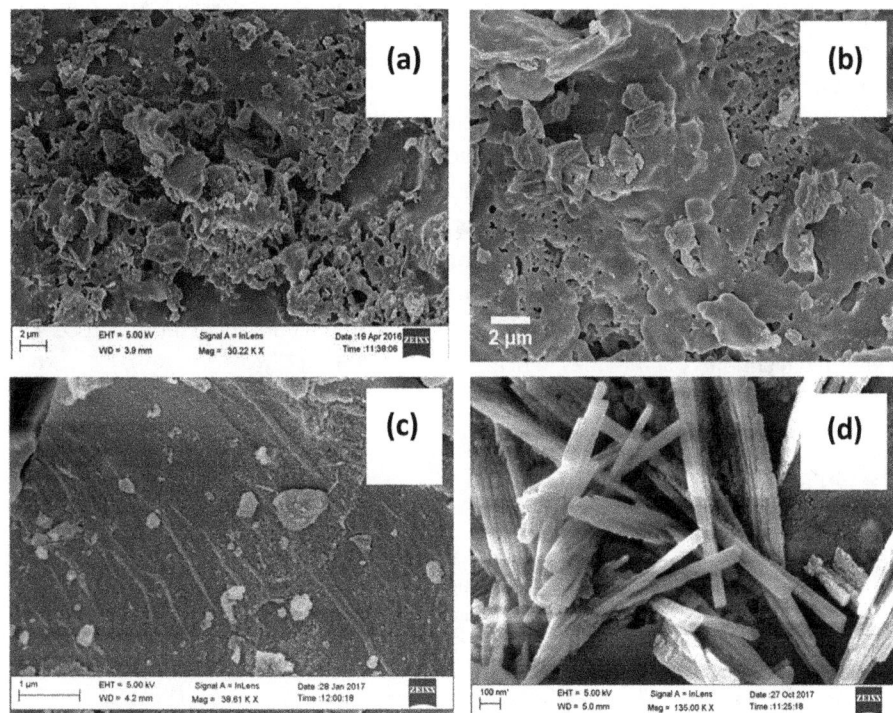

Figure 14.6: FESEM images of (a) clay based ceramic nodules obtained on firing at 1000 °C (Reproduced from ref. [55] Copyright 2017 with the permission of Elsevier), (b) clay based flat ceramic membrane fired at 900 °C (Reproduced from ref. [64] Copyright 2019 with the permission of Springer), (c) nano-aluminum oxyhydroxide deposited carbon (NAOC) coated clay based flat ceramic barrier (Reproduced from ref. [56] Copyright 2019 with the permission of Springer) and (d) crystalline phase of the clinkers prepared by substitution of clay portion of raw mixture by nodular ceramic waste obtained after fluoride adsorption.

3 Conclusion

Although one of the cheapest ways of mitigation of Fluoride contamination of ground water is adsorption yet across the world the method is skeptically received as post adsorption sludge disposal is a serious environmental threat. However, utilizing siliceous common agricultural wastes like paddy husk ash or common plastic potters clays or fired ceramic barriers derived from such potters clays adsorbents, which can be easily surface modified with aluminum oxide nano centers derived from hydrolytic aluminium oligomer are not only a simpler route to develop superior fluoride adsorbents also it can provide good supplementary cementitious materials or raw material component. They can be handled by both common 'lime-silica' reactions or through design of raw mix component for high temperature burning in kilns to give high quality clinkers at a reduced firing temperature with reduced energy consumption and consequent reduced green house gas emission. Thereby the waste sludge can be converted to a secondary economic raw material and the economic and environmental benefits obtained from the process if transmitted to the rural water users can make the water affordable to them at a cheaper rate. The, ease with which they can be reacted with lime and stabilized can facilitate their temporary immobilization at the water treatment plant site as a measure for intermediate stabilization for ultimate transport to a safe central repository as par environmental regulatory authorities safety protocols. The intermediate stabilized adsorbents then can used for the preparation of raw mix for ultimate burning in high temperature kilns. The efforts are underway with the authors to lift the process to a matured technology sometimes in near future.

References

[1]   WHO (World Health Organization) Guidelines for Drinking-Water Quality, 2011, 4th ed.
[2]   BIS (Bureau of Indian Standards) Specification for Drinking Water IS 10500: 2012, New Delhi, India. Available: http://cgwb.gov.in/Documents/WQ-standards.pdf.
[3]   EU (European Union) Council, Council Directive 98/83/EC of 3 November 1998 on the Quality of Water Intended for Human Consumption.
[4]   WHO (world health organization), guideline for drinking water quality, World Health Organization, Geneva, 2004.
[5]   Thakre D, Rayalu S, Kawade R, Meshram S, Subrt J, Labhsetwar N, Magnesium incorporated bentonite clay for de-fluoridation of drinking water, J. Hazard Mater. 2010, 180(1–3), 122–30.
[6]   Yadav AK, Kaushik CP, Haritash AK, Kansal A, Neetu R, J Hazard Mater 2006, 128 289–293.
[7]   M, Gourich B, Essadki AH, Vial C, Delmas H, Defluoridation of Morocco drinking water by electrocoagulation/electroflottation in an electrochemical external-loop airlift reactor, Chem. Eng. J. 2009, 148, 122–131.
[8]   Hu CY, Lo SL, Kuan WH, Effect of co-exiting anions on fluoride removal in electro coagulation process using aluminium electrodes, Water Res. 2003, 37, 4513–4523.
[9]   Meenakshi S, Viswanathan N, Identification of selective ion-exchange resin for fluoride sorption, J. Colloid Interface Sci. 2007, 308, 438–450.
[10] Chubar N, New inorganic (an) ion exchangers based on Mg–Al hydrous oxides: (Alkoxide-free) sol–gel synthesis and characterisation, Journal of Colloid and Interface Science, 2011, 357,198–209.
[11] Liu J, Xu Z, Li X, Zhang Y, Zhou Y, Wang Z, Wang X, An Improved process to prepare high separation performance PA/PVDF hollow fiber composite nano filtration membranes. Sep. Purif. Technol. 2007, 58, 53–60.
[12] Geethamani CK, Ramesh ST, Gandhimathi R, Nidheesh PV, Fluoride sorption by treated fly ash: kinetic and isotherm studies. J. Mater. Cycles Waste Manag. 2013, 15(3),381–392.

[13] Sailaja BK, Bhagawan D, Himabindu V, Cherukuri J, Removal of fluoride from drinking water by adsorption onto Activated Alumina and activated carbon. IJERA.2015, 5, 19–24.

[14] Paudyal H, Inoue K, Kawakita H, Ohto K, Kamata H, Alam, S, Removal of fluoride by effectively using spent cation exchange resin. J. Mater. Cycles Waste Manag. 2017, 20(2),975–984.

[15] Lanas SG, Valiente M, Aneggi E, Trovarelli A, Tolazzi M, Melchior A, Efficient fluoride adsorption by mesoporous hierarchical alumina microspheres, RSC Adv. 2016, 6, 42288.

[16] Nawlakhe WG, Kulkami DN, Pathak BN, Bulusu KR, De-fluoridation of water by Nalgonda technique, Indian J Environ Health. 1975, 17, 26–65.

[17] Mondal P, George S, A review on adsorbents used for defluoridation of drinking water water". Rev. Environ. Sci. Biotechnol. 2015, 14, 195–210.

[18] Abo Markeb A, Alonso A, Sánchez A, Fon X, Adsorption process of fluoride from drinking water with magnetic core-shell Ce-Ti@Fe$_3$O$_4$ and Ce-Ti oxide nanoparticles, Sci. Total Environ. 2017, 598, 949–958.

[19] Kameda T, Oba J, Yoshioka T, Recyclable Mg-Al layered double hydroxides for fluoride removal: Kinetic and equilibrium studies, J Hazard Mater 2015, 300, 475–482.

[20] Sadik N, Mountadar M, Sabbar E, Defluoridation by calcined layered double hydroxides synthesized from seawater, J. Mater. Environ. Sci. 2014, 6 (8) 2239–2246.

[21] Khichar M, Kumbhat S, Defluoridation-A review of water from aluminium and alumina based compound, Int. J. Chem. Stud. 2015, 2(5): 04–11.

[22] Cai J, Zhao X., Zhang Y, Zhang Q, Pan B, Enhanced fluoride removal by La-doped Li/Al layered double hydroxides. J Colloid Interface Sci. 2018, 509, 353–359.

[23] Teagarden DL, Kozlowski JF, White JL, Hem SL, Aluminum chlorohydrate I: Structure studies, J. Pharm. Sci. 1981, 70, 758–761.

[24] Bi S, Wang C, Cao Q, Zhang C, Studies on the mechanism of hydrolysis and polymerization of aluminum salts in aqueous solution: correlations between the "Core-links" model and "Cage-like" Keggin-Al13 model. Coordin Chem Rev. 2004, 248(5–6), 441–455.

[25] Goswamee RL, Poellmann H, XRD study of thermal stability of hydroxyl-aluminium chloride, Indian Journal of Chemistry. 1998, 37A, 561–563.

[26] Coppel CP, Jennings HY, Reed MG. (1973). Field Results From Wells Treated With Hydroxy-Aluminum. Journal of Petroleum Technology, 1973, 25(09),1108–1112.

[27] Sarmah S, Saikia J, Bordoloi D, Goswamee RL, Surface modification of paddy husk ash by hydroxyl-alumina coating to develop an efficient water defluoridation media and the immobilization of the sludge by lime-silica reaction, J. Environ. Chem. Eng. 2017, 5, 4483–4493.

[28] Fan X, Parker DJ, Smith MD, Adsorption kinetics of fluoride on low cost materials. Water Res. 2003, 37, 4929–4937.

[29] Jia Y, Wang H, Zhao X, Liu X, Wang Y, Fan Q, Zhou J, Kinetics, isotherms and multiple mechanisms of the removal for phosphate by Cl-hydrocalumite, Appl.Clay Sci. 2016,129, 116–121.

[30] Gerente C, Lee VKC, Le Cloirec P, McKay G, Application of chitosan for the removal of metals from wastewaters by adsorption – mechanisms and models review. Crit. Rev. Environ. Sci. Technol. 2007, 37, 41–127.

[31] Qiu H, Lv Lu L, Pan BC, Zhang QJ, Zhang WM, Zhang QX, Critical review in adsorption kinetic models, J Zhejiang Univ Sci A., 2009, 10(5):716–724.

[32] Weber WJ, Morris JC, Kinetics of adsorption on carbon from solutions. J. Sanit. Eng. Div. Am. Soc. Civ. Eng. 1963, 89, 31–60.

[33] Ho YS, McKay G, Pseudo-second order model for sorption processes, Process Biochemistry, 1999, 34,451–465.

[34] Ponsot I, Falcone R, Bernardo E, Stabilization of fluorine-containing industrial waste by production of sintered glass-ceramics. Ceram. Int., 2013, 39(6): 6907–6915.

[35] Sarmah S, Saikia J, Bordoloi DK, Kalita PJ, Bora JJ, Goswamee RL, Immobilization of fluoride in cement clinkers using hydroxyl-alumina modified paddy husk ash based adsorbent, J Chem Technol Biotechnol. 2018, 93, 533–540.

[36] Ismail Z Z, Abdel Kareem H N, Sustainable Approach for Recycling Waste Lamb and Chicken Bones for Fluoride Removal From Water Followed by Reusing Fluoride-Bearing Waste in Concrete. Waste Manag. 2015, 45, 66–75.

[37] Rao SM, Reddy BVV, Lakshmikanth S, Ambika NS, Re-use of fluoride contaminated bone char sludge in concrete. J. Hazard. Mater. 2009, 166, 751–756.

[38] Baeza-Brotons F, Garces P, Paya J, Saval MJ, Portland cement systems with addition of sewage sludge ash. Application in concretes for the manufacture of blocks. J. Clean. Prod. 2014, 82, 12–124.

[39] Zhan BJ, Poon CS, Study on feasibility of reutilizing textile effluent sludge for producing concrete blocks. J. Clean. Prod. 2015, 101, 174–179.

[40] BS EN 196-2, British Standards Institution: London, 1995.

[41] MacPherson DR, Forbrich LR, Determination of Uncombined Lime in Portland Cement: The Ethylene Glycol Method. Ind. Eng. Chem. Anal. Ed. 1937, 9, 451–453.

[42] Odler I, Maula SA, Structure and Properties of Portland Cement Clinker doped with CaF_2, J Am Ceram Soc. 1980, 63(11–12), 654–659.

[43] Piekos R, Paslawska S, Leaching characteristics of fluoride from coal fly ash. Fluoride, 1998, 31, 188.

[44] Sengupta P, Saikia NJ, Borthakur PC, Bricks from petroleum ETP sludge: Properties and Environmental Characteristic, J. Env. Eng., Amer. Soc. Of Civil Eng. (ASCE), 2002, 128, 1090–1094.

[45] Loganathan P, Vigneswaran S, Kandasamy J, Naidu R, Defluoridation of drinking water using adsorption processes. J. Hazard. Mater. 2013, 248–249, 1–19.

[46] Gao S, Sun R, Wei Z, Zhao H, Li H, Hu F, Size-dependent defluoridation properties of synthetic hydroxyapatite. J. Fluorine Chem. 2009,130, 550–556.

[47] Velumani P, Senthikumar S, Production of Sludge-Incorporated Paver Blocks for Efficient Waste Management. J Air Waste Manag Assoc. 2018, 68, 626–636.

[48] Liu W T, Li K C, Application of Reutilization Technology to Calcium Fluoride Sludge From Semiconductor Manufacturers. J Air Waste Manag Assoc. 2011, 61, 85–91.

[49] Serjun V Z, Mladenovic A, Mirtic B, Meden A, Scancar J, Milacic R, Recycling of Ladle Slag in Cement Composites: Environmental Impacts. Waste Manag. 2015, 43, 376–385.

[50] Chen N, Zhang Z, Feng C, Zhu D, Yang Y, Sugiura N, Preparation and characterization of porous granular ceramic containing dispersed aluminum and iron oxides as adsorbents for fluoride removal from aqueous solution. J. Hazard. Mater. 2011, 186, 863–868.

[51] Rodriguez N H, Granados R J, Blanco–Varela M T, Cortina J L, Martinez–Ramirez S, Marsal M, Guillem M, Puig J, Fos C, Larrotcha E, Flores J, Evaluation of a Lime-Mediated Sewage Sludge Stabilisation Process. Product Characterisation and Technological Validation for Its Use in the Cement Industry. Waste Manag. 2012, 32, 550–560.

[52] Singh M, Kapur P C, Pradip, Preparation of alinite based cement from incinerator ash. Waste Manag. 2008, 28, 1310–1316.

[53] Garcés P, Pérez Carrión M, García–Alcocel E, Payá J, Monzó J, Borrachero M V, Mechanical and physical properties of cement blended with sewage sludge ash. Waste Manag. 2008, 28, 2495–2502.

[54] Chen Q Y, Tyrer M, Hills C D, Yang X M, Carey P, Immobilisation of Heavy Metal in Cement-Based Solidification/Stabilisation: A Review. Waste Manag. 2009, 29, 390–403.

[55] Saikia J, Sarmah S, Ahmed T H, Kalita P J, Goswamee R L, Removal of toxic fluoride ion from water using low cost ceramic nodules prepared from some locally available raw materials of Assam, India, J. Environ. Chem. Eng. 2017, 5, 2488–2497.

[56] Saikia J, Goswamee R L, Use of carbon coated ceramic barriers for adsorptive removal of fuoride and permanent immobilization of the spent adsorbent barriers. SN Applied Sciences 2019, 1, 634.

[57] Menéndez E, Glasser F P, Aldea B, Andrade C, Zimmermann Y C, Effects of the incorporation of aluminum fluoride mineralizers in Portland cement clinker phases. National Council for Cement and Building materials. 2016.

[58] Maheswaran S, Kalaiselvam S, Karthikeyan S K S S, Kokila C, Palani G S, β-Belite cements (β-dicalcium silicate) obtained from calcined lime sludge and silica fume. Cem. Concr. Compos. 2016, 66, 57–65.

[59] Misra K C, Borthakur P C, Alite Formation in the Rice Husk Ash-CaCO₃ System: Influence of Mineralizers on Formation. T. Indian Ceram. Soc. 1985, 44, 101–105.

[60] Chen N, Zhang Z, Feng C, Sugiura N, Li M, Zhu D, Chen R, Sugiura N, An excellent fluoride sorption behavior of ceramic adsorbent. J. Hazard. Mater. 2010, 183, 460–465.

[61] Chen N, Feng C, Li M, Fluoride removal on Fe–Al-impregnated granular ceramic adsorbent from aqueous solution. Clean Technol. Environ. Policy, 2014, 16, 609–617.

[62] Tor A, Danaoglu N, Arslan G, Y., Removal of Fluoride From Water by Using Granular Red Mud: Batch and Column Studies. J. Hazard. Mater. 2009, 164, 271–278.

[63] Nijhawan A, Butler E C, Sabatini D A, Macroporous hydroxyapatite ceramic beads for fluoride removal from drinking water. J. Chem. Technol. Biotechnol. 2017, 92 (8), 1868–1875.

[64] Saikia J, Sarmah S, Bora J J, Das B, Goswamee R L, Preparation and characterization of low cost flat ceramic membranes from easily available potters' clay for dye separation. Bull. Mater. Sci. 2019, 42, 104.

[65] He J, Cai X, Chen K, Li Y, Zhang K, Jin Z, Meng F, Liu N, Wang X, Kong L, Huang X, Liu J, Performance of a novelly-defined zirconium metal-organic frameworks adsorption membrane in fluoride removal. J. Colloid Interface Sci. 2016, 484, 162–172.

Part 4: **Residues from mining**

Andreas Kamradt
Chapter 15
Characterization and mineral processing options of "Kupferschiefer"-type low-grade black shale ore from mining dumps in Central Germany

Abstract: Dumps in the former Mansfeld mining district, Central Germany, consisting of waste rocks and low-grade ore were investigated lithologically and geochemically focussing the metal content. One of the most important steps to perform the assessment of the economic potential addresses a methodology for a representative sampling of the heterogeneous stockpiled black shale ore. Investigations were concentrated in particular on two flat dumps, which were deposited between 1870 to 1930. Low-grade black shale ore from one of the dumps showed a metal content of 0.5-0.6% Cu, 0.56-0.6% Pb and 0.69-0.75% Zn, from which technical bulk samples were extracted for mineral processing test studies. Chemical analyses of particle size classes revealed that the majority (> 70%) of the base metal content is contained in coarse-sized particles (> 1.6 cm), which is a prerequisite for the successful application of automated sensor-based sorting techniques. Elution and extraction analyses demonstrated that a low metal elutability in general emanates from dump material and just marginally weathering took place since the deposition period of low-grade ore on dump, although a slight increase of sulphide oxidation was determined for dump material with longer deposition time.

A holistic approach to the utilization of black shale low-grade ore considers extensive processing applications comprising alternative methods and enhanced technical protocols in sensor-based sorting, comminution, flotation as well as acid and bioleaching. The use of sensor-based sorting shows that in combination with a novel X-ray fluorescence scanner a 100 % increase (doubling) of the copper content in the pre-sorted black shale fraction can be achieved. Additionally, alternative comminution methods enhance the liberation of the extremely finely dispersed sulphides in the low-grade ore, which has a positive effect on the yield of the downstream flotation. The combination of energy-saving extraction methods comprising modified approaches of sulphide flotation and bioleaching could be considered as a connected copper processing line in general, in which an alternative recovery of copper from black shales with a complex ore mineralogy was experimentally confirmed and

Andreas Kamradt, Economic Geology & Petrology Research Unit, Institute of Geosciences and Geography, Martin Luther University Halle Wittenberg, von-Seckendorff-Platz 3, D-06120 Halle, Germany

https://doi.org/10.1515/9783110674941-015

characterized by advanced analysing techniques such as automated mineralogy. Thus, adapted mineral processing and the implementation of hydrometallurgical extraction are capable to increase the economic potential of low-grade ores and the recovery of valuable metals as well as improve the low efficiency and reduce the environmental impact of conventional technical applications.

Keywords: black shale, mining dumps, copper extraction, representative sampling, automated mineralogy, sulphide flotation, energy-saving comminution, acid- / bioleaching, XRF-based sorting, sequential extraction

1 Introduction

The growing demand of metals in the global economy is opposed by shortage of raw materials in respect to low-priced market availability. This is countered by saving measures and minimizing raw material use on the one hand and by using existing mineral resources, such as mining residues, on the other hand. A striking advantage is the availability of secondary mining resources, which are usually preceded by cost-intensive ore extraction in primary ore deposits and thus considerably reduces production expenditure. Many long-term mining districts host considerable amounts of mined ore and gangue material dumped during operation in some cases deposited in different times back up to the era of industrialization in 1850s. Mining dumps deposited before the technology jump around 1950 contain often portions of ore, which was not processed due to economic and technical reasons related to the low-grade mineralization of the ore at the time of deposition. The advantages for today's processing of these low-grade ores are obvious due to the lack of costs for exploration and extraction and additionally the use of already existing infrastructure in order to be supplied to the mineral processing facilities.

Past extensive research on mining dumps were generally focussed on assessing the environmental impact. Relatively few investigations have been directed to an objective that links economic processing and raw material extraction with simultaneous recycling of the tailings. However, an innovative approach of mineral processing can offer technical as well as economic and ecological alternatives for the processing of dump material and to the production of value metals such as Cu, Ag, Zn and Pb, but also largely avoiding new waste flows [1].

This chapter will delineate strategies to characterize low-grade black shale ore on mining dumps and addresses low-cost processing options for the beneficiation of copper and by-products. The main part of investigation and results shown hereinafter have been realized and achieved by a joint research project aiming for the extraction of metals and mineral products from dumped residues of the former mining industry in the Mansfeld area, Central Germany, and represented a part of the BMBF-funding priority "r2 – Innovative Technologies for Resource Efficiency – Raw Material Intensive

Production Processes" (2009–2013). The interdisciplinary and holistic approach was focussed on the economic extraction of valuable metals while simultaneously considering sustainable recycling of tailings and process waste. A main outcome of the research association was the development of a holistic innovative process including the global utilization of the Mansfeld "Kupferschiefer" mining dumps as a case study, from which the mineralogical geochemical characterization of dump material and the application of processing methods are presented largely.

1.1 Distribution of black shale copper deposits

Sediment-hosted copper deposits associated with black shale are globally distributed as shown in Figure 15.1 and include renowned mining districts such as the Central African Copperbelt and the European Kupferschiefer and deposits, such as White Pine (USA), Udokan (Russia) and Dzhezkazgan (Kazakhstan). The European Kupferschiefer-type deposits contain considerable base metal reserves partly associated with increased contents of precious metals (including Cu, Ni, Zn, Pb, Ag, Au, Pt, and Pd) that potentially represent strategic and critical raw materials for Europe. The economic importance of black shale-hosted deposits illustrates the ranking of the copper resource contained in the Polish Kupferschiefer-type deposits of the Lubin-Sieroszowice mining district that represents the fourth largest copper (26.4 Mt Cu) and first silver (0.11 Mt) resource worldwide. 2.6 Mt of 3.4 Mt Cu and 0.14 Mt of 0.18 Mt Ag were extracted by long lasting mining from deposits on the German territory predominantly mined in the former Mansfeld-Sangerhausen mining district until the mine closure in 1990 [2–4].

Extensive industrial mining of black shale hosted copper has been taken place over the last 150 years in the Sangerhausen-Mansfeld mining district and 100 years in the Central African Copperbelt, which has left behind huge amounts of waste dumps, partially considered to be low-grade ore economically not exploitable at the time of deposition. In recent times, the initially termed low-grade ore of black shale-associated deposits represents copper resources often exceeding the cut-off grade of recent copper mining operations. However, low-grade ores of "Kupferschiefer"-type mines consist predominantly of fine-grained, organic carbon-rich black shale with extremely fine-grained disseminated base metal sulphide mineralization, which has proven to be extremely difficult to process. This is one of the main reasons for the material to still exist on mine dumps. Similar subeconomic low-grade ore and mineralized waste dumps left by processing of black shale-hosted copper ores can be found worldwide, e.g., the Polish Kupferschiefer district around Lubin, the African Copperbelt of Zambia and Zaire, and the deposits of Dzhezkazgan, Central Kazakhstan [1, 5, 6].

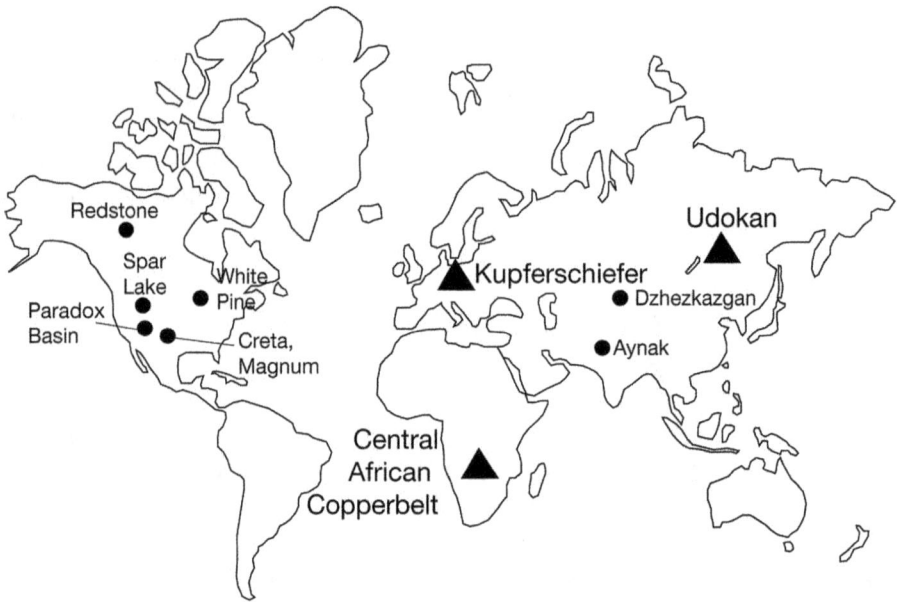

Figure 15.1: Global distribution of important black shale-associated copper deposits
(●) and mining districts (▲).

1.2 Characteristics of black shale ores

The occurrence of Kupferschiefer-type ore with economic potential is linked to sediment-hosted copper deposits of Permian age, in which enrichment zones of ore minerals occur in varying extend over three lithological layers comprising black shale
(Kupferschiefer) as well as footwall sandstone and hanging wall carbonate rock. Kupferschiefer-type ore mineralization is polymetallic and characterized by the presence
of base metal and complex sulphides variable in the spatial distribution within the
stratigraphic horizons as well as their relative proportion.

Lithological, the Permian black shale is a dark anthracite to dark grey finely
laminated carbonaceous marl (Figure 15.2), mainly composed of quartz, carbonates
(calcite/dolomite), feldspars, clays (mainly illite), organic matter (kerogen type II)
and varying portions of Cu-, Pb-, Zn-and Fe-sulphides, depending on the ore zone.

A striking feature of the black shale is the increased and partially extremely
high organic carbon content that exceeds at least 0.5 wt.% and usually varies from
2 to 15 wt.%. The metal-bearing sulphides either occur dispersed as small-sized particles (Figure 15.3) or form stratiform, up to mm-sized layers, more rarely crosscut-
ting veins. The common sulphide assemblage associated with Kupferschiefer-type
mineralization contains chiefly bornite, chalcopyrite, chalcocite, galena, sphalerite
and pyrite. The main part of the ore mineralization occurs as fine disseminated

sulphide particles within the layered rock fabric, which often consists of intimately intergrowths of several sulphides or sulphides with gangue minerals.

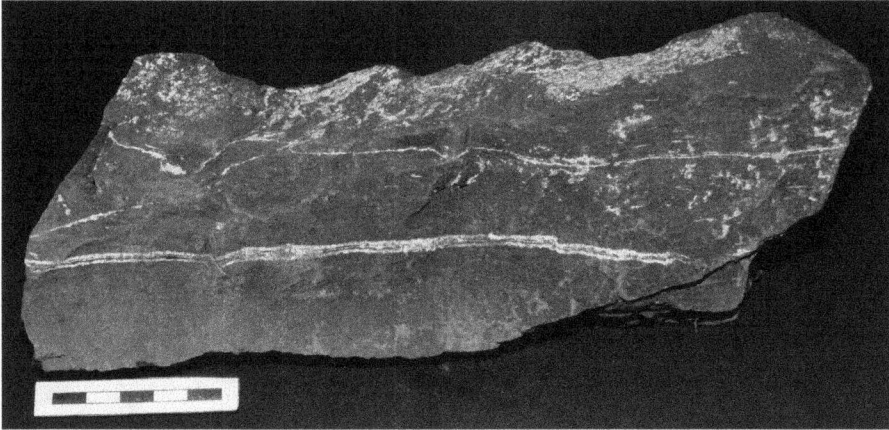

Figure 15.2: Photograph of black shale hand specimen from a mine dump showing locally secondary coatings on the dark grey-anthracite rock surface due to partial oxidation of sulphides along bedding-parallel veins. (Scale bar indicating cm-scale).

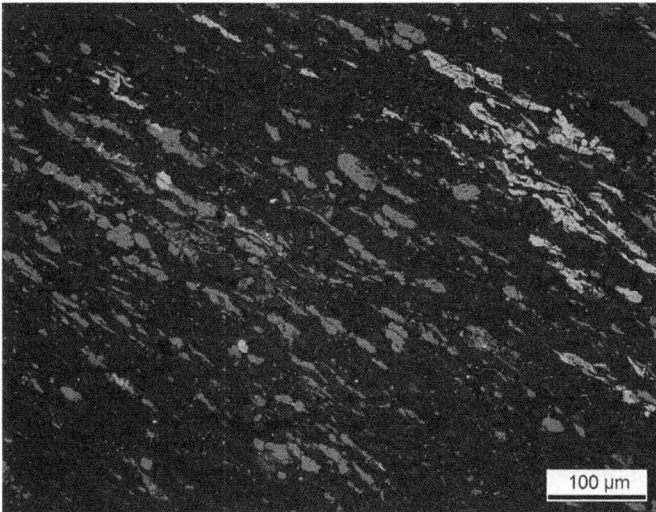

Figure 15.3: Microphotograph (reflected light) showing chalcopyrite (yellow) and bornite (brownish) along the stratification of the black shale host rock.

1.3 Mining history and related legacies of the Mansfeld-Sangerhausen mining district

The mining of the Kupferschiefer-type black shale-hosted copper ore occurred over a timespan of nearly 800 years and has formed significant legacies of mining and processing residues within the landscape of the Mansfeld-Sangerhausen mining district. Until the closure of the last mining shaft, 109 Mt of Kupferschiefer ore were mined to produce mainly copper and silver, as well as by-products such as zinc, lead, cobalt, nickel, rhenium and gold [3, 7]. Mining legacies caused by the mining of black shale ore and subsequent extraction of metal commodities comprise numerous historic small-scale dumps, tabular dumps piled up from the beginning of the industrial mining age and huge conical dumps, which mark the youngest remnants of the local mining history from the socialist era, as well as tailing ponds of process residues.

Artisanal mining followed the northern boundaries of both districts, where the black shale layer outcrops, but can be tracked equally on the western rim of the Mansfeld basin (see Figure 15.4). These small-scale mining shafts reached only few meters depth and left unexceptional wall rocks heaps of 10 to 20 meters diameter, because of the very selective mining during the late Middle Age (1200–1600). With increasing quantities of mined Kupferschiefer ore, the new shafts were sunk further basin inwards corresponding with increasing mining depths and volumes of waste rocks deposited on increasingly larger heaps/dumps (mining period 1600–1800). Improvement of mining and extraction techniques resulted in mining depths reaching over 100 metres with a corresponding rise in volumes of wall rocks that had to be excavated during the sinking of the deeper shafts.

From about 1850 a new era of mining in the Mansfeld district took place and marks the change to the application of industrial mining methods. As a consequence, the mining depths and the size of the dumps rose further, due to increasing adit roof heights and the extraction of thicker portions of the metal-bearing black shale bed towards the hanging wall limestone. Low-grade ore, uneconomic at that time, was stored mostly at a certain part of the top of the tabular dumps and represent nowadays an easily accessible resource. Large scale mining of Kupferschiefer starting from the technology jump after the Second World War resulted in large conical dumps in both, the Mansfeld and Sangerhausen district, with heights of up to 150 metres with a 300 m diameter, piled up in the Mansfeld area on top of tabular dumps from the former mining era.

The region of the former Mansfeld-Sangerhausen mining district hosts extensive legacies from the historic mining activities that represent a substantial raw material potential. Several studies of the Mansfeld-Sangerhausen area in respect to the inventory of mining legacies and their environmental impact have been carried out in the 1990s commissioned by the German Federal Environment Agency [9, 10].

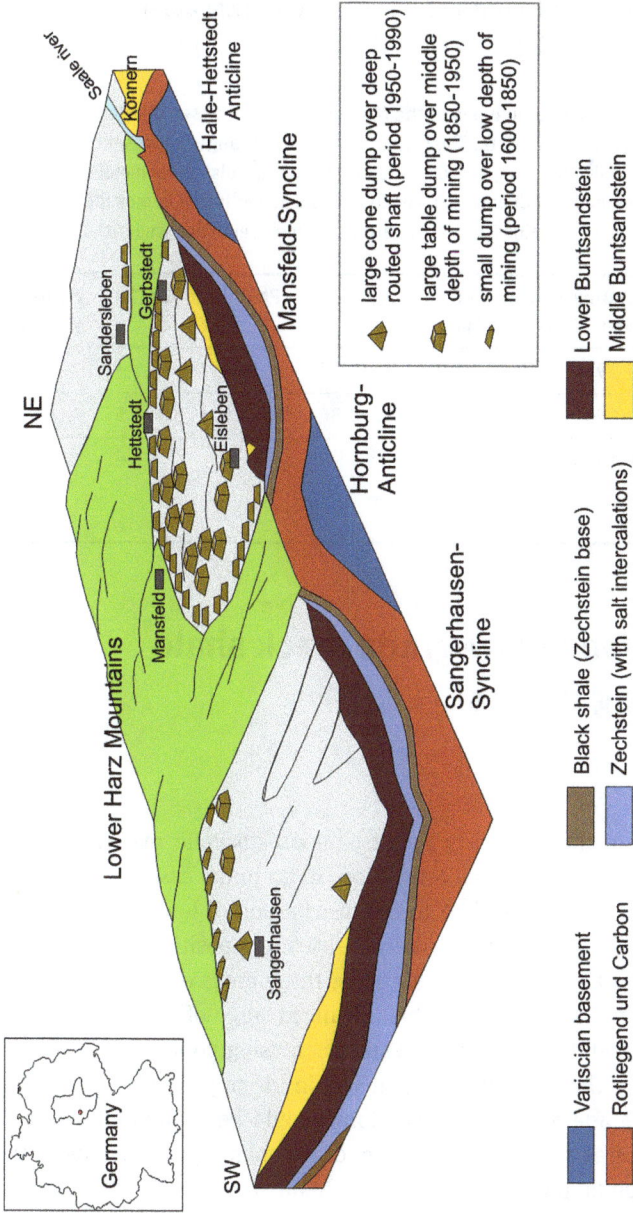

Figure 15.4: Block diagram illustrating the geological setting in the Mansfeld-Sangerhausen mining district, Central Germany, and the position of different dump types within Rotliegend subbasins. Modified after [8].

Table 15.1 summarizes the main features of tabular and conical dumps of the former Mansfeld-Sangerhausen mining district.

Table 15.1: General features of mining dump types of the Mansfeld-Sangerhausen district. Information summarized from [9] and own data. The given metal content ranges represent averaged values collected from individual dumps. Procedure for the data acquisition of metal contents on dumps is described in the beginning of Chapter 2.1. For cone and small artisanal dumps metal content data are not available, but they mainly consist of barren, not metal enriched wall rocks.

Type	Number	Operation period	Low-grade ore [10^6 m^3]	Cu [%]	Pb [%]	Zn [%]	Wallrock [10^6 m^3]
Cone dumps	6	1950–1990	< 0.1–	–	–	–	32.2
Table dumps	11	1850–1930	2.9	0.3–0.8	0.6–3.5	0.3–4.5	22.9
Small artisanal heaps	> 100	1400–1800	< 0.1	–	–	–	0.4

2 Characterization of low-grade black shale ore from mining dumps

2.1 Sampling

Representative sampling of the extremely heterogeneous stockpile material represents an enormous challenge for resource estimation. In an initial assessment phase dump bodies should be mapped according to the distribution of low-grade ore and wall rock within the dump. Due to the spatially restricted deposition of low-grade black shale ore, largely on top of wall rock sections, these areas can be delineated and subsequently investigated in detail. In this advanced stage of resource estimation, the screening of the metal content of hand specimen using portable XRF (pXRF) can be useful to determine the metal distribution within the upper level of mapped areas dominated by low-grade ore. Sample locations should be created following a grid of at least 10 × 10 metres and the upper 20 cm of weathered material should be removed. On sampling points, pXRF screening on 10 hand specimens should be carried out with at least five pXRF readings of each hand specimen and averaged. The pXRF screening delivers a first estimation of the metal concentration range for each sampling point and further a rough assessment of the metal distribution within the sampling grid.

A more representative acquisition of data was achieved by a sampling technique carried out along trenches of at least 1–2 metre depth provided by an excavator (Figure 15.5). Bulk samples of 40–60 kg were collected in regular distance and

Figure 15.5: Construction of a ramp on a flat dump section of the "Fortschritt" shaft dump by an excavator.

sieved on site down to 6.3 cm sieve size in order to split larger hand specimen from the finer fraction (Figure 15.6). Sieve fractions and hand specimen define one bulk sample, which ensures that the whole spectrum of differently ore-bearing rock fragments including finer constituents are sampled representatively. Sieve analyses have shown that large-sized rock samples (10–30 cm) represent the main mass portion of the bulk sample. Several of these larger rock pieces from a bulk sample were selected for geochemical analyses. Therefore, they were cut into 300–400 g subsamples and subsequently milled, homogenized and representatively split for analysis by pXRF and additionally by ICP-MS whole rock analysis at a certified laboratory. ICP-MS data were also used to check accuracy of pXRF-readings on powdered samples and to form a correction factor for pXRF readings chiefly of base metals. The fraction < 7 cm bulk sample was sieved to 11 standardized sieve size classes, which are split representatively by a sample divider, milled and also analysed.

Since the majority of the bulk sample is dominated by larger rock pieces, the base metal content of a bulk sample is controlled by its larger constituents, which underlines the importance of a suitable sorting technique in a mineral processing line. The mass-balanced sum of geochemical data of the larger individual samples together with the sieve fractions reveals the base metal content of a bulk sample.

Figure 15.6: Sampling an exploration bulk sample on the "Fortschritt I" shaft dump using a screening sieve to prepare particle size fractions.

A further approach to determine the distribution of metals in deeper levels of a dump can be realized by drilling. Due to the lumpy feature of dump material, the selection of a suitable drilling technique is limited. To protect the borehole from collapsing due to the lose cohesion of the dump rock fragments, the bore hole is tubed tracking the drilling progress. Thus, sample material can be provided metre by metre using the auger drilling technique (see Figure 15.7). The drill metres are packed in buckets and subsequently split in two fractions by a 1.6 cm size sieve. The finer fraction is quartered again and classified by a standardized sieve set. Similar to the geochemical bulk samples, sieve classes were milled, split and analysed by pXRF as well as partially by ICP-MS.

Big bag-sized bulk samples for mineral processing test work were taken with an excavator providing sample material of 3 t, 1.5 t and a further of 1 t. The run-off-mine ore (avg. 12 cm size) from the "Fortschritt I" shaft dump (see Figure 15.8) was comminuted stepwise prior to usage for mineral processing test work (see Chapter 3.2).

Figure 15.7: Detail of sampling by auger drilling.

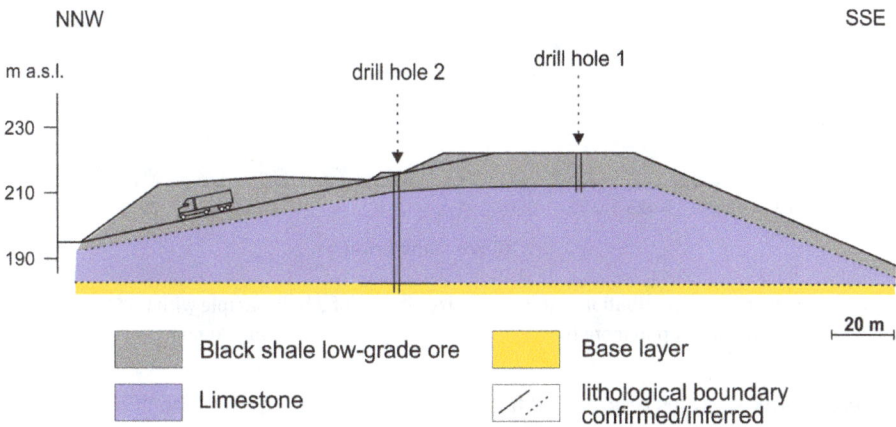

Figure 15.8: Cross section of the "Fortschritt I" shaft table dump indicating drilling locations and distribution of black shale low-grade ore and limestone (confirmed and proposed) within the dump.

2.2 Metal distribution and representativity of sampling techniques

In order to adapt process options and adjust technical parameters for the efficient recovery of value minerals it is indispensable to know the quantitative occurrence of the desired element to be extracted and its mineralogical deportment.

Geochemical analysis of sieve size classes from bulk samples showed that the finer grain size portion within the sieve classes contained systematically increasing base metal concentrations, exemplarily shown for Cu in Figure 15.9. Additionally, it can be ascertained that the base metal enrichment in the lower grain size classes as shown in Figure 15.10 just marginally influences the total distribution within the bulk sample due to low percentage of the fine fraction in the total bulk sample. Figure 15.9 reveals additionally an important fact, the grain size fraction > 70 mm hosted around 50% of the total Cu from the whole bulk sample, which indicates that a sorter system can be applied to produce a preconcentrate prior to subsequent enrichment techniques such as flotation or gravity separation. Another important finding is that the finer size fractions (−125 µm) showed increased Cu contents up to 2.5%, which constituted a very subordinate portion of the bulk sample representing just 5% of the total sample. Furthermore, granulometric data showed the total weight of a total bulk sample is dominated by 90% of rock components in the grain size fraction > 1 cm.

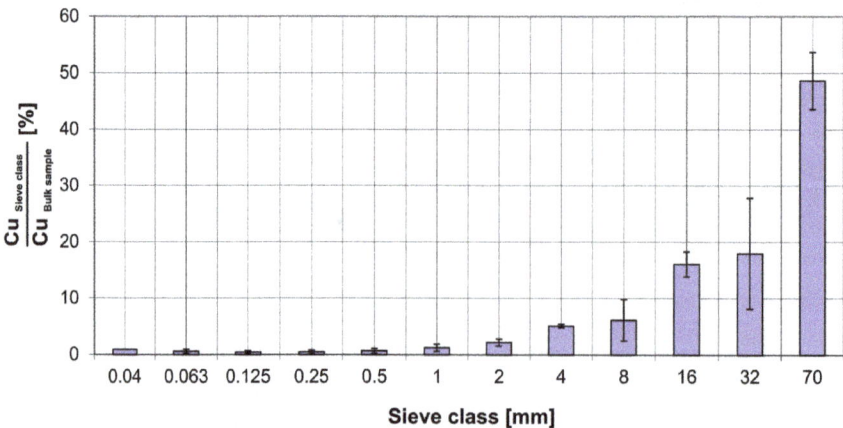

Figure 15.9: The copper distribution within grain size classes of a bulk sample with 0.78% Cu shows a predominance (up to more than 50%) in coarse-grained fraction > 70 mm.

Different bulk samples originating from the low-grade ore part of the "Fortschritt I" shaft table dump were analysed regarding their base metal content (see Table 15.2). They comprised a smaller manually taken exploration bulk sample at the scale of tens of kilogram (Sample 1 in Figure 15.10) and larger technical taken bulk samples at ton scale (Sample 2 and 3 in Figure 15.10).

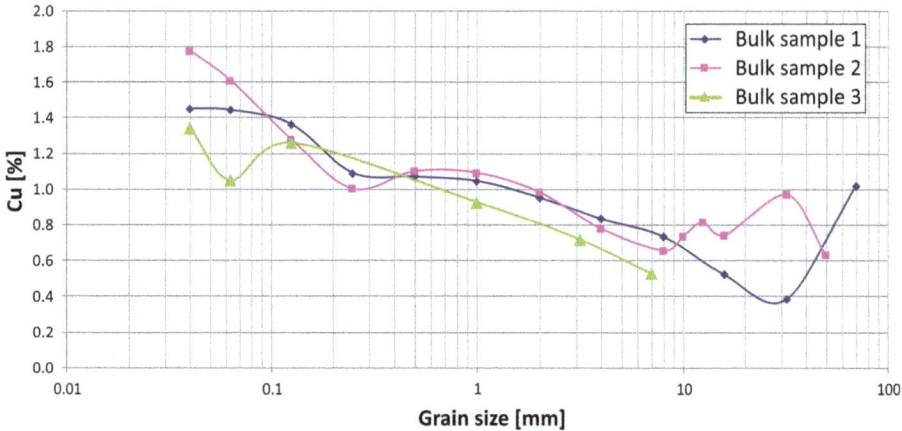

Figure 15.10: Distribution of the copper content in different bulk sample in respect to the distribution of grain size classes.

The total content of base metals for exploration bulk samples is compiled and averaged from the base metal concentration of each grain size class, whereas for the larger technical bulk samples aliquots were split and geochemically analysed.

As Table 15.2 indicates, the Cu and Zn content of manually taken bulk samples correlate precisely with the contents of technically taken bulk samples and confirms the base metals values, which shows that a representative sampling of lumpy, heterogeneous low-grade dump ore can be achieved.

Table 15.2: Base metal content of different bulk sample from the "Fortschritt" shaft dump show a concordance for Cu, whereas Zn and Pb deviates more.

Bulk sample	Cu [%]	Zn [%]	Pb [%]
1 (42 kg)	0.78	0.96	0.50
2 (2 t)	0.78	0.93	0.72
3 (1.5 t)	0.78	0.95	0.79

The investigation of the metal content of individual grain size classes of exploration bulk samples revealed that certain grain size classes of a bulk sample correspond with the total copper content and that they can be used to define the lower and upper limits of base metal concentrations in a bulk sample. Therefore, it was ascertained that the base metal content of the size fraction 4–8 mm is just below that of the whole bulk sample, whereas the size fraction 2–4 mm showed slightly increased base metal concentrations compared to the whole bulk sample. Thus, two consecutive grain size classes of bulk samples can be used to determine the metal content of a sample location, on which just two subsamples have to split on-site using a 2-, 4- and 8-mm mesh.

Based on these findings, the extent of sieving and analysis of numerous grain size classes can be reduced to two individual sieve classes, whereby the time-consuming over all analysis of bulk samples is substantially minimized. Depending to the type of geochemical analysis carried out on the sieve size classes either by pXRF or lab geochemical assays, the optimized sampling of sieve size classes characterizing the maximum and minimum content of base metals can be very effective and cost-saving. Even data return can be sped up by on-site data acquisition using appropriate sieve meshes and pXRF and furthermore larger areas of a dump can be investigated more detailed by sieved exploration samples. This procedure additionally allows a rapid and established estimation of a narrow concentration range of base metals. On basis of the analysis of sieve fractions of 5 bulk samples and 15 sieved exploration samples, the base metal range for the "Fortschritt I"-dump can be stated as follows:
- lower base metal content: >0.50% Cu, 0.56% Pb, 0.69% Zn and
- upper base metal content: <0.60% Cu, 0.60% Pb, 0.75% Zn.

Technical bulk sample (> 1 t) can deliver more representative data due to their large volume but that is dependent on an appropriate splitting method. The comparison of the base metal content of an exploration bulk sample (Sample 2 in Figure 15.10, Table 15.2) and a technical bulk sample (Sample 1 in Figure 15.10, Table 15.2) revealed a good agreement especially for Cu with 0.78% as well as for Zn (0.96% vs. 0.93%) but with some divergence for Pb (0.5% vs. 0.72%). A further technical bulk sample (Sample 3 in Figure 15.10, Table 15.2) confirmed base metal concentrations for Cu and Zn as well as shows closer accordance to the Pb content in the technical bulk Sample 2 (0.79% vs. 0.72%). In particular, the identical copper values demonstrate impressively the reproducibility of data raised by the analysis of exploration bulk samples on that heterogeneous dump material and thus prove the effectiveness of this sampling method used.

Table 15.3 shows geochemical analyses of several comminution products of a further technical bulk sample (1 t) carried out by international acknowledged and certified laboratories.

The results show that data is reproducible and that the technical bulk sample (1 t) contains a copper content of 0.54–0.60%, which corresponds with the estimation of the base metal content of low-grade ore determined by exploration sieve samples. Generally, metal concentrations determined among different comminution products of the bulk sample show minor deviations (< 5%) as Table 15.3 indicates.

On the opposite, metal content of individual hand specimen of low-grade black shale ore fluctuated from virtually barren samples to samples containing up to 12%

Table 15.3: Concentration of base and trace metals in comminution products of low-grade black shale ore (feed size: $F_{80} = 12$ cm) determined by ICP-MS by two different international approved laboratories. (MDL = minimum detection limit).

			1. crusher stage cone crusher		final product ball mill	
Grain size			−12 mm		−0.1 mm	
laboratory			1	1	2	2
Method			Aqua regia – ICP/MS			
Element	Unit	MDL				
Cu	PPM	0.1	5387	5551	5980	6010
Pb	PPM	0.1	7671	7388	7680	8350
Zn	PPM	1	9778	9706	10100	10400
Ag	PPM	0.1	>100	>100	236	227
Au	PPB	0.5	505	690	n.a.	n.a.
Co	PPM	0.2	58	54	50	48
Ga	PPM	0.5	14	13	<50	<50
Mo	PPM	0.1	85	85	90	95
Ni	PPM	20	143	141	130	109
Sb	PPM	0.1	5	5	<25	30
V	PPM	8	391	413	159	159

Cu, which was higher than previously reported for high-grade ore [11]. Exploitable and economic Cu-content of Kupferschiefer- type ore ranges from > 3.5% Cu for high grade ore and 2.5–3.5% Cu for moderate mineralized ore, whereas a Cu-content of < 2.5% undercut the economic and technical requirements. Additionally, Kupferschiefer-type ores also contain elevated to anomalously high concentrations of several minor and trace elements, which were also industrially extracted in the past and recently. Their concentration range in low-grade ores is generally lower, but for some trace metals still increased or equal to high-grade ore. It was shown that Mo concentrations of low-grade ores can exceed the concentration in high-grade ores, which also points to the limited significance of the published concentration ranges as pointed out in [1]. Partly elevated trace metal concentration as shown for Ag, Mo, Ni and V in Table 15.3 occur mainly as substitute elements in sulphides or appears to be associated with the increased amount of organic matter in black shales. In this context, the main Ag carriers are bornite, chalcocite and galena, which was previously reported for both, Polish and German deposits [3, 12]. Other trace metals (Co, Ni, Mo, and Se) can be contained in significant amounts in widespread pyrite that occurred as late diagenesis-formed, tiny framboidal aggregates or larger scale pyrite crystals related to hydrothermal-induced massive replacement mineralization. However, especially Ni and V can be enriched in organic porphyrin compounds [13–16].

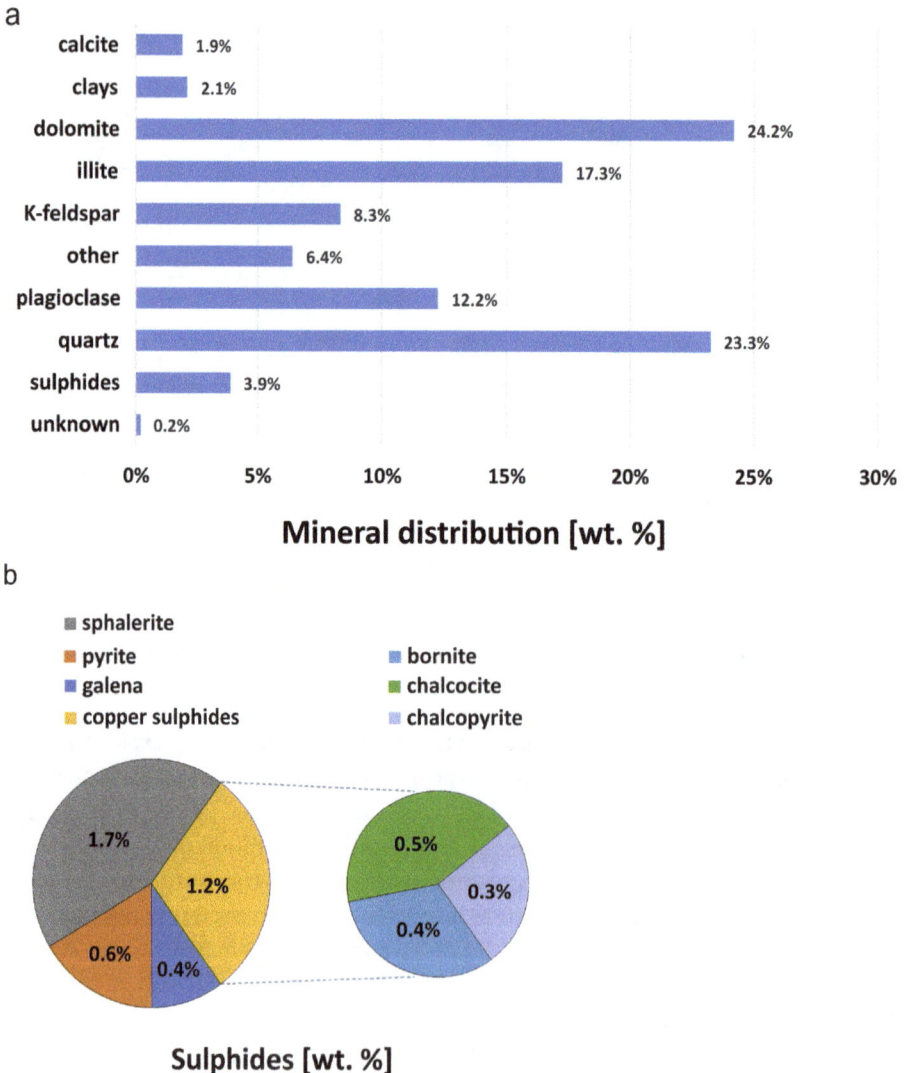

Figure 15.11: Mineral distribution diagrams presenting a) the main mineral phases and mineral classes and b) the content and proportion of copper and of base metal sulphides. The low-grade black shale ore consists mainly of carbonates, illite and quartz.

An EDX-based mineral distribution analysis, depicted in Figure 15.11a, carried out on a sample split of the ball mill product originating from the technical bulk sample in Table 15.3 shows that the low-grade black shale ore is dominated by carbonate minerals (27%; 24.5% dolomite), quartz (23.8%), feldspar minerals (20.5%; 12.5% plagioclase and 8.0% K-feldspar) and illite (17.7%). Additionally, it contained 4.0% of minor mineral phases comprising mainly gypsum, iron oxyhydroxides as well as rutile and an organic matter portion of 2.4% analysed independently by LECO analysis. The bulk sample contained 3.9% sulphides but due to the low content of copper, copper-bearing sulphides occurred in low concentrations and comprised chalcocite (0.5%), bornite (0.4%) and chalcopyrite (0.3%) (Figure 15.11b). Among the portion of sulphide minerals, the amount of sphalerite was comparedly increased with 1.7%, whereas pyrite (0.6%) and galena (0.4%) also showed low contents in the ball mill product of the bulk sample [17].

2.3 Metal mobilisation and weathering

Mining dumps of the former Mansfeld mining district in Central Germany containing low-grade ore were mainly deposited during the production period between 1850–1950, which means an exposition to weathering by environmental conditions of at least 70 years and in many cases far longer. Two drill holes sunk on the mining dump of the older "Theodor" shaft (deposition ceased 1890) and one drill hole sunk on the tabular dump of the younger "Fortschritt I" shaft (deposition ceased 1940) (see Figures 15.7, 15.8) were investigated to trace potential metal mobility of black shale-hosted ores as well as limestone that generally represents the main volume of the dumps due to its stratigraphic hanging wall position within the deposit and the necessity of removal to extract the ore. On dumps, limestone waste rocks commonly underlie areas of separately dumped low-grade ore. The change from black shale-dominated to limestone-dominated drill metres was determined at 5–6 metres depth for the "Theodor" shaft drill holes and 7–8 metre depth for the "Fortschritt I" drill hole.

Regarding the base metal content shown in Figure 15.12a, it can be generally ascertained that the drill material of the "Theodor" dump was increased in Pb (up to 1.2%) and Zn (up to 2.4%), whereas Cu content was higher in drill metres of the "Fortschritt I" dump (up to 0.9%). The base metal content was determined by pXRF-readings (corrected by ICP-MS-analysed references) on the 2–4 mm sieve fraction, which has been proven to be the representative grain size fraction regarding the average content of the total material of a drill metre. The difference of the base metal content among the drill holes of the "Theodor" dump is generally noticeable in the upper three metres. In drill hole 2, Cu values vary slightly from 0.1 to 0.2%, whereas in drill hole 1 Cu shows a maximum of 0.6% in the first drill metre and steadily decreasing values up to 5 metres depth from 0.4 to 0.2% Cu. Generally, low Cu contents were determined in the underlying limestone-dominated drill metres, but an exception-ally increased Cu content of 0.8% was found in the "Fortschritt I" drill hole at metre 11. Pb contents of 0.6–0.9% were slightly changing in the latter, whereas both drill holes from the "Theodor" shaft dump showed increased Pb-values slightly differing from 0.9–1.2% within the black shale-dominated part. Zn contents (1.8–2.4%) were likewise especially increased compared to the "Fortschritt I" drill hole, in which Zn contents fluctuated between 0.7 and 1.5% in the black shale-dominated upper 8 drill metres. Limestone-dominated drill metres contain generally lower contents of Pb and Zn, whereas the decline is particularly remarkable in the drill holes of the "Theodor" shaft dump by a decrease from 2.1 to 0.9% Zn and 1.0 to 0.5% Pb.

To investigate the portion of water-soluble constituents, the drilling material was tested in respect to the elutability of elements with wider emphasis on metal ions. For S4-Elution tests, 3 aliquots (each 20 g) of the 4 mm sieve fraction of the examined drill metres were eluted for 24 h with 200 ml deionized water while placed in a shaker according to [18]. Filtered eluates were investigated in regards to pH, conductivity and metal inventory using ICP-OES.

As Figure 15.12b indicates, eluates of the drill metres of the "Theodor" shaft dump are substantially more acidic especially in the upper part up to 3 m in depth, whereas for drill hole 2 the pH continuously decreases from 7.7 to 7.2 in a depth of 5 m. The pH decreased from 7.7 to 7.4 in the upper part of drill hole 1 (1–3 m) and increased con-stantly for eluates from deeper drill metres to a more alkaline pH of 8.1. Eluates of tested drill metres of the "Fortschritt I" dump showed generally slight pH fluctuations between 7.9 and 8.2, although eluates of the upper black shale-dominated and gener-ally limestone-dominated drill metres show pH above 8, whereas eluates from black shale-dominated drill metres from 2 to 7 metres depth are between pH 7.9 and 8.

Conductivity data, shown in Figure 15.12c, indicates that increased values corre-late with a pH decrease. Generally, the conductivity was significantly lower in eluates of the drill metres from the "Theodor" shaft dump and ranged between 258–1002 µS/cm for drill hole 1 and 358–1305 µS/cm for drill hole 2, whereas conductivity values of eluates from "Fortschritt I" dump drill metres were almost twice as high ranging be-tween 1141 and 2257 µS/cm. Most increased conductivity values were determined in two limestone-dominated drill metres (10 and 11 metre depth), although the overlying limestone-dominated part showed the lowest conductivity. Generally, the conductiv-ity was low in eluates of the first drill metre and increased for black shale-dominated drill metres with varying values. Increased conductivity within the course of the bore holes often correlated with increased Zn contents in drill metres.

Eluates gained by S4-Elution were analysed in respect to content of selected metals (As, Cd, Cu, Fe, Mn, Mo, Pb, V, Zn) and S by ICP-OES. Figure 15.12d shows, most of the metals showed low concentration comparable to Cu or even lower (0.01–0.05 mg/L), whereas the concentration of dissolved Zn was increased up to three orders of magnitude (0.02–12.38 mg/L) and S contents by up to 5 orders of magni-tude (9.72–673.22 mg/L). While dissolved Cu in eluates of drill metres of all bore

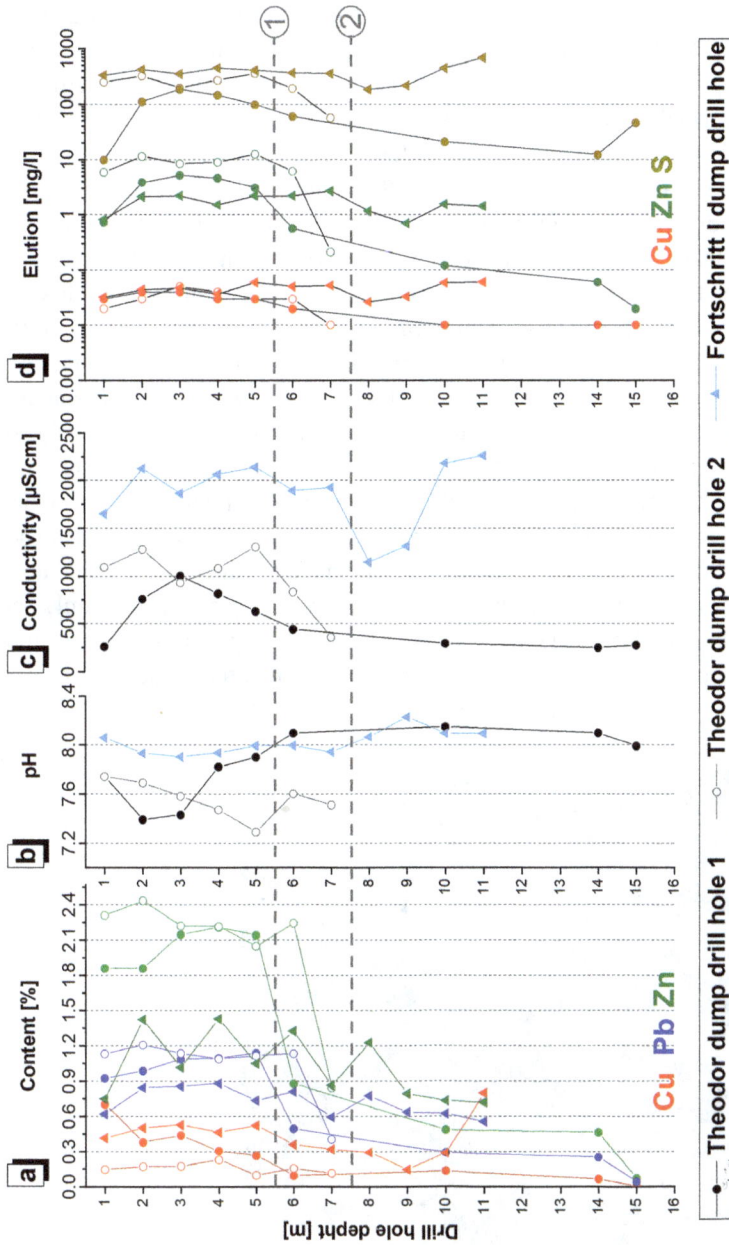

Figure 15.12: Depth profile from drill holes on the "Theodor" shaft dump and "Fortschritt" shaft dump for a) base metal content in the drill metres, b) pH and c) conductivity of elution solutions, as well as d) content of dissolved ions displayed exemplarily for Cu, Zn and S. Note that the lithological change from upper black shale- to lower limestone-dominated part (① Theodor / ② Fortschritt shaft dump) of the dump is generally indicated by a partly strong decreasing metal content of the drill material, but also in an increasing pH of the elution solution.

holes had similar concentration ranges, strong difference in the content of dissolved Zn and S ions are clearly notable. The content of eluted Zn from drillings of the "Theodor" shaft dump was comparably increased and correlated especially in the upper black shale-dominated section with a markedly increased content of Zn analysed in the feed material. For S in turn, eluates of the "Fortschritt I" shaft material contained the most increased values, whereas dissolved S in eluates of drill hole 2 from the "Theodor" shaft dump were similar increased chiefly in the black shale-dominated section but drill hole 1 showed generally lower contents of S in the eluates.

It can be noted that the ions strength largely controls the conductivity of eluates. Furthermore, increased conductivity values within the drilled profile coincide with lower pH values determined in particular at upper black shale-dominated drill metres. It is obvious that an increase of conductivity in eluates is mainly driven by the amount of dissolved ions extracted by S4-Elution and largely coincides with the amount of dissolved S and Zn. The amount of dissolved base metal ions is clearly linked to base metal content in the drilled dump material. Thus, an increased amount of available and elutable Zn is apparent, which is largely caused by the occurrence of widespread secondary Zn-hydroxide coatings on dump rock fragments, whereas the content of eluted S is mainly caused by dissolution of secondary formed gypsum commonly occurring in association with secondary amorphous Cu-Pb-Zn-hydroxide coatings and minerals such as brochantite, malachite (Figure 15.13), serperite, langite and others that occupy surfaces of dump rock fragments (see Figure 15.14).

Figure 15.13: Tiny botryoidal malachite (< 100 μm) forming mineral coatings onto surfaces of black shale hand specimen from low-grade dump material.

Figure 15.14: Secondary mineral association of mainly Cu-Zn hydroxysulphates such as langite (blue, $Cu_4(SO_4)(OH)_6 \cdot 2H_2O$) and serpierite (greenish white, $Ca(Cu,Zn)_4(SO_4)_2(OH)_6 \cdot 3H_2O$) as well as amorphous coatings.

Stability fields of these secondary phases suggests a formation at a pH range from pH 3–8 documenting the neutralisation of acid metal-bearing weathering solutions derived from sulphide weathering/oxidation (brochantite < pH 6 > malachite, $Zn(OH)_2$ > pH 8) and field observations point to an in-situ formation of secondary Cu minerals and precipitation of secondary Zn phases along seepage run offs.

A sequential extraction procedure adapted from [19] was slightly modified and applied to investigate the type of chemical bonding of mainly base metals and their distribution within the drilling profiles. Aliquots of 0.1 g from the 2–4 mm sieve and milled fraction of selected drill metres were treated by a leaching protocol modified according to common constituents of black shale rocks, considering the increased organic carbon content. A sequence of five leaching fractions attacking the pulverized samples from pH 7 to pH 2 were used to dissolve specific ions or digest defined mineral groups. The selected sequential leaching steps aimed to the dissolution of the mineral inventory by the five leaching fractions is shown in Table 15.4.

After filtration, all extracted solutions were acidified to pH < 2 by addition of suprapure HCl. Major elements (Ca, Fe, Mg, Al) were analysed by ICP-OES and minor elements (Cu, Pb, Zn, Mo, V) by ICP-MS.

Figure 15.15 show exemplarily the chemical fixation of Cu and Zn in various depth metres of the drill holes from the "Theodor" and "Fortschritt" shaft dump. The distribution of leaching fractions indicates that the portion of Cu and Zn related to

Table 15.4: Sequential leaching fractions and corresponding leaching solutions.

Leaching step	Leaching solution	Quantity	Time
L1: Exchangeable cations (clay minerals, LDH)	1 M $C_2H_7NO_2$ (pH = 7)	10 ml	5 h
L2: Carbonate fraction	1 M $C_2H_3NaO_2$ (pH = 5)	50 ml	5 h
L3: Easily reducible elements (Dissolution of Fe- and Mn-hydroxides)	0.1 M [NH_3OH]Cl + 25% CH_3COOH (pH = 2)	20 ml	12 h
L4: Organic and sulphide fraction	30% H_2O_2 + 0.01 M HNO_3 (pH = 2)	30 ml	10 h
L5: Residual fraction (silicates etc.)	HF-HCl-HNO_3 acid pressure digestion	–	–

fraction L4/5 (sulphides) increased with depth especially in black shale-dominated drill metres, whereas the portion of Cu and Zn extracted by fractions L1, L2 and L3 decreased steadily with depth. Generally, the lowest proportion of leached Cu was found in the extraction fraction L3 (easily reducible elements, Fe- and Mn-OOH's). Here, the values, as shown in Figure 15.15, are in the single-digit range (< 7%), but the fixation to the L4/5-fraction (sulphides) represented the largest proportions. However, the proportion of Cu extracted by leaching step L4/5 (sulphides) was generally increased within the whole drill section of the "Theodor" shaft dump compared to samples from the drill hole of the "Fortschritt" shaft dump.

Proportions of Zn related to extraction steps L1, L2 and L3 decreased continuously in the range from 2 to 6/7 m depth and showed a slight increase at the transition from black shale- to limestone-dominated drill metres, which indicates that Zn and also Cu occurred in more easily soluble mineral phases, such as hydroxides, sulfate and carbonate minerals.

As with Cu, the sulfidic bond is the dominant type of Zn fixation. Furthermore, the carbonate and cation-exchangeable fixation form was determined by sequential extraction to be substantially increased. Such chemical fixation includes, for example, amorphous zinc hydroxides, which can be found quite often as white coatings on black shale rock fragments on dump. The proportion of Cu and Zn extracted by leaching steps L1 and L2 was generally decreasing with depth, whereas the higher leaching rates were noticed for L1. A renewed increase can be noted at the transition from black shale- to limestone-dominated drill metres, which clearly indicates an increased abundance of secondary mineral phases more susceptible to dissolution by less acidic extractants.

Results from the investigations determining chemical bonding type of base metals and water elutability were combined in order to draw conclusion and to assess the degree of weathering of the examined table dumps.

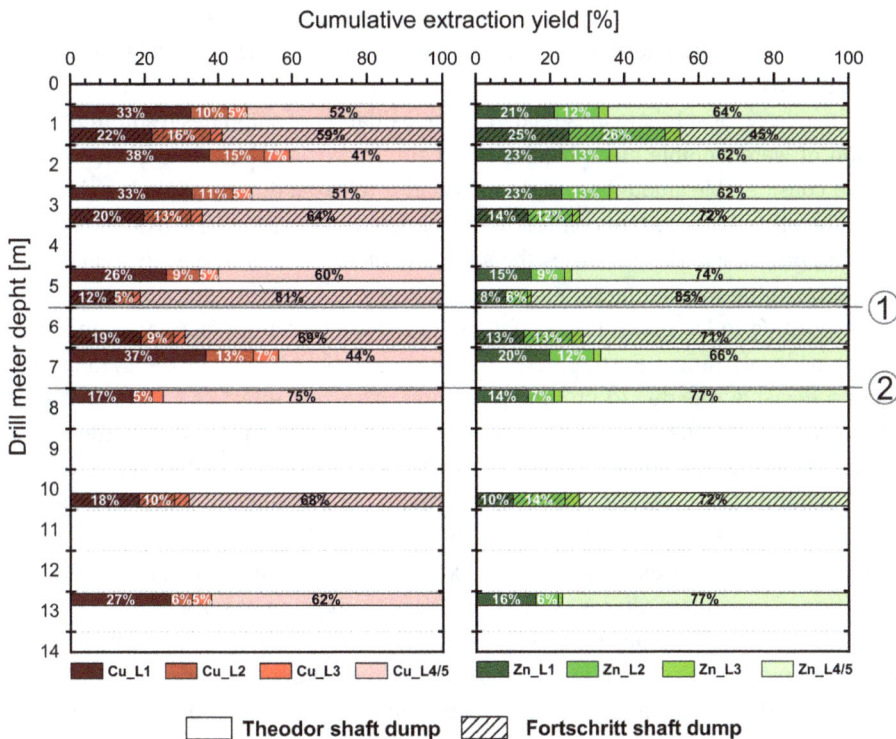

Figure 15.15: Chemical fixation of Cu and Zn regarding to a sequential extraction protocol in Table 15.4 for selected samples within drill holes from "Theodor" and "Fortschritt" shaft dumps. (lithological change from upper black shale- to lower limestone-dominated part ① Theodor / ② Fortschritt shaft dump).

It may be stated, of the three base metals, Zn is the one with the highest mobility. Zn is relatively soluble and elution curves of the investigated drill holes correlate with each other. Rose [20] stated increased zinc solubility occurs at a pH of 7 under oxidizing conditions and that zinc sulfate occurs up to a pH of 8.6. The investigated bore hole sections showed a vertical shift from the middle to lower black shale-dominated drill metres due to the increased solubility or mobility of Zn, which is manifested by Zn-rich precipitation of commonly amorphous zinc hydroxide onto rock fragments. Furthermore, conductivity data correlates very strongly with the eluted Zn contents as well as the pH decrease of the eluates. Generally, low and high pH values can have a mobilizing effect on Zn ions, which is substantiated by the wide range (up to pH 8) of Zn mobilization shown by Brookins [21].

Although sulphur is not one of the main elements investigated, it still plays a decisive role in the transport and mobility of metals. A slight variation of the pH results in an increase or rapid decrease in the S content of the eluates, which was particularly noticeable in the black shale-dominated parts. Additionally, S has a significant influence on the conductivity and with regards to the solubility of base metals the following relationship that controls the conductivity can be considered: $S \gg Zn > Cu/Pb$.

The release of sulphate ions related to the oxidation of sulphides result in the formation of water-soluble metal-bearing sulphate, hydroxisulphate as well as hydroxides. A considerable amount of sulphurous acid is released during sulphide oxidation. Here, the pH is an important controlling factor that accelerates sulfide oxidation at decreasing pH, which also increases the solubility of the sulphides and the metal concentration in the weathering solution.

According to Langmuir [22], sulfate anions released during oxidation are stable over a wide pH range. Furthermore, the presence of Fe^{3+} ions in the system also plays a decisive role due to proton release and oxidizing effect of Fe^{3+} ions [23]. In black shale rocks, the excessive supply of Fe^{3+} ions are generated by weathering of pyrite, chalcopyrite and Fe-containing sphalerite. The presence of dolomite or calcite, common carbonate minerals in the Permian black shale and limestone, causes a buffering of the acidic leachates from pH 1–2 to a slight acidic or neutral pH range [24]. The result of this buffering and the excess of SO_4^{2-} in the system allow the precipitation of secondary phases such as brochantite, gypsum or serpierite.

An important external parameter is the location of the mining dumps to the west Harz mountains, which is considered to be part of Central German arid region. It is affected by its geomorphological position in the rain shadow of the Harz mountains and characterized by very low annual precipitation (< 500 mm) [25, 26]. Less frequent but intense rain falls in the summer season limit the transport and dissolution processes in the dump body extremely. Heavy rain falls are the main reason for the transport of water and oxygen within the mining dumps. With an initial pH of 6.4 [27], the penetrating seepage waters have a weakly acidic character which result partly in a mobilization and relocation of soluble metals in the upper area of the dump body, whereas Cu can be regarded to be immobile and remains in the upper part of the dump. Due to the further penetration of the seepage water into the interior of the dump (approx. below 3 m in depth), an increasing buffering of acidic seepage water taking place through the presence and availability of not yet consumed/dissolved calcite and dolomite. However, Mibus [28] determined a neutral to weakly basic pH for leachates of drill material from a further dump ("Zirkel" shaft) investigated more detailed, which corresponds to the pH values obtained in the elution experiments generally and especially for "Fortschritt" dump material.

Sequential extraction leaching data revealed an increased sulphide content in the deeper part of black shale-dominated drill metres, which are more preserved due to the carbonates' buffer capacity increasing with depth. For base metals, the following mobility sequence can be observed: $Cu < Pb \ll Zn$. Furthermore, there is only a very minimal retention capacity for seepage water within the dump body due to the high pore space capacity, so that the majority of the seepage water is discharged rapidly draining in channel-like pathways, which has been proven by investigations around and on dumps in the region [29–31]. Mibus [28] classified the seepage water in dumps of the Mansfeld region according to the prevailing chemical inventory and concludes a $CaSO_4$-enhanced type in the study area, whereas investigations by Dunger [29] revealed high permeability rates and low storage capacities are to be expected in dump bodies containing large rock fragments. Small components are more likely to be subject to erosion or gravitationally displacement by seepage water into deeper areas of the dump [32]. However, grain size analyses of the dump material have shown that the main part of this material correspond to medium to coarse gravel grain size, which supports very low storage capacities for seepage water. According to a slightly increased portion of the fine-grained fraction in the depth range of 2–3 m from the top of the dump it can be assumed that seepage water has longer retention times where it enhances the release of metal ions by the dissolution of secondary mineral phases that often occurred accumulated in filled cavities between larger rock fragment. The formation of secondary mineral precipitation is triggered by buffering of acidic metal-bearing solutions, which corresponds with observations and sequential extraction data.

Generally, it can be concluded, only a small proportion (approx. 15%) of the base metal content is present in a mobilizable or easily mobilizable bound form comprising mainly secondary sulphates and hydroxides formed from sulphide oxidation. The content of base and trace metals in the eluates is rather low. A further limiting factor is the neutral to weakly basic pH value of seepage water, which largely inhibits the mobilisation of metals by sulphide oxidation. Thus, parameters mentioned, such as neutral pH, low metal content in leachates and metal precipitation as well as increased buffer capacity by carbonates cause a very slow sulfide oxidation in the system, since catalysts such as sulfide oxidizing bacteria do not exist. It can be also ascertained that varying deposition times leave no significant progress in the oxidation state of the sulphide assemblage, although it is recognizable that a longer retention time of low-grade ore deposited on dump causes a slight increased degree of weathering as shown for the older "Theodor" shaft dump in this study.

3 Processing options to recover copper and by-products from low-grade black shale ore

The efficient and economic processing of black shale ore has always been a challenge in the former Sangerhausen-Mansfeld mining district as well as in the current exploitation of Polish Kupferschiefer-type ores. Thus, novel or improved mineral processing methods that are more efficient in the application of low-grade ores are in high demand.

3.1 Sensor-based sorting

Sorting represents an early processing method on coarser feed streams in mining operations to separate them into two or more products generally termed concentrate and tailings. Mass reduction and ore enrichments are essential contributions to initiate the economic processing of an ore. According to material-specific properties present in low-grade ores from dumps, an upstream sorting process, which produces a copper-enriched concentrate and tailings, is a prerequisite for the economic beneficiation by extraction techniques. The following properties can be used for sorting experiments: colour, base metal content and mineralogical composition of the host rock. In detail, it has been shown that, on the one hand, colour differences between rock fragments containing increased or low ore content exist, which is indicated by greenish weathering coatings on highly mineralised black shale rock fragments. Secondly, the difference in light intensity is related to the different lithotypes occurring on dumps. Clearly, the light, quartz-rich sandstone differs from the dark black shale as well as brownish limestone. However, a copper enrichment by optical sorting can be achieved but a mere separation of the lithological members in sorting stream may have an effect on subsequent processing techniques. Current colour scanners detect minor differences in colour, brightness, reflection, or transparency.

Test work investigating various sorting principles were carried out on black shale low-grade ore mainly from the "Fortschritt" shaft dump by the Department of Mineral Processing of the Technical University Aachen (RWTH) and results are compiled hereinafter from [33].

Optical sorting tests of black shale low-grade ore based on colour differences by secondary copper coatings showed that a copper enrichment from originally 0.7 to 1.1% can be achieved. This corresponds to an almost 50% Cu recovery. A stronger upgrade to 1.9% Cu yielded a significantly lower loss of 25% Cu with a 25% mass reduction in the concentrate. The corresponding tailings were characterized by a low Cu grade of 0.3%.

Within the scope of the investigation on sensor-based sorting of Kupferschiefer-type black shale ores, sorting tests were carried out for the first time on an industrial X-ray fluorescence (XRF) sorter "TiTech x-tract XRF" by TOMRA Sorting Solutions. The design of the XRF sorter does not differ fundamentally from the usual design of belt sensor sorters (see Figure 15.16). In addition to the feed periphery (screen classifier, conveyor belt and vibrating chute) and the sorting mechanism (air pressure tank and nozzle bar), the X-ray source and the energy emission detector are located above the conveyor belt.

Figure 15.17 shows the sorting results of three test runs according to mass and copper distribution. The copper content served as a threshold value for the sorting criteria below which a rock piece is assigned to the tailings. The grain size of the rock pieces was between 20 and 50 mm; which corresponds to the mean grain size

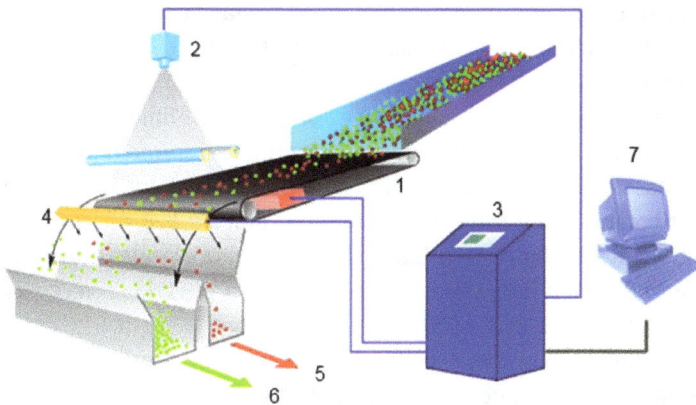

Figure 15.16: Generalised design of a sensor-based sorter from [34]. Numbered units refer to (1) material conditioning, (2) scanner, (3) data processing unit, (4) mechanical sorting device, (5) rejected product, (6) accepted product, (7) control unit.

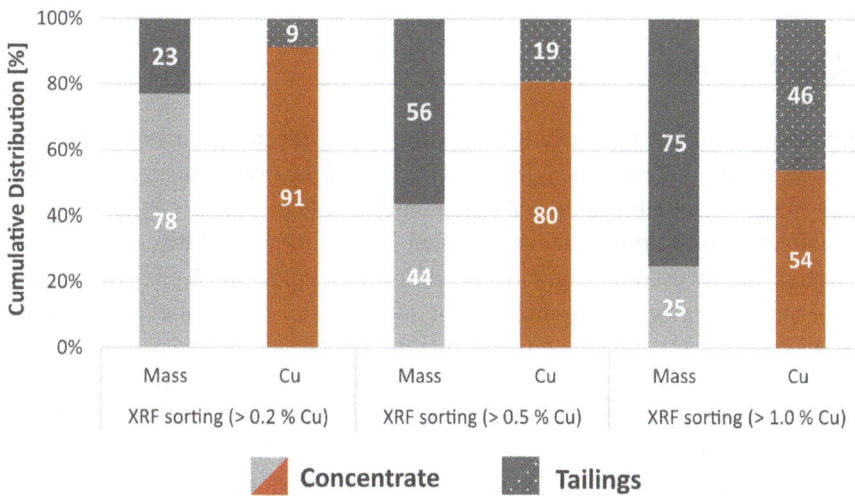

Figure 15.17: Mass and Cu recovery (wt.%) in concentrates and tailings by the application XRF-based sorting using different copper content thresholds.

of frock fragments from low grade ore dump material. The selected threshold values chosen for test runs were 0.2%, 0.5% and 1% copper content.

Results at a low threshold value of 0.2% Cu showed that Cu loss in the tailings was less than 10%. The zinc content in the concentrate could also be slightly enriched. Using a threshold value of 0.5% Cu, more than 50% of the mass were sorted to the tailings, which had an average copper content of 0.26%. In this configuration, the Cu loss increased to 20%. The highest threshold of 1% meant that only 25% of the feed mass was allocated to the concentrate. Although the average copper content in the concentrate is over 2% Cu, the loss of Cu increased significantly to 46% regarding the initial copper content in the feed.

Sorting tests using mixed ore feed (sandstone, black shale, limestone) from Rudna mine, Poland, carried out by a combined colour and near infrared sorter on different grain size classes (12.5–20 mm, +20 mm) have shown that 20–30% of the feed prior to the first comminution stage can be reduced at a Cu loss of less than 5–10%. Sorting approaches to produce a black-shale-rich concentrates from the mixed ore feed (prior to flotation line) yield 45–60% shale and 17–35% Cu recovery with sorting concentrates of 1.8–3.1% Cu [35].

X-ray fluorescence sorting provides excellent results for copper enrichment to gain a processable sorter concentrates. Depending on the test settings, either a high-grade concentrate can be produced while accepting extensive process losses (tailings) or a pre-separation of the tailings can be achieved by producing a coarse grain fraction with the lowest possible copper content that can be discharged at an early stage within the processing line.

3.2 Comminution

Comminution, comprising crushing and grinding, is the most cost- and energy-consuming step in mineral processing. Nearly 50% of the overall operating costs in mineral processing plants are caused by comminution to prepare ore feed for subsequent extraction techniques such as flotation [36–38]. Consequently, a significant reduction of energy used for particle size reduction at equal or advanced liberation of target minerals is desired to improve the overall balance of mineral processing routes.

The whole comminution process comprised a multi-stage comminution line. The feed was pre-crushed using two cone crushers (gap size 1: 2 cm and gap size 2: 5 mm), the first interposed by a vibrating screen of 12 mm. The final grain size of the product was achieved by two-stage milling using a screening ball mill (screen size: 3.15 mm and 90 µm). Comminution products are used for different mineral processing tests e.g. froth flotation and leaching. Aliquots of the final ball mill product were detailed mineralogically and geochemically analysed.

Comminution as well as comminution tests on low-grade black shale ore were exclusively carried out in the pilot plant facilities of the UVR-FIA GmbH, Freiberg/Saxony, Germany. Bulk samples of up to 2 tons were treated by a comminution line using a primary jaw crusher (Pulverisette 1, model II, Fritsch GmbH, Germany), a cone crusher (type Symons, size 1, FIA Freiberg, Germany) and a screening ball mill (SKM 400, FIA Freiberg, Germany), furthermore vibrating screens and riffle splitter were interposed to single comminution steps. Figure 15.18 illustrates the multistep comminution line, which delivered a final comminution product generated by a screening ball mill with a grain size of –100 µm, but also exemplary tests producing a finer final grain size of –40 µm have been executed.

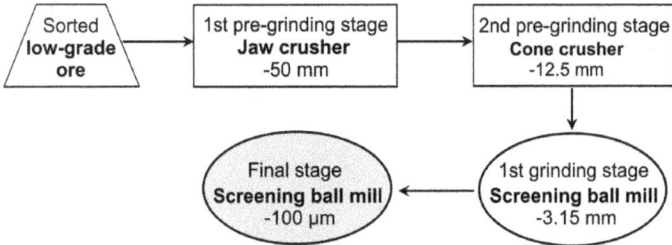

Figure 15.18: Comminution route applied to low-grade black shale ore in order to generate a final ball mill product of –100 µm grain size.

In addition, low-grade black shale ore was comminuted in grinding tests using a vertical roller mill (Loesche mill LM4.5 operated in both the overflow and with internal sieving mode, see [39, 40]). Roller mill products were compared to ball mill products in regards to particle size distribution, liberation of value minerals and their flotation performance.

To estimate the energy required to grind raw materials, grindability tests according to Bond [41] were carried out. The Bond work index has proven to be a widely used and important criteria for the evaluation of grinding results in laboratory, semi-industrial and operational scale as well as for the design of mills and grinding media fillings. To determine the Bond working index W_i, an intermittently operating circuit of ball mill and classifier is modelled [42]. The results are summarized in Table 15.5. The Bond work index for grinding low-grade black shale ore was 12.8 kWh/t. Therefore, a specific energy consumption to gain a product –100 µm by grinding in a ball mill of 13.1 kWh/t (wet operated mill) and 17.1 kWh/t (dry operated mill), respectively, would be necessary. However, the dry operating vertical roller mill consumes 7.2 kWh/t and produces, compared to ball mill product, an equal if not an improved product in respect to grain size distribution and sulphide liberation [1, 43].

Table 15.5: Bond-Indices for lithotypes occurring on black shale mining dumps.

Lithotype	Work index W_i [kWh/t]
Black shale	12.8
Limestone	13.8
Sandstone	17.0

In order to investigate the influence of the breakage principle (shearing, abrasion) caused by different mill types, granulometric and liberation data of final mill products with a grain size of d99 = 40 and 100 μm, respectively, generated by a screening ball mill (laboratory scale) and a vertical roller mill (pilot scale) were compared.

Figure 15.19 shows the particle size distribution of ball mill products (−40 μm and −100 μm) and equivalent roller mill products. The granulometric distribution indicates that roller mill products contained generally 5–10% lower portion of fines < 5 μm compared to the mill product.

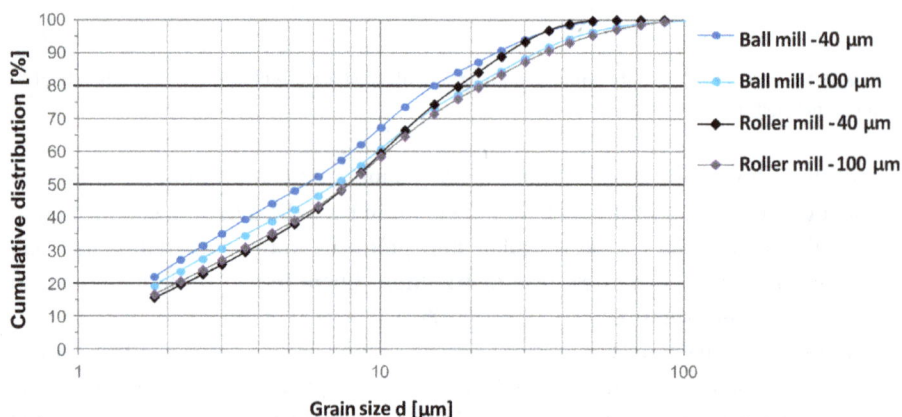

Figure 15.19: Comparison of the grain size distribution of grinding products of ball mill and vertical roller mill in different final grain sizes (−40 and −100 μm).

Liberation analyses of ball mill and vertical roller mill products revealed that a significant difference in the degree of liberation of copper sulphides was not ascertained. However, the proportion of specific gangue minerals interlocked with copper sulphides differed at a notable extent. Illite and carbonate minerals, mainly dolomite and some calcite, form over 40% of the main mineral constituents of the black shale feed and thus, they represented potential mineral phases containing copper sulphide grains. The portion of copper sulphide grains intergrown with dolomite/calcite was considerably increased in the vertical roller mill product compared to the ball mill product, whereas the latter contained copper sulfides substantially more intergrown with illite (Figures 15.20, 15.21). The proportion of calcite/dolomite in particles, which consisted of 20–70% copper sulphide, was substantially increased in the product of the vertical roller mill compared to the ball mill product. In contrast, the illite/clay content of particles containing 25–85% copper sulphide was significantly increased in the ball mill product. Particle size reduction in ball mills based on breakage failure caused by impact and attrition forces works most efficiently on brittle minerals such

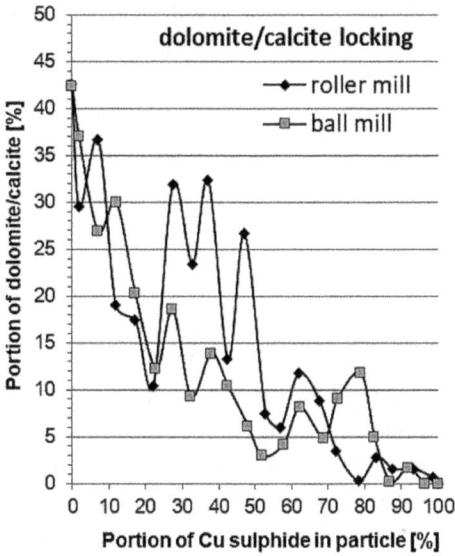

Figure 15.20: Percentual locking of copper sulphides with carbonates (dolomite/calcite) in roller and ball mill products.

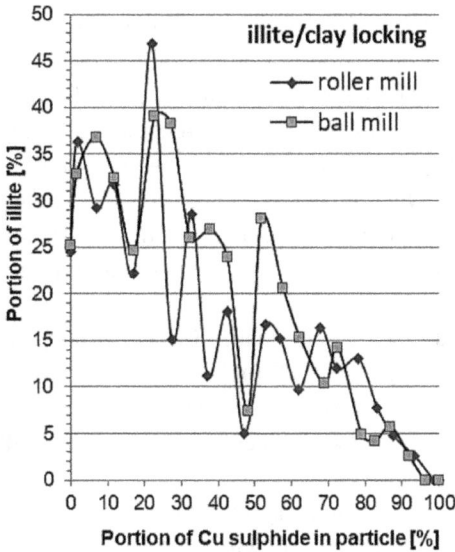

Figure 15.21: Percentual locking of copper sulphides with illite/clay in roller and ball mill products.

as carbonate, while comminution by roller milling applies mainly shearing forces, which can obviously detach anisotropic sheeted clay mineral from sulphide grains more easily.

By comparing comminution products of the ball and vertical roller mill it can be concluded that the roller mill consumed 40% less specific energy than the conventional ball mill when producing a product $P_{80} = -100$ µm at improved liberation of sulphide grains. An energy-saving contribution to the processing of copper ores, which has been already shown by other studies, e.g. on a chalcopyrite ore [44] or different iron oxide ores [39, 45].

3.3 Flotation

The physico-chemical separation process of flotation is the most important and most commonly applied mineral processing technique used for the beneficiation of copper ores. The mineral processing of black shale ore is a challenge due to its fine-grained intergrowth of valuable minerals and its special carbonaceous and argillaceous composition. Because of this fact numerous tests were carried out in recent decades. From an early stage, flotation issues caused by the mineralogical composition and organic constituents of black shales were recognized [46, 47]. Especially intergrowths of organic matter, clay minerals and copper sulphides were considered to interfere the selective flotation of base metal sulphides mainly of bornite, chalcocite, chalcopyrite, galena, pyrite, and sphalerite. Thus, the depression of organic compounds, pyrite, galena and sphalerite at simultaneous enhanced recoverability of copper sulphides poses the greatest challenge to the improved flotation of black shale ores.

Flotation tests were carried out in the technical centre by the UVR-FIA GmbH, Freiberg, Germany. Results of investigations conducted by UVR-FIA compiled in the following paragraphs are detailed reported in [33] (figure 15.23). Numerous flotation experiments were aimed to increase the recovery of copper and comprised agglomeration flotation, bitumen flotation, oil flotation, carrier flotation, froth flotation in agitator cells and column flotation. Recovery data showed that the froth flotation in mechanical flotation cells achieved the highest recovery yields.

In order to determine the optimal grain size of the feed, different screening ball mill products with varying mesh sizes (-71, -100, -125 µm) were subjected to a standard flotation using xanthate as collector and Flotanol H54 as frother at pH 8.8. Figure 15.22 shows that the highest Cu recovery and enrichment factor was achieved with a ball mill product -100 µm.

Froth flotation tests were carried out at lab-scale using different sized DENVER flotation machines (1.2, 2.7, 4.5-litre-cells) with a solid content in the flotation feed of about 30%. In order to produce a larger amount of a flotation product, tests were conducted with a semi-technical 4-staged flotation bank.

A pH of 10 was adjusted with sodium hydroxide, polyacrylate was added as dispersing agent and xanthate was used as sulfide mineral collector. A mixture of alcohol

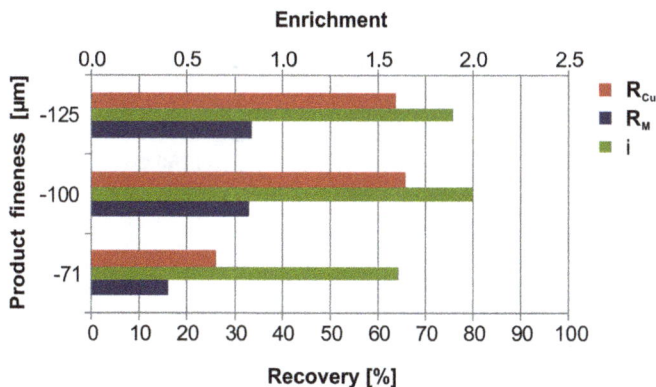

Figure 15.22: Standard sulfide flotation results obtained by the use of different product grain sizes of low-grade black shale feed. (from [33]).

and polyethylene glycol ether acted as the frother. Froth flotation experiments with mechanical flotation cells were carried out in chronological steps as listed below:

- 0–5. min: suspend the solids (25 wt.%) and adjust the desired pH value,
- 5. min: addition of the dispersant,
- 7. min: addition of the collector,
- 9. min: addition of the frother,
- 14.–19. min: air supply and froth extraction.

In preparation for a multi-stage flotation experiment on a semi-technical scale, flotation tests consisting of a standard sulphide flotation (SSF) and a subsequent three-staged cleaner flotation (CF) was first carried out on a laboratory scale (see Table 15.6).

Figure 15.23 shows the recovery yields and the enrichment factor (i) of the 4-stage lab-scale flotation test, Table 15.6 lists test conditions applied in the individual flotation steps. The copper recovery refers to an optical pre-sorted feed (−50 mm) comminuted to a final grain size −75 μm by ball milling and a copper content of 2%. The froth product of the last cleaning flotation had a copper content of 9.4%, which corresponds to an enrichment factor of 4.6 at 70% Cu recovery.

In a semi-technical approach using a flotation bank (volume: 5 × 8 Litre cells), 80 kg of run of mine-black shale low-grade ore (not pre-sorted) were treated by the same flotation protocol and yield in a froth concentrate with 2.4% Cu, which corresponded to a similar enrichment factor of 4 considering a Cu content of 0.6% in the feed.

To investigate the influence of the liberation state of copper sulphide in comminution products generated by different grinding techniques (ball mill, roller mill), feed material with different final grain sizes (d99 = −40 or −100 μm) was treated by standard sulphide flotation. 11 products from the roller mill and the two ball mill products were individually flotated applying a standard sulphide flotation protocol.

Table 15.6: Experimental setting of multi-stage flotation test in laboratory scale (from [33]).

Stage	Flotation time [min]	Flotation conditions	Reagents
Conditioning	3	30 wt.% feed −75 µm	300 g/t xanthate, 150 g/t Flotanol H54
Standard sulphide flotation	15	300 l/h, 1200 min^{-1} 4.5 l-Denver cell	
1. Cleaner flotation	12	150 l/h, 1100 min^{-1} 2.7 l-Denver cell	
Conditioning	3		500 g/t sodium silicate (water glass)
2. Cleaner flotation	10	100 l/h, 1000 min^{-1} 1.2 l-Denver cell	
3. Cleaner flotation	10	150 l/h, 1000 min^{-1} 1.2 l-Denver cell	

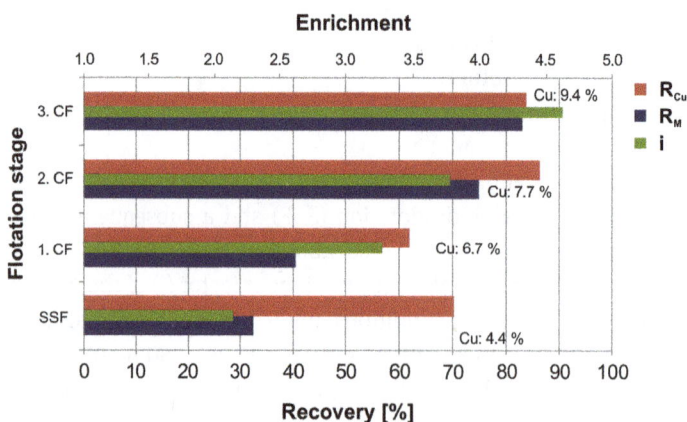

Figure 15.23: Recovery data of lab-scale multi-stage flotation test of a −75 µm ball mill product from low-grade black shale ore consisting of a standard sulphide flotation (SSF) stage and reflotated by three downstream cleaner flotation (CF) stages. (from [33]).

Ball mill and roller mill products from which flotation concentrates showed the highest Cu recovery were also used for agglomeration flotation tests. The best results for standard sulphide and agglomeration flotation tests of mill products −100 µm are summarized in Figure 15.24 and indicate that standard sulphide flotation of the ball mill product achieved an increased Cu recovery, whereas the roller mill product was flotated more selectively showing a considerable higher enrichment factor (2.4 vs 1.6). A difference notable also by the use of a more fine-grained feed (–40 μm). Agglomeration flotation achieved the best flotation characteristics using the roller mill product, which is manifested by increased recovery and selectivity, whereas agglomeration flotation of the finer ball mill product (–40 μm) led to less recovery and selectivity compared to the more coarse-grained product (–100 μm).

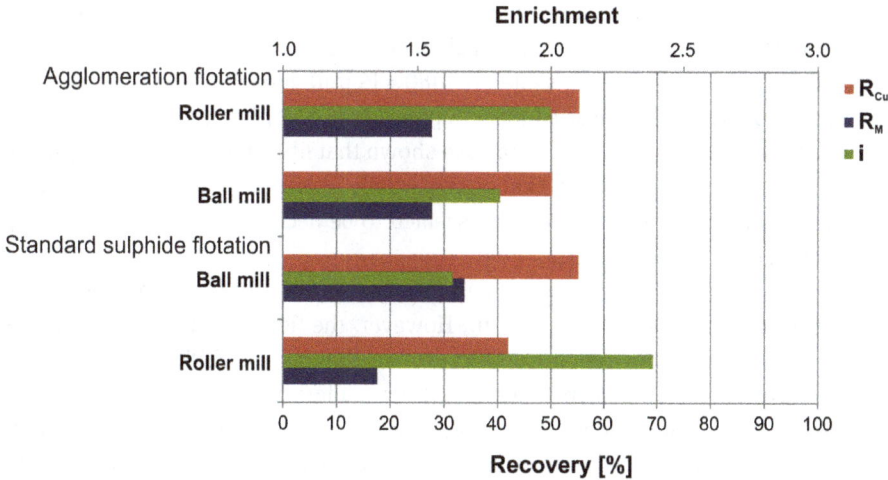

Figure 15.24: Comparison of recovery data from agglomeration and standard sulphide flotation tests carried out on –100 μm products generated by roller and ball mill. (from [33]).

As shown in Chapter 3.2, the liberation of copper sulphides in roller and ball mill products were almost equal. Obviously, varying degrees of liberation is therefore not the determining factor for different flotation properties. Mineral Liberation Analyses (MLA) performed on Scanning Electron Microscope (SEM) deliver basically modal mineralogy and liberation/exposition data of target minerals regarding intergrowths with other minerals on the base of back scatter electron imaging of polished sections. MLA-data of flotation concentrates, and tailings revealed that flotation products from roller mill feed –100 μm contained an increases proportion of poorly liberated copper sulphides compared to concentrates from the corresponding ball mill feed. Additionally, the portion of completely liberated copper sulphides was higher compared to the flotation product from ball mill feeds. The generally increased recovery of copper sulphides in flotation products from roller mill feed, in particular of completely liberated copper sulphides was also achieved by flotation of –40 μm feed. Further, MLA-data were analysed regarding mineralogical interlocking associations of copper sulphide-bearing particles. The assumption that heavy metal ions originating from other base metal sulphides (galena, sphalerite) intergrown with copper sulphides could interfere with flotation yields could not be verified because of missing differences in the amount of such intergrowths. Finally, different proportions of gangue minerals intergrown with copper sulphides affected their recoverability. Copper sulphides were less intergrown with illite/clays, but more interlocked with dolomite/calcite in the roller mill feed. This suggests that intergrowths of copper sulphides with clay minerals hinders their recovery as recognized in the flotation tests using ball mill products.

Comparative flotation tests of grinding products from a ball mill and a roller mill showed a tendency towards a more selective flotation of copper sulphides using roller mill products independently from the grain size of the feed. In comparison of agglomeration flotation tests, the best flotation characteristics were achieved by flotation of an extremely fine milled black shale ore (d80 = 15 μm) generated by the roller mill, whereas a coarser feed grain size (d80 = 75 μm) produced by both, roller and ball mill impaired the recoverability. MLA-data have shown that liberation of copper sulphides was coequal, but rather the combination of partially liberated copper sulphide minerals intergrown with illite/ clay minerals seemed to be decisive for the lower recovery by flotation of ball mill products.

Basically, it is technically feasible to produce an 10% Cu flotation concentrate from Mansfeld low-grade ore (0.6% Cu). However, the flotation yield of 70% Cu recovery is insufficient for economic efficiency so that the flotability of copper sulphides needs to be gained in order to extract a marketable copper concentrate, which meets the requirements of internationally operating smelters. Otherwise, an alternative beneficiation of copper, zinc and silver can be realized by leaching extraction methods.

In the industrial application of the flotation of black shale-containing ore in concentrator plants of the Polish Kupferschiefer-type deposits for the last 30 years, a partly doubled black shale fraction in the mixed ore (sandstone, dolomite, black shale) was processed especially in Rudna and Polkowice concentrators. However, it was reported that finely disseminated copper sulphides hosted in the black shale feed are considerably enriched in tailings of the 1st stage cleaning flotation, which result in serious problems in the recovery of copper [48, 49]. Drzymala et al. [50] showed by modelling of grade-recovery curves that the maximal theoretical copper content in concentrates from black shale ore feed was generally significantly below of that from sandstone feed. However, black shale ore is highly enriched in Cu and associated metals compared to sandstone ore, which is easily processable, but corresponding metal contents are contained in considerably larger tonnages and enlarge process capacities and costs. Due to that, organic carbon is regarded to interfere with the recovery of fine-grained copper sulphide particles. Principally, lab- to industrial scale approaches to diminish organic carbon content prior to flotation were undertaken and finally ended up with the segregation in pre-concentrates, one poor and a second rich in organic matter [49].

3.4 Acid leaching

Currently, sulphide copper is beneficiated mainly by pyrometallurgical processes worldwide, although the production demands a high energy consumption and causes a considerable environmental impact. Copper concentrates processed in smelters contain usually about 30% Cu unless the sulphur content essentially contributes to an autothermic process to lower the tremendous operating costs. However, this is not applicable to concentrates from low-grade black shale ores due to the low sulphur content. Concentrates of Kupferschiefer-type ores are currently processed industrially only at KGHM Polska Copper S.A., Poland, where the pyrometallurgical route produces marketable copper and wrought products. Recently, copper contents of Polish concentrates are decreasing as reported exemplarily for Lubin flotation concentrates, which had 15.9% Cu in 2009 and decreased to 12.9% Cu in 2013. This was accompanied by a significant increase of the lead content in Lubin concentrates (2009: 2.7%, 2013: 5.4%), which is unwanted regarding the processing in the smelter [48].

Hydrometallurgical processes are though suitable for materials containing only a small amount of valuable metals and hydrometallurgical extraction aims to precipitate a marketable metal or compound. Leaching test series using black shale ore and related concentrates were carried out by the IME Process Metallurgy and Metal Recycling Institute of the Technical University (RWTH) Aachen, Germany. Results presented in the following paragraphs have been previously published [1, 33].

Several copper concentrates containing 2–10% Cu, 3–6% Zn, 1–2% Pb and up to 800 ppm Ag were investigated regarding the most efficient extractability using different leaching agents. These copper contents are relatively low for standard concentrates, so that the economic efficiency of extraction processes to be selected plays a special role and has to cause low costs in plant construction and operation. Besides copper, the extraction of silver can result in a significant increase in value.

The aim of the leaching test series was to develop an optimum process route with regards to the following parameters: solvents and concentration, temperature, leaching time, solid/liquid ratio as well as the necessity of thermal pre-treatment or the addition of oxidants.

A first test series carried out on an averaged ore grade black shale (2.3% Cu) served to select a suitable solvent and to investigate the influence of thermal pre-treatment due to a pyrometallurgical conditioning like roasting that might be recommendable.

A second series of tests was carried out with a copper concentrate (9.8% Cu) extracted from low-grade ore by a 4-stage flotation, whereas a third series of tests investigated the suitability of nitric acid, which is a residue from the production of silicon wafers. The series of tests was also carried out in parallel with HCl, which on the one hand served the purpose of comparison and on the other hand was intended to show the influence of hydrochloric acid on leaching of silver.

All leaching tests have been carried out on laboratory scale. A beaker was prepared with the appropriate amount of solvent and solid, which was heated using an adjustable heating plate. The solution was constantly mixed by a magnetic stirrer so that homogeneity and a constant temperature profile over the filling level can be ensured. The temperature was kept constant at 450 or 900 °C ± 1 °C. For cementation of copper, the well-established precipitation by addition of iron was applied.

Sulfuric or hydrochloric acids are known to be most efficient for treating sulfide copper ores. H_2SO_4, HCl, H_2O and NH_3 have been used as solvents during the experiments to leach a copper concentrate containing 9.8% Cu. The extraction process has been carried out for 24 h at 50 °C. The used material had three modifications: the original concentrate and two concentrates which have been roasted at different temperatures (450 °C and 900 °C, see Figure 15.25). Although roasted materials have achieved a higher yield in the dissolution of Cu, the use of unroasted material might be preferred due to the energy amount used for roasting processes, if it is not an autothermic process. However, the dissolution of silver seems to be only possible after roasting at 900 °C.

As shown in Figure 15.25, the highest concentration of copper in the solution is achieved by the use of sulphuric or hydrochloric acid, but a previous roasting had a decisive influence on the strength of dissolution. However, since the hydrochloric acid (4 mol/l) had a much higher concentration than the sulphuric acid (2 mol/l), the latter was preferred. With unroasted material, which is indicated in Figure 15.25 on columns without frame, a significantly lower leaching yield can be achieved. Roasting temperatures of 450 °C and 900 °C were previously determined by differential thermo analysis (DTA) on black shale ore. At 450 °C, the thermal treatment is expected to convert the sulphides into sulphates, whereas at 900 °C complete oxidation is realized. Thus, it can be stated that sulphates are more soluble than oxides, but water and NH_3 as a basic solvent are not suitable for leaching the material.

A promising alternative to reduce the costs for the leaching agent can be a reuse of a nitric acid solution, which accumulates as a waste product in industrial processes such as the etching of wafers. Leaching with used HNO_3 obtained highest yields on unroasted feed and up to 90% of the contained Cu can be dissolved using this alternative leaching agent. Figure 15.26 shows the highest possible yields of various unroasted ores and concentrates marked by different copper contents. The comparison of the solvents shows clear variations, but it is not obvious which solvent should be preferred. Although the use of the waste acid (HNO_3) achieves the highest yield, in comparison to the use of conventional HNO_3 it is clearly recognizable that it is not the nitric acid itself that contributes to the high dissolution of copper sulphides. It is more likely that other components of the waste acid such as HF, which is contained at approx. 1% in the waste nitric acid enhanced the advanced dissolution of sulphides.

The disadvantage of leaching with the prepared nitric acid is a strongly exothermic reaction during leaching, which produces nitrous gases. These gases are characterized by their harmful effects on the environment and health and this is why the acid should not be used or costly health and industrial safety measures have to be

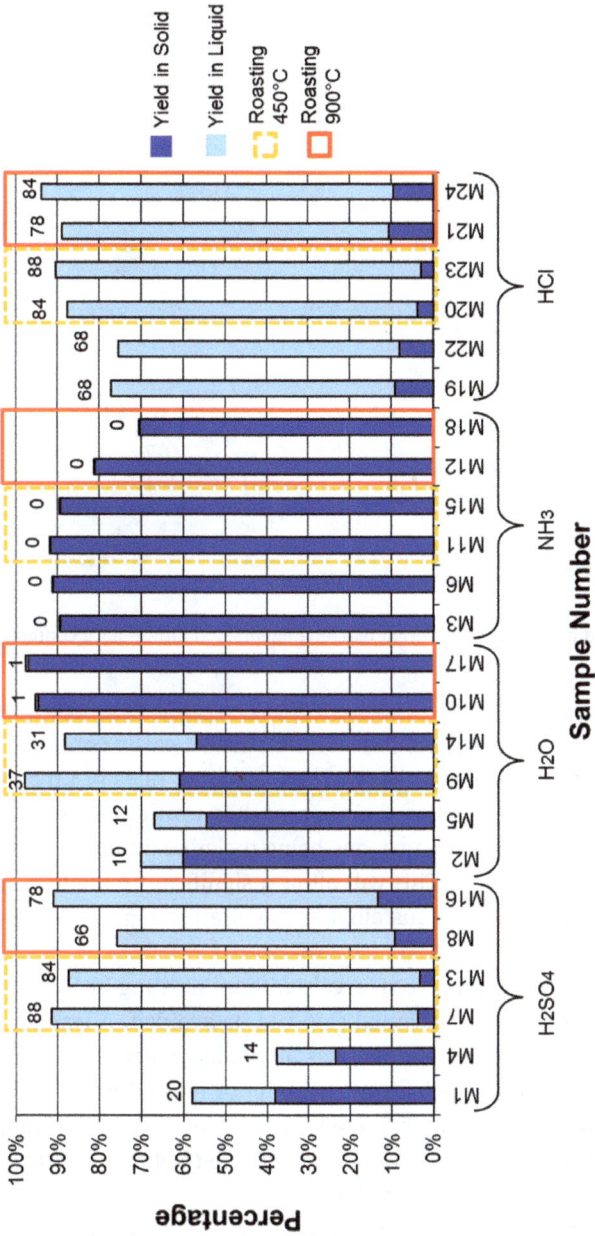

Figure 15.25: Mass balance of Cu in different solvents with varying pre-treatment of the feed material. (from [1]).

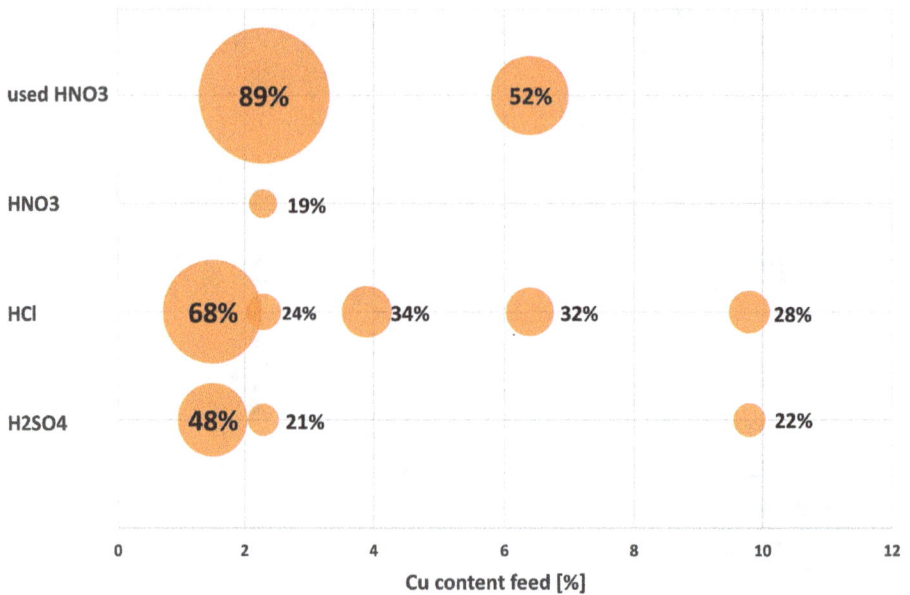

Figure 15.26: Cu recovery achieved by the use of different leaching agents and different Cu content in the feed. (modified from [33]).

implemented. However, test series using HNO_3 waste acid showed partly contradictory results, which showed partly high and also very low leaching rates of Cu and Zn, thus can hardly be conclusively evaluated. This approach is worth following up on due to the combination of two waste streams, one being the mine dump material and the other being the waste nitric acid. This will allow a significant reduction of acid costs and – simultaneously – a more sustainable use of all raw materials involved in the process.

A further series of tests has been carried out largely without roasting in order to examine if a comparatively cost-intensive thermal step can be avoided. A leaching time of 5 days with the addition of ferrous sulphate as an oxidising agent can produce approximately the same result as the leaching using prior roasted black shale feed. However, due to the strongly varying extraction yields achieved by leaching unroasted feed and long-lasting extraction times, a thermal pre-treatment to transfer sulphides in sulphates could amplify leaching kinetics.

To continue the concept of bringing together different waste streams, a cementation process should take place, which is robust, cost-saving, and can easily be established on a small scale. Cementation is based on the different standard electrode potential of elements. Iron scrap or zinc dust can be used as precipitation agent for Cu.

Black shale-bearing feed materials containing comparable Cu-contents (2.7%) are tailings from the 1st cleaner stage of the Lubin flotation plant, Poland, which were investigated regarding a hydrometallurgical treatment to improve to efficiency of the overall processing by extraction technologies considering additionally troublesome middlings. Copper-bearing minerals such as predominantly bornite and chalcocite in the latter were sufficiently leached by a two-stage leaching first with sulfuric acid and second with oxygenated sulfuric acid at an increased leaching temperature (90 °C). The first leaching step removes carbonate minerals, common constituents of the feed, and liberates valuable minerals, whereas in the second leaching step the oxidized sulfuric acid was used to extract Cu, Co, Ni and As but Pb and Ag remained in the leaching residue [51, 52]. Middlings subjected to sulfuric acids leaching represented flotation feed characterized by improved flotation kinetics of Cu and Ag-containing value minerals and increased recovery rates [53].

3.5 Bioleaching

As conventional copper extraction via smelting is extremely energy-consuming, (bio)-hydrometallurgy in combination with electrowinning represents a low-cost and environmental-friendly and energy-saving extractive method to win copper products. Studies on the processing of low-grade Kupferschiefer ore and dump material by bioleaching started in the 1970s in the laboratory and pilot scale percolator experiments to simulate heap bioleaching and have been relaunched recently [1]. Following bioleaching tests achieving higher recovery rates, in several joint research projects resulting in improved bioleaching protocols were implemented on black shale ore and related flotation concentrates [54, 55].

While improved recovery rates were just notable for Zn by bioleaching of low-grade black shale ores, a commonly more advanced dissolution of base metal sulphides is clearly apparent by bioleaching of concentrates. It is generally known that the increased content of base metal sulphides provided by flotation concentrates from black shale feed establishes the base for a substantial enhanced leaching efficiency feasible by bioleaching in comparison to sterile control leaching.

Bioleaching tests with flotation concentrates gained from low-grade black shale ore were not conducted, but a comparable concentrate (10.0% Cu) extracted by flotation of black shale feed from the Rudna Mine, Poland, has a reasonable similarity to the concentrate extracted by multi-stage flotation from Mansfeld low-grade black shale ore, which achieves a Cu content of 9.8%. Compositionally, both black shales are similar in terms of content and type of gangue minerals (carbonates, quartz, illite) and a bornite-chalcocite-dominated copper sulphide association [17].

The bioleaching tests were conducted in 2 litre stirred batch bioreactors at 42 °C with 10% (w/v) solids load (−100 μm ball mill product) using a consortium of acidophilic, autotrophic mineral-oxidizing bacteria kept in basal salt medium [56]. In a first step, the pH of the suspension was adjusted by adding H_2SO_4 until the pH stayed below 1.8. The bioreactors were sparged with air at 120 l/h. Each experiment was carried out in triplicate. Control leaching tests were carried out with the same set up but under sterile conditions. EDX-based mineral distribution and geochemical analyses were conducted on leaching residues.

The sulphide assemblage of the Rudna black shale copper concentrate formed a fraction of 29% and was dominated by chalcocite (11.0%), bornite (6.7%) and galena (8.5%), as shown in Figure 15.27. The proportion of sulphides in the bioleaching residues (7.4%) was more than the half of that in the sterile control leaching residues (15.9%).

While a far advanced dissolution of copper sulphides was achieved by bioleaching, the sterile control residues contained a larger portion of not dissolved copper-bearing sulphides. Accordingly, traces of chalcocite (0.6%) and chalcopyrite (0.3%) remained in the bioleaching residues, whereas 2.6% bornite, 2.6% chalcocite and 1.7% chalcopyrite, was found in the latter indicating a relative enrichment in comparison to the sterile control residues. In addition, covellite (1.9%) was detected in the sterile control residues and was obviously formed during the leaching process. Furthermore, the sterile control residues contained 1.1% idaite (Cu_3FeS_4) that had a lower content in the feed material, whereas it was absent in the bioleaching residues. Idaite might represents an oxidation product of bornite. Galena, which was one of the major sulphides in the feed material, was more intense dissolved by the sterile control leaching so that 4.8% remained in the residues, whereas bioleaching residues contained 5.4% galena. The leached portion of galena is supposed to be the prerequisite for the formation of lead sulphate during the leaching process.

The mineral distribution revealed that anglesite-like lead sulfate precipitations occurred in both, but in the sterile control residues contained a lower amount (3.0%) and the bioleaching residues contained slightly more (4.6%). However, due to the partly extremely fine-grained nature of these leaching related precipitations, a part of the lead sulphate particles might not be detected due to the selected detection area limit (5 µm) of the EDX-based particle analysis. According to the global mineral mass balance, gangue minerals such as quartz and minerals of the feldspar-clay group were especially enriched in the bioleaching residues. Thus, bioleaching residues contained 52.0% clay and feldspar minerals and 13.8% quartz. Opposite to that, the sterile control residues contained 42.1% minerals belonging to the feldspar-clay group, 12.6% quartz and was additionally enriched in calcium sulphate (gypsum, 13.2%,) and iron sulphate (1.0%, mainly jarosite-like composition). The bioleaching residues accommodated a similar quantity of gypsum (9.9%), but only traces of iron sulphate minerals (0.1%).

According the varying contents of single metals in sterile control and bioleaching residues, the recovery from the feed material was different. Table 15.7 indicates that the recovery of Cu and Fe was nearly complete using bioleaching, but Pb was enriched in both, bioleaching and sterile leaching residues, whereas the initial low content of Zn in the concentrate was leached completely. Almost complete leaching rates by bioleaching were also ascertained for As and Co, but the leaching efficiency

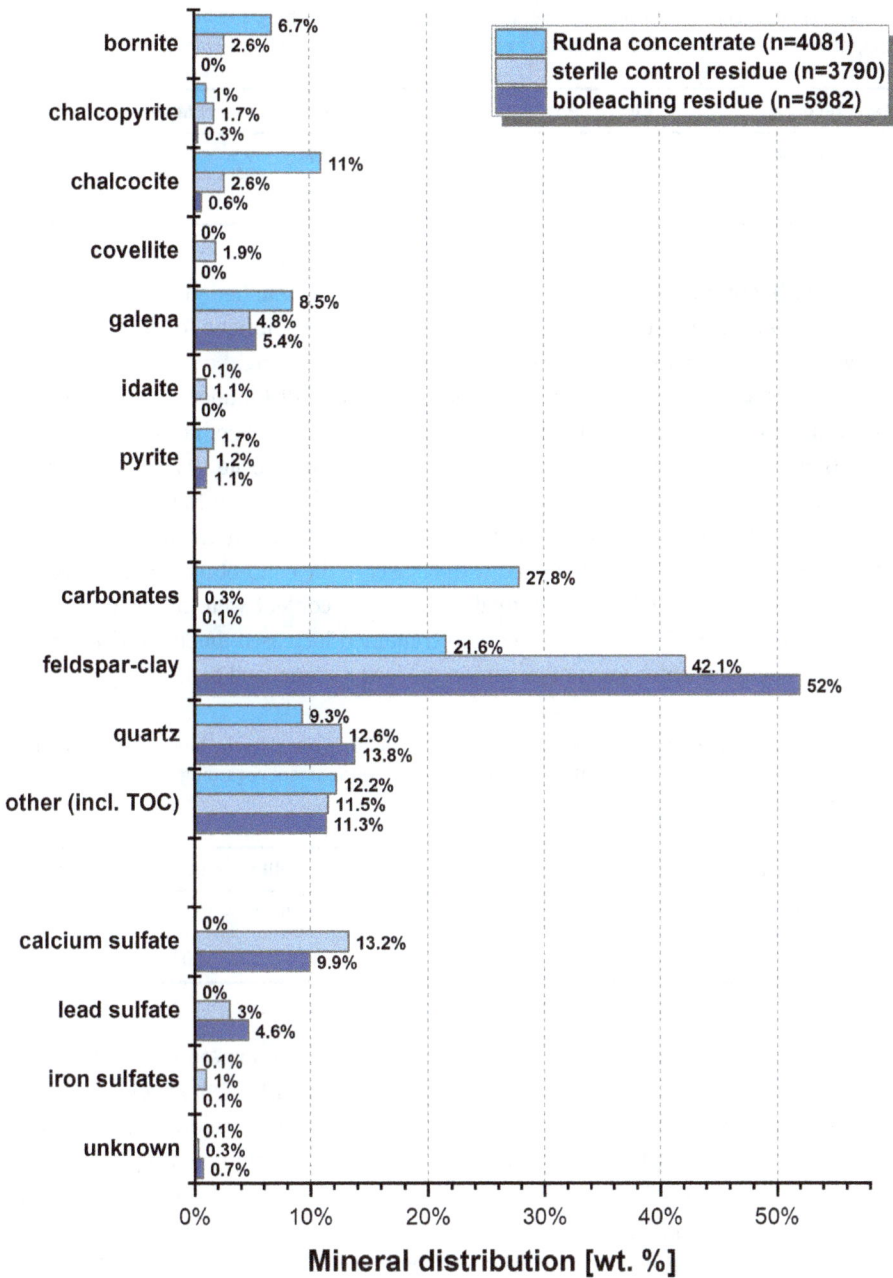

Figure 15.27: Bar chart displaying the distribution of sulphides as well as gangue-type minerals and mineral groups in a copper concentrate and its bioleaching and sterile control leaching residues. (from [16]).

Table 15.7: Recovery rates of selected metals by sterile control leaching and bioleaching tests. (Negative values indicate an enrichment regarding the content in the feed. (from [17]).

Metal recovery	Cu	Pb	Fe	Ag	As	Co	Mo	Ni	V
Sterile control leaching	39%	−3%	45%	3%	56%	39%	21%	74%	44%
Bioleaching	97%	−16%	84%	−3%	85%	92%	50%	68%	35%

of the sterile control leaching tests stood far behind and reached just 30–50% of the metal recovery achieved by bioleaching. Exceptions were noticed for Ni and V; both showed a slight increased recovery by sterile control leaching.

Leaching yields for sulphides determined by mineral distribution analyses are displayed in Table 15.8. Bioleaching achieved generally higher recovery rates, expect for galena. Bornite was leached completely and most of the chalcocite (94%), whereas the leachability of sulphides by sterile control leaching was limited, so that 45% of bornite remained in the residue. Additionally, the content of chalcopyrite in the sterile control leaching residues was slightly increased with respect to the concentrate indicating an enrichment. However, two-third of the total chalcopyrite content were recovered by bioleaching. Sterile control residues furthermore contained increased amounts of idaite and secondary covellite, which was not present in the feed material and bioleaching residues.

Table 15.8: Leaching recovery proportion of base metal sulfides dissolved by bioleaching as well as sterile control leaching tests. (Negative values indicate an enrichment regarding the content in the feed. (from [17]).

Mineral recovery	bornite	chalcopyrite	chalcocite	covellite	idaite	pyrite	galena
Sterile control leaching	61%	−64%	77%	−190%	−124%	28%	43%
Bioleaching	100%	69%	94%	0%	100%	37%	37%

The leaching efficiency for pyrite was low in both leaching experiments, but nearly one-third was extracted by bioleaching compared to a nearly 10% lower dissolution achieved by sterile control leaching tests. The dissolution of galena was generally hindered, with a slightly better dissolution by sterile control leaching (43%) than by bioleaching (30%), but dissolved lead ions were transferred into lead sulphate precipitates that remained in both, bioleaching and sterile control residues.

It was shown that increased contents of base metal sulphides provided by flotation concentrates of black shale feed revealed substantial deviations in the leaching performance of bioleaching in comparison to sterile control leaching. The exceptionally increased Cu-recovery (97%) caused by a more advanced dissolution of mostly base metal sulphides is clearly apparent by bioleaching of flotation concentrates, in which the leaching performance exceeds that of sterile control leaching substantially.

Additionally, in particular Co and Mo showed significantly increased recovery rates by bioleaching caused by the dissolution of pyrite, which was recognized to be the host of both trace metals.

Bioleaching tests on a copper concentrate similar to that produced by multi stage flotation of low-grade black shale ore indicate that an alternative processing of low-grade black shale ores is feasible. An increased copper recovery in respect to acid leaching can substantially contribute to an integrated concept using new low-cost, low-energy processing technology for the intelligent metal recovery from Kupferschiefer-type low-grade black shale ores.

4 Conclusions

Dumps in the former Mansfeld mining district, Central Germany, containing black shale low-grade ore were first mapped lithotype-specifically and the metal content was analysed by portable XRF. Investigations were concentrated in particular on two special flat dumps, which were deposited in the period from 1870 to 1930. The "Theodor" shaft dump, one of the oldest industrial mine dumps in the region, and the flat dump part of the "Fortschritt I" shaft, which is one of the youngest flat dumps in the mining district. For research purposes, the dump material was sampled on a ton scale for the mineral processing tests, on the other hand the flat dumps were sampled for the purpose of geochemical and mineralogical characterisation by means of hand specimen, trenching (bulk samples 40–50 kg) and drilling samples.

Mineralogical and geochemical results clarify the highly variable metal content of the stockpiled low-grade ore. The Cu-content fluctuates from virtually barren samples to individual samples with up to 12% Cu. Based on 21 'bulk samples', consisting of numerous sub-samples, it was possible to set limits of the average base metal content of the black shale low grade ore on the "Fortschritt I" dump: 0.5–0.6% Cu, 0.56–0.6% Pb, 0.69–0.75% Zn. The majority (> 70%) of the base metal content is contained in coarse-sized particles (> 1.6 cm), which is a prerequisite for the successful application of automated sensor-based sorting techniques. Low-grade ore deposited separately consist predominately of shale (approx. 80 wt.%) and subordinate of limestone (approx. 18%), whereas the occurrence of other rock types is negligible (< 2%).

Detailed investigations on drillings of the "Theodor" and "Fortschritt I" shaft dump focussed on charactering the degree of weathering of the low-grade ore. The metal inventory and data from pH and conductivity probe of water-eluates of the drilling samples revealed a low mobility for most metals but increased elutability for Zn and S. Sequential leaching extraction data showed that generally only low proportions of the primary sulphide ore minerals have been converted into easily soluble secondary phases (< 10%), which include copper hydroxi-sulphates and carbonates, widespread gypsum and some iron hydroxides. The results revealed that low-grade ore on dump bodies were hardly changed by meteorological influences in their potential weathering period of more than 100 years in some cases, and thus the recoverability of the valuable metals, e.g. by extensive sulphide oxidation, is barely affected. It could be demonstrated that the age of deposition of the low-grade ore just marginally affected the weathering state and resulted in low elutability in general and a slight increase of sulphide oxidation for dump material deposited for longer.

The processing of the black shale ore has always posed a major challenge, as the particularly finely disseminated ore minerals in organic-rich host rocks cause major processing problems. Thus, a main focus was the development or adaptation of targeted and efficient processing methods aiming to the extraction of predominantly Cu and subordinately Ag as a beneficial by-product.

The use of sensor-based sorting had shown that a significant enrichment of Cu in a pre-sorted run-of-mine ore can be achieved. The lithological separation of sandstone, black shale and limestone fragments from dump material was most efficient using NIR-based sorting and allowed the highest selectivity. However, the combination with a novel X-ray fluorescence scanner caused a 100% increase (doubling) of the copper content in the pre-sorted black shale low-grade ore.

After a first-stage crushing of run-of-mine ore by jaw crusher, the pre-sorted low-grade ore was milled to the required final flotation grain size using an established sieving ball mill. Alternatively, large-scale tests were carried out with a vertical roller mill, which is currently used primarily in the cement industry. According to the respective specification to mill 99% of the feed material (d_{99}) to the particle size range -40 or -100 μm, it can be noted that the ball mill product contains 14% more fines than the vertical roller mill product for the -40 μm product, which contributes negatively to the flotation. In the upper specification range of 100 μm the grain distribution of both grinding products are nearly equivalent. However, 43% of the feed were milled by roller mill comminution to the final grain size which was 11% more than achieved by ball milling. Apart from a better grinding behaviour with regards to the desired target grain size, the more energy-efficient use of the vertical roller mill saves according to the specific Bond work index almost 50% of the energy input compared to the ball milling. In addition, the use of vertical roller mill technology enhances the liberation of the extremely finely dispersed sulphides in the low-grade ore, which has a positive effect on the yield of the downstream flotation.

An upstream bitumen flotation failed due to excessive copper losses in the bitumen concentrate. Therefore, a four-stage flotation (a standard sulphide flotation with three cleaning stages) using pre-sorted material (2.2% Cu) extracted a copper concentrate with 9.4% Cu at a Cu recovery of 70%. A corresponding semi-technical test in the pilot plant with non-pre-sorted low-grade ore (0.6% Cu) could only enrich Cu to 2.4% in the flotation product. However, a four-fold enrichment of Cu was achieved for both multi-stage flotation tests, which shows that the flotation protocol is generally applicable. Comparative flotation tests of grinding products from the ball mill and the roller mill showed a slightly more selective flotation of roller mill products by the improved recovery of copper minerals.

The downstream processing of copper flotation concentrates into metal products by a hydrometallurgical leaching test series showed the most efficient recovery of copper with the usage of sulfuric and hydrochloric acid. To increase the leaching yields, feed material has been treated by roasting at 450 and 900 °C prior to leaching application. It was shown that feed material roasted at 450 °C converted into sulphates and could be dissolved more easily than oxides generated by roasting at 900 °C. Due to the high additional energy demand of roasting, leaching test on unroasted feeds were preferred. The use of nitric waste acid from the photovoltaic industry as a cost-effective and resource-saving solvent achieved the highest Cu recovery yield (89%) of unroasted flotation concentrate, but it caused a strong exothermic reaction with the release of harmful nitrous gases. For the technical implementation, a 3-stage leaching cascade with an upstream roasting step can be implemented.

A further alternative extraction method applicable to flotation concentrates of low-grade black shale ore has been proven by bioleaching. Residues of bioleaching tests conducted on a black shale-derived concentrate were investigated by mineral distribution analyses focussing residual sulphides and process-related formed metal sulphates. Generally, the recovery of Cu was vastly enhanced by bioleaching compared to sterile control leaching. A particularly high copper recovery rate of 98% was achieved due to the low portion of chalcopyrite in the copper sulphide assemblage. As it is commonly known that chalcopyrite is often subjected to incomplete dissolution and thus hinders the total Cu recovery by bioleaching, Cu extraction from chalcopyrite-poor feeds can exceed the efficiency of acid leaching considerably. Thus, the usage of an energy-saving bio-hydrometallurgical extraction of Cu and other value metals opens up a wider replacement of pyrometallurgical Cu production.

References

[1]   Kamradt A, Borg G, Schaefer J, Kruse S, Fiedler M, Romm P, et al. An Integrated Process for Innovative Extraction of Metals from Kupferschiefer Mine Dumps, Germany. Chemie Ing Tech 2012;84:1694–703. https://doi.org/10.1002/cite.201200070.

[2]   Borg G, Piestrzynski A, Bachmann GH, Puttmann W, Walther S, Fiedler M. An Overview of the European Kupferschiefer Deposits. SEG Spec Publ 2012;16:455–86.

[3]   Knitzschke G. Metall- und Produktionsbilanz für die Kupferschieferlagerstätte im südöstlichen Harzvorland. In: Jankowski G, editor. Zur Geschichte des Mansfelder Kupferschieferbergbaus, Clausthal-Zellerfeld: GDMB-Informationsgesellschaft mbH; 1995, p. 270–84.

[4]   Mudd GM, Weng Z, Jowitt SM. A Detailed Assessment of Global Cu Resource Trends and Endowments. Econ Geol 2013;108:1163–83. https://doi.org/10.2113/econgeo.108.5.1163.

[5]   Bellenfant G, Guezennec AG, Bodénan F, D'Hugues P, Cassard D. Reprocessing of mining waste: Combining environmental management and metal recovery? Mine Clos 2013:571–82.

[6] Magwaneng RS. Recovery of Copper from Mine Tailing and Complex Carbonaceous Sulfide Ore by Flotation and High – Pressure Leaching. PhD-thesis, Akita University, Japan, 2019.

[7] Rappsilber I, Stedingk K, König S, Heckner J, Thomae M. Geologisch-montanhistorische Karte Mansfeld -Sangerhausen 1 : 50 000 – Geotourismus in den Kupferschieferrevieren. Landesamt Für Geologie Und Bergwesen Sachsen-Anhalt (LAGB), Halle (Saale): 2007.

[8] Wagenbreth O, Steiner W. Bau und Bildungsgeschichte der Landschaften in der DDR. Geol. Streifzüge, Berlin, Heidelberg: Springer; 1990, p. 21–178. https://doi.org/10.1007/978-3-662-44728-4_3.

[9] Arge TÜV Bayern/L.U.B. Umweltsanierung Mansfelder Land. Abschlußbericht zum Forschungs- und Entwicklungsvorhaben "Umweltsanierung Mansfelder Land." Eisleben: 1991.

[10] Sanierungsverbund Mansfeld e.V. Theisenschlamm, Manuskriptsammlung zum Fachkolloquium Theisenschlamm am 07.12.1993. In: Mansfeld S e. V, editor. Sanierungsverbund e.V. Mansfeld im Auftrag des Minist. für Umwelt und Naturschutz des Landes Sachsen-Anhalt, 1993.

[11] Hammer J, Junge F, Rösler HJ, Niese S, Gleisberg B, Stiehl G. Element and isotope geochemical investigations of the Kupferschiefer in the vicinity of "Rote Fäule", indicating copper mineralization (Sangerhausen basin, G.D.R.). Chem Geol 1990;85:345–60. https://doi.org/10.1016/0009-2541(90)90012-V.

[12] Oszczepalski S. Origin of the Kupferschiefer polymetallic mineralization in Poland. Miner Depos 1999;34:599–613. https://doi.org/10.1007/s001260050222.

[13] Püttmann W, Fermont WJJ, Speczik S. The possible role of organic matter in transport and accumulation of metals exemplified at the Permian Kupferschiefer formation. Ore Geol Rev 1991;6:563–79. https://doi.org/10.1016/0169-1368(91)90047-B.

[14] Sawlowicz Z. Organic Matter and its Significance for the Genesis of the Copper-Bearing Shales (Kupferschiefer) from the Fore-Sudetic Monocline (Poland). Bitumens Ore Depos., 1993. https://doi.org/10.1007/978-3-642-85806-2_23.

[15] Czechowski F. Metalloporphyrin composition and a model for the early diagenetic mineralization of the Permian Kupferschiefer, SW Poland. In: Glikson M, Mastalerz M, editors. Org. Matter Miner. Therm. Alteration, Hydrocarb. Gener. Role Metallog., 2000. https://doi.org/10.1007/978-94-015-9474-5_12.

[16] Szubert A, Sadowski Z, Gros CP, Barbe JM, Guilard R. Identification of metalloporphyrins extracted from the copper bearing black shale of Fore Sudetic Monocline (Poland). Miner Eng 2006. https://doi.org/10.1016/j.mineng.2005.11.007.

[17] Kamradt A. Mineralogical and textural analysis of processing products, tailings and residues from Mid-European Kupferschiefer-type black shale ores. Martin-Luther-Universität Halle-Wittenberg, 2019. https://doi.org/http://dx.doi.org/10.25673/31727.

[18] DIN 38414-4:1984-10. German standard methods for the examination of water, waste water and sludge; sludge and sediments (group S); determination of leachability by water (S 4) 1984.

[19] Tessier A, Campbell PGC, Bisson M. Sequential Extraction Procedure for the Speciation of Particulate Trace Metals. Anal Chem 1979. https://doi.org/10.1021/ac50043a017.

[20] Rose AW. Mobility of copper and other heavy metals in sedimentary environments. Geol Assoc Canada Spec Pap 1989;36:97–110.

[21] Brookins DG. Eh-pH Diagrams for Geochemistry. 1988. https://doi.org/10.1007/978-3-642-73093-1.

[22] Langmuir D. Aqueous Environmental Geochemistry. Prentice Hall; 1997.

[23] Rimstidt JD, Chermak JA, Gagen PM. Rates of Reaction of Galena, Sphalerite, Chalcopyrite, and Arsenopyrite with Fe(III) in Acidic Solutions, 1993. https://doi.org/10.1021/bk-1994-0550.ch001.

[24] Langmuir D. Use of laboratory adsorption data and models to predict radionuclide releases from a geological repository: A brief history. Mater. Res. Soc. Symp. – Proc., 1997.

[25] Zinke G. Die natürlichen Verhältnisse des Einzugsgebietes der Bösen Sieben unter Berücksichtigung der Mansfelder Seen. Der Süsse See – Das Blaue Auge des Mansfelder Landes, Staatliches Amt für Umweltschutz Halle (Saale); 1993, p. 8–12.

[26] Döring J. Zu den Klimaverhältnissen im östlichen Harzvorland. Hercynia 2004.

[27] Matheis G, Jahn S, Marquardt R, Schreck P. Mobilization of heavy metals in mining and smelting heaps, Kupferschiefer district, Mansfeld, Germany. Chron La Rech Minière 1999;534:3–12.

[28] Mibus J. Column experiments with heap material of Kupferschiefer mining and thermodynamic interpretation. Karlsruhe: 2002.

[29] Dunger C. Sickerwasser- und Schadstoffbewegung aus ausgewählten Bergehalden Sachsens und Sachsen-Anhalts. Technical University Bergakademie Freiberg, 1998.

[30] Schmidt G. Umweltbelastung durch Bergbau – Der Einfluss der Halden des Mansfelder Kupferschieferbergbaus auf die Schwermetallführung der Böden und Ge-wässer im Einzugsgebiet Süßer See. Hallesche Stud Zur Geogr 2000;3:117.

[31] Mibus J. Geochemische Prozesse in Halden des Kupferschieferbergbaus im südöstlichen Harzvorland. Technical University Freiberg, 2001.

[32] Wiggering H. Verwitterung auf Steinkohlenbergehalden: Ein erster Schritt von anthropo-technogenen Eingriffen zurück in den natürlichen exogen-geodynamischen Kreislauf der Gesteine. Zeitschrift Der Dtsch Geol Gesellschaft 1986;137:431–46.

[33] Borg G, Du Bois M, Friedrich B, Morgenroth H, Wotruba H, Kamradt A, et al. Gewinnung von Metallen und mineralischen Produkten aus deponierten Reststoffen der ehemaligen Montanindustrie im Mansfelder Gebiet : Schlussbericht. Halle (Saale): 2013. https://doi.org/10.2314/GBV:785965777.

[34] Wotruba H. Sensor sorting technology – Is the minerals industry missing a chance? IMPC 2006 – Proc. 23rd Int. Miner. Process. Congr., 2006.

[35] Grotowski A, Witecki K. Research on the Possibility of Sorting Application for Separation of Shale and/or Gangue from the Feed of Rudna Concentrator. E3S Web Conf., 2017. https://doi.org/10.1051/e3sconf/201712301004.

[36] Curry JA, Ismay MJL, Jameson GJ. Mine operating costs and the potential impacts of energy and grinding. Miner Eng 2014;56:70–80. https://doi.org/10.1016/j.mineng.2013.10.020.

[37] Herbst JA, Lo CY, Flintoff B. Size Reduction and Liberation. In: Fuerstenau MC, Han KN, editors. Princ. Miner. Process., Littleton, Colorado: Society for Mining, Metallurgy, and Exploration, Inc. (SME); 2003, p. 61–118.

[38] Wills BA, Napier-Munn T. Introduction. Wills' Miner. Process. Technol., Elsevier; 2005, p. 1–29. https://doi.org/10.1016/B978-075064450-1/50003-5.

[39] Reichert M, Gerold C, Fredriksson A, Adolfsson G, Lieberwirth H. Research of iron ore grinding in a vertical-roller-mill. Miner Eng 2015;73:109–15. https://doi.org/10.1016/j.mineng.2014.07.021.

[40] Schaefer HU. LOESCHE vertical roller mills for the comminution of ores and minerals. Miner Eng 2001. https://doi.org/10.1016/S0892-6875(01)00133-9.

[41] Bond FC. The third theory of comminution. Min Eng 1952;4:484–94.

[42] Bond FC. Grinding ball size selection. Min Eng 1958;10:592–5.

[43] Kamradt A. Comminution as an economic efficient key for the improvement of metal recovery from low-grade Kupferschiefer-type black shale -hosted copper ore from the Mansfeld mining district, Germany. In: André-Mayer AS, Cathelineau M, Muchez P, Pirard E, Sindern S, editors. Miner. Resour. a Sustain. world. Proc. 13th Bienn. SGA Meet. 2015, vol. 4, Nancy: 2015, p. 1373-1376.

[44] Altun D, Gerold C, Benzer H, Altun O, Aydogan N. Copper ore grinding in a mobile vertical roller mill pilot plant. Int J Miner Process 2015. https://doi.org/10.1016/j.minpro.2014.10.002.

[45] Boehm A, Meissner P, Plochberger T. An energy based comparison of vertical roller mills and tumbling mills. Int J Miner Process 2015;136:37–41. https://doi.org/10.1016/j.minpro.2014.09.014.

[46] Aletan G. Untersuchungen zur Schwimmaufbereitung bituminöser Kupfererze vom Typ Niedermarsberg und Mansfeld. PhD-thesis, Technical University (Bergakademie) Clausthal, 1932.

[47] Babiński W, Madej W, Bortel R. Erfüllungsbericht betreffend den 1. Teil der Untersuchungen über die Entwicklung eines Aufbereitungsverfahrens für Kupfererze aus Feld 1 der Lagerstätte. Internal Report, Forschungsinstitut Für Aufbereitung (FIA), Freiberg: 1978.

[48] Chmielewski T, Konieczny A, Drzymala J, Kaleta R. Development concepts for processing of Lubin- Glogow complex sedimentary copper ore. Proc. XXVII Int. Min. Process. Congr. IMPC 2014, Santiago, Chile: 2014, p. 20–4.

[49] Konieczny A, Pawlos W, Krzeminska M, Kaleta R, Kurzydlo P. Evaluation of organic carbon separation from copper ore by pre-flotation. Physicochem Probl Miner Process 2013;49: 189–201. https://doi.org/10.5277/ppmp130117.

[50] Drzymala J, Kowalczuk PB, Oteng-Peprah M, Foszcz D, Muszer A, Henc T, et al. Application of the grade-recovery curve in the batch flotation of Polish copper ore. Miner Eng 2013;49: 17–23. https://doi.org/10.1016/j.mineng.2013.04.024.

[51] Chmielewski T. Non-oxidative leaching of black shale copper ore from Lubin mine. Physicochem Probl Miner Process 2007;41:323–35.

[52] Luszczkiewicz A, Chmielewski T. Acid treatment of copper sulfide middlings and rougher concentrates in the flotation circuit of carbonate ores. Int J Miner Process 2008;88:45–52. https://doi.org/10.1016/j.minpro.2008.06.003.

[53] Konopacka Ż, Łuszczkiewicz A, Chmielewski T. Effect of non-oxidative leaching on flotation efficiency of lubin concentrator middlings. Physicochem Probl Miner Process 2007;41: 275–89.

[54] Spolaore P, Joulian C, Gouin J, Morin D, D'Hugues P. Relationship between bioleaching performance, bacterial community structure and mineralogy in the bioleaching of a copper concentrate in stirred-tank reactors. Appl Microbiol Biotechnol 2011;89:441–8. https://doi.org/10.1007/s00253-010-2888-5.

[55] Hedrich S, Joulian C, Graupner T, Schippers A, Guézennec AG. Enhanced chalcopyrite dissolution in stirred tank reactors by temperature increase during bioleaching. Hydrometallurgy 2018;179:125–31. https://doi.org/10.1016/j.hydromet.2018.05.018.

[56] Hedrich S, Guézennec A-G, Charron M, Schippers A, Joulian C. Quantitative Monitoring of Microbial Species during Bioleaching of a Copper Concentrate. Front Microbiol 2016;07. https://doi.org/10.3389/fmicb.2016.02044.

Tim Rödel, Stefan Kiefer, Gregor Borg

Chapter 16
Rare-earth elements in phosphogypsum and mineral processing residues from phosphate-rich weathered alkaline ultramafic rocks, Brazil

Abstract: Abstracts: Phosphogypsum is regarded today mainly as a by-product or waste generated during industrial processes such as the production of phosphoric acid for fertilizer. During the process of sulphuric acid leaching of apatite-rich concentrates from weathered alkaline ultramafic rocks large volumes of gypsum and accessory minerals are produced. Recently nine phosphogypsum samples from Catalão, Brazil have been investigated as a potential secondary source for rare-earth elements (REEs). Identifying the minerals hosting REE, the mineral composition and modal abundance are key in evaluating the economic potential of phosphogypsum as a source for critical metals. A combination of detailed petrographic investigations, SEM-based mineral distribution analyses, EPMA and geochemical analyses using ICP-MS/AES was successfully applied to identify REE carrier minerals and the fraction of associated REE.

The analysed phosphogypsum samples are mainly composed of euhdral gypsum crystals constituting around 93% of the total mass. Accessory minerals identified include quartz, octahedral REE-bearing, (Ca-Al-) fluorides, monazite, celestine, Fe-oxides, ilmenite, barite, pyrochlore, and baddeleyite. Apart from gypsum most minerals also occur in the weathered phosphate-rich rocks and are therefore carried over throughout the sulphuric acid leaching process of apatite concentrates. Monazite has been identified as the most important carrier for REE in phosphogypsum. The mineral mainly consists of Ce, La, Nd and P. The total rare-earth content in the mineral amounts to a mean 57 %. Furthermore the Th concentration, a major contaminant, are comparably low.

Acknowledgements: The Federal Ministry of Education and Research has funded our research project CaMona (033R187C) within the framework of the CLIENT II funding scheme and research collaboration. Jörg Reichert and Marco Fiedler of Ceritech AG are thanked for providing sample logistics, practical know-how and numerous fruitful scientific discussions. CMOC Brazil, the operator of Catalão Mine is thanked for providing access to sample material and visits to the mine and processing facilities. The reviewers are thanked for their input which imroved the quality of the text.

Tim Rödel, Martin Luther University, Institute of Geosciences and Geography, Von Seckendorffplatz 3, 06120 Halle, Germany

Stefan Kiefer, Friedrich Schiller University, Institute of Geosciences, Burgweg 11, 07749 Jena, Germany

Gregor Borg, Martin Luther University, Institute of Geosciences and Geography, Von Seckendorffplatz 3, 06120 Halle, Germany

https://doi.org/10.1515/9783110674941-016

The mean abundance of monazite in phosphogypsum amounts to 0.6wt%. Overall monazite hosts 50% to 60% of the total REE in phosphogypsum, while the remaining 50% are locked in gypsum and Ca-Al-fluorides. Therefore only a fraction of the geochemically available total rare-earth oxide content of 0.6% in phosphogypsum is likely to be recoverable. A good complete particle liberation of 60% to 80% and a grain size range of 15mμm to 50μm is promising for a further beneficiation of monazite by means of physical separation prior to winning the REE. Based on the data at hand phosphogypsum from the Catalão region in Brazil could potentially provide a significant supply of REE as a secondary resource.

Keywords: rare-earth elements, phosphogypsum, monazite, quantitative, mineral composition, gypsum, geochemistry, EPMA, AMICS, Brazil

1 Rare-earth elements and phosphogypsum

The 17 rare-earth elements (REEs) consist of metallic elements from the lanthanide series including scandium and yttrium. Their uncommon accumulation as minable natural ore deposits is in stark contrast to their ubiquitous distribution throughout the earth's crust. This becomes obvious when comparing average crustal abundance of the REEs in the upper crust with the abundances of other metals consumed by the human society. Cerium for example is the most common REE with a crustal abundance of 63 ppm, which is significantly higher than for instance the crustal abundance of the mass commodity copper with 28 ppm [1]. Generally, high concentrations of REEs in natural rocks are rarely observed although a number of occurrence and classified deposits are known worldwide [2, 3]. The most important rare-earth minerals such as bastnaesite, monazite, or xenotime are hosted by igneous rocks such as carbonatites or pegmatites and hydrothermal veins [3, 4]. Weathered or redistributed equivalents of such primary rock or mineralisation types can also contain considerable amounts of residual or compositionally transformed rare earth minerals [5]. During weathering and supergene alteration of primary REE-bearing rocks, these elements can become redistributed or absorbed to so called ion-adsorption clays [6]. Technical constraints in mining and particularly in beneficiation of REEs from natural primary and secondary deposits drastically limit the economically recoverable amount of REE further. One major geochemical and mineralogical problem is the association of REE deposits with uranium or thorium minerals and the resulting environmental challenges when processing these ores and treating and storing the residues [7].

In 2017, China was the main producer of REEs, constituting over 80% of the annual world production [8] and being almost the sole supplier, followed by Australia [9]. Ten other countries contribute additional but small proportions to the worldwide production. The range of the global supply is therefore strongly restricted, a situation which can be exploited economically, and may cause political dissents that were witnessed during the REE crisis between 2009 and 2013.

REEs have been classified as critical raw materials in consequence to these supply risks, low "end-of- life recycling rates" and their high economic and technical value [10]. Especially their application in so-called future and clean technologies such as wind turbines, electric vehicles or energy-efficient lighting, an increasing demand is to be expected in the next decades [9].

Diversifying of supply from the presently limited REE sources is therefore crucial for stable economic development in manufacturing countries. Therefore, the exploration and exploitation of new primary deposits is needed. Additionally, the identification and utilisation of secondary sources for REEs are also investigated. Recycling of end-of-life products from the technosphere currently receives most attention [11]. The beneficiation of REEs as by-products of primary ores or recovery of REEs from industrial and mining residues can comprise significant economic potentials [12]. Using secondary material streams as potential sources for REEs can help to improve the overall supply. This can therefore increase resource efficiency in the sense of a sustainable circular economy.

In our present study, the residues from phosphate fertilizer production and phosphate rock acid leaching are being reviewed for their mineralogical and geochemical composition with respect to the contained REE-bearing mineral phases. Special emphasis is put on phosphogypsum as a by-product generated during the industrial wet process of phosphoric acid production from weathered carbonatite ores from Brazil. Technically and chemically, this process is a reaction of solid calcium phosphate with sulphuric acid to form phosphoric acid and solid calcium sulfate, i.e. gypsum. Common impurities are phosphates, fluorides, sulfates, trace metals, and radioactive elements [13, 14]. These are commonly inherited from the feed material of primary phosphate ore. Approximately five tonnes of phosphogypsum are produced for every tonne phosphoric acid. Estimations for a worldwide production reach from 100 to 280 million tons of phosphogypsum per year [15, 16]. However only 15% of the production is recycled to building materials, agricultural fertilisers, or agricultural soil conditioners [14].

A substantial amount of REEs is bound to phosphogypsum and it can therefore be considered as an important potential secondary resource [17–20]. Although several studies have been carried out recently, no economically viable technology has yet been developed to recover REEs from phosphogypsum. Most research focussed on hydrometallurgical processing and direct leaching of bulk phosphogypsum material on a laboratory scale [19, 21–29]. Generally, all known methods for REE recovery by leaching with strong sulphuric acid provide a low degree of REE recovery rate of only 30–40% [30–32]. Only few publications refer briefly to the mineralogical composition of the phosphogypsum and thus the understanding of the REE-fixation in specific mineral phases is limited [14, 27, 33–35]. The dihydrate gypsum or the hemihydrate bassanite might be accompanied by fluorine-bearing minerals such as chukhrovite $Ca_3Ce_{0.75}Y_{0.25}Al_2(SO_4)\ F_{13}\cdot10(H_2O)$ or meniaylovite $Ca_4AlSi(SO_4)F_{13}\cdot12(H_2O)$ as well as iron oxides and silicate minerals. Gypsum is generally regarded as the main host for REE in phospho-gypsum, structurally incorporating REEs released during apatite leaching [23, 30, 36].

Recently, the German start-up company Ceritech AG developed the concept of pre-concentration of REE-minerals from phosphogypsum by physical separation followed by hydrometallurgical leaching. After reviewing several phosphogypsum sources worldwide, Ceritech AG investigated material from the Chapadão Mine of the Catalão I deposit in the state of Goiás, Brazil. Currently the phosphate mine and fertilizer plant at Chapadão Mine is being operated by Copebrás Indústria Ltda, a subsidiary of CMOC International, who is based in São Paulo. Brazil produces an estimated total of 5.4 million tonnes phosphogypsum each year [37] and the residues from Catalão are almost exclusively used as soil conditioners. Total REE concentrations in three Brazilian phosphogypsum sites are reported with 3,200 ppm to 5,670 ppm [38].

In the collaborative project "CaMona", funded by the German Federal Ministry of Education and Research (BMBF), Ceritech AG, Technical University Clausthal, Federal University of Catalão, and Martin Luther University Halle-Wittenberg investigated the innovative use of phosphogypsum as a new and sustainable source of additional REE supply. A hydrometallurgical process for the production of a mixed-REE carbonate from REE-mineral concentrates has also been developed [39]. Ceritech AG and Martin Luther University Halle-Wittenberg conducted the first detailed petrological study of the mineral composition and geochemical characterisation of REE-bearing mineral phases of the Catalão phosphogypsum. Additionally, material streams from primary phosphate ore to phosphate beneficiation have also been investigated for their material properties as these parameters control the REE recovery from the phosphogypsum [13, 14].

In the following a brief introduction to the geology of Catalão I, the alkaline-ultramafic complex in Brazil, and the processing of the primary phosphate ore is presented.

The primary phosphate ore, as feed of the Catalão phosphate beneficiation and fertilizer plant, consists of the deeply weathered part of an alkaline intrusive complex. Geologically this complex is part of the Late Cretaceous Alto Paranaíba Igneous Province (APIP), consisting of multi-stage carbonatite-bearing ultramafic complexes, intruded into Late-Proterozoic metamorphic rocks of the Araxá Group [40]. The circular Catalão I plutonic complex (Figure 16.1) covers roughly 27 km^2 and is composed of a carbonatite dominated core, surrounded by pyroxenite, glimmerite, serpentinized peridotite, and minor nepheline syenite [41]. The core zone consists of several cross-cutting bodies of phoscorite, nelsonite, phlogopitite but also calcitic and dolomitic carbonatites [42]. The different phoscorite series P1 to P3 are the main hosts for the phosphorous-rich ores. Additionally, rocks of the P2 and P3 intrusive series hosts Nb-rich mineralization whereas the late dolomitic carbonatites contain REE mineralization [43, 44]. The phoscorite series and carbonatite rocks have been intensively weathered and transformed, leading to residual accumulation of more weathering-resistant minerals, and REE-fractionation in the weathering profile [45, 46]. Similar supergene processes are known from alkaline intrusive

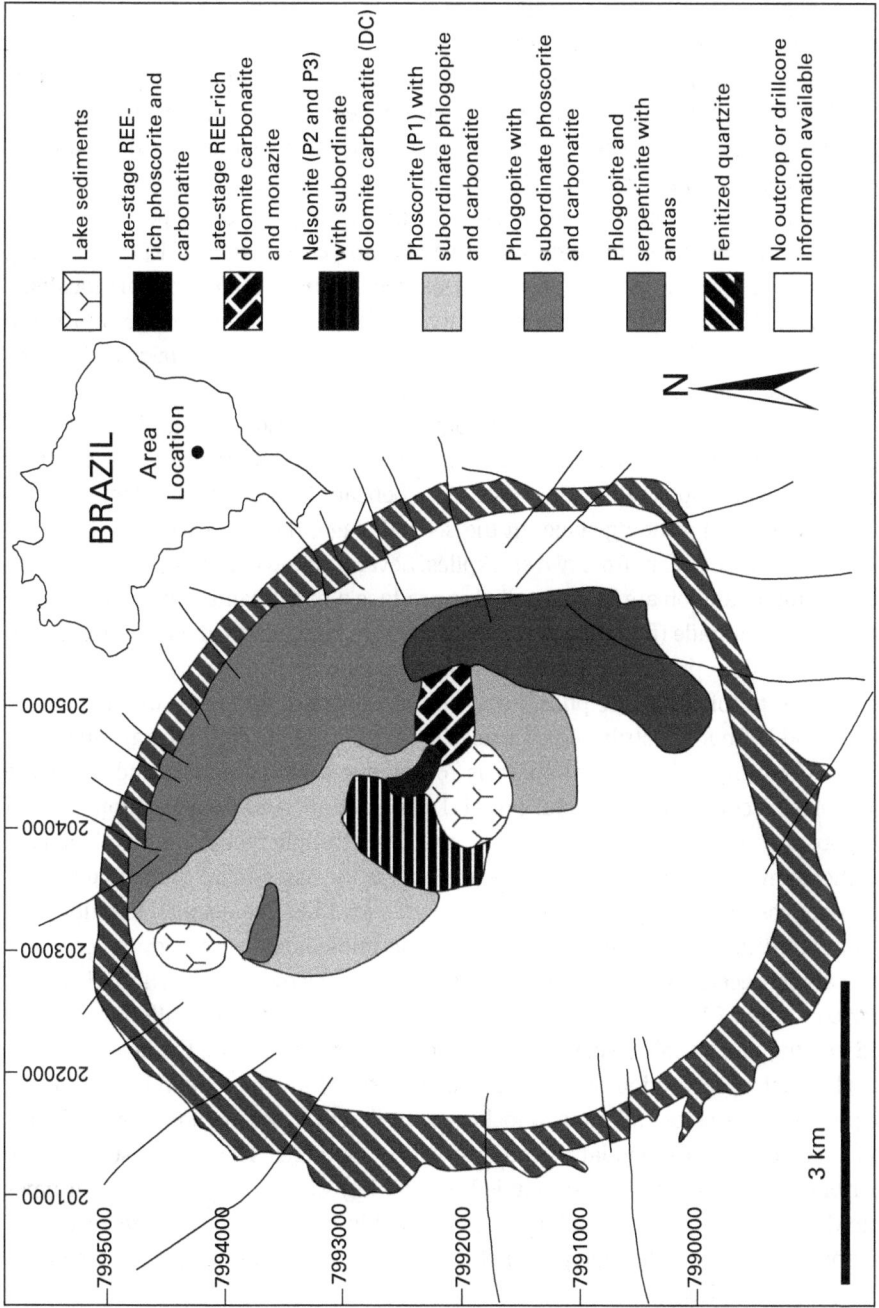

Figure 16.1: Geological sketch map of the circular Catalão I alkaline-ultramafic complex, Brazil (Cordeiro et al. 2011, adapted from Brod et al. 2004).

complexes like Mt. Weld in Australia or other occurrences [5, 47]. Apatite, gorceix-ite, monazite, and pyrochlore are regarded as the main hosts of REE-enrichment in weathered residues of the alkaline complexes. At Catalão I Chapadão Mine, only the main phosphate ore zone, situated between the lateritic overburden and the fresh rock is mined. The REE-mineralisation of the larger area of the Catalão I igeneous complex has been reviewed by several authors but the publications are commonly limited to specific areas such as the Corrego do Garimpo REE-deposit [48, 49]. In the ore an estimated total of 1.12 Mt rare earth oxides (TREOs), grading at an average content of 7.6% rare earth oxides has been stated for this localised occurrence. The REE-minerals were mainly concentrated in the fines and process-ing was deemed difficult [48]. Another focus has been on niobium-rich domains and tailings from niobium enrichment [50, 51].

The current process of the phosphogypsum production on site is practically achieved in three stages: The first step is mining of the primary phosphate ore from the weathered portions of the Chapadão Mine beneath lateritic overburden, followed by a crushing and grinding stage. In the second stage, the ore is fed into the phos-phate beneficiation plant from two stockpiles. Several stages of milling, magnetic sep-aration, and floatation are being applied to produce a final apatite concentrate. In the last stage, high-grade (fluor)apatite concentrates are being leached with concentrated sulphuric acid to produce phosphoric acid, gypsum, and minor hydrofluoric acid. Subsequent to this stage the phosphogypsum is deposited and rests for one year for curing and is finally distributed on trucks to farms to act as agricultural soil condi-tioner. The aim of the current study is to achieve a better understanding of the geochemical composition, modal mineral distribution, and textural properties of the phosphogypsum,associated minerals and other sample typesfrom the phosphate beneficiation processes. Focus of these analyses is the assessment of the potential of phosphogypsum as a potential secondary source for REE. The detailed identification of the quantitative mineral composition of the phosphogypsum and its accessory minerals is necessary to determine the technically recoverable portion of the to-tally contained REE. Only monazite is considered as an economically recoverable REE-bearing mineral phase based on the concept of concentrating the heavy mineral trough physical separation prior to hydrometallurgical REE extraction [39, 52]. Parti-cle properties like size and mineral locking with other phases are important informa-tion together with the modal abundance of the minerals. The SEM-based mineral distribution analyses have been applied to a variety of REE- bearing material types from Catalão to allow a calibration of the method in conjunction with the geochemi-cal analyses. Investigated samples include primary phosphate ore, apatite concen-trates, and phosphogypsum.

2 Materials and methods

Nine samples of phosphogypsum have been taken in several batches. Additionally, sixteen samples have been taken during phosphate beneficiation including primary phosphate ore, apatite concentrates, and intermediate products.

Four phosphogypsum samples have been taken fresh from the conveyor belt after the sulphuric acid leaching stage. Two samples have been taken from the stockpile. The sample mass ranges from 500 g to 1000 g. Three samples represent bulk samples acting as a representative base for mineral processing tests. Weights measure 350 kg, 2 t and 23 t respectively. All samples have been homogenised, split, and dried at 40 °C. Macroscopically, phosphogypsum is a yellowish to beige, powdery fine crystalline material consisting of minute minerals.

Geochemical analyses have been performed on powdered homogenised splits of ~15 g sample material. Analyses were carried out at the certified commercial lab ALS Minerals in Loughrea, Ireland. The chosen analytical package includes 14 elements by inductively coupled plasma atomic emission spectroscopy (ICP-AES) for major elements and 30 element mass spectroscopy (ICP-MS) for trace elements on lithium borate fused discs, followed by acid digestion. The Geochemical Data Kit plugin for R was used for plotting geochemical compositional data [53].

The chemical composition of gypsum, Ca-Al-fluorides, and monazite was determined by electron probe micro analysis (EPMA) using a JEOL JXA-8230 at the University of Jena, Germany.

The operating conditions were set to an accelerating voltage of 15 kV and a beam current of 15 nA. The wavelength dispersive X-ray spectrometers were used to measure the elements and X-ray lines of Na (Kα), F (Kα), Si (Kα), Mg (Kα), Al (Kα), Y (Lα), Th (Mα), Sr (Lα), Ba (Lα), K (Kα), P (Kα), S (Kα), Ca (Kα), Ti (Kα), Cl (Kα), Fe (Kα), Mn (Kα), La (Lα), Ce (Lα), and Nd (Lα). To improve the count-rate statistics the counting times were 60s on the peak and 30s on the background for La, Ce, Nd and Th and 20s on each peak and 10s on the background for the other elements. The standard specimens used for calibration were Al_2O_3 for Al, MgO for Mg, orthoclase for K, wollastonite for Si, albite for Na, rutile for Ti, Fe_2O_3 for Fe, rhodonite for Mn, apatite for Ca, P and F, halite for Cl, barite for Ba and S, celestine for Sr and doted glass-standards (10%) for Y, Th, La, Ce, and Nd. Peak overlap correction was used to avoid interference between the lines of Fe and Th, F and P, Ce and Th, Al and Ba, Si and Nd, F and Ce as well as F and Fe. The detection limits given in Table 16.1 are calculated from the peak and background counts, the measurement time, the beam current, and the standard material concentration.

For the determination of chemical composition of the monazites a wavelength dispersive X-ray spectrometer was used to measure the elements and X-ray lines of Na (Kα), F (Kα), Si (Kα), Mg (Kα), Al (Kα), Y (Lα), Th (Mα), Sm (Lα), Sr (Lα), Ba (Lα), P (Kα), S (Kα), Ca (Kα), Ti (Kα), Cl (Kα), Fe (Kα), Mn (Kα), La (Lα), Ce (Lα), Pr (Lα), Nd (Lα), Eu (Lα), and Gd (Lα). To improve the count-rate statistics the counting times were 60s on the peak and 30s on the background for Sm, La, Ce, Pr, Nd, Eu and Gd and 20s on each peak and 10s on the background for the other elements. The standard specimens used for calibration were Al_2O_3 for Al, MgO for Mg, wollastonite for Si, albite for Na, rutile for Ti, Fe_2O_3 for Fe, rhodonite for Mn, apatite for Ca, P and F, halite for Cl, barite for Ba and S, celestine for Sr and doted glas-standards (10%) for Y, Th, La, Ce, Nd, Sm, Pr, Eu, and Gd. Peak overlap correction was used to avoid interference between the lines of Fe and Th, F and P, Ce and Th, Al and Ba, Si and Nd, F and Ce, Eu and Nd, Gd and Ce, Gd and La as well as F and Fe. The detection limits are given in Table 16.1.

When dealing with the statistical accuracy of our analytical data presented in this study it is important to note that analytical values with a precision of two decimal places are calculated averages, i.e. these are statistical accuracies of averages calculated from large numbers of EPMA measurements. However, these two decimal place precisions are not giving the accuracy of individual single spot EPMA analyses. It is also important to note that the analytical data of this paper is for scientific purposes only and not suitable for mineral resource estimations as in prefeasibility studies or bankable feasibility studies.

Powder X-ray diffraction (XRD) for qualitative phase analyse have been performed, using a Panalytical X'pert Pro MPD with PIXcel[1D] detector. The following parameters were used: step size 0.013 °2θ, 30 seconds per step, 5 to 65 °2θ observation angle, Cu-K α1 ($\lambda = 1.5406$ Å) target, 45 kV and 40 mA, soller collimator 0.04 rad, divergence slit 1/8°, anti-scatter screen 1/4°.

Scanning electron microscopy with energy dispersive X-ray spectroscopy (SEM-EDX) has been performed on a Hitachi TM4000 + with 15 kV acceleration voltage using a 30 mm High-Sensitivity 4-segment back-scatter-electron (BSE) detector. The hardware is coupled to the Bruker software package **A**dvance **M**ineral **I**dentification and **C**haracterization **S**ystem (AMICS). Automated quantitative mineral analyses were performed on petrographic sections and grain mounts. This sophisticated software package is based on the Mineral Liberation Analyser (MLA) algorithm used in FEI systems [54]. AMICS uses SEM-BSE greyscale images for segmentation of areas with similar brightness (Figure 16.2). A single point analysis is performed for each segment with a minimum size of 10 μm^2. The collected EDX spectra are subsequently compared and matched with entries from a user customised database. Neighbouring identical database matches are combined to grains and agglomerated to particles as intergrowths with different mineral spectra. Database entries most importantly include spectral information, elemental composition and density. Theoretical mixed spectra can be calculated to match mineral series or fine intergrowth. The calculation of a modal mineral abundance is based on the spectral match and the total pixel area of the assigned segment. The spatial relation of successfully classified mineral spectra is used to calculate grain size, mineral liberation, or mineral associations.

For comparison, SEM-EDX based mineral distribution analyses were also performed using a JEOL JSM 6300 with XFlash 5010 detector with an acceleration

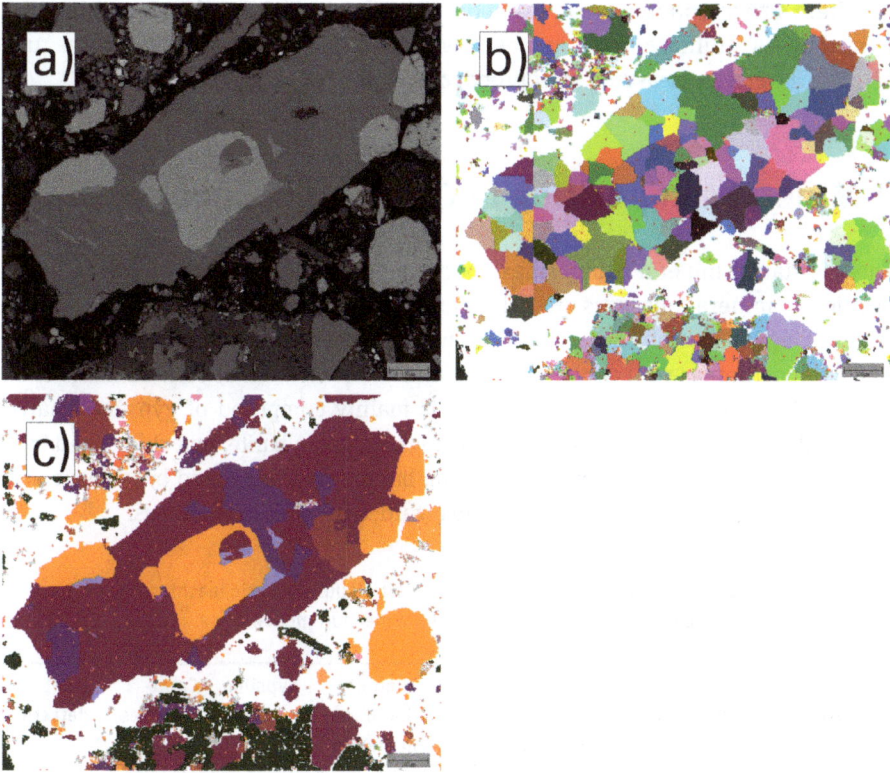

Figure 16.2: SEM-BSE image (a), segmentation (b) and classification (c) using AMICS SEM-EDX based automated mineralogy. Image width 700 μm.

voltage of 20kV. The hardware is coupled to a Bruker Quantax 200 Esprit 2.1.2 Feature Analysis Tool. Additionally, analyses on six samples using **Q**uantitative **E**valuation of **M**inerals by **SCAN**ning electron microscopy (QEMSCAN®) have been performed for comparison at the certified commercial lab ALS Global Metallurgy in Perth.

Morphological observations of phosphogypsum were performed using unconsolidated powder on 9.5 mm × 5 mm aluminium stubs. Petrographic samples were prepared as granular polished sections on 46 × 27 mm glass slides. Epoxy and sample material were mixed and spread carefully to avoid particle density segregation. Several samples have been mixed in a 1:5 ratio with fine-grained graphite powder (2–15 μm) to allow for optimal particle separation and the application of particle liberation and particle size analyses using AMICS. Particle liberation analysis is calculating the amount of free mineral surface based on neighbouring identified mineral pixels. Samples from phosphate beneficiation have been milled in a planetary ball mill combined with 125 μm mesh to allow simultaneous sizing of the material and to prevent over-milling.

Grain size distribution analyses in the range between 0.3 and 400 µm were performed by laser diffraction granulometry with a CILAS 920 granulometer. Three runs of 0.15 g have been measured in isopropanol solution for each sample.

3 Mineral composition of Catalão phosphogypsum

The qualitative mineral composition, particle properties, and the elemental composition of mineral phases is crucial for assessing the processability of the sample material. Special emphasis is placed on mineral intergrowth, shape, size, and the REE-composition as well as the presence and concentration of contaminants such as F or Th. The Catalão phosphogypsum is mainly composed of gypsum, quartz, (Ca-Al-) fluorides, monazite, celestine, Fe-oxides, ilmenite, barite, pyrochlore, and baddeleyite (ZrO_2). Monazite, Fe-oxides, ilmenite, barite, pyrochlore, and baddeleyite also occur in the primary phosphate ore and are regarded as inherited minerals.

Table 16.1: Mean composition of monazite, before and after sulphuric acid leaching, gypsum and Ca-Al-fluorides with the number of representative EPMA measurements.

mean composition (%)	monazite after leaching n = 221	monazite before leaching n = 132	Limit of detection monazite	gypsum n = 97	Ca-Al-fluorides n = 84	Limit of detection
Na_2O	0.00	0.00	0.05	0.02	0.14	0.05
F	1.43	0.50	0.12	0.07	18.94	0.14
SiO_2	0.57	0.84	0.04	0.06	6.28	0.05
MgO	0.04	0.07	0.02	0.00	0.00	0.03
Al_2O_3	0.18	0.09	0.02	0.05	14.08	0.03
Y_2O_3	0.25	0.24	0.08	0.02	0.19	0.08
ThO_2	0.42	0.28	0.21	0.06	0.06	0.22
Sm_2O_3	0.95	1.10	0.08			
SrO	3.07	2.74	0.08	0.12	0.74	0.09
BaO	1.16	1.62	0.10	0.01	0.01	0.07
K_2O				0.01	0.01	0.03
P_2O_5	25.38	24.83	0.05	0.49	0.35	0.05
SO_3	1.57	0.42	0.03	53.58	8.50	0.06
CaO	3.59	3.67	0.03	40.71	43.87	0.03

Table 16.1 (continued)

mean composition (%)	monazite after leaching n = 221	monazite before leaching n = 132	Limit of detection monazite	gypsum n = 97	Ca-Al-fluorides n = 84	Limit of detection
TiO$_2$	0.04	0.03	0.04	0.01	0.01	0.04
Cl	0.05	0.07	0.02	0.01	0.01	0.02
FeO	0.30	1.10	0.06	0.18	0.02	0.06
MnO	0.00	0.20	0.07	0.01	0.00	0.07
La$_2$O$_3$	15.31	14.39	0.07	0.04	1.35	0.08
Ce$_2$O$_3$	25.17	25.07	0.07	0.06	1.63	0.07
Pr$_2$O$_3$	2.65	2.68	0.09			
Nd$_2$O$_3$	8.87	9.82	0.09	0.06	1.90	0.11
Eu$_2$O$_3$	0.28	0.31	0.06			
Gd$_2$O$_3$	3.80	3.95	0.09			
Total	95.1	94.0		95.6	98.1	
TREO	57.0	57.3		0.2	4.9	

3.1 Gypsum

Microscopically, all phosphogypsum samples exhibit a well-developed euhedral, authigenic crystalline texture consisting of rhombic gypsum crystals (Figure 16.3). The tabular shape originates from a perfect cleavage along [010] plane and preferred growth along a/c-axis. Most of the mineral particles occur as solitary individuals but rarely, v-shaped twinning on [100] contact or x-shaped penetration twins were observed. Similar mineral properties have been described from other phosphogypsum deposits [14, 34, 35].

Qualitative phase analyses with powder XRD was performed on two samples. Prominent XRD reflexes are mainly caused by gypsum [010] and only subordinately by bassanite (hemihydrate). Comparable diffraction patterns have been published for synthetic calcium sulfates from Greek and Moroccan phosphogypsum and gypsum from flue gas desulfurisation [33, 35, 55].

The elemental composition of the gypsum has been determined with 97 individual EPMA point measurements (Table 16.1). SO$_3$ (53.58%) and CaO (40.71%) are the main components, whereas the elemental oxide total sum is 95.57%. Concentrations for REEs are below the individual limit of detection. Therefore, the data is regarded as qualitative or semi-quantitative. For a reliable REE-concentrations, laser ablation

Figure 16.3: SEM-BSE overview image of representative Catalão phosphogypsum, formed newly during mineral processing. The material consists of euhedral gypsum crystals and relictic bright mineral phases comprising monazite, barite, and celestine.

ICP-MS analyses would be necessary. Total rare earth oxides (TREO) amount to an estimated 0.17% consisting in turn of 37.8% Nd, 36.9% Ce and 25.3% La. Considerable amounts of the total REE concentrations are therefore bound to gypsum as the main mineral constituent of the phosphogypsum samples. These REE concentrations are regarded as trapped and can only be recovered using bulk leaching processes [19, 21, 23, 31, 32], which is currently economically not viable.

Diffraction granulometry data indicate a narrow range of the mean grain diameter, generally well below 120 μm (Figure 16.4). The D80 varies between 90 to 109 μm and D50 between 52 to 62 μm. SEM-EDX based mineral distribution analyses revealed similar granular properties. The mean grain diameter is in accordance with the mean grain size range of 45 to 250 μm reported for phosphogypsum from Florida [56]. Individual gypsum crystals can have a length of up to 500 μm but with a very limited width and height (thickness) range. The aspect ratio of length to width, is between 1:3 to 1:6 whereas the length to height ratio varies between 1:10 to 1:30.

3.2 Common accessory minerals

More than twelve accessory minerals have been identified using detailed SEM analysis. Most of these accessory minerals have a modal abundance of less than 1% and commonly represent "heavy minerals" containing most of the metals and trace

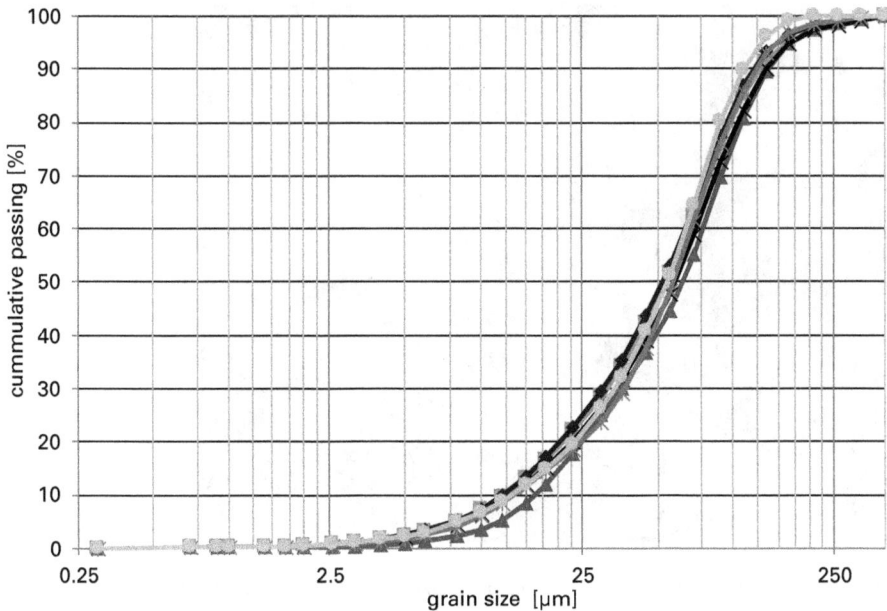

Figure 16.4: Laser diffraction granulometry data of six different Catalão phosphogypsum samples from conveyor belt and stockpile. The grain size distribution is similar for all analysed samples.

elements such as the REE, but also Nb, Ba, Sr, and Zr. Most of the accessory minerals are inherited from the primary phosphate rock whereas some have formed newly during acid leaching of the apatite concentrates.

3.2.1 Monazite

Most important for the resource potential of the Catalão phosphogypsum is the REE-phosphate mineral monazite [57, 58]. In daylight monazite appears as yellowish to slight pale orange mineral. Individuals are characterised by high SEM-BSE brightness and collomorphous, botryoidal, or spherulitic mineral shapes (Figure 16.5). Mean grain size diameters are generally below 50 µm with a P80 of 30 to 35 µm and a P50 between 18 and 24 µm. Larger grain size diameters occur mainly as aggregates of multiple intergrown botryoidal particles. The majority of monazite occurs as liberated single particles or aggregates of several intergrown bulged, rounded minerals. Rarely monazite appears as clusters of parallel tabular radial plates (rosettes) forming botryoidal aggregates. Euhedral short prismatic crystals, which are common in fresh igneous rocks were not observed. Partial mineral locking with quartz, Sr-sulfates, fluorite-related minerals, to lesser extent with ilmeno-rutile, Fe-oxides, or surficial overgrown by minute gypsum laths is readily observed (Figure 16.5, Figure 16.6).

Figure 16.5: SEM-BSE image of collomorphous monazite (Mon) intergrown with euhedral Ca-Al-fluorides (CaAlF) together with gypsum (Gp). Imaging mode: shadow 1.

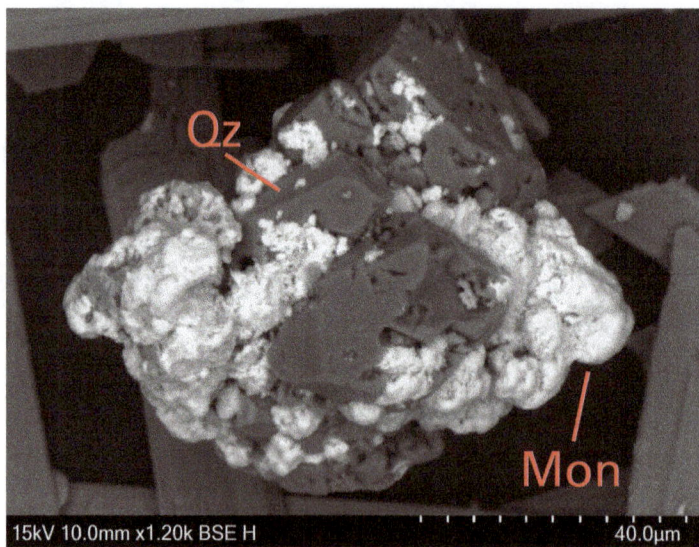

Figure 16.6: SEM-BSE image of botryoidal monazite (Mon) intergrown with euhedral quartz (Qz).

The overall modal abundance of monazite in Catalão phosphogypsum is not higher than 1%. Conventional mineralogical phase identification of accessory minerals in partly oriented gypsum using powder XRD is challenging. Qualitative analyses were therefore performed on monazite obtained by physical separation from phosphogypsum andconcentration to more than 50%. Ce-dominant monazite can be identified based on best-fit main XRD reflexes alongside gypsum, quartz, Fe-oxides. The hydrated REE-phosphate rhabdophane, which occurs commonly at other monazite occurrences has not been identified at Catalão [59]. Semi-quantitative SEM-EDX analyses of monazite revealed a complex elemental composition with Ce, P, La, and Nd and minor additional, Sr, S, Ba, Ca, and F. The critical contaminant Th only occurs sporadically at detectable concentration levels. The mean composition of monazite from Catalão phosphogypsum was calculated based on 221 EPMA point measurements on more than 80 single particles (Table 16.1). The analysed minerals are dominated by P_2O_5 (25.38%), Ce_2O_3 (25.17%), La_2O_3 (15.31%) and Nd_2O_3 (8.87%). The oxides for Gd (3.80%), Pr (2.65%), Sm (0.95%), and Eu (0.28%) have also been determined and TREO concentrations amount to 57.31%. The relative REE contents are: Ce 44.1%, La 26.8%, Nd 15.6%, Gd 6.8%, Pr 4.6%, Sm 1.7% and Eu 0.5%. The mean La/Nd ratio of 2.67 is low and comparable to early stage carbonatite-related monazite [60]. Rarely, unusually La- and Ce-rich minerals as well as some Nd-poor minerals have been measured (Figure 16.7).

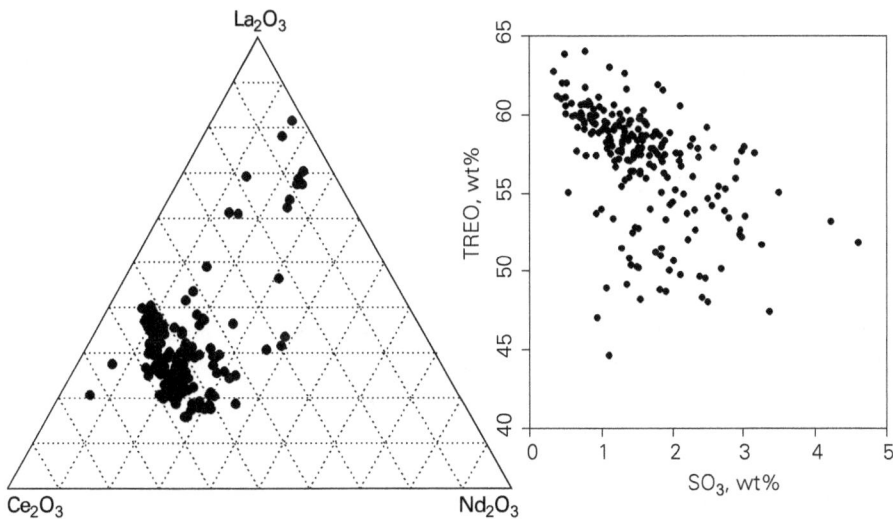

Figure 16.7: Ternary Ce-La-Nd oxide composition (left) and TREO vs. SO3 content (right) for Catalão monazite after sulphuric acid leaching. Calculations based on wt%.

Minor contents of other element oxides commonly are CaO (3.59%), SrO (3.07%), and BaO (1.16%). These are probably due to the monazite cation exchange capability as pointed out by Toledo et al. [61]. Peculiar are elevated concentrations of SO_3 (1.57%) and F (1.43%). The mean total sum of all oxides is 95.09% indicating low concentrations of water. Some monazite has probably been partly hydrated as proposed for monazite from the weathered Mt. Weld deposit in Australia [5]. The water bearing REE-phosphate rhabdophane is absent in the diffraction analyses.

Mean ThO_2 contents are comparably low, amounting to a mean of 0.4%. The maximum concentrations in single measurements are 2.0%. In some cases, ThO_2 has not been detected in the analysed particles. Generally, ThO_2 in monazite from carbonatite related deposits is considered to be low (<2%) compared to monazite from metamorphic, felsic intrusive, or sedimentary origin [62, 63]. The coupled substitution of Ce^{3+} by $[Ca^{2+} (Sr^{2+}) + Th^{4+}]$ in the monazite-cheralite solid solution could explain the divers elemental composition of Catalão monazite [64, 65]. Additional $[Ce^{3+} + P^{4+}]$ substitution by $[Si^{4+} + Th^{4+}]$ is possible, called huttonite substitution [64, 66]. It was not possible to derive any of these substitution mechanisms based on the microprobe analyses of monazite from Catalão phosphogypsum. Only SO_3 and TREO express a negative correlation, possibly as a result from sulphuric acid leaching and loss of TREO from the structure (Figure 16.7).

Monazite from the primary and weathered calc-alkaline rocks from Catalão has previously been analysed [49, 61]. Poor crystallinity, similar variable elemental composition, and morphological characteristics have been described. Low TREO and P_2O_5 concentrations and elevated SrO and CaO have also been reported. The analysed monazite La/Nd ratio of 2.67 is comparably low and similar to monazite analysed previously but not nearly as low as previously reported for Catalão carbonatites [61, 67]. The La/Nd ratio for carbonatite-hosted monazite from other deposits worldwide may vary between 3 and 7 as summarised by Chen et al. [63]. Low ratios comparable to Catalão monazites have also been reported from the weathered monazite occurrence at Mt. Weld [5]. Additionally, the chondrite normalised REE-patterns of the analysed monazite particles have a similar and less steep character (Figure 16.8) [61]. A negative Ce anomaly as indicated by some analyses for Mt. Weld is not present in the majority of Catalão phosphogypsum monazite EPMA measurements. The presence of the anomaly was previously interpreted as an indication for a oxidizing formation environment [63].

The majority of the monazite from the Catalão phosphogypsum seems to represent relicts of primary minerals or supergene re-precipitates from the weathered REE-bearing ultramafic igneous rocks. Therefore, monazites from the stockpile of primary phosphate ore, apatite concentrate, and tailings were investigated to allow a comparison of monazite before and after sulphuric acid leaching. Morphologically, all monazite commonly shows botryoidal collomorphous shapes with concentric growth zones. Additionally, fine-grained acicular crystals and radial bundles of monazite intergrown with and overgrown by apatite, Fe-oxides, gorceixite, or quartz were observed. Fine-crystalline monazite present in samples before acid leaching is rare or almost absent in phosphogypsum either due to partial or total H_2SO_4-dissolution of smaller crystals. Apart from that the monazite particle size is similar before and after the leaching process.

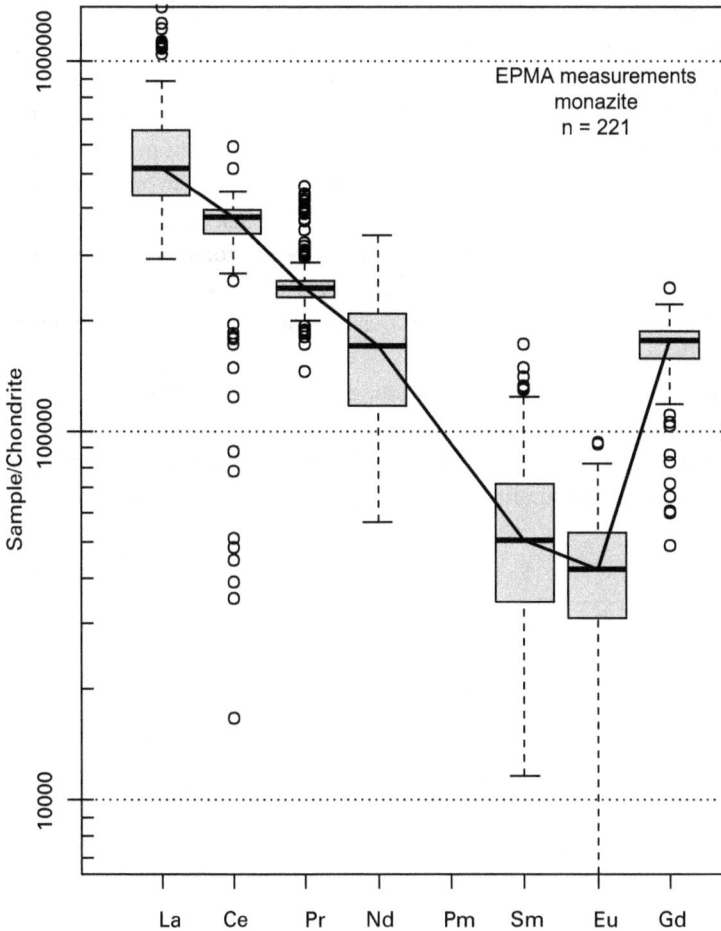

Figure 16.8: Chondrite-normalised (Anders and Grevesse 1989) REE composition of monazite from Catalão visualized as boxplots based on 221 EPMA measurements.

In total, 127 EPMA point measurements have been performed on multiple particles before acid leaching to identify compositional differences (Figure 16.9). Generally, the elemental composition of the monazite before and after acid leaching of the phosphate ore is relatively similar. Differences in concentrations might be explained by the overall geochemical variation of monazite and the limited number of microprobe analyses performed. Only the F and SO_3 contents show a significant two- or threefold increase from primary to treated monazites. Elevated SO_3 contents for processed monazite have been previously described in literature but substitution mechanisms are not known yet [63]. A substitution of TREO by SO_3 is indicated by the negative correlation of these components.

Elevated fluorine concentrations most likely result from the acid leaching of the primary fluoroapatite and therefore from the release of fluorine into the fluid. This solution enriched in fluorine in turn interacts with residual minerals such as monazite. Apatite of the Catalão complex contains 3.6% F and therefore represents an important carrier mineral of this element [67]. The contents of ThO_2 in monazite after acid leaching are 50% higher compared to the concentrations prior to leaching. SiO_2 and FeO concentrations decreased by 30 to 70% during the leaching process. Striking however is the marked similarity of the TREE-concentrations before and after acid leaching. Relative changes vary between 1 and 10% with a slight increase in La and decrease in Nd after acid leaching. The overall composition of the different REEs however remains very similar.

In sum, based on our mineralogical, textural, and compositional results, monazite in phosphogypsum is consequently regarded as inherited from the original phosphate ore. Due to the interaction with sulphuric acid and elements in the processing solution (e.g. from disintegrated apatite) a slight chemical transformation of monazite must be assumed.

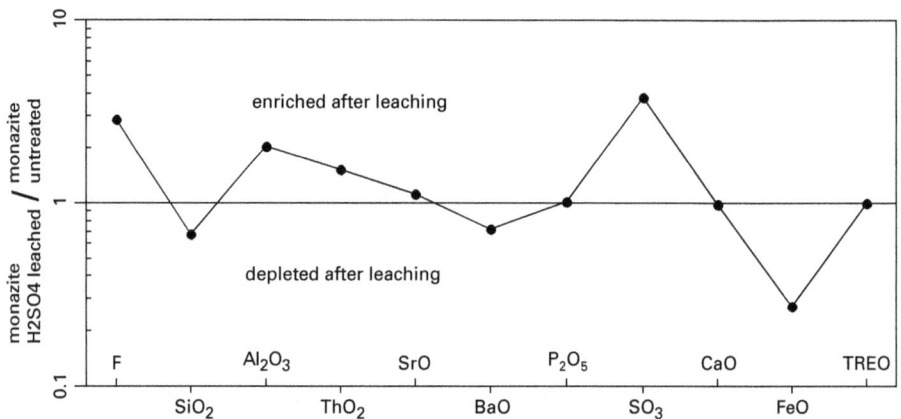

Figure 16.9: Elemental composition of monazite from phosphogypsum (after H_2SO_4 leaching) normalized against monazite from phosphate rock before acid leaching. Enriched and depleted elements are represented as points above or below a value of 1.

3.2.2 Calcium-Alumino-Fluorides (Ca-Al-Fluorides)

Peculiarly euhedral isometric octahedral shaped crystals can be observed in all phosphogypsum samples (Figure 16.10, Figure 16.11). They commonly occur as solitary, liberated crystals and subordinately as penetration twins or intergrown with gypsum on the [010] face of the gypsum crystals. Rarely the euhedral minerals occur as aggregates of several octahedrons cemented by Sr-sulfates. Most of the octahedrons have crystal sizes smaller than 100 μm, a P80 of 30 and 48 μm and a P20 between 21 and 30 μm.

Figure 16.10: SEM-BSE image of euhedral gypsum (Gp) overgrown by amorphous crusts of celestine (Cln) and intergrown with perfectly octahedral Ca-Al-fluorides (CaAlF).

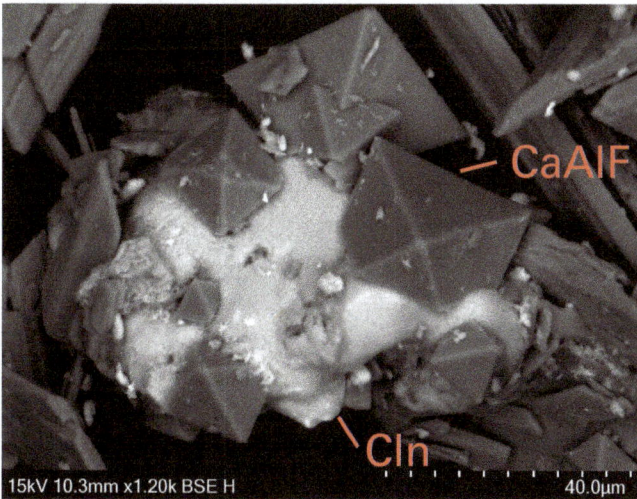

Figure 16.11: SEM-BSE image of octahedral Ca-Al-fluoride (CaAlF) crystals cemented by crusts of anhedral celestine (Cln).

Qualitative elemental analyses by EDX indicate a predominance of Ca and F alongside Al, S, Si, to lesser extent Sr and P as well as minor REE concentrations. Generally, the geochemical composition is complex and a simple identification is complicated. Fluoride-bearing phases have been described before as precipitates from apatite leaching in phosphogypsum from other sources [68, 69] and the cubic minerals chukh-rovite(Ce) $Ca_3Ce_{0.75}Y_{0.25}Al_2(SO_4)F_{13} \cdot 10(H_2O)$ and meniaylovite $Ca_4AlSi(SO_4)F_{13} \cdot 12(H_2O)$ were identified [33, 70–72]. The minerals above were described as newly formed during the process.

For Catalão phosphogypsum the mean elemental composition of octahedral calcium-aluminium-fluorides (Ca-Al-fluorides) has been established based on 84 EPMA point measurements of more than 50 crystals (Table 16.1). The mean composition is dominated by CaO (43.87%), F (18.94%), Al_2O_3 (14.08%), SO_3 (8.50%), and SiO_2 (6.28%). As shown in Figure 16.12, the overall variability of the measured geochemical composition of the minerals is low. SiO_2 concentrations correlate positively with SO_3 and CaO but negatively with Al_2O_3 contents. Other oxides like SrO (0.74%), P_2O_5 (0.35%), and Y_2O_3 (0.19%) occur in low concentrations only. The total oxides sum up to 98.1% and therefore only minor amounts of water or carbonates are probable. TREO concentrations range up to 4.89% and the Ca-Al-fluoride minerals should therefore be considered an important host mineral for the REE in the Catalão phosphogypsum (Figure 16.12). Among the analysed REE, Nd is the most abundant with 39%, followed by Ce (33%), and La (27%) (Figure 4.2.4.6). These minerals have supposedly formed syngenetically with the gypsum during sulphuric acid leaching of the phosphate rock because of the close textural relationship, the occurrence intergrown with gypsum (Figure 16.12), and absence in phosphate ores. Comparably low mineral densities of the similar minerals chukrovite (2.3 g/cm^3) and meniaylovite (2.7 g/cm^3), together with the small crystal sizes of these minerals will

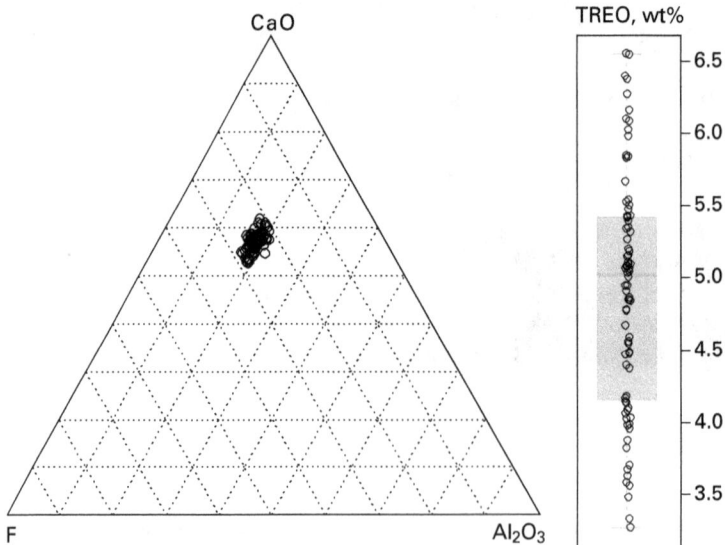

Figure 16.12: Ternary F-CaO-Al2O3 composition (wt%) and TREO boxplot (wt%) of the Catalão Ca-Al-fluorides based on EPMA measurements.

be a hindrance for the recovery of REE from calciumfluorides by physical separation and are therefore considered as economically not feasible. Altough similar, the mineralogical identification on the basis of the elemental composition is not satisfactory as both chukhrovite (Ce) or meniaylovite contain substantial quantities of water of 21.8% and 27.9% respectively, which are not probable for the Catalão Ca-Al-fluorides due to the good total sum., F contents in the measured minerals are considerably lower with 18.94% compared to the theoretic content by formula of 29.95% (chukhrovite) or 31.88% (meniayovite). The composition of chukhrovite though has certain variability, i.e. cation exchange of REE and Ca and this mineral might therefore accommodate a variety of cations [73].

3.2.3 Further accessory minerals

A variety of rare minerals have been found in Catalão phosphogypsum by detailed scanning electron microscopy. These were identified on the basis of spectral features (semi quantitative EDX) and morphology. In order of their modal abundance these are quartz, primary and secondary celestine, Fe-oxides, ilmenite, barite, baddeleyite, Ba-pyrochlore, relictic apatite, and very rarely zirconolite.

Quartz is the most common accessory mineral alongside Ca-Al-fluorides. The silicate commonly occurs as medium-sized particles with mean particle size ranges between 20 μm to 80 μm. Quartz generally occurs as solitary crystals or surficial overgrown by other minerals. Peculiar is the intergrowth of quartz and spherulitic monazite. Euhedral to subhedral prismatic crystal faces of quartz can be observed. Partly, quartz surfaces exhibit locally small cavities. The quartz is a mineral, typically inherited from the locally quartz-rich primary phosphate ore. Fine cavities in the surfaces of quartz crystals might represent etchings originating from sulphuric acid treatment of the material or dissolved minerals that were intergrown with or occurred as inclusions within the quartz (see Figure 16.6).Celestine commonly occurs intergrown with quartz. Based on EDX analysis the Sr-sulfates can contain minor amounts of Ba or REE, Ca, and even F. Celestine exhibits morphological characteristics and mineral shapes similar to the monazite and collomorphous and botryoidal particles and crusts are common. Although the BSE brightness is typically slightly lower the celestine can be easily mistaken for monazite. Celestine commonly occurs as crusts, overgrowing gypsum, quartz, and Ca-Al-fluoride (Figure 16.11). Celestine might also form minute acicular crystal aggregates of fine needles (<4 μm) overgrowing other minerals. The latter celestine probably formed syngenetically together with gypsum during sulphuric acid leaching. Fe-oxides occur as compact, shard-like particles, aggregates of subhedral radial slender prisms, and spongy porous rounded anhedral masses. The observed features are an indication for the presence of several different mineral phases, including goethite, haematite, and magnetite. Magnetite and goethite have been identified by XRD in heavy mineral concentrates produced from phosphogypsum. These minerals are inherited from the primary phosphate ore. The primary alkaline intrusive rocks contain significant Fe_2O_3 concentrations of 15 to 70% [43]. Own investigations of the primary phosphate ore revealed Fe_2O_3 concentrations of 38 to 44% in the feed material. During mineral processing, most Fe-oxides are removed by high and low intensity magnetic separation and flotation and only minor quantities of Fe-minerals pass through into the phosphogypsum product.

The Ti-Fe-oxide ilmenite found in phosphogypsum is most likely inherited from primary phosphate ore analogous to the Fe-oxide minerals. Most ilmenites show compact anhedral, shard-like mineral shapes. Rarely, subhedral thick tabular particles with partly rhombohedral mineral faces can be observed (Figure 16.13). The majority of the ilmenite occurs as liberated single particles. The EDX-spectrum not only hosts energy peaks of Ti, Fe, and O but also Mg and rarely Nb. These elements have been reported previously in ilmenite from the Catalão I igneous rocks and the Mg content has been used to classify different generations of igneous rocks [67].

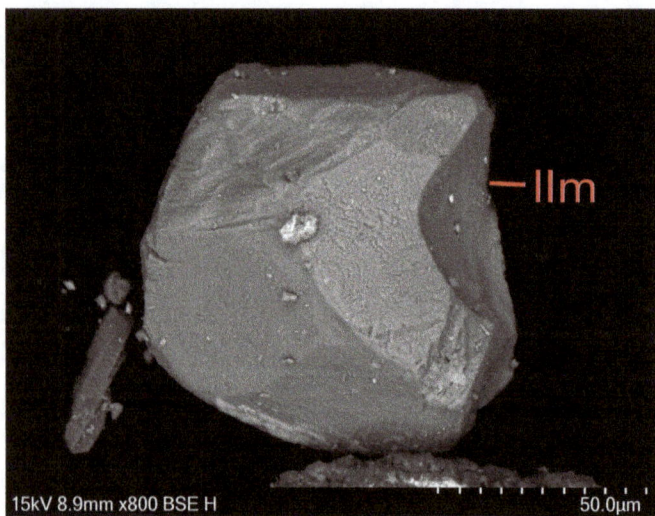

Figure 16.13: SEM-BSE images of ilmenite (Ilm) observed in Catalão phosphogypsum.

Barite, baddeleyite, pyrochlore, and zirconolite occur in very low abundances and have been identified by scanning electron microscopy. These mineral phases were undetectable in XRD phase analyses of heavy mineral concentrates from the Catalão phosphogypsum. Barite is the densest mineral found in the phosphogypsum and occurs as compact, sometimes angularly fractured particles (Figure 16.14). EDX spectra show variable Sr-contents in the barite that resulting in lower BSE brightness of some particles. The Zr-oxide baddeleyite and the important Nb-mineral Ba-pyrochlore have BSE grey scale values similar to that of barite. Even though these minerals only occur

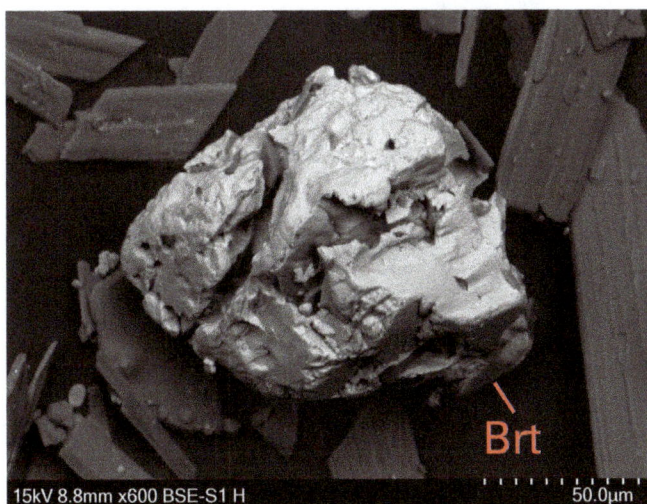

Figure 16.14: SEM-BSE images of barite (Brt) observed in Catalão phosphogypsum.

in accessory quantities, they both contribute significantly to the overall composition with their respective characteristic elements. Baddeleyite occurs exclusively as compact, fractured and liberated anhedral crystals. The only other Zr-bearing phase is a Zr-Ca-Ti mineral, most likely zirconolite or zirkelite. Based on EDX analysis, these rare minerals might contain low concentrations (≪1%) of REE. They have not been considered in our current REE deportment analysis, due to their small. The Nb-Ca-Ti-rich pyrochlore commonly occurs as anhedral or euhedral minerals. Crystal faces show octahedral or tetrahedral shapes. Pyrochlore typically exhibits oscillatory concentric compositional zoning, due to varying Ba and Sr concentrations. The zonation apparently developed along octahedral growth boundaries. In some parts of the Catalão alkaline complex pyrochlore is regarded as a major Nb resource [42, 43, 50]. Nb is currently mined and extracted at the Catalão II Mine and also produced as a by-product from Catalão I. TREO (La + Ce + Nd) concentrations in the pyrochlore structure vary between 3.5 and 6% [42]. Due to its low overall abundance and minor concentrations it is not considered to have an important influence on the elemental deportment of REE. Rarely, apatite relicts are observed as massively corroded crystals overgrown by gypsum and anhedral collomorphous fluorite.

4 Quantitative material composition

In the following chapter the bulk geochemical composition and the total abundance of the relevant minerals are presented. Based on modal abundance and mineral composition data the deportment of the REE during the mineral processing at Catalão are further calculated.

The bulk geochemical composition of the analysed phosphogypsum almost represents the stoechiometric composition of the mineral gypsum. Therefore, the geochemical composition of the phosphogypsum with respect to the major elements is simple and the variation of the analysed material is low. Trace element concentrations of REE, Zr, Nb, Ba, or Th originate primarily from the contained accessory minerals.

CaO concentrations vary between 29.9 and 32.7%. The loss on ignition value (LOI) is between 20.8 and 21.4% and most probably reflects the stoichiometric H_2O content in gypsum (20.93%) [74]. Only one sample exhibits lower LOI values after drying at 105 C and therefore water loss. SiO_2 contents range between 1.47 and 2.51% whereas Fe_2O_3 (0.52–0.85%), TiO_2 (0.18–0.28%) and Al_2O_3 (0.05–0.28%) concentrations are very low and result from low respective abundances of quartz, Fe-oxide, and ilmenite. P_2O_5 concentrations are low (0.87–0.55%) and reflect relictic apatite, monazite, and minor amounts bound to gypsum. Low SrO (0.6–0.7%) and BaO (0.29–0.42%) concentrations originate from celestine, gypsum, and barite. Zr, Nb, and Th occur in trace concentrations only with mean concentrations of 549 ppm, 369 ppm, and 76 ppm respectively. The majority of Zr and Nb is most probably hosted by baddeleyite, Nb-bearing ilmenite, and pyrochlore.

The mean TREO concentration of 0.63% (analysed by ICP-MS) in Catalão phosphogypsum are relatively high and thus explains the economic interest for recovery. Published TREO concentrations at other occurrences range between 0.03 and 0.6% [12, 17, 21, 24, 27, 30].

Phosphogypsum from other deposits, Cubatão and Uberaba in Brazil, have TREO contents of ~0.4 to ~0.6% and are thus comparable to the concentrations at Catalão. The chondrite normalised REE spectra for Catalão phosphogypsum exhibit a steep curve and strong enrichment of the light REE, La to Sm (LREE) compared to heavy REE, Eu to Lu (HREE).

The general REE pattern shows close resemblance with weathered and altered primary rock types previously described for Catalão I [46]. This has been further supported by own analysis performed on run of mine phosphate ore from stockpile (Figure 16.15). Although the REE concentrations are 40% higher in the phosphate ore the distribution of the REE closely resembles the phosphogypsum REE pattern. Analogous to the mineral monazite the corresponding REE signature is inherited by phosphogypsum from the primary phosphate bearing host rocks. The slight negative Ce anomaly present in monazite elemental composition and bulk phosphogypsum likely indicates oxidizing environments during formation or alteration of the mineral [63, 75].

Generally, the the plot of the modal abundance of monazite and of the total available TREO from phosphogypsum, run of mine phosphate ore, apatite concentrates and intermediate products shows a good positive correlation trend (Figure 16.16). Therefore, samples with higher monazite contents carry also higher concentrations of TREO.

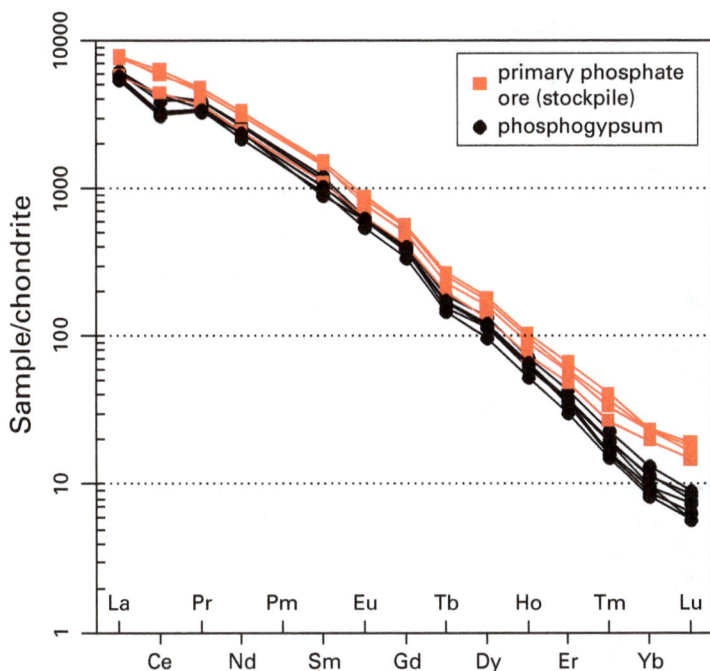

Figure 16.15: Chondrite-normalized REE-patterns for the primary phosphate ore and the processed phosphogypsum from Catalão.

This is due to monazite being the main host mineral for REE in all sample types. There are several other mineral phases occurring in different abundance hosting REE like gypsum, apatite, gorceixite, pyrochlore, and celestine explaining the partly non-ideal linear trend of the correlation. The intermediate products represent tailings from magnetic separation and carry a high abundance of monazite but a comparably low concentration of apatite. Our own investigation has shown, that apatite acts as an important REE carrier (mean 0.58% TREO). It's high or low abundance in some sample groups therefore highly impacts the correlation of monazite abundance and geochemical TREO content.

The AMICS mineral distribution analyses are performed by back-calculated geochemistry based on mineral abundance, theoretical density and composition. The quality of the used analytical approach and therefore reliability of the mineral distribution analyses a sufficient if the results from lab geochemistry and back calculated elemental assays match well. The composition for the major REE-bearing phases (monazite, apatite, gypsum, gorceixite, and Ca-Al-fluorides) has been determined with EPMA. Calculated geochemical results are subsequently compared to laboratory-based geochemical results (Figure 16.17). Apatite concentrates and primary phosphate ore samples have been chosen to illustrate the results because these

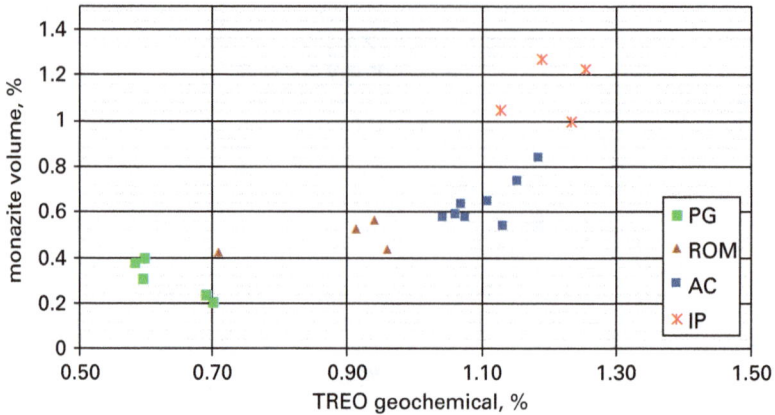

Figure 16.16: Correlation of the monazite mineral content in Catalão phosphogypsum with the TREO concentration. The data is based on AMICS mineral distribution analyses and geochemical bulk TREO concentration for different sample types of the Catalão phosphate beneficiation. PG = phosphogypsum, ROM = run of mine phosphate ore, AC = apatite concentrate, IP = intermediate products (tailings).

sample groups cover a wide range of different mineral groups and show a large difference in the abundance of REE bearing minerals like monazite or apatite. A high quality and adequate statistical basis is achieved if the results are in good accordance [51, 76, 77]. Generally, the results from geochemical analyses and back-calculated assays based on SEM-based quantitative mineralogy with AMICS show a low deviation of less than 30%. Especially for elements in trace concentration, a higher deviation can be expected to due smaller tolerance for error in sample preparation and analyses. Some elements like Nb exhibit an extremely poor correlation of the results, most likely due to not defined bonding of the element to ilmenite. The results for the major REE La, Ce, and Nd are in good agreement and deviation is mostly less than 10%. Therefore, the calculation of the bulk geochemical composition based on SEM-EDX based mineral distribution analyses is sufficiently accurate. Consequently, SEM-EDX mineral abundance studies are thus a suitable scientific method to describe the quantitative mineralogy of such phosphogypsum samples.

In the following the quantitative mineral distribution of phosphogypsum based on mineral distribution analysis with AMICS is presented. The composition of phosphogypsum is dominated by the mineral gypsum sensu stricto (93.2 wt% $\sigma \pm 1.8\%$), although a slight under-representation can be assumed due to flat tabular mineral shape and segregation effects during sample preparation (Figure 16.18). Monazite as the main REE host mineral has a mean abundance of 0.59 wt% ($\sigma \pm 0.14\%$). The relatively high variation between different samples is due to the varying distribution of monazite in the primary phosphate ore. Monazite is inherited from the weathered phosphate rock and thus largely reflects the abundance of monazite in the feed

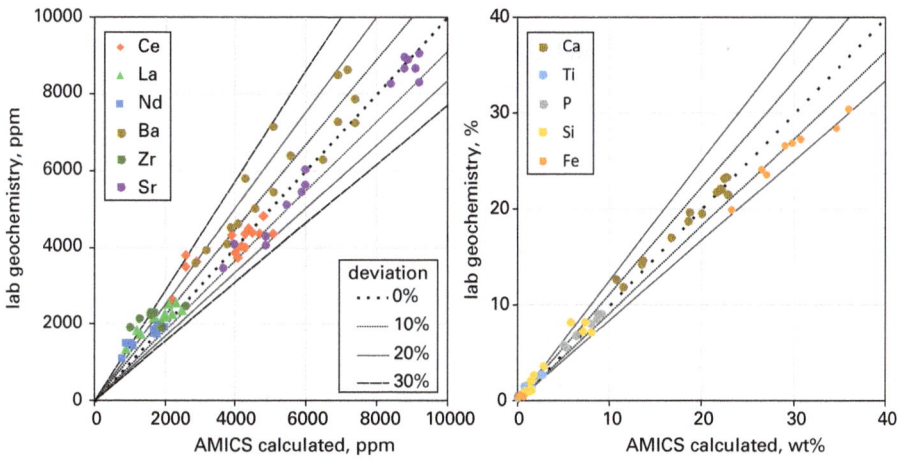

Figure 16.17: Lab geochemistry (ICP-MS/OES,XRF) versus back-calculated geochemistry derived from quantitative mineral distribution analyses (AMICS) of primary phosphate ore and apatite concentrates.

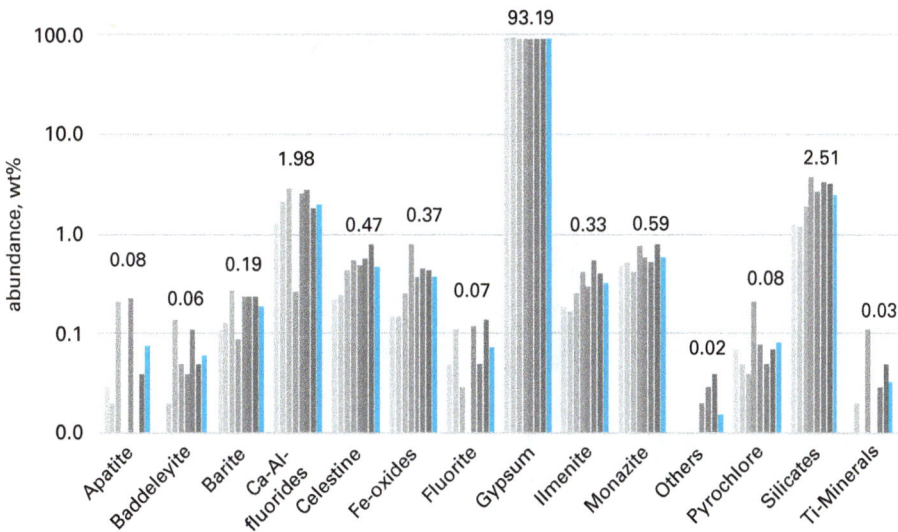

Figure 16.18: Quantitative mineral distribution of seven phosphogypsum samples and mean value (blue bar).

material for the phosphate beneficiation. Additionally, the heavy mineral monazite can segregate by density or can show nugget-effects during sample preparation. Ca-Al-fluorides have a highly variable abundance of mean 1.98 wt% ($\sigma \pm 0.93\%$). The origin of this unsystematic variation is not clear. Ore samples from stockpile and freshly sampled from the conveyor belt exhibit both, high and low abundance.

Silicate-group minerals with quartz as the major mineral phase have a mean abundance of 2.51 wt% ($\sigma \pm 1.1\%$) and are therefore the most common minerals along gypsum and Ca-Al-fluorides. All other minerals occur in quantities of less than 1% such as celestine (0.47%), Fe-oxides (0.37%), ilmenite (0.33%), and barite (0.19%). Pyrochlore (0.08% $\sigma \pm 0.06\%$) and baddeleyite (0.06% $\sigma \pm 0.05$) occur as accessory mineral fractions only and exhibit a strong variation throughout the sample range. Analogous to monazite these variations originate from mineralogical differences in the feed material.

The monazite liberation has been analysed using AMICS and QUEMSCAN analyses (Figure 16.19). A major fraction of 80 wt% to 63 wt% monazite occurs completely liberated. QUEMSCAN analyses classified a further 20 wt% to 30 wt% as high-grade and low-grade middlings. Only 9 wt% to 6 wt% are locked in or with other minerals, based on QUEMSCAN analyses. Based on AMICS analyses, less than 1% of the monazite is completely locked by other particles and 95% of the particles exhibit a liberation better than 80%. This correlates well with observations made during SEM microscopy, where most of the monazite occurs as single minerals or aggregates of

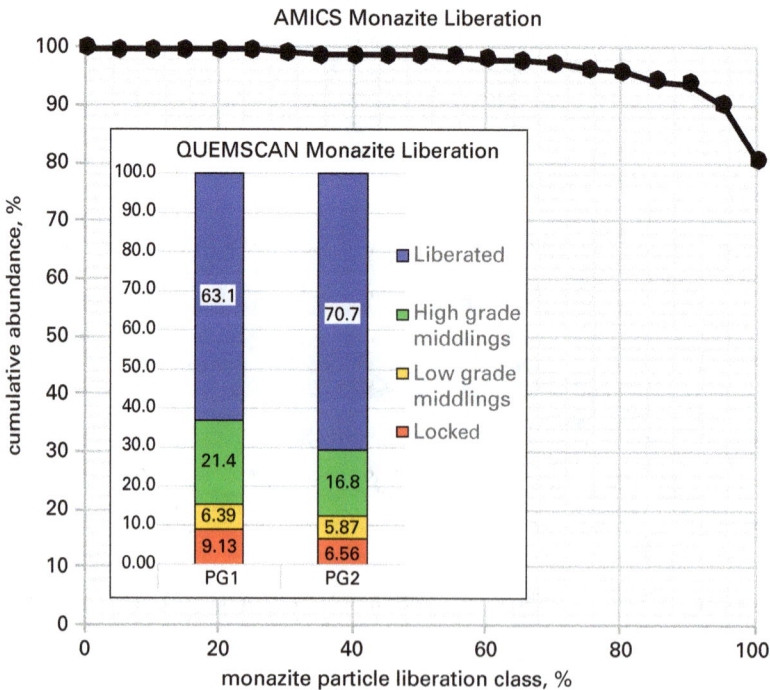

Figure 16.19: Good monazite particle liberation in Catalão phosphogypsum, calculated with AMICS and QUEMSCAN. 80% to 60% of the monazite particles in the analysed phosphogypsum occur fully liberated.

multiple collomorphous particles. Limited intergrowth with quartz and Ca-Al-fluorides is only rarely observed.

The deportment of the REE is also calculated with AMICS. This allows a more detailed assessment of the minerals hosting the REE in the phosphogypsum. The calculation is based on the composition for each of the associated minerals (Figure 16.20), its total abundance (4.2.4.4), and density. Only the three main REE (Ce, La, Nd) are considered in the deportment assessment. Based on the identified mineral phases four main carriers for REE in phosphogypsum have been identified: monazite, gypsum, and Ca-Al-fluorides (Figure 16.21). Some REE are hosted by an intergrowth of monazite and celestine. Due to the fine grain sizes this group is classified as one REE-hosting particle group. Of the REE, totally contained in the bulk phosphogypsum from Catalão, some 60% of Ce, 54% of La and 34% of Nd is bound to monazite. Ce, La, and Nd all have a high standard deviation of 8.2 to 8.7%, which predominantly originates from the comparatively high variation of monazite total abundance in the samples. Another major and thus critical host for the REE at Catalão is the gypsum itself. The mineral hosts 43% of the total Nd, 27% of the La, and 23% of the Ce. Due to the high abundance of gypsum, the standard deviation in the analysed samples for the REE deportment is comparably low with 3.1% to 4.6%. The third REE hosting phase in the bulk phosphogypsum is Ca-Al-fluoride containing 21% Nd, 15% La and 12% Ce. Deportment is therefore lower but similarly distributed as calculated for the gypsum. Minor contents of REE are bound to fine grained celestine, which occurs intricately intergrown with monazite.

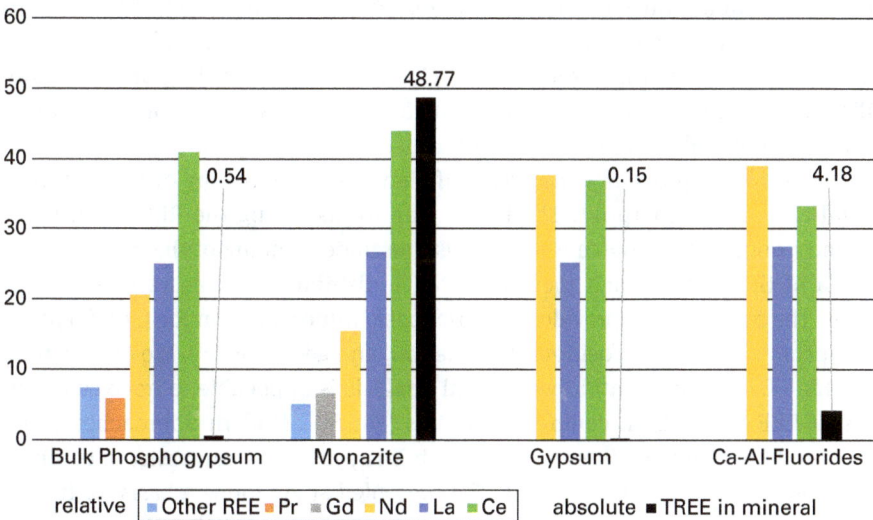

Figure 16.20: REE composition of Catalão phosphogypsum, monazite, Ca-Al-fluorides, and mineral gypsum.

MEAN REE DEPORTMENT PHOSPHOGYPSUM

☐ Ce   ☐ La   ☐ Nd   ■ Total Ce+La+Nd Normalized

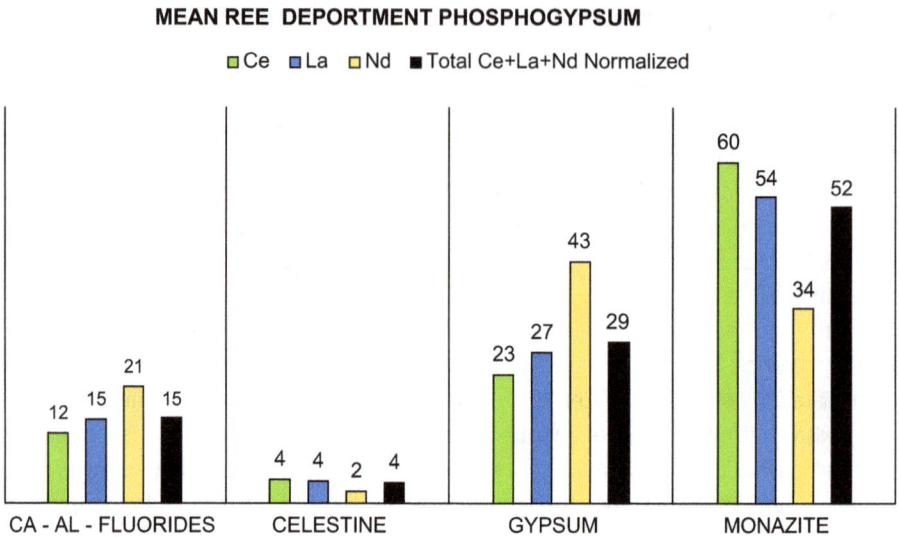

CA - AL - FLUORIDES   CELESTINE   GYPSUM   MONAZITE

Figure 16.21: REE deportment in phosphogypsum from Catalão, based on mineral distribution analyses using AMICS automated mineralogy and compositional data from EPMA.

5 Summary and conclusion

Phosphogypsum represents today a large potential secondary source for REE with up to 280 million tonnes being globally produced annually as mining waste. To date, only 15% are recovered and used but large quantities are still deposited on landfills or discharged into the oceans [13, 14, 78, 79]. Recycling and therefore utilization of previously unused waste material streams may create additional economic potential for the phosphate fertiliser industry by recovering REE as a by-product and improve raw materials efficiency and reduce the overall environmental and carbon footprint. The initial step for assessing the REE potential of any phosphogypsum resource is to establish an understanding of the mineralogical fixation of the elements and their quantitative distribution in the various mineral phases. The present study provides essential basic information on the Catalão phosphogypsum mineral composition, REE fixation, and geochemistry. We demonstrate that SEM-EDX-based quantitative mineral analysis is applicable to trace compositions of REE minerals in complex materials [54, 80]. Modern approaches using SEM-based liberation analyses coupled with X-ray micro CT can assist in confirming previously acquired datasets [81]. The presented study forms a basis for further work that might improve knowledge on secondary resource beneficiation, e.g. using predictive geometallurgy [82].

Detailed petrographic investigations revealed a peculiar mineral composition of the Catalão phosphogypsum, consisting of euhedral tabular gypsum, octahedral

Ca-Al-fluorides, partly cavernous and possibly etched subhedral quartz, Fe-oxides, subhedral ilmenite, cryptocrystalline to amorphous celestine or Sr-sulfates, barite, baddeleyite, pyrochlore, and apatite relicts. Monazite is the major REE-bearing mineral (besides the gypsum itself), occurring as collomorphous particles or aggregates rarely intergrown with quartz, Ca-Al-fluorides, or celestine. The monazite composition is complex and dominated by Ce, La, Nd, and P_2O_5. The TREO in analysed particles amount to 57% and low ThO_2 contents of approximately 0.42%. The La/Nd ratio is low and thus comparable to monazite in weathered REE-bearing rocks from Catalão or to those from Mt. Weld, Western Australia [5, 61]. The monazite is predominantly inherited from weathered phosphate rock, based on the major similarities of monazite before and after leaching with H_2SO_4. Only minor changes in the monazite composition are expressed as enrichment in F and SO_3 that occurred during acid leaching of apatite. The REE available in the phosphogypsum are therefore highly dependent on the monazite grade in the primary phosphate ore, feed to the phosphate beneficiation plants. Tabular gypsum and octahedral Ca-Al-fluoride crystals both form during phosphate rock leaching, simultaneously adsorb substantial TREO contents (0.17% and 4.88% respectively) from dissolved REE-bearing fluoroapatite. Gypsum constitutes the bulk phase of all samples with a total abundance of more than 92%. Even though the TREO contents are relatively low, the absolute concentrations of REE hosted by the mineral are significant but preliminary tests deemed the recovery of these REE uneconomic [17, 19, 21, 23, 25, 26, 31, 32]. Regarding the EPMA data used in this study it is important to note that the REE concentrations in gypsum are relatively low and, in some cases, close or even below the individual detection limit. The data and concentrations reported in this article must therefore be considered as semi-quantitative. The data is suitable for metallogenic modelling purposes as well as for geometallurgical interpretations but should not be used for detailed economic evaluation such as pre-feasibility studies or feasibility studies.

Monazite accounts for 50 to 60% of the total Rare Earth Elements (TREEs) in Catalão phosphogypsum, based on the combined results from EPMA and mineral distribution analysis. Consequently up to 50% of TREEs are locked by gypsum and Ca-Al-fluorides and are therefore considered economically non-recoverable.

The monazite from Catalão phosphogypsum is well suited for further beneficiation of the mineral using physical separation techniques before isolation of the REE [3, 39, 63]. Key properties are:
a) comparably high abundance of the mineral in the bulk material (0.6 wt%),
b) low ThO_2 contents (0.4 wt%) and low (<1%) accessory mineral abundance (ilmenite, Fe-oxides) and the absence of acid-consuming carbonates,
c) very good to almost complete particle liberation (60–80%), rare intergrown textures, and compact mineral shapes,
d) similar particle size distribution of monazite with grain sizes generally between 15 to 50 μm, thus avoiding an additional comminution stage,

e) no necessity for direct or additional mining operation and no technical interference with the phosphate fertilizer process,

Our study thus demonstrates that monazite from Catalão phosphogypsum could potentially play an important role for expanding the REE supply from waste materials. The combination of detailed petrographic investigations, SEM-based mineral distribution, EPMA and geochemical analyses was successfully applied to identify REE carrier minerals and the fraction of associated REE. Especially the geometallurgical methods of SEM-EDX based mineral distribution analyses using AMICS has proven to be scientifically reliable and economically relatively fast and thus feasible. Transferability of the workflow employed in this study to asses other global phosphogypsum occurrences largely depends on the in-situ mineralogical material composition and REE-fixation.

References

[1] Rudnick RL, Gao S. Composition of the continental crust. In: Rudnick RL, editor. Treatise on Geochemistry. Oxford: Elsevier; 2003, p. 1–64.

[2] Orris GJ, Grauch RI. Rare earth element mines, deposits, and occurrences: U.S. Geological Survey, Open-File Report 02-189; 2002.

[3] Krishnamurthy N, Gupta C. Extractive metallurgy of rare earths. Boca Raton, London, New York: CRC Press Taylor & Francis Group; 2016.

[4] Weng Z, Jowitt SM, Mudd GM, Haque N. A detailed assessment of global rare earth element resources: opportunities and challenges. Economic Geology 2015;110(8):1925–52.

[5] Lottermoser BG. Rare-earth element mineralisation within the Mt. Weld carbonatite laterite, Western Australia. Lithos 1990;24(2):151–67.

[6] Kanazawa Y, Kamitani M. Rare earth minerals and resources in the world. Journal of Alloys and Compounds 2006;408–412:1339–43.

[7] Haque N, Hughes A, Lim S, Vernon C. Rare earth elements: overview of mining, mineralogy, uses, sustainability and environmental impact. Resources 2014;3(4):614–35.

[8] U.S. Geological Survey. Mineral commodity summaries. Reston, USA: U.S. Geological Survey; 2019.

[9] Zhou B, Li Z, Chen C. Global potential of rare earth resources and rare earth demand from clean technologies. Minerals 2017;7(11):203.

[10] European Commission. Communication from the commission to the European Parliament, the Council, the European Economic and Social Committee and the Committee of the regions. 2014th ed. Brussels; 2014.

[11] Voncken JHL. The Rare Earth Elements: An Introduction. 1st ed. Cham, s.l.: Springer International Publishing; 2016.

[12] Binnemans K, Jones PT, Blanpain B, van Gerven T, Pontikes Y. Towards zero-waste valorisation of rare-earth-containing industrial process residues: A critical review. J. o, Cleaner Production 2015;99:17–38.

[13] Rutherford PM, Dudas MJ, Samek RA. Environmental impacts of phosphogypsum. Science of The Total Environment 1994; 149 (1–2): 1–38.

[14]   Tayibi H, Choura M, López FA, Alguacil FJ, López-Delgado A. Environmental impact and management of phosphogypsum. Journal of environmental management 2009;90(8):2377–86.
[15]   Yang J, Liu W, Zhang L, Xiao B. Preparation of load-bearing building materials from autoclaved phosphogypsum. Construction and Building Materials 2009;23(2):687–93.
[16]   Parreira AB, Kobayashi ARK, Silvestre OB. Influence of Portland Cement Type on Unconfined Compressive Strength and Linear Expansion of Cement-Stabilized Phosphogypsum. J. Environ. Eng. 2003;129(10):956–60.
[17]   Liang H, Zhang P, Jin Z, DePaoli D. Rare earths recovery and gypsum upgrade from Florida phosphogypsum. Minerals & Metallurgical Processing 2017;34(4):201–6.
[18]   Sinha S, Abhilash, Meshram P, Pandey BD. Metallurgical processes for the recovery and recycling of lanthanum from various resources – A review. Hydrometallurgy 2016;160:47–59.
[19]   Lokshin EP, Tareeva OA, Elizarov IR. Agitation leaching of rare earth elements from phosphogypsum by weak sulfuric solutions. Theo. Found. o. Chem. Eng. 2016;50(5):857–62.
[20]   Lokshin EP, Tareeva OA. Production of high-quality gypsum raw materials from phosphogypsum. Russ J Appl Chem 2015;88(4):567–73.
[21]   Al-Thyabat S, Zhang P. In-line extraction of REE from Dihydrate (DH) and HemiDihydrate (HDH) wet processes. Hydrometallurgy 2015;153:30–7.
[22]   Al-Thyabat S, Zhang P. REE extraction from phosphoric acid, phosphoric acid sludge, and phosphogypsum. Mineral Processing and Extractive Metallurgy 2015;124(3):143–50.
[23]   Lokshin EP, Tareeva OA, Elizarova IR. Sorption of rare-earth elements from phosphogypsum sulfuric acid leaching solutions. Theo. Found. o. Chem. Eng. 2015;49(5):773–8.
[24]   Valkov AV, Andreev VA, Anufrieva AV, Makaseev YN, Bezrukova SA, Demyanenko NV. Phosphogypsum Technology with the Extraction of Valuable Components. Procedia Chemistry 2014;11:176–81.
[25]   Kulczycka J, Kowalski Z, Smol M, Wirth H. Evaluation of the recovery of Rare Earth Elements (REE) from phosphogypsum waste – case study of the WIZÓW Chemical Plant (Poland). Journal of Cleaner Production 2016;113:345–54.
[26]   Preston JS, Cole PM, Craig WM, Feather AM. The recovery of rare earth oxides from a phosphoric acid by-product. Part 1: Leaching of rare earth values and recovery of a mixed rare earth oxide by solvent extraction. Hydrometallurgy 1996;41(1):1–19.
[27]   Rychkov VN, Kirillov EV, Kirillov SV, Semenishchev VS, Bunkov GM, Botalov MS et al. Recovery of rare earth elements from phosphogypsum. Journal of Cleaner Production 2018;196:674–81.
[28]   Todorovsky D, Terziev A, Milanova M. Influence of mechanoactivation on rare earths leaching from phosphogypsum. Hydrometallurgy 1997; 45 (1–2): 13–9.
[29]   Walawalkar M, Nichol CK, Azimi G. Process investigation of the acid leaching of rare earth elements from phosphogypsum using HCl, HNO3, and H2SO4. Hydrometallurgy 2016;166: 195–204.
[30]   Habashi F. The recovery of the lanthanides from phosphate rock. J Chem Tech Biotechnol 1985;35(1):5–14.
[31]   Lokshin EP, Tareeva OA, Elizarova IP. A study of the sulfuric acid leaching of rare-earth elements, phosphorus, and alkali metals from phosphodihydrate. Russ J Appl Chem 2010;83 (6):958–64.
[32]   Lokshin EP, Tareeva OA, Elizarova IP. Processing of phosphodihydrate to separate rare-earth elements and obtain gypsum free from phosphates and fluorides. Russ J Appl Chem 2011;84 (9):1461–9.
[33]   Papageorgiou F, Godelitsas A, Xanthos S, Voulgaris N, Nastos P, Mertzimekis TJ et al. Characterization of phosphogypsum deposited in Schistos remediated waste site (Piraeus,

Greece). In: Merkel BJ, editor. Uranium – past and future challenges: Proce. o. t. 7th Intern. Conf. o. Uranium Mining a. Hydrogeol. Cham: Springer; op. 2015, p. 271–280.

[34] Rajkovic M, Toskovic D. Phosphogypsum surface characterisation using scanning electron microscopy. Acta per tech 2003(34):61–70.

[35] Koukouzas N, Vasilatos C. Mineralogical and chemical properties of FGD gypsum from Florina, Greece. J. Chem. Technol. Biotechnol. 2008;83(1):20–6.

[36] Dutrizac JE. The behaviour of the rare earth elements during gypsum (CaSO 4 ·2H 2 O) precipitation. Hydrometallurgy 2017;174:38–46.

[37] Mazzilli B, Palmiro V, Saueia C, Nisti MB. Radiochemical characterization of Brazilian phosphogypsum. Journal of environmental radioactivity 2000;49(1):113–22.

[38] Le Bourlegat FM, Saueia CHR, Mazzilli BP, Fávaro DIT. Metals concentration in phosphogypsum and phosphate fertilizers produced in Brazil using INAA: INAC 2009 – International Nuclear Atlantic Conference. Rio de Janeiro: Instituto de Pesquisas Energeticas e Nucleares (Brazil) Brazil; 2009.

[39] Brückner L, Elwert T, Schirmer T. Extraction of Rare Earth Elements from Phospho-Gypsum: Concentrate Digestion, Leaching, and Purification. Metals 2020;10(1):131.

[40] Brod JA, Junqueira-Brod TC, Gibson SA, Thompson RN, Hildor J, Moraes LCD et al. The Kamafugite-Carbonatite Association in the Alto Paranaíba Igneous Province (APIP) Southeastern Brazil. Revista Brasileira de Geociências 2017;30(3):408–12.

[41] Gibson SA, Thompson RN, Leonardos OH, Dickin AP, Mitchell J G. The Late Cretaceous Impact of the Trindade Mantle Plume: Evidence from Large-volume, Mafic, Potassic Magmatism in SE Brazil. Journal of Petrology 1995;36(1):189–229.

[42] Cordeiro PFdO, Brod JA, Palmieri M, Oliveira CG de, Barbosa ESR, Santos RV et al. The Catalão I niobium deposit, central Brazil: Resources, geology and pyrochlore chemistry. Ore Geology Reviews 2011;41(1):112–21.

[43] Cordeiro PFO, Brod JA, Dantas EL, Barbosa ESR. Mineral chemistry, isotope geochemistry and petrogenesis of niobium-rich rocks from the Catalão I carbonatite-phoscorite complex, Central Brazil. Lithos 2010;118(3):223–37.

[44] Ribeiro CC. Geologia, geometalurgia, controles e gênese dos depósitos de fósforo, terras raras e titânio do complexo carbonatítico Catalão I GO. Brasilia, Universidade de Brasilia: Ph.D. thesis; 2008.

[45] Imbernon RAL, Oliveira SMB, Figueiredo A. Concentração dos ETR nos produtos de alteração intempérica do complexo alcalino-carbonatítico de Catalão I. GO: Boletim Geociências Centro-Oeste 1994;17:25–8.

[46] Maria Barros de Oliveira S, Aparecida Liguori Imbernon R. Weathering alteration and related REE concentration in the Catalão I carbonatite complex, central Brazil. J. o. S. Am. Earth Sc. 1998;11(4):379–88.

[47] Walter A-V, Nahon D, Flicoteaux R, Girard JP, Melfi A. Behaviour of major and trace elements and fractionation of REE under tropical weathering of a typical apatite-rich carbonatite from Brazil. Earth and Planetary Science Letters 1995;136(3):591–602.

[48] Tassinari MML, Kahn H, Ratti G. Process mineralogy studies of Corrego do Garimpo REE ore, catalao-I alkaline complex, Goias, Brazil. Minerals Engineering 2001;14(12):1609–17.

[49] Neumann R. Caracterização tecnológica dos potenciais minérios de terras-raras de Catalão I, GO. São Paulo, Universidade de São Paulo: Ph.D. thesis; 1999.

[50] Cordeiro PFdO. Petrologia e metalogenia do depósito primário de nióbio do complexo carbonatítico-foscorítico de Catalão I, GO. Brasilia, Universidade de Brasília: M.Sc. thesis; 2009.

[51] Pereira L, Birtel S, Möckel R, Michaux B, Silva AC, Gutzmer J. Constraining the Economic Potential of By-Product Recovery by Using a Geometallurgical Approach: The Example of Rare Earth Element Recovery at Catalão I, Brazil. Econ Geol 2019.

[52] Jordens A, Cheng YP, Waters KE. A review of the beneficiation of rare earth element bearing minerals. Minerals Engineering 2013;41:97–114.

[53] Janoušek V, Farrow CM, Erban V. Interpretation of Whole-rock Geochemical Data in Igneous Geochemistry: Introducing Geochemical Data Toolkit (GCDkit). Journal of Petrology 2006;47 (6):1255–9.

[54] Gu Y. Automated Scanning Electron Microscope Based Mineral Liberation Analysis An Introduction to JKMRC/FEI Mineral Liberation Analyser. JMMCE 2003;02(01):33–41.

[55] El Zrelli R, Rabaoui L, Daghbouj N, Abda H, Castet S, Josse C et al. Characterization of phosphate rock and phosphogypsum from Gabes phosphate fertilizer factories (SE Tunisia): high mining potential and implications for environmental protection. Environmental science and pollution research international 2018;25(15):14690–702.

[56] May A, Sweeney JW. Assessment of Environmental Impacts Associated with Phosphogypsum in Florida. In: Kuntze RA, editor. The chemistry and technology of gypsum. Philadelphia, Pa: American Society for Testing and Materials; 1984, 116–139.

[57] Ni Y, Hughes JM, Mariano AN. Crystal chemistry of the monazite and xenotime structures. American Mineralogist 1995; 80 (1–2): 21–6.

[58] Boatner LA. Synthesis, Structure, and Properties of Monazite, Pretulite, and Xenotime. Reviews in Mineralogy and Geochemistry 2002;48(1):87–121.

[59] Mitchell RS, Swanson S, Crowley JK. Mineralogy of a deeply weathered perrierite-bearing Pegmatite, Bedford County, Virginia. Southeast. Geol. 1976;18(37–47).

[60] Wall F, Zaitsev AN. Rare earth minerals in Kola carbonatites. In: Wall F, Zaitsev AN, editors. Phoscorites and carbonatites from mantle to mine: The key example of the Kola alkaline province. London: Mineralogical Society; 2004, p. 341–373.

[61] Toledo MCM de, Olivera SMB de, Fontan F, Ferraril VC, Parseval P de. Mineralogia, morfologia e cristaloquímica da monazita de Catalão I (GO, Brasil). Revista Brasileira de Geociências 2004;34(1):135–46.

[62] Mariano AN, Mariano A. Rare Earth Mining and Exploration in North America. Elements 2012;8(5):369–76.

[63] Chen W, Honghui H, Bai T, Jiang S. Geochemistry of Monazite within Carbonatite Related REE Deposits. Resources 2017;6(4):51.

[64] Forster H-J, Harlov DE, Milke R. Composition and Th-U-total Pb ages of huttonite and thorite from Gillespie's Beach, South Island, New Zealand. The Canadian Mineralogist 2000;38(3):675–84.

[65] Foerster H-J. The chemical composition of REE-Y-Th-U-rich accessory minerals in peraluminous granites of the Erzgebirge-Fichtelgebirge region, Germany; Part I, The monazite -(Ce)-brabantite solid solution series. American Mineralogist 1998; 83 (3–4): 259–72.

[66] Foerster HJ, Harlov DE. Monazite-(Ce)-huttonite solid solutions in granulite-facies metabasites from the Ivrea-Verbano Zone, Italy. Mineralogical Magazine 1999;63(4):587–94.

[67] Ribeiro CC, Brod JA, Junqueira-Brod TC, Gaspar JC, Petrinovic IA. Mineralogical and field aspects of magma fragmentation deposits in a carbonate–phosphate magma chamber: evidence from the Catalão I complex, Brazil. J. o. S. Am. Earth Sc. 2005; 18 (3–4): 355–69.

[68] Lehr JR, Frazier AW, Smith JP. Phosphoric Acid Impurities, Precipitated Impurities in Wet-Process Phosphoric Acid. J. Agric. Food Chem. 1966;14(1):27–33.

[69] Coates RV, Woodard, G. D. Similarity between "Chukhrovite" and the Octahedral Crystals found in Gypsum in the Manufacture of Phosphoric Acid. Nature 1966;212(5060):392.

[70] Walenta K. Chukhrovit-(Ce) und Rhabdophan-(Ce) aus der Grube Clara bei Oberwolfach im mittleren Schwarzwald. Chem. Erde 1979;38:331–9.

[71] Matta S, Stephan K, Stephan J, Lteif R, Goutaudier C, Saab J. Phosphoric acid production by attacking phosphate rock with recycled hexafluosilicic acid. International Journal of Mineral Processing 2017;161:21–7.

[72] M. Mathew, S. Takagi, K. R. Waerstad, A. W. Frazier. The crystal structure of synthetic chukhrovite, Ca4AlSi(SO4)F13 · 12H2O. American Mineralogist 1981; 66 (3–4): 392–7.

[73] Vignola P, Hatert F, Bersani D, Diella V, Gentile P, Risplendente A. Chukhrovite-(Ca), Ca4.5Al2 (SO4)F13·12H2O, a new mineral species from the Val Cavallizza Pb-Zn-(Ag) mine, Cuasso al Monte, Varese province, Italy. Euro. J. o. Min. 2012;24(6):1069–76.

[74] Anthony JW, Bideaux RA, Bladh KW, Nichols MC. Handbook of Mineralogy: Volume V – Borates, Carbonates, Sulfates. Tuscon, Arizona: Mineral Data Publishing.

[75] Tostevin R, Shields GA, Tarbuck GM, He T, Clarkson MO, Wood RA. Effective use of cerium anomalies as a redox proxy in carbonate-dominated marine settings. Chemical Geology 2016;438:146–62.

[76] Mackay DAR, Simandl GJ, Ma W, Redfearn M, Gravel J. Indicator mineral-based exploration for carbonatites and related specialty metal deposits – A QEMSCAN® orientation survey, British Columbia, Canada. Journal of Geochemical Exploration 2016;165:159–73.

[77] Pooler R, Dold B. Optimization and Quality Control of Automated Quantitative Mineralogy Analysis for Acid Rock Drainage Prediction. Minerals 2017;7(1):12.

[78] Prasad MNV. Resource Potential of Natural and Synthetic Gypsum Waste. In: Prasad MNV, Shih K, editors. Environmental materials and waste: Resource recovery and pollution prevention. Amsterdam [u.a.]: Elsevier AP; 2016, p. 307–337.

[79] Papastefanou C, Stoulos S, Ioannidou A, Manolopoulou M. The application of phosphogypsum in agriculture and the radiological impact. Journal of environmental radioactivity 2006;89(2):188–98.

[80] Fandrich R, Gu Y, Burrows D, Moeller K. Modern SEM-based mineral liberation analysis. International Journal of Mineral Processing 2007; 84 (1–4): 310–20.

[81] Miller JD, Lin C-L, Hupka L, Al-Wakeel MI. Liberation-limited grade/recovery curves from X-ray micro CT analysis of feed material for the evaluation of separation efficiency. International Journal of Mineral Processing 2009;93(1):48–53.

[82] van den Boogaart, K. G., Tolosana-Delgado R. Predictive Geometallurgy: An Interdisciplinary Key Challenge for Mathematical Geosciences. In: Daya Sagar BS, Cheng Q, Agterberg F, editors. Handbook of Mathematical Geosciences: Fifty Years of IAMG. Cham: Springer International Publishing; Springer; 2018, p. 673–686.

Kássia L. L. Marinho, Bruno A. M. Figueira, Dorsan S. Moraes,
Oscar J. C. Fernandez, Marcondes L. da Costa

Chapter 17
The Mn oxides tailing from Amazon Region as low-cost raw material to synthesis of shigaite-type phase

Abstract: The work herein describes the synthesis of layered double hydroxide with shigaite structure from Mn tailings found in the Kalunga dam (Carajás Mineral Province, Brazil). The characterization results showed that the tailings were chemically formed by MnO (54,015 wt.%), Al_2O_3 (12,454 wt.%), Fe_2O_3 (9,061 wt.%) and SiO_2 (7,003 wt.%). Mineralogically, they were composed by birnessite, cryptomelane, lithiophorite, kaolinite, gibbsite and anatase. After the development of the synthesis process, Mn tailings were converted into shigaite without impurities with the ratio $Mn^{2+}:Al^{3+}= 2:1$, temperature of 85 °C and 24 h of reaction time. The lamellar product was characterized by diagnostic FTIR bands in the 4000–400 cm^{-1} range, thermal stability above 180 °C, hexagonal plate morphology around 5 μm and 0.365 nm lattice fingers. The results demonstrated that the use of abundant low-cost starting material was adequate to obtain a rare phase belonging to the manganous layered double hydroxide.

Keywords: Mn residues, transformation, layered double hydroxide, shigaite

Acknowledgments: Authors thank to Brazilian National Council for Scientific and Technological Development (CNPQ-420169/2016-4; 305015/2016-8; and 442871/2018-0), Coordination for the Improvement of Higher Education Personnel (CAPES 88881.160695), Centro de Microscopia (UFMG) and LAMIGA Laboratories (IG/UFPA) by financial and technical support.

Kássia L. L Marinho, Bruno A. M. Figueira, Universidade Federal do Oeste do Pará, Programa de Pós Graduação em Sociedade, Ambiente e Qualidade de Vida, Santarém-Pará, Brazil
Dorsan S. Moraes, Universidade Federal do Pará, Instituto de Geociências, Belém-Pará, Belém-PA, Brazil
Oscar J. C. Fernandez, Instituto Federal do Pará, Programa de Pós Graduação em Engenharia de Materiais, Belém-PA, Brazil
Marcondes L. da Costa, Universidade Federal do Pará, Programa de Pós-Graduação em Geologia e Geoquímica, Belém-Pará, Brazil

https://doi.org/10.1515/9783110674941-017

1 Introduction

Mn mining tailings from Azul Manganese Deposit in Carajás, Amazon Region, are products of medium-grade ores beneficiation processes, often of fine particles (<100 μm) and high MnO content (>35 Wt. %), composed of manganese oxyhydroxides (cryptomelane, birnessite, todorokite, lithiophorite, nsutite and pyrolusite) as mineral ores, aluminosilicates (kaolinite and muscovite), hematite and goethite as ganga. The tailings are stored in open air dams causing serious environmental problems [1–3].

During the last decade, considerable attention has been given to the characterization or conversion of tailings into products such as catalysts, adsorbents for organic pollutants, ion exchangers, nanocomposites, zeolites, bricks, geopolymers and supplementary cementitious materials [4–13]. In addition, the Mn tailings from Azul Mine have been reported as an excellent raw material for the obtention of octahedral layer manganese oxide and molecular sieve, as well as cathodic material of o-LiMn$_2$O$_3$ [2, 3]. Nonetheless, no work has been published in what regards the use of this material in the synthesis of layered double hydroxide (LDH) with a shigaite structure. Thus, this study decided to reach this goal. According to [14–17], the shigaite-type phase may be employed as the only material in catalysis, environmental remediation, cement hydration and emulsifiers in oil/water emulsions. Shigaite consists of oxycations sheets interspersed with oxyanions sheets. It is a rare clay belonging to the family of manganous layered double hydroxides, nominally $[Me^{2+}_{1-x} Me^{3+}_{x} (OH)_2]^{x+} [A^{r-}_{x/r} \cdot nH_2O]^{x-}$, where M^{2+} is occupied by Mn^{2+}, Me^{3+} by Al^{3+}, Fe^{3+}, Cr^{3+}, Mn^{3+} and Ga^{3+} at octahedral sites (brucite-like layers) and A^{r-} by SO_4^{2-} or other interlayer exchanger anions such as Cl^-, ClO_4^-, NO_3^-, CO_3^{2-} and organic molecules [18–22].

The aim of this work was the synthesis of shigaite, using for that Mn tailings from the Carajás Mineral Province (Amazon Region) as a low-cost starting material through co-precipitation route and control of the synthesis parameters (ratio Mn^{2+}:Al^{3+}, temperature and time).

2 Experimental

2.1 Synthesis by co-precipitation route

Mn tailing represented by sample MnTai was employed as raw material for transformation into shigaite-type material. Firstly, around 50 g of sample was extracted, dried at room temperature, pulverized and homogenized. Optimized hydrothermal synthesis route conditions is presented in Figure 17.1. Approximately 1 g of MnTai was added to HCl: H2O (1:1) solution and heated at 100 °C to obtain a rich solution in Mn^{2+} (SOLMn). The stoichiometric ratio of Mn^{2+}/Al^{3+} (2:1) ideal for obtaining shigaite was reached adding 1.75 g of $Al_2(SO_4)_3{*}6H_2O$ (Sigma Aldrich) to the solution (SOLMn). Then, approximately 3.5 mol.L^{-1} of NaOH solution was slowly added under vigorous stirring at 85 °C and pH ~ 11.2 for 24 h. Finally, the obtained product was filtered, washed thoroughly several times with deionized water and dried at room temperature.

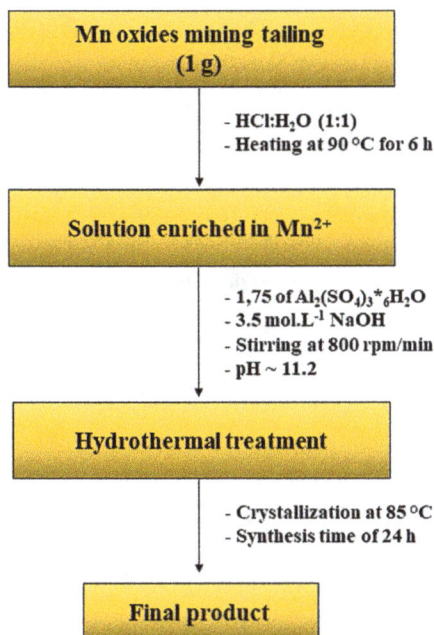

Figure 17.1: Flowchart of synthesis of shigaite-type material.

2.2 Chemical and mineral characterization

The chemical characterization of the tailings sample was performed by X-rays fluorescence using the Panalytical spectrometer, Axios Minerals, and the mineralogical one was performed by X-rays diffraction (XRD) in a Burker diffractometer, model D2Phaser, with copper CuKα = 1.5406 Å) at 30 kV and 10 mA. The XRD measurement was also used to characterize the synthesis products. In addition, the final shigaite product was analyzed by FT-IR, using for that a Bruker infrared spectrometer, Vertex 70 model, while the thermal behavior was monitored by TG-DTG-DSC curves using a thermal analyzer from Thermal Sciences, model PL-ST, at a heating rate of 15 °C min^{-1} to 1050 °C, which employed synthetic air as a gas carrier. The morphological characterization of shigaite was performed by a scanning electron microscope (SEM) from Veja Tescan, which was coupled to semi-quantitative chemical analysis by X-ray dispersive energy (EDX) under a cold field emission gun environment, operating at 20 kV and 10 µA. Additional characterization of the microstructure was made in a high-resolution transmission electron microscope (HR-TEM) from Tecnai-G2TF20 operating at 200 KV to obtain photomicrographs and selected-area electron diffraction (SAED) of the layered phase.

3 Results and discussion

The XRD pattern of the MnTai sample is shown in Figure 17.2, in which it can be observed that the tailings were mainly composed of Mn minerals, such as cryptomelane (PDF 044-1386), lithiophorite (PDF 041-1378) and birnessite (PDF 043-1456). Other minerals such as kaolinite (PDF 029-1488), hematite (PDF 033-0664), gibbsite (PDF 033-0018) and anatase (PDF 021-1272) were also observed. The chemical composition of MnTai was analyzed and showed that it was predominantly formed by MnO (54,015 wt.%), Al_2O_3 (12,454 wt.%), Fe_2O_3 (9,061 wt.%), SiO_2 (7,003 wt.%), besides MgO (0.315) and LOI (15.220 wt.%). These chemical and mineralogical data revealed that the Mn tailings used in this work were appropriate for the process of Mn-LDH synthesis.

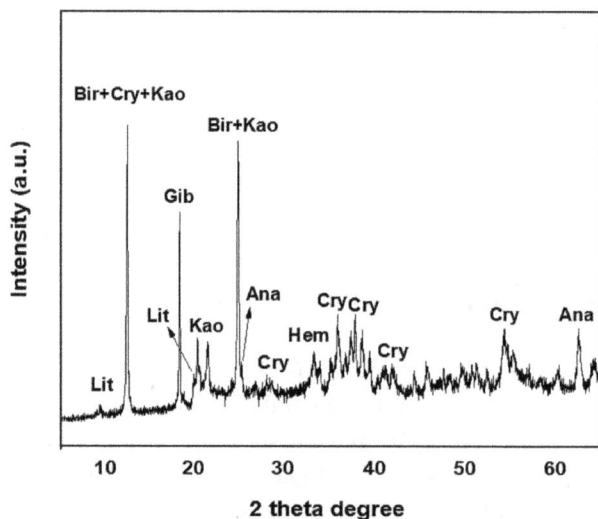

Figure 17.2: X-ray diffraction pattern for whole Mn tailings sample (MnTai). (Lit: lithiophorite; Bir: birnessite; Cry: cryptomelane; Kln = kaolinite; Gib = gibbsite; Ana = anatase; Hem = hematite).

Figure 17.3 shows DRX patterns of products obtained in this work that investigated the influence of the $Mn^{2+}:Al^{3+}$ ratio in the shigaite synthesis in 24 h at 85 °C. It was possible to observe that the Mn tailings were completely converted into shigaite (PDF 01−086-1398) in the 1.5, 2, 3 and 4 ratios, being observed greater crystallinity and ordering in the layers where the $Mn^{2+}:Al^{3+}= 2$ ratio was applied. These results were in good agreement with studies related to the synthesis of shigaite that used other sources of Mn in their synthetic routes [14–22]. An interlayer distance of 10.631 Å at the $Mn^{2+}Al^{3+} = 1.5$ ratio was also observed, which varied on average to ~11 Å with an increase in the ratio $Mn^{2+}:Al^{3+}$ to 4. The DRX patterns of the products synthesized with the $Mn^{2+}:Al^{3+} = 5$ and 8 revealed the formation of a new product with interlamellar distance of 7 Å with low ordering of layers and degree of crystallinity.

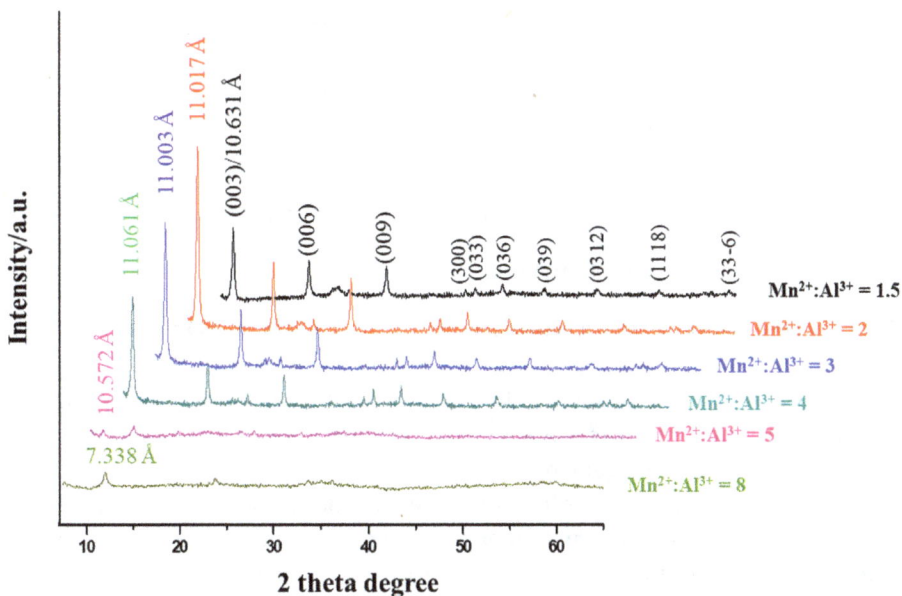

Figure 17.3: XRD patterns of products synthesized at 85 °C during 2 days at distinct $Mn^{2+}:Al^{3+}$ (1.5, 2, 3, 4, 5 and 8) ratios.

On the other hand, products synthetized at different temperatures (5, 25, 85, 150 e 190 °C) showed different phases after the XRD analysis (Figure 17.4). The DRX pattern at 5 °C showed the presence of a typically amorphous phase, exhibiting two broad peaks with low definition and intensity at 16 and 22 ° (2 theta). At 25 °C, the XRD pattern presented well-defined peaks of around 8.43, 16.44, 20.06, 24.58, 32.92, 33.93, 36.90, 41.42 and 46.92° (2 theta), related to the shigaite planar lattices (006), (113), (024), (300), (217), (306), (134) and (3012) with an inter-lamellar spacing of 10.556 Å. As the hydrothermal treatment increased to 85 °C, the intensity peaks corresponding to the basal planes (003) and (006) have also increased, as well as the inter-lamellar spacing to 11.064 Å. At higher temperatures (150 and 190 °C), the shigaite XRD patterns disappeared and a new phase was formed, which became more crystalline at 190 °C. Thus, these results showed that the ideal temperature for shigaite formation is below 90 °C.

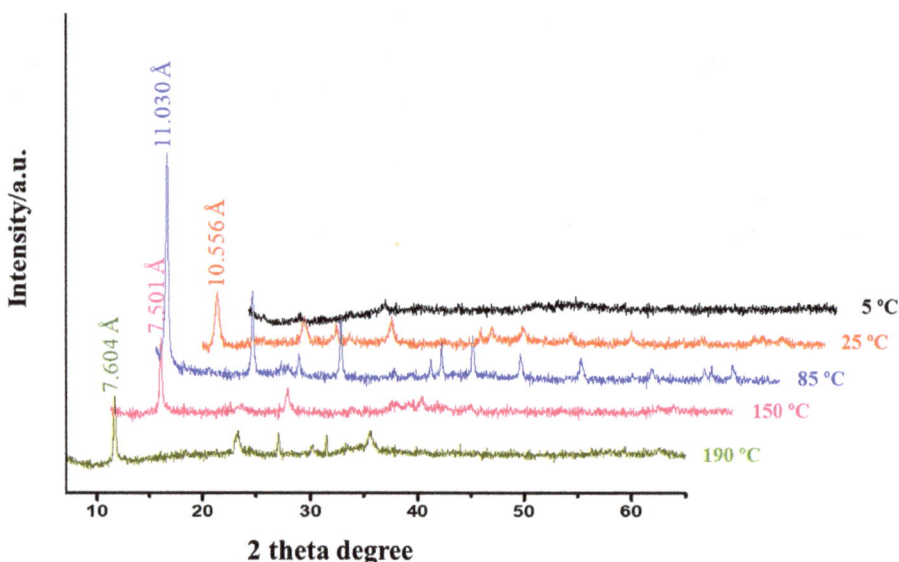

Figure 17.4: XRD patterns of the synthesized products obtained at 5, 25, 85, 150 and 190 °C.

The degree of crystallization of the products synthesized at 85 °C was evaluated at different running times and proved to be very sensitive (Figure 17.5). After 24 h, the XRD pattern of the shigaite phase with the highest crystallinity and ordering of the octahedral layers was identified. The same XRD profile of the lamellar phase remained the same during the hydrothermal time of 48 h, but it showed a lower degree of crystallinity. In a shorter time (6 h), it was not possible to observe the formation of any crystalline phase, whilst in a longer time (~96 h), a lamellar product was formed, however with an inter-lamellar spacing of 7.341 Å, remaining stable up to 168 h of hydrothermal treatment, despite the appearance of impurities at 27 and 30 ° (2 theta). Thus, the ideal crystallization running time for shigaite was 24 h at 85 °C.

The previously obtained shigaite morphology showing high crystallinity and no impurities was investigated by scanning electron microscopy (SEM-EDS) and high-resolution transmission electron microscopy (HR-TEM), as shown in Figure 17.6.

The results revealed that the lamellar phase presented a typical double lamellar hydroxide morphology that consisted of hexagonal plate crystals with a size variation ranging from 0.5 a 2.5 μm (Figure 17.6a), similar to what has already been reported for the morphology (SEM) of shigaite and its synthetic analogue [18–25], but with a slightly smaller crystal size. The chemical composition (Figure 17.6b) obtained by EDX indicated the presence of Mn, Al, Fe, Na, O, S, Cl and C, which could be related to the ionic species Mn^{2+}, Al^{3+}, Fe^{3+}, Na^+, SO_4^{2-}, Cl^- and CO_3^{2-}, in accordance with the chemical formula of the shigaite mineral. It is important to draw attention to the presence of Fe^{3+} ions originating from Mn tailings that may be in the

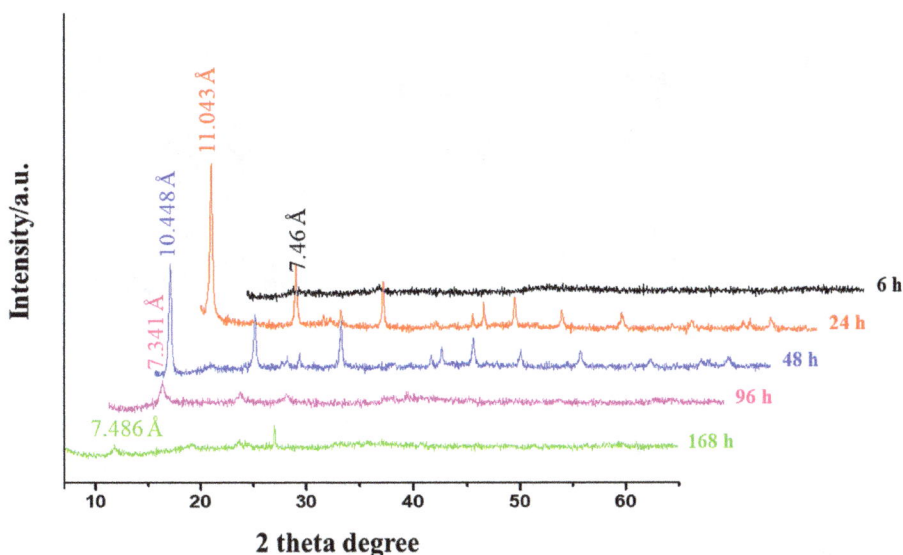

Figure 17.5: XRD patterns of the samples obtained for 6 h, 24 h, 48, 96 and 168 h at 85 °C.

octahedral sites in the layers of the brucite sheet of the lamellar structure. In addition, the ratio $Mn^{2+}/Al^{3+} + Fe^{3+} = 2.55$ was reasonably close to the stoichiometric formula of shigaite $[Mn_6Al_3(OH)_{18}]^{3+}$ $[(SO_4)_2Na \cdot _{12}H_2O]^{3-}$ [14]. The microstructure analysed by HR-TEM corroborated with the SEM data by clearly showing a morphology of hexagonal or pseudo hexagonal plates parallel to each other. The length of these hexagonal plates ranged from 150 to more than 500 nm (Figure 17.6c). The calculated network spacing fringes were ~0.365 nm, which corresponded to the shigaite plane (009) (Figure 17.6d), suggesting that the crystals grow along the 00 *l* direction. The inset in Figure 17.6d corresponded to the SAED standard, confirming the polycrystalline nature of the material and corroborating the XRD data [24].

The FTIR spectrum in the 4000–400 cm^{-1} range of the shigaite obtained at 85 °C for 24 h is shown in Figure 17.7. The broad and intense band at 3437 cm^{-1} may be correlated with the υ-OH stretching vibrations linked to the interlayer anionic species (SO_4^{2-}, CO_3^{2-} and Cl). The next band visualized at 1661 cm^{-1} referred to the H_2O molecules between the layers greatly linked to the sulphate and hydrogen ions. The band at 1455 cm^{-1} were assigned to the CO_3^{2-} asymmetric stretching vibration. Stretching vibrations at around 1195, 1145, 1105 (υ_3), 955 (υ_1) and 605 cm^{-1} (υ_4) were attributed to asymmetric S-O bonds of distorted tetrahedral SO_4^{2-} in Na-shigaite. Additionally, the bands of the FTIR spectrum at 770, 620, 535 and 421 cm^{-1} could be well correlated to the O-M-O vibration modes (M = $Mn^{2+,}$ Al^{3+}, Fe^{3+}) [15, 25–28]. Additionally, the FTIR spectrum bands at 770, 620, 535 and 421 cm^{-1} could be well correlated to the O-M-O vibration modes ((M = Mn^{2+}, Al^{3+}, Fe^{3+}) [15, 25–28].

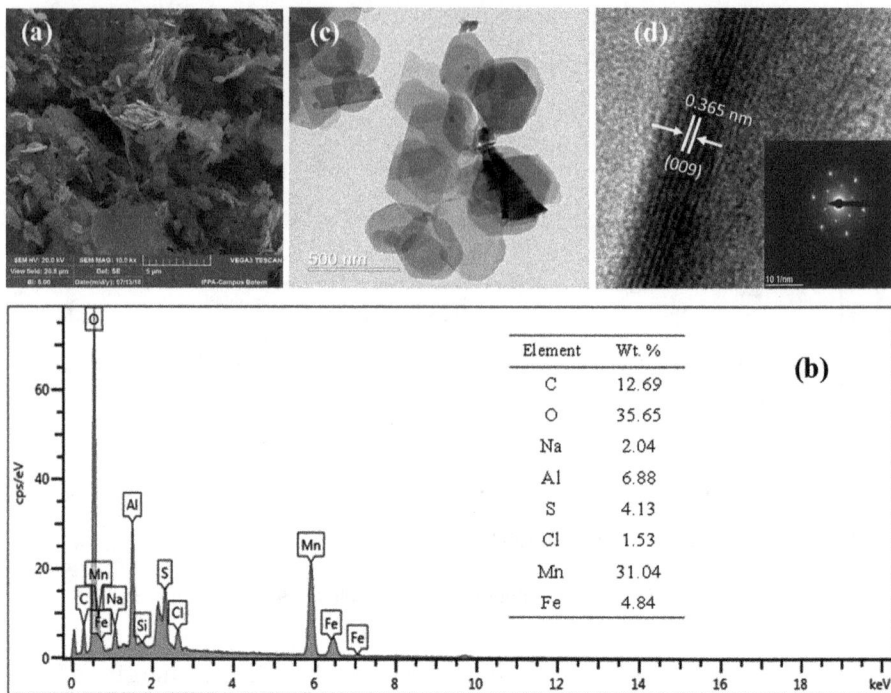

Figure 17.6: SEM photomicrograph (a), EDS analysis (b), HR-TEM micrograph (c) and SAED pattern (d) of shigaite.

Element	Wt. %
C	12.69
O	35.65
Na	2.04
Al	6.88
S	4.13
Cl	1.53
Mn	31.04
Fe	4.84

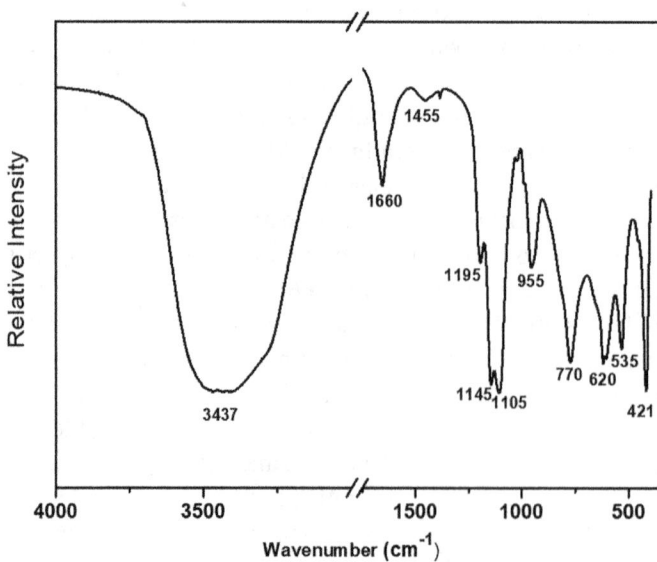

Figure 17.7: FT-IR spectrum of the final product synthesized at 85 °C for 24 h.

The thermal behaviour of the product previously characterized was monitored by TG-DTG-DSC curves and XRD analysis (Figure 17.8). An initial mass loss of 11.45% between room temperature and 175 °C (Figure 17.8a) was identified. In this temperature

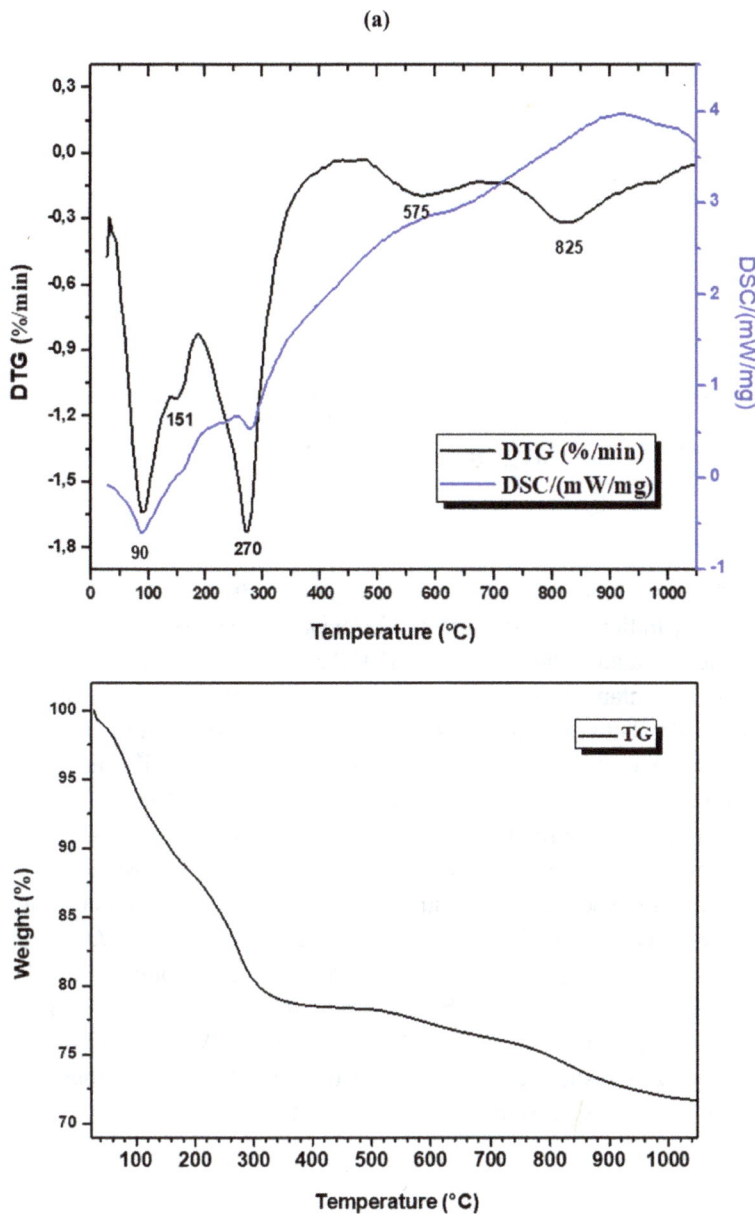

Figure 17.8: TG-DTG-DSC curves of the final product synthesized at 85 °C for 24 h (a); and thermal behavior of sample monitored by XRD measurements between 25–250 °C (b).

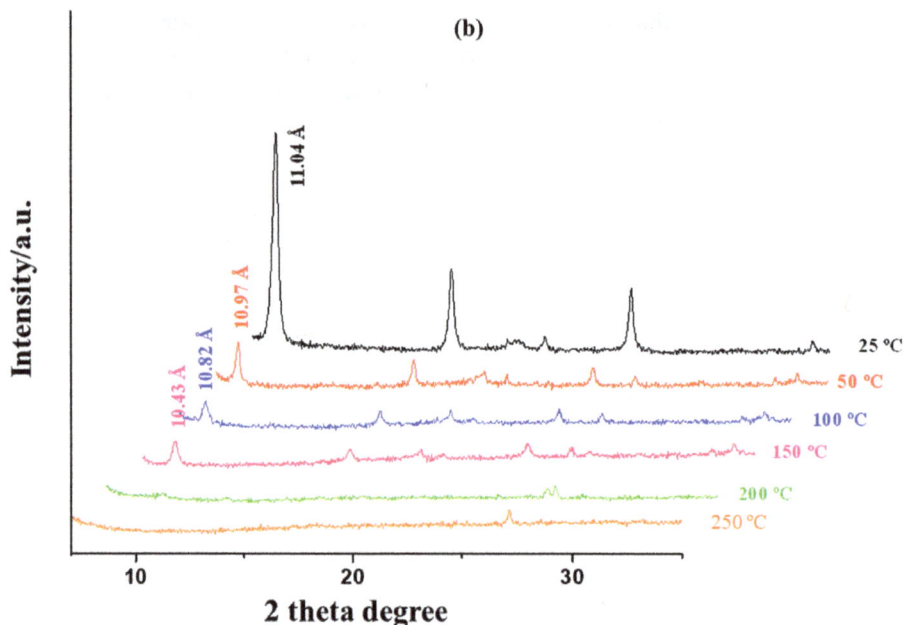

Figure 17.8 (continued)

range, the DTG-DSC curves showed two endothermic peaks at 90 and 150 °C, indicating a continuous dehydration of the shigaite interlayer water. The XRD patterns of the shigaite samples heat-treated at 25, 50, 100 and 150 °C confirmed the reduction of the interlayer space and maintenance of the lamellar structure (Figure 17.8b). A continuous weight loss of 8.45% between 175 and 270 °C because of the loss of water remaining in the interlayer space and dehydroxylation of the $[Mn_6Al_3(OH_{18})]^{3+}$ sheet was observed. At 200 °C, the XRD pattern showed only traces of shigaite. The total loss of mass from the interlayer dehydration process (19.79%) was slightly greater than the calculated theoretical loss (18.92%). The weight loss of 7.47% and intense peak between 230 and 345 °C represented the collapse of shigaite and its conversion into amorphous phase, which was confirmed by XRD analysis (Figure 17.8b). At higher temperatures, the endothermic peaks around 575 and 825 °C corroborated with the thermal events already reported for the shigaite structure [14, 25–28], revealing its transformation into $MnAl_2O_4$ and MnO phases. The monitoring by XRD analysis of the sample heat-treated at 250 °C showed a typical pattern of amorphous phase, thus suggesting that the limit stability of the structure was at 200 °C.

4 Conclusion

The synthesis of hexagonal shigaite-like material using Mn tailings from the Azul Mine dam in Carajás Province Mineral was successfully achieved through a co-precipitation chemical process and hydrothermal treatment. The synthesis parameters (Mn^{2+}/Al^{3+} ratio, temperature and synthesis time) strongly affected the final product of Mn layered double hydroxide, which showed an average particle size between 150 and 500 nm. Thermal stability above 150 °C has been reported. Furthermore, the results showed that an undesirable product from the Mn oxide ore beneficiation industry could be converted into an interesting double lamellar hydroxide, a potential value-added product.

References

[1]   Verendra S, Tarun C, Sunil K T. A Review of Low-Grade Manganese Ore Upgradation Processes. Miner. Procc. and Extractive Metall Rev, 2019, 1–2.
[2]   Bruno A M Figueira, Romulo S Angelica, Marcondes L da Costa, Herbert Poellmann, Karla S. Conversion of different Brazilian manganese ores and residues into birnessite -like phyllomanganate, Applied Clay Science, 2013, 86, 54–58.
[3]   Kamilla C M, Bruno A M Figueira, Thays C C L, Oscar J C F, Pio C G, Jose M R M. Hydrothermal synthesis of o-LiMnO$_2$ employing Mn mining residues from Amazon (Brazil) as starting material. Material Letters: X, 2019, 1000123.
[4]   Pereira M J, Lima F F, Lima R M F. Calcination and characterisation studies of a Brazilian manganese ore tailing. International Journal of Mineral Processing, 2014, 131, 26–30.
[5]   Changxin L, Hong Z, Shuai W, Jianrong X, Zhenyu Z. Removal of basic dye (methylene blue) from aqueous solution using zeolite synthesized from electrolytic manganese residue. Journ. Ind. And Eng. Chem, 2015, 23, 344–352.
[6]   Jirong L, Yan S, Ping H, Yaguang D, Wei Z, Tian C Z, Dongyun D. Using Electrolytic Manganese Residue to prepare novel nanocomposite catalysts for efficient degradation of Azo Dyes in Fenton-like process. Chemosphere, 2020, 252, 12487.
[7]   Yaguang W, Shuai G, Xiaoming Liu, Binwen T, Emile M, Na Z. Preparation of non-sintered permeable bricks using electrolytic manganese residue: environmental and NH$_3$-N recovery benefits. Journ. Hazard. Mat. 2019, 378, 120768.
[8]   Dengquan W, Qiang Qiang W, Junfeng X. Reuse of hazardous electrolytic manganese residue: Detailed leaching characterization and novel application as a cementous material. Resources, Conservation and Recycling, 2020, 154, 104645.
[9]   Jia L, Ying L, Xiangke J, Peng S, Jianxin L, Lage W, Tian C Z. Electrolytic manganese dioxide residue based autoclaved bricks with Ca(OH)2 and thermal-mechanical activated K-feldspar additions. Construction and Building Materials, 2020, 230, 116848.
[10] Jia L, Peng S, Jiaxin L, Ying L, Hengpeng Y, Li S, Dongyun D. Synthesis of electrolytic manganese residue-fly ash based geopolymers with high compressive strength. Construction and Building Materials, 2020, 248, 118489.
[11] Xiunan C, Yanjuan Z, Huayu H, Zuqiang H, Yanzhen Y, Xingtang L, Yuben Q, Jing L. Journal of cleaner production. 2020, 258, 120741.

[12] Koivula R, Pakarinen J, Sivenius M, Sirola K, Harjula R, Paatero E. Use of hydrometallurgical wastewater as a precursor for the synthesis of cryptomelane -type manganese dioxide ion exchange material. Separation and purification technology, 2009, 70, 53–57.

[13] Jiancheng S, Renlong L, Haiping W, Zuohua L, Xiaolong S, Changyuan T. Adsorption of methylene blue on modified electrolytic manganese residue: Kinetics, isotherm, thermodynamics and mechanism analysis. Journ. of the Taiwan Institute of Chemical Eng., 2018, 82, 351–359.

[14] Uwe Koenig & Herbert Poellmann Synthesis, Properties and Characterisation of Manganeous Layered Double Hydroxides Using In Situ X-ray Techniques. Mat. Sci. Forum, 2004, 443–444, 307–310.

[15] Lilian F M A, Rilton A F, Fernando W. K-shigaite -like layered double hydroxide particles as Pickering emulsifiers in oil/water emulsions. Applied Clay Sci., 2020, 193, 105660.

[16] Herbert Poellmann & Oberste-Padtberg R. Manganese in High Alumina. ICCC Durban, S. 2002–2011, Vol. 4, (2003).

[17] Herbert Poellmann & Oberste-Padtberg R. Manganese in High Alumina cement (HAC). International Conference on Calcium Aluminate Cements, Edinburgh, Scotland 16.-19. Juli, S. (2001).

[18] Uwe Koenig & Herbert Poellmann Properties and synthesis of alkali-substituted shigaite. 21st General Meeting of IMA South Africa, S. 15, (2014)

[19] Uwe Koenig & Herbert Poellmann Synthesis and in situ X-ray investigation of manganeous Layered Double Hydroxides (LDHs) intercalated with inorganic anions. 21st General Meeting of IMA South Africa, S. 14, (2014)

[20] Uwe Koenig & Herbert Poellmann Synthesis, Properties and Characterisation of Manganeous Layered Double Hydroxides Using In Situ X-Ray Techniques. EPDIC 8, S. 307–310), (2002)

[21] Uwe Koenig & Herbert Poellmann Properties and synthesis of alkali-substituted shigaites. Organizează Simpozionul: Mediul-Cercetare, Protectie si Gestiune. Universitäta "Babes – Bolyai"
Cluj-Napoca, S. 60. (2002).

[22] Uwe Koenig & Herbert Poellmann Synthesis, Properties and Characterisation of Manganeous Layered Double Hydroxides Using in Situ X-ray Techniques. Epdic-8, Upsala, S.150, (2002).

[23] Pring A, Slade P G, Birch W D. Shigaite from Iron Monarch, South Australia. Mineralogical Magazine, 1992, 56, 417–419.

[24] Anne R S, Neffer A G Gomez, Suelen C. Silva, Fernando W. Layered Double Hydroxides with the Composition $Mn/Al-SO_4-A$ (A = Li, Na, K; Mn: Alca. 1:1) as Cation Exchangers. J. Braz. Chem. Soc, 2019, 30, 1807–1813.

[25] Neffer A G G, Anne R S, Fernando W. Layered double hydroxides with the composition $[Mn_6Al_3 (OH)_{18}]$ $[(HPO_4^{2-})_2A^+].yH_2O$ (A^+ = Li, Na or K) obtained by topotactic exchange reactions. Applied Clay Sci., 2020, 193, 105658.

[26] Sumio A, Hidetoshi H, Hiroaki U, Satoshi T, Eiichi N. Synthesis and Thermal Decomposition of Mn-Al Layered Double Hydroxides. Journal of Solid State Chemistry, 2002, 167, 152–159.

[27] Anne R S, Fernando W. Converting Mn/Al layered double hydroxide anion exchangers into cation exchangers by topotactic reactions using alkali metal sulfate solutions. Chem Comm, 2019, 55, 7824–7828.

[28] Anne R S, Neffer A G Gomez, Fernando W. Thermogravimetric analysis of layered double hydroxides intercalated with sulfate and alkaline cations $[M_6^{2+}Al_3(OH)_{18}][A^+(SO_4)_2].12H_2O$ (M^{2+} = Mn, Mg, Zn; A^+ = Li, Na, K). J. Therm. Anal. and Calorimetry, 2020, 140, 1715–1723.

Leonardo Boiadeiro Ayres Negrão, Herbert Pöllmann,
Marcondes Lima da Costa

Chapter 18
Eco-cements out of Belterra Clay: An extensive Brazilian bauxite overburden to produce low-CO$_2$ eco-friendly calcium sulphoaluminate based cements

Abstract: Among the strategies for CO$_2$ reduction in the field of sustainable low-CO$_2$ cementitious materials, some new cement types, instead of ordinary Portland cements (OPC), are discussed. Calcium-sulphoaluminate (CSA) or ye'elimite dominating cement types, like Belite-Calcium-Sulphoaluminate (BCSA), Belite-Calcium-Sulphoaluminate-ferrite (BCSAF or BYF), Belite-Ternesite-Calcium-Sulphoaluminate (BTCSA or BTY), have comparable properties to OPC, but with the advantages of much less CO$_2$ emission during their production and 15% enlowered clinkering temperature. Nevertheless, CSA large-scale production is limited due to the expensiveness of bauxite and its need as a raw material for the aluminum industry. Therefore, the promising new CSA cement types need alternative Al-rich raw materials. Belterra Clay (BTC), an alumina-rich clay overburden on the bauxites of Brazilian Amazon, can be considered as an important raw material for the production of CSA eco-cement types. The wide distribution of BTC in the Amazon region, its high alumina contents, simple mineralogy and wide distribution direct on the surface makes it an easy-to-exploit and encouraging raw material to produce CSA cement types. Preliminary results, using Belterra Clay from Rondon do Pará in Eastern Amazon/Brazil, show the formation of different CSA clinkers with approximately 35% of ye'elimite at 1250°C. Belite, ternesite, ferrite, Fe-perovskite and other minor phases can also be present in variable contents. The clinkers show fast hydration when mixed to gypsum. The produced CSA binders save about 30% of CO$_2$ emissions in comparison to OPC production due to mineral formation. Less energy consumption is expected due to 200°C lower clinkering temperatures and easier clinker grindability. The use of Belterra Clay, a mining overburden, to produce CSA cement types is highlighted.

Leonardo Boiadeiro Ayres Negrão, Institute of Geosciences and Geography, Mineralogy/ Geochemistry, Martin Luther University Halle-Wittenberg, Von-Seckendorff-Platz 3, 06120 Halle, Germany, e-mail: boiadeiro.negrao@gmail.com
Herbert Pöllmann, Institute of Geological sciences, mineralogy/geochemistry, Martin-Luther-University Halle, Halle (Saale), Germany
Marcondes Lima da Costa, Institute of Geosciences, Universidade Federal do Pará, Belém-PA, Brazil

https://doi.org/10.1515/9783110674941-018

Keywords: Belterra Clay; calcium sulphoaluminate; Eco-cement; ye'elimite; belite; CO_2 reduction; ternesite; bauxite; residue.

The following cement notation will be used in this chapter: C = CaO; S = SiO_2; A = Al_2O_3; F = Fe_2O_3; s = SO_3; H = H_2O; T = TiO_2

1 Introduction

Cements are a needful building material in the construction industry and thanks to the rising world population and urbanization it increases demand will last far in the future (Figure 18.1). One of the main challenges in the cement industry has been the reduction of CO_2 emissions, as this sector is responsible for roughly 8% of the global CO_2 output [1, 2]. Such significant emissions highlighted the importance to create new approaches for the sector considering the 2 °C scenario of the Paris Agreement negotiated in 2015. The main CO_2-mitigation strategies by the cement sector include the optimization of industrial processes, use of alternative (and less carbon-intensive) fuels, the addition of calcined clays in the clinker, and the progressive use of alternative binders to the world-used Ordinary Portland Cement (OPC) [2, 3]. The research on new alternative binders has a particular concern, considering that only the clinkerization process (limestone calcination) is responsible for up to 70% of CO_2 generated in the OPCs production.

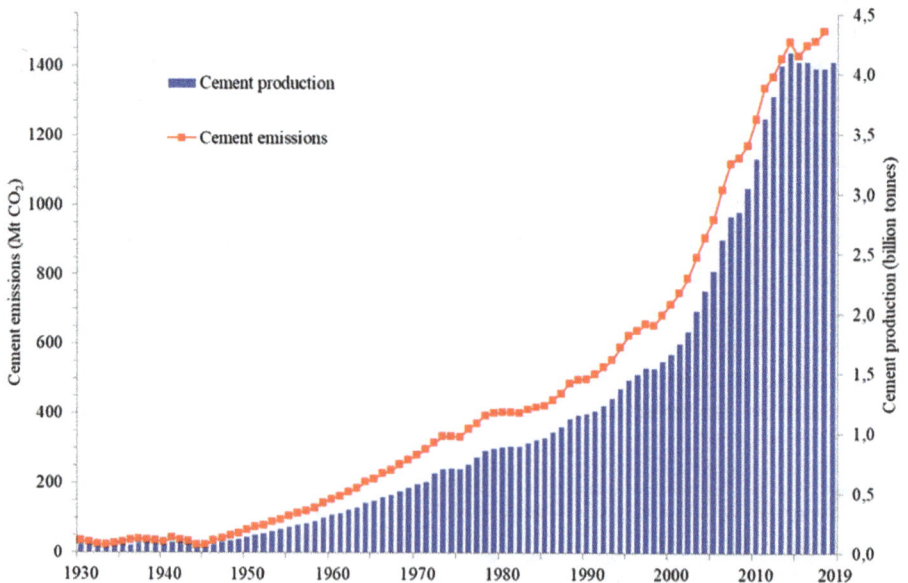

Figure 18.1: Global cement production and its CO_2 emissions from 1930 to 2019 [1, 4, 5].

Belterra Clay (BTC) is a widespread clay covering most of the bauxite deposits of the Brazilian Amazon region [6–9]. The material lies on the top of plateaus held by mature lateritic profiles, has been ground to agricultural (mostly soya bean) and cattle pasture expansions in Eastern Amazon, and investigated for the production of red ceramic [10, 11]. BTC's high alumina contents and related mineralogical composition set it as a promissory raw material for the upcoming new generation of low-CO_2 binders.

Calcium Sulphoaluminate based cements (CSA), including the belite-CSA (BCSA) types, are of special interest as they can be produced using lower $CaCO_3$ amounts in the kiln mix, under reduced temperatures (up to 300 °C lower than OPC's) and are, therefore, easier to grind [12–15]. Only these features stand CSA as one of the most promissory low-CO_2 cements with considerably lower CO_2 emissions when compared to OPC's (Figure 18.2). CSA's faster hardening, higher strength development, slight expansion, and lower alkalinity are among further special advantages [16, 17].

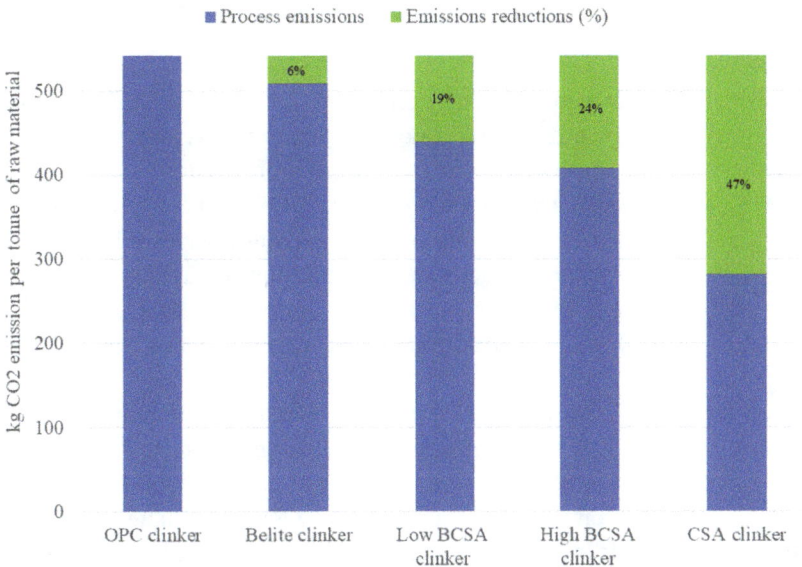

Figure 18.2: CO_2 emissions per binding material of selected clinkers. Source: Gartner and Sui [18].

To form the main CSA phase ye'elimite (C_4A_3s), limestone (CaO source) is mixed with calcium sulfate and alumina-rich raw materials, the most obvious being bauxite. The high prices of bauxite have though limited large CSA production throughout the years, with a notable larger production only in China during the seventies [19]. In order to overcome the high bauxite prices limitation, several works on the use of industrial wastes and byproducts to produce CSA-based cements have been reported [17, 20–23]. The composition of such byproducts is obviously different of bauxite, and the frequent iron and silica impurities result in clinkers with ye'elimite as the main hydration component of CSA, but also with high contents of belite (C_2S), ferrite (C_4AF), and several other minor phases. BCSA (i.e. BYF, belite-ye'elimite-ferrite) cements [18], for example, can be generally understood as a member of the extensive family of CSA-based cements, with ye'elimite as the main early-hydration phase, and high amounts of less reactive belite.

BTC's large uniform distribution, high alumina contents, known mineral assemblage, and clayey to silty fineness rises it as a possible raw material for the close-future cement industry. Preliminary investigations have shown that this material can be used as raw material for CSA-based cements. In this chapter, we present the basis of these studies and some preliminary results.

2 Belterra Clay – an overview

2.1 Distribution

Bauxite deposits in the Amazon region are commonly covered by yellowish to reddish clays that can reach up to 25 m depth, known as Belterra Clay, a term attributed by Sombroek (1966) after he observed similar clays, but not related to bauxites in the region of Belterra, in the Brazilian lower Amazon. BTC is widespread on the top of extensive plateaus in Amazonia, mainly covering bauxite reserves (Figure 18.3). Some of these reserves configure large mines in operation, including the one in Porto Trombetas-PA (Mineração Rio do Norte-MRN), in Juruti-PA (ALCOA), and in Paragominas-PA (HYDRO), while others are still in the research phase as in Rondon do Pará-PA.

Large amounts of BTC are relocated during the bauxite strip mining method (Figure 18.4) when BTC together with all other overburden material (e.g. forest, topsoil, and part of the lateritic profile) need to be removed to access the bauxite ore. The huge amounts of BTC on the bauxites are decisive for the bauxite mining, as thicker BTC packages result in more overburden relocation and, consequently extra mining cost. After bauxite mining, the disturbed BTC lays on the surface until the mined areas are rehabilitated.

The extent of the BTC in the Amazon region and its association with bauxite deposits fomented a range of studies to explain the genesis of BTC [6, 7, 24–30]. Although its geological origin is controversial, Belterra Clay is generally related to latosols (e.g. oxisols in the USDA soil taxonomy, or ferrasols in the World Reference Base for Soil Resources). The BTC close relation with the underneath laterite-bauxitic profile is clearly in the field and further attested by mineralogical and chemical analysis, suggesting that BTC origin is related to intense weathering processes under tropical environment, possibly of the underneath laterite-bauxitic profile, followed by local transport and deposition [6, 8, 31].

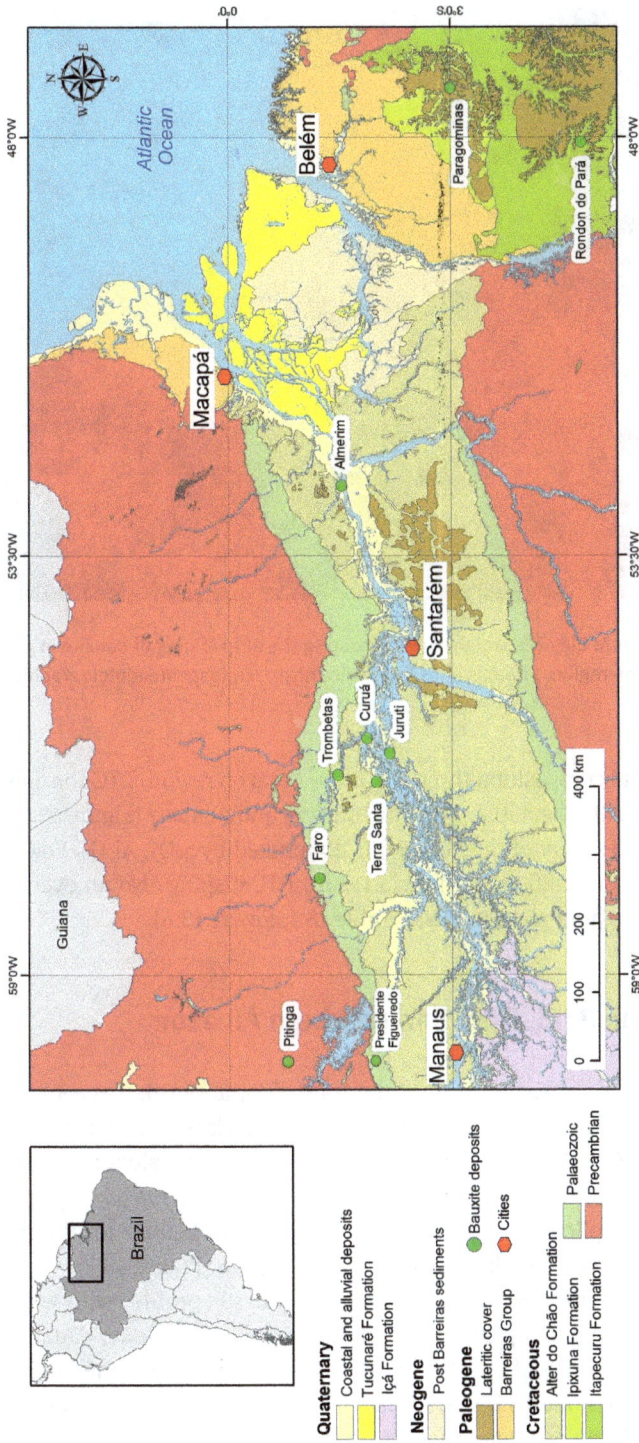

Figure 18.3: Simplified geological framework of the eastern Amazon with the main bauxite deposits, all covered with Belterra Clay [8]. Licensed under CC BY 4.0 https://creativecommons.org/licenses/by/4.0/.

Figure 18.4: Relocation of Belterra Clay (yellowish material) during the strip mining of bauxite in Porto Trombetas, lower Amazon region. Photo: Carlos Penteado, http://carlospenteadofotographia. blogspot.com/.

With a uniform mineral composition throughout the Amazon region, BTC consists basically of kaolinite, gibbsite, hematite, goethite, quartz, and anatase (Figure 18.5). Consequently, the main chemistry of the material is dominated by SiO_2, Al_2O_3, Fe_2O_3, and TiO_2, besides the loss on ignition (LOI). As a result, BTC display also an overall close chemical composition from place to place (Table 18.1, Figure 18.6).

2.2 BTC in Rondon do Pará, Brazilian easthern Amazon

Rondon do Pará, a county in the eastern part of the state of Pará, Brazil, was set as a prospective bauxite producer in the last decade due to its world-class bauxite reserves. These reserves of at least 642 Mt of bauxite (grading 42.7% available alumina), integrate the larger Paragominas Bauxitic District in the region [32]. During the ore exploration campaigns, three large pilot mines (i.e. trenches) were performed for initial ore evaluation, exposing up to 13 m thick BTC layers in the region (Figure 18.7).

The BTC in Rondon do Pará is on the top of laterite-bauxitic profiles with well-defined horizons, from bottom to top composed of: bauxitic clay, compact bauxite, ferroaluminous crust, dismantled ferroaluminous crust, ferruginous spheroliths, and a horizon with bauxitic nodules (Figure 18.8). The BTC is reddish brown at the base to yellow in the top, fairly homogeneous with no visible sedimentary structures,

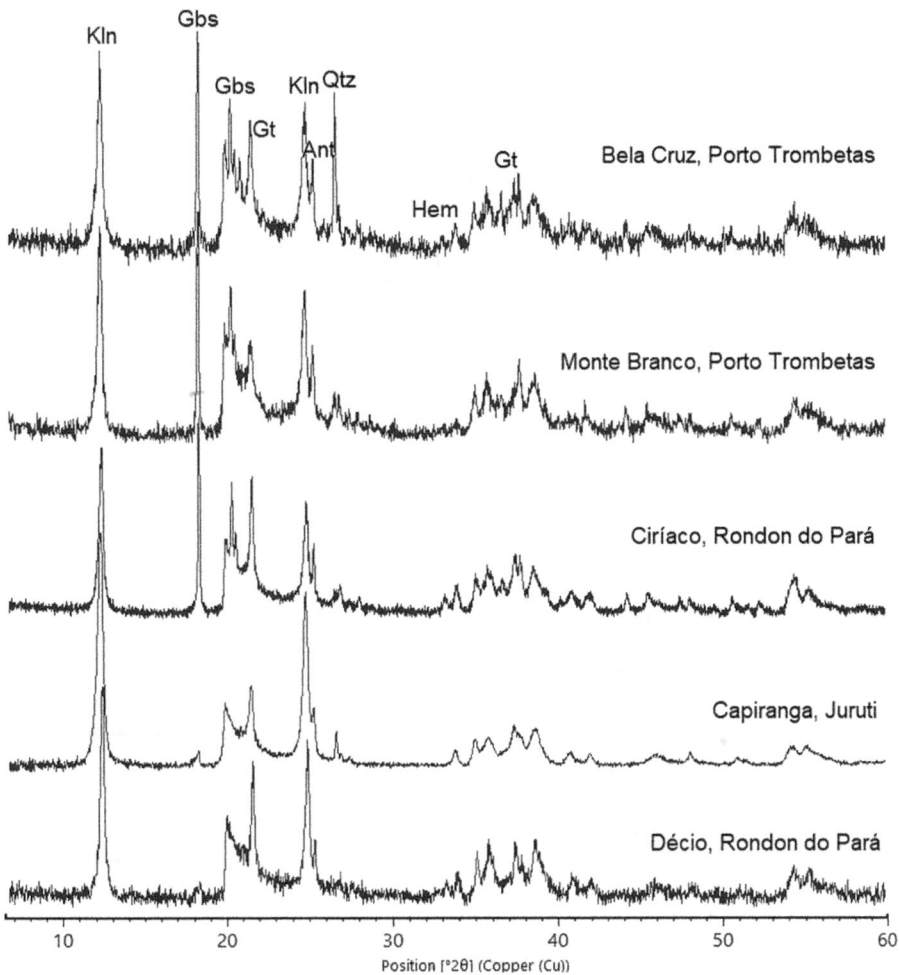

Figure 18.5: Comparison of X-ray diffractograms of BTC (all from the bottom layers) from different localities within the Brazillian Amazon region. Kln: kaolinite; Gbs: gibbsite, Gt: goethite; Ant: anatase; Hem: hematite; Qtz: quartz.

silt to clayey granulometry, and with bauxite fragments at its base [8]. The material is fairly continuous, with an average thickness close to 10 m (Figure 18.9).

As in other localities throughout Amazon, the BTC in Rondon do Pará shares the same mineralogy of the underlying bauxite, with only different mineral proportions. The main mineral phase is notably kaolinite, representing around 75% of the BTC composition in two pilot mines (Décio and Branco) and close to 60% in the pilot mine Ciríaco, where the more aluminous composition is partially explained by relatively higher gibbsite contents, reaching up to 20% (Figure 18.10). The main iron-rich phase is goethite, with contents close to 15%, followed by some few

Table 18.1: Chemical composition of BTC in different localities throughout the Brazilian Amazon.

City	Locality	Depth	Al_2O_3	SiO_2	Fe_2O_3	TiO_2	LOI	Total
Rondon do Pará	Branco	0.8	33.7	35.9	12.5	2.4	15.1	99.7
		5.5	34.1	35.8	12.7	2.5	14.6	99.7
		10.5	33.8	35.2	13	2.6	15.1	99.7
	Ciríaco	1.0	36.4	31.6	12.2	2.6	17	99.8
		5.0	37.1	30.2	12.5	2.7	17.2	99.7
		12	38.6	27.7	12.9	2.7	17.7	99.7
	Décio	0.8	33.3	36.3	12	2.5	15.2	99.3
		4.5	34.1	36.1	12.4	2.5	14.7	99.8
		10	34.2	35.6	13.1	2.6	14.4	99.8
Juruti	Capiranga	1.5	34.1	36.4	9	2.9	17.4	99.8
		9.5	32.9	35	8.1	2.7	21.1	99.7
		16	35.4	37.8	8.2	2.9	15.5	99.7
Paragominas	PA 256	0.5	36.4	38.1	5.9	2.1	17	99.5
		1.8	40.4	35.2	3.9	2.4	17.7	99.6
Oriximiná	Jabuti	1.0	40.1	26.5	6.6	3.2	22.9	99.2
		2.5	43.1	24.6	5.9	3.2	22.6	99.3
		3.5	45.1	22.1	5.7	3.3	23.2	99.3
	Cruz Alta	–	40.8	27	8.4	3	19.9	99.9

Source: Rondon do Pará [8]; Juruti [31]; Oriximiná [26].

hematite 1–2%, richer in the deepest parts of BTC. Anatase (~3%) is the main Ti-carrier, with a relatively homogeneous distribution, and quartz occurs as residual grains, normally not reaching more than 1% [8].

Microscopically (Figure 18.11), the BTC is composed of nanometric (~250 nm) pseudo-hexagonal kaolinite, goethite, and anatase with undistinguished morphology. BTC seems fairly homogeneous concerning its chemical composition, mineralogical and textural aspects in the three pilot mines [8].

3 Synthesis of CSA-based clinkers out of Belterra Clay

CSA cements contain ye'elimite (C_4A_3s) as basic calcium aluminum sulfate [19] and the main hydraulic phase. Further common components occurring together might include, but are not restricted to: dicalcium silicate in the natural mineral analog structures of larnite (β-C_2S) or flamite (α-C_2S); tetracalcium iron aluminate (C_2A,F) as brownmillerite; calcium sulfosilicate (C_5S_2s) as ternesite; calcium aluminates (C_3A, CA, $C_{12}A_7$); and calcium silicoaluminates (C_2AS as gehlenite, and CAS_2). In the cement industry C_2S is commonly referred to as belite, and $C_2(A,F)$ as ferrite.

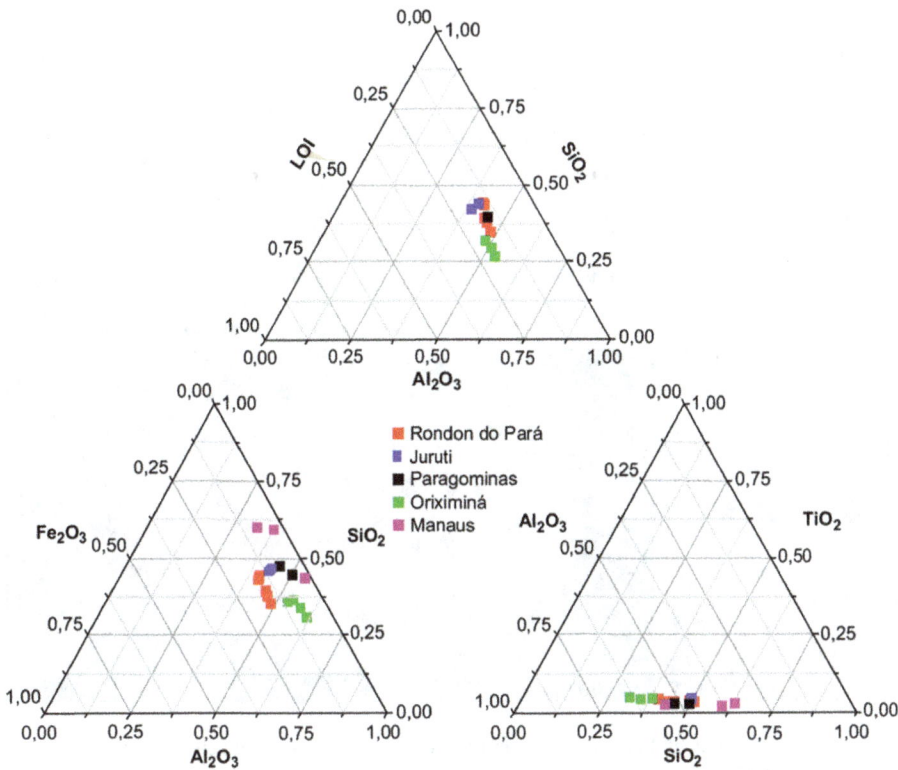

Figure 18.6: Similar chemistry composition of Belterra Clay from different locations within the Amazon region [8]. Licensed under CC BY 4.0 https://creativecommons.org/licenses/by/4.0/.

Here we use the natural mineral analog notation for these phases to better differentiate the present polymorph.

Limestone is the main source of CaO due to its high contents of calcite, $CaCO_3$; gypsum, $CaSO_4 \cdot 2H_2O$, gives CaO and SO_3; normally bauxite is used as a source of SiO_2 and Al_2O_3. Bauxites are essentially composed of aluminum hydroxide minerals such as gibbsite, $Al(OH)_3$ and/or boehmite, $AlO(OH)$, frequently with low silica contents, as kaolinite $Al_2[Si_2O_5/(OH_4)]$, and associated iron oxyhydroxides commonly represented by goethite, $FeO(OH)$, and hematite, Fe_2O_3.

The high alumina/silica ratios of bauxite in the raw meal promote the formation of ye'elimite in restriction of calcium silicoaluminates such as CAS_2, undesirable in CSA clinkers as it has no hydraulic activity. Berrio et al. [37] investigated the importance of the alumina to silica equilibrium in the production CSA cements, observing that clinkers produced (at 1250 °C) with ratios lower than 2 had CAS_2 contents exceeding ye'elimite. On the other hand, the 2-ratio limit should not be strict, as complex chemical phase interactions when using secondary raw materials can favor the formation of C_2S.

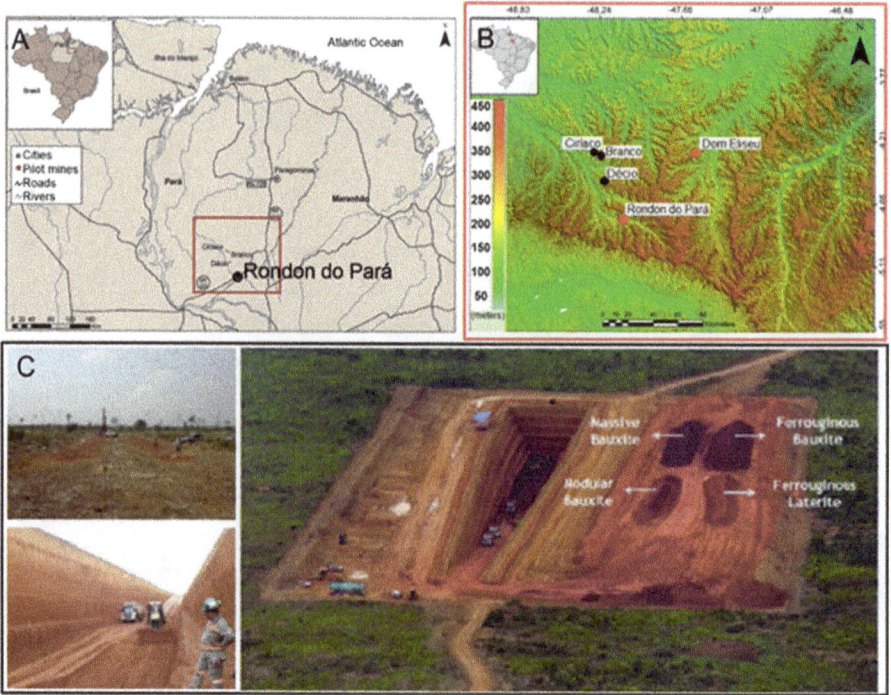

Figure 18.7: A–Location of Rondon do Pará and respective pilot mines conducted for bauxite evaluation. B: SRTM image of the area showing the bauxitic plateaus where BTC occurs on the top. C– Aerial and in situ photographs of the pilot mine excavations [33].

Figure 18.8: A: Belterra Clay expositions at Rondon do Pará. B: underlying lateritic bauxite profile (LP) and its summarized description (left) in the Ciríaco pilot mine. Source: [8]. CC BY 4.0 https://creativecommons.org/licenses/by/4.0/.

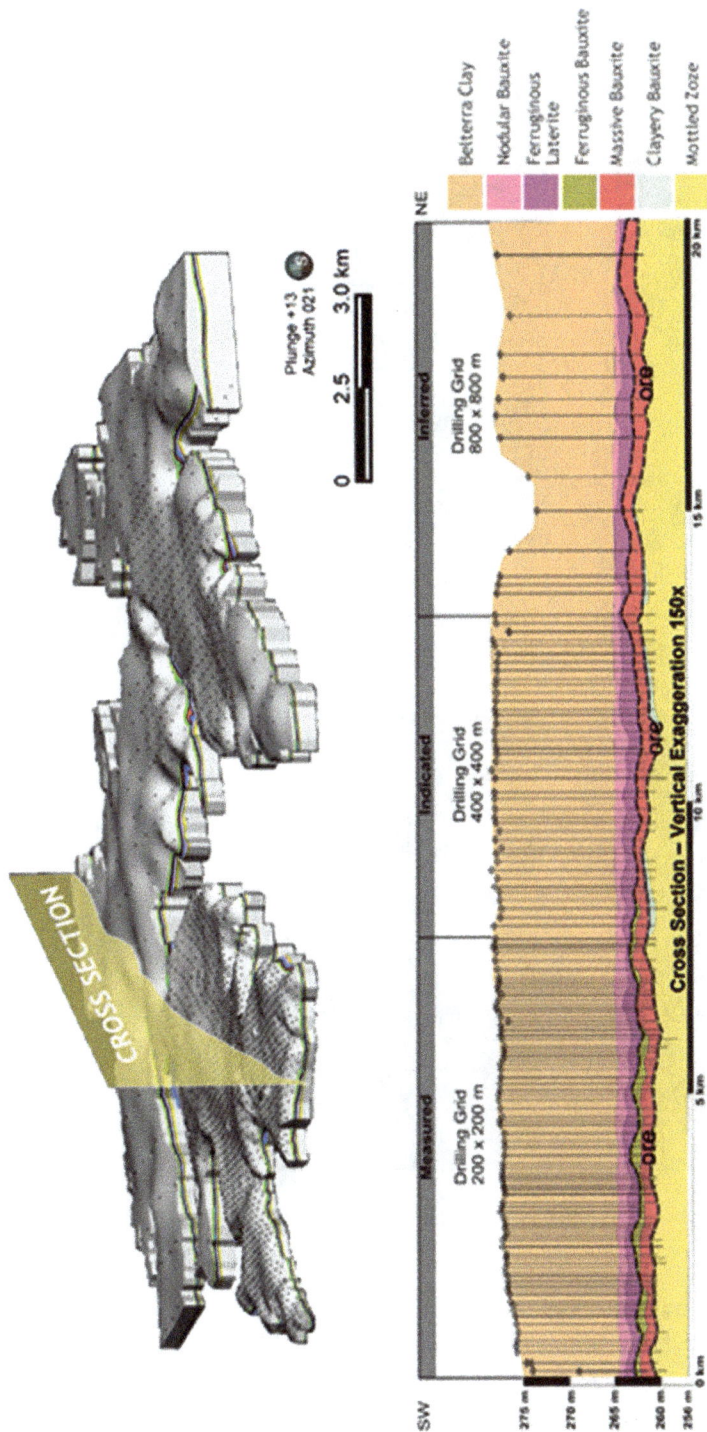

Figure 18.9: Cross-section of a plateau in the bauxite exploration areas in Rondon do Pará showing the continuous thick Belterra Clay covering the bauxite ore [34].

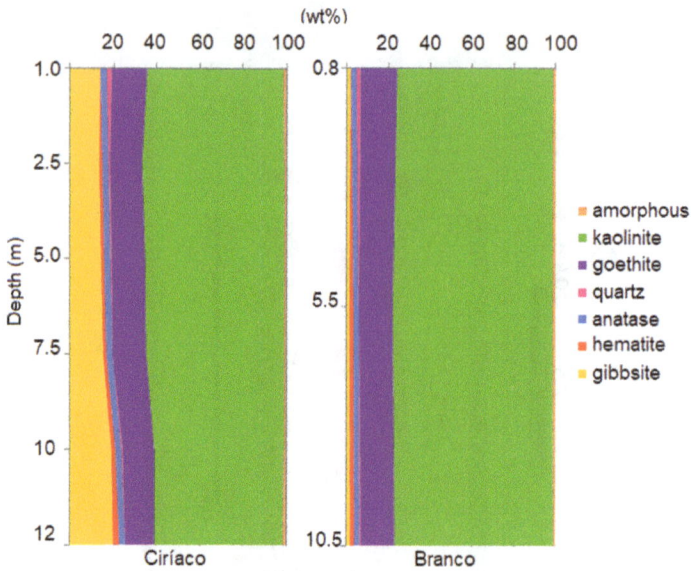

Figure 18.10: Mineral composition of BTC in the pilot mines Ciríaco and Branco, determined by combined X-ray diffraction and stoichiometry X-ray fluorescence calculations [8].

Figure 18.11: Micromorphological aspect of Belterra Clay, formed by nanocrystals. Sample from Décio pilot mine.

Diverse secondary sources of reactive silica and alumina have been tested for the production of such binders, including coal fly ashes [20, 38], steelmaking slags [39], kaolin and bauxite byproducts [22], red mud and alumina powders [40]. The use of alumina-rich by-products is of particular interest as partial or total substitutes for the expensive bauxite.

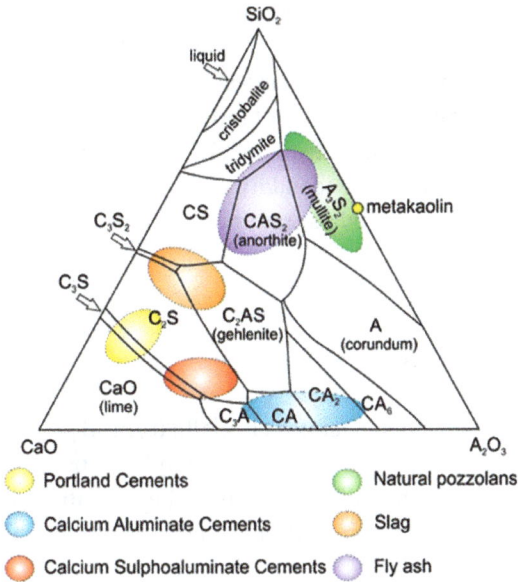

Figure 18.12 legend items:
- Portland Cements
- Calcium Aluminate Cements
- Calcium Sulphoaluminate Cements
- Natural pozzolans
- Slag
- Fly ash

Figure 18.12: Cement-related system $CaO-Al_2O_3-SiO_2$ with the simplified phase compositions, and approximated chemical compositions of some cementitious materials [35, 36].

When using BTC as substitute raw material for bauxite, the relative high Al_2O_3 contents of BTC (~35–40%) coming from gibbsite and kaolinite can be addressed to form ye'elimite. C_2S is expected by the additional SiO_2 from BTC's kaolinite. The iron contents can form C_2A,F, and/or be partially retained in the structure of perovskite (CT), expected after the low TiO_2 in the meal (Table 18.2). Iron is known to decrease the sintering temperature, and thus the quality of the raw material must not necessarily be poor in iron.

Table 18.2: Main phases expected in the clinkers using BTC.

Phase	ye'elimite	larnite	brownmillerite	ternesite	gehlenite	perovskite
Composition	C_4A_3s	β-C_2S	C_4AF	C_5S_2s	C_2AS	CT

Despite all cements containing ye'elimite as the main phase are known as "CSA cements", the occurrence of belite, ternesite, and/or ferrite in considerably higher amounts can change the properties of these binders. Therefore an extensive family of CSA with members described as CSA-belite, CSA-ternesite, or CSA-ferrite, according to the second most abundant phase, are frequently common. Similarly, BYF cements are used to describe clickers with a relative abundance of phases such as belite > ye'elimite > ferrite [18].

A wide possibility of mixing ratios of Belterra Clay with other natural raw materials (e.g. limestone and gypsum) is possible under different experimental designs, varying temperature, and/or sintering time. Sintering temperatures varying from 1000 °C to 1350 °C are reported in the formation of similar binders. The broad possibility of combinations can be handled using statistical modeling software to optimize the experimental work [22]. Other available secondary raw materials can be further added to the mixtures, as low-grade bauxites, red mud, or other bauxite mining residues.

3.1 Hydration of CSA-based cements

Upon hydration, the main CSA phase ye'elimite generates ettringite ($C_6As_3H_{32}$) [41]. Further addition of lime (CaO) and/or calcium sulfates controls the hydration behavior of CSA cements. When lime or calcium sulfate is absent, the hydration of pure ye'elimite is likely to first form monosulfate ($C_3A \cdot Cs \cdot \underline{12}H$), microcrystalline Al $(OH)_3$, and only minor amounts of ettringite [42]:

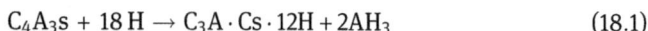

$$C_4A_3s + 18\,H \rightarrow C_3A \cdot Cs \cdot 12H + 2AH_3 \tag{18.1}$$

Calcium sulfates, belonging or added to CSA clinker (in hydrous or anhydrous forms) are known to increase ettringite contents in such systems (Figure 18.13), by favoring its formation according to the following equation:

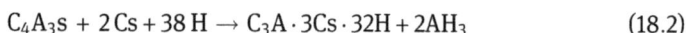

$$C_4A_3s + 2\,Cs + 38\,H \rightarrow C_3A \cdot 3Cs \cdot 32H + 2AH_3 \tag{18.2}$$

When lime is added, calcium hydroxide reacts with ye'elimite, also forming larger amounts of ettringite [41]:

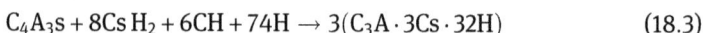

$$C_4A_3s + 8Cs\,H_2 + 6CH + 74H \rightarrow 3(C_3A \cdot 3Cs \cdot 32H) \tag{18.3}$$

Depending on the conditions, ettringite formation regulates some properties of CSA cements. For instance, ettringite formed without lime is non-expansive and responsible for early strengths cementitious materials [3], whereas it is expansive when formed in the presence of lime [43].

The composition of the hydrate assemblage can also vary considerably depending on the contents of belite present in the system. Here, belite hydration depends on its polymorph, as α'_H-belite is considered more reactive than β-belite (i.e. larnite). Straetlingite (C_2ASH_8), C-S-H gel, and portlandite (CH) are attributed to the hydration of belite with aluminum hydroxide (from eqs. (18.1) and (18.2)), according to the reactions [42, 44]:

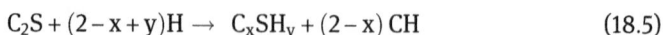

$$C_2S + AH_3 + 5H \rightarrow C_2ASH_8 \tag{18.4}$$

$$C_2S + (2-x+y)H \rightarrow C_xSH_y + (2-x)\,CH \tag{18.5}$$

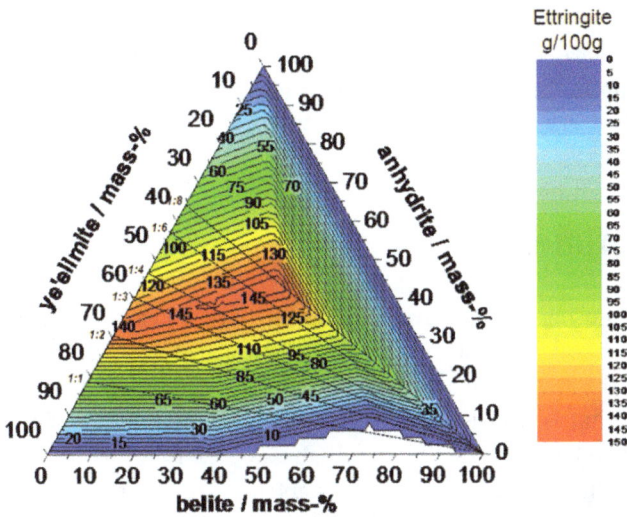

Figure 18.13: Modelled ettringite ternary diagram of the ye'elimite-belite–anhydrite hydration system. Ye'elimite/anhydrite molar ratios are represented by the lines crossing the diagram. Source: Winnefeld and Lothenbach (2016). Licensed under CC BY 4.0 https://creativecommons. org/licenses/by/4.0/ Permission.

Ternesite ($C_5S_2\bar{s}$), initially considered an unreactive phase, shows a very slow hydration reaction [45]. Nevertheless [46], observed that this phase contributed more than belite to the early strength development of particular CSA cements. According to these authors, the hydration of CSA clinkers with 5 and 12% of ternesite resulted in, respectively, 28 and 69% of reactivity for this phase after 48 h. The restricted terne-site's hydraulic reactivity is enhanced in the presence of $[Al(OH)_4]^-$, as SO_4 from ter-nesite dissolution enhances the formation of both ettringite and monosulphate [47]. Montes et al. [48] studied the activation of ternesite by calcium aluminates, including ye'elimite, and concluded that ternesite not only hasted the hydration of the ternesite-aluminate mixtures but also altered the hydration products by forming straetlingite and monosulfoaluminate.

Ferrite $[C_2(A,F)]$, i.e. brownmillerite, hydration is slower than that of ye'emilite and influenced by its Al/Fe ratio, by the calcium sulfate and belite concentrations in the solution [49]. These complex interactions make it difficult to predict the final products of ferrite hydration, as these might include the stabilization of Fe-rich et-tringite and formation of Fe-Si-hydrogarnets [44, 50, 51]. Katoite ($C_3(A,F)SH_4$), for example, usually appears only in the much later hydration stages [3]:

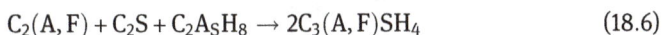

$$C_2(A, F) + C_2S + C_2A\bar{s}H_8 \rightarrow 2C_3(A, F)SH_4 \tag{18.6}$$

The distinct hydration paths of ye'elimite, belite, ternesite and ferrite, the most com-mon CSA phases, result in complex hydration reactions of the CSA pastes. The initial formation of ettringite from ye'elimite controls almost entirely the early strength gain of these types of cement, whereas the further reactions will depend on complex interactions among the present phases, as well as available sulfate and/or basic aluminum in solution.

4 Design of experiments

Formulation of new binders using atypical raw materials might rely on extreme large experimental trials until an ideal compositional mixture is defined. The use of Bogue calculations [52] to predict the formed clinker phases is helpful, but often tricky when using non-pure reagents. That is because Bogue equations use the amount of the main oxides of the raw materials to predict the main phases present in the clinker, but do not consider solid solutions [53]. When further adapted to CSA cements, Bogue computations consider the phase assemblage $C\text{-}C_2S\text{-}C_2A,F\text{-}C_4A_3s\text{-}Cs$ [54], while other possible previously mentioned phases (e.g. C_5S_2s, C_2AS, CT) are neglected. For CSA-based cements, recent new insights on the system C-S-A-s-F, investigated at 1250 °C, made it possible to predict a more accurate and realistic phase assemblage for CSA-based clinkers produced at this temperature [55]. Nevertheless, the stability of TiO_2 (present in BTC) is still not completely understood in these systems. Therefore, a definition of a "specific receipt" will rely on the specific chemical composition of the raw materials, as well as on the chosen temperature and amount of time during sintering, often turning reliable predictions very difficult.

Design of Experiments (DoE) has been used to reduce the number of trials in the research of new binder compositions [22, 56] and in the investigation of different cements [57–60]. DoE is a very useful tool to deal with variable parameters by computing all observations and trends with minimal experimental work [61]. The influence of individual and combined factors is taken into account to predict a maximum or a minimum performance of a target result. In the investigation of new binder formulations, the factor can be referred to as the raw materials used in the meal, the different possible firing temperatures, and the sintering time. A specific clinker phase is normally chosen as a response variable (target), to be maximized or minimized using DoE (Figure 18.14).

Diverse software solutions, many using "black box" processes, are available to conduct DoE [62]. DoE can be performed in steps depending on the experimental phase. The software STAVEX (AICOS Technologies®), for example, distinguishes three successive design stages: screening; modeling (linear model), and optimization (quadratic model) [63]. The successive stages claim to reduce the number of factors that influence the response variable, identifying an optimum after the last step. Although, care must be taken when working in such interfaces, as the mathematical software calculations do not consider the complex chemical interactions

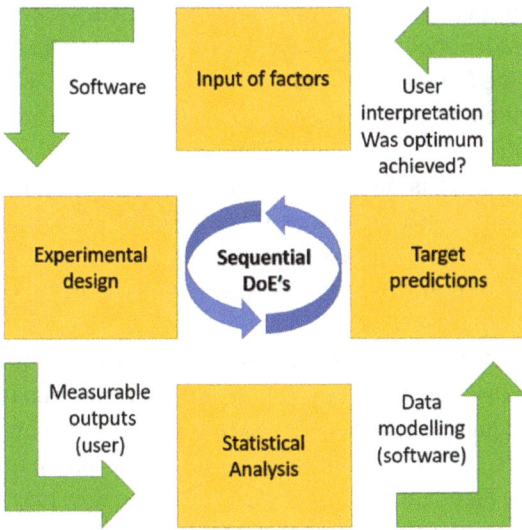

Figure 18.14: Structure conduction of sequential Design of Experiments.

that occur during the clinkerization process. Still, DoE can be very useful to reduce the experimental work and help to achieve raw meal compositions at least close to the ideal one [22].

5 Ongoing research and preliminary results

As previously shown, BTC can present slightly different alumina contents depending on its gibbsite and kaolinite amounts (Figure 18.15). A sample from Ciríaco, in Rondon do Pará, Brazil, was used in the first trial of experiments aiming the production of CSA-based cements using BTC, gypsum, and calcite. DoE was performed to reduce the number of trials in the view of the software STAVEX.

A total of 32 experiments from BTC, gypsum, and calcite mixtures were produced after two planned sequential modelings. The raw meal compositions, the temperature, and the sintering time were adjusted to maximize ye'elimite in the produced clinkers, as well as to reduce the calcite consumption in the raw meal and consequently increase the use of BTC. 1250 °C sintering temperature, under one hour, was defined as optimal after testing the initial temperature range varying from 1100 °C to 1300 °C and under one and two hours used in the experiments.

Mixing raw compositions in the range of 35–43.5% of BTC, 43.5–50% of calcite, and 8–13% of gypsum resulted in clinkers with maximum optimized contents of ye'elimite varying from 30 to 35% when produced at 1250 °C (Figures 18.16, 18.17).

Further phases include larnite, ternesite, brownmillerite, gehlenite, anhydrite, and perovskite, besides amorphous/non-identified phases by the XRD-Rietveld method.

The main phases ye'elimite, belite (i.e. larnite), and ternesite, made it possible to classify such clinkers as CSA-belite and CSA-ternesite, depending on its second most abundant phase, after ye'elimite.

The micromorphology of the clinkers is dominated by nano-sized ye'elimite and larnite, edgy but in general with rounded shapes (Figure 18.18). Ternesite forms prismatic crystals that can reach few microns in the ternesite-rich clinkers (Figure 18.19). Brownmillerite is represented by crystals with rhombohedral-like shapes within ye'elimite and larnite aggregates.

The main hydration of the produced clinkers, without additions (pure), occurs in the first 40 hours (Figure 18.20). Ettringite, kuzelite, strätlingite, and hemicarboaluminate are the main hydration products, as observed by XRD analysis. Unlike hydration behavior among the clinkers can be partially addressed to their different fineness and phase composition. The predominance of a single main heat flow maxima is related to the hydration of the main ye'elimite phase to form ettringite, while larnite, ternesite, and brownmillerite have a very limited hydraulic behavior under the investigated hydration time.

Gypsum was added to the most representative produced clinkers (i.e. SD2-OP and SD2-6) to optimize their setting time. The gypsum additions accelerated the hydration of the clinkers (Figure 18.21), reducing the induction periods and increasing contents of ettringite in the systems, as predicted in eq. (18.2) (Table 18.3). Five percent of gypsum additions anticipated the main heat flow maxima to 15 h in the clinker SD2-OP and 11.5 h in SD2-6. The ten percent additions, on the other hand, resulted in overall longer hydration.

The hydrated clinkers show ettringite occurring both as large prisms (close to 10 μm) and as needle-shaped crystals (Figure 18.22). Larnite occurs as irregular and surface (corroded) fragments, and platy monosulphate (kuzelite) is observed in the systems without gypsum additions (Figure 18.22).

6 Concluding remarks

Belterra Clay is an available and extensive alumina-rich clay cover on the bauxite deposits in the Brazilian Amazon region, so far not used by the industry and commonly considered a drawback during surface bauxite mining. The chemical and mineralogical composition of BTC indicates it can be used in the production of calcium sulphoaluminate based cements, that are promissory binder candidates for the upcoming cement industry, as it generates much less CO_2 than the currently ordinary Portland cements.

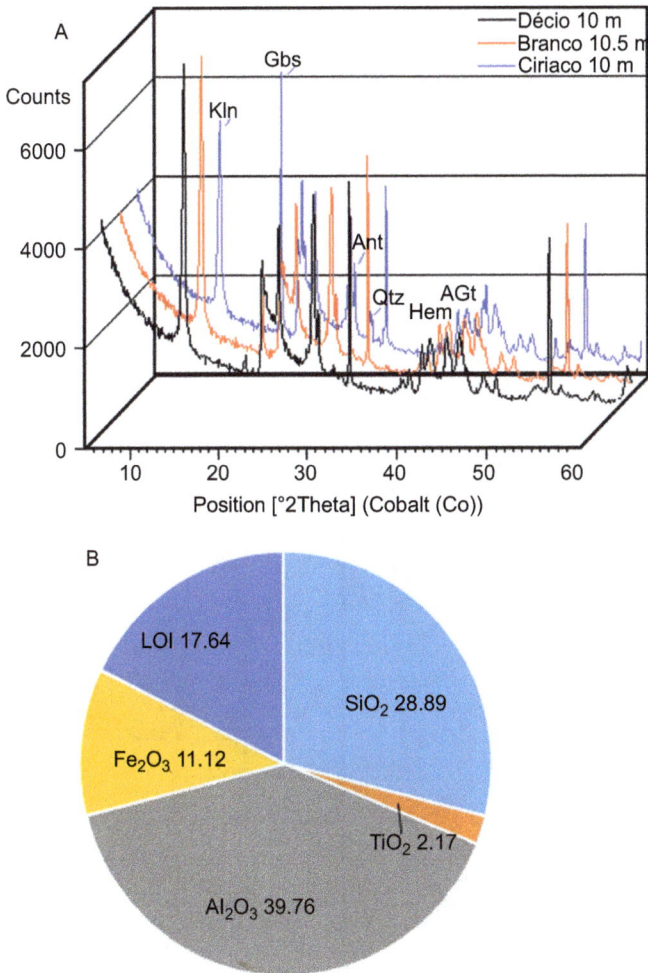

Figure 18.15: A: Comparison of the X-ray Diffractograms of the BTC in the pilot mines Ciríaco, Branco, and Décio in Rondon do Pará. Kln: kaolinite, Gbs: gibbsite, AGt: Al-goethite, Qtz: quartz, Ant: anatase. B: Chemical composition and loss on ignition (LOI) of the used BTC from the Ciríaco pilot mine.

Previous results confirm that BTC can be used for CSA-based cements, after the production of at least two main types of clinkers: CSA-ternesite and CSA-belite. The hydration of these binders occurred within 24 h when mixed with 5% of gypsum, a comparable time to the hydration of OPCs.

At least 30% less CO_2 is generated by the produced clinkers when compared to OPC's. Further savings on energy due to softer grinding and lower sintering temperature, as well as on raw material costs are among the advantages of these clinkers. The results encourage further researches by using BTC from other localities, as well as an investigation on the hydration paths of the hydrated systems.

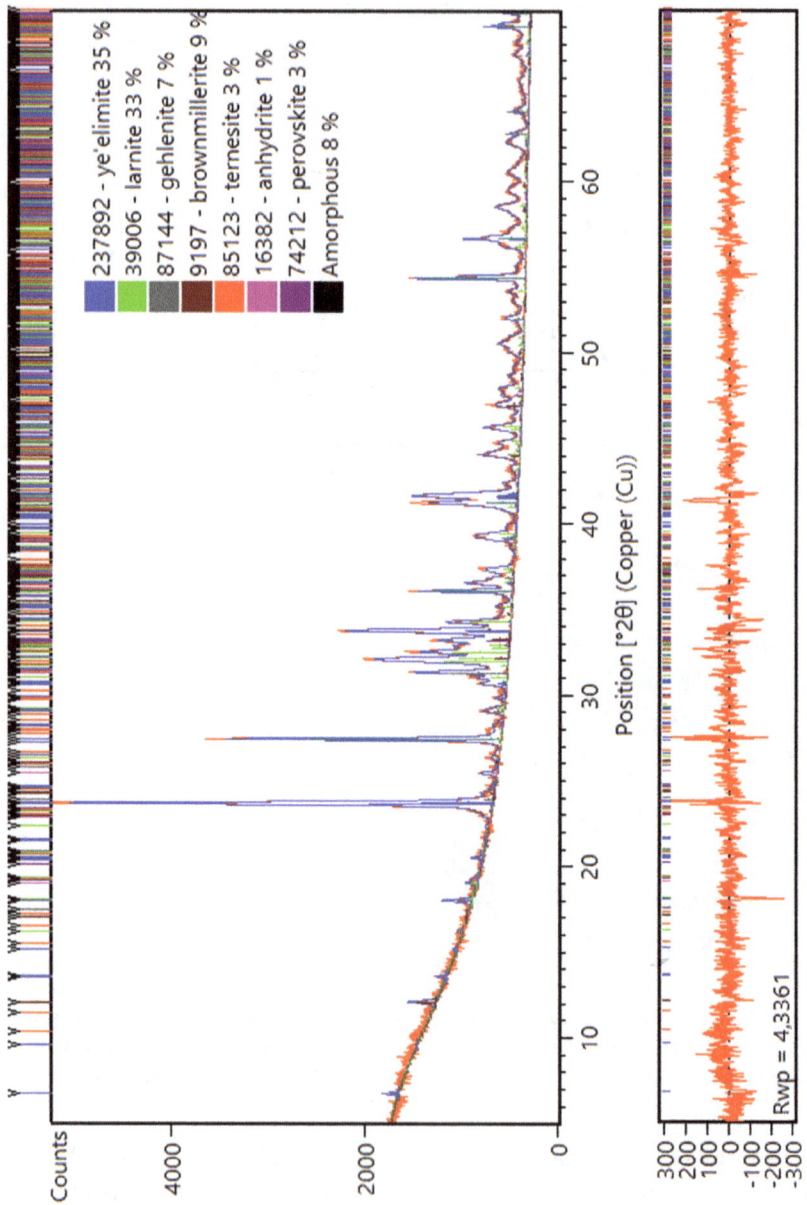

Figure 18.16: Rietveld refined diffractogram and difference plot of the clinker SD2-6, produced at 1250 °C.

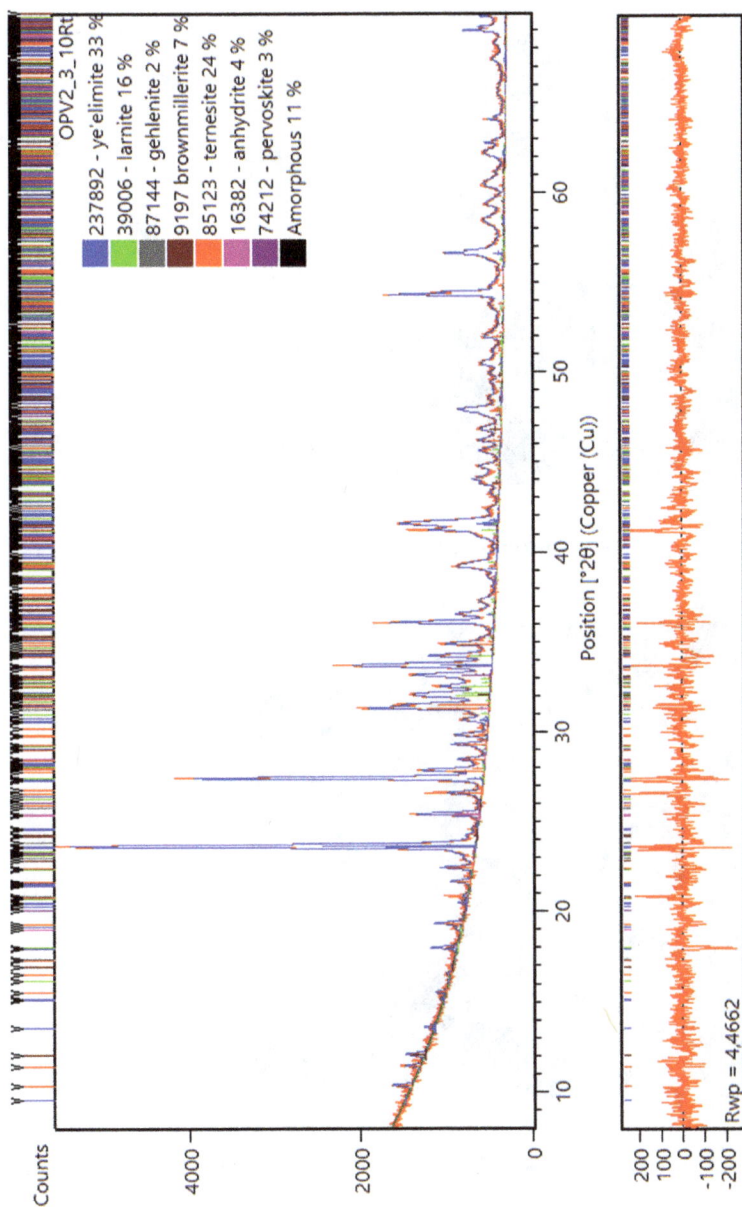

Figure 18.17: Rietveld refined diffractogram and difference plot of the clinker SD2-OP, produced at 1250 °C.

Figure 18.18: Ye'elimite crystallites in the clinker SD2-OP.

Figure 18.19: Micromorphology of the clinker SD2-OP showing prismatic ternesite and smaller rounded ye'elimite crystals.

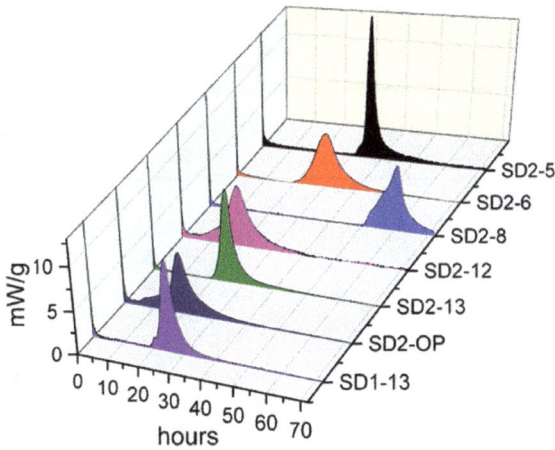

Figure 18.20: Isoperibolic heat flow curves showing the hydration behavior of the produced clinkers at 1250 °C. Water/cement ratio = 1. The specific surface area of the clinkers is as followed, determined by BET analysis: SD1-13 2601 cm^2/g; SD2-OP 3040 cm^2/g; SD2-13 3260 cm^2/g; SD2-12 2236 cm^2/g; SD2-8 2735 cm^2/g; SD2-6 2396 cm^2/g; SD2-5 2560 cm^2/g.

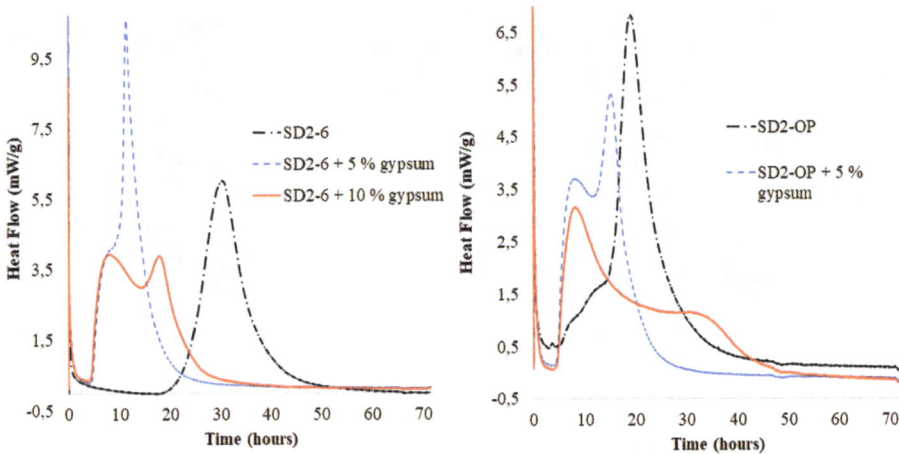

Figure 18.21: Isoperibolic heat flow curves after the hydration of clinkers SD2-OP and SD2-6, with gypsum additions. W/C = 1, after 72 h.

Table 18.3: Phase and amorphous contents (in wt.%) of the hydration products quantified by Rietveld analysis.

Phase (%)	SD2-6			SD2-OP		
	pure	+5%Gp	+10%Gp	pure	+5%Gp	+10%Gp
ettringite	7	19	36	16	27	42
larnite	28	20	21	13	10	13
ternesite	8	0	2	19	15	17
kuzelite	11	7	4	10	4	2
strätlingite	4	6	3	1	0	0
brownmillerite	3	2	2	3	2	3
perovskite	2	2	2	3	2	2
gehlenite	6	5	4	0	0	0
hemicarboaluminate	16	5	4	6	4	0
ye'elimite	0	1	2	1	1	3
amorphous	14	33	21	28	34	19
GOF	1	1	1	1	1	2
Rwp	7	7	8	8	8	9
Total	100	100	100	100	100	100

Figure 18.22: Micromorphology of the hydrated systems after 72 h using a w/c = 1. A: prismatic and needle-shaped ettringite, clinker SD2-OP, with 10% of added gypsum. B: platy monosulphate aggregates and larnite fragments in the clinker SD2-6, without gypsum additions. Images acquired at Centre for Material Analysis/ZWL in Lauf, Germany.

Acknowledgements: We thank the Brazilian CAPES foundation through the grant 88881.199654/2018-01 to the author Leonardo Boiadeiro Ayres Negrão. The authors also gratefully acknowledge the company NEXA Resources (former Votorantim Metais) for the Belterra Clay samples and, the Zentrum für Werkstoffanalytik at Lauf for the SEM analyzes.

References

[1] R.M. Andrew, Global CO2 emissions from cement production, 1928–2018, Earth Syst. Sci. Data. 11 (2019) 1675–1710. doi:10.5194/essd-11-1675-2019.

[2] IEA, Technology Roadmap Low-Carbon transition in the Cement Industry, 2018. doi:10.1007/1-4020-0612-8_961.

[3] A. Naqi, J.G. Jang, Recent progress in green cement technology utilizing low-carbon emission fuels and raw materials: A review, Sustain. (2019). doi:10.3390/su11020537.

[4] U.S Geological Survey, Mineral commodity summaries 2020, Reston, VA, 2020. doi:10.3133/mcs2020.

[5] T.D. Kelly, G.R. Matos, Historical Statistics for Mineral and Material Commodities in the United States, U.S. Geol. Surv. Data Ser. 140. (2014).

[6] A.M.C. Horbe, M.L. da Costa, Lateritic crusts and related soils in eastern Brazilian Amazonia, Geoderma. 126 (2005) 225–239. doi:10.1016/j.geoderma.2004.09.011.

[7] W. Truckenbrodt, B. Kotschoubey, Argila da Belterra – Cobertura Terciária das bauxitas amazônicas, Rev. Bras. Geociências. 11 (1981) 203–208. doi:10.25249/0375-7536.1981203208.

[8] L.B.A. Negrão, M.L. da Costa, H. Pöllmann, The Belterra Clay on the bauxite deposits of Rondon do Pará, Eastern Amazon, Brazilian J. Geol. 48 (2018) 473–484. doi:10.1590/2317-4889201820180128.

[9] L.B.A. Negrão, M.L. da Costa, H. Pöllmann, A. Horn, An application of the Rietveld refinement method to the mineralogy of a bauxite -bearing regolith in the Lower Amazon, Mineral. Mag. 82 (2018) 413–431. doi:10.1180/minmag.2017.081.056.

[10] I.A.R. Barreto, M.L. da Costa, Viability of Belterra clay, a widespread bauxite cover in the Amazon, as a low-cost raw material for the production of red ceramics, Appl. Clay Sci. (2018). doi:10.1016/j.clay.2018.06.010.

[11] I.A.R. Barreto, M.L. da Costa, Sintering of red ceramics from yellow Amazonian latosols incorporated with illitic and gibbsitic clay, Appl. Clay Sci. (2018). doi:10.1016/j.clay.2017.11.003.

[12] G. Álvarez-Pinazo, A. Cuesta, M. García-Maté, I. Santacruz, E.R. Losilla, A.G. De la Torre, L. León-Reina, M.A.G. Aranda, Rietveld quantitative phase analysis of Yeelimite-containing cements, Cem. Concr. Res. 42 (2012) 960–971. doi:10.1016/J.CEMCONRES.2012.03.018.

[13] J. Beretka, R. Cioffi, M. Marroccoli, G.L. Valenti, Energy-saving cements obtained from chemical gypsum and other industrial wastes, Waste Manag. 16 (1996) 231–235. doi:10.1016/S0956-053X(96)00046-3.

[14] M.C. Martín-Sedeño, A.J.M. Cuberos, Á.G. De la Torre, G. Álvarez-Pinazo, L.M. Ordónez, M. Gateshki, M.A.G. Aranda, Aluminum-rich belite sulfoaluminate cements: Clinkering and early age hydration, Cem. Concr. Res. (2010). doi:10.1016/j.cemconres.2009.11.003.

[15] P. Mehta, Investigations on energy-saving cements, World Cem. Technol. (1980).

[16] H. Pöllmann, Calcium aluminate cements – Raw materials, differences, hydration and properties, Rev. Mineral. Geochemistry. (2012). doi:10.2138/rmg.2012.74.1.

[17] H. Pöllmann, S. Stöber, Investigations on commercial and synthetic calciumsulfoaluminate cements, in: 36th Int. Conf. Cem. Microsc. 2014, 2014.

[18] E. Gartner, T. Sui, Alternative cement clinkers, Cem. Concr. Res. 114 (2018) 27–39. doi:10.1016/J.CEMCONRES.2017.02.002.

[19] J. Péra, J. Ambroise, New applications of calcium sulfoaluminate cement, Cem. Concr. Res. (2004). doi:10.1016/j.cemconres.2003.10.019.

[20] P. Arjunan, M.R. Silsbee, Della M. Roy, Sulfoaluminate-belite cement from low-calcium fly ash and sulfur-rich and other industrial by-products, Cem. Concr. Res. 29 (1999) 1305–1311. doi:10.1016/S0008-8846(99)00072-1.

[21] N. Ukrainczyk, N. Franković Mihelj, J. Šipušić, Calcium sulfoaluminate eco-cement from industrial waste, in: Chem. Biochem. Eng. Q., 2013.

[22] S. Galluccio, T. Beirau, H. Pöllmann, Maximization of the reuse of industrial residues for the production of eco-friendly CSA -belite clinker, Constr. Build. Mater. (2019). doi:10.1016/j.conbuildmat.2019.02.148.

[23] H. Pöllmann, M.L. da Costa, R.S. Angélica, Sustainable Secondary Resources from Brazilian Kaolin Deposits for the Production of Calcined Clays, in: K. Scrivener, A. Favier (Eds.), Calcined Clays Sustain. Concr., Springer Netherlands, Dordrecht, 2015: pp. 21–26.

[24] W.G. Sombroek, Amazon soils. A reconnaissance of the soils Region, the Brazilian Amazonitle, Wageningen University, 1966.

[25] P.L.C. Grubb, Genesis of bauxite deposits in the lower Amazon Basin and Guianas coastal plain, Econ. Geol. 74 (1979) 735–750. doi:10.2113/gsecongeo.74.4.735.

[26] W. Truckenbrodt, B. Kotschoubey, W. Schellmann, Composition and origin of the clay cover on North Brazilian laterites, Geol. Rundschau. (1991). doi:10.1007/BF01803688.

[27] B.I. Kronberg, W.S. Fyfe, B.J. McKinnon, J.F. Couston, B.S. Filho, R.A. Nash, Model for bauxite formation: Paragominas (Brazil), Chem. Geol. (1982). doi:10.1016/0009-2541(82)90008-0.

[28] Y. Tardy, étrologie des latérites et des sols tropicaux, {M}asson, 1993. http://www.documentation.ird.fr/hor/fdi:38818.

[29] B. Kobilsek, Y. Lucas, Morphologic and petrographic study of a bauxitic formation in Amazonia, district of Juruti, State of Para, Brazil, Sci. Geol. – Bull. 41 (1988) 71–84.

[30] W. Abouchami, K. Näthe, A. Kumar, S.J.G. Galer, K.P. Jochum, E. Williams, A.M.C. Horbe, J.W.C. Rosa, W. Balsam, D. Adams, K. Mezger, M.O. Andreae, Geochemical and isotopic characterization of the bodélé depression dust source and implications for transatlantic dust transport to the Amazon basin, Earth Planet. Sci. Lett. (2013). doi:10.1016/j.epsl.2013.08.028.

[31] M.L. da Costa, G. da S. Cruz, H.D.F. de Almeida, H. Poellmann, On the geology, mineralogy and geochemistry of the bauxite -bearing regolith in the lower Amazon basin: Evidence of genetic relationships, J. Geochemical Explor. 146 (2014) 58–74. doi:10.1016/J.GEXPLO.2014.07.021.

[32] S.B. De Oliveira, M.L. da Costa, H.J. Dos Prazeres Filho, The lateritic bauxite deposit of Rondon do Pará: A new giant deposit in the Amazon Region, Northern Brazil, Econ. Geol. 111 (2016) 1277–1290. doi:10.2113/econgeo.111.5.1277.

[33] H. Prazeres Filho, S. Oliveira, L. Molinari, J. Belther, The rediscovery of Rondon do Para, the last giant world-class bauxite deposit in an attractive geography, in: 33th ICSOBA – International Committee for Study of Bauxite, Alumina & AluminiumAt: Dubai, United Arab Emirates, 2015.

[34] C. Gatti, The Allumina Issue and the Bauxite Project in Rondon, Pará, VII Brazilian Symp. Miner. Explor. (2016). http://www.adimb.com.br/simexmin2016/palestra/auditorio_sao_joao_delrey_16/15h00CarlosGatti.pdf (accessed October 4, 2020).

[35] D.E. Macphee, E.E. Lachowski, 3 – Cement Components and Their Phase Relations, in: P.C. Hewlett (Ed.), Lea's Chem. Cem. Concr. (Fourth Ed., Fourth Edi, Butterworth-Heinemann, Oxford, 1998: pp. 95–129. doi:https://doi.org/10.1016/B978-075066256-7/50015-1.

[36] B. Lothenbach, K. Scrivener, R.D. Hooton, Supplementary cementitious materials, Cem. Concr. Res. (2011). doi:10.1016/j.cemconres.2010.12.001.

[37] A. Berrio, C. Rodriguez, J.I. Tobón, Effect of Al2O3/SiO2 ratio on ye'elimite production on CSA cement, Constr. Build. Mater. 168 (2018) 512–521. doi:10.1016/j.conbuildmat.2018.02.153.

[38] R. Schmidt, H. Pöllmann, Quantification of calcium sulpho-aluminate cement by Rietveld analysis, Mater. Sci. Forum. (2000).

[39] D. Adolfsson, N. Menad, E. Viggh, B. Bjo, Steelmaking slags as raw material for sulphoaluminate belite cement, Adv. Cem. Res. 7605 (2007) 147–156. doi:10.1680/adcr.2007.19.4.147.

[40] M.L. Pace, A. Telesca, M. Marroccoli, G.L. Valenti, Use of industrial byproducts as alumina sources for the synthesis of calcium sulfoaluminate cements, Environ. Sci. Technol. 45 (2011) 6124–6128. doi:10.1021/es2005144.

[41] I. Odler, Cements containing calcium sulfoaluminate, in: A. Bentur, S. Miindess (Eds.), Spec. Inorg. Cem., London, 2000: pp. 69–87.

[42] F. Winnefeld, B. Lothenbach, Phase equilibria in the system $Ca_4Al_6O_{12}SO_4 - Ca_2SiO_4 - CaSO_4 - H_2O$ referring to the hydration of calcium sulfoaluminate cements, RILEM Tech. Lett. 1 (2016) 6. doi:10.21809/rilemtechlett.2016.5.

[43] P.K. Mehta, Mechanism of expansion associated with ettringite formation, Cem. Concr. Res. (1973). doi:10.1016/0008-8846(73)90056-2.

[44] G. Álvarez-Pinazo, I. Santacruz, M.A.G. Aranda, Á.G. De La Torre, Hydration of belite - ye'elimite -ferrite cements with different calcium sulfate sources, Adv. Cem. Res. (2016). doi:10.1680/jadcr.16.00030.

[45] S. Skalamprinos, G. Jen, I. Galan, M. Whittaker, A. Elhoweris, F. Glasser, The synthesis and hydration of ternesite, $Ca_5(SiO_4)_2SO_4$, Cem. Concr. Res. (2018). doi:10.1016/j.cemconres.2018.06.012.

[46] F. Bullerjahn, M. Zajac, M. Ben Haha, CSA raw mix design: effect on clinker formation and reactivity, Mater. Struct. Constr. (2015). doi:10.1617/s11527-014-0451-z.

[47] M. Ben Haha, F. Bullerjahn, M. Zajac, On the reactivity of ternesite, in: Proc. 14th Int. Congr. Chem. Cem. Beijing, China, 2015. doi:10.1248/cpb.6.638.

[48] M. Montes, E. Pato, P.M. Carmona-Quiroga, M.T. Blanco-Varela, Can calcium aluminates activate ternesite hydration? Cem. Concr. Res. (2018). doi:10.1016/j.cemconres.2017.10.017.

[49] X. Huang, F. Wang, S. Hu, Y. Lu, M. Rao, Y. Mu, Brownmillerite hydration in the presence of gypsum: The effect of Al/Fe ratio and sulfate ions, J. Am. Ceram. Soc. 102 (2019) 5545–5554. doi:10.1111/jace.16384.

[50] E. Gartner, G. Walenta, V. Morin, P. Termkhajornkit, I. Baco, J.-M. Casabonne, Hydration of a Belite-CalciumSulfoaluminate-Ferrite cement: AetherTM, in: 13th Int. Congr. Chem. Cem. Madrid, Spain, 2011.

[51] J. Wang, Hydration mechanism of cements based on low-CO_2 clinkers containing belite, ye'elimite and calcium alumino-ferrite, 7th Int. Symp. Cem. Concr. (2010).

[52] H.F.W. Taylor, Modification of the Bogue calculation, Adv. Cem. Res. (1989). doi:10.1680/adcr.1989.2.6.73.

[53] P. Stutzman, A. Heckert, A. Tebbe, S. Leigh, Uncertainty in Bogue-calculated phase composition of hydraulic cements, Cem. Concr. Res. (2014). doi:10.1016/j.cemconres.2014.03.007.

[54] J. Majling, J. Strigáč, D.M. Roy, Generalized Bogue computations to forecast the mineralogical composition of sulfoaluminate cements based on fly ashes, Adv. Cem. Res. (1999). doi:10.1680/adcr.1999.11.1.27.

[55] I. Galan, T. Hanein, A. Elhoweris, M.N. Bannerman, F.P. Glasser, Phase compatibility in the system $CaO-SiO_2-Al_2O_3-SO_3-Fe_2O_3$ and the effect of partial pressure on the phase stability, Ind. Eng. Chem. Res. (2017). doi:10.1021/acs.iecr.6b03470.

[56] L. Senff, A. Castela, W. Hajjaji, D. Hotza, J.A. Labrincha, Formulations of sulfobelite cement through design of experiments, Constr. Build. Mater. (2011). doi:10.1016/j.conbuildmat.2011.03.032.

[57] M.H. Moreira, A.P. Luz, A.L. Christoforo, C. Parr, V.C. Pandolfelli, Design of Experiments (DOE) applied to high-alumina calcium aluminate cement-bonded castables, Ceram. Int. (2016). doi:10.1016/j.ceramint.2016.08.079.

[58] A. Mohan, K.M. Mini, Strength and durability studies of SCC incorporating silica fume and ultra fine GGBS, Constr. Build. Mater. (2018). doi:10.1016/j.conbuildmat.2018.03.186.

[59] A.J. Moseson, D.E. Moseson, M.W. Barsoum, High volume limestone alkali-activated cement developed by design of experiment, Cem. Concr. Compos. (2012). doi:10.1016/j.cemconcomp.2011.11.004.

[60] R. Kosalram, G.S. Vijayabhaskara, M. Anoop, Mix Design of Fly Ash based concrete using DOE Method, Int. J. Trends Eng. Technol. (2015).

[61] J. Antony, Design of Experiments for Engineers and Scientists, 2003. doi:10.1016/B978-0-7506-4709-0.X5000-5.

[62] O. Rynne, M. Dubarry, C. Molson, D. Lepage, A. Prébé, D. Aymé-Perrot, D. Rochefort, M. Dollé, Designs of experiments for beginners – A quick start guide for application to electrode formulation, Batteries. (2019). doi:10.3390/batteries5040072.

[63] W. Seewald, Y.L. Grize, Stavex: A sophisticated but user-friendly expert system for the design and analysis of experiments, Ther. Innov. Regul. Sci. (1997). doi:10.1177/009286159703100227.

Index

https://doi.org/10.1515/9783110674941-019

www.ingramcontent.com/pod-product-compliance
Lightning Source LLC
Chambersburg PA
CBHW060941210326
41598CB00031B/4696